MANUAL OF GEOSPATIAL SCIENCE AND TECHNOLOGY

Second Edition

MANUAL OF GEOSPATIAL SCIENCE AND TECHNOLOGY

Second Edition

Edited by
JOHN D. BOSSLER

Associate Editors
JAMES B. CAMPBELL
ROBERT B. McMASTER
CHRIS RIZOS

CRC Press
Taylor & Francis Group
Boca Raton London New York

CRC Press is an imprint of the
Taylor & Francis Group, an **informa** business

CRC Press
Taylor & Francis Group
6000 Broken Sound Parkway NW, Suite 300
Boca Raton, FL 33487-2742

First issued in paperback 2019

© 2010 by Taylor and Francis Group, LLC
CRC Press is an imprint of Taylor & Francis Group, an Informa business

No claim to original U.S. Government works

ISBN-13: 978-1-4200-8733-8 (hbk)
ISBN-13: 978-0-367-88421-5 (pbk)

Library of Congress Cataloging-in-Publication Data

Manual of geospatial science and technology / editors, John D. Bossler ... [et al.]. -- 2nd ed.
    p. cm.
"A CRC title."
Includes bibliographical references and index.
ISBN 978-1-4200-8733-8 (hard back : alk. paper)
    1. Geographic information systems. 2. Remote sensing. 3. Global Positioning System.
I. Bossler, John D. II. Title.

G70.212.M287 2010
910.285--dc22
                                          2009051402

Visit the Taylor & Francis Web site at
http://www.taylorandfrancis.com

and the CRC Press Web site at
http://www.crcpress.com

Contents

PART I   Prerequisites

PART II   Global Positioning System

PART III   Remote Sensing

PART IV   Geographic Information Systems

Preface

The *Manual of Geospatial Science and Technology* is written for those who are involved in setting up a "GIS project," but who have only limited knowledge when it comes to completing this task. In order to satisfy this need, the manual is divided into five parts: Prerequisites, Global positioning system, Remote sensing, Geographic information systems, and Applications. If the reader is somewhat familiar with these topics, it is suggested that he or she skip Part I and proceed to the part of the manual that interests him or her the most. If the reader is unfamiliar with the three main topics, appropriate chapters in Part I, along with the references provided, are highly recommended. The integration of these three technologies is best found in Part V.

The editor and the associate editors believe that this edition is a considerable improvement over the first edition of the manual. Almost every chapter of the first edition has been edited, revised, or completely rewritten. The index has been improved and expanded by several hundred words. Each figure from the first edition that has been used here has been reviewed, refined, resized, and/or redrawn. Dozens of new figures have been added. Part I, covering the basics, has undergone the least change, which is understandable because basics are not subject to change. Below is a discussion of specific changes in all the parts of the manual.

Part I: The chapter on computer basics has been dropped. We thought that every user of the manual would have excellent and easily available material covering this subject. Moreover, most practitioners in today's business and engineering circles have basic knowledge about computers.

The chapter on datums and geodetic reference systems (Chapter 3) was revised significantly because there have been many changes in the international scientific community regarding the formulation and distribution of these data.

In the other parts of the manual, many chapters refer to statistics and least squares. Therefore, we decided to add a chapter devoted to basic statistics and least squares solutions. All of the other chapters in Part I were either edited lightly or were not edited at all.

Part II: This part has nearly been completely rewritten, but it has the same basic structure as the previous edition. The rewritten chapters are simply a reflection of the rapid changes that have occurred in the technology, applications, and usage of GPS. With regard to the technology, the use of non-U.S. satellite systems has expanded rapidly. That is why many of the chapters address the Global Navigation Satellite System (GNSS) rather than just the U.S.-based system, GPS. Almost every chapter is impacted by the inclusion of the Russian Federation system, GLONASS, the European space agency system, GALILEO, and other such systems. The number of satellites available to a receiver on earth changes many aspects of satellite positioning technology, e.g., accuracy, time of positioning, and cost. The second edition provides state-of-the-art information on this important technology.

Part III: As in the case with GPS, remote sensing has experienced major changes during the last eight years. During this time, remote sensing technology has exploited advances in computer technology, and the timing of this volume has been nearly perfect to capture these developments. The massive data sets produced by satellite-observing platforms can be used efficiently and effectively because of advances in computer technology. Further, remote sensing imagery is now incorporated into applications that are directly accessible by the public, and that are imbedded into everyday commercial products used throughout the world. One of the most important nascent remote sensing (or photogrammetric) technologies that has matured during the period between both editions is Light Detection and Ranging (LIDAR), which is covered in Chapter 23, one of the most significant chapters in the manual. We are not aware of any presentation of this material that is clearer and better presented than that provided in Chapter 23. The manual is worth buying just for this chapter! The structure of Part III has changed, and now has many helpful references including Web sites, tools for decision making, etc.

Part IV: The field of Geographic Information Science/Systems has grown tremendously over the last eight years. In addition to the previously mentioned new sources of data, topics such as spatiotemporal data modeling, mobile GIS technologies, Web mapping, geovisualization, and GIS and society have emerged as new areas of inquiry and are included in this edition. Chapter 29 on data models now addresses, more specifically, time as part of the GIS. A new chapter that shows the relationship between geographic information systems and society addresses issues such as privacy, legal concerns, and the emerging field of public participation GIS. Chapter 31 on cartography has new material on the expanding field of geovisualization, which looks at cartography as a dynamic process.

Part V: The applications of the three nascent technologies covered in this manual have grown enormously. The reader will find that the three technologies are integrated far more tightly than they were eight years ago and the usage has been creative and far-reaching. A new chapter on applications in the utility industry, which was inappropriately missing in the first edition, provides a wealth of information about how utilities have positively exploited this technology. The applications in state government have been fleshed out and provide insight into this arena.

John D. Bossler
Robert B. McMaster
Chris Rizos
James B. Campbell

Acknowledgments

This edition of the *Manual of Geospatial Science and Technology* was authored by 53 individuals. I want to thank each and every one of them for their significant contribution. Their impressive bios are listed in the section "Authors." The monetary remuneration for writing a chapter in this manual is not sufficient motivation to contribute their valuable time and energy. Their motivation came from believing that they were providing a service to the community of users of geospatial data. For this service, they deserve our sincere thanks. The three associate editors, Chris Rizos for Part II (Global positioning system), Jim Campbell for Part III (Remote sensing), and Bob McMaster for Part IV (Geographic information systems), deserve our special thanks because they had to coordinate, organize, and edit all the chapters in their part of the manual in addition to writing a large number of chapters themselves. The staff at Taylor & Francis provided patient advice throughout the 16 months that spanned the development of this manual.

It is my opinion that one of the most important factors in the acceptance and use of any technical document of this kind depends on the quality of the figures or illustrations. Once again, I called on my old friend Robert (Bud) Hanson, who worked with me in the National Oceanic and Atmospheric Administration (NOAA), to take charge of the development of the figures in this manual. This was one of my best decisions! Some of the 218 figures that are placed throughout the manual have been taken from the first edition. In such cases, they have been reviewed, resized, or redrawn from scratch. Nearly all of the new figures were redrawn using the author's submittal as a draft. They are, and I believe you will agree, superb! In addition, as with the first edition, Bud served as a valuable mentor throughout the project.

Editors

John D. Bossler is a consultant; he is also Professor Emeritus at The Ohio State University, specializing in GIS, GPS, and remote sensing. He retired from the positions of professor and director of the Center for Mapping at The Ohio State University, Columbus. Previously, he was director of both the National Geodetic Survey and the Coast and Geodetic Survey. He has authored over 100 papers in the fields mentioned.

Robert B. McMaster is a professor of geography and a vice provost and dean of undergraduate education at the University of Minnesota. He received his BA (cum laude) from Syracuse University, New York in 1978 and his PhD in geography and meteorology from the University of Kansas, Lawrence in 1983. He currently serves as the president of the University Consortium for Geographic Information Science, Alexandria, Virginia and is a member of the National Research Council's Mapping Science Committee.

Chris Rizos is a professor at the School of Geomatic Engineering, the University of New South Wales (UNSW), Sydney, Australia, and is the head of the School of Surveying and Spatial Information Systems, UNSW. He has been researching precise GPS/navigation technologies and applications since the mid-1980s. He is currently the vice president of the International Association of Geodesy. He is the author of over 400 journal and conference papers, and several books relating to GPS, surveying, and geodesy.

James B. Campbell has devoted his career to applications of aerial survey and remote sensing to studies of land use, geomorphology, and soils. Since 1997, he has served as a codirector of Virginia Tech's Center for Environmental Analysis of Remote Sensing. He is the author of numerous technical articles and several books. His recent research addresses the human impact upon coastal geomorphology and the history of aerial survey. He currently is a professor in the Department of Geography, Virginia Tech, Blacksburg, Virginia.

Contributors

Jan A. N. van Aardt specializes in system state assessment of forestry and natural resources using spectral (imaging spectroscopy) and structural (LIDAR) remote sensing approaches. He has published his findings in various peer-reviewed journals. Jan worked in the academic and private (Council for Scientific and Industrial Research [CSIR], Pretoria, South Africa) sectors before joining the faculty at the Rochester Institute of Technology (RIT), New York. He currently is an associate professor in the Laboratory for Imaging Algorithms and Systems (LIAS) in RIT's Center for Imaging Science.

Hasanuddin Z. Abidin is a professor in the Geodesy Research Division, Faculty of Earth Science and Technology, Institute of Technology Bandung (ITB), Indonesia. Hasan is head of the Geodesy Research Division at ITB, where he conducts research in a range of geodetic problems using GPS technology and InSAR. These include volcano and land subsidence monitoring, active fault study, geodynamics, and GPS meteorology. Hasan is one of the foremost GPS experts in Indonesia. He has authored many papers and has written two books on GPS and one book on satellite geodesy.

Thomas R. Allen is an associate professor of geography at East Carolina University, Greenville, North Carolina. His research interests are in remote sensing, especially change detection, and GIS modeling and visualization of environmental processes, particularly in biogeographic, geomorphic, and environmental health research.

John D. Althausen is a mission officer for Lockheed Martin's IEC program. He has been actively involved in satellite imagery analysis for 20 years in both the academic (the University of South Carolina, the University of South Florida, and other universities) and private (ERDAS, Inc. and Lockheed Martin Corp.) sectors. His analysis of satellite imagery has been published in numerous professional remote sensing journals and he has presented his research at national and international conferences. He currently is a systems engineer, and a senior staff member at LMC, Gaithersburg, Maryland.

John C. Antenucci is an engineer, planner, management consultant, executive, and author. He is also the founder and current president of PlanGraphics, Inc., a firm that specializes in the design and implementation of geographic information system and related spatial technologies. He has served as president and director of the Geospatial Information & Technology Association.

Robin Antenucci has more than 20 years experience in GIS project planning and management, photogrammetric mapping, and remote sensing and design and development of GIS applications and databases. She served as senior technical consultant for the Convergent Group and as consultant at PlanGraphics, Inc. She currently is the director of sales and marketing, Frankfort Tourism Commission.

Marc Armstrong is a professor and chair of geography, an interim director of the School of Journalism and Mass Communication, and an administrative fellow at the University of Iowa. He holds a PhD in geography from the University of Illinois. Marc's research interests include collaborative decision-making, geocomputation, geovisualization, and cyberinfrastructures. He has served as the editor of the *International Journal of Geographic Information Science*.

Robert Chris Barnard is a certified photogrammetrist and consultant in remote sensing. He currently works for Science Applications International Corporation (SAIC) where he supports the Department of Homeland Security by assisting in the development of policy and strategy for the use of geospatial information. Prior to this, Barnard was the director of business development for EarthData International, a global provider of aerial mapping and remote sensing services. Barnard is a graduate of the American University, Washington, District of Columbia.

David Bennett is an associate professor of geography at the University of Iowa, where he received his PhD in geography. His research interests are in applications of GIScience in complex human–environment decision systems, and GIS and biophysical modeling. He currently serves as an associate editor of cartography and geographic information science.

Joel Campbell has spent nearly 20 years in the geospatial market as both a user and purveyor of geospatial systems and data. He has consulted for many leading GIS companies throughout the world with regard to their business, sales, and marketing strategies. Campbell has lectured internationally on the topic of geospatial and remote sensing systems. Currently, he serves on the External Advisory Board for the Center for Geospatial Information Technology (www.cgit.vt.edu), a research center at the Virginia Polytechnic Institute and State University, Blacksburg.

Russell G. Congalton is an expert in the practical application of remote sensing and spatial data analysis for natural resource management. He has developed the most widely adopted methods for assessing the accuracy of maps derived from remotely sensed images and has written numerous articles and books on this topic. He currently is a professor in the Department of Natural Resources and the Environment, James Hall, the University of New Hampshire, Durham.

Dmitry Ershov is a graduate student in the Department of Geography at the University of Iowa. He obtained his undergraduate degree from the Department of Geography and Geo-Ecology at St. Petersburg State University, Russia, where his focus was geodetic data preprocessing. His undergraduate thesis was "Development of software for the purpose of processing of tachometric measurements of architectural buildings." His current interest is on the applications of hyperspectral imagery. He currently works at the GeoTREE Center in the Department of Geography, the University of Northern Iowa, Cedar Falls.

Joe Evjen is the chief of the National Geodetic Survey's Standards and Applications branch. He has enjoyed surveying from mountaintop triangulation recoveries to charting wrecks offshore. A 1990 honors graduate from the University of Florida's surveying and mapping program, Joe helped manage the High Accuracy Reference Network (HARN) and Federal Base Network (FBN) survey projects and now focuses on improving National Geodetic Survey (NGS) products and guidelines.

Charles D. Ghilani is a professor of engineering in the surveying engineering program at The Pennsylvania State University, University Park. He received his PhD in civil and environmental engineering from the University of Wisconsin-Madison in 1989. He has been involved as an educator for 40 years and has presented workshops related to statistics and least squares to surveyors across the country. He has written numerous papers and is the author of *Adjustment Computations*: *Surveying Measurement Analysis*, 4th edition, and *Elementary Surveying*: *An Introduction in Geomatics*, 12th edition.

Debarchana Ghosh is a PhD candidate in the Department of Geography at the University of Minnesota, Minneapolis. Her research focuses on the methodology and application of geographic information science and spatial statistics to understand the spatial distribution of complex health outcomes and human–environment interactions.

Michael F. Goodchild is a professor of geography at the University of California, Santa Barbara (UCSB). He received his PhD from McMaster University, Hamilton, Canada and was on the faculty of geography at the University of Western Ontario, London, Canada before moving to UCSB in 1988. He is former director of the National Center for Geographic Information and Analysis and a member of the National Academy of Sciences.

Dorota A. Grejner-Brzezinska is a professor in the Department of Civil and Environmental Engineering and Geodetic Science, The Ohio State University, Columbus. Dorota's current research interest is in integrated multisensory data acquisition systems, which involves kinematic positioning with GPS, GPS/INS integration, personal navigation, terrain-based navigation technologies, machine learning techniques, and robust estimation techniques. In the last 15 years, she has been active in basic research as well as engineering and scientific applications of GPS.

Ayman Fawzy Habib is a teaching professor at the University of Calgary, Canada and has research interests that span terrestrial and aerial mobile mapping systems, modeling the perspective geometry of nontraditional imaging scanners, automatic matching and change detection, the automatic calibration of low-cost digital cameras and integrating photogrammetric data with other sensors/datasets (e.g., GPS/INS, multi- and hyper-spectral sensors, and LIDAR). Dr. Habib is the recipient of numerous awards, which include the Talbert Abrams Grand Awards from the American Society of Photogrammetry and Remote Sensing.

Kathryn Hail is the marketing and communications manager for the Geospatial Information & Technology Association, and has worked with Mary Ann Stewart to compile the Geospatial Technology Report since 2006. Hail's background includes work in marketing, journalism, and public relations. In 2005, she earned her master of arts degree in mass communication from Texas State University, San Marcos in 2005.

Travis E. Hardy is a consultant specializing in GIS, geospatial technology, and federal information technology policy. Hardy has focused on helping his customers analyze both system-level technical requirements as well as broader policy and governance requirements. He has published in the *Military Geospatial Technology* magazine and is a recognized professional in the geospatial industry. Currently, Travis provides full-time technical and management support services to the Department of Homeland Security's Geospatial Management Office in Washington, D.C.

Francis Harvey is an associate professor at the University of Minnesota. His research interests include semantic interoperability, spatial data infrastructures, geographic information and sharing, and critical GIS. He serves on the editorial board of the *International Journal for Geographical Information System*, *Cartographica*, *GeoJournal*, and the *URISA Journal*. He edits the GIS section of the *Geography Compass* with coeditor Elizabeth Wentz. His book, *A GIS Primer*, has recently been published by Guildford Press (2008).

William (Bill) Henning is a registered professional land surveyor with over 40 years of active experience in all phases of surveying technology. He has helped plan, construct, process, adjust, and manage height modernization geodetic networks for countywide projects in the United States. Bill is a past president of the American Association for Geodetic Surveying (AAGS) and a fellow at the ACSM/ AAGS. He is currently employed by NOAA's NGS as a senior geodesist supporting methodology for real time positioning with state, national, and international organizations.

Pierre Héroux is employed by the Geodetic Survey Division, Natural Resources Canada (NRCan). Pierre leads the GNSS technology team at the Geodetic Survey Division, NRCan. Over the last 25 years, he has been involved with the development of services and applications for precise positioning using space-based technologies. He is particularly interested in facilitating access to a common global geodetic reference frame to enable the integration of geospatial data at the highest level of accuracy in support of Earth sciences.

Stephen Hilla is a member of the Geosciences Research Division at NOAA's National Geodetic Survey where he is responsible for developing and maintaining GNSS software for use in field processing, online processing, and precise orbit determination. He is the coeditor of the GPS Toolbox column in the journal *GPS Solutions*.

Gérard Lachapelle is a professor in the Department of Geomatics Engineering, the University of Calgary, Canada. Gerard has been involved with satellite and ground-based navigation systems since 1980, and has eight years of GPS development experience in the industry. He has developed a number of precise GPS positioning methods and has coauthored several GPS software packages and GNSS signal processing techniques. He has been the recipient of some 30 awards for his contributions. Since 1988, he has managed a highly successful GPS research and development program at the University of Calgary.

Rongxing (Ron) Li is the Lowber B. Strange professor of civil engineering, professor of geoinformation and geodetic science, and director of the mapping and GIS laboratory in the Department of Civil and Environmental Engineering and Geodetic Science, The Ohio State University, Columbus. Dr. Li is a participating scientist of the Mars Exploration Rover and the Lunar Reconnaissance Orbiter missions, and a member of the Mars Exploration Program Geodesy and Cartography Working Group. He has received numerous academic and professional awards from NASA, International Society of Photogrammetry and Remote Sensing (ISPRS), American Society of Photogrammetry and Remote Sensing (ASPRS), and Chinese Professional in Geographic Information Science (CPGIS).

Chun Liu's research focus is on spatial data quality, image data processing, and integrated geographical data and GPS measurements for engineering applications. His research results have been published in the *Journal of Navigation* and other journals covering the topics of surveying and GIS. He currently is an associate professor in the Department of Surveying and Geo-Informatics, Tongji University, Shanghai, People's Republic of China.

Wolfgang Lück is a technology manager at the Council for Scientific and Industrial Research (CSIR)'s Satellite Applications Centre (SAC). He has extensive experience in operational land cover assessment and algorithm development for automated image processing using a wide range of optical and structural remote sensing systems. He is a remote sensing system specialist and is responsible for the identification and implementation of new technologies and the development of new production chains at SAC. He currently is a technology manager at CSIR's SAC.

Melanie Lück-Vogel is an environmental remote sensing expert. She specializes in environmental monitoring at different spatial scales using a variety of multispectral remote sensing data sets. She is the author of various papers and conference contributions in this field. She currently is a senior researcher, Remote Sensing, CSIR, Pretoria, South Africa.

Steven M. Manson is an associate professor of geography at the University of Minnesota. He received his PhD in geography from Clark University, Worcester, Massachusetts. Dr. Manson combines environmental research, social science approaches, and geographic information science to understand changing urban and rural landscapes in the United States and Mexico. He was a NASA Earth System Science Fellow and is a NASA New Investigator in Earth–Sun System Science.

Carolyn J. Merry's current research focuses on land cover change mapping and water quality mapping using data from various satellite systems. The remote sensing products related to land cover and water quality are used in watershed and water quality engineering models. Her research articles have been published in several journals of remote sensing. She currently is a professor and chair in the Department of Civil and Environmental Engineering and Geodetic Science, The Ohio State University, Columbus.

David Mockert is the director of local and state practice for GeoAnalytics and is responsible for leading local and state professional and services teams. Prior to joining GeoAnalytics, Mockert was the geographic information officer (GIO) and deputy chief information officer (CIO) for the State of Wisconsin. He held the position as the CIO for the City of Indianapolis/Marion County, Indiana. His private sector experience includes project management and technical consulting.

John J. Moeller is a senior principal engineer for geospatial intelligence with Northrop Grumman Corporation. Prior to this, he was the staff director for the Federal Geographic Data Committee and was a founding member of the GSDI Association. He is currently employed by Northrop Grumman Corporation.

Joel Morrison is a professor Emeritus in the Department of Geography at The Ohio State University, Columbus. He received his PhD from the University of Wisconsin, where he served as a faculty member and chair of the Department of Geography. He has also worked for the United States Geological Survey, Bureau of the Census, and was the director of the Center for Mapping at The Ohio State University, Columbus. He currently is a coeditor for the twentieth century volume of the *History of Cartography* project.

Ian Muehlenhaus is currently a PhD student at the University of Minnesota and an instructor/laboratory director in the Department of Geography at the University of Wisconsin, River Falls. He received his MA from The Pennsylvania State University, University Park. His research interests include geovisualization, cartographic design, and political geography/cartography.

Zorica Nedović-Budić earned her PhD at the University of North Carolina, Chapel Hill. Her research interests focus on the implementation of GIS in local governments and the building of spatial data infrastructures. She served on the University Consortium for Geographic Information Science (UCGIS) and Urban and Regional Information System Association (URISA) board of directors, and has published extensively in related geospatial technology journals and books. She currently is a professor at the University of Illinois at Urbana-Champaign.

Mark E. Reichardt is the president and chief executive officer of the Open Geospatial Consortium, Inc. (OGC), a voluntary consensus standards organization that provides a global collaborative forum for the development, testing, and promotion of international standards for geospatial interoperability. Reichardt has overall responsibility for the consortium's operations. He joined OGC after a 20 year career managing geospatial production and technology development programs within the U.S. federal government, including the U.S. Department of Defense, the U.S. Federal Geographic Data Committee, and Vice President Gore's Partnership for Reinventing Government.

Sam Ryan is a manager in electronics and informatics in the Canadian Coast Guard, Ottawa, Canada. Sam is the manager of electronics and informatics in the Integrated Technical Services Directorate of the Canadian Coast Guard. Most recently, he has been involved with the implementation of the long range identification and tracking system for the Canadian Coast Guard.

Vincent V. Salomonson's career extends over 40 years performing research using spaceborne remote sensing instruments. He has served as the NASA Landsat 4 and 5 project scientist and the Earth Observing System (EOS) MODIS Science team leader. His publication record shows over 130 publications in scientific journals, conference proceedings, and NASA reports. He is currently at the Departments of Meteorology and Geography, the University of Utah, Salt Lake City.

Dru Smith has been the chief geodesist of the National Geodetic Survey (NGS) since 2005. Prior to 2005, he served as a research geodesist for NGS, working on geoid modeling and ionospheric modeling. He has published extensively, and is a member of the International Association of Geodesy and the American Geophysical Union. He has also served as the vice president of the American Association for Geodetic Surveying.

J. Woodson Smith's research interests include multimodal and alternative transportation, and the spatial analysis of historical records and data. He is at present studying the early development of Kentucky's system of roads. He currently is a consultant at PlanGraphics, Inc.

Richard Snay currently serves as chief of the spatial reference system division at NOAA's National Geodetic Survey. From 1998 to 2007, he was the project manager of the U.S. Continuously Operating Reference Station (CORS) system. He served as a visiting scientist at Stuttgart University, Germany and with the Southern California Earthquake Center, Los Angeles. Dr. Snay has published over 50 papers, including articles on geodetic reference systems, crustal motion, and the Global Positioning System.

Rebecca Somers has more than 25 years experience helping government agencies, companies, and nonprofit organizations develop GIS programs, publications, and instructional resources. She has published dozens of articles, columns, monographs, and book chapters on GIS planning, development, and management. She is also a prominent GIS workshop and course instructor and speaker. She is the president of Somers-St. Claire GIS Management Consultants.

Mary Ann Stewart is a professional civil engineer with a business, Mary Ann Stewart Engineering LLC, focused on GIS implementations for government and utilities. Her focus includes market research in the utility industry, return on investment analysis for utilities and government, and the analysis of landbase issues for geospatial projects. Stewart is a former IT manager for an investor-owned utility.

Ramanathan Sugumaran's research focuses on the application of remote sensing and GIS to water quality modeling, heat loss detection, land cover mapping, and other environment-related problems. He is the author of over 50 journal and conference papers relating to GIS and remote sensing areas. He currently is an associate professor in the Department of Geography, University of Northern Iowa, Cedar Falls.

Howard Veregin is the director of geographic information services and the editor of *Goode's World Atlas* at Rand McNally, a large commercial map publisher. Prior to this, he was an associate professor of geography at the University of Minnesota–Twin Cities. He received his PhD in geography from the University of California, Santa Barbara. During his academic career he published numerous professional articles on spatial data quality.

Matthew Voss is a remote sensing researcher specializing in object-based image analysis, hyperspectral remote sensing, and applications of LIDAR. His current areas of interest are vegetation remote sensing and issues relating to crop management. He currently is a remote sensing scientist, the STORM Project and GeoTREE Center, Department of Geography, University of Northern Iowa, Cedar Falls.

Yong Wang's research interests include the studies of responses and variations of shorelines and coastal wetlands to changes of environments and climate, and to sea level rise; landcover types and landuse changes caused by nature and human disturbance; microwave canopy backscatter numerical modeling; retrieving forest physical parameters by model inversion; and geographic information sciences and analyses, and image processing and analysis. He currently works in the Department of Geography, East Carolina University, Greenville, North Carolina.

Timothy A. Warner's research focuses on the spatial properties of images, high-resolution images, LIDAR, and land cover change detection. He is a coeditor of the Letters section of the *International Journal of Remote Sensing*, and a member of the editorial board of *Geography Compass*. He was awarded the 2006 Outstanding Contributions Award from the Association of American Geographers (AAG) Remote Sensing Specialty Group (RSSG). He currently is a professor of geology and geography, West Virginia University, Morgantown.

May Yuan is the Edith Kinney Gaylord presidential professor and Brandt professor of atmospheric and geographic sciences, as well as being the director of the Center for Spatial Analysis at the University of Oklahoma, Norman. May's research interest is in temporal GIS, geographic representation, spatiotemporal information modeling, and applications of geographic information technologies to dynamic systems. Her research projects center on representation models and developing algorithms for spatiotemporal analysis.

Part I

Prerequisites

1 An Introduction to Geospatial Science and Technology

John D. Bossler

CONTENTS

1.1 ABOUT THIS BOOK

This book is written for professionals working in the private sector, various levels of government, and government agencies that are faced with, or require knowledge of, the tasks involved in performing an area-wide mapping, inventory, and data conversion and analysis project. Such projects are commonly referred to as geographic information system projects.

The book covers the three sciences and technologies needed to accomplish such a project—the Global Positioning System (GPS), remote sensing, and Geographic Information Systems (GIS). These three disciplines are sometimes called the geospatial information sciences. This book covers the basic mathematics and physics necessary to understand how these disciplines are used to accomplish mapping, inventory, and data conversion. Examples of the basics covered in Part I of this book include coordinate systems, coordinate transformations, datums, electromagnetic radiation, atmospheric disturbances, basic statistics, and data adjustment.

The book addresses the background, how-to, and frequently asked questions. Questions such as "What GPS equipment should I use?," "What accuracy is required?," "What scale should I use?," "How do I buffer this river?," "What satellite data are available?," and "What spectral bands can I obtain, and which should I use?" are typical of the questions addressed in this book. It contains 39 chapters written by 54 authors, most of whom are widely known experts in the field. The authors and editors hope that the book will make decision making easier and better for the professionals using this book.

1.2   GEOSPATIAL DATA

Features shown on maps or those organized in digital databases that are tied to the surface of the earth by coordinates, addresses, or other means are collectively called geospatial data. These data are also called *spatial* or *geographic* data. Almost 80% of all data are geospatial data. A house whose address is provided or a geodetic control monument with its latitude and longitude are examples of geospatial data. An example of data that is not geospatial is a budget for an organization.

Geospatial data can be acquired by digitizing maps, by traditional surveying, or by modern positioning methods using GPS. Geospatial data can also be acquired using remote sensing techniques, that is, by airborne or spaceborne platforms. After such data are acquired, they must be organized and utilized. A GIS serves that purpose admirably by providing organizing capability through a database and utilizing query capability through sophisticated graphics software. The reader should now understand the intent of this book—to help understand the process involved in modern map making and information age decision making.

If we define the process needed to make decisions using geospatial data as (1) acquiring data, (2) analyzing and processing data, and (3) distributing data, then this book provides significant material about the first two steps.

The book does not address the issue of distributing geospatial data.

1.3   SPATIAL DATA INFRASTRUCTURE

A large number of futurists believe that we are probably in the middle of the information age. The ramifications of this assertion require us to collect, process, manage, and distribute geospatial and other data. In the United States, we have defined a *process* (*not* an organization!) called the National Spatial Data Infrastructure (NSDI), which is comprised of the people, policies, information, technology, and institutional support needed to utilize geospatial data for the enhancement of society (NAS/NRC, 1993). The term NSDI has been generalized to SDI, which includes global data infrastructures. For an in-depth discussion of SDIs, see Chapter 37. This book describes the three nascent information technologies that are central to the NSDI process because they allow us to *acquire* digital information and to *process* and *analyze* these data.

The NSDI and in particular the technologies described in this book, along with the Internet, are now essential for an extremely wide variety of applications. Today, one of the most important applications is probably the management of our environment. Other important applications include sustainable economic development, taxation, transportation, public safety, and many more. While we have attempted to describe these three technologies in a broad manner, we have deliberately focused our description of these technologies on natural resource questions, land use planning, and other issues concerned with the land and its use, because in our opinion, there is significant "GIS growth" associated with land issues, and we believe that there is a need for educational material in this arena.

1.4 RELATIONSHIP OF GEOSPATIAL SCIENCE TO OTHER DISCIPLINES

The foundation of the geospatial sciences, especially those topics covered in this book, is mathematics, computer science, physics, and engineering. Depending on the depth and breadth of study, the practitioner may find biology, cartography, geodetic science, geography, geology, and surveying very helpful.

Table 1.1 shows the basic educational needs when working with each of the topics covered in this book. The table is highly subjective, and all topics require some knowledge of all the educational subjects shown. However, if the practitioner wants to acquire fundamental *background* information about the topics discussed in this book, Table 1.1 provides a reasonably general guidance. For continuing education purposes, most universities offer courses in these fundamentals and many provide specific courses in GPS, remote sensing, and GIS. Consider, for example, the universities represented by the editor and the three associate editors of this book. They all offer courses in these subjects.

1.5 THREE IMPORTANT GEOSPATIAL TECHNOLOGIES

This book covers the topics of GPS, remote sensing, and GIS. It is important, therefore, that we discuss each of these briefly here and refer the reader to the individual part of this book covering this material and to other writings.

1.5.1 GPS/GNSS

It is probably true that no other technology has affected the surveying, mapping, and geodesy professions as profoundly as GPS. The purpose of this book is to introduce practitioners to the principles of this technology so that they can understand its capability, accuracy, and limitations. In addition, it is necessary to understand the implications of the improvements to GPS, and the impacts of the proliferation of other satellite-based positioning systems on users. During the second decade of the twenty-first century, we will enter the era of multiple Global Navigation Satellite Systems (GNSS). GNSS is an umbrella term that will be increasingly used. GNSS includes

TABLE 1.1
Educational Background Needed for Geospatial Science and Technology

Topic	Educational Needs in Approximate Order of Relevance
GIS	Computer science, geography, cartography, surveying, operations research
GPS	Geodetic science, mathematics, physics, surveying, celestial mechanics
Remote sensing	Physics, mathematics, computer science, engineering, geology/biology

the modernized GPS, the Russian GLONASS, the European Union's Galileo, and the Chinese COMPASS navigation satellite constellations and services.

The single-most powerful feature related to GPS (and GNSS, in general), which is not true of traditional surveying techniques, is that its use does not require a line of sight between adjacent surveyed points. This factor is very important in understanding the impact that GPS/GNSS has had (and will continue to have) on the geospatial communities.

GPS has been used by the surveying and mapping community since the late 1970s, when only a few hours of satellite coverage was available. It was clear even then that centimeter-level accuracy was obtainable over very long baselines (hundreds of kilometers). In the early 1980s, users of GPS faced several problems: the cost of GPS receivers; poor satellite coverage, which resulted in long lengths of time at each survey location; and poor user–equipment interfaces. Today, instantaneous measurements with centimeter-level accuracy over tens of kilometers and with one part in 10^9 accuracy over nearly any distance greater than 10 km can be made. The cost of "surveying and mapping-level" receivers in 2008 ranged from $10,000 to $30,000.

Traditional land surveying is increasingly being accomplished using GPS because of the increase in user-friendliness. This trend toward the use of GPS for surveying and mapping has enhanced the volume of survey receiver sales, and will accelerate as multi-constellation GNSS becomes a reality.

The standard surveying and mapping activities that are enhanced by GPS/GNSS include surveying for subdivisions, highways, planned communities, etc. These will no doubt continue to be important applications. However, the use of real-time GPS/GNSS means that this technology will be increasingly used for machine guidance and control (vehicles in open-cut mining, construction, agriculture, etc.), deformation monitoring (civil structures such as bridges, dams, tall buildings, etc.), and as the crucial technology for direct georeferencing of terrestrial, airborne and satellite-borne imaging, and scanning (NAS/NRC, 1995).

1.5.2 REMOTE SENSING

The practice of remote sensing began almost 100 years ago, when, during World War I, the marriage of the airplane and the camera provided the capability to image the earth's surface from the sky. Within a few decades, this technology developed into an operational capability based on cameras tailored for use in aircraft, specialized films, and photogrammetric instruments that could extract accurate spatial measurements from aerial imagery. By the 1970s, such images could be acquired in multiple regions of the electromagnetic spectrum, using new imaging technologies, satellite platforms, and digital analyses. Today, remote sensing technologies have been integrated with those of GPS and GIS to provide systems with exceptional positional accuracy and analytical capabilities.

We can now call upon a family of varied remote sensing instruments that provide a broad range of capabilities that, directly or indirectly, contribute to the operations of industry, commerce, agriculture, and government. Imagery acquired by satellite systems designed for broadscale observations offer global views that document the impact of earth surface processes, land use changes, and climatic patterns. At the opposite extreme, high-resolution imagery provide the detail necessary to plan urban systems, monitor agricultural productivity, assess the impact of natural disasters, guide public safety operations, and support efficient operation of private businesses. Additional capabilities are offered by intermediate-resolution systems, such as the LANDSAT-class satellite systems that routinely acquire wide views at intermediate levels of detail to depict (for example) hydrologic processes, natural resource patterns, and the growth of urban regions.

Even a casual user of the Internet is familiar with the systems such as Google Earth that use aerial imagery as backgrounds for applications devoted to navigation, real-estate analysis, marketing, health care planning, and public safety operations. Such innovations have brought remotely sensed imagery from the exclusive realm of the specialist into the broader province of public access and applications. Further, the Internet has opened access to image archives, rapid delivery of imagery, and access to technical information and training materials that expand access to the field of remote sensing.

Our purpose here is to inform the reader of the capabilities of remote sensing analysis; provide the background to acquire, manage, and understand remotely sensed imagery; and to highlight the interrelationships between remote sensing, GPS, and GIS technologies. Although the analysis of imagery requires specialized software and expertise, the basic image background forms a valuable resource for nonspecialists.

1.5.3 GIS

Prior to the 1960s, there was no efficient way to manage natural or man-made resources associated with large areas of land. During the 1960s and 1970s, computer capability increased significantly and graphics capability became feasible and even commonplace. Today GIS are used extensively to perform spatial analysis related to many resources. Examples of the applications supporting natural resource management, equitable taxation, environmental monitoring, and civil infrastructure are numerous and are discussed in Parts IV and V. To understand how a GIS can be used to benefit society, consider a demographic analysis of recent crime increases. The areas of increased crime can be tabulated, displayed graphically, and covered with helpful overlays, and then increased police patrols can be dispatched accordingly.

Fundamentally, a marriage of database management systems with graphics capability, GIS is designed to allow for changes in the activities of individuals, in the processes of organizations and changes in the data. Therefore, they are able to serve the complete spectrum of individuals, from political appointee to supervisor to police officer. It is now an indispensable tool for policy makers as well as technicians.

GIS has continued to develop over the past decade, with the advancement in the core areas of spatial data acquisition, spatial data models focusing on object-based approaches, spatial analysis and modeling, and rapid expansion in methods for visualizing both geographical data and processes. Particular advancements have been made in several areas, including mobile GIS and "ubiquitous mapping," continued work on understanding scale and the creation of multiscale databases, and a new area called public participation GIS (PPGIS), which emerged from basic research and in geographic information science/systems. PPGIS has nurtured research in the democratization of GIS for grassroots organizations and community groups. Some examples are local in nature whereas others, such as Google Earth, allow for access to spatial data at the global scale. This global utilization provides us with a potential for significant impacts on society.

Mobile mapping has been made possible with the rapid advances in technology, including the ability to put GPS chips in cell phones, PDAs, and other handheld devices. Commensurate with this has been the development of spatial databases that can be used by such technologies, thus spatially enabling most of our society. An example is the proliferation of vehicle navigation systems. Research in the development of multiscale databases and cartographic generalization has also made rapid progress, with new approaches for generalization and several operational systems that allow for true multiscale databases.

REFERENCES

National Research Council, National Academy of Sciences (NAS/NRC). *Toward a Coordinated Spatial Data Infrastructure for the Nation.* National Academy Press, Washington, DC, 1993.

National Research Council, National Academy of Sciences (NAS/NRC). *The Global Positioning System, A Shared National Asset.* National Academy Press, Washington, DC, 1995.

2 Coordinates and Coordinate Systems

John D. Bossler

CONTENTS

In the following sections, the fundamentals of coordinates and coordinate systems are described. A description of the most important and commonly used coordinate system associated with Global Positioning System (GPS), Geographic Information Systems (GIS), and Remote Sensing (RS) usage is provided.

2.1 RECTANGULAR COORDINATES

Consider two straight lines *OX* and *OY* intersecting at a right angle (Figure 2.1). On each line, we construct a scale using the point of intersection *O* as a common origin. Positive numbers are generally shown to the right, on horizontal line *OX* and upward on *OY*. Consider the point *P* in the plane of the two axes. The distance of the point *P* from the y-axis (line *OY*) is called the abscissa of the point, denoted by *x*, and its distance from the x-axis (line *OX*) is called the ordinate and is denoted by *y*. In this case, the abscissa and ordinate together are called the *rectangular coordinates* of the point *P*. In Figure 2.1, *P* has coordinates (−2, 4), the abscissa, $x = -2$, and the ordinate, $y = +4$. The distance *d* is equal to

$$\sqrt{x^2 + y^2} = \sqrt{4 + 16} = \sqrt{20}$$

by trigonometry. The distance between any two points in general is equal to

$$\sqrt{(x_j - x_i)^2 + (y_j - y_i)^2}$$

where the subscripts *j* and *i* refer to point *j* and point *i*, respectively.

9

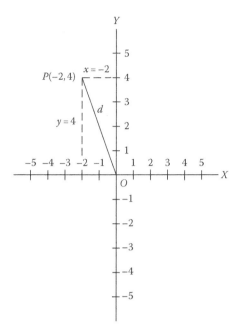

FIGURE 2.1   Rectangular coordinate system.

2.2   POLAR COORDINATES

Rather than locating a point by its distances from two intersecting lines, we may locate it from its distance and direction from a fixed point. Consider a fixed point O, which is called the pole and a fixed line OX, which we call the polar axis (Figure 2.2).

Suppose now that an angle θ has been generated by the rotation of a line from initial coincidence with the polar axis and let P be a point on the terminal side of the angle. If we denote OP by ρ, the polar coordinates of P are given by ρ and θ, that is, $P(\rho, \theta)$. Theta, θ, is called the vectorial angle and ρ the radius vector. If θ has been generated by a counterclockwise rotation, it is usually regarded as positive; if it has been generated by a clockwise rotation, it will be regarded as negative. The radius vector will be regarded as positive or negative as the point P lies in the direction determined by the vectorial angle or backward through the origin. If theta is allowed to take on any positive or negative value, greater or less than 360°, a point can have an infinity of polar coordinates; for example, (3, 60°), (−3, 240°), and (3, −300°) all represent the same point.

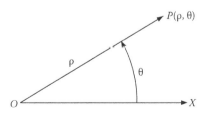

FIGURE 2.2   Polar coordinate system.

2.3 TRANSFORMATION BETWEEN POLAR AND RECTANGULAR COORDINATES

If the polar axis coincides with the positive x-axis of a rectangular system of coordinates, the pole being at the origin, the relationship between the two coordinate systems is readily obtained using trigonometry (see Figure 2.3), as follows:

$$x = \rho \cos \theta, \quad y = \rho \sin \theta, \qquad (2.1)$$

$$\rho = \pm\sqrt{x^2 + y^2}, \quad \theta = \tan^{-1} y/x \qquad (2.2)$$

2.4 RECTANGULAR COORDINATES IN THREE DIMENSIONS

To locate a point in space, three coordinates are necessary. Consider three mutually perpendicular planes that intersect in lines OX, OY, and OZ, as shown in Figure 2.4. These lines are mutually perpendicular. The three planes are called the coordinate planes (xy plane, xz plane, and yz plane); the three lines are called the coordinate axes (x-axis, y-axis, and z-axis), and the point O is the origin. A point P (x, y, z) may be located by providing the orthogonal distances from the coordinate planes to the point P.

Again, the quantities x, y, and z are the rectangular coordinates of the point P. The positive and negative directions of the axes are somewhat arbitrary. Most mathematical textbooks use a left-handed coordinate system, but practitioners in the geospatial community use right-handed systems.

Definition: A rectangular coordinate system in three dimensions that has the positive directions on the three axes (x, y, and z) is defined in the following way: If the thumb of the right hand is imagined to point in the positive direction of the z-axis and the forefinger in the positive direction of the x-axis, then the middle finger, extended at

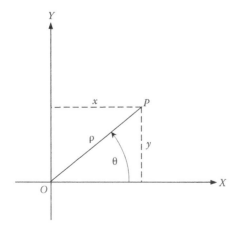

FIGURE 2.3 Relationship between rectangular and polar coordinates.

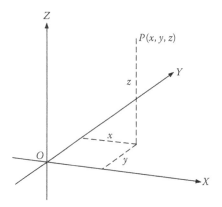

FIGURE 2.4   Rectangular coordinates in 3D.

right-angles to the thumb and forefinger, will point in the positive direction of the
y-axis (Geodetic Glossary 1986).

If the coordinate system is left-handed, the middle finger will point in the negative
direction of the y-axis.

The use of GPS and other geometric procedures has fostered the use of rectangular
coordinate systems for geospatial endeavors.

2.5   SPHERICAL COORDINATES

A point in space can also be located by the length ρ of its radius vector from the origin, the angle ϕ, which this vector makes with the xy-plane and the angle λ from the
x-axis to the projection of the radius vector in the xy-plane. The coordinates (ρ, ϕ, λ)
are called spherical coordinates (see Figure 2.5).

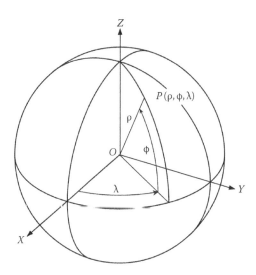

FIGURE 2.5   Spherical coordinate system.

The relationship between spherical and rectangular coordinates is easily derived from the figure as follows:

$$x = \rho \cos\phi \cos\lambda \tag{2.3}$$

$$y = \rho \cos\phi \sin\lambda \tag{2.4}$$

$$z = \rho \sin\phi \tag{2.5}$$

The reader may recognize the angles ϕ, λ as the familiar latitude and longitude used on globes, maps, etc.

2.6 ELLIPSOIDAL COORDINATES

A reasonable approximation of the surface of the earth is found by rotating an ellipse that is slightly flattened at the poles, around its shortest (minor) axis. Such a figure is called an ellipsoid of revolution or simply an ellipsoid. It is defined by its semimajor axis a and semiminor axis b (see Figures 2.6 and 2.7).

The equation governing the ellipsoid obtained by rotating Figure 2.6 about axis b is

$$\frac{x^2 + y^2}{a^2} + \frac{z^2}{b^2} = 1 \tag{2.6}$$

The above equation follows directly from the definitions of an ellipse (Zwillinger 1996, Swokowski 1998) and an ellipsoid with two of its three orthogonal axes equal in length. To facilitate the development of additional equations and relationships, it is convenient to define the following quantities:

$$f = \frac{a - b}{a}, \quad e = \sqrt{\frac{a^2 - b^2}{a^2}} = \sqrt{2f - f^2} \tag{2.7}$$

where
  f is defined as the flattening
  e is defined as the first eccentricity

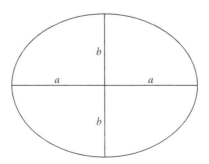

FIGURE 2.6 Ellipse.

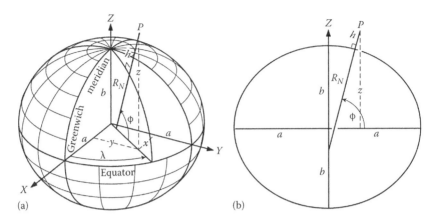

FIGURE 2.7   Ellipsoidal coordinate system.

Textbooks in geodesy develop the equations relating rectangular coordinates to ellipsoidal coordinates (Bomford 1980, Torge 1980), and this development will not be repeated here. The following equations refer to the right-handed coordinate system shown in Figure 2.7a. Figure 2.7b is a cross section of the ellipsoid shown in Figure 2.7a.

$$x = (R_N + h)\cos\phi\cos\lambda \tag{2.8}$$

$$y = (R_N + h)\cos\phi\cos\lambda \tag{2.9}$$

$$z = \left(\frac{b^2}{a^2} R_N + h\right)\sin\phi \tag{2.10}$$

where

$$R_N = \frac{a^2}{(a^2\cos^2\phi + b^2\sin^2\phi)^{1/2}} = \frac{a}{(1 - e^2\sin^2\phi)^{1/2}} \tag{2.11}$$

R_N is the radius of curvature in the prime vertical (Torge 1980, p. 50). The GRS 80 reference ellipsoid used in the North American Datum, NAD 83, has values of a and b as

$$a = 6,378,137.0\,\text{m}, \quad b = 6,356,752.314\,\text{m}$$

This reference ellipsoid is now used throughout the world.

2.7   STATE PLANE COORDINATE SYSTEMS (UNITED STATES)

Using the curved reference ellipsoid is awkward and time-consuming for carrying out mapping, surveying, and GPS projects, so most countries use plane coordinate systems for those activities. The equations and concepts in Section 2.1 can then be used.

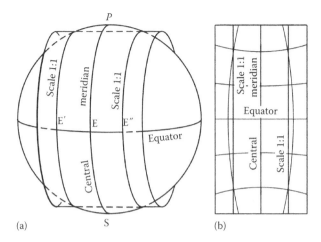

FIGURE 2.8   Transverse Mercator projection.

In the United States, the individual states are of such size that one or two plane coordinate systems can be used to cover an entire state, which allows plane surveying to be used over relatively great distances while maintaining the simplicity of plane rectangular x and y coordinates. With such a system, azimuths are referred to grid lines that are parallel to one another. Two types of projections are used in developing the state plane coordinate systems: the transverse Mercator projection and the Lambert conformal projection. The transverse Mercator projection employs a cylindrical surface (see Figure 2.8). The distortions in this system occur in an east–west direction; therefore, the projection is used for states, such as New Hampshire, that have relatively short east–west dimensions. The Lambert conformal projection uses a conical surface, the axis of which coincides with the ellipsoid's axis of rotation and which intersects the surface of the ellipsoid along two parallels of latitude that are approximately equidistant from a parallel lying at the center of the area to be projected (see Figure 2.9). The distortions occur in a north–south direction, and

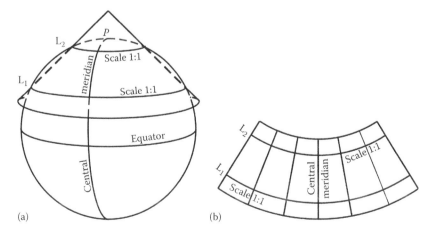

FIGURE 2.9   Lambert conformal projection.

therefore the projection is used for states, such as Tennessee, with relatively short north–south dimensions. One exception to this rule exists. The state of Alaska uses an oblique transverse Mercator.

2.8 UNIVERSAL TRANSVERSE MERCATOR

Because of its popularity worldwide and its use by geospatial practitioners, a brief discussion of this projection is appropriate. The universal transverse Mercator (UTM) is simply a transverse Mercator projection to which specific parameters, such as central meridians, have been applied. The UTM covers the earth between latitudes 84° north and 80° south. The earth is divided into 60 zones each generally 6° wide in longitude. The bounding meridians are therefore evenly divisible by 6° and zones are numbered from 1 to 60 proceeding east from the 180th meridian from Greenwich, with minor exceptions. Tables and other easy-to-use data and information for the UTM are widely available (Snyder 1987).

The equations governing state plane coordinates, both the Lambert conformal and the transverse Mercator, are provided in Chapter 4. Several books are valuable as references. To cover the background of the systems, tables of projection systems, and their use in surveying, see Moffitt and Bossier (1998). To understand the development of the mapping equations, the reader is referred to Snyder (1987).

REFERENCES

Bomford, G., *Geodesy* (4th edn.), Clarendon Press, Oxford, U.K., 1980.

Geodetic Glossary, *National Geodetic Survey*, United States Governmental Printing Office, Washington, DC, 1986.

Moffitt, F. and Bossier, J., *Surveying* (10th edn.), Addison Wesley Longman, Menlo Park, CA, 1998.

Snyder, J.P., *Map Projections—A Working Manual*, United States Government Printing Office, Washington, DC, 1987.

Swokowski, E.W., *Calculus and Analytic Geometry* (2nd alternate edn.), PWS Kent Publishing Co., Boston, MA, 1998.

Torge, W., *Geodesy*, Walter de Gruyter, New York, 1980.

Zwillinger, D. (ed.), *Standard Mathematical Tables and Formulae* (30th edn.), CRC Press, New York, 1996.

3 Datums and Geospatial Reference Systems

John D. Bossler and Richard Snay

CONTENTS

This chapter introduces and defines the concepts of *datums* and *geospatial reference systems*. It also discusses various applications involving datums and geospatial reference systems and, in tabular form, provides a list of the major datums used in the world as of this writing.

A datum, in general, is any quantity or set of quantities that may serve as a referent or basis for the calculation of other quantities.

3.1 GEODETIC DATUMS

According to the Geodetic Glossary (1986), a geodetic datum is "a set of constants specifying the coordinate system used for geodetic control, i.e. for calculating coordinates of points on the earth." Geodetic datums have also been referred to as geospatial reference systems. Datums may be classified as geometric, geopotential or some combination thereof. A purely geometric datum is concerned only with those spatial relationships between terrestrial points that are independent of the Earth's gravity field, for example, (a) the straight-line distance between two points, (b) the length of the geodesic connecting two points on an ellipsoidal surface, and (c) the angles formed by the sides of a triangle. Geopotential datums, on the other hand, address those spatial relationships that relate to the Earth's gravity field.

3.1.1 HORIZONTAL DATUMS

If surveying, mapping, and geodetic activities are carried out over large areas, the use of plane coordinates becomes impractical. Therefore, for geodetic purposes, especially, the earth is considered as a whole body. Prior to about 1650, the earth was assumed to be spherical in shape. However, the results of using more accurate measuring instruments, combined with increased knowledge of earth physics, yielded the fact that an ellipsoid of revolution better approximated the figure of the earth (see Section 2.6).

During the period from about 1700 to 1850, numerous measurements were made along arcs of meridians to verify the ellipsoidal assumption and then to determine the flattening of the ellipsoid. An accounting of these activities can be found in Torge (1980). Once the ellipsoidal figure of the earth was accepted, it was logical to adopt an ellipsoidal coordinate system. Then, the problem was to establish these coordinates, that is, to assign coordinates to a monument on the surface of the earth.

In the days before the "deflection of the vertical" (Geodetic Glossary 1986) was known or understood, astronomic positions were used to establish these coordinates. Later, in the pre-satellite era, using classical geodetic techniques, the purely geometric horizontal coordinates, geodetic latitude and longitude, defined a horizontal datum. Current horizontal datums, consequently, belong in the class of geometric datums. For the most part, horizontal datums have been defined on a country-by-country basis, but sometimes, neighboring countries have adopted a common horizontal datum. This was usually accomplished on a country-by-country basis. A few examples of such datums are the North American Datum (NAD) 1927, European Datum (ED) 1950, Japanese Geodetic Datum (JGD) 2000, Australian Geodetic Datum (AGD) 1966, and South American Datum (SAD) 1969.

At least eight quantities are needed to form a complete datum—three to specify the location of the origin of the coordinate system, three to specify the orientation of the coordinate system, and two to specify the reference ellipsoid. Moritz (1980) provides a discussion of recent determinations of the size of the reference ellipsoid.

The question of how many local horizontal datums defined under the classical definition are in use in the world may never be answered correctly. A very good attempt to list the datums in use today can be found in DIA (1971).

However, the problem with this list is that a number of datum names are either generic or extremely localized; in such cases, it becomes practically impossible to associate them with a country or any region, for example, the Red Bridge Datum. In some cases, there are multiple names listed for the same datum, for example, Leigon Pillar (GCS No. 1212), Accra, and Ghana, and the listed names may not strictly relate to a geodetic datum or may refer to the "approximate" versions of the same datum.

Due to "localized" time-saving readjustments, many local datums have been developed inadvertently within or in parallel to the original datums. These "pseudo" datums then exist concurrently with other recognized coordinates, which are different for the same control points. Unless the two sets of coordinates indicate systematic biases and/or show very large differences, it would not be possible for users to suspect their existence. An interesting case is when two or more countries define datums with the same ellipsoid and name, but use different defining parameters. Fortunately, this case is easy to identify and the chances of any mix up are almost nil.

3.1.1.1   Horizontal Datums with Multiple Ellipsoids

In some cases, a local datum already defined by an ellipsoid was later redefined using a different ellipsoid. Two such datums are the Old Hawaiian Datum and the Timbalai Datum 1948.

In the case of the Old Hawaiian Datum, the two ellipsoids, namely, Clarke 1866 and International 1924, were assumed to be tangent at an arbitrarily selected point. Then, polynomials were developed to establish transformations between the two coordinate sets based on the two reference ellipsoids.

3.1.2   Vertical Datums

Traditionally, vertical datums have been members in the class of geopotential datums, because geospatial professionals and others are concerned with determining the direction in which water will flow. Important applications of vertical datums include concerns about flooding and crop irrigation. If the Earth were a perfectly spherical, nonrotating solid body of homogeneous density, then a droplet of water on its surface would remain at rest. If a mountain, however, was suddenly placed on the surface of this spherical body, then the droplet would start flowing toward this mountain, because of the resulting change in the gravitational field. Hence, geospatial professionals have introduced several different measures of "height" that are designated to quantify the concept that water will tend to flow from a higher height to a lower height. The Helmert orthometric height of a terrestrial point, for example, is the distance along the curve of the plumb line from the geoid to this point, where the geoid is defined as the equipotential surface of the Earth's gravity field which best fits, in a least squares sense, the Mean Sea Level (MSL) (Geodetic Glossary 1986). Other popular concepts related to heights are discussed in a series of articles by Meyer et al. (2006) and in books such as that written by Heiskanen and Moritz (1967).

In the past, geodesists could not determine the geoid accurately enough over land areas for it to serve as a reliable reference surface. Hence, historically, the MSL was used as the vertical datum. The sea level was (and still is) monitored and measured at Tidal Bench Marks (TBM) for a minimum period of 18.67 years to compute the MSL. Today, determinations of the geoid by many scientists in different countries around the world have yielded adequate accuracies so that the geoid can be used as a vertical datum over land areas (Lemoine et al. 1997).

In the past, such vertical datums were defined separately from the horizontal datum—even in the same country. A few examples are the National Geodetic Vertical Datum (NGVD) 1929, the North American Vertical Datum (NAVD) 1988, the Genoa 1942, and the Bluff 1960.

The reference system for heights used in North America today is the NAVD 88. The NAVD 88 heights are the result of a mathematical (least squares) adjustment of the vertical control in North America. Over 500,000 permanent benchmarks are included in the datum. The datum surface is an equipotential surface that passes through a point on the International Great Lakes Datum. For further information about the NAVD 88, see Zilkoski et al. (1992).

3.2 GEOSPATIAL REFERENCE SYSTEMS

Soon after the launch of the first man-made satellite in 1957, geodesists defined geodetic coordinates with respect to a 3D rectangular coordinate system whose origin is located at the center of mass (geocenter) of the earth. Using a mean earth (best fitting) ellipsoid, these coordinates can be converted to geodetic latitude, longitude, and height above the ellipsoid using the inverse of Equations 2.8 through 2.10. Figure 2.7a and b shows the relationship between rectangular coordinates and geodetic latitude and longitude.

The concept of a 3D geospatial reference system is closely related to a datum in that it belongs to the class of purely geometric datums. A 3D geospatial reference system reflects the evolution of a datum given changes in technology. The introduction of Doppler and Global Positioning System (GPS) technologies led the way to a more global, rather than local, system. More importantly, these technologies demanded that the definitions of the coordinate system(s) be well defined since the accuracy of GPS, for example, is 1 part in 10^8 of the distance between points. This implies that the orientation, size, and scale of the (reference) coordinate system should be even more accurate.

Therefore, the modern 3D geospatial reference system can be viewed as an evolved (updated) datum.

3.2.1 MODERN GEOMETRIC REFERENCE SYSTEMS

The practitioner in the geospatial sciences in most developed countries in the world will be faced with the possibility of using several reference systems. Therefore, it is important that the differences between these possibilities be discussed. North America will be used as an example. The fundamental concept behind the three reference systems that are discussed is nearly identical. However, the implementation (realization) of these coordinate systems is different (Snay and Soler 1999). The reason for this is straightforward. In order to implement these systems, measurements had to be obtained. Even if the scientists involved used the same instruments, times of observations, etc., they would not achieve the same results. The measurements would be contaminated by small errors and hence different results would be obtained.

The three reference systems that are in use in North America are the North American Datum of 1983 (NAD 83), the World Geodetic System of 1984 (WGS 84), and the International Terrestrial Reference System (ITRS). The practical result of the above discussion is that the coordinates of a point on the ground would be different in each of the above-mentioned systems. The magnitude of this difference is as much as several meters; however, the difference in the distances between two points in each system would be quite small. In general, the differences between quantities that tend to be independent of the coordinate system will be very small.

For many applications in mapping and Geographic Information Systems (GIS) development, the coordinate differences between these systems may not be important. For geodetic control work, nearly all differences are important.

3.2.1.1 Reference System Definition

Two broad aspects of the three reference systems will be discussed: the reference ellipsoid and the orientation of the system. The *reference ellipsoid* for each of the above reference systems is shown in Table 3.1.

The orientation of a reference system is complicated. This is because the rotation and wobble of the earth, at the level of centimeters, is somewhat irregular. For purposes of this book, the *z*-axis of the rectangular coordinate system is aligned with the North Pole. However, because of the small irregularities of the earth's motions and the increased accuracies of the instrumentation used to measure the motion, there is a need for something more precise than "North Pole." This refinement is the International Reference Pole (IRP) as defined by an organization known as the International Earth Rotation and Reference Services (IERS). The IERS is headquartered in Paris, France. The *x*-axis of the right-handed system passes through the point of zero longitude located on the plane of the equator. The zero longitude is also defined by the IERS. This meridian passes through Greenwich, England, although this fact has more meaning historically today than it does scientifically.

The origin of the coordinate system is at the geocenter, and this has probably been accomplished to centimeter-level accuracy. It should be pointed out that the landmasses of the earth are in a state of constant motion and that even the center of mass moves slightly with respect to the solid earth. These motions are at the centimeter- or millimeter-per-year level and are of interest, generally speaking, only to earth scientists. To them, not only are the coordinates important, but so are the velocities of the points. In Chapter 7, the coordinates obtained using GPS technology are discussed using the term Earth-Centered-Earth-Fixed (ECEF) coordinates. The three systems discussed in this chapter are ECEF systems. Figure 3.1 describes the coordinate system utilized by all three of the geodetic reference systems that are described above. The interested reader is referred to Kouba and Popelar (1994) and McCarthy (1996) for further details.

TABLE 3.1
Reference Ellipsoid Parameters for Three Reference Systems

Reference System[a]	Reference Ellipsoid Semimajor Axis (m)	Reference Ellipsoid Semiminor Axis (m)	Flattening, $f(f = (a - b)/a)$
NAD 83	6,378,137.0	6,356,752.314	1/298.257222101
WGS 84	6,378,137.0	6,356,752.314	1/298.257223563
ITRS	6,378,136.49	6,356,751.750	1/298.25645

[a] The semimajor axis and the flattening are given quantities. The semiminor axis is computed from them and rounded to the nearest millimeter.

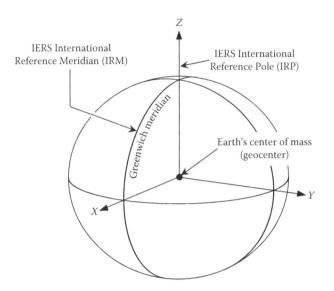

FIGURE 3.1   Geodetic reference system.

3.2.1.2   3D Geospatial Reference Systems Used in North America

It is important to understand the context of the general usage of each of these systems. The NAD 83 system is used by surveyors, geodesists, mappers, remote sensing users, and those involved in GIS. Generally, these professionals occupy, with various instrumentation, the NAD 83 geodetic reference stations whose coordinates are known in the NAD 83 system. They make additional measurements that are computed using the NAD 83 system.

The ITRS is the most rigorously defined 3D geospatial reference system. The IERS has introduced new realizations of the ITRS every few years by using measurements from various space-based observational systems including GPS, Very Long Baseline Interferometry (VLBI), Satellite Laser Ranging (SLR), and Doppler Orbitography by Radiopositioning Integrated on Satellite (DORIS). The ITRS was the first global reference frame to address crustal motion by assigning a 3D constant velocity, as well as 3D positional coordinates referred to a designated reference date, to each of its reference stations. The ITRS velocity field is currently defined by the condition that the average horizontal motion over the entire surface of the Earth is to be zero.

Because of the rigor that has gone into defining the ITRS, this reference system has become the standard to which other 3D geospatial reference systems are being related. Indeed, Canada and the United States jointly agreed to relate the current NAD 83 realization to the ITRS via a mathematical transformation (Soler and Snay 2004). Hence, the NAD 83 coordinates can be rigorously converted to corresponding ITRS coordinates and *vice versa*.

The U.S. Department of Defense (DoD) introduced WGS 84 in the mid-1980s. Originally, WGS 84 was designed to agree closely with NAD 83, as their first realizations were based on the TRANSIT (Doppler) satellite positioning system. In the early 1990s, SLR and GPS observations revealed that the origins of both NAD 83 and WGS 84, as had been deduced from the Doppler observations, were located more than 2 m from the true geocenter. Hence, DoD has redefined WGS 84 such that its subsequent realizations are each consistent with the most recent ITRS realization available at the time when each new WGS 84 realization is introduced. WGS 84 is extremely popular around the world because of the many applications of GPS. GPS uses the WGS 84 system.

There are a few important summary points to make. All three reference systems are very similar. Distances between points as determined in each of these references systems differ by parts in 10^8. All three will probably be revised on a periodic basis for decades to come.

3.2.2 Transformations between Systems

This general topic is covered in detail in Chapter 4, where both traditional datum transformations and 3D transformations are discussed. In the case of North America, the National Geodetic Survey (NGS) in the United States has developed transformation software that is freely available to all via the Internet. The URL is http://www.ngs.noaa.gov/TOOS/Htdp/Htdp.shtml. This software enables users to transform individual positions entered interactively or a collection of positions entered as a formatted file. Also, if users expect to transform only a few positions, they may exercise the Horizontal Time Dependent Positioning (HTDP) program interactively from this web page. A discussion of this issue can be found in the series of articles published in the Professional Surveyor (Snay and Soler 1999). Similar software and transformation procedures are available for many other developed countries.

The translations and rotations for all possible realizations of all three modern coordinate systems would result in dozens of sets of transformation values since each system has been almost continuously updated over the past decade. If the latest values were provided, they would be out of date by the time this book is published. Therefore, the practitioner should consider contacting the following individuals/ agencies for the latest and best information:

For WGS 84

Dr. Stephen Malys
National Geospatial-Intelligence Agency
Basic and Applied Research Office, Mail Stop: P-126
12310 Sunrise Valley Drive
Reston, Virginia 20191
Stephen.Malys@nga.mil

For NAD 83

Dr. Richard Snay
National Geodetic Survey
National Ocean Service
1315 East-West Highway
Silver Spring, Maryland 20910
Richard, Snay@noaa.gov

For IERS
Dr. Zuheir Altamimi
Institut Geographique National
6 & 8 Avenue Blaise Pascal
77455 Champs-sur-Marne, Paris, France
altamimi@ensg.ign.fr

3.3 APPLICATIONS

Both datums and reference systems are used to determine new coordinates of points needed for boundary determination, navigation, construction, natural resource management, and myriad other applications. Because datums may be classified into geometric and geopotential datums, they are discussed accordingly.

3.3.1 GEOMETRIC DATUMS

Both local and global geometric datums are used throughout the world. The typical procedure in surveying and geodesy is to occupy a monument with suitable equipment, such as a GPS receiver, and make observations. From these observations, coordinates of new points can be computed. There is an origin and an orientation associated with each datum but these are transparent to most practitioners. The coordinate values, azimuths to nearby points, origin, etc. can (usually) be obtained from the appropriate government agency in the country where the work is carried out.

In ocean areas, the horizontal network is simply *extended* for use in these areas. The ocean network is often sparse. An implication of this sparseness is that it generally leads to inaccurate positioning over these areas.

In their effort to provide safe marine and air navigation, the International Hydrographic Organization (IHO) in 1983 and the International Civil Aviation Organization (ICAO) in 1989 have mandated the use of WGS 84 as a common geodetic system for global usage. Under this requirement, all nautical and aeronautical charts will be converted and civil airports worldwide surveyed in WGS 84.

As the use of the GPS in surveying and in automatic navigation became accepted for universal usage, the WGS 84 system was the most logical choice as a global geodetic system. In the interim, especially for marine navigation and nautical charts, local geodetic datums require transformation to WGS 84.

Due to the paucity of required common geodetic data, datum transformations could only be developed for a subset of the DIA's list. Table 3.2 lists one such subset (NIMA 1997).

3.3.2 GEOPOTENTIAL DATUMS

As mentioned earlier, the geoid is the ideal surface for referencing heights or elevations. However, from a practical point of view, the MSL, as determined from tide gages, is usually utilized. The non-geodesist should be aware of the fact that elevations or heights are more complicated than they appear initially. For example, water does not necessarily flow from a point that has a "higher" GPS-derived-ellipsoid height to a

TABLE 3.2
Geodetic Datums/Reference Systems Related to the World Geodetic System 1984 (through Satellite Ties)

Local Geodetic Datum	Associated Ellipsoid
Adindan	Clarke 1880
Afgooye	Krassovsky 1940
Ain el Abd 1970	International 1924
American Samoa 1962	Clarke 1866
Anna I Astro 1965	Australian National
Antigua Island Astro 1943	Clarke 1880
Arc 1950	Clarke 1880
Arc 1960	Clarke 1880
Ascension Island 1958	International 1924
Astro Beacon "e" 1945	International 1924
Astro DOS 71/4	International 1924
Astro Tern Island (FRIG) 1961	International 1924
Astronomical Station 1952	International 1924
Australian Geodetic 1966	Australian National
Australian Geodetic 1984	Australian National
Ayabelle Lighthouse	Clarke 1880
Bellevue (IGN)	International 1924
Bermuda 1957	Clarke 1866
Bissau	International 1924
Bogota Observatory	International 1924
Campo Inchauspe	International 1924
Canton Astro 1966	International 1924
Cape	Clarke 1880
Cape Canaveral	Clarke 1866
Carthage	Clarke 1880
Chatham Island Astro 1971	International 1924
Chua Astro	International 1924
Coordinate System 1937 of Estonia	Bessel 1841
Corrego Alegre	International 1924
Dabola	Clarke 1880
Deception Island	Clarke 1880
Djakarta (Batavia)	Bessel 1841
DOS 1968	International 1924
Easter Island 1967	International 1924
European 1950	International 1924
European 1979	International 1924
Fort Thomas 1955	Clarke 1880
Gan 1970	International 1924
Geodetic Datum 1949	International 1924
Graciosa Base SW 1948	International 1924
Guam 1963	Clarke 1866

(*continued*)

TABLE 3.2 (continued)
Geodetic Datums/Reference Systems Related to the
World Geodetic System 1984 (through Satellite Ties)

Local Geodetic Datum	Associated Ellipsoid
GUX I Astro	International 1924
Hjorsey 1955	International 1924
Hong Kong 1963	International 1924
Hu-Tzu-Shan	International 1924
Indian	Everest
Indian 1954	Everest
Indian I960	Everest
Indian 1975	Everest
Indonesian 1974	Indonesian 1974
Ireland 1965	Modified airy
ISTS 061 Astro 1968	International 1924
ISTS 073 Astro 1969	International 1924
Johnston Island 1961	International 1924
Kandawala	Everest
Kerguelen Island 1949	International 1924
Kertau 1948	Everest
Kusaie Astro 1951	International 1924
L. C. 5 Astro 1961	Clarke 1866
Leigon	Clarke 1880
Liberia 1964	Clarke 1880
Luzon	Clarke 1866
Mahe 1971	Clarke 1880
Massawa	Bessel 1841
Merchich	Clarke 1880
Midway Astro 1961	International 1924
Minna	Clarke 1880
Montserrat Island Astro 1958	Clarke 1880
M'Poraloko	Clarke 1880
Nahrwan	Clarke 1880
Naparima, BWI	International 1924
North American 1927	Clarke 1866
North American 1983	GRS80
North Sahara 1959	Clarke 1880
Observatorio Meteorologico 1939	International 1924
Old Egyptian 1907	Helmert 1906
Old Hawaiian	Clarke 1866
Oman	Clarke 1880
Ordnance Survey of Great Britain 1936	Airy 1830
Pico de las Nieves	International 1924
Pitcairn Astro 1967	International 1924
Point 58	Clarke 1880
Pointe Noire 1948	Clarke 1880

TABLE 3.2 (continued)

Geodetic Datums/Reference Systems Related to the World Geodetic System 1984 (through Satellite Ties)

Local Geodetic Datum	Associated Ellipsoid
Porto Santo 1936	International 1924
Provisional South American 1956	International 1924
Provisional South Chilean 1963	International 1924
Puerto Rico	Clarke 1866
Qatar National	International 1924
Qornoq	International 1924
Reunion	International 1924
Rome 1940	International 1924
S-42 (Pulkovo 1942)	Krassovsky 1940
Santo (DOS) 1965	International 1924
Sao Braz	International 1924
Sapper Hill 1943	International 1924
Schwarzeck	Bessel 1841
Selvagem Grande 1938	International 1924
Sierra Leone 1960	Clark 1880
S-JTSK	Bessel 1841
South American 1969	South American 1969
South Asia	Modified Fischer 1960
Timbalai 1948	Everest
Tokyo	Bessel 1841
Tristan Astro 1968	International 1924
Viti-Levu 1916	Clarke 1880
Voirol 1960	Clarke 1880
Wake-Eniwetok 1960	Hough 1960
Wake Island Astro 1952	International 1924
Zanderij	International 1924

"lower" GPS-derived-ellipsoid height. This is because ellipsoid heights are geometric quantities that do not account for gravitational effects. Therefore, in the United States, geodesists have developed a series of models that enable the conversion between a GPS-derived-ellipsoid height, denoted by h, and its corresponding orthometric height, denoted by H. The conversion is based on the approximation $H \sim h - N$, where N denotes the height of the geoid above the ellipsoid. Each model provides an estimate for N as a function of geodetic latitude and longitude. For the conterminous United States, the latest in this series of models is called GEOID03 and it is designed specifically to convert between NAD 83 ellipsoid heights and NAVD 88 orthometric heights. Because both NAD 83 and NAVD 88 contain systematic distortions, GEOID03 has been designed to compensate for these distortions. That is, GEOID03 does not provide the value of N for the actual geoid, but for some "hybrid" geoid. Interested readers are referred to http://www.ngs.noaa.gov/GEOID for details.

The normal procedure in determining geopotential heights is to occupy a reference station that has a known geopotential height and then make appropriate observations to points whose geopotential heights are to be determined. Such observations have traditionally been performed with leveling instrumentation, yielding extremely accurate heights when the distances between points are less than a few kilometers. Leveling, however, is labor intensive and, consequently, relatively expensive. Since the mid-1990s, geopotential professionals have increasingly turned to GPS technology to determine ellipsoid heights for points relative to known ellipsoid heights at other points. These professionals have then used a geoid model, such as GEOID03, to convert the newly derived ellipsoid heights into geopotential heights. The use of GPS technology has proven extremely cost-effective relative to leveling when the heights need to be transferred are over tens of kilometers and/or when accuracy requirements are less stringent. It should be noted that within the vast number of civil applications, the *difference* between geopotential heights is the needed quantity, and not the actual heights, for example, in determining the proper flow of liquid in a pipe. Hence, GPS technology may never replace leveling for such applications.

In coastal and ocean areas, the vertical datum, or a quantity related to it, is used as the water level on which the depths shown on nautical charts are based. Although a controversy still lingers concerning the use of the Mean Low Water (MLW) and the Mean Lower Low Water (MLLW) for a datum, it appears to have been resolved by the use of Lowest Astronomic Tide (LAT), (IHO 1998). The definitions and procedures needed to compute or model the tidal surfaces may vary between different agencies within the same country. One result of the varying definitions for the zero reference is that the bathymetric data from two different sources and ocean areas will not "match" (Kumar 1996).

REFERENCES

DIA, Mapping, charting, and geodesy support systems, data elements and related features, Technical Manual DIAM 65-5. Defense Intelligence Agency, Washington, DC, 1971.

Federal Register, 60(157), August 15, 1995.

Geodetic Glossary, National Geodetic Survey, United States Governmental Printing Office, Washington, DC, 1986.

Heiskanen, W. A. and Moritz, H., *Physical Geodesy*, W. H. Freeman & Co., San Francisco, CA, 1967.

IHO, Publication M-3, Resolutions of the International Hydrographic Organization, International Hydrographic Bureau, Monaco, 1998.

Kouba, J. and Popelar, J., Modern reference frame for precise satellite positioning and navigation, *Proceedings of the International Symposium on Kinematic Systems in Geodesy*, Geomatics and Navigation (FIS 94), Banff, Canada, August 1994, pp. 79–86.

Kumar, M., Time-invariant bathymetry: A new concept to define and survey using GPS, *PACON International Proceedings*, Honolulu, HI, 1996.

Lemoine, F. G., Kenyon, S. C., Trimmer, R., Factor, J., Pavlis, N. K., Klosko, S. M., Chin, D. S., Torrence, M. H., Pavlis, E. C., Rapp, R. H., and Olson, T. R., EGM96 the NASA GSFC and NIMA joint geopotential model. NASA Technical Memorandum, 1997.

McCarthy, D. D., IERS conventions 1996, IERS Technical Note, 21, Paris, France, 1996.

Meyer, T. H., Roman, D. R., and Zilkoski, D. B., What does height really mean? *Surveying and Land Information Science*, 66(2), 127–160 and 66(3), 165–183, 2006.

Moritz, H., Geodetic reference system 1980, *Bulletin Geodesique*, 54(3), 1395–40, 1980.

NIMA, *Department of Defense World Geodetic System 1984, its Definition and Relationships with Local Geodetic Systems (3rd edn.)*, National Imagery and Mapping Agency TR 8350.2, Bethesda, MD, 1997.

Snay, R. and Soler, T., *Professional Surveyor*, 19(10), and sequential series, 1999.

Soler, T. and Snay, R., Transforming positions and velocities between the International Terrestrial Reference Frame of 2000 and the North American Datum of 1983, *ASCE Journal of Surveying Engineering*, 130(2), 49–55, 2004.

Torge, W., *Geodesy*, Walter de Gruyter, New York, 1980.

Zilkoski, D., Richards, J., and Young, G., Results of the general adjustment of the North American Vertical Datum of 1988, *Surveying and Land Information Systems*, 53(3), 133–149, 1992.

4 Coordinate Transformations

Ayman Fawzy Habib

CONTENTS

4.1 INTRODUCTION

This chapter explains the concepts and the mechanics behind transforming coordinates between different systems. A right-handed coordinate system is always assumed. First, the concepts of two- and three-dimensional translation and rotation will be explained. Then, the seven-parameter transformation will be addressed along with some of its applications in geodetic activities. Finally, the general concepts and computational procedures for some map projections will be outlined.

4.2 TRANSLATION

The goal of this section is to establish the relationship between the coordinates of a given point and its coordinates in another coordinate systems that is shifted from the original system (i.e., they are parallel to each other with a shift in the origin).

4.2.1 Two-Dimensional Translation

In this section, two-dimensional coordinate systems are addressed. Assume that the two coordinate systems under consideration are (xy) and $(x'y')$. The second coordinate system $(x'y')$ is displaced through shifts x_T and y_T from the y- and x-axes of the first coordinate system (Figure 4.1). The mathematical relationship between the coordinates of a given point (P) with respect to these coordinate systems can be determined through visual inspection of Figure 4.1, which leads to the following equations:

$$x_p = x_T + x'_p$$

$$y_p = y_T + y'_p$$

(4.1)

where
   (x_p, y_p) are the coordinates of point P with respect to the xy-coordinate system
   (x'_p, y'_p) are the coordinates of point P with respect to the $x'y'$-coordinate system

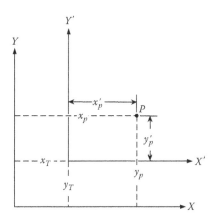

FIGURE 4.1 Two-dimensional translation.

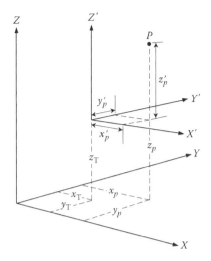

FIGURE 4.2 Three-dimensional translation.

4.2.2 THREE-DIMENSIONAL TRANSLATION

Three-dimensional coordinate systems are dealt with in this section. Figure 4.2 illustrates the two coordinate systems, (xyz) and $(x'y'z')$. The origin of the $(x'y'z')$ is shifted with x_T, y_T, and z_T, relative to the origin of the first coordinate system (xyz). The coordinates of a certain point (P) with respect to these coordinate systems can be related to each other according to

$$\begin{bmatrix} x_p \\ y_p \\ z_p \end{bmatrix} = \begin{bmatrix} x_T \\ y_T \\ z_T \end{bmatrix} + \begin{bmatrix} x'_p \\ y'_p \\ z'_p \end{bmatrix} \tag{4.2}$$

where

(x_p, y_p, z_p) are the coordinates of point P with respect to the xyz-coordinate system
(x'_p, y'_p, z'_p) are the coordinates of point P with respect to the $x'y'z'$-coordinate system

4.3 ROTATION

The goal of this section is to establish the relationship between the coordinates of a given point with respect to two coordinate systems that are not parallel to each other and share the same origin (no shift is involved). One of the coordinate systems can be obtained from the other through a sequence of rotation angles.

4.3.1 TWO-DIMENSIONAL ROTATION

In this section, two-dimensional coordinate systems are considered. The coordinate system $(x'y')$, in Figure 4.3, can be obtained from the xy-coordinate system through

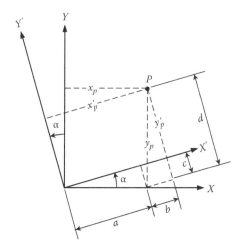

FIGURE 4.3 Two-dimensional rotation.

a counterclockwise rotation angle. The objective is to derive the mathematical relationship between the coordinates of a given point (P) with respect to these coordinate systems. Once again, by inspecting Figure 4.3 and using some trigonometric relationships, one can write

$$\begin{bmatrix} x'_p \\ y'_p \end{bmatrix} = \begin{bmatrix} a+b \\ -c+d \end{bmatrix} = \begin{bmatrix} \cos\alpha x_p + \sin\alpha y_p \\ -\sin\alpha x_p + \cos\alpha y_p \end{bmatrix} \tag{4.3}$$

where
(x_p, y_p) are the coordinates of point P with respect to the xy-coordinate system
(x'_p, y'_p) are the coordinates of point P with respect to the $x'y'$-coordinate system

Equation 4.3 can be rewritten in a matrix form as follows:

$$\begin{bmatrix} x'_p \\ y'_p \end{bmatrix} = \begin{bmatrix} \cos\alpha & \sin\alpha \\ -\sin\alpha & \cos\alpha \end{bmatrix} \begin{bmatrix} x_p \\ y_p \end{bmatrix} \tag{4.4}$$

A matrix is simply a notation representing a series of equations. For a more detailed explanation about rotation matrices and their properties, the reader can refer to Section 4.3.2.

4.3.2 Three-Dimensional Rotation

Now, three-dimensional coordinate systems are investigated. Assume that the two coordinate systems (xyz) and ($x'y'z'$) are under consideration. These two coordinate systems can be related to each other through a sequence of rotation angles (i.e., one

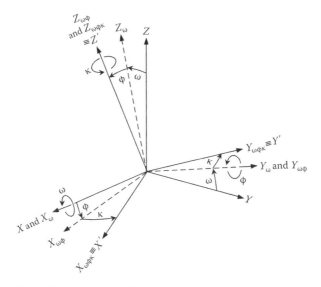

FIGURE 4.4   Three-dimensional rotation.

coordinate system can be obtained by rotating the other system around the x-, y-, and z-axes, respectively). In Figure 4.4, the xyz- and the $x'y'z'$-coordinate systems can be made parallel by rotating the former system as follows:

- Primary rotation angle (ω) around the x-axis
- Secondary rotation angle (ϕ) around the y-axis
- Tertiary rotation angle (κ) around the z-axis

One should note that the rotation order could be arbitrarily chosen. The above-mentioned order is just an example of how to make the involved coordinate systems parallel. In this section, the relationship between the coordinates of a given point (P), with respect to these coordinate systems, as a function of the rotation angles is established.

4.3.2.1   Primary Rotation (ω) around the x-Axis

Assume that the xyz-coordinate system has been rotated through an angle (ω) around the x-axis yielding the $x_\omega y_\omega z_\omega$-coordinate system (Figure 4.5). The mathematical relationship between the coordinates of a given point with respect to these coordinate systems can be seen in the following equation:

$$\begin{bmatrix} x_\omega \\ y_\omega \\ z_\omega \end{bmatrix} = \begin{bmatrix} 1 & 0 & 0 \\ 0 & \cos\omega & \sin\omega \\ 0 & -\sin\omega & \cos\omega \end{bmatrix} \begin{bmatrix} x \\ y \\ z \end{bmatrix} = R_\omega \begin{bmatrix} x \\ y \\ z \end{bmatrix} \qquad (4.5)$$

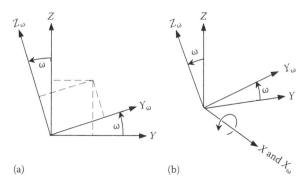

FIGURE 4.5   Primary rotation (ω) around the x-axis.

4.3.2.2   Secondary Rotation (ϕ) around the y_ω-Axis

After the primary rotation (ω), assume that the $x_\omega y_\omega z_\omega$-coordinate system has been rotated with an angle (ϕ) around the y_ω-axis yielding the $x_{\omega\phi} y_{\omega\phi} z_{\omega\phi}$-coordinate system (Figure 4.6). The mathematical relationship between the coordinates of a given point with respect to these coordinate systems, as a function of the rotation angle ϕ can be seen in the following equation:

$$\begin{bmatrix} x_{\omega\phi} \\ y_{\omega\phi} \\ z_{\omega\phi} \end{bmatrix} = \begin{bmatrix} \cos\phi & 0 & -\sin\phi \\ 0 & 1 & 0 \\ \sin\phi & 0 & \cos\phi \end{bmatrix} \begin{bmatrix} x_\omega \\ y_\omega \\ z_\omega \end{bmatrix} = R_\phi \begin{bmatrix} x_\omega \\ y_\omega \\ z_\omega \end{bmatrix} \tag{4.6}$$

4.3.2.3   Tertiary Rotation (κ) around the $z_{\omega\phi}$-Axis

Finally, assume that the $x_{\omega\phi} y_{\omega\phi} z_{\omega\phi}$-coordinate system has been rotated with an angle (κ) around the $z_{\omega\phi}$-axis yielding the $x_{\omega\phi\kappa} y_{\omega\phi\kappa} z_{\omega\phi\kappa}$-coordinate system (Figure 4.7). The mathematical relationship between the coordinates of a given point with respect to these coordinate systems, as a function of the rotation angle κ can be seen in the following equation:

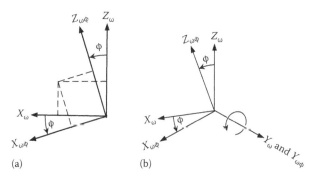

FIGURE 4.6   Secondary rotation (ϕ) around the y_ω-axis.

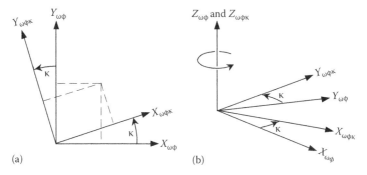

FIGURE 4.7 Tertiary rotation (κ) around the $z_{\omega\phi}$-axis.

$$\begin{bmatrix} x_{\omega\phi\kappa} \\ y_{\omega\phi\kappa} \\ z_{\omega\phi\kappa} \end{bmatrix} = \begin{bmatrix} \cos\kappa & \sin\kappa & 0 \\ -\sin\kappa & \cos\kappa & 0 \\ 0 & 0 & 1 \end{bmatrix} \begin{bmatrix} x_{\omega\phi} \\ y_{\omega\phi} \\ z_{\omega\phi} \end{bmatrix} = R_\kappa \begin{bmatrix} x_{\omega\phi} \\ y_{\omega\phi} \\ z_{\omega\phi} \end{bmatrix} \qquad (4.7)$$

Assuming that the $x_{\omega\phi\kappa}y_{\omega\phi\kappa}z_{\omega\phi\kappa}$-coordinate system is parallel to the $x'y'z'$-coordinate system and by substituting Equations 4.5 and 4.6 into Equation 4.7, one gets the following equation:

$$\begin{bmatrix} x' \\ y' \\ z' \end{bmatrix} = R_\kappa R_\phi R_\omega \begin{bmatrix} x \\ y \\ z \end{bmatrix} = R(\omega,\phi,\kappa) \begin{bmatrix} x \\ y \\ z \end{bmatrix} \qquad (4.8)$$

The product of the three matrices ($R_\kappa R_\phi R_\omega$) in Equation 4.8 is referred to as the rotation matrix (R). It is a function of the rotation angles ω, ϕ, and κ and gives the relationship between the coordinates of a given point with respect to the involved coordinated systems.

Remarks on the rotation matrix (R)

1. The rotation matrix is an orthogonal matrix. This means that its inverse will be equal to its transpose (i.e., $R^{-1} = R^T$). This stems from the fact that the magnitude (norm) of any column or row is unity (REA, 1997). In addition, the dot product of any two rows or columns is zero.
2. The elements of the rotation matrix R will change as a result of changing the rotation order (i.e., $R_{\omega,\phi,\kappa} \neq R_{\phi,\omega,\kappa}$).
3. A positive rotation angle is one that is counterclockwise when looking at the coordinate system with the positive direction of the rotation axis pointing toward us.

4.4 THREE-DIMENSIONAL SIMILARITY TRANSFORMATION (SEVEN-PARAMETER TRANSFORMATION)

In Sections 4.2 and 4.3, the influence of rotation and translation between two coordinate systems was discussed. The scales along the axes of the involved coordinate systems were assumed to be equal. The impact of having uniform change in the scale will be considered in this section. Considering three rotation angles (one around each axis) and three translations (one in the direction of each axis) and a uniform scale change, a total of seven parameters must be involved in the transformation between the coordinate systems (Figure 4.8). This transformation is called a seven-parameter transformation, or sometimes a linear conformal transformation in three dimensions, or simply a three-dimensional similarity transformation. The mathematical model will be explained in Section 4.4.1, followed by some geodetic and photogrammetric applications.

4.4.1 MATHEMATICAL MODEL

If (x, y, z) and (x', y', z') are the rectangular coordinates of a given point before and after the transformation, the three-dimensional similarity transformation can be given by

$$\begin{bmatrix} x' \\ y' \\ z' \end{bmatrix} = \begin{bmatrix} x_T \\ y_T \\ z_T \end{bmatrix} + SR(\omega, \phi, \kappa) \begin{bmatrix} x \\ y \\ z \end{bmatrix} \tag{4.9}$$

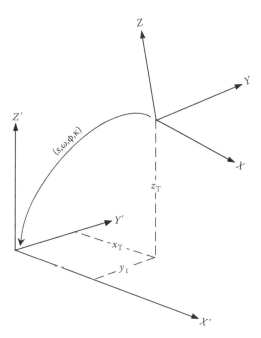

FIGURE 4.8   Three-dimensional similarity transformation.

where

S is the scale factor

(x_T, y_T, z_T) are the involved translations

$R(\omega, \phi, \kappa)$ is a three by three matrix that rotates the xyz-coordinate system to make it parallel to the $x'y'z'$-system

Equation 4.9 can be sequentially explained as follows:

- The xyz-coordinate system is made parallel to the $x'y'z'$-system through a multiplication with the rotation matrix $R(\omega, \phi, \kappa)$.
- The uniform change in the scale is compensated for through the multiplication with the scale factor S.
- Finally, the three translations x_T, y_T, and z_T are applied in the x', y', and the z' directions, respectively.

4.4.2 APPLICATIONS

To illustrate the importance of the three-dimensional similarity transformation, its use in geodetic and photogrammetric applications is explained.

4.4.2.1 Collinearity Equations

Photogrammetry can be defined as the art and science of deriving three-dimensional positions of objects from imagery (Kraus 1993). The main objective of photogrammetry is to inverse the process of photography. When the film inside the camera is exposed to light, light rays from the object space pass through the camera's perspective center (the lens), until it hits the focal plane (film) producing images of the photographed objects. The main mathematical model that has been in use in the majority of photogrammetric applications is the collinearity equations. The collinearity equations mathematically describe the fact that the object point, the corresponding image point, and the perspective center lie on a straight line (Moffitt and Mikhail 1980) (Figure 4.9).

The collinearity equations can be derived using the concept of the three-dimensional similarity transformation (Figure 4.10).

The reader should note that the focal plane is shown in the negative position (above the perspective center) in Figure 4.9, but it is shown in the positive position (below the perspective center) in Figure 4.10. Image coordinate measurements are accomplished on the positive, usually called the diapositive. The diapositive is used in all the photogrammetric mathematical derivations since it maintains directional relationship between points in the object and image spaces. The directional relationship is not maintained between the object space and the negative (i.e., the negative is an inverted perspective representation of the object space).

Before deriving the collinearity equations, the following coordinate systems need to be defined (Figure 4.10):

- The ground coordinate system (XYZ): this can be a local coordinate system associated with the area being photographed (e.g., state plane coordinate system and height).

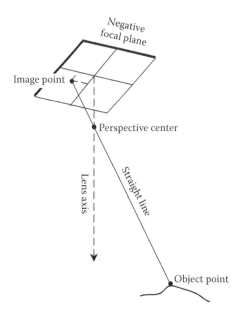

FIGURE 4.9   The principle of the collinearity condition.

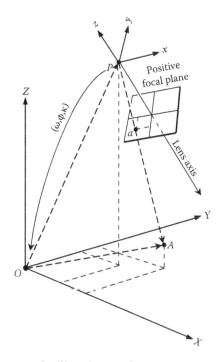

FIGURE 4.10   The concept of collinearity equations.

- The image coordinate system (*xyz*): this coordinate system is associated with the camera body. All the measurements in the image plane are referenced to this coordinate system.

The three angles, ω, ϕ, and κ, describe the rotational relationship between the image and the ground coordinate systems. Now, let us define the following points:

- The origin of the ground coordinate system will be denoted by (*O*)
- The perspective center of the lens is denoted by (*P*)
- (*A*) and (*a*) will denote the object point and the corresponding image point, respectively

The vector (one-dimensional array, REA, 1997) connecting the origin of the ground coordinate system and the object point is given as

$$\vec{V}_{OA} = \begin{bmatrix} X_A \\ Y_A \\ X_A \end{bmatrix} \tag{4.10}$$

One should note that this vector is referenced relative to the ground coordinate system. The vector connecting the origin of the ground coordinate system to the perspective center is defined by

$$\vec{V}_{OP} = \begin{bmatrix} X_P \\ Y_P \\ Z_P \end{bmatrix} \tag{4.11}$$

Once again, this vector is defined relative to the ground coordinate system. Finally, the vector connecting the perspective center to the image point is given by

$$\vec{v}_{Pa} = \begin{bmatrix} x_a \\ y_a \\ -c \end{bmatrix} \tag{4.12}$$

where (*c*) is the principal distance (the normal distance between the perspective center and the image plane). $(x_a, y_a, -c)$ are the coordinates of the image point (*a*) relative to the image coordinate system. The vector connecting the perspective center and the image point can be transformed into the ground coordinate system through a multiplication with the rotation matrix $R(\omega, \phi, \kappa)$ as follows:

$$\vec{V}_{Pa} = R(\omega, \phi, \kappa) \begin{bmatrix} x_a \\ y_a \\ -c \end{bmatrix} \tag{4.13}$$

The vector connecting the perspective center and the object point under consideration can be obtained by multiplying Equation 4.13 with a scale factor, S (Figure 4.10):

$$\vec{V}_{PA} = SR(\omega, \phi, \kappa) \begin{bmatrix} x_a \\ y_a \\ -c \end{bmatrix} \tag{4.14}$$

The three vectors in Equations 4.10, 4.11, and 4.14 form a closed polygon. Therefore, the following equation can be written:

$$\vec{V}_{OA} = \vec{V}_{OP} + \vec{V}_{PA}$$

$$\begin{bmatrix} X_A \\ Y_A \\ Z_A \end{bmatrix} = \begin{bmatrix} X_P \\ Y_P \\ Z_P \end{bmatrix} + SR(\omega, \phi, \kappa) \begin{bmatrix} x_a \\ y_a \\ -c \end{bmatrix} \tag{4.15}$$

Note the similarity between Equations 4.15 and 4.9. One should also note that these vectors are all referenced to the ground coordinate system. Rearranging the terms in Equation 4.15 and using the orthogonality property of the rotation matrix (refer to Section 4.3.2) this equation can be rewritten as follows:

$$\begin{bmatrix} x_a \\ y_a \\ -c \end{bmatrix} = 1/SR^{\mathrm{T}}(\omega, \phi, \kappa) \begin{bmatrix} X_A - X_P \\ Y_A - Y_P \\ Z_A - Z_P \end{bmatrix} \tag{4.16}$$

Equation 4.16 is one form of the collinearity equations. The classical and traditional form of the collinearity equations can be obtained by dividing the first two lines in Equation 4.16 by the third line (thus, the scale factor S is eliminated), which leads to

$$x_a = -c \frac{r_{11}(X_A - X_P) + r_{21}(Y_A - Y_P) + r_{31}(Z_A - Z_P)}{r_{13}(X_A - X_P) + r_{23}(Y_A - Y_P) + r_{33}(Z_A - Z_P)}$$
$$\tag{4.17}$$
$$y_a = -c \frac{r_{12}(X_A - X_P) + r_{22}(Y_A - Y_P) + r_{32}(Z_A - Z_P)}{r_{13}(X_A - X_P) + r_{23}(Y_A - Y_P) + r_{33}(Z_A - Z_P)}$$

4.4.2.2 Datum-to-Datum Transformation

As mentioned in Chapter 2, an ellipsoid of revolution, or simply a reference ellipsoid, is a reasonable approximation of the earth's surface. The shape (first two parameters) and the position of the reference ellipsoid in space are defined by the following parameters:

- The semimajor axis (a)
- The semiminor axis (b), eccentricity (e), or flattening (f)

- The position of the center of the reference ellipsoid relative to the earth's center of gravity
- The rotational relationship between the ellipsoid axes and the natural coordinate system associated with the earth (as defined by the earth's axis of rotation and the equatorial plane)

The geodetic latitude (ϕ), the geodetic longitude (λ), and the ellipsoidal height (h) define the position of any point (P) in space. These quantities are specific for a certain reference ellipsoid. If, for any reason, the geodetic datum is changed—that is, the reference ellipsoid and its position—then the geodetic coordinates (ϕ, λ, h) will also change. This problem is simply a transformation of coordinates. In this section, the mathematical derivations of this type of coordinate transformation are not explained. Rather, alternative approaches to solving this problem are described. For a detailed description, the reader can refer to Heiskanen and Moritz (1996).

First approach (direct approach): In this approach, differential equations that relate changes in the geodetic coordinates of a given point to changes in the geodetic datum are formulated, as shown in the following equation:

$$a\delta\phi = \sin\phi\cos\lambda\delta x_o + \sin\phi\sin\lambda\delta y_o - \cos\phi\delta z_o + 2a\sin\phi\cos\phi\delta f$$

$$a\cos\phi\delta\lambda = \sin\lambda\delta x_o - \cos\lambda\delta y_o \qquad\qquad (4.18)$$

$$\delta h = -\cos\phi\cos\lambda\delta x_o - \cos\phi\sin\lambda\delta y_o - \sin\phi\delta z_o - \delta a + a\sin^2\phi\delta f$$

where ($\delta\phi$, $\delta\lambda$, δh) are the changes in the geodetic coordinates of a given point due to changes of (δx_o, δy_o, δz_o, δa, δf) in the position, semimajor axis, and the flattening of the reference ellipsoid, respectively. Note that in Equation 4.18, the axes of the reference ellipsoids are assumed to be parallel (no rotation angles are involved).

Second approach (indirect approach): In the previous approach, a mathematical function that directly relates the change in the geodetic coordinates to the changes in the parameters of the corresponding reference ellipsoid was shown. In the following approach, this transformation can be carried out sequentially:

- Transform the geodetic coordinates into Cartesian geocentric coordinates using the shape parameters of the initial reference ellipsoid (refer to Section 2.6, Equations 2.8 through 2.10).
- Apply three-dimensional similarity transformation to the geocentric coordinates that were determined in the previous step. The similarity transformation should compensate for the change in the position and the orientation between the new and the original reference ellipsoids.
- Transform the new Cartesian geocentric coordinates into geodetic coordinates using the shape parameters of the new reference ellipsoid.

One should note that both direct and indirect approaches are expected to yield almost identical results. However, some approximations in the first approach are incorporated to simplify the transformation formulae. These approximations might lead to relatively less accurate results compared to the second approach.

4.5   MAP PROJECTIONS AND TRANSFORMATIONS

A map projection is a systematic representation of all or part of the surface of a round body, especially the earth, on a plane (Snyder 1987). The transformation from a round surface to a plane cannot be accomplished without distortion. If the map covers a large area of the earth's surface—for example, a continent—distortions will be visually apparent. If the region covered by the map is small, distortions might be just measurable using many projections, yet it can be serious with other projections.

The characteristics that are normally considered when choosing a map projection can be listed as follows:

- Area: Some map projection schemes are designed to be equal areas. This means that equal areas on the map correspond to equal areas on the earth's surface.
- Shape: Map projections can be conformal. This means that relative local angles about every point on the map are shown correctly. One important result of conformality is that the local scale in every direction around any point is constant. Since local angles are correct, meridians intersect parallels at right angles on a conformal projection. One should note that no map can be both conformal and equal area.
- Scale: No map projection can show the scale correctly throughout the whole map. Rather, the map projection can be designed in such a way that guarantees uniform scale along one or more lines. Some map projections show true scale along every meridian.
- Direction: Conformal projection gives the relative local directions correctly at any given point. In this type of projection, the directions or azimuths of all points on the map are shown correctly with respect to the center.
- Special characteristics: Some map projections offer special characteristics such as lines of constant directions and are shown as straight lines. Some other projections show the shortest distance between two points as straight lines.

Developable surfaces are ones that can be transformed to a plane without distortion. There are three types of developable surfaces that can be used for map projection, namely, the cylinder, the cone, and the plane. All three surfaces are variations of the cone. A cylinder is a limiting form of a cone with an increasingly sharp apex. As the cone becomes flatter, its limit is a plane.

A cylinder, whose axis is coincident with the earth's polar axis, can be wrapped around the globe with its surface touching the equator. Alternatively, it can intersect the globe (i.e., be secant to) at two parallels. In either case, the meridians may be projected onto the cylinder as equidistant straight lines perpendicular to the equator. In addition, the parallels can be drawn as lines parallel to the equator. A cylindrical projection with straight meridians and straight parallels results when the cylinder is cut along one of the meridians and unrolled. The Mercator projection is the best-known example of a cylindrical projection.

Similarly, a cone can be placed over the globe with its apex along the polar axis of the earth and its surface either touching the earth's surface at a particular parallel of latitude, or intersecting the earth's surface (i.e., secant to) at two parallels (Figure 2.9). In this case, the meridians can be projected onto the cone as equidistant straight lines radiating from the apex. The parallels, on the other hand, can be drawn as lines around the circumference of the cone in planes perpendicular to the earth's axis. When the cone is cut along a meridian, unrolled, and laid flat, the meridians remain straight lines and the parallels will appear as circular arcs centered at the apex. The angles between the meridians are shown smaller than the true angles. This type of projection is known as conic or conical map projection.

A plane tangent to the earth at one of its poles is the basis for polar azimuthal projections. The meridians are projected as straight lines radiating from a point. The meridians are spaced at their true angles instead of smaller angles of the conic projections. The parallels of latitude will appear as complete circles, centered at the pole.

The concepts outlined above might be modified in two ways and still produce cylindrical, conic, or azimuthal projections.

1. As discussed earlier, the cylinder or cone may be secant to or cut the globe at two parallels (standard parallels) instead of being tangent at just one. This conceptually provides two standard parallels. In addition, the plane may cut the globe at any parallel instead of touching a pole.
2. The axis of the cylindrical or cone can have a direction different from that of the earth's axis of rotation. Similarly, the plane may be tangent to any point other than the pole. This modification leads to oblique, transverse, and equatorial projections. In this case, meridians and parallels of latitude are no longer straight lines or arcs of circles.

In the following sections, the projection formulae are listed in a manner suitable for computer computations. All the angular elements are assumed to be in radians.

4.5.1 TRANSVERSE MERCATOR PROJECTION

The main characteristics of the Transverse Mercator projection can be summarized as follows:

- It is a cylindrical (transverse) projection.
- It is a conformal projection.
- The central meridian, the meridians 90° away from the central meridians and the equator will appear as straight lines.
- Other meridians and parallels will appear as complex curves.
- The central meridian and the two nearly straight lines equidistant from and almost parallel to the central meridian will have true scale.
- This projection is mainly used for quadrangle maps at scales from 1:24,000 to 1:250,000.

4.5.1.1 Geodetic to *xy*-Transverse Mercator (Forward Formulae)

Given the reciprocal of the flattening ($RF = 1/f$) and the semimajor axis (a) of the reference ellipsoid, together with the latitude and the longitude of the origin of the rectangular coordinates (ϕ_0, λ_0), and the scale factor along the central meridian, one can compute the following constants:

$$f = 1.0/RF \tag{4.19}$$

$$e^2 = 2f - f^2 \tag{4.20}$$

$$e'^2 = e^2/(1-e^2) \tag{4.21}$$

$$p' = (1.0 - f)a \tag{4.22}$$

$$e_n = (a - p')/(a + p') \tag{4.23}$$

$$A = 1.5e_n + 9.0/16.0e_n^3 \tag{4.24}$$

$$B = 0.93750e_n^2 - 15.0/32.0e_n^4 \tag{4.25}$$

$$C = -35.0/48.0e_n^3 \tag{4.26}$$

$$U = 1.5e_n - 27.0/32.0e_n^3 \tag{4.27}$$

$$V = 1.31250e_n^2 - 55.0/32.0e_n^4 \tag{4.28}$$

$$W = 151.0/96.0e_n^3 \tag{4.29}$$

$$R = a(1.0 - e_n)(1.0 - e_n^2)(1.0 + 2.25e_n^2 + 225.0/64.0e_n^4) \tag{4.30}$$

$$M_o - \phi_o + A \sin(2.0\phi_o) + B \sin(4.0\phi_o) + C \sin(6.0\phi_o) \tag{4.31}$$

$$S_0 = \text{scale factor } RM_o \tag{4.32}$$

For a given (ϕ, λ), one can compute the *xy*-transverse Mercator rectangular coordinates as follows:

$$M = \phi + A \sin(2.0\phi) + B \sin(4.0\phi) + C \sin(6.0\phi) \tag{4.33}$$

$$S = RM \text{ scale factor} \tag{4.34}$$

$$T_N = \sin \phi / \cos \phi \tag{4.35}$$

$$T_S = T_N^2, \tag{4.36}$$

$$ETS = e'^2 \cos^2 \phi \tag{4.37}$$

$$L = (\lambda - \lambda_o) \cos \phi \tag{4.38}$$

$$RN = (\text{scale factor } a) / \sqrt{1.0 - e^2 \sin^2 \phi} \tag{4.39}$$

$$A_2 = 0.5RNT_N \tag{4.40}$$

$$A_4 = (5.0 - T_s + \text{ETS}(9.0 + 4.0\text{ETS}))/12.0 \tag{4.41}$$

$$A_6 = \{61.0 + T_s(T_s - 58.0) + \text{ETS}(270.0 - 330.0T_s)\}/360.0 \tag{4.42}$$

$$A_1 = -RN \tag{4.43}$$

$$A_3 = (1.0 - T_s + \text{ETS})/6.0 \tag{4.44}$$

$$A_5 = \{5.0 + T_s(T_s - 18.0) + \text{ETS}(14.0 - 58.0T_s)\}/120.0 \tag{4.45}$$

$$A_7 = \left\{ 61.0 - 479.0T_s + 179.0T_s^2 - T_s^3 \right\}/5040.0 \tag{4.46}$$

$$y = S - S_o + A_2L^2[1.0 + L^2(A_4 + A_6L^2)] + \text{false northing} \tag{4.47}$$

$$x = \text{false easting} - A_1L[1.0 + L^2\{A_3 + L^2(A_5 + A_7L^2)\}] \tag{4.48}$$

Note that the false easting and false northing are used to avoid negative coordinates.

4.5.1.2 *xy*-Transverse Mercator into Geodetic Coordinate Transformation (Inverse Formulae)

Given (x, y) rectangular coordinates, the corresponding geodetic coordinates need to be computed. First, compute the following constants:

$$f = 1.0/RF \tag{4.49}$$

$$e^2 = 2f - f^2 \tag{4.50}$$

$$e'^2 = e^2/(1 - e^2) \tag{4.51}$$

$$p' = (1.0 - f)a \tag{4.52}$$

$$e_n = (a - p')/(a + p') \tag{4.53}$$

$$C_2 = -1.5e_n + 9.0/16.0e_n^3 \tag{4.54}$$

$$C_4 = 15.0/16.0e_n^2 - 15.0/32.0e_n^4 \tag{4.55}$$

$$C_6 = -35.0/48.0e_n^3 \tag{4.56}$$

$$C_8 = 315.0/512.0e_n^4 \tag{4.57}$$

$$U_0 = 2.0(C_2 - 2.0C_4 + 3.0C_6 - 4.0C_8) \tag{4.58}$$

$$U_2 = 8.0(C_4 - 4.0C_6 + 10.0C_8) \tag{4.59}$$

$$U_4 = 32.0(C_6 - 6.0C_8) \tag{4.60}$$

$$U_6 = 128.0C_8 \tag{4.61}$$

$$C_1 = 1.5e_n - 27.0/32.0e_n^4 \tag{4.62}$$

$$C_3 = 21.0/16.0e_n^2 - 55.0/32.0e_n^4 \tag{4.63}$$

$$C_5 = 151.0/96.0e_n^3 \tag{4.64}$$

$$C_7 = 1097.0/512.0e_n^4 \tag{4.65}$$

$$V_0 = 2.0(C_1 - 2.0C_3 + 3.0C_5 - 4.0C_7) \tag{4.66}$$

$$V_2 = 8.0(C_3 - 4.0C_5 + 10.0C_7) \tag{4.67}$$

$$V_4 = 32.0(C_5 - 6.0C_7) \tag{4.68}$$

$$V_6 = 128.0C_7 \tag{4.69}$$

$$R = a(1.0 - e_n)(1.0 - e_n^2)(1.0 + 2.25e_n^2 + 225.0/64.0e_n^4) \tag{4.70}$$

$$M_o = \phi_o + \sin \phi_o \cos \phi_o(U_0 + U_2 \cos^2 \phi_o + U_4 \cos^4 \phi_o + U_6 \cos^6 \phi_o) \tag{4.71}$$

$$S_o = \text{scale factor } RM_o \tag{4.72}$$

Then, one can proceed to compute the corresponding geodetic coordinates as follows:

$$M = (y - \text{false northing} + S_o)/(R \text{ scale factor}) \tag{4.73}$$

$$F = M + \sin M \cos M(V_0 + V_2 \cos^2 M + V_4 \cos^4 M + V_6 \cos^6 M) \tag{4.74}$$

$$T_N = \sin F/\cos F \tag{4.75}$$

$$\text{ETS} = e'^2\cos^2 F \tag{4.76}$$

$$RN = a \text{ scale factor}/\sqrt{1 - e^2 \sin^2 F} \tag{4.77}$$

$$Q = (x - \text{false easting})/RN \tag{4.78}$$

$$B_2 = -0.5T_N(1.0 + \text{ETS}) \tag{4.79}$$

$$B_4 = -\left(5.0 + 3.0T_N^2 + \text{ETS}\left[1.0 - 9.0T_N^2\right] - 4.0\text{ETS}^2 \right)/12.0 \tag{4.80}$$

$$B_6 = \left(61.0 + 45.0T_N^2 \left[2.0 + T_N^2\right] + \text{ETS}\left[46.0 - 252.0T_N^2 - 60T_N^4\right]\right)/360.0 \quad (4.81)$$

$$B_1 = 1.0 \quad (4.82)$$

$$B_3 = -(1.0 + 2.0T_N^2 + \text{ETS})/6.0 \quad (4.83)$$

$$B_5 = \left(5.0 + T_N^2 [28.0 + 24.0T_N^2] + \text{ETS}[6.0 + 8.0T_N^2]\right)/120.0 \quad (4.84)$$

$$B_7 = -\left(61.0 + 662.0T_N^2 + 1320.0T_N^4 + 720.0T_N^6\right)/5040.0 \quad (4.85)$$

$$\phi = F + B_2 Q^2 (1.0 + Q^2 [B_4 + B_6 Q^2]) \quad (4.86)$$

$$L = B_1 Q(1.0 + Q^2 [B_3 + Q^2 \{B_5 + B_7 Q^2\}]) \quad (4.87)$$

$$\lambda = L/\cos F + \lambda_o \quad (4.88)$$

4.5.2 LAMBERT (CONIC) PROJECTION

The main characteristics of the Lambert projection can be summarized as follows:

- Lambert projection is conic.
- Lambert projection is conformal.
- Parallels are unequally spaced arcs of concentric circles, more closely spaced near the center of the map.
- Meridians are equally spaced radii of the same circles. Thus, the parallels will cut the meridians at a right angle.
- The scale will be true along the standard parallels.
- This projection is used for regions and countries with predominant east–west extent.
- The pole in the same hemisphere as the standard parallels will appear as a point. The other pole will be at infinity.

Now, the mathematical formulas to transform geodetic into xy-Lambert coordinates (forward equations) and the inverse formulas for transforming xy-Lambert into geodetic coordinates will be listed. In both cases, the following parameters that describe the reference datum and the location of the projection surface (the cone) are given:

- The semimajor axis of the reference ellipsoid (a)
- The first eccentricity (e)
- The latitudes ϕ_1 and ϕ_2 of the standard parallels
- The latitude and the longitude ϕ_o and λ_o of the origin of the rectangular coordinates

Given these parameters, one can compute the following constants:

$$m = \frac{\cos \phi}{\sqrt{1.0 - e^2 \sin^2 \phi}} \qquad (4.89)$$

Substituting ϕ_1 and ϕ_2 yields m_1 and m_2

$$t = \frac{\tan(\pi/4 - \phi/2)}{\left(\dfrac{1.0 - e \sin \phi}{1.0 + e \sin \phi} \right)^{0.5e}} \qquad (4.90)$$

ϕ_0, ϕ_1, ϕ_2 yield t_0, t_1, and t_2

$$n = \frac{\log m_1 - \log m_2}{\log t_1 - \log t_2} \qquad (4.91)$$

$$F = \frac{m_1}{n t_1^n} \qquad (4.92)$$

$$R_0 = a F t_0^n \qquad (4.93)$$

4.5.2.1 Geodetic to xy-Lambert Coordinate Transformation (Forward Equations)

Here, the geodetic coordinates (ϕ, λ) of any point are given. The goal is to compute the corresponding rectangular coordinates. The computational procedure can be listed as follows:

$$t = \frac{\tan(\pi/4 - \phi/2)}{\left(\dfrac{1.0 - e \sin \phi}{1.0 + e \sin \phi} \right)^{0.5e}} \qquad (4.94)$$

$$R = a F t^n \qquad (4.95)$$

$$\theta = n(\lambda - \lambda_0) \qquad (4.96)$$

$$x = R \sin \theta + \text{false easting} \qquad (4.97)$$

$$y = R_0 - R \cos \theta + \text{false northing} \qquad (4.98)$$

4.5.2.2 xy-Lambert to Geodetic Coordinate Transformation (Inverse Equations)

Here, the rectangular coordinates are given for the purpose of computing the corresponding geodetic coordinates. The computation procedure should be as follows:

$$x_t = x - \text{false easting} \tag{4.99}$$

$$y_t = y - \text{false northing} \tag{4.100}$$

$$r = \frac{n}{|n|}\sqrt{x_t^2 + (R_0 - y_t)^2} \tag{4.101}$$

$$\Theta = \tan^{-1}\left(\frac{x_t}{R_0 - y_t}\right) \tag{4.102}$$

$$T = \left(\frac{r}{aF}\right)^{1.0/n} \tag{4.103}$$

$$\phi = \pi/2.0 - 2.0\tan^{-1}\left(T\left(\frac{1.0 - e\sin\phi}{1.0 + e\sin\phi}\right)^{e/2.0}\right) \tag{4.104}$$

To compute the latitude ϕ in Equation 4.104, one has to iterate starting from an initial value (e.g., $\phi = \pi/2.0 - 2.0\tan^{-1}(T)$). Finally, the longitude can be computed as follows:

$$\lambda = \Theta/n + \lambda_0 \tag{4.105}$$

4.5.3 OBLIQUE MERCATOR PROJECTION

The main characteristics of the Oblique Mercator Projection can be summarized as follows:

- It is an oblique cylindrical projection
- It is a conformal projection
- The two meridians 180° apart are straight lines
- Other meridians and parallels are complex curves
- The scale will be true along the chosen central line
- Scale becomes infinite 90° from the central meridian
- It is mainly used for areas with greater extent in an oblique direction

Given the following parameters

- Semimajor axis (a) and the first eccentricity (e) of the reference ellipsoid
- The central scale factor
- The latitude and the longitude of the selected center of the map (ϕ_0, λ_c)
- The azimuth (α_c) of the central line as it crosses the latitude (ϕ_0) measured east of north

One can start by computing the following constants:

$$B = \sqrt{1.0 + e^2 \cos^4 \phi_o /(1.0 - e^2)} \tag{4.106}$$

$$A = aB \text{ scale factor } \sqrt{1.0 - e^2} \Big/ (1.0 - e^2 \sin^2 \phi_o) \tag{4.107}$$

$$t_o = \tan (\pi/4.0 - \phi_o/2.0) \Big/ \{(1.0 - e \sin \phi_o)/(1.0 + e \sin \phi_o)\}^{0.5e} \tag{4.108}$$

$$D = B\sqrt{(1.0 - e^2)} \Big/ \left\{ \cos \phi_o \sqrt{1.0 - e^2 \sin^2 \phi_o} \right\} \tag{4.109}$$

$$F = D + \frac{\phi_o}{|\phi_o|} \sqrt{D^2 - 1.0} \tag{4.110}$$

$$E = Ft_o^B \tag{4.111}$$

$$G = 0.5(F - 1.0/F) \tag{4.112}$$

$$\gamma_o = \sin^{-1}\{\sin (\alpha_c)/D\} \tag{4.113}$$

$$\lambda_o = \lambda_c - \sin^{-1}(G \tan (\gamma_o))/B \tag{4.114}$$

$$u_o = \frac{\phi_o}{|\phi_o|} A \Big/ B \tan^{-1}\left\{ \sqrt{(D^2 - 1.0)} \Big/ \cos(\alpha_c) \right\} \tag{4.115}$$

$$v_o = 0.0 \tag{4.116}$$

4.5.3.1 Geodetic to *xy*-Oblique Mercator Coordinate Transformation (Forward Equations)

Given the geodetic coordinates of one point (ϕ, λ), one can compute the corresponding rectangular coordinates as follows:

$$t = \tan (\pi/4.0 - \phi/2.0)/\{(1.0 - e \sin \phi)/(1.0 + e \sin \phi)\}^{0.5e} \tag{4.117}$$

$$Q = E/t^B \tag{4.118}$$

$$S = 0.5(Q - 1.0/Q) \tag{4.119}$$

$$T = 0.5(Q + 1.0/Q) \tag{4.120}$$

$$V = \sin (B\{\lambda - \lambda_o\}) \tag{4.121}$$

$$U = (-V \cos \gamma_o + S \sin \gamma_o)/T \tag{4.122}$$

$$v = A \log \left\{ \frac{1.0 - U}{1.0 + U} \right\} \Big/ \{2.0B\} \tag{4.123}$$

$$u = A/B \tan^{-1}\{[S \cos \gamma_o + V \sin \gamma_o]/\cos [B(\lambda - \lambda_o)]\} \tag{4.124}$$

$$x = v \cos \alpha_c + u \sin \alpha_c + \text{false easting} \tag{4.125}$$

$$y = u \cos \alpha_c - v \sin \alpha_c + \text{false northing} \tag{4.126}$$

4.5.3.2 *xy*-Oblique Mercator to Geodetic Coordinate Transformation (Inverse Equations)

Given the rectangular coordinates (x, y), the corresponding geodetic coordinates can be computed as follows:

$$x_r = x - \text{false easting} \tag{4.127}$$

$$y_r = y - \text{false northing} \tag{4.128}$$

$$v = x_r \cos \alpha_c - y_r \sin \alpha_c \tag{4.129}$$

$$u = y_r \cos \alpha_c + x_r \sin \alpha_c \tag{4.130}$$

$$Q' = 2.718281828^{-Bv/A} \tag{4.131}$$

$$S' = 0.5(Q' - 1.0/Q') \tag{4.132}$$

$$T' = 0.5(Q' + 1.0/Q') \tag{4.133}$$

$$V' = \sin(Bu/A) \tag{4.134}$$

$$U' = (V' \cos \gamma_o + S' \sin \gamma_o)/T' \tag{4.135}$$

$$t = E\Big/\sqrt{[(1.0 + U')/(1.0 - U')^{1.0/B}} \tag{4.136}$$

$$\phi = \pi/2.0 - 2.0 \tan^{-1}(t[(1.0 - e \sin \phi)/(1.0 + e \sin \phi)]^{0.5e}) \tag{4.137}$$

One can iterate for ϕ using an initial estimate ($\phi = \pi/2.0 - 2.0 \tan^{-1}t$). Finally, the longitude can be computed as follows:

$$\lambda = \lambda_o - \tan^{-1}\{(S' \cos \gamma_o - V' \sin \gamma_o)/\cos (Bu/A)\}/B \tag{4.138}$$

The above-mentioned projections are employed in developing the state plane coordinate system in the United States. The Lambert Conformal projection is used for states with dominant east–west extent. The Transverse Mercator projection is used for states with dominant north–south extent. The Oblique Mercator projection is used for states with oblique extent. The type of map projection and the associated constants (e.g., the standard parallels, the latitude and the longitude of the origin of the map, the scale factor, false easting, false northing, etc.) for each state can be seen in Snyder (1987). PC-Software for state plane coordinate transformations can be found in Moffitt and Bossler (1998).

REFERENCES

Heiskanen, W. and Moritz, H., *Physical Geodesy* (Reprint), Institute of Physical Geodesy, Technical University, Graz, Austria, 1996.
Kraus, K., *Photogrammetry, Fundamentals and Standard Processes* (vol. 1), Dummler, Bonn, Germany, 1993.
Moffitt, F. and Bossler, J., *Surveying* (10th edn.), Addison Wesley Longman, Menlo Park, CA, 1998.
Moffitt, F. and Mikhail, E., *Photogrammetry* (3rd edn.), Harper & Row, New York, 1980.
REA (Staff of Research and Education Association), *Handbook of Mathematical, and Engineering Formulas, Tables, Functions, Graphs, Transforms*, Research and Education Association, Piscataway, NJ, 1997.
Snyder, J. P., *Map Projections—A Working Manual*, U.S. Government Printing Office, Washington, DC, 1987.

5 Basic Electromagnetic Radiation

Carolyn J. Merry

CONTENTS

This chapter serves as an introduction to the basic concepts of electromagnetic radiation. Much of the material discussed in this chapter is taken from Avery and Berlin (1992), Jensen (2007), and Lillesand and Kiefer (2004). The reader is encouraged to consult these three textbooks for additional technical information.

Remotely sensed data of the earth are collected by special sensors that have the capacity to record electromagnetic energy, acoustic energy (sound), or variations in force (gravity, magnetic). However, the focus of this chapter will be on the electromagnetic wavelengths that are recorded by aircraft or satellite sensors. Passive remote-sensing records electromagnetic radiation that is reflected or emitted naturally from an object and normally depends on energy from the sun. In contrast, active remote sensing depends on electromagnetic energy that is generated by the sensor itself and then the sensor records the reflected energy from the object. Examples of active remote sensing are radar and sonar systems.

5.1 ELECTROMAGNETIC ENERGY

The world we live in is full of energy. For example, light, heat, electricity, and sound are some of the forms of energy. One source of energy important to remote sensing is electromagnetic energy. The main source of electromagnetic energy received by the earth is light from the sun. Electromagnetic energy travels at the speed of light,

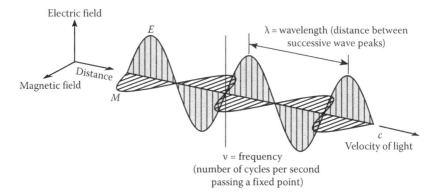

FIGURE 5.1 Electromagnetic wave. (Adapted from Lillesand, T.M. and Kiefer, R.W., *Remote Sensing and Image Interpretation*, 5th edn., John Wiley & Sons, New York, 2004.)

which is about 299,792,458 m/s (about 186,000 miles/s). Electromagnetic energy is composed of two parts—an electrical wave and a magnetic wave that vibrate at right angles to each other (Figure 5.1). We measure the electromagnetic wave energy in terms of wavelength, λ, which is the length from the top of one wave to the top of the next; amplitude, which is the height of the wave; and frequency (ν), which is the number of waves that pass through a specified point each second.

Light-energy waves belong to a family called the electromagnetic spectrum. The electromagnetic spectrum includes visible rays, light that our eyes can detect, and invisible rays, such as gamma rays, x-rays, ultraviolet (UV) waves, infrared (IR) radiation, microwaves, and television and radio waves. As previously mentioned, electromagnetic energy (light waves) travels at the same speed—the speed of light—but their wavelengths are different, and light waves have different effects on earth materials. Special instruments are used to record certain wavelengths of electromagnetic energy in the form of images, which we call remote-sensing imagery.

5.2 ELECTROMAGNETIC SPECTRUM

The electromagnetic spectrum defines the entire region of wavelengths. Normally, the spectrum is divided into arbitrary regions with a beginning and an ending wavelength. This interval is commonly referred to as a spectral band, channel, or region (Figure 5.2, see plate 1). The spectral regions, which are approximate, are further defined in Table 5.1. The spectral regions are normally called UV, visible, near IR, middle IR, thermal IR, and microwave. Below the UV region are the short wavelengths of gamma waves measured in nanometers (nm) and above the microwave region are the long wavelengths of radio waves measured in centimeters. However, these two wavelength regions are rarely used in terrestrial remote sensing because the earth's atmosphere absorbs the energy in these wavelength regions.

The UV region ranges from 0.03 to 0.4 μm (micrometers). The UV (literally meaning "above the violet") region borders the violet portion of the visible wavelengths. These are the shortest wavelengths practical for remote sensing. The UV can be divided into the far UV (0.01–0.2 μm), middle UV (0.2–0.3 μm), and near

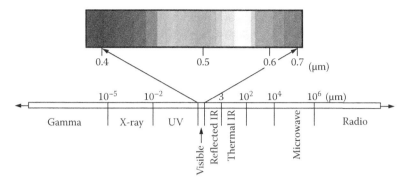

FIGURE 5.2 (**See color insert following page 426.**) Spectral regions of the electromagnetic spectrum.

TABLE 5.1
Spectral Regions

Spectral Region	Wavelength
Gamma rays	<0.03 nm
X-rays	0.03–30 nm
UV region	0.03–0.4 μm
Far UV	0.01–0.2 μm
Middle UV	0.2–0.3 μm
Near UV (photographic UV band)	0.3–0.4 μm
Visible region	0.4–0.7 μm
Visible blue	0.4–0.5 μm
Visible green	0.5–0.6 μm
Visible red	0.6–0.7 μm
Reflected near-IR region	0.7–3 μm
Photographic IR	0.7–1.3 μm
Middle IR	1.5–1.8 μm, 2.0–2.4 μm
Thermal IR (far IR)	3–5 μm, 8–14 μm (below ozone layer), 10.5–12.5 μm (above ozone layer)
Microwave	0.1–100 cm
Radar region	0.1–100 cm
K	0.8–2.4 cm
X (3.0 cm)	2.4–3.8 cm
C (6 cm)	3.8–7.5 cm
S (8.0 cm, 12.6 cm)	7.5–15.0 cm
L (23.5 cm, 25.0 cm)	15.0–30.0 cm
P (68 cm)	30.0–100.0 cm
Radio	>100 cm

UV (0.3–0.4 μm). The sun is a natural source of UV radiation. However, wavelengths shorter than 0.3 μm are unable to pass through the atmosphere and reach the earth's surface due to atmospheric absorption by the ozone layer. Only the 0.3–0.4 μm wavelength region is useful for terrestrial remote sensing.

The most common region of the electromagnetic spectrum used in remote sensing is the visible band, which spans from 0.4 to 0.7 μm (Figure 5.2, see plate 1). These limits correspond to the sensitivity of the human eye. Blue (0.4–0.5 μm), green (0.5–0.6 μm), and red (0.6–0.7 μm) represent the additive primary colors—colors that cannot be made from any other color. Combining the proper proportions of light that represent the three primary colors can produce all colors perceived by the human eye. Although sunlight seems to be uniform and homogeneous in color, sunlight is actually composed of various wavelengths of radiation, primarily in the UV, visible, and near-IR portions of the electromagnetic spectrum. The visible part of this radiation can be shown in its component colors when sunlight is passed through a prism, which bends the light through refraction according to wavelength (Figure 5.3).

Most of the colors that we see are a result of the reflection and absorption of wavelengths from objects on the earth's surface. For example, chlorophyll pigments in healthy vegetation naturally absorb the blue and red wavelengths of white light, as these wavelengths are used in photosynthesis, and reflect more of the green wavelengths. Thus, vegetation appears as green to our eyes. Snow is seen as white, since all wavelengths of visible light are scattered. Fresh basaltic lava appears black, as all wavelengths are absorbed and, therefore, no wavelengths of light are being reflected back to the human eye.

The IR band includes wavelengths between the red light (0.7 μm) of the visible band and microwaves at 1000 μm. Infrared literally means "below the red" because it is adjacent to red light. The reflected near-IR region (0.7–1.3 μm) is used in black-and-white IR and color IR–sensitive film. The middle IR includes energy at wavelengths ranging from 1.3 to 3 μm. Middle IR energy is detected using electro-optical sensors.

The thermal (far) IR band extends from 3 to 1000 μm. However, due to atmospheric absorption, the wavelength regions of 3–5 and 8–14 μm are typically used for remote-sensing studies. The thermal IR region is directly related to the sensation of heat. Heat energy, which is continuously emitted by the atmosphere and all features on the earth's surface, dominates the thermal IR band. Optical-mechanical scanners

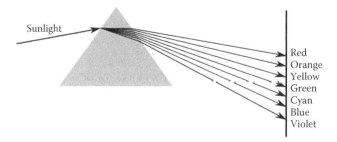

FIGURE 5.3   White light passed through a prism will generate a spectrum of colors.

and special vidicon systems are typically used to record energy in this part of the electromagnetic spectrum.

The microwave region is from 1 mm to 1 m. The radio wavelengths are the longest waves. Microwaves can pass through clouds, precipitation, tree canopies, and dry surficial deposits. There are two types of sensors that operate in the microwave region. Passive microwave systems detect natural microwave radiation that is emitted from the earth's surface. Radar (*ra*dio *d*etection *a*nd *r*anging) systems—active remote sensing—propagate artificial microwave radiation to the earth's surface and record the reflected component. Typical radar systems are listed in Table 5.1.

In summary, the spectral resolution of most remote-sensing systems is described in terms of their bandwidths in the electromagnetic spectrum.

5.3 ENERGY TRANSFER

Energy is the ability to do work, and in the process of doing work, energy is moved from one place to another. The three ways that energy moves include conduction, convection, and radiation.

Conduction is when a body transfers its kinetic energy to another body by physical contact. For example, when frying your eggs in a pan, heat is transferred from the electric coils on the stove to the frying pan. Another example is when you place your spoon in a hot cup of coffee. The spoon becomes hot because of the transfer of energy from the hot liquid to the spoon.

Convection is the transfer of kinetic energy from one place to another by physically moving the bodies. For example, convection occurs when warmer air rises in a room, setting up currents. Heating of the air near the ground during the early morning hours sets up convection currents in the atmosphere. Likewise, water in a lake will turn over during the fall, moving the cooling water on the surface to the bottom, mixing with the warmer water. Convection occurs in liquids and gases.

Radiation is the transfer of energy through a gas or a vacuum. Energy transfer from the sun to the earth by electromagnetic radiation takes place in the vacuum of space.

There are two models that are used to describe electromagnetic energy—the wave model and the particle model.

5.3.1 WAVE MODEL

The process of light movement is nearly impossible to see. Most physicists believe light energy travels in waves, like water. The light energy is carried along in very tiny ripples, which are much smaller than the actual waves in water. If you think of a cork in a pond, the waves make the cork move up and down, but the cork does not move in the direction of the water wave. In a similar manner, the light energy waves vibrate up and down at right angles like the cork, while the light energy moves along the wave.

Electromagnetic radiation is generated when an electrical charge is accelerated. The wavelength (λ) depends on the time that the charged particle is accelerated and is normally measured from peak to peak in µm, nm, or cm. The frequency of the electromagnetic radiation depends on the number of accelerations per second and is measured as the number of wavelengths that pass a point for a given unit of time. For example, an electromagnetic wave that sends one crest every second is said to have a frequency of one cycle per second or 1 hertz (Hz).

The relationship between λ and ν is (Boleman 1985)

$$c = \lambda\nu \tag{5.1}$$

where c is defined as the speed of light in a vacuum.

The frequency is then defined as

$$\nu = \frac{c}{\lambda} \tag{5.2}$$

and the wavelength is defined as

$$\lambda = \frac{c}{\nu} \tag{5.3}$$

From these two equations, the frequency is inversely proportional to wavelength, which means that the longer the wavelength, the lower the frequency or, the shorter the wavelength, the higher the frequency. As electromagnetic radiation passes from one substance to another, the speed of light and wavelength will change while the frequency remains the same.

All earth objects—water, soil, rocks, and vegetation—and the sun emit electromagnetic energy. The sun is the initial source of electromagnetic energy that can be recorded by many remote-sensing systems; radar and sonar systems are exceptions, as they generate their own source of electromagnetic energy.

A blackbody is defined as an object that totally absorbs and emits radiation at all wavelengths. For example, the sun is a 6000 K blackbody (Figure 5.4). The total emitted radiation from a blackbody (M_λ) is measured in watts per m^2 (W/m^2) and is proportional to the fourth power of its absolute temperature (T), measured in Kelvin:

$$M_\lambda = \sigma T^4 \tag{5.4}$$

where σ is the Stefan–Boltzmann constant of 5.6697×10^8 W/m^2/K^4.

This relationship is known as the Stefan–Boltzmann law. The law states that the amount of energy emitted by an earth object (or the sun) is a function of its temperature. The higher the temperature, the greater the amount of radiant energy emitted by the object.

In addition to computing the total amount of energy, the dominant wavelength of this radiation can be calculated as

$$\lambda_{max} = \frac{k}{T} \tag{5.5}$$

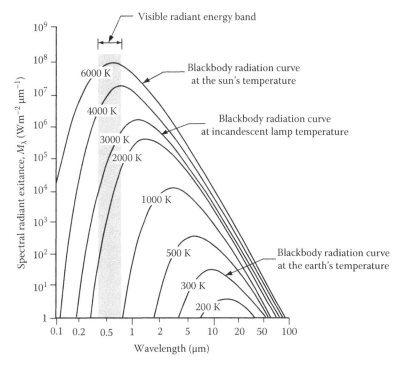

FIGURE 5.4 Spectral energy distribution from representative blackbodies. (Adapted from Lillesand, T.M. and Kiefer, R.W., *Remote Sensing and Image Interpretation*, 5th edn., John Wiley & Sons, New York, 2004.)

where

k is a constant of 2898 μm K

T is the absolute temperature, K

This is known as Wien's displacement law. For example, we can calculate the dominant wavelength of the radiation for the sun at 6000 K and the earth at 300 K (27°C or 80°F) as

$$\lambda_{max} = 2898 \, \mu m \, \frac{K}{6000 \, K}$$

$$\lambda_{max} = 0.483 \, \mu m$$

and

$$\lambda_{max} = 2898 \, \mu m \, \frac{K}{300 \, K}$$

$$\lambda_{max} = 9.66 \, \mu m$$

Thus, the sun's maximum wavelength of radiation is 0.48 μm, located in the visible spectrum, which are the wavelengths our eyes are sensitive to. The earth's peak wavelength at 9.66 μm is in the thermal region. As this relationship demonstrates, as the temperature of an object increases, its dominant wavelength (λ_{max}) shifts toward the shorter wavelengths of the electromagnetic spectrum.

5.3.2 PARTICLE MODEL

Electromagnetic energy can also be described as a particle model by considering that light travels from the sun as a stream of particles. Einstein found that when light interacts with matter, the light has a different character and behaves as being composed of many individual bodies called photons that carry particle-like properties, such as energy and momentum. Physicists view electromagnetic energy interaction with matter by describing electromagnetic energy as discrete packets of energy or quanta.

An atom is made up of electrons—negatively charged particles—that move around a positively charged nucleus. The electron is kept in orbit around the nucleus by the interaction of the negative electron around the positive neutron. For the electron to move up to a higher level, work has to be performed, and energy needs to be available to move the electron up a level. Once an electron jumps up to a higher level, that is, the electron becomes excited, radiation is emitted. This radiation is a single packet of electromagnetic radiation—a particle unit of light called a photon. Another word for this is a quantum. In its orbital path around the nucleus, the electron makes a quantum leap or jump.

Planck described the nature of radiation and proposed the quantum theory of electromagnetic radiation. A relationship between the frequency of radiation, as described by the wave model, and the quantum is (Boleman 1985)

$$Q = h\nu \tag{5.6}$$

where
    Q is the energy of a quantum, in joules (J)
    h is the Planck constant (6.626×10^{-34} J/s)
    ν is the frequency of radiation

If we take Equation 5.3 and multiply it by 1 or h/h, then

$$\lambda = \frac{hc}{h\nu} \tag{5.7}$$

By substituting Q for $h\nu$ from Equation 5.6, the wavelength, λ, associated with a quantum of energy is

$$\lambda = \frac{hc}{Q}$$

or rearranging

$$Q = \frac{hc}{\lambda}$$

From this, the energy of a quantum is inversely proportional to its wavelength. In other words, the longer the wavelength, the lower its energy content. This inverse relationship is important to remote sensing. The relationship indicates that it is more difficult to detect longer wavelength energy being emitted, such as the thermal IR or microwave wavelengths, than those at shorter or visible wavelengths. In fact, remote-sensing sensors typically "look at" larger pixel sizes at thermal wavelengths or in the microwave region, so that there is enough electromagnetic energy to record. You also experience this phenomenon by getting sunburn from UV rays when outdoors too long—there is a lot more energy at these shorter wavelengths.

5.4   ENERGY–MATTER INTERACTIONS IN THE ATMOSPHERE

Electromagnetic energy travels from the sun to the earth. In the vacuum of space, not very much happens to electromagnetic energy. However, once the electromagnetic energy hits the earth's atmosphere, then the speed, the wavelength, the intensity, and the spectral distribution of energy may change. Four types of interactions are common—transmission, scattering, absorption, and reflection.

5.4.1   TRANSMISSION

Transmission occurs when the incident radiation passes through the medium without any measurable changes. Essentially, the matter is transparent to the radiation. However, when electromagnetic energy encounters materials of different densities, such as air or water, then refraction or the bending of light occurs when the light passes from one medium to another. The refraction of light occurs because the media are of different densities and the speed of electromagnetic energy will be different in each medium. The index of refraction, n, is a measure of the optical density of a substance. This index is the ratio of the speed of light in a vacuum, c, to the speed of light in a second medium, c_n:

$$n = \frac{c}{c_n} \tag{5.8}$$

The speed of light in a material can never reach the speed of light in a vacuum; therefore, the index of refraction is always greater than one. Light travels slower in water than air. For example, the index of refraction for the atmosphere is about 1.0002926 and for water it is about 1.33.

Snell's law can be used to describe refraction. This law states that for a given frequency of light, the product of the index of refraction and the sine of the angle between the ray and a line normal to the interface at the two media is constant:

$$n_1 \sin \theta_1 = n_2 \sin \theta_2 \tag{5.9}$$

The amount of refraction is a function of the angle θ made with the vertical, the distance involved (the greater the distance of the atmosphere, the more the changes in density), and the density of the air involved (usually air is more dense near sea level than at higher elevations).

5.4.2 SCATTERING

Three types of scattering are important in remote sensing—Rayleigh, Mie, and nonselective scattering. Particles in the atmosphere cause the scattering of the electromagnetic energy. Scattering occurs when radiation is absorbed and then reemitted by atoms or molecules. The direction of the scattering is impossible to predict.

Rayleigh scattering—molecular scattering—occurs when the effective diameter of the particle is many times smaller (<0.1) than the wavelength of the incoming electromagnetic energy. Usually air molecules, such as oxygen and nitrogen, cause scattering. Rayleigh scattering occurs predominantly in the upper 4.5 km of the atmosphere. The amount of scattering is inversely related to the fourth power of the wavelength. For example, UV light at $0.3\,\mu m$ is scattered approximately 16 times more ($(0.6/0.3)^4 = 16$) and blue light at $0.4\,\mu m$ is scattered about 5 times more ($(0.6/0.4)^4 = 5.06$) than red light at $0.6\,\mu m$. Rayleigh scattering causes the blue sky that we are used to seeing on clear, bright sunny days—the shorter violet and blue wavelengths are more efficiently and preferentially scattered than the longer orange and red wavelengths. In the morning or in the evening, the sun's light is passing through a longer path length. As a consequence, the red and orange wavelengths are hardly scattered at all, which makes for the red sunrises and red sunsets.

Mie scattering—nonmolecular scattering—occurs in the lower 4.5 km of the atmosphere. The particles that are responsible for Mie scattering are roughly equal in size to the incoming wavelength. The actual size of particles ranges from 0.1 to 10 times (0.1 mm to several mm in diameter) the wavelength of the incident energy and are typically dust and other particles. Mie scattering is greater than Rayleigh scattering. Pollution particles—smoke and dust—in the air contribute even more to the red sunrises and sunsets.

Nonselective scattering takes place at the lowest portions of the atmosphere. The particles are roughly greater than 10 times the incoming wavelength. Since this type of scattering is nonselective, all wavelengths are scattered and appear as white light. Water droplets and ice crystals causing up clouds and fog are examples of nonselective particle scattering, making clouds and fog to appear white.

The reason why scattering is important to remote sensing is that scattering can reduce the information content of imagery. Image contrast is lost, making it difficult to tell objects apart from one another. Filters on cameras have to be used to eliminate or filter out wavelengths that would normally cause scattering.

5.4.3 ABSORPTION

Absorption is the process where radiant energy is absorbed and converted into other forms of energy. This can take place in the atmosphere or on the earth's surface. Absorption bands are a range of wavelengths in the electromagnetic spectrum, where the radiant energy is absorbed by a substance, such as water (H_2O), carbon dioxide (CO_2), oxygen (O_2), ozone (O_3), or nitrous oxide (N_2O). This is important for remote sensing because in these absorption bands, there is no energy available to be sensed by a remote-sensing instrument. The visible spectrum—0.4–0.7 μm—does not absorb all of the incident energy, but rather transmits the energy. This portion of the electromagnetic spectrum is known as an atmospheric window.

Chlorophyll in plants absorbs blue and red wavelengths for use in photosynthesis. This characteristic is used to map vegetation types. Also, water is an excellent absorber of energy for most wavelengths of light. Minerals important for geologic purposes also have unique absorption properties that are used for identification purposes.

5.4.4 REFLECTION

Reflection occurs when the electromagnetic radiation "bounces off" an object. The incident radiation, the reflected radiation, and the vertical to the surface where the angles of incidence and reflection are measured, all lie in the same plane (Figure 5.5). Two types of reflecting surfaces are possible—specular and diffuse surfaces (Figure 5.5). Specular reflection occurs when the surface is essentially smooth and the angle of incidence is equal to the angle of reflection. The average surface profile height is several times smaller than the wavelength of the electromagnetic radiation striking the surface. For example, a calm water body will act like a near-perfect specular reflector.

If the surface has a large surface height relative to the wavelength of the incoming electromagnetic radiation, the reflected light rays will go in many different directions—this is known as diffuse radiation. Examples would include white powders, white paper,

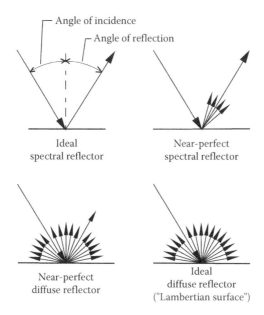

FIGURE 5.5 Specular versus diffuse reflectance. (Adapted from Lillesand, T.M. and Kiefer, R.W., *Remote Sensing and Image Interpretation*, 5th edn., John Wiley & Sons, New York, 2004.)

or other similar materials. A perfectly diffuse surface occurs when the reflecting light rays leaving the surface are constant for any angle of reflectance. This is known as an ideal surface—a Lambertian surface (Schott 1997).

5.5 ENERGY–MATTER INTERACTIONS WITH THE TERRAIN

An understanding and knowledge of energy–matter interactions of radiant energy is essential for accurate image interpretation and to extract biophysical information from remote-sensing imagery. This understanding and knowledge is the key and focus of remote-sensing research.

The amount of radiant energy absorbed, reflected, or transmitted through an object per unit time is called the radiant flux (Φ) and is measured in watts (W) (Figure 5.6). The radiation budget equation for incoming and outgoing radiant flux to a surface from any angle in a hemisphere can be stated as

$$\Phi_{i\lambda} = \rho_\lambda + \tau_\lambda + \alpha_\lambda \tag{5.10}$$

where
  $\Phi_{i\lambda}$ is the total amount of incoming radiant flux at a specific wavelength (λ)
  ρ_λ is the amount of energy reflected from the surface
  τ_λ is the amount of energy transmitted through the surface
  α_λ is the amount of energy absorbed by the surface

Hemispherical reflectance (ρ_λ) is defined as the ratio of the radiant flux reflected from a surface to the radiant flux incident to it:

$$\rho_\lambda = \frac{\Phi_{\text{reflected}}}{\Phi_{i\lambda}} \tag{5.11}$$

Hemispherical transmittance (τ_λ) is defined as the ratio of the radiant flux transmitted through a surface to the radiant flux incident to it:

$$\tau_\lambda = \frac{\Phi_{\text{transmitted}}}{\Phi_{i\lambda}} \tag{5.12}$$

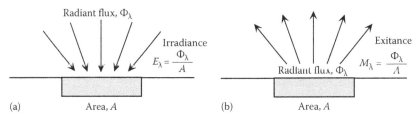

FIGURE 5.6   Radiant flux density. (After Jensen, J.R., *Remote Sensing of the Environment: An Earth Resource Perspective*, 2nd edn., Prentice Hall, Upper Saddle River, NJ, 2007.)

Hemispherical absorptance (α_λ) is defined by the following relationship:

$$\alpha_\lambda = \frac{\Phi_{absorbed}}{\Phi_{i\lambda}} \qquad (5.13)$$

or

$$\alpha_\lambda = 1 - (\rho_\lambda + \tau_\lambda) \qquad (5.14)$$

These three equations imply that radiant energy must be conserved. These radiometric quantities are useful to produce general statements about the spectral reflectance, absorptance, and transmittance of earth terrain features. A percentage reflectance ($\rho_{\%\lambda}$) can be obtained by taking Equation 5.11, multiplying by 100, to get

$$p_{\%\lambda} = \frac{\Phi_{reflected}}{\Phi_{i\lambda}} \times 100 \qquad (5.15)$$

This equation is used commonly in remote-sensing research to describe the spectral reflectance characteristics of various terrain features. Figure 5.7 shows several typical spectral reflectance curves. These curves do not provide any information about the absorption or transmittance characteristics of the terrain features. However, since the remote-sensing sensors—cameras and multispectral scanners—only record reflected energy, this information is still very valuable to the remote-sensing interpreter and forms the basis for identifying and assessing objects on the ground.

These hemispherical reflectance, transmittance, and absorptance radiometric quantities do not provide information on the exact amount of energy reaching an area on the ground from a given direction or about the exact amount and direction of energy

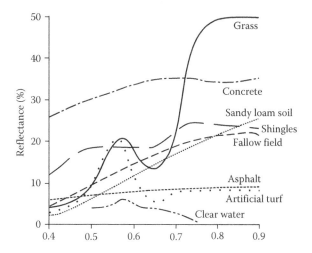

FIGURE 5.7 Typical spectral reflectance curves. (After Jensen, J.R., *Remote Sensing of the Environment: An Earth Resource Perspective*, 2nd edn., Prentice Hall, Upper Saddle River, NJ, 2007.)

leaving this same ground area. It is important to refine the radiometric measurement techniques so that more precise radiometric information can be extracted from remotely sensed data. Additional radiometric quantities will be introduced next.

Imagine an area $1 \times 1\,m^2$ in size that is bathed in radiant flux (Φ) for specific wavelengths (Figure 5.6). The average radiant flux density—the irradiance, E_λ—will be the amount of radiant flux divided by the area of this flat surface in Watts per square meter:

$$E_\lambda = \frac{\Phi_\lambda}{A} \qquad (5.16)$$

The amount of radiant flux leaving—the exitance, M_λ—will be the energy leaving per unit area, A, of this flat surface in Watts per square meter:

$$M_\lambda = \frac{\Phi_\lambda}{A} \qquad (5.17)$$

Radiance is the most precise radiometric measurement used in remote sensing. Radiance for a given wavelength—L_λ—is the radiant flux leaving a surface in a given direction ($A \cos \theta$) and solid angle (Ω) per unit area:

$$L_\lambda = \frac{\dfrac{\Phi_\lambda}{\Omega}}{A \cos \theta} \qquad (5.18)$$

Ideally, energy from the earth's surface will not become scattered in this solid angle field of view and "contaminate" the radiant flux from the ground. However, there will most likely be scattering in the atmosphere and from nearby areas of the earth's surface that contribute to this energy in the solid angle field of view. As a result, atmospheric transmission, scattering, absorption, and reflection—described earlier—will influence the radiant flux before the energy is recorded by the remote-sensing system.

The discussion in this chapter has focused on the propagation of energy through the atmosphere and the interaction of this energy with terrain features. Detection of this electromagnetic energy is then performed either by using a photographic process (aerial photography) or by an electronic process (video cameras or multispectral scanners). Details of these two processes will be covered in Part III.

In summary, an understanding of basic electromagnetic energy concepts is important when using remote-sensing imagery. In this way, the nature of interactions that take place from when the electromagnetic energy leaves the source, travels through the atmosphere, illuminates the terrain feature, and then is finally recorded by the remote-sensing instrument can be understood. Additional information on remote sensing—sensors, processing the data, hardware and software, accuracy—are covered in much more detail in Chapters 16 through 24.

REFERENCES

Avery, T. E. and Berlin, G. L., *Fundamentals of Remote Sensing and Airphoto Interpretation* (5th edn.). MacMillan, New York, 1992.

Boleman, J., *Physics: An Introduction*. Prentice Hall, Englewood Cliffs, NJ, 1985.

Jensen, J. R., *Remote Sensing of the Environment: An Earth Resource Perspective* (2nd edn.). Prentice Hall, Upper Saddle River, NJ, 2007.

Lillesand, T. M. and Kiefer, R. W., *Remote Sensing and Image Interpretation* (5th edn.). John Wiley & Sons, New York, 2004.

Schott, J. R., *Remote Sensing—The Image Chain Approach*. Oxford University Press, New York, 1997.

6 Data Analysis

Charles D. Ghilani

CONTENTS

6.1 INTRODUCTION

Professionals in the geospatial industries collect large quantities of data. No matter the level of care taken in collecting these data, they will contain error. This chapter is about the analysis of observations and their errors.

Observations are defined as measurements made to determine unknown quantities. Observations may be either *direct* or *indirect*. Direct measurements are the application of a device or apparatus to determine an unknown quantity. The application of a tape to determine the length and the width of a tabletop are examples of direct measurements. Direct measurements always contain some errors. Assume, for example, that the tape used to determine the length of the tabletop is graduated to the nearest 0.1 ft, and the length is determined to be 6.12 ft, where the hundredths of a foot is estimated by the observer. If this same observer used another tape graduated to the nearest 0.01 ft, the length may be observed as 6.116 ft. If the observer continued to obtain more precise tapes, the observations for the length of the table could continue to be refined. These are examples of errors caused by the refinement of the instrument. We would also see error if we repeatedly used the same tape to determine the length of the table due to the placement of the tape, the amount of tension on the tape, the temperature of the tape at the time of the reading, and the operator's ability to estimate the reading of the tape.

Indirect observations are determined by applying mathematics to direct measurements. For example, the area of the aforementioned tabletop is the product of the length and the width. Thus, as shown in Figure 6.1, the area of the tabletop contains errors from both the length and width measurements. These errors are propagated into the error of the computed area. The resulting area will have the error in the shaded region. The computation of area is an indirect observation that has a resulting uncertainty due to *error propagation*.

FIGURE 6.1   Error in the area of tabletop.

Sources of errors can be categorized as *natural, instrumental,* or *personal.* For example, the previously discussed tape will change length due to its expansion or contraction caused by changes in temperature. This is an example of a natural error. Temperature, pressure, and humidity often affect direct observations in surveying. In satellite surveying, the total free-electron count in the ionosphere is usually the largest natural source of error. Natural errors can often be modeled mathematically to reduce their impact on observations. At other times, proper field procedures can minimize the effect of natural errors. For example, it is known that lines of sight will refract differently in different atmospheric conditions. Since the amount of refraction is dependent on the length of the sight, this source of error is minimized in differential leveling by keeping sight distances approximately equal in length (Ghilani and Wolf, 2008).

Examples of instrumental errors include inherent errors in the instrumentation and the calibration status of the instrument. For example, tapes with more precise graduations will typically yield more precise observations when used properly. However, no matter what the level of graduation, the manufacturing process will also create errors in observations. Thus, instruments are periodically *calibrated* to ensure that they are functioning properly and in some cases to ensure that mechanical connections have maintained proper alignment. For example, the level bubble of a differential leveling instrument can change with respect to the horizontal plane of the instrument. Surveyors typically check the level bubbles after each setup. They also perform a specific calibration procedure known as a horizontal collimation test periodically. Instruments that are found to contain too much error for the required accuracy of the job are repaired and brought back to within the manufacturer's specifications (Ghilani and Wolf, 2008).

Personal errors arise due to limitations in human senses. The size of errors can be affected by a person's ability to see and their manual dexterity. Several factors can influence the size of personal errors including environmental conditions, insects, and the operator's state of alertness.

Errors can also be categorized as either *systematic* or *random.* Systematic errors, which are also called *biases,* are errors that follow some mathematical or physical law. They often bias the solution to one side or the other of the true value. Often these errors can be removed or reduced in size by proper field procedures or by the application of a mathematical model. For example, if an oven thermometer reads 50° too high, the cook will simply set the oven to a 50° higher temperature to get the proper conditions for baking. Likewise, if a calibrated 100 ft tape is found to be 0.012 ft too long under field conditions, the resulting measurements will be corrected by −0.012 ft per every 100 ft measured with this tape. That is, a 323.56 ft observation will be corrected to

$$\text{Length corrected } = 323.56 - 323.56 \left(\frac{0.012}{100} \right) = 323.52 \text{ ft}$$

In this instance, a mathematical formula has been used to remove the systematic error.

Random errors are all the errors that remain after removing systematic errors and any mistakes from the observations. In general, they are the result of human and instrumental imperfections. An example of a random error is one's ability to read an instrument. For example, when determining the length of the table, repeated readings of 6.12, 6.11, 6.13, 6.07, and 6.17 ft may occur. Assuming systematic errors are properly taken into account, the variation in these observations is due to random errors. In general, random errors in surveying observations follow the laws of probability and conform to the normal distribution curve (see Section 6.3). That is, they are impossible to avoid, tend to be small in size, and are equally distributed in sign about the true value.

One last observation problem that must be defined is a *blunder*. Blunders are commonly known as *mistakes*. Blunders are not classified as errors. They are often caused by the confusion or carelessness of the observer. To isolate mistakes, professionals will observe more than the minimum amount of observations needed to establish geometric closures. For example, to determine the elevation of a stake, a surveyor need only perform differential leveling from one benchmark* to the stake. However, if any mistake occurred in the leveling process, it would result in an incorrect elevation on the stake. To check this elevation, additional observations are made to determine the observed elevation of the stake from another benchmark. After an observed elevation from the second benchmark is established, it is checked for agreement with the previous value. These additional observations are known as *redundant* observations. Redundant observations are used to check observations. For example, assume that we have six equations with three unknown parameters. Using any three of these equations, we could solve for the unknown parameters. Thus the remaining three equations are redundant, but can be used to check the computations. When dealing with systems of equations, these redundant equations are commonly referred to as *degrees of freedom*.

6.2 BASIC STATISTICS

In surveying, a *population* of observations tends to infinity, and thus it is impossible to acquire all the possible observations. To determine the length of a table, we may take only three observations. These three observations represent a *sample* of

---

* A benchmark is a location with known and published elevation.

the entire population of observations for the length of the table. From this sample of observations, we then make inferences about the population. For example, the *true value* for the length of the table is the arithmetic mean of all the observations in the population. Since we can never obtain all the observations for the population, the true value is never known. However, the arithmetic mean of the sample is used as an estimate for this true value. For a general review of basic mathematical statistics, the reader is referred to Mendenhall and Sincich (2006).

Assume that we obtain n observations, l_1 *through* l_n, for the length of a table. The sample mean $\bar{l}$ for the length of the table is

$$\bar{l} = \frac{l_1 + l_2 + \cdots + l_n}{n} \tag{6.1}$$

where the bar over l indicates that it is the mean of sample data. Since the sample mean is taken from a representative sample of the population, some variation between the sample mean and the population mean can be expected.

The mean is one measure of stating the central tendency of data. Other measures for the central tendency include the *median* and the *mode*. The median is the middle value of the data when it is arranged in numerical order. If the data set has an even number of elements, it is then the average of the two values nearest to the middle. The mode is the most commonly occurring data value. A sample data set can have more than one mode.

A sample of observations can be analyzed graphically or numerically. For example, a large sample of observations of the same value may be plotted on a graph so that the tendencies of the observations can be observed. Thus, the variations in the computed coordinates of a point can be plotted to determine the central locus of the coordinates and the amount of variation between the coordinates for the point. This is often done in Global Positioning Systems (GPS), where the number of observations may range in the hundreds and would be difficult to analyze individually. Therefore, it is more common to perform these analyses numerically.

Some common numerical methods for analyzing data include the *range* or *dispersion* of the data, *discrepancies* between the individual observations, and the *mean* and the *variance* of the observations. The range of the data is the difference between the largest and smallest values when the data is ordered numerically. The mean of a data set was defined in Equation 6.1 and is an estimate for the true value of the population. The most commonly used discrepancy is known as the *residual*, which is the difference between the mean and an observation. For example, the residual v_i for observation l_i is

$$v_i = \bar{l} - l_i \tag{6.2}$$

where $\bar{l}$ is previously defined as the mean of the sample of observations. Since observations tend to follow the normal distribution (Ghilani, 2006), we expect the size of the residuals to be small in size and distributed equally about the mean.

The *standard deviation* for a sample of data is a measure of variation in the sample data and is defined as

$$S = \sqrt{\frac{v_1^2 + v_2^2 + \cdots + v_n^2}{n-1}} = \sqrt{\frac{\sum v_i^2}{n-1}} \tag{6.3}$$

where $\sum v_i^2$ is the notation for the sum of the squared residuals, that is, $\sum v_i^2 = v_1^2 + v_2^2 + \cdots + v_n^2$. The square of the standard deviation is known as the variance of the sample and is determined as

$$S^2 = \frac{v_1^2 + v_2^2 + \cdots + v_n^2}{n-1} = \frac{\sum v_i^2}{n-1} \tag{6.4}$$

The standard deviation and the variance of the data are estimates of their equivalent values in the population from which the sample was taken. The standard deviation and the variance for a sample of data are measures of the *precision* of the data. The precision is a measure of the repeatability of the data. This term is often confused with *accuracy*, which is a measure of the absolute nearness of the data to the true value for the population. Since we never know the true value for the population, we never really know the accuracy of our data.

The *standard deviation of the mean* provides a range for the population mean based on a sample mean and sample standard deviation. The standard deviation of the mean is

$$S_y = \sqrt{\frac{v_1^2 + v_2^2 + \cdots + v_n^2}{n(n-1)}} = \sqrt{\frac{\sum v_i^2}{n(n-1)}} = \frac{S}{\sqrt{n}} \tag{6.5}$$

Example 6.1

Assume that the data 6.12, 6.11, 6.13, 6.07, 6.17, 6.11, 6.09, 6.12, 6.16, and 6.14 ft were collected for the length of a table. What are the range, the mean, the standard deviation, the variance, and the standard deviation of the mean for the data?

Solution

The data arranged in numerical order is 6.07, 6.09, 6.11, 6.11, 6.12, 6.12, 6.13, 6.14, 6.16, and 6.17 ft. Thus, the range for this sample of data is $6.17 - 6.07 = 0.10$ ft.

The mean of the data is

$$\bar{l} = \frac{6.12 + 6.11 + 6.13 + 6.07 + 6.17 + 6.11 + 6.09 + 6.12 + 6.16 + 6.14}{10}$$

$$= \frac{61.2}{10} = 6.122 \text{ ft.}$$

Its median value is the average of the fifth and sixth elements in the numerically ordered data. In this sample, it is 6.12. This data set has two values that appear twice, which are 6.11 and 6.12. Thus, this data set is bimodal.

The residuals and their squared values are

Obs	$v = \bar{l} - l_i$	v^2
6.12	0.002	0.000004
6.11	0.012	0.000144
6.13	−0.008	0.000064
6.07	0.052	0.002704
6.17	−0.048	0.002304
6.11	0.012	0.000144
6.09	0.032	0.001024
6.12	0.002	0.000004
6.16	−0.038	0.001444
6.14	−0.018	0.000324
Σ	0.000	0.00816

The standard deviation for the data is $S = \sqrt{0.00816 / (10 - 1)} = \pm 0.030$ ft.

By Equation 6.4, the variance is $S^2 = 0.000907$.

The standard deviation of the mean is $S_{\bar{y}} = \pm 0.03/\sqrt{10} = \pm 0.009$ ft.

6.3　NORMAL DISTRIBUTION

As stated earlier, the data collected by geospatial professionals generally conforms to the normal distribution shown in Figure 6.2. The curve ranges from negative to positive infinity and is asymptotic to the x axis, the peak being at the population mean. The graph shows the probability of the occurrence of a residual. As can be seen in

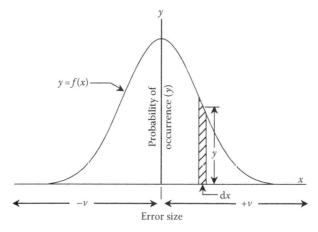

FIGURE 6.2　Plot of the normal distribution curve.

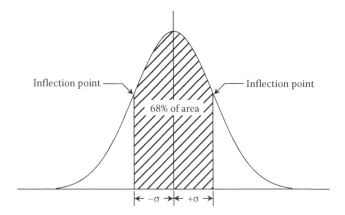

FIGURE 6.3 Points of inflection on normal distribution curve.

the figure, the mean with a residual of zero has the highest probability of occurrence. Also note that large residuals have the lowest probabilities and thus seldom occur. This means that small errors occur more frequently than large errors. Points of inflection occur where the concavity of a curve changes. As shown in Figure 6.3, the points of inflection occur at one standard deviation away from the curve's central value, its mean (Ghilani, 2006). This represents about 68.27% of the total area under the curve. Thus, about 68% of the data should be within one standard deviation of the mean for a data set. From Example 6.1, we see that about 68% of the data should be within the range determined from 6.12 ± 0.03, that is, 68% of the data should be between 6.09 and 6.15. Ranges that are equally distributed about the mean are often called *confidence intervals*. In this case, we have a one-sigma confidence interval. By observation, we see that 7 of the 10 values are within this one-sigma confidence interval. This represents 70% of the sample data.

As stated previously, the area under the normal distribution curve represents the probability of occurrence. The area that goes from negative infinity to some critical *t* value is determined as

$$F_x(t) = \int_{-\infty}^{t} \frac{1}{\sigma\sqrt{2\pi}} e^{-x^2/2\sigma^2} dx \tag{6.6}$$

Figure 6.4 depicts the area determined by this integral. Since Equation 6.6 cannot be integrated in closed form, a table of critical values is derived for a population having a mean of 0 and variance of 1. This table is typically available in introductory statistics books. The tabular values are used as multipliers to obtain other *probable errors*. Table 6.1 lists some of these multipliers, their probable errors, and their common names. For example, a multiplier of 2 is commonly used to determine the range for the *two-sigma level* of confidence. Thus from Example 6.1, the two-sigma level for the data is

$$\bar{l} \pm 2.00S = 6.122 \pm 2(0.030)$$
$$6.062 < l < 6.182$$

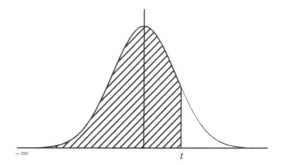

FIGURE 6.4   Area under curve defined by integral in Equation 6.6.

TABLE 6.1
Commonly Used Probable Errors

Probable Error (%)	Multiplier	Common Name
50	0.6745σ	Linear error probable
68.3	1σ	One-sigma error
90	1.6449σ	90% confidence level
95	1.9599σ	95% confidence level
95.4	2σ	Two-sigma level
99.7	3σ	Three-sigma level

This means that 95.4% of the data should be between 6.062 and 6.182. By observation, we see that all of the data is within this range. Often geospatial professionals use the two-sigma multiplier to ferret out blunders in their data. This means that approximately 5% of the time an observation with a large random error rather than a blunder will be discarded. This is why mistakes or blunders are sometimes called *outliers*, since they could be mistake-free observations with large random errors.

6.4   SAMPLE VERSUS POPULATION

As previously discussed, statistics computed from samples can vary with the elements in the sample and the size of the sample. Thus, it is possible to determine the length of the table as 6.12 ft at one observation session and later determine it to be 6.13 ft. This variation in the sample means and variances is the subject of statistics. For example, if only the odd observations in the Example 6.1 had been used, a mean and a variance of 6.13 and 0.00155, respectively, would be computed. Since the mean and the variance computed from samples of data are only estimates for their equivalent population values, we need to determine the accuracy of these estimates. To do this, the t, χ^2, and F distributions are used (Ghilani, 2006). These distributions list the probabilities for the sample means, variances, and the ratio of two sample variances, respectively, and can typically be found at the end of introductory statistics books.

In Section 6.3, the multipliers from the normal distribution were used to create a two-sigma confidence interval. However, these multipliers were based on the normal distribution, which is based on an infinite number of observations. As stated earlier, geospatial professionals collect samples of data. Depending on the size of the sample more or less confidence can be placed on the sample mean as a good estimate for the population mean. For example, we tend to place more confidence on the mean derived from a sample of 30 elements than we would on a sample mean based on only 3 elements even if both samples had the same variance. The t distribution is used to develop confidence intervals for the population mean. For example, a sample set with only two redundant observations would have a t multiplier of 4.30 when building a 95% confidence interval for the data as compared to the 1.96 used from the normal distribution in Section 6.3 (Ghilani, 2006). The t multiplier for a data set having 25 redundant observations would have a multiplier of 2.06. Table 6.2 lists a sample of multipliers used when developing 95% confidence intervals for varying numbers of redundant observations. From this, we see that as the number of redundant observations increases, the t multiplier approaches that of the normal distribution. In fact, statisticians state that it is appropriate to use the normal distribution t multiplier when the number of redundant observations is over 30. We can use this number (30) as guidance in various measurement processes.

In Example 6.1, there were 10 observations yielding 9 redundant observations. From Table 6.2, we see that the t multiplier used in the example should have been 2.26 rather than 1.96. Using 2.26 yields a larger 95% confidence interval and allows for error in the sample mean and standard deviation. In this example, it yields the same results as the normal distribution t multiplier of 1.96 since all of the data was within the interval based on the normal distribution value. Geospatial professionals often use 2 as the multiplier accepting the fact that the resultant confidence interval will be smaller, and thus an occasional acceptable observation will be discarded as an outlier.

The χ^2 distribution is used to develop confidence intervals for the variance. One application of this is called the *goodness of fit* test in a least-squares adjustment. In establishing a stochastic model, the geospatial professional typically sets the population variance to a value of 1. Following the adjustment of the data, a sample

TABLE 6.2
Critical t Values for Developing 95%
Confidence Intervals Varying
Numbers of Redundant Observations

Redundancies	t Value	Redundancies	t Value
1	12.71	10	2.23
2	4.30	15	2.13
3	3.18	20	2.09
5	2.57	25	2.06
9	2.26	30	2.04

TABLE 6.3

Critical χ^2 Values for Developing 95% Confidence Intervals Varying Numbers of Redundant Observations

Redundant Observations	Lower Bound	Upper Bound	Redundant Observations	Lower Bound	Upper Bound
1	0.00098	5.023	10	6.262	20.48
3	0.216	9.348	15	9.591	27.49
5	0.831	12.83	20	13.12	34.17
9	2.700	19.02	30	16.79	46.98

variance, called the *reference variance*, is computed for the overall adjustment. The size of the reference variance demonstrates how well the observations fit the functional and stochastic models. For a weighted adjustment, a reference variance of 1 is expected. If the confidence interval for the population variance based on the adjustment's values does not contain 1, then we either have blunders in the data, a poor stochastic model, or both (Ghilani, 2006).

The χ^2 distribution is used to develop a confidence interval for the population variance based on a sample reference variance and the number of redundant observations in the adjustment. If the confidence interval contains 1, the adjustment is said to *pass* the goodness of fit test. As stated previously, this test can fail if there are blunders in the data or errors in the stochastic model. The sample χ^2 value is computed as

$$\chi^2 = \frac{rS^2}{\sigma^2} \tag{6.7}$$

where
  r is the number of redundant observations in the adjustment
  S^2 is the adjustment's reference variance
  σ^2 is the population's variance, which is typically set to 1 in the stochastic model

Unlike the normal and t distributions, the χ^2 distribution is not symmetric. Thus, the bounds of the confidence interval must be computed independently based on the critical values listed in a table of χ^2 critical values. Table 6.3 list the χ^2 distribution critical values for a 95% confidence interval based on varying degrees of freedom. A complete table of χ^2 critical values can be found in most introductory statistics books.

Example 6.2

A weighted least-squares adjustment (see Section 6.5) having nine redundant observations has a reference variance of 2.32. Does this adjustment pass the goodness of fit test at a 95% level of confidence? If not, what could be done to correct the adjustment?

SOLUTION

The adjustment's χ^2 value is $\chi^2 = (9(2.32)/1) = 20.88$. From Table 6.3, we see that the 95% confidence interval for an adjustment with nine redundant observations is

$$2.70 < \chi^2 < 19.02$$

It can be seen that the computed value for χ^2 of 20.88 is not within the 95% confidence interval for a value computed with nine degrees of freedom. Thus, it can be stated that the adjustment's reference variance of 1 is not within the bound of the 95% confidence interval. This is caused by either a blunder in the observations or poor estimates for variances of the observations that were used to build the stochastic model for the adjustment. Since the computed χ^2 value is greater than the upper limit of the confidence interval, we can state that there was either a blunder in the data or some of the estimated variances for the observations were too small (Ghilani, 2006).

The F distribution is used to compute confidence intervals for the ratio of two sample variances. This is often used in the geospatial industries to compare two adjustments that used different geodetic or survey control.* For example, the differential leveling survey shown in Figure 6.5 ties into three benchmarks A, B, and C into a station U with unknown elevation. The arrows in the drawing indicate the direction of the leveling operation. Any of the six lines could be used to determine the elevation of U. This means that there are five redundant observations shown in the figure. Since any of the three benchmarks may contain a blunder, an adjustment is performed using only benchmark A's known elevation. Stations B and C are treated as unknown stations and also included in the adjustment along with station U. Using only the minimum number of control to solve the adjustment is known as a *minimally constrained*[†] adjustment. Following the minimally constrained adjustment, a second adjustment is performed that includes the stations B and C as control. The second adjustment results in a second sample reference variance being computed. The only difference in the two adjustments is the inclusion of the benchmark elevations for B and C in the second adjustment. If all

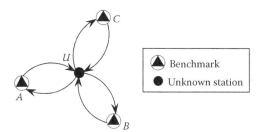

FIGURE 6.5 Differential leveling survey.

---

* Controls are used in adjustments to fix the observations in space. For example, in differential leveling, a minimum of one station with known elevation is used to establish elevations of unknown stations. In plane surveying, a station with known coordinates and a line with known or fixed direction is used to fix the survey in position and rotation in space.

† A minimally constrained adjustment is sometimes called a *free-network* adjustment.

TABLE 6.4

95% Confidence Interval Multipliers for Error Ellipse Axes with Varying Numbers of Redundant Observations

Redundant Observations	Multiplier c	Redundant Observations	Multiplier c
1	19.97	10	2.86
2	6.16	15	2.71
3	4.37	20	2.64
4	3.72	30	2.58
5	3.40	60	2.51

three control elevations for A, B, and C agree with the observations, and if the observations do not contain any systematic errors, the two reference variances should agree with each other, that is, their ratio should be statistically equal to 1. The computed F value is

$$F = \frac{\text{larger sample variance}}{\text{smaller sample variance}} \tag{6.8}$$

The F value computed from Equation 6.8 is compared against a critical value from an F distribution table. If the computed F is greater than the tabulated value, the test is said to fail. This means that either one or more of the benchmark elevations do not agree or there are systematic errors in the observations that need to be removed. Since the F distribution tables are extensive, the reader is referred to the references at the end of this chapter for critical values from the F distribution tables (Ghilani, 2006).

Another common use of the F distribution is as a multiplier for increasing the probability of computed error ellipses (see Section 6.6) from a least-squares adjustment. Table 6.4 contains multipliers derived from the F distribution tables for varying numbers of redundant observations in the adjustment. For example, assume that the error ellipse for a station has a semimajor axis length of 0.034 ft from an adjustment with 10 redundant observations. An error ellipse with a 95% level of confidence would have a length of 2.86 × 0.034, or 0.097 ft.

6.5 LEAST-SQUARES ADJUSTMENTS

Since all observations have errors, they seldom meet geometric closures. In order to force geometric closures and aid in the discovery of blunders, geospatial data are often adjusted. Since random errors follow normal distribution theory, they are adjusted accordingly. As a graduate student, Gauss used the method known as least squares in 1794. The method of least squares was first published by Legendre in 1805 and is commonly used to adjust observations now. With the least-squares method, observations are adjusted according to the normal distribution theory and yield the most probable values for the unknowns based on the sample set of data. However, before the advent of computers, the simultaneous solution of multitudes of equations was tedious and difficult. Thus, several approximate methods, such as the compass rule, were developed. With the availability and the computing power of the personal computer, least-squares adjustment of observations has become an integral part of the office process.

The least-squares method involves the simultaneous solution of a set of *observation equations*. That is, for each observation an equation is written in terms of the known and unknown parameters and the error in the observation. For example, in the case of differential leveling, the observation equation relates the unknown or known elevations of stations to the observed differences in elevation. The residual error, υ, is included in each equation to yield a consistent set of equations. This is expressed mathematically as

$$\text{Elevation}_j \ - \ \text{Elevation}_i \ = \Delta\text{Elev}_{ij} + \upsilon \tag{6.9}$$

where
> Elevation_i and Elevation_j are the known or unknown elevations of the stations I and J, respectively
> ΔElev_{ij} is the observed difference in elevation between stations I and J
> υ is the amount of residual error in the observation after the adjustment

The system of equations in matrix form is expressed as

$$AX = L + V \tag{6.10}$$

where
> A is the matrix of coefficients for the unknown parameters
> X is the vector of unknown parameters
> L is the vector of observations
> V is the vector of residual error

This matrix system typically involves more observations than unknown parameters and thus has redundant observations. It is solved using the least-squares method as

$$X = \left(A^{\mathsf{T}}WA\right)^{-1} A^{\mathsf{T}}WL = N^{-1}A^{\mathsf{T}}WL = Q_{xx}A^{\mathsf{T}}WL \tag{6.11}$$

where
> A and L are defined in Equation 6.10
> X contains the *most probable values* for the unknowns
> W is a matrix of weights for each observation

It should be noted here that $A^{\mathsf{T}}WA$ is the so-called *normal* matrix, which is typically denoted as N, and its inverse is a *cofactor* matrix, which is typically denoted as Q_{xx}.

One of the advantages of the least-squares method is the ability to weight observations proportionately to their estimated variances. Doing this allows the adjustment to place more error in the observations with the higher variances (Ghilani, 2006). This weighting process is known as the *stochastic* model for the adjustment. For independent observations, individual weights are computed as

$$w_i = \frac{\sigma_0^2}{\sigma_i^2} = \frac{1}{\sigma_i^2} \tag{6.12}$$

where
    w_i is the weight of the ith observation
    σ_i^2 is the estimated variance of the ith observation
    σ_0^2 is the a priori reference variance for the adjustment, which, as shown in Equation 6.12, is typically assigned a value of 1

For each observation, σ_i is estimated and a square matrix of weights is developed. However, in some cases, the W matrix is replaced by the identity matrix. In this case, the adjustment is said to be unweighted. The typical format for the weight matrix having independent observation is

$$W = \begin{bmatrix} w_1 & 0 & \cdots & 0 \\ 0 & w_2 & & \vdots \\ \vdots & & \ddots & 0 \\ 0 & \cdots & 0 & w_m \end{bmatrix} \tag{6.13}$$

where $w_1, w_2, \ldots, w_m$ are the weights of the observations 1 through m as defined by Equation 6.12.

In some instances, the observations are dependent on each other. This is especially true in the case of indirect observations. For example, most GPS softwares employ a two-step solution process. In the first adjustment, the individual baseline vectors are computed from the pseudorange observations between the satellites and receivers. The result of this adjustment creates dependencies between the baseline vector components. Following these adjustments, baseline vectors are simultaneously adjusted to compute the coordinates of the unknown stations. In this second adjustment, the baseline vector components are indirect observations. That is, they are computed from direct observations. The baseline vector components have dependencies, which are depicted in the weight matrix by nonzero off-diagonal elements. Thus, the weight and *covariance* matrices for a baseline vector are

$$W = \Sigma^{-1} = \begin{bmatrix} \sigma_{XX} & \sigma_{XY} & \sigma_{XZ} \\ \sigma_{XY} & \sigma_{YY} & \sigma_{YZ} \\ \sigma_{XZ} & \sigma_{YZ} & \sigma_{ZZ} \end{bmatrix}^{-1} \tag{6.14}$$

where

W is the weight matrix for the adjustment

Σ is the covariance matrix

σ_{XX}, σ_{YY}, and σ_{ZZ} are the variances for the changes in the X, Y, and Z coordinates, respectively

σ_{XY}, σ_{XZ}, and σ_{YZ} are the covariance elements depicting the dependencies in the baseline vector components

When nonzero covariance elements are present in the weight matrix, it implies that the unknowns are *correlated*; that is, a change in one unknown will directly affect the computation of the other unknowns that have nonzero elements.

Example 6.3

Benchmarks A, B, and C in Figure 6.5 have elevations of 100.00, 124.16, and 108.76 ft, respectively. The observed elevation differences and estimated standard deviations are

SOLUTION

Following Equation 6.9, the six observation equations are

$U - 100.00 = 12.61 + v_{ua}$

$100.00 - U = -13.09 + v_{au}$

$U - 124.16 = -11.32 + v_{ub}$

$124.16 - U = 9.88 + v_{bu}$

$U - 108.76 = 4.87 + v_{cu}$

$108.76 - U = -4.27 + v_{uc}$

Line	ΔElev (ft)	S (ft)
AU	12.61	0.56
UA	-13.09	0.56
BU	-11.32	0.76
UB	9.88	0.76
CU	4.87	0.64
UC	-4.27	0.64

There are six observation equations with only one unknown. Thus, there are five redundant observations since any of the six equations could be used to solve for the elevation of U. However, the most probable value for U is obtained by using the least-squares method. The observation equations written in matrix form are

$$AX = \begin{bmatrix} 1 \\ -1 \\ 1 \\ -1 \\ 1 \\ -1 \end{bmatrix} [U] = \begin{bmatrix} 12.61 + 100.00 \\ -13.09 - 100.00 \\ -11.32 + 124.16 \\ 9.88 - 124.16 \\ 4.87 + 108.76 \\ -4.27 - 108.76 \end{bmatrix} + \begin{bmatrix} v_{au} \\ v_{ua} \\ v_{bu} \\ v_{ub} \\ v_{cu} \\ v_{uc} \end{bmatrix} = L + V$$

Using Equations 6.12 and 6.13, the stochastic model for the adjustment is

$$
W = \begin{bmatrix}
\dfrac{1}{0.56^2} & 0 & 0 & 0 & 0 & 0 \\
0 & \dfrac{1}{0.56^2} & 0 & 0 & 0 & 0 \\
0 & 0 & \dfrac{1}{0.76^2} & 0 & 0 & 0 \\
0 & 0 & 0 & \dfrac{1}{0.76^2} & 0 & 0 \\
0 & 0 & 0 & 0 & \dfrac{1}{0.64^2} & 0 \\
0 & 0 & 0 & 0 & 0 & \dfrac{1}{0.64^2}
\end{bmatrix}
$$

$$
= \begin{bmatrix}
3.19 & 0 & 0 & 0 & 0 & 0 \\
0 & 3.19 & 0 & 0 & 0 & 0 \\
0 & 0 & 1.73 & 0 & 0 & 0 \\
0 & 0 & 0 & 1.73 & 0 & 0 \\
0 & 0 & 0 & 0 & 2.44 & 0 \\
0 & 0 & 0 & 0 & 0 & 2.44
\end{bmatrix}
$$

Applying Equation 6.11, the least-squares solution for the most probable value for U given this set of observations is

$$
X = (14.72297)^{-1} 666.28907
$$
$$
= 113.176
$$

Thus, based on this set of observations, the most probable value for the elevation of U is 113.176 ft. The adjusted elevation differences and residuals for each observation are computed using Equation 6.9, yielding

From	To	ΔElev	v
A	U	13.18	0.566
U	A	−13.18	−0.086
B	U	−10.98	0.336
U	B	10.98	1.104
C	U	4.42	−0.454
U	C	−4.42	−0.146

Note in this adjustment that observation UA has the smallest residual and UB has the largest residual. This agrees with the variances that were used prior to the adjustment. That is, UA had the smallest variance and thus the largest weight while UB had the largest variance and thus the lowest weight. This demonstrates how weights control the distribution of the error during the adjustment, typically placing more error in the observations having the largest variance.

6.6 POST-ADJUSTMENT STATISTICS

After a least-squares adjustment, several statistical values can be determined, which are used as estimates of the accuracy of the results. The reference variance discussed in Section 6.4 for an adjustment is computed as

$$S_0^2 = \frac{\sum w_i v_i^2}{m-n} = \frac{V^T W V}{m-n} \tag{6.15}$$

where
   S_0^2 is the reference variance for the adjustment based on the sample of data
   w_i and v_i are the weights and residuals of the observations
   m is the number of independent observation equations
   n is the number of unknowns

As shown in Equation 6.15, the sample reference variance can also be computed using the residual matrix V and weight matrix W as defined in Equations 6.10 and 6.11. In Example 6.3, the reference variance for the adjustment is

$$S_0^2 = \frac{\begin{bmatrix} 0.566 & -0.086 & 0.336 & 1.104 & -0.454 & -0.146 \end{bmatrix} W \begin{bmatrix} 0.566 \\ -0.086 \\ 0.336 \\ 1.104 \\ -0.454 \\ -0.146 \end{bmatrix}}{6-1}$$

$$= \frac{3.90958}{5} = 0.7812$$

where W was defined previously. Another commonly computed statistic is the *standard deviation of unit weight* so called since the weights were based on an a priori reference variance of 1. In this example, the standard deviation of unit weight S_0 is $\sqrt{0.7812} = 0.88$.

Repeating that matrix $A^T W A$ in Equation 6.11 is the normal matrix N, its inverse, Q_{xx}, multiplied by the reference variance for the adjustment creates the covariance matrix, S_{xx}, for the unknown parameters (Ghilani, 2006). In Example 6.3, the covariance matrix for the adjustment is 0.7812[0.06792], which is [0.05306]. The square roots of the diagonal elements of covariance matrix for the unknown parameters yield the standard deviations for the unknowns. In this simple example, the standard deviation for the adjusted elevation of U is $\sqrt{0.05306} = \pm 0.23$ ft. One of the biggest advantages of the least-squares method over other adjustment techniques is the ability to determine not only the most probable values for the unknowns but also the estimated error in these values.

Other statistical values derived from a least-squares adjustment include the covariance matrix for the adjusted observations S_{ll}, which is computed in matrix form as

$$S_{ll} = S_0^2 A N^{-1} A^T = S_0^2 A Q_{xx} A^T = S_0^2 Q_{ll} \tag{6.16}$$

where

N and Q_{xx} are defined in Equation 6.11

Q_{ll} is the covariance matrix for the adjusted observations

The covariance matrix for the adjusted elevation differences in Example 6.3 is

$$S_{ll} = 0.7812 \begin{bmatrix} 1 \\ -1 \\ 1 \\ -1 \\ 1 \\ -1 \end{bmatrix} [0.06792] \begin{bmatrix} 1 & -1 & 1 & -1 & 1 & -1 \end{bmatrix}$$

$$= \begin{bmatrix} 0.05306 & -0.05306 & 0.05306 & -0.05306 & 0.05306 & -0.05306 \\ -0.05306 & 0.05306 & -0.05306 & 0.05306 & -0.05306 & 0.05306 \\ 0.05306 & -0.05306 & 0.05306 & -0.05306 & 0.05306 & -0.05306 \\ -0.05306 & 0.05306 & -0.05306 & 0.05306 & -0.05306 & 0.05306 \\ 0.05306 & -0.05306 & 0.05306 & -0.05306 & 0.05306 & -0.05306 \\ -0.05306 & 0.05306 & -0.05306 & 0.05306 & -0.05306 & 0.05306 \end{bmatrix}$$

The diagonal of this matrix are the variances for the adjusted elevation differences. Each diagonal element is assigned to each adjusted observation in the same order as they were entered into the adjustment. Thus, the 1,1 element of S_{ll} is the variance for the observation going from A to U, the 2,2 element is the variance for the observation UA, and so on. Since this example had only one unknown, the six adjusted observations have the same standard deviation as the adjusted elevation for U. This atypical example yields a standard deviation for each adjusted elevation of ± 0.23 ft. In a least-squares problem involving multiple unknowns, the variances for each adjusted observation will vary according to the errors in the computed unknowns typically.

In some adjustments, the maximum error at a station having multiple unknowns is desired. For example, in a horizontal survey, the standard deviations in the unknowns are in the cardinal directions of the coordinate axes. However, as shown in Figure 6.6, the maximum error at the station will be in some other alignment typically. The reason for this is the correlation between the x and y coordinates (Ghilani, 2006). Figure 6.7 depicts the standard error rectangle as determined by the standard deviations in the coordinates. The t angle is the amount of rotation of the y axis that is

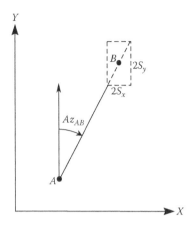

FIGURE 6.6 Region defined by standard deviations in station's coordinates.

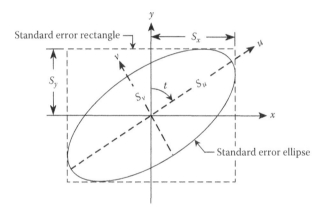

FIGURE 6.7 Components of the standard error ellipse and its bounding standard error rectangle.

necessary to remove the correlation between the coordinates. This rotation determines the direction of the u axis at the station and the v axis, which is the direction of minimum error at the station. This correlation can be removed at each station by determining the amount of rotation, t, at the station that is in the direction of the maximum error. The equations for determining the t angle and the sizes of the S_u and S_v axes are

$$\tan 2t = \frac{2q_{xy}}{q_{yy} - q_{xx}}$$

$$S_u = S_0 \sqrt{q_{xx} \sin^2 t + 2q_{xy} \cos t \sin t + q_{yy} \cos^2 t} \qquad (6.17)$$

$$S_v = S_0 \sqrt{q_{xx} \cos^2 t - 2q_{xy} \cos t \sin t + q_{yy} \sin^2 t}$$

where

S_0 is defined by Equation 6.15

q_{xy} is the covariance element of the station's submatrix

q_{xx} and q_{yy} are the diagonal elements of the station's submatrix with respect to the station's x and y coordinates

The individual elements of the error ellipse are from the Q_{xx} matrix defined in Equation 6.11.

The probability levels of S_u and S_v as computed in Equation 6.17 are somewhere between 35% and 39% (Ghilani, 2006). This is known as the *standard error ellipse*. To increase the probability of the error ellipse, a multiplier can be determined using the F distribution tables and

$$c = \sqrt{2F_{\alpha,2,\text{number of redundant observations}}} \qquad (6.18)$$

where

c is the multiplier

α is defined as 1—level of probability

the critical value from the F distribution table is determined with two degrees of freedom in the numerator and the number of redundant observations in the adjustment for the denominator. Table 6.4 lists some multipliers for an error ellipse at a 95% level of confidence. In this table, α is determined as $1-0.95$, which equals 0.05. The lengths of the semimajor and semiminor axes of the error ellipse at some other level of probability are computed as

$$S_{u\%} = cS_u$$
$$S_{v\%} = cS_v \qquad (6.19)$$

Example 6.4

A least-squares adjustment is performed on a network of four stations having stations R, S, T with unknown coordinates. The standard deviation of unit weight as defined by Equation 6.14 is 1.648. The adjustment had 10 redundant observations. The inverse of the normal matrix, Q_{xx}, and its corresponding X matrix is

$$Q_{xx} = \begin{bmatrix} 0.0000 & 0.0053 & 0.0002 & 0.0041 & 0.0006 & 0.0033 \\ 0.0053 & 2.8514 & 0.1007 & 2.2151 & 0.3453 & 1.7761 \\ 0.0002 & 0.1007 & 2.6825 & -0.5296 & 0.3112 & -1.1707 \\ 0.0041 & 2.2151 & -0.5296 & 3.9658 & 1.0957 & 3.4216 \\ 0.0006 & 0.3453 & 0.3112 & 1.0957 & 3.1200 & 1.4329 \\ 0.0033 & 1.7761 & -1.1707 & 3.4216 & 1.4329 & -6.0716 \end{bmatrix} \times 10^{-4} \quad X = \begin{bmatrix} x_R \\ y_R \\ x_S \\ y_S \\ x_T \\ y_T \end{bmatrix}$$

What are the standard deviations and the standard error ellipse semimajor- and semiminor-axes lengths for station S? Compute these values at the 95% level of confidence.

SOLUTION

Before responding to the problem at hand, note the correlation between the x and y coordinates for the stations as evidenced by the covariance (off-diagonal) nonzero elements in the Q_{xx} matrix. These correlations imply that if any single coordinate is changed, the other coordinates must also change.

To do this problem, we must isolate the submatrix that corresponds to the unknown x and y coordinates for station S. We see in the X matrix that these coordinates are in the third and fourth rows. Thus, the submatrix we need is at the intersection of rows 3 and 4 and columns 3 and 4 in the Q_{xx} matrix. This submatrix is

$$\begin{bmatrix} 2.6825 & -0.5296 \\ -0.5296 & 3.9658 \end{bmatrix} \times 10^{-4}$$

The standard deviation in the x and y coordinates for station S are

$$S_x = S_0\sqrt{q_{xx}} = 1.648\sqrt{0.00026825} = \pm0.027 \text{ ft}$$

$$S_y = S_0\sqrt{q_{yy}} = 1.648\sqrt{0.00039658} = \pm0.033 \text{ ft}$$

These standard deviations are at 68% probability. To increase probability to 95%, we use the multiplier of 2.23 from Table 6.2 yielding values of

$$S_{x,95\%} = tS_x = 2.23(0.027) = \pm0.060 \text{ ft}$$

$$S_{y,95\%} = tS_y = 2.23(0.033) = \pm0.073 \text{ ft}$$

Using Equation 6.17, the standard error ellipse for station S is

$$\tan 2t = \frac{2(-0.00005296)}{0.00039658 - 0.00026825} = -0.82537$$

$$t = 160.232°$$

$$S_u = S_0\sqrt{q_{xx}\sin^2 t + 2q_{xy}\cos t \sin t + q_{yy}\cos^2 t}$$

$$= 1.648\sqrt{0.0000307 - 0.0000337 + 0.0003512}$$

$$= \pm0.034 \text{ ft}$$

$$S_v = S_0\sqrt{q_{xx}\cos^2 t - 2q_{xy}\cos t \sin t + q_{yy}\sin^2 t}$$

$$= 1.648\sqrt{0.0002376 - 0.0000337 + 0.0000454}$$

$$= \pm0.026 \text{ ft}$$

S_u and S_v are the lengths of the semimajor and semiminor axes, respectively, with the semimajor axis being rotated about 160° from the y axis. To get these values at a 95% level of probability, the standard error ellipse axes lengths are multiplied by the appropriate c value found in Table 6.4. Since this adjustment has 10 redundant observations, we see that the c multiplier is 2.86 so that the 95% error ellipse has semimajor and semiminor axes lengths of ±0.097 ft and ±0.074 ft, respectively. Notice in this example that the semimajor axis S_u is greater than S_y and that the semiminor axis S_v is less than S_x.

6.7   SUMMARY

Using statistics and least squares, geospatial professionals can analyze their observations, adjust these values to find the most probable solution to any set of equations, and compute post-adjustment statistics to aid in further analysis and understanding about the precisions of their data. In fact, it may be stated that statistics are the cornerstone around which all geospatial professionals manage their observations.

REFERENCES

Baarda, W. 1967. *Statistical Concepts in Geodesy*. Netherlands Geodetic Commission, Delft, the Netherlands.

Baarda, W. 1968. *A Testing Procedure for Use in Geodetic Networks*. Netherlands Geodetic Commission, Delft, the Netherlands.

Foster, R. 2003. Uncertainty about positional uncertainty. *Point of Beginning* 28(11): 40.

Ghilani, C. D. 2003a. Statistics and adjustments explained—Part 1: Basic concepts. *Surveying and Land Information Science* 63(2): 62.

Ghilani, C. D. 2003b. Statistics and adjustments explained—Part 2: Sample sets and reliability. *Surveying and Land Information Science* 63(3): 141.

Ghilani, C. D. 2003c. Statistics and adjustments explained—Part 3: Error propagation. *Surveying and Land Information Science* 64(1): 23.

Ghilani, C. D. and P. R. Wolf. 2006. *Adjustment Computations: Spatial Data Analysis*. John Wiley & Sons, Inc., Hoboken, NJ.

Ghilani, C. D. and P. R. Wolf. 2008. *Elementary Surveying: An Introduction to Geomatics*. Prentice Hall, Upper Saddle River, NJ.

Mendenhall, W. and T. Sincich. 2006. *Statistics for Engineering and the Sciences,* 5th edn. Pearson Education, Upper Saddle River, NJ.

Mikhail, E. M. 1976. *Observations and Least Squares*. University Press of America, Inc., Washington, DC.

Mikhail, E. M. and G. Gracie. 1981. *Analysis and Adjustment of Survey Measurements*. Van Nostrand Reinhold, New York.

Moffit, F. and J. Bossler. 1997. *Surveying*. Addison Wesley Longman, Menlo Park, CA.

Pope, A. J. 1976. The statistics of residuals and the detection of outliers. NOAA Technical Report NOS 65 NGS 1. U.S. Department of Commerce, Washington, DC.

Schwarz, C. R. 2005. The effects of unestimated parameters. *Surveying and Land Information Science* 65(2): 87.

Tan, W. 2002. In what sense a free net adjustment? *Surveying and Land Information Science* 62(4): 251.

Part II

Global Positioning System

7 Introducing the Global Positioning System

Chris Rizos

CONTENTS

The NAVSTAR Global Positioning System (GPS) is a satellite-based positioning, navigation, and timing (PNT) system designed, financed, deployed, and operated by the U.S. Department of Defense (PNT 2009), and for which applications have burgeoned since the system was declared fully operational in 1995. The attractions of GPS as a PNT technology include:

- High positioning accuracy, ranging from meters down to the millimeter level.
- The capability of determining velocity and time, to an accuracy commensurate with position.
- No intervisibility of GPS ground stations is required for high-accuracy positioning.

- Results are obtained with reference to a single, global datum.
- Position information is provided in three dimensions.
- Signals are available to users anywhere on or above the earth—in the air, in space, on the ground, or at sea.
- There are no user charges.
- Relatively low-cost user hardware.
- An all-weather system, available 24 h a day.

GPS has also revolutionized the fields of geodesy, surveying, and mapping, commencing with its introduction to the civilian community in the early 1980s. Indeed, among the first users were geodetic surveyors who applied GPS to the task of surveying primary control networks that form the basis of all map data and digital databases. Today, around the world, GPS is unchallenged as the quintessential technology for such geodetic applications. However, as a result of user equipment and algorithmic innovations over the last two decades, GPS technology is increasingly addressing the precise positioning needs of cadastral, engineering, environmental, planning, and geographic information system (GIS) surveys, as well a range of precise land, air, and marine navigation applications.

Over the next 5–10 years, several other satellite-based navigation systems will be deployed. These will augment the current (and upgraded) GPS, providing users with access to significantly more satellite signals than is currently the case. Nowadays the phrase *Global Navigation Satellite System* (GNSS) is used as an umbrella term for all current and future global, satellite-based, radio-navigation systems. Although GPS is currently the only fully operational GNSS, the Russian Federation's GLONASS is undergoing replenishment and will be fully operational by 2010, it is planned that the European Union's GALILEO will be deployed and be operational by 2013, and China's COMPASS is also likely to join the "GNSS Club" by the middle of the next decade. The focus of Part II of this book, however, is on the current GPS, with occasional reference to the incomplete GLONASS constellation and how it is used in combination with GPS in specialist GPS+GLONASS receivers. The acronym "GNSS" will therefore be mostly used in Chapter 15, where the future of GPS/GNSS will be discussed.

7.1 BACKGROUND

The development work on GPS commenced within the U.S. Department of Defense in 1973, and *full operational capability* was declared on July 17, 1995—the milestone reached when 24 satellites were transmitting navigation signals. The objective was to design and deploy an all-weather, 24 h, global, satellite-based navigation system to support the positioning requirements of the U.S. armed forces and its allies. For a background to the development of the GPS, the reader is referred to Parkinson (1994). GPS was intended to replace the large number of navigational systems already in use, and great importance was placed on the system's reliability and survivability. Therefore, a number of stringent conditions had to be met:

- Suitable for all military platforms: aircraft (jet to helicopter to Unmanned Aerial Vehicle (UAV)), ships (of all sizes), land (vehicle-mounted to hand-held), and space-based vehicles (missiles and satellites).
- Able to handle a wide range of platform dynamics.
- A real-time positioning, velocity, and time (PVT) determination capability to an appropriate level of accuracy.
- The positioning results were to be available on a single, global, geodetic datum.
- The highest accuracy was to be restricted to the military user.
- Resistant to jamming, whether intentional and unintentional.
- Incorporating redundancy mechanisms to ensure the survivability of the system.
- A passive positioning system that did not require transmission of signals by the user.
- Able to provide the positioning service to an unlimited number of users.
- Use low-cost, low-power user hardware.
- A replacement for the transit satellite system, as well as other terrestrial navigation systems.

What was unforeseen by the system designers was the power of commercial product innovation, which has added significantly to the versatility of the GPS, but in particular as a system for *precise positioning*. For example, GPS is able to support a variety of positioning and measurement modes in order to simultaneously address the requirements of a wide range of users; from those satisfied with navigational accuracies (of the order of 10 m or so—dekameters), to those demanding very high (even sub-centimeter) positioning accuracies. *GPS has now so penetrated certain application areas that it is difficult to imagine life without it!* Part II of this book is not intended to be a comprehensive textbook on the GPS technology and its applications. Excellent general references to the engineering aspects of GPS are Kaplan and Hegarty (2006) and Parkinson and Spilker (1996). Texts dealing extensively with the high-precision GPS surveying techniques include Leick (2004) and Hofmann-Wellenhof et al. (2008).

At about the same time GPS was declared fully operational, the then Soviet Union deployed 24 GLONASS satellites. The abbreviation GLONASS is derived from the Russian "Global'naya Navigatsion-naya Sputnikovaya Sistema" (Hofmann-Wellenhof et al. 2008). Following the dissolution of the Soviet Union, the Russian Federation initially struggled to find sufficient funds to maintain GLONASS, and the number of functioning satellites steadily decreased to less than 10. However, the Russian Federation is *rebuilding* GLONASS (Section 15.5.1).

Any discussion of the GPS technology and its applications starts with the identification of the three components (Figure 7.1):

- *The space segment:* satellites and transmitted signals
- *The control segment:* ground facilities carrying out the task of satellite tracking, orbit computation, satellite clock behavior and system monitoring, telemetry, and supervision necessary for routine operations
- *The user segment:* the applications, equipment, and computational techniques that are available to users

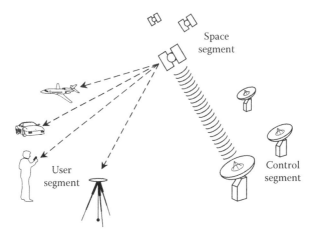

FIGURE 7.1 GPS elements.

7.1.1 SPACE SEGMENT

The space segment consists of the constellation of spacecraft—and the signals that are broadcast by them—which allow users to determine PVT. The basic functions of the GPS satellites are to

- Receive and store data uploaded by the control segment
- Maintain accurate time by means of an onboard atomic clock
- Transmit information and signals to users on two microwave frequencies

Several constellations of GPS satellites have already been deployed, and several more are planned. The first *experimental* satellite of the so-called Block I constellation was launched in February 1978. The last of this 11 satellite series was launched in 1985. The *operational* constellation of GPS satellites, the so-called Block II and Block IIA satellites, were launched from 1989 onward. The "Block IIR" satellites are the *replenishment* constellation, with the first launched in 1997. There are at present 12 orbiting "Block IIR" satellites. Under the *GPS Modernization* program (see Section 15.4), some of the original "Block IIR" series were *modernized* with new civilian and military signals, and these eight satellites are now known as the "Block IIR-M" series. The "Block IIF" *follow-on* satellite series are planned for launch from 2010 onward with similar enhancements as the "Block IIR-M" satellites, as well as transmitting a third civilian frequency. Note that all plans to date call for a 24 satellite constellation size. That is, there are 24 orbital "slots," and extra satellites beyond the nominal 24 occupy the same slots as functioning satellites, and do not necessarily improve availability in challenging signal environments such as urban, mountainous, or forested areas. The status of the current GPS satellite constellation, and such details as the launch and official commissioning date of each GPS satellite, the orbital plane and position within the plane, the satellite

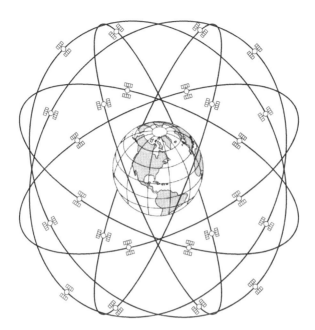

FIGURE 7.2   The GPS constellation "birdcage" showing the nominal 24 orbiting satellites.

I.D. number(s), etc., can be obtained from the U.S. Coast Guard Navigation Center (NavCen 2009).

At an altitude of approximately 20,200 km, a nominal constellation of 24 GPS satellites, located in six orbital planes inclined at about 63° to the equator (Figure 7.2), is sufficient to ensure that there will be *at least four satellites visible*, at any unobstructed site on the earth, at any time of the day. (At the time of writing, the GPS constellation consists of 31 satellites transmitting navigation signals.) As GPS satellites are in nearly circular orbits:

- Their orbital period is approximately 11 h 58 min, so that each satellite makes two revolutions in one "sidereal" day (the period taken for the earth to complete one rotation about its axis with respect to the stars).
- At the end of a sidereal day, the satellites are again over the same location on the earth.
- Reckoned in terms of a solar day (24 h in length), the satellites are in the same position in the sky about 4 min earlier each day.

The satellite visibility at any point on the earth, and for any time period, can be computed using mission planning tools provided with standard GPS surveying software as well as satellite visibility tools on the Internet.

A GPS satellite may be above an observer's horizon for many hours, perhaps 6–7 h or more in the one pass. At various times of the day, and at various locations on the surface of the earth, the number of satellites and the length of time they are above an observer's horizon will vary. Although at certain times of the day there may be as many as 14 satellites visible simultaneously, there are nevertheless occasional periods of degraded satellite coverage, made worse if there are sky obstructions preventing the signals of some of the satellites above the horizon not being trackable—though naturally their frequency and duration will increase if some of the satellites fail. "Degraded satellite coverage" is typically defined in terms of the magnitude of the *Dilution of Precision* (DOP) value, a measure of the quality of *receiver-satellite geometry* (see Section 9.3.3). The higher the DOP value, the poorer is the satellite geometry with respect to the user on the ground.

Each orbiting GPS satellite—of the Block IIA/IIR/IIR-M generations—transmits unique navigational signals centered on two L-band frequencies of the electromagnetic spectrum: L1 at 1575.42 MHz and L2 at 1227.60 MHz. (Two signals at different frequencies permit the ionospheric delay effect on the signal raypaths to be estimated—see Section 9.2.1.1—thus improving measurement accuracy.) At these two frequencies, the signals are highly directional and can be reflected or blocked by solid objects. Clouds are easily penetrated, but the signals may be blocked by foliage (the extent of this is dependent on a number of factors, such as the type and the density of leaves and branches, and whether they are wet or dry), or reflected by surfaces (causing multipath—Section 8.5.2). The current GPS satellite signals have the following components:

- Two L-band *carrier waves*
- *Ranging codes* modulated on the carrier waves using a CDMA (Code Division Multiple Access) scheme
- *Navigation message*

The primary function of the ranging codes is to permit the *signal transit time* (from the satellite to the receiver) to be determined. The transit time when multiplied by the speed of light then gives a measure of the receiver-satellite "range" or distance. In reality, the measurement process is more complex and the measurement is contaminated by a variety of biases and errors (Langley 1991b, 1993)—see mathematical model in Section 8.6. The *navigation message* contains the satellite orbit (or ephemeris) information, satellite clock error parameters, and the pertinent general system information necessary for real-time navigation to be performed. Although for positioning and timing the function of the GPS signal is quite straightforward, the stringent performance requirements of GPS are responsible for the complicated nature of the GPS signal structure. Table 7.1 summarizes the GPS requirements and their corresponding implications on the signal characteristics.

7.1.2 CONTROL SEGMENT

The control segment consists of facilities necessary for satellite health monitoring, telemetry, tracking, command and control, and satellite orbit and clock error computations. The U.S. Air Force operates six ground facility stations—Hawaii, Colorado Springs, Florida, Ascension Island, Diego Garcia, and Kwajalein—which perform the following functions:

TABLE 7.1

GPS Requirements and the Nature of GPS Signal

System Requirements	Implication on GPS Signals
GPS has to be a multiuser system	Signals can be simultaneously observed by unlimited numbers of users
	Accomplished by one-way measurement to passively listening users
	Signal has to have a relatively wide spatial coverage
GPS has to provide real-time positioning and navigation capability for the users	At a certain epoch, signals from several satellites have to be simultaneously observed by a single user
	Each signal has to have a unique code, so the receiver can differentiate different signals coming from different satellites
	Signal has to provide data for the user to estimate its range to the observed satellite in real time
	Signal has to enable time delay measurement by the user
	Signal has to provide the ephemeris data in real time to the user
	Ephemeris data is included in a broadcast message
GPS has to serve both military and civilian users	Signals have to provide two levels of accuracy for time delay measurements
	One code for the military and a different set of codes for civilian users
	Signal has to carry the two codes
	Signal has to support the "AS" policy, in which the military code has to be encrypted to prevent unauthorized use
GPS signal has to be impervious to jamming	Requires a unique code structure
	Uses spread spectrum modulation technique
GPS can be used for precise positioning	Provide range measurements at two frequencies, allowing for the compensation of the ionospheric refraction effect
	Require carrier waves with centimeter wavelengths

- All six stations are *monitor stations*, equipped with GPS receivers to track the satellites. The resultant tracking data is sent to the Master Control Station (MCS). An additional 10 globally distributed monitor stations operated by the National Geospatial-Intelligence Agency (NGA) have recently been incorporated into an upgraded control segment.
- Colorado Springs (Shriever AFB) is the *MCS*, where the tracking data are processed in order to compute the satellite ephemerides (or coordinates) and satellite clock error parameters. It is also the station that initiates all operations of the space segment, such as spacecraft maneuvering, signal encryption, satellite clock-keeping, and so on.
- Four of the stations (Colorado Springs, Ascension Is., Diego Garcia, and Kwajalein) are *Upload Stations* through which data is telemetered to the satellites. There are an additional eight upload stations in the Air Force Space Command Network.

Each of the upload stations views all of the satellites at least once per day. All satellites are therefore in contact with an upload station several times a day, and new navigation messages as well as command telemetry can be transmitted to GPS satellites on a regular basis. The computation of each satellite's ephemeris (Section 8.3.1) and the determination of each satellite's clock errors (Section 8.3.2) are the most important tasks of the control segment. The first is necessary because GPS satellites function as "orbiting control stations" and their coordinates must be known to a high accuracy, while the latter permits a significant distance measurement bias to be reduced.

The product of the orbit computation process at the MCS is each satellite's *predicted ephemeris*, expressed in the reference system most appropriate for positioning: an *earth-centered-earth-fixed* (ECEF) reference system known as WGS84 (Chapter 3). The accuracy with which the orbit is predicted is typically at the few meters level. The behavior of each GPS satellite clock is monitored against *GPS Time*, as maintained by an ensemble of atomic clocks at the MCS, aligned in turn to the master clocks of the U.S. Naval Observatory. The satellite clock *bias*, *drift*, and *drift-rate* relative to GPS time are determined at the same time as the estimation of the satellite ephemeris. The measured satellite clock error is made available to all GPS users via clock error coefficients in a polynomial form broadcast in the navigation message. However, what is available to users is really a *prediction* of the clock behavior into the future. Due to random deviations—even cesium and rubidium oscillators are not entirely predictable—the deterministic models of satellite clock error are only accurate to about a few nanoseconds for the best-performing satellites, and this accuracy degrades with the age of the satellite clocks. This is not precise enough for range measurements that must satisfy the requirements of centimeter-level accuracy GPS positioning. Operational and data processing techniques therefore have had to be developed to account for this *residual* range bias that remains after correction for the broadcast satellite clock error.

7.1.3 USER SEGMENT

This is the component of the GPS with which users are most concerned—the space and control segments are largely transparent. Of interest is the range of GPS user applications, equipment, positioning strategies, and data processing techniques that are now possible. The engine of commercial GPS product development is, without doubt, the *user applications*. New applications are being continually identified, each with its unique requirements in terms of accuracy, reliability, operational constraints, user hardware, form factor and power consumption, data processing algorithms, latency of GPS results, and so on. As a result, the GPS user equipment has undergone tremendous development, and continues to this day. In this context, the GPS *equipment* refers to the combination of hardware, software, and operational procedures or requirements. Chapter 9 discusses the various measurement models and data processing strategies; Chapter 10 introduces the hardware issues; Chapter 11 describes various GPS techniques; and Chapters 12 and 13 deal with the planning and field operations of relevance to precise GPS positioning.

While military R&D has concentrated on achieving a high degree of miniaturization, modularization, and reliability, the commercial equipment manufacturers have, in addition, sought to bring down costs and to develop features that *enhance* the capabilities of the positioning system. Civilian users have, from the earliest days of GPS availability, demanded increasing levels of performance, in particular higher accuracy, improved reliability, and faster results. This is particularly true of the *surveyor*, seeking levels of accuracy several orders of magnitude higher than that of the *navigator*. In some respects, the GPS user equipment development is being driven by the precise positioning applications—in much the same way that automotive technology often benefits from car racing. Another major influence on the development of the GPS equipment has been the increasing variety of civilian applications. Although it is possible to categorize positioning applications according to many criteria, the most important from the perspective of geospatial applications are:

- *Accuracy*, which leads to a differentiation of the GPS user equipment and techniques into several subclasses
- *Timeliness*, whether GPS results are required in real time, or may be derived from post-mission data processing
- *Dynamics*, distinguishing between static receiver positioning, and those applications in which the receiver is moving (or in the so-called kinematic mode)

The different GPS positioning modes and data processing strategies are all essentially designed to account for biases (or systematic errors) in GPS measurements to different levels of accuracy (Sections 7.3, 8.6, 9.3, and 9.4). In this regard, there are two aspects of GPS that fundamentally influence the entire user segment—the user equipment, data processing techniques, and operational (field) procedures. They are:

1. The type of *measurement* that is used for the positioning solution (Sections 8.2 and 8.4). There is, on the one hand, the basic satellite-to-receiver range measurement with a precision typically at the few meter level. However, for high-accuracy applications carrier phase measurements must be used. These have measurement noise at the millimeter level, but require more complex data processing in order to obtain centimeter-level positioning accuracy.
2. The *mode of positioning*, whether it is based on single-receiver techniques, or in terms of defining the position of one receiver *relative* to another that is located at a known position. Relative positioning is the standard mode of operation if accuracies higher than a few meters are required, although the actual accuracy depends on a number of other factors as well.

7.1.3.1 Absolute Positioning

In this mode of positioning, the reference system must be rigorously defined and maintained, and total reliance is placed on the *integrity* of coordinated points that realize the datum. In general, the coordinate origin of the coordinate system is the geocenter, and the axes of the system are defined in a conventional manner as in ECEF systems such as WGS84 and *International Terrestrial Reference System* (ITRS)

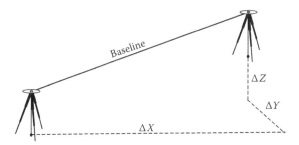

FIGURE 7.3   The baseline linking two simultaneously observing GPS receivers.

(Chapter 3; and Section 7.2). Satellite single-point positioning (SPP) (Section 9.3) is the process by which *given* the position vector of the satellite (in the global system) and a set of measurements from one ground tracking station to the satellite (or satellites) being tracked, the position vector of the ground station is determined.

Some space geodesy technologies can determine the absolute position of a ground station to a very high accuracy, as for example the *Satellite Laser Ranging* technique. However, the coordinates of a GPS receiver *in an absolute sense* are determined to a much lower accuracy than the precision of the measurements themselves, because it is not possible to fully account for the effects of measurement biases. Combining data from two GPS receivers is an effective way of eliminating or mitigating the effects of unmodeled measurement biases, as discussed in the following text.

7.1.3.2   Relative Positioning

Conceptually, the relative position is the difference between the two position vectors (in the *global system*), expressed in a local reference system with origin at one of the ground stations. Most of the error in absolute position are common to both sets of coordinates (due to similar biases on all simultaneous GPS measurements), and hence largely cancel from the *baseline components*—the vector linking the reference receiver to the user receiver. In this case, the positioning accuracy approaches that of the measurement precision itself.

There are different ways in which such differential positioning can be implemented using GPS. Data processing techniques such as those implemented for GPS surveying are essentially concerned with the determination of the baseline vector (Figure 7.3) (Section 9.4).

7.2   ISSUE OF GPS DATUMS

Chapter 3, introduced the concept of datums and geodetic systems. In this section, the geodetic reference systems are discussed from the viewpoint of GPS positioning.

7.2.1   WGS84 Datum

The World Geodetic System 1984 (WGS84) is defined and maintained by the U.S. NGA as a *global geodetic datum* (WGS84 2000). It is the datum to which all GPS positioning information is referred by virtue of being the reference datum of the broadcast GPS satellite ephemerides (Langley 1991a). The *realization* of the WGS84 satellite datum is the set of coordinates of the monitor stations within the GPS control segment. They fulfill the same function as national control benchmarks; that is, they provide the means by which a position can be related to a datum.

The relationships between WGS84 (as well as other global datums) and local geodetic datums have been determined empirically (WGS84 2000), and transformation models have been developed. Reference systems are periodically redefined, for various reasons, and the result is generally a small refinement in the datum definition, and a change in the numerical values of the coordinates of benchmarks. However, with dramatically improving tracking accuracies another phenomenon impacts on datum definition and its maintenance: *the motion of the tectonic plates across the earth's surface* (or "continental drift"). This motion is measured in centimeters per year, with the fastest rates being over 10cm/year. Nowadays this motion can be monitored and measured to sub-centimeter accuracy, on a global annual-average basis. In 1994, the GPS reference system underwent a subtle change to WGS84(G730) to bring it into alignment with the same system as used by the *International GNSS Service* (IGS) to generate its precise GPS ephemerides. Another small change was made in 1996, to WGS84(G873) (WGS84 2000).

7.2.2 THE INTERNATIONAL TERRESTRIAL REFERENCE FRAME

Since the mid-1980s geodesists have been using GPS to measure crustal motion, and to define more precise satellite datums. The latter were essentially by-products of the sophisticated data processing, which included the computation of the GPS satellite orbits. These surveys required coordinated tracking by GPS receivers spread over a wide region during the period of GPS survey campaigns. Little interest was shown in these alternative datums until the network of tracking stations evolved into a *global one* that was maintained on a *permanent basis*, and the scientific community initiated a *project to define and maintain a datum at the highest level of accuracy*.

In 1991, the International Association of Geodesy established the first of its space geodesy services, the "International GPS Service for Geodynamics" (nowadays the acronym "IGS" stands for the "International GNSS Service"), to promote and support activities such as the maintenance of a permanent network of GPS tracking stations, and the continuous computation of the satellite ephemerides, satellite clock error, ground station coordinates, earth orientation parameters, and other quantities (IGS 2009). Routine operations commenced at the beginning of 1994 and the network now consists of several hundred GPS tracking stations located around the world. The precise orbits of the GPS satellites (and other products) are available from the IGS at no charge, via the Internet, with varying delays—though some products are predicted into the future.

The definition of the reference frame in which the coordinates of the IGS tracking stations are expressed and periodically redetermined is the responsibility of the International Earth Rotation and Reference Systems Service (IERS 2009). The reference system is known as the ITRS, and its definition and maintenance is dependent on a suitable combination of Satellite Laser Ranging, Very Long Baseline Interferometry, and GPS coordinate results (although nowadays it is the GPS system that provides most of this data) —see Section 3.2.1. Every so often a new combination of precise tracking results is performed, and the resulting new coordinates of SLR, VLBI, and GPS tracking stations constitute a new *International Terrestrial Reference Frame* (ITRF) or "ITRF datum," which is referred to as "ITRF*yy*," where *yy* is the computation year identifier. A further characteristic that distinguishes the ITRS series of datums is that the definition consists of not only the station coordinates but also their *velocities* (due to the continental and regional tectonic motion). Hence, it is possible to determine station coordinates within the datum, say ITRF2005 (ITRF2005 2009), at some *epoch* such as the year 2010, by applying the velocity information and predicting the coordinates of the station at any time into the future (or the past). For example, the WGS84(G730) reference frame is identical to that of ITRF91 at epoch 1994.0.

7.3   THE PERFORMANCE OF GPS

As far as users are concerned, there are a number of *measures of performance*. For example, how many observations are required to assure a certain level of accuracy is one measure that is important for survey-type applications. The less time required to collect observations, the more *productive* is the GPS, because productivity is closely related to the number of points that can be surveyed per day. Another measure of performance might be the maximum distance between two GPS receivers that would still assure a certain level of accuracy. However, the most common measure of performance is the positioning *accuracy*.

7.3.1   Factors Influencing GPS Accuracy

Biases and *errors* affect all GPS measurements. GPS biases have one of the following characteristics:

1. Affect all measurements made by a receiver by an equal (or similar) amount
2. Affect all measurements made to a particular satellite by an equal (or similar) amount
3. Unique to a particular (receiver-satellite) range or carrier phase observation

7.3.1.1   Biases and Errors

Their combined magnitude will affect the accuracy of the positioning results. Errors may be considered synonymous to internal instrument noise or *random errors*. Biases, on the other hand, may be defined as being those measurement errors that cause *true ranges* to be different from *measured ranges* by a "systematic amount," such as, for example, all distances being measured either too short, or too long.

In the case of GPS, a very significant bias was *selective availability* (SA), a policy of the U.S. government imposed on March 25, 1990 and finally revoked on May 1, 2000 (OoP 2000). SA was a bias that caused all distances from a particular satellite, at an instant in time, to be in error by up to several tens of meters. The magnitude of the SA-induced bias varied from satellite-to-satellite, and over time, in an unpredictable manner. The policy *Anti-Spoofing* (AS), on the other hand, although not a signal bias, does affect positioning accuracy as it prevents civilian users access to the second GPS signal frequency (L2) on Block IIA and IIR satellites. Measurements on two frequencies simultaneously is the best means by which a significant measurement bias—the ionospheric refraction delay—can be accounted for (Section 9.2.1.1).

Biases must be accounted for in the measurement model used for data processing if high accuracy is sought. There are several sources of bias with different characteristics of magnitude, periodicity, satellite or receiver dependency, and so on. Biases may have physical bases, such as the atmospheric effects on signal propagation, but may also enter at the data processing stage through imperfect knowledge of constants, for example, any "fixed" parameters such as the satellite orbit, station coordinates, etc. *Residual biases* may therefore arise from incorrect or incomplete observation modeling, and hence it is useful to assemble under the heading of "errors" all random measurement process effects, as well as any unmodeled biases that remain after any data reduction.

7.3.1.2 Absolute and Relative Positioning

There are two GPS positioning modes that are fundamental to considerations of (a) *bias propagation* into (and hence the impact on the accuracy of) GPS results and (b) the *datum* to which GPS results refer. The first is *absolute or point positioning*, with respect to a datum such as WGS84 or the ITRF, and is often referred to as SPP. As the satellite coordinates are essential for the computation of user position, any error in these values (as well as the presence of other biases) will directly affect the quality of the position determination. The satellite-receiver geometry will also influence the error propagation into the GPS positioning results (see Section 9.3).

Higher accuracies are possible if the relative position of two GPS receivers, simultaneously tracking the same satellites, is computed (Section 9.4). Because many errors will affect the absolute position of two or more GPS users to almost the same extent, these errors largely cancel when the *differential or relative positioning* mode is used. There are different implementations of relative positioning procedures but all share the characteristic that the position of the GPS receiver of interest is derived *relative* to another fixed *reference* receiver whose absolute coordinates are assumed to be known. One of these implementations, based on differencing the carrier phase data from the two receivers, is the standard mode for precise GPS techniques (Section 8.6.2).

7.3.1.3 Other Factors Influencing Accuracy

GPS accuracy is also dependent on a host of other *operational and algorithmic factors*:

- Whether the user is *moving* or *stationary*. Clearly repeat observations at a static benchmark permit an improvement in precision due to the effect of averaging over time. A moving GPS receiver does not offer this possibility and the accuracy is dependent on single-epoch processing.

- Whether the results are required in *real-time*, or if *post-processing* of the data is possible. The luxury of post-processing the data permits more sophisticated modeling of the GPS data in order to improve the accuracy and the reliability of the results.
- The level of measurement noise has a considerable influence on the quality of GPS results. Low measurement noise would be expected to result in comparatively high accuracy. Hence *carrier phase* measurements are the basis for high-accuracy techniques (Section 8.4), while *pseudorange* measurements are used for comparatively low-accuracy applications (Section 8.2).
- The degree of *redundancy* in the solution as provided by extra measurements, which may be a function of the number of tracked satellites as well as the number of observables (e.g., carrier phase and pseudorange measurements made on L1 and L2 signals).
- The algorithm type may also impact on GPS accuracy (although this is largely influenced by the observable being processed and the mode of positioning). In the case of carrier phase-based positioning, to ensure centimeter-level accuracy it is crucial that a so-called ambiguity-fixed solution be obtained (Section 9.4.3).
- "Data enhancements" and "solution aiding" techniques may be employed. For example, the use of carrier phase-smoothed pseudorange data, external data from Inertial Navigation Systems (and other such devices), additional constraints, etc.

7.3.2 ACCURACY VERSUS POSITIONING MODE

Figure 7.4 illustrates the positioning accuracies associated with the different GPS positioning modes (accuracies are quoted as two-sigma values, that is, 95% confidence level). The following comments may be made to this figure:

- The top half refers to SPP and the lower half to the relative positioning mode.
- The basic SPP services provided by the U.S. Department of Defense are the *Standard Positioning Service* (SPS) and the *Precise Positioning Service* (PPS), both intended for single-epoch positioning (NavCen 2009).
- There is a large range of horizontal SPS and PPS accuracy possible due to a variety of factors:
  - 100 m level accuracy SPS positioning when SA was on during the 1990s, *as a result of an artificial degradation of the system.*
  - 5–15 m level accuracy of SPS positioning without SA, representing the current natural accuracy ceiling when using basic navigation-type GPS receivers, because of the difficulty in accounting for the ionospheric bias in the single-frequency C/A-code measurements.
  - 10%–20% improvement is possible using dual-frequency GPS receivers.
  - 2–10 m level accuracy PPS positioning, using dual-frequency P-code pseudorange measurements.

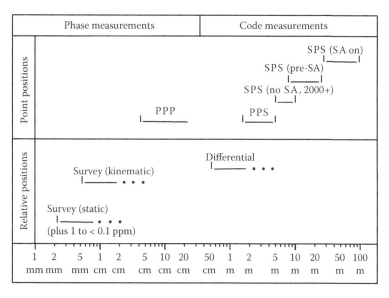

FIGURE 7.4 GPS accuracies and positioning modes.

- Dual-frequency GPS, coupled with the high-accuracy satellite clock and ephemeris data provided by the IGS, can deliver at least a twofold improvement in basic SPS accuracy.
- Surprisingly, the averaging of SPS results for up to 60 min at a single benchmark does not significantly improve positioning accuracy, with studies typically indicating an improvement of the order of 10%–15% compared to single-epoch solutions.
- The carrier phase-based procedures are typically only applied in the relative positioning mode for most engineering, surveying, and geodetic applications, and the *relative* position accuracy is usually expressed in terms of *parts per million* ("ppm"—e.g., 1 cm error in 10 km).
- Carrier phase-based positioning may be in the single-epoch mode (as is necessary for "kinematic" positioning), or takes advantage of the receiver being static in order to collect data over an *observation session*.
- *Precise Point Positioning* is possible using carrier phase data, with accuracies better than a decimeter currently possible if the observation session is long enough and the receiver is stationary.
- The accuracy of carrier phase-based positioning techniques is a function of baseline length, the number of observations, the length of observation session, whether ambiguities have been fixed to their integer values or not, and others.
- In all cases, the vertical accuracy is about two to three times worse than the horizontal positioning accuracy.

The *resolution of the carrier phase ambiguities* is central to precise carrier phase-based positioning in many surveying and engineering applications (Section 9.4.2), and requires the determination of the correct number of *integer* wavelengths in the carrier measurement of satellite-to-receiver distance (or linear combinations of the measurements from a pair of receivers to a pair of satellites—see Section 8.6.2).

It should be emphasized that GPS was originally designed to provide accuracies of the order of a *dekameter* (10 m) or so in the SPP mode, and is optimized for real-time operations. All other innovations to improve this basic accuracy capability must be viewed in this context.

7.4 HIGH-PRECISION GPS POSITIONING

GPS is having a profound impact on society. *It is estimated that the worldwide market for GPS products and services in 2008 was about US$30 billion.* Market surveys suggest that the greatest growth is expected to be consumer markets such as in-vehicle applications, GPS-enabled cellular phones, and portable GPS for outdoor recreation and similar activities. These are expected to ultimately account for more than 80% of the GPS market in volume. The penetration of GPS into many applications (and in particular into consumer devices) helps make the processes and products of geospatial information technology more and more a part of the mainstream *information society*. However, in the following chapters, the focus will be on the surveying and mapping disciplines, and how GPS is now an indispensable tool for geospatial professionals.

7.4.1 GPS in Support of Geospatial Applications

In this book, the authors have adopted a very broad definition of "GPS surveying," encompassing all applications where coordinate information is sought in support of mapping or geospatial applications. In general, such applications:

- Are of *comparatively high accuracy*. This is, of course, a subjective judgement, but in general the phrase "high accuracy" implies a level of coordinate precision much higher than that originally intended of GPS. As GPS is a navigation system designed to deliver dekameter-level SPP accuracy, the accuracy threshold for *surveying* may be arbitrarily set at the submeter level, while *mapping* accuracies may be satisfied by differential GPS (DGPS) techniques that can deliver accuracies at the few meter level. *In this book, "GPS surveying" will be considered synonymous with carrier phase-based positioning.*
- Require the use of *unique observation procedures, measurement technologies, and data analysis*. In fact, the development of distinctive field procedures, specialized instrumentation, and sophisticated software is the hallmark of GPS surveying.
- Do *not require positioning information "urgently."* Navigation, on the other hand, is concerned with the safe passage of vehicles, ships, and aircraft, and hence demands location information in *real time*.
- Permit *post-processing of data* to obtain the highest accuracy possible, although increasingly real-time GPS surveying is the norm.

- Have as their raison d'être the *production of a digital map*, or the establishment of a *network of coordinated points*, which support traditional tasks of the surveying discipline, as well as new applications such as GIS database generation and engineering machine guidance.

In the case of *land surveying applications*, the characteristics of GPS satellite surveying are:

1. The points being coordinated are in general *stationary*.
2. Depending on the accuracy sought, GPS *data are collected over some observation session*, ranging in length from a few seconds to several hours, or more.
3. Restricted to the *relative positioning mode of operation*.
4. In general (depending on the accuracy sought), the measurements used for the data reduction are those made on the satellites' *L-band carrier waves*.
5. Generally associated with the *traditional surveying and mapping functions*, but accomplished using GPS techniques in less time, to a higher accuracy (for little extra effort), and with greater efficiency.

A convenient approach is to adopt a geospatial applications classification on the basis of accuracy requirements. For purposes of discussion, four classes can be identified on this basis:

- Scientific surveys (category A): better than 1 ppm
- Geodetic surveys (category B): 1–5 ppm
- General surveying (category C): lower than 5 ppm, centimeter-level
- Mapping/geolocation (category D): better than 1 m

Category A primarily consists of those surveys undertaken in support of precise engineering, deformation analysis, global geodesy, and geodynamic applications. Category B includes geodetic surveys undertaken for the establishment, the densification, and the maintenance of control networks. Category C encompasses lower accuracy surveys, primarily to support engineering and cadastral applications, sensor georeferencing, geophysical prospecting, etc., and increasingly for machine guidance applications. Category D includes all other general purpose geolocation surveys intended to coordinate objects or features for map production and GIS data capture (Chapter 25). Users in the latter two categories form the majority of the GPS user community. Categories A and B users may provide the "technology-pull" impetus for the development of new instrumentation and processing strategies, which may ultimately be adopted by the categories C and D users. Note, this classification scheme is entirely *arbitrary*, and does not relate to any specification of "order" or "class" of survey as may be defined by national or state survey agencies.

7.4.2 Using GPS in the Field

With respect to category D users (using the pseudorange-based techniques), the planning issues, as well as the field and office procedures, are not as stringent as for the GPS surveying users. Hence most of the attention will be focused on carrier phase-based techniques. Some comments to the *operational aspects* of GPS surveying (categories A, B, and C described earlier):

- Survey planning considerations are derived from
  - The nature and the aim of the survey project—*as for conventional surveys*
  - The unique characteristics of GPS, and in particular no requirement for receiver intervisibility—*a simplification in survey design*
  - The number of points to be surveyed, the resources at the surveyor's disposal, and the strategy to be used for propagating the survey—*a logistical challenge*
  - Prudent survey practice, requiring redundant and check measurements to be incorporated into the network design
- Field operations are characterized by requirements for
  - Clear skyview
  - Setup of antennas over ground marks
  - Simultaneous operation of two or more GPS receivers
  - Common data collection over some observation session (if in static mode)
  - Deployment of GPS hardware to new stations
- Field validation of data collected, in order to
  - Verify sufficient common data collected at all sites operating simultaneously
  - Verify quality of data to ensure that acceptable results will be obtained
  - Where data dropout is high, or a station has not collected sufficient data, reoccupation may be necessary
- Office calculations
  - To obtain GPS solutions for single sessions or baselines
  - To combine the baseline results into a network solution
  - To incorporate external information (e.g., local control station coordinates), and hence modify the GPS-only network solution
  - To transform the GPS results (if necessary) to the local geodetic datum, and to derive orthometric (or sea-level referenced) heights
  - To verify the accuracy and the reliability of the GPS survey (including cases where real-time techniques are used)

The GPS project planning and field operation issues are discussed in Chapters 12 and 13.

7.4.3 GPS COMPETITIVENESS

GPS needs to be competitive with other terrestrial techniques of surveying. Several criteria for judging the utility of GPS can be identified:

1. Cost benefit: *issues such as the capital cost of equipment, ongoing operational costs, data processing costs, development, training, and maintenance costs.* This can be best measured according to *productivity.* The direct cost of a GPS survey (not including equipment and training costs) can be estimated during the planning phase. It needs to be established whether competing technologies offer lower costs.
2. Ease-of-use: *issues such as servicing, the timeliness of results, and the expertise of users.* Experience indicates that the primary factors affecting servicing are those of distance to the servicing agents, their technical expertise, and their customer service. To ensure high-quality results in a reasonable time (real-time operations may not be required), it is important that all personnel be adequately trained.
3. Accuracy: *obviously related to the class of application.* The level of accuracy sought will directly influence many other factors, such as the type of instrumentation, the technique to be used, the sophistication of the software, and the cost of survey, field operations, etc.
4. External factors *such as the availability of satellite ephemerides and other performance constraints such as superior GPS networks to connect into, base station operation, etc.*

GPS *complements* the traditional electronic distance measurement (EDM)-theodolite ("total station") techniques for routine surveying activities. Indeed traditional techniques are likely to continue playing the dominant role for some time to come, unless the conditions for survey are ideal from a GPS point of view. In that case, the "real-time kinematic" (RTK) technique will be favored. For mapping of features, for GIS-type applications, GPS is the ideal low-to-moderate accuracy point coordination tool. Finally, for high-precision (geodetic) positioning, particularly over long distances, GPS is without peer.

REFERENCES

Hofmann-Wellenhof, B., Lichtenegger, H., and Wasle, E., 2008, *GNSS Global Navigation Satellite Systems: GPS, GLONASS, Galileo, and More*, Springer Verlag, Wien, Austria, New York, ISBN 978-3-211-73012-6, 516p.

IERS, 2009, International earth rotation and reference systems service, Web page http://www.iers.org, accessed March 10, 2009.

IGS, 2009, International GNSS service, Web page http://www.igs.org, accessed March 10, 2009.

ITRF2005, 2009, International Terrestrial Reference Frame 2005, Web page http://itrf.ensg.ign.fr/ITRF_solutions/2005/ITRF2005.php, accessed March 10, 2009.

Kaplan, E. and Hegarty, C.J. (eds.), 2006, *Understanding GPS: Principles and Applications*, 2nd edn, Artech House, Norwood, MA, ISBN 1-58053-894-0, 570p.

Langley, R.B., 1991a, The orbits of GPS satellites, *GPS World*, 2(3), 50–53.

Langley, R.B., 1991b, Time, clocks, and GPS, *GPS World*, 2(10), 38–42.

Langley, R.B., 1993, The GPS observables, *GPS World*, 4(4), 52–59.

Leick, A., 2004, *GPS Satellite Surveying*, 3rd edn., John Wiley & Sons, New York, ISBN 0-471-05930-7, 435p.

NavCen, 2009, U.S. Coast Guard Navigation Center, Web page http://www.navcen.uscg.gov/gps/, accessed March 10, 2009.

Office of the U.S. President's Press Secretary (OoP), 2000, Statement by the president regarding the United States' decision to stop degrading global positioning system accuracy, May 1, see Web page http://www.ngs.noaa.gov/FGCS/info/sans_SA.

Parkinson, B.W., 1994, GPS eyewitness: The early years, *GPS World*, 5(9), 32–45.

Parkinson, B.W. and Spilker, J.J. (eds.), 1996, *Global Positioning System: Theory and Applications*, Vols. I and II, American Institute of Aeronautics and Astronautics, Inc., Washington, DC, ISBN 1-56347-106-X, 793p.

PNT, 2009, National Space-Based PNT (Positioning, Navigation & Timing) Coordination Office, Web page http://pnt.gov, accessed March 10, 2009.

Wells, D.E., Beck, N., Delikaragolou, D., Kleusberg, A., Krakiwsky, E.J., Lachapelle, G., Langley, R.B. et al., 1986, *Guide to GPS Positioning, Canadian GPS Associates*, Fredericton, NB, Canada, 600p.

WGS84, 2000, *Department of Defense World Geodetic System 1984: Its Definition and Relationships with Local Geodetic Systems*, 3rd edn., NIMA Tech. Rept. TR8350.2, http://earth-info.nga.mil/GandG/publications/tr8350.2/wgs84fin.pdf, accessed March 10, 2009.

8 Fundamentals of GPS Signals and Data

Hasanuddin Z. Abidin

CONTENTS

8.1 GPS SIGNAL CHARACTERISTICS

8.1.1 SIGNAL STRUCTURE

As of the time of writing this chapter, Global Positioning System (GPS) satellites transmit signals at two frequencies, designated as L1 and L2, on which three binary modulations are impressed: the C/A-code, the P(Y)-code, and the broadcast (or navigation) message, as depicted in Figure 8.1. L1 is the principal GPS carrier signal with a frequency of 1575.42 MHz, and is modulated with the P(Y)-code, the C/A-code, and the navigation message. The second signal, L2, is transmitted at a frequency of 1227.60 MHz and in the case of the Block IIA/IIR series of satellites is modulated

115

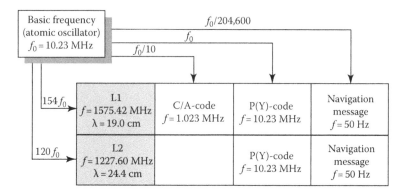

FIGURE 8.1   Current structure of the GPS L1 and L2 satellite signals.

only with the P(Y)-code and the navigation message. In the case of the *modernized* GPS satellites, commencing with the Block IIR-M satellites, the L2 frequency is also modulated by the L2C-code (Table 8.5 and Section 15.4).

The precision P(Y)-code has a bit rate of 10.23 MHz, while the coarse acquisition C/A-code and navigation message have bit rates of 1.023 MHz and 50 Hz, respectively. The P(Y)-code is an encryption of the known P-code by a code sequence referred to as the "W-code" (resulting in the so-called Y-code), and can only be used for navigation by the U.S. Department of Defense and other authorized users. This encryption of the P-code was imposed on January 31, 1994, for all satellites under the so-called *anti-spoofing* (AS) policy. The characteristics of signal modulations are summarized in Table 8.1. A detailed explanation of the characteristics of GPS signals can be found in such engineering texts as Kaplan and Hegarty (2006) and Parkinson and Spilker (1996).

All signals transmitted by GPS satellites are right-hand polarized, coherently derived from a basic frequency of 10.23 MHz (Figure 8.1), and are transmitted within a bandwidth of about 20.46 MHz (at both center frequencies) by a shaped-beam antenna array on the nadir-facing side of the satellite. Each satellite transmits a different C/A-code and a unique 1-week long segment of the P(Y)-code (and this segment is set back to zero each week at midnight (0 h UT from Saturday to Sunday). Such a modulation scheme is known as *Code Division Multiple Access* (CDMA). The transmitted power levels are 23.8 and 19.7 dBW for the P(Y)-code signals on L1 and L2, respectively, and 28.8 dBW for the C/A-code signal (Langley 1998). (*Decibel*, abbreviated as dB, is a unit for the logarithmic measure of the relative power of a signal, hence a 3 dB increase in the strength of a signal corresponds to a doubling of the power level. dBW indicates the actual power of a signal compared to a reference of 1 W.) The strength of the electromagnetic wave decreases during propagation primarily as a function of the transmitter–receiver distance; thus the GPS signal arriving at the receiver antenna is very weak, with roughly the same strength as the signal from geostationary TV satellites. However, the carefully designed GPS signal structure allows the use of small antennas, as opposed to comparatively large TV dishes.

TABLE 8.1
Characteristics of the GPS Signal Modulations

Component	C/A-Code	P(Y)-Code	Navigation Message
Bit generation rate	1.023 Mbps (10^6 bits/s)	10.23 Mbps	50 bps
Bit length	≈293 m	≈29.3 m	≈5950 km
Repetition rate	1 ms	1 week	N/A
Code type	37 unique codes (pseudorandom gold codes)	37 one-week segments (PRN code)	N/A
Carrier frequency	L1	L1, L2	L1, L2
Expected minimum signal strength at the user receiver, referenced to 0 dBic antenna[a]	−160 dBW	−163 dBW (on L1), −166 dBW (on L2)	N/A
Fundamental characteristics	Easy to acquire due to its short period and low cross-correlation between different PRN codes; thus easy to rapidly distinguish among signals arriving simultaneously from multiple satellites. Acquired before P(Y)-code so as to provide the timing information necessary to acquire the more complex P(Y)-code	More accurate and jam resistant than C/A-code. Not available to civilian users. Resistant to mutual interference when signals are received simultaneously from multiple satellites. Difficult to acquire, that is, receiver correlator must be timed to within roughly one P-code chip to allow correlation	Provides satellite health status, time, ephemeris data, various correction terms, and handover words that tell the receiver where to start the search for the P(Y)-code. Total transmission time is 30 s

[a] dBic describes antenna gain in dB with respect to a circularly polarized isotropic radiator (a hypothetical ideal reference antenna).

Since the GPS satellite and receiver move with respect to each other, the signal arriving at the antenna is *Doppler-shifted*. Consequently, the received frequency differs from the transmitted frequency by an amount that is proportional to the radial velocity of the satellite with respect to the receiver. Signals from different satellites therefore arrive at the receiver antenna with different Doppler-shifted frequencies. For example, assuming a stationary receiver in the satellite orbit plane, the maximum frequency shift of about ±6 kHz due to the Doppler effect would be at the epoch of the local horizon crossing (maximum satellite–receiver distance about 25,783 km), where the radial velocity is a maximum (about ±0.9 km/s). Naturally, there is no Doppler frequency shift at the point of closest approach (the zenith, at an approximate distance of 20,100 km).

8.1.2 SIGNAL COVERAGE

The GPS signal is transmitted by the satellite toward the earth in the form of a *signal beam*, as illustrated in Figure 8.2 (Parkinson and Spilker 1996). The figure shows that the signal illuminates not only the Earth's surface (where the signal can be used for positioning and navigation) but above it as well, as long as the user is within the main beam of the GPS signal, but outside the earth's shadow. This is useful for a variety of space applications.

In order to achieve high accuracy, positioning by the GPS is typically performed in the *differential* mode (Section 7.1.3). In this mode, two GPS receivers observe the same satellites, at the same time, and at the *relative* position (or baseline vector) between the two receivers is estimated. If one considers the spatial coverage of the GPS signal from a satellite as depicted in Figure 8.2, then the distances between two points on the earth's surface that can still view the same satellite as a function of the observation *mask angle* is indicated in Figure 8.3. For example, even with an

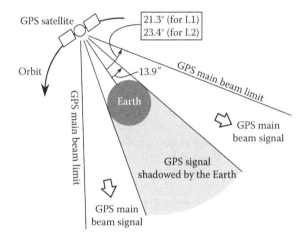

FIGURE 8.2   Spatial coverage of the GPS signal beam.

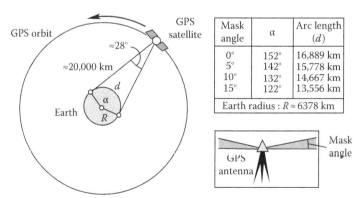

Mask angle = minimum elevation of observed satellites

FIGURE 8.3   Visibility of the same GPS satellite from two points on the Earth's surface.

elevation mask angle of 15° (the local elevation angle below which the satellite signals will not be tracked), two receivers on the earth's surface separated as far apart as about 13,500 km can still view the same GPS satellite. However, both receivers "see" the signal very near the horizon and, in general, since the signal will be obstructed by either topography or objects around the antenna, receiver separations will be much less than those shown in Figure 8.3. For most surveying/mapping applications, maximum receiver separations are of the order of tens of kilometers (and generally less), while for continental geodesy applications receiver separations may be up to several thousands of kilometers.

8.2 CODES AND PSEUDORANGE MEASUREMENTS

8.2.1 PSEUDORANDOM NOISE CODES

There are two pseudorandom noise (PRN) codes, which are modulated on the signals transmitted by a GPS satellite: the P(Y)-code and the C/A-code (Table 8.1). The two main functions of these codes are (a) to provide time delay measurements so the user can determine the distance from the receiver's antenna to the observed satellite (either code could be used, but the P(Y)-code provides a more precise distance estimate than the C/A-code) and (b) to help the receiver in differentiating the incoming signals from different satellites. These codes are sequences of binary values (zeros and ones), and although the sequence *appears to be random*, each code has a unique structure generated by a mathematical algorithm (Figure 8.4). One version of the code is generated within the satellite, and the identical code sequence is replicated within the receiver if it is known (of course, that is not the case for the P(Y)-code for civilian receivers). Two such identical codes will only be aligned at *zero lag* (i.e., when the sequence of one code is time-shifted to the instant when all the zeros and ones in that code match the sequence in the other code). Because the codes are generated by either the satellite clock or the receiver clock, they are in fact a means of representing the time defined by the respective clock. If the two clocks—satellite and receiver—are synchronized to the same time system, then the clock times can be compared, and the difference is a measure of the time taken by a signal to travel from the satellite to the receiver. From this, the distance can be derived.

As indicated in Table 8.1, each C/A-code is a unique sequence of 1023 binary numbers, which repeats itself every 1 ms. Each binary bit of the C/A-code is generated at

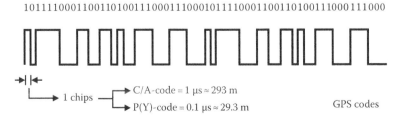

FIGURE 8.4   Example of a sequence of binary numbers of a PRN code.

the rate of 1.023 MHz and has a duration or "length" of approximately 1 μs (or about 293 m in units of length).

In contrast to the C/A-code, the P(Y)-code consists of a much longer binary sequence of 2.3547×10^{14} binary numbers, and its pattern will not repeat itself until after 266 days. The P(Y)-code is generated at a rate 10 times faster than the C/A-code, that is, 10.23 MHz. This means each bit has a duration of approximately 0.1 μs (or a "length" of about 29 m).

8.2.2 DETERMINING SATELLITE–RECEIVER RANGE

By acquiring the P(Y)-code, or the C/A-code, the observer can measure the distance or the *range* to the satellite. The basis for obtaining this range is the so-called code-correlation technique whereby the incoming code from the satellite is correlated with a replica of the corresponding code generated inside the receiver, as depicted by Figure 8.5. The time shift (d*t*) required to align the two codes is, in principle, the time required by the signal carrying the code to travel from the satellite to the receiver. Multiplying d*t* with the speed of light results in a measure of the range. This range is referred to as a *pseudorange* because it is still biased by the time offset (or mis-synchronization) between the satellite clock and the receiver clock used to measure the time delay.

As a rule of thumb, the precision of a pseudorange measurement is about 1% of its code length (or *resolution*). The nominal precision of P(Y)-code pseudorange is therefore about 0.3 m, and for the C/A-code pseudorange it is about 3 m. The P(Y)-code pseudorange measurement is also more resistant to the effects of multipath and jamming/interference. Moreover, since the P(Y)-code is modulated on both the L1 and L2 signals, the user can obtain pseudorange measurements at both frequencies, that is, the P(Y)-L1 pseudorange and the P(Y)-L2 pseudorange, so that by combining these two measurements it is possible to derive a new pseudorange observable that is not affected by the ionospheric delay bias (Section 9.2.1.1). However, due to the implementation of the AS policy, only authorized users can gain access to the P(Y)-code directly to make pseudora,nge measurements using the code-correlation technique. Civilian receivers must employ special signal processing techniques to make dual-frequency measurements (Section 10.2.4). Apart from the receivers used for GPS surveying applications, where dual-frequency measurements are a prerequisite for obtaining fast centimeter-level accuracy coordinates, most civilian GPS receivers

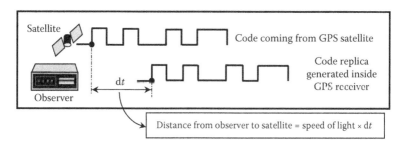

FIGURE 8.5   The principle of obtaining a range measurement using PRN codes.

intended for navigation applications only observe the C/A-code pseudorange, and hence are referred to as *single-frequency navigation receivers* (Section 10.4).

8.3 NAVIGATION MESSAGE

Besides the ranging codes, GPS signals also are modulated with the *navigation message*. This contains information such as the satellite's orbital data (or ephemeris), satellite almanac data, satellite clock error parameters, satellite health and constellation status, ionospheric model parameters for single-frequency users, and the offset between the GPS and the UTC (*Universal Time Coordinated*) time systems. The content of the navigation message is continuously updated by the GPS control segment (Section 7.1.2) and broadcast to the users by the GPS satellites.

8.3.1 BROADCAST EPHEMERIS

The most important data contained within the navigation message is the *broadcast ephemeris*. This ephemeris is in the form of Keplerian orbital elements and their perturbations, as listed in Table 8.2 (Seeber 2003). The geometric visualization of these parameters is given in Figure 8.6. The coordinates of the satellite in the WGS84

TABLE 8.2
Content of GPS Clock and Broadcast Ephemeris Messages

Time parameters

t_{oe}	Reference time for ephemeris parameters (s)
t_{oc}	Reference time for clock parameters (s)
a_0, a_1, a_2	Polynomial coefficients for satellite clock correction, that is, representing the bias (s), drift (s/s), and drift-rate (s/s^2) components
IOD	Issue of data (arbitrary identification number)

Satellite orbit parameters

$\sqrt{a}$	Square root of the semimajor axis (m$^{1/2}$)
e	Eccentricity of the orbit (dimensionless)
i_0	Inclination of the orbit at t_{oe} (semicircles)
Ω_0	Longitude of the ascending node at t_{oe} (semicircles)
ω	Argument of perigee (semicircles)
M_0	Mean anomaly at t_{oe} (semicircles)

Orbital perturbation parameters

Δn	Mean motion difference from computed value (semicircles/s)
$\dot{\Omega}$	Rate of change of right ascension (semicircles/s)
i_{dot}	Rate of change of inclination (semicircles/s)
C_{us} and C_{uc}	Amplitude of the sine and cosine harmonic correction terms to the argument of latitude (rad)
C_{is} and C_{ic}	Amplitude of the sine and cosine harmonic correction terms to the inclination angle (m)
C_{rs} and C_{rc}	Amplitude of the sine and cosine harmonic correction terms to the orbit radius (m)

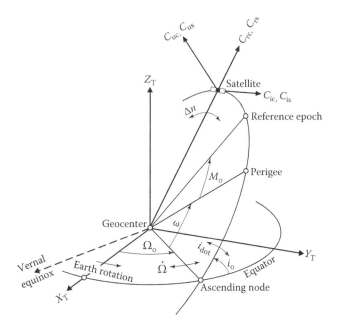

FIGURE 8.6   Geometric visualization of the GPS broadcast ephemeris parameters.

datum (Section 7.2.1) at every observation epoch can be computed using the algorithm given in Seeber (2003).

8.3.2   BROADCAST SATELLITE CLOCK ERROR MODEL

The satellite clock error at any time t can be computed using the following model
(broadcast by the navigation message):

$$dt^j = \alpha_0 + \alpha_1(t - t_{oc}) + \alpha_2(t - t_{oc})^2 \tag{8.1}$$

where α_0, α_1, and α_2 are the clock offset, the clock drift-rate, and half the clock drift
acceleration at the reference clock time t_{oc} (time of clock), respectively. Equation 8.1
provides a good prediction for the satellite clock behavior because GPS satellites use
atomic clocks (cesium or rubidium oscillators), which have a stability of about 1 part
in 10^{13} over a period of 1 day. The satellite clock correction computed from Equation
8.1 has a *residual* error of the order of a few meters (error remaining after correcting
for this clock error model). (However, when *selective availability* was "on" during
the 1990s this error was of the order of 20–25 m.)

8.4   CARRIER WAVES AND CARRIER PHASE MEASUREMENTS

The main function of GPS carrier waves L1 and L2 is, as the name implies, to
"carry" the PRN codes and navigation messages to the receiver. The codes and navigation messages are modulated onto the carrier waves using the biphase shift key

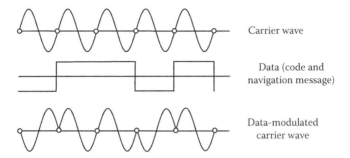

Carrier wave

Data (code and navigation message)

Data-modulated carrier wave

FIGURE 8.7 Biphase shift key modulation of the GPS carrier wave.

(BPSK) modulation technique, as illustrated in Figure 8.7 (Hofmann-Wellenhof et al. 2008). When the data (code and navigation message) value for a binary bit is 0 and it changes to a value of 1 (or vice versa, 1–0) then the carrier phase is shifted by 180°. When there is no change in the value of the adjacent bits, then there is no change in the phase.

Phase measurements made on the carrier waves can also be used to derive very precise range measurements to the satellites. For precise GPS applications, the carrier phase measurements must be used instead of pseudoranges.

8.4.1 Ambiguous Carrier Phase

The carrier phase measurement is the difference between the reference phase signal (which has the form of a sine wave) generated by the receiver and the phase of the incoming GPS signal (after stripping away the code and navigation message modulations), and is in fact the *beat signal*. The measurement of phase by a GPS receiver is complicated by the fact that the receiver cannot measure directly the complete range from the receiver to the satellite. At the initial epoch of signal acquisition (t_0) only the *fraction* of the cycle of the beat phase can be measured by the receiver. (One cycle corresponds to one *wavelength* of the sine wave that is the carrier signal without modulation, which for the L1 carrier wave is about 19 cm and for L2 is approximately 24 cm.) The remaining *integer* number of cycles to the satellite cannot be measured directly. This unknown integer number of cycles is usually termed the "initial cycle (phase) ambiguity" (N), or simply the "ambiguity." Subsequently, the receiver only counts the number of integer cycles that are being tracked. As long as there is no *loss-of-lock* of tracking of the satellite, the value of N remains a *constant*.

The phase measurement at a certain epoch t_i, that is, $\varphi(t_i)$, is therefore equal to the fractional phase (Fr) at that epoch augmented with the number of full cycles "counted" (Int) since t_0:

$$\varphi(t_i) = \mathrm{Fr}(\varphi(t_i)) + \mathrm{Int}(\varphi; t_0, t_i) + N(t_0) \tag{8.2}$$

Equation 8.2 shows how the GPS carrier phase measurement (when expressed in length units) is not a true or absolute distance from receiver to the satellite as in the

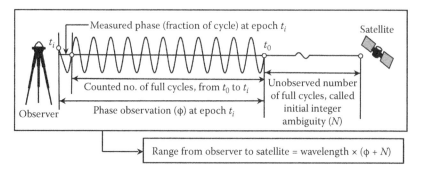

FIGURE 8.8   Phase (carrier) range determination using GPS phase measurement.

case of pseudorange, but is an *ambiguous* range. There is an unobserved part of the range caused by the initial cycle ambiguity of the phase, as illustrated in Figure 8.8. In order to convert this ambiguous range into a true range, the cycle ambiguity N has to be estimated. If the value of the integer cycle ambiguity can be correctly estimated, then the resultant *carrier-range* will be transformed into a very accurate range measurement (at the few millimeter level), and can be used for high-accuracy positioning (Section 9.4.3). Correctly estimating the value of the integer cycle ambiguity may be a challenge under certain circumstances. This computation is generally referred to as *ambiguity resolution* (Section 9.4.2).

8.4.2   UNAMBIGUOUS CARRIER-RANGE

From Equation 8.2, it can be seen that the first two terms on the left-hand side of the equation will change because the satellite is moving. If the two terms are expressed as

$$\Delta\varphi_i = \mathrm{Fr}(\varphi(t_i)) + \mathrm{Int}(\varphi; t_0, t_i) \tag{8.3}$$

then the geometrical interpretation of this temporal change in carrier phase (range) is illustrated in Figure 8.9. Note that this representation assumes that there is a continuous lock on the signal. If loss-of-lock on the signal results, due to any cause, then the second term on the right-hand side of Equation 8.3 (the integer "count") is corrupted and a *cycle slip* has occurred. Another interpretation of the cycle slip is that the ambiguity term in Equation 8.2 is not the same constant as at initial signal lock-on. The value of N has changed (or "jumped") after the epoch of the cycle slip by a quantity that is equal to the integer number of cycles in the cycle slip. Cycle slip "repair" is a crucial operation in carrier phase data processing, as it ensures that the ambiguity term is a constant for the entire observation session. Remember, the value of N must be determined and, if possible, "resolved" to its likeliest integer value if the ambiguous carrier phase measurements made during an entire observation session are to be converted to accurate carrier-range observables. Note that there is a different ambiguity for each satellite measurement—different satellites and different frequencies.

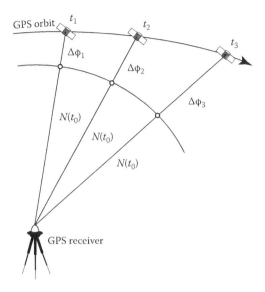

FIGURE 8.9   Geometrical interpretation of carrier-range and cycle ambiguity.

8.5 GPS SIGNAL PROPAGATION

In its propagation from a satellite to the receiver's antenna, the signal has to travel through several layers of the atmosphere, as illustrated in Figure 8.10. Due to the refraction inside the tropospheric and ionospheric layers, the speed and the direction of the signal will be affected, and in turn the measured pseudorange and carrier phase will be *biased*. In addition, the signal propagation may be influenced by other phenomena such as intentional or accidental interference due to multipath and jamming.

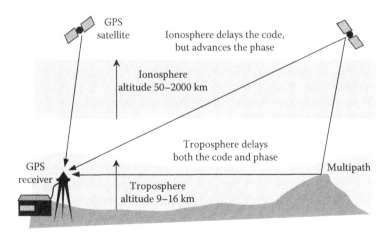

FIGURE 8.10   Propagation effects on the GPS signal.

8.5.1 Ionospheric and Tropospheric Delay

The ionosphere will affect the speed, the frequency, the direction, and the polarization of GPS signals, as well as cause amplitude and phase scintillation (Parkinson and Spilker 1996). By far the largest effect will be on the speed of the signals, which will directly affect the carrier-range and pseudorange measurements to the satellite. The ionosphere will *delay* the code measurement (the pseudorange) and *advance* the signal phase (the carrier-range), by the same amount. The thickness of the ionosphere is several hundreds of kilometers, and is generally considered to start from about 50 km above the surface of the earth.

The slant range error caused by the ionosphere can be up to about 100 m, and its magnitude is dependent on factors such as the time of day (the ionosphere is generally quieter at night), the latitude of the observer, the time of the year, the period within the 11 year sunspot cycle (the highest ionospheric activity for the current cycle is expected in the years 2010–2011), the elevation angle to the satellite, and the frequency of the signal. The estimated maximum rate-of-change of ionospheric propagation delay is about 19 cm/s, which corresponds to about 1 cycle/s on L1 (Parkinson and Spilker 1996). Very rapidly changing ionospheric conditions can cause losses of signal lock, especially on the L2 frequency (due to suboptimal tracking by civilian receivers under AS—Section 10.2.4). When dual-frequency observations are available, a linear combination of measurements made on L1 and L2 can be constructed that eliminates the first-order ionospheric delay (Section 9.2.1.1). In the case of single-frequency users, the broadcast navigation message contains parameters of an approximate model for the ionospheric delay, and although it is preferable to having no model, only about half of the true delay is typically represented by such a model under active ionospheric conditions.

The troposphere will also affect the speed and the direction of the GPS signals, as well as cause some attenuation of the signal. As with the ionosphere, the largest effect will be on the speed of the signal and, in this case, the troposphere will *delay* both the code and phase measurements by the same amount. The magnitude of the tropospheric delay is in the range 2–25 m, a minimum in the vertical (or zenith) direction and becoming larger as the elevation angle of the signal reduces down to the horizon. Both the wet and dry components of the troposphere contribute to the delay, with the wet component being approximately 10% the magnitude of the dry component. The troposphere is only about 10 km thick, and hence signals reaching high-altitude points will have less troposphere to travel through, and will therefore experience less tropospheric delay.

8.5.2 Multipath Effects

In addition, the GPS signal can experience *multipath*, a phenomenon in which a signal arrives at the antenna via two or more paths, that is, a direct signal and one or more reflected signals (Figure 8.10). The signals will interfere with each other and the resultant signal will cause an error in the pseudorange and carrier phase measurements. The range errors caused by multipath will vary depending on the relative geometry of the satellites, reflecting surfaces, and the GPS antenna, and also on the properties and dynamics of the reflecting surfaces. In the case of the carrier phase, the multipath effect can be up to about 5 cm, that is, 0.25 cycle of the wavelength (Hofmann-Wellenhof et al. 2008), while for the pseudoranges it can be as high as a few tens of meters. Multipath is considered one of the limiting factors for improving the GPS carrier phase-based positioning "productivity." That is, the presence of multipath disturbance impacts on the reliability of ambiguity resolution, particularly when observation sessions are very short (as is the case of "rapid static" and kinematic "on-the-fly" GPS surveying techniques—Section 11.1).

It should also be noted here that in order to be used for positioning, the signals transmitted by the satellites have to reach the GPS antenna. If the satellite signal is blocked by the topography, foliage, or other structures, then measurements cannot be made (Figure 8.11). GPS positioning therefore cannot be performed in tunnels, indoors, or underwater. In areas with many high-rise buildings or tall trees, which can obstruct the signal reception from the satellites (and cause a lot of cycle slips), it can also be expected that positioning with GPS will be a less optimal technique than in relatively open sky areas. There are a few strategies that can be used to overcome the signal obstruction problem (Section 7.4.2). The obvious one is to choose a location for the antenna site that has the clearest visibility to the sky. In such cases, where the observation site has been fixed beforehand, elevating the antenna above the obstructing objects, for example above the tree canopy, can help. If it is permitted, cutting the trees around the observation side is another option. Finally, the siting of antennas near sources of microwave radiation, such as communication towers, radio, and TV broadcasting stations, or cellular phone towers, is to be avoided wherever possible so as to minimize the chances of accidental jamming of the GPS signals.

The effects of the above-mentioned propagation biases and errors, along with others such as satellite ephemeris and clock errors, will contaminate derived position, velocity, and time information. Hence accounting for such effects is always necessary in order to ensure high-quality results. Chapter 9 introduces the basic mathematical

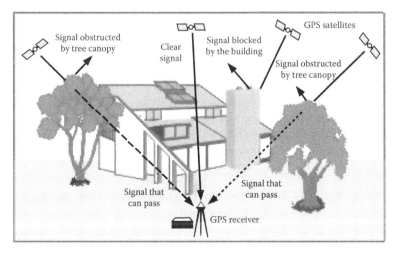

FIGURE 8.11 Obstructions and interference to GPS signals.

observation models for pseudorange and carrier phase used in GPS data processing. The most effective strategy for eliminating or mitigating measurement biases is to take advantage of the fact that observations are made to several GPS satellites, from two or more receivers (in relative positioning mode), at the same time. *Differencing* GPS measurements is the means by which complex data-processing models (which must explicitly include all the different GPS signal biases and errors) can be simplified in order to yield high-accuracy results, even in the presence of so many measurement biases.

8.6 DIFFERENCING OF GPS DATA

The pseudorange and carrier phase data are related to the receiver coordinates (x, y, z) and other variables through the following mathematical relations (in units of meters):

$$P_i^j = \rho_i^j + d_{\text{trop}i}^j + d_{\text{ion}\,qi}^j + c \cdot (dt_i - dt^j) + mp_i^j + \varepsilon p_i^j \qquad (8.4)$$

$$L_i^j = \rho_i^j + d_{\text{trop}\,i}^j - d_{\text{ion}\,qi}^j + c \cdot (dt_i - dt^j) + \lambda_q N_{qi}^j + mc_i^j + \varepsilon c_i^j \qquad (8.5)$$

where the subscript refers to the receiver identifier i and the signal frequency q and the superscript refers to the satellite identifier j:

P_i^j is the pseudorange at frequency f_q ($q = 1, 2$) (in m)
L_i^j is the carrier phase at frequency f_q ($q = 1, 2$) (in m) = $\lambda_q \cdot \varphi_i^j$
φ_i^j is the carrier phase at frequency f_q ($q = 1, 2$) (in units of cycles)
λ_i^j is the carrier wavelength at frequency f_q ($q = 1, 2$) (in m)
ρ_i^j is the geometric range between the receiver q and the satellite
$d_{\text{trop}\,i}^j$ is the range bias effect caused by tropospheric delay
$d_{\text{ion}\,qi}^j$ is the range bias effect caused by ionospheric delay at frequency f_q
dt_i, dt^j are the clock offset for the receiver and satellite clocks, respectively (in s)
c is the speed of electromagnetic radiation in a vacuum (in m/s)
mp_i^j, mc_i^j are the multipath disturbance on P_i^j and L_i^j observables, respectively
N_1, N_2 are the cycle ambiguities of the L1 and L2 signals (integer cycles)
$\varepsilon p_i^j, \varepsilon c_i^j$ are the noise of the P_i^j and L_i^j observables

The pseudorange and carrier phase measurements can be differenced by the data processing software to create new range data combinations with specific properties. The differencing of GPS data can be performed in several modes. Depending on the way in which the differencing is carried out, *one-way* (OW), *single-difference* (SD), *double-difference* (DD), and *triple-difference* (TD) data can be constructed. OW data is the basic range data from one receiver to one satellite at one frequency, as represented by Equations 8.4 and 8.5. Different combinations of OW pseudorange and carrier phase measurements to the same satellite, from the same receiver, are possible. Some of these combinations have beneficial characteristics for cycle slip editing, ambiguity resolution, or coordinate parameter estimation, making dual-frequency instrumentation essential for all high-accuracy applications.

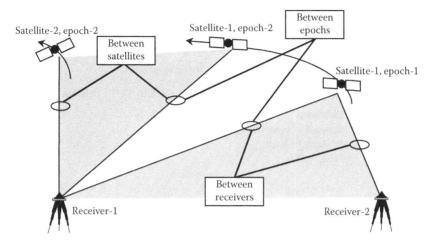

FIGURE 8.12 Data differencing modes.

GPS data differencing can be carried out between different receivers, different satellites, or different epochs, as illustrated in Figure 8.12, as well as between frequencies (e.g., L1–L2). It should be noted that between-receiver and between-satellite differencing are performed among the measurements made at the same observation epoch.

There are several consequences of the data differencing process:

- differencing can eliminate or reduce the effects of many biases, and
- differencing will reduce the quantity of data, however
- differencing will introduce mathematical correlations among the data, and
- differencing will increase the noise level of the resulting data

8.6.1 SINGLE-DIFFERENCE DATA

SD data is the difference between two OW measurements. There are three types of SD data, namely, *between-receiver* SD, *between-satellite* SD, and *between-epoch* SD. Each type of SD differencing will have different effects, as summarized in Table 8.3. Between-receiver differencing of measurements involving the same satellite signal (subscript j), for example, will eliminate the effects of satellite clock errors, and will reduce the effects of tropospheric and ionospheric biases, depending on the distance between the receivers. A model for the between-receiver phase SD (receiver subscripts i and m) in units of meters is

$$\lambda(\varphi_i^j - \varphi_m^j) = (\rho_i^j - \rho_m^j) + (d_{\text{trop } i}^j - d_{\text{trop } m}^j) - (d_{\text{ion } i}^j - d_{\text{ion } m}^j)$$

$$+ c \cdot (dt_i - dt_m) + \lambda(N_i^j - N_m^j) + (mc_i^j - mc_m^j) + (\varepsilon c_i^j - \varepsilon c_m^j) \quad (8.6)$$

Between-satellite differencing will eliminate the effects of receiver clock errors (satellite superscripts j and l):

TABLE 8.3

Characteristics of Errors and Biases in Single-Differenced Data

Errors/Biases	Single-Differenced Data		
	Between-Receiver (Δ)	Between-Satellite (∇)	Between-Epoch (δ)
Satellite clock	Eliminated	—	—
Ephemeris	Reduced (depends on the distance between receivers)	—	—
Receiver clock	—	Eliminated	—
Ionospheric	Reduced (depends on the distance between receivers)	Reduced (depends on the angular separation between satellites)	Reduced (depends on the time interval between epochs)
Tropospheric	Reduced (depends on the distance between receivers)	Reduced (depends on the angular separation between satellites)	Reduced (depends on the time interval between epochs)
Cycle ambiguity	—	—	Eliminated (if no cycle slips)
Noise	Increased by sqrt2	Increased by sqrt2	Increased by sqrt2

$$\lambda(\varphi_i^j - \varphi_i^l) = (\rho_i^j - \rho_i^l) + (d_{\text{trop}\,i}^j - d_{\text{trop}\,i}^l) - (d_{\text{ion}\,i}^j - d_{\text{ion}\,i}^l)$$
$$+ c \cdot (dt^j - dt^l) + \lambda(N_i^j - N_i^l) + (mc_i^j - mc_i^l) + (\varepsilon c_i^j - \varepsilon c_i^l) \qquad (8.7)$$

In the case of between-epoch differencing, the cycle ambiguity of the phase observation will be eliminated when there is no cycle slip between the two epochs, and depending on the time interval between epochs.

It should be noted that in performing differencing between satellites, when more than two satellites are involved, then there are several strategies for selecting which combinations of satellites should be differenced.

8.6.2 Double-Difference Data

DD data is the differencing result of two SD data, or four OW measurements. Three DD data can be distinguished: *receiver–satellite* DD, *satellite–epoch* DD, and *receiver–epoch* DD. The geometry of the standard DD data used in GPS data processing is depicted in Figure 8.13, and the characteristics of the differencing process are summarized in Table 8.4.

Combining Equations 8.6 and 8.7 leads to the following model of the carrier phase DD where all clock errors have been eliminated:

$$\lambda(\varphi_i^j - \varphi_m^j - \varphi_i^l + \varphi_m^l) = (\rho_i^j - \rho_m^j - \rho_i^l + \rho_m^l) + (d_{\text{trop}\,i}^j - d_{\text{trop}\,m}^j - d_{\text{trop}\,i}^l + d_{\text{trop}\,m}^l)$$
$$- (d_{\text{ion}\,i}^j - d_{\text{ion}\,m}^j - d_{\text{ion}\,i}^l + d_{\text{ion}\,m}^l) + \lambda(N_i^j - N_m^j - N_i^l + N_m^l)$$
$$+ (mc_i^j - mc_m^j - mc_i^l + mc_m^l) + (\varepsilon c_i^j - \varepsilon c_m^j - \varepsilon c_i^l + \varepsilon c_m^l) \qquad (8.8)$$

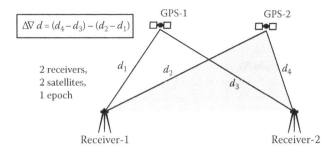

FIGURE 8.13 Receiver–satellite DD.

TABLE 8.4
Characteristics of Errors and Biases in DD Data

Errors/Biases	DD Data		
	Receiver–Satellite ($\Delta\nabla$)	Receiver–Epoch ($\Delta\delta$)	Satellite–Epoch ($\nabla\delta$)
Satellite clock	Eliminated	Eliminated	—
Ephemeris	Reduced (depends mainly on the distance between receivers)	Reduced (depends mainly on the distance between receivers)	—
Receiver clock	Eliminated	—	Eliminated
Ionospheric	Reduced (depends mainly on the distance between receivers)	Reduced (depends mainly on the distance between receivers)	Reduced (depends on the time interval between epochs and the distance between satellites)
Tropospheric	Reduced (depends mainly on the distance between receivers)	Reduced (depends mainly on the distance between receivers)	Reduced (depends on the time interval between epochs and the distance between satellites)
Cycle ambiguity	—	Eliminated (if no cycle slips)	Eliminated (if no cycle slips)
Noise	Increased by 2	Increased by 2	Increased by 2

It should be noted that the receiver–satellite DD phase data is the standard observable used for precise positioning with GPS (Section 9.1.2). In this case, however, the cycle ambiguity still needs to be estimated and "resolved."

8.6.3 TRIPLE-DIFFERENCE DATA

TD data is the difference between two DD data, or four SD data, or eight OW measurements. Regardless of the differencing sequences between DD data, only one TD data is obtained, that is, receiver–satellite–epoch TD, or simply TD. In GPS surveying, the TD phase data is usually used for automatic cycle slip detection and repair, and for estimating the approximate baseline vector to be used as a priori values in the baseline computation process (Section 9.4).

8.7   NEW GPS SIGNALS

As part of GPS *modernization* (Section 15.4), two new civilian signals will be added: first at the existing L2 frequency and the second at the frequency 1176 MHz, known as L5 (Divlis 1999, van Dierendonck et al. 2000, Fontana et al. 2001). The modernization of GPS signals will occur in stages (Figure 8.14).

The modernization of GPS signals will have tremendous positive impacts on navigation, positioning, and surveying applications using GPS. It will increase the availability, the accuracy, and the reliability of GPS. Increased availability is a direct consequence of the introduction of the new GPS signals shown in Table 8.5. (The availability will be increased even further in the case of multi-constellation Global Navigation Satellite System [GNSS] positioning.) With the original Block II/IIA/IIR

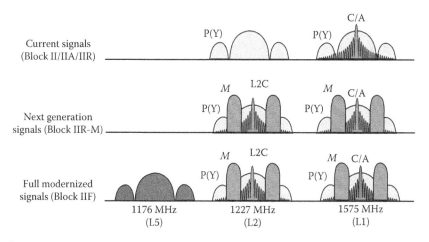

FIGURE 8.14   Modernization stages of GPS signals.

TABLE 8.5
Availability of New GPS Signals on the Block II Series of Satellites

Code Signal	Chipping Rate (Mchips/s)	Carrier Frequency (MHz)	Availability		
			Block IIR	Block IIR-M	Block IIF
L1 C/A	1.023	1575.42 (L1)	✓	✓	✓
L1 P(Y)	10.23	1575.42 (L1)	✓	✓	✓
L1 M	5.150	1575.42 (L1)		✓	✓
L2C	1.023	1227.60 (L2)		✓	✓
L2 P(Y)	10.23	1227.60 (L2)	✓	✓	✓
L2 M	5.150	1227.60 (L2)		✓	✓
L5C	10.23	1176.45 (L5)			✓

satellites, three GPS codes are available, that is, L1 C/A, L1 P(Y), and L2 P(Y). Block IIR-M satellites will transmit six GPS codes, with the addition of L2C (L2 civil), L1 M (military), and L2 M codes. The Block IIF (and Block III) satellites will transmit new codes on the L5 frequency. The subsequent Block III satellite generation (not shown in Figure 8.14 or Table 8.5) will transmit all these signals, and in addition a new L1 civilian code. Accuracy and reliability will be improved due to more rapid and more certain ambiguity resolution, even as the length of the baseline increases to many tens of kilometers.

REFERENCES

Divlis, D.A., 1999, Finally, a second signal decision, *GPS World*, 10(2), 16–20.

Fontana, R.D., Cheung, W., and Stansell, T., 2001, The modernized L2 civil signal leaping forward in the 21st century, *GPS World*, 12(9), 28–34.

Hofmann-Wellenhof, B., Lichtenegger, H., and Wasle, E., 2008, *GNSS Global Navigation Satellite Systems: GPS, GLONASS, Galileo, and More*, Springer Verlag, Wien, Austria/ New York, ISBN 978-3-211-73012-6, 516pp.

Kaplan, E. and Hegarty, C.J. (eds.), 2006, *Understanding GPS: Principles and Applications*, 2nd edn., Artech House, Norwood, MA, ISBN 1-58053-894-0, 570pp.

Langley, R.B., 1998, A primer on GPS antennas, *GPS World*, 9(7), 50–54.

Leick, A., 2004, *GPS Satellite Surveying*, 3rd edn., John Wiley & Sons, Hoboken, NJ, ISBN 0-471-05930-7, 435pp.

Parkinson, B.W. and Spilker, J.J. (eds.), 1996, *Global Positioning System: Theory and Applications*, Vol. I and II, American Institute of Aeronautics and Astronautics, Inc., Washington, DC, ISBN 1-56347-106-X, 793pp.

Seeber, G., 2003, *Satellite Geodesy*, 2nd edn., Walter de Gruyter, Berlin, Germany/New York.

Van Dierendonck, A.J. and Hegarty, C., 2000, The new L5 civil GPS signal, *GPS World*, 11(9), 64–71.

9 GPS Positioning Models for Single Point and Baseline Solutions

Chris Rizos and Dorota A. Grejner-Brzezinska

CONTENTS

This chapter briefly describes the observation models for the Global Positioning System (GPS) measurements for the two most common position estimation techniques: (a) single-point positioning (SPP) using pseudorange data and (b) baseline determination using double-differenced (DD) carrier phase data. For a detailed treatment of these models, and the computational procedures used in GPS data processing, the reader is referred to such texts as Leick (2004), Hofmann-Wellenhof et al. (2008), and Strang and Borre (1997). The last of these references is particularly useful for those interested in the algorithms themselves as it supplies sample MATLAB® code for SPP and baseline determination.

9.1 THE GPS MEASUREMENT MODELS

9.1.1 PSEUDORANGE OBSERVATION

A GPS receiver uses the satellite-generated pseudorandom noise (PRN) codes to perform a correlation between the incoming satellite PRN code and the receiver-generated replica of the same C/A-code (Section 8.2). This process produces an observation of the (code-derived) range or, equivalently, the signal *transmit time* t^j of the received signal at time t_i. Through this process the propagation delay $(t_i - t^j)$ is measured, which is scaled by the speed of light to generate the pseudorange observable (Langley 1993):

$$P_i^j = c \cdot (t_i - t^j) \tag{9.1}$$

The reason this observable is referred to as a "pseudorange" is because the receiver and satellite clock errors affect the measurement of propagation delay, and therefore the receiver–satellite range or distance, in a *systematic* manner. (The satellite clock controls the signal generation, and the receiver clock controls the code's replica generation.) In addition, the pseudoranges are also affected by the ionospheric and tropospheric delays, multipath, and receiver noise, and can be modeled for either the L1 or the L2 observable by Equation 8.4. However, a simplified observation model is generally adopted for pseudorange-based SPP:

$$P_i^j = \sqrt{[X^j(t) - x_i]^2 + [Y^j(t) - y_i]^2 + [Z^j(t) - z_i]^2} + c \cdot dt_i \tag{9.2}$$

All terms have been defined in Section 8.6.

9.1.2 DOUBLE-DIFFERENCED CARRIER PHASE OBSERVABLE

Although code-range measurement technology has advanced substantially during the last decade, the most precise observable is the carrier phase measurement φ_i^j. This observable is equal to the *difference* between the phase φ_i of the receiver-generated carrier wave at signal reception time, and the phase φ^j of the satellite-generated carrier phase at transmission time. Ideally the carrier phase observable would be equal to the total number of full carrier wavelengths plus the fractional cycle between the transmitting satellite antenna and the receiver antenna at any instant. However, the GPS receiver cannot distinguish one carrier cycle from another but it can measure the fractional phase when it locks onto the satellite, and then to keep track (or "count") of the phase changes thereafter. As a result, the initial phase is *ambiguous* by a quantity N_i^j, that is, by an unknown integer number of cycles (Equation 8.2 and Figure 8.9).

The basic observable for carrier phase-based GPS positioning is *not* the one-way measurement model, but rather it is the DD observable involving simultaneous measurements to two satellites and two receivers, as represented by Equation 8.8. The principal advantage of such an observable is the elimination of the receiver and satellite clock errors. Assuming that the DD atmospheric biases are negligible, and only considering the parameters of interest, the following simplified observation model is obtained (in units of meters):

$$\lambda \cdot \Delta\nabla\varphi_{im}^{jl} = \Delta\nabla\rho_{im}^{jl} + \lambda \cdot \Delta\nabla N_{im}^{jl} \tag{9.3}$$

There are now four coordinate triplets involved: two coordinate sets for the receivers i and m, as well as two sets of coordinates for satellites j and l. *What if the ionospheric bias is not negligible?* One option is to form the "ionosphere-free" observable from dual-frequency measurements (Section 9.2.1.1).

9.2 PREPARING FOR DATA PROCESSING

There are several issues regarding GPS data processing including:

- The degree of preprocessing of the data
- The parameter estimation technique used
- The data combinations used in processing
- The parameterization of the observations
- The quality control procedures used to evaluate the solutions
- The options and capabilities of the data processing software

The reader is referred to, for example, the texts of Leick (2004) and Hofmann-Wellenhof et al. (2008) for details. In this chapter, only some brief remarks will be made with regards to some of the above-mentioned issues, with particular attention being paid to *ambiguity resolution* (*AR*) for carrier phase data processing.

9.2.1 COMBINATIONS OF DUAL-FREQUENCY GPS DATA

GPS data (either pseudorange or carrier phase) at the L1 and L2 frequencies can be linearly combined in different ways.

9.2.1.1 The Ionosphere-Free Combinations

A first-order approximation for the ionospheric delay d_{ion} (in units of meters) at frequency q is (Leick 2004)

$$\frac{d_{ion_q}}{c} \approx (1.35 \times 10^{-7}) \frac{STEC}{f_q^2} \tag{9.4}$$

where
   f_q is the frequency (in Hz)
   c is the speed of light
   STEC is the slant total electron content of a column of ionosphere condensed onto
      a disk (in units of free electrons per square meter)

Note that the higher the frequency, the smaller the ionospheric delay, and therefore the delay on the L1 signal is less than that on the L2 signal. The relationship between the ionospheric delays on the two GPS frequencies can be expressed as

$$f_1^2 \cdot d_{ion_1} = f_2^2 \cdot d_{ion_2} \tag{9.5}$$

Hence the L2 ionospheric delay is approximately 1.647 times that on L1 ($1.647 \approx f_1^2/f_2^2$). Equation 9.5 can be simplified (and expressed in units of cycles of the f_q frequency, noting that $\lambda = c/f$) by dropping some terms and leaving off the subscripts and superscripts:

$$\varphi_q = \frac{f_q}{c} \cdot \rho - d_{ion_q} + N_q \tag{9.6}$$

Constructing Equation 9.6 for both the L1 and L2 measurements, then multiplying each equation by the associated frequency, and finally subtracting the two equations yields the following relation:

$$f_1 \cdot \varphi_1 - f_2 \cdot \varphi_2 = \frac{f_1^2 - f_2^2}{c} \cdot \rho - \frac{1}{c}(f_1^2 \cdot d_{ion_1} - f_2^2 \cdot d_{ion_2}) + f_1 \cdot N_1 - f_2 \cdot N_2 \tag{9.7}$$

where the second term on the right hand side of Equation 9.7 is zero, due to the relation at Equation 9.5. In order to combine the L1 and L2 phase observations, which are in units of cycles (of different wavelengths for L1 and L2), they have to be converted to the same units, for example, scaling by the L1 frequency:

$$\frac{f_1(f_1 \cdot \varphi_1 - f_2 \cdot \varphi_2)}{f_1^2 - f_2^2} = \frac{f_1}{c} \cdot \rho + \frac{f_1(f_1 \cdot N_1 - f_2 \cdot N_2)}{f_1^2 - f_2^2} \tag{9.8}$$

which yields the following expressions for the "ionosphere-free" L1 phase measurement:

$$\varphi_{1ion-free} = \alpha_1 \cdot \varphi_1 + \alpha_2 \cdot \varphi_2 \tag{9.9}$$

$$\varphi_{1ion-free} = \varphi_1 + \frac{f_2}{f_1} \cdot \varphi_2$$

$$\varphi_{1ion-free} - \frac{f_1}{c} \cdot \rho + \alpha_1 \cdot N_1 + \alpha_2 \cdot N_2$$

where
$\alpha_1 = f_1^2/(f_1^2 - f_2^2) \approx 2.546$
$\alpha_2 = -f_1 f_2/(f_1^2 - f_2^2) \approx -1.984$

Alternatively, if Equation 9.7 were scaled by the L2 frequency, the ionosphere-free L2 phase measurement would have been obtained, having the same form as Equation 9.9 except that α_1 would be replaced by β_1, and α_2 replaced by β_2, where $\beta_1 = f_1 f_2/(f_1^2 - f_2^2) \approx 1.984$ and $\beta_2 = -f_2^2/(f_1^2 - f_2^2) \approx -1.54$. The ambiguity term for the ionosphere-free combination is no longer an integer, being $N_{1\text{ion-free}} \approx 2.546 N_1 - 1.984 N_2$ (when expressed in units of L1 cycles) or $N_{2\text{ion-free}} \approx 1.984 N_1 - 1.54 N_2$ (when expressed in units of L2 cycles). By a similar process, an expression for the ionosphere-free pseudorange combination can be obtained:

$$P_{\text{ion-free}} = \frac{f_1^2 \cdot P_1 - f_2^2 \cdot P_2}{f_1^2 - f_2^2} \tag{9.10}$$

9.2.1.2 Linear Combinations of GPS Data: General Form

Other linear combinations of GPS phase measurements have been found to be useful for AR (Hofmann-Wellenhof et al. 2008). A general form for the linear combination of GPS data at two frequencies can be developed (with its cycle ambiguity still an integer):

$$\varphi_{n,m} = n \cdot \varphi_1 + m \cdot \varphi_2 \tag{9.11}$$

where

 n and m are integers
 φ_1 and φ_2 are the phase measurements made on the L1 and L2 carrier waves

(Such expressions can be generalized for three frequencies, for example, L1–L2–L5.) The ambiguity of the combined observable $N_{n,m}$ is related to the cycle ambiguity of the L1 and L2 signals, N_1 and N_2, through the following equation:

$$N_{n,m} = n \cdot N_1 + m \cdot N_2 \tag{9.12}$$

For example, the *widelane* combination is defined by $n = 1$ and $m = -1$, and the effective wavelength of this linear combination is 86 cm ($=[n \cdot f_1 + m \cdot f_2]/c$). The processing of this observable (and the resolution of its ambiguity) is useful for long-range positioning (i.e., long baseline between reference receiver and user) or for rapid AR. However, on the negative side, the noise of the combined observable is usually increased, and the covariance matrix (see the following text) of the differenced observables becomes non-diagonal.

9.2.2 THE LEAST-SQUARES SOLUTION

It is assumed that Equations 9.2 and 9.3 express the relationship between the "true" values of the observations and the model parameters. The mathematical basis of least-squares solution requires the definition of two models: (a) the *functional model* and (b) the *stochastic model*. The former is based on the observation models represented by, for example, Equation 9.2 or 9.3. The latter is tantamount to describing the accuracy of the measurement and its statistical properties. The reader should refer to Chapter 6 for an introduction to this material.

Because least-squares estimation assumes a linear mathematical model, and both the pseudorange and carrier phase observations are *nonlinear* with respect to both the receiver and satellite position vectors—see Equation 8.6—then Equations 9.2 and 9.3 must be *linearized* before proceeding. This is done by using approximate values of the unknown receiver coordinate parameters, so that the estimated parameters become the *corrections* to these initial values, not the values themselves. The least-squares estimation problem is usually described in terms of matrices (represented here by symbols that are underlined), for which the general solution is (Strang and Borre 1997)

$$\underline{\Delta x} = (\underline{A}^T \underline{\Sigma}^{-1} \underline{A})^{-1} \underline{A}^T \underline{\Sigma}^{-1} \underline{\Delta P} \tag{9.13}$$

where
  $\underline{\Delta x}$ is the matrix containing the estimable parameters
  $\underline{A}$ is the "design matrix" containing the partial derivatives of the processed observable with respect to the estimable parameters
  $\underline{\Delta P}$ is the matrix representing the residual quantities: observations minus the "calculated observations" (generated using the approximate values of the estimable parameters)
  $\underline{\Sigma}$ is the "covariance matrix," and is also the inverse of the so-called weight matrix of the observations
  $\underline{A}^T \underline{\Sigma}^{-1} \underline{A}$ is the "normal equation matrix"

The "covariance matrix of the solution" is $\underline{\Sigma}_s = (\underline{A}^T \underline{\Sigma}^{-1} \underline{A})^{-1}$, which is the inverse of the normal equation matrix.

"Quality" information can be derived from the covariance matrix of the solution, such as the standard deviations of the parameters, and their correlations, and is useful for statistical testing. Assuming that the observations are all independent (and hence the weight matrix is diagonal), each with the same standard deviation σ, then a simplified expression for this matrix is

$$\underline{\Sigma}_s = \sigma^2 (\underline{A}^T \underline{A})^{-1} \tag{9.14}$$

In general, the solution to a nonlinear problem must be iterated by updating the approximate values of the estimable parameters using the current solution for $\underline{\Delta x}$. Then the process is repeated until a convergent solution is obtained (generally within two to three iterations). If the functional and stochastic models are correct, then the observation residuals in the $\underline{\Delta P}$ matrix should be small and randomly distributed.

9.3 SINGLE-POINT POSITIONING

Although geodesists and surveyors typically use two (or more) GPS receivers in the relative positioning mode (Section 7.1.3.2), the primary objective of the GPS is to allow positioning anywhere, under any weather conditions, 24 h a day, with a single receiver. This mode of positioning is nevertheless important for positioning features for inventory/Geographical Information System (GIS) mapping purposes (Section 13.2.3). Single receiver positioning is generally referred to as "single-point positioning" or "absolute positioning" (Section 7.1.3.1), and can be performed using pseudoranges alone, or (under certain circumstances) carrier phase measurements, or both phase and pseudoranges. Only the former will be discussed here.

9.3.1 PSEUDORANGE-BASED POSITIONING

The following comments may be made to the solution procedure based on Equation 9.2 (Langley 1991):

- There are four parameters that must be estimated: the 3D coordinates of the receiver and the receiver clock error, requiring simultaneous observations from receiver i to four satellites ($j = 1...4$) for a unique solution.
- If pseudorange measurements to more than four satellites are made, a least-squares solution procedure is used to derive the *optimal* solution.
- An estimate of the satellite clock error is computed using parameters contained within the satellite navigation message (Equation 8.1), and applied as a *correction* to the pseudorange observation P, hence no satellite clock parameters need to be estimated. (Since *selective availability* [SA] was switched off on May 1, 2000, this clock error estimate is accurate to a few meters or less of equivalent range.)
- GPS receivers have inbuilt models for the ionospheric and tropospheric biases. Furthermore, if dual-frequency pseudorange observations are available, a new observable that is a combination of L1 and L2 measurements can be generated, which is "ionosphere-free" (Equation 9.10).
- Any errors in the known coordinates of the satellite (X, Y, Z) will impact directly on the accuracy of the solution. Hence ephemeris errors, together with the satellite clock error model and multipath error, are one of the limiting factors to SPP accuracy.
- The *reference datum* in which the receiver coordinates are expressed is that of the known (fixed in Equation 8.6) satellite coordinates, which, if derived from the broadcast ephemeris, is WGS84 (Section 7.2.1).
- Such a solution can be obtained instantaneously (as more than four satellites are visible at the same time if the skyview is clear). Hence the user's GPS receiver may be moving, and each independent solution is a "snapshot" of the receiver's location at a particular epoch. However, SPP solutions can also be computed *post-mission*, in which case more precise models of the satellite clock error and the satellite ephemeris may be used (see Section 14.4.2).

The factors influencing the final positioning accuracy may be identified from Equation 9.5:

1. The measurement noise (or random) error, represented by σ
2. The magnitude of any unmodeled biases, such as satellite ephemeris and clock, multipath, and atmospheric biases, that may also be absorbed into a larger value for σ
3. The factors that govern the magnitude of the elements of the design matrix $\underline{A}$

A combination of (1) and (2) is referred to in GPS terminology as the "user range error," while the factor at (3) is defined in terms of a quantity known as "dilution of precision."

9.3.2 USER RANGE ERROR

If the standard deviation σ of the pseudoranges represents *both* the measurement and modeling errors, and it is assumed that these errors are all independent, so that the quantity σ is equal to the square root of the sum of the variances of all these errors. If these errors include the internal receiver noise, the residual satellite clock error (remaining after the model is applied), ephemeris errors, atmospheric errors, and multipath, then the resulting standard deviation σ is known as the User Range Error.

For the standard positioning service (SPS—see Section 7.3.2), the user range (URE) was of the order of 25 m with SA on. However, with SA turned off the SPS, the URE value has dropped to below the 5 m level, and is now dominated by the ionospheric and multipath errors. For the precise positioning service (i.e., the service available to military users), using dual-frequency receivers, the URE value is of the order of 1–2 m.

9.3.3 DILUTION OF PRECISION

An important parameter as far as a GPS user is concerned is the accuracy of the estimated position. *How much have the pseudorange measurement and modeling errors affected the estimated position?* To answer this question it is necessary to study the covariance matrix of the solution Σ_s. If it is assumed that all the pseudoranges have the same standard deviation, then Equation 9.14 is the expression defining the relationship between the solution accuracy and the measurement precision.

The 4×4 covariance matrix $\Sigma_s = \sigma^2 (\underline{A}^T \underline{A})^{-1}$ has 10 independent (nonidentical) elements, and is a combination of the satellite geometry (i.e., the matrix $\underline{A}$), and the quality of the pseudorange measurements (including both random and systematic components). In order to reduce this information to just one number, dependent only on the satellite geometry, one can take the square root of the trace of the matrix, divided by the standard deviation of each pseudorange measurement (Strang and Borre 1997, Hofmann Wellenhof et al. 2008). If URE is substituted for σ, the resulting quantity is known as the "Geometric Dilution of Precision" (GDOP):

$$\text{GDOP} = \frac{\sqrt{\text{trace}(\Sigma_s)}}{\text{URE}} = \frac{\sqrt{\sigma_x^2 + \sigma_y^2 + \sigma_z^2 + \sigma_{clk}^2}}{\text{URE}} = \frac{\sqrt{\sigma_e^2 + \sigma_n^2 + \sigma_u^2 + \sigma_{clk}^2}}{\text{URE}} \quad (9.15)$$

where

    $\sigma_x, \sigma_y, \sigma_z$, and σ_{clk} are the standard deviations of the x, y, z components and receiver clock solution (taken from the diagonal elements of the covariance matrix), respectively

    $\sigma_e, \sigma_n, \sigma_u$ are the standard deviations of the estimated position along the east, north, and up directions

GDOP *factor* is a dimensionless quantity that is independent of the measurement errors, and dependent only on receiver–satellite geometry.

An estimate of the overall solution error therefore is

$$\sigma_s = \sqrt{\text{trace}(\Sigma_s)} = \text{URE} \cdot \text{GDOP} = \sqrt{\sigma_x^2 + \sigma_y^2 + \sigma_z^2 + \sigma_{\text{clk}}^2} = \sqrt{\sigma_e^2 + \sigma_n^2 + \sigma_u^2 + \sigma_{\text{clk}}^2} \qquad (9.16)$$

Note that the pseudorange standard deviation is *multiplied* by the GDOP scaling factor, and because this scaling factor is usually greater than unity it *amplifies* the pseudorange error, or "dilutes" the precision of the position determination. Sometimes it is desirable to examine the quality of the 3D position itself, or that of its horizontal and vertical components separately. In such cases only the corresponding elements of the covariance matrix need be used. For example, using only the first three diagonal elements $\sigma_x^2, \sigma_y^2, \sigma_z^2$ the "Position Dilution of Precision" (PDOP) can be obtained:

$$\text{PDOP} = \frac{\sqrt{\sigma_x^2 + \sigma_y^2 + \sigma_z^2}}{\text{URE}} = \frac{\sqrt{\sigma_e^2 + \sigma_n^2 + \sigma_u^2}}{\text{URE}} \qquad (9.17)$$

The last term in Equations 9.15 through 9.17 is included because the trace of the covariance matrix is unaffected when transforming the x, y, z coordinates to the local e, n, u coordinates. Hence it is also possible to determine the "Horizontal Dilution of Precision" (HDOP) and the "Vertical Dilution of Precision" (VDOP):

$$\text{HDOP} = \frac{\sqrt{\sigma_e^2 + \sigma_n^2}}{\text{URE}} \qquad (9.18)$$

$$\text{VDOP} = \frac{\sqrt{\sigma_u^2}}{\text{URE}} = \frac{\sigma_u}{\text{URE}} \qquad (9.19)$$

Typically more observed satellites, more evenly distributed across the sky, yields smaller Dilution of Precision (DOP) values, and hence more precise (and generally more accurate) position solutions. Also, in general, the HDOP values are between 1 and 2 and the VDOP values are larger than the HDOP values, indicating that the vertical positions are less precise (and less accurate) than the horizontal position components (Section 7.3.2).

Although DOP values are of particular interest for SPP applications, they also have a role in indicating the quality of a position determined using certain carrier phase-based techniques. These are the techniques based on very short static periods such as the *kinematic* procedures (Section 11.1.3). Such techniques are sensitive to receiver–satellite geometry, and DOP values are a convenient way of expressing this relationship. Mission planning software can be used to compute the DOP values at a certain location, for a specific instant in time, or for a span of time. Because the orbits of the GPS satellites repeat in such a way that they rise and set each day (but by about 4 min earlier from 1 day to the next), then the DOP values will vary on a predictable daily basis.

9.4 BASELINE DETERMINATION

The objective of relative positioning is to estimate the 3D *baseline* vector between the known (generally assumed stationary) and the unknown point or points (Figure 7.3). Although different processing schemes are possible, using either pseudorange or carrier phase data, only the carrier phase-based techniques capable of centimeter-level positioning accuracy will be discussed here. Many of the systematic errors in carrier phase measurements are completely (or nearly so) eliminated by taking DD combinations of the simultaneous phase observations recorded by receivers at the known and unknown points (Figure 8.13). Equation 9.3 is the standard mathematical model for the DD implemented in commercial GPS software. However, when it is compared with the more rigorous model at Equation 8.8 it is clear that all systematic biases due to the clocks, multipath, and atmospheric signal delay are assumed to have canceled. The degree of cancellation of the atmospheric biases is dependent on the baseline length, since these common errors tend to decorrelate with increasing receiver separation (i.e., increasing baseline length). Clock biases can indeed be assumed to have canceled completely. Multipath is more problematic, as it is a station environmental effect that affects each receiver differently, and when present (at either end of the baseline) will degrade the baseline solution. Hence all efforts should be made to ensure that the observation environment is as *multipath-free* as possible, or that the receiver's electronics and antenna are able to mitigate as much of the multipath effects as possible.

9.4.1 USING DOUBLE-DIFFERENCED CARRIER PHASE OBSERVABLES

There are a number of comments that may be made regarding the observable model in Equation 9.3:

- There are two classes of estimable parameters: the 3D receiver coordinates and the DD ambiguities.
- The *integer* nature of the ambiguities is preserved despite the measurement double-differencing.
- The ambiguities are assumed to be *constants* for as long as measurements are made, hence *all* data contributes to the estimation of these parameters. If, for whatever reason, the signal tracking to a satellite is interrupted, then the ambiguities involving that satellite before the break are *not equal* to the ambiguities after the break. That is, a *"cycle slip"* has occurred.

- One set of coordinates in the observation model are held fixed in the solution, while the other receiver's coordinates are estimated, requiring that the pair of GPS receivers be deployed with one at a *reference* station of known location and the second receiver at the station whose coordinates are to be determined.
- In principle, the second receiver can be in motion, necessitating the computation of a new set of coordinates at each epoch. This mode of positioning is referred to as *kinematic* positioning (Section 11.1.3).
- The entire computational process may be carried out in *post-mission* mode, after the recorded measurements from both receivers are brought together within the baseline determination software; or in *real time*, if the reference station measurements are transmitted to the second receiver (see Section 11.2).

When a least-squares solution is implemented to estimate the coordinate and ambiguity parameters, the main challenges are:

1. How to reliably estimate *both* the coordinate and ambiguity parameters
2. How to ensure that the errors and biases in the observable model remain *small* (compared to the wavelength of the carrier signal)
3. How to take advantage of the *integer* nature of the ambiguities

In general, (1) requires that observations are collected by the pair of GPS receivers over some *observation session* whose length varies according to a number of factors, principally the distance between the receivers (or the baseline length), the accuracy required, and the GPS technique used (see Section 12.1). Factor (2) typically requires that the baseline length be kept below a few tens of kilometers, and that no cycle slips are present in the DD observables. (If cycle slips do occur, then a preprocessing procedure must be used to "repair" the data so that the ambiguity parameters can be modeled as constants for an entire observation session.) Factor (3) is crucial to high-accuracy carrier phase-based GPS positioning techniques.

9.4.2   RESOLVING AMBIGUITIES

What is AR? It is a means of improving the accuracy of carrier phase-based GPS positioning using a mathematical process by which the ambiguous carrier phase measurements (Section 8.4.1) are converted to unambiguous "carrier-range" measurements (Section 8.4.2). In other words, the *integer* value ambiguity term in Equations 8.2, 8.5, 8.7, 8.8, and 9.3 is determined, and applied as a correction to the measured carrier phase, and in effect creates an accurate pseudorange-type observable (from Equation 9.3):

$$\Delta\nabla L_{im}^{jl} = \Delta\nabla\rho_{im}^{jl} \qquad (9.20)$$

where $\Delta\nabla L_{im}^{jl}$ is the DD carrier-range (in m).

The starting point for the AR process is the real-valued estimates of the DD ambiguities $\Delta\nabla N_{im}^{jl}$ derived from a least-squares solution using the observation model at Equation 9.3. If these estimates are accurate then AR can be attempted with some confidence of success. In the early days of GPS positioning, 1 h or more of carrier phase measurements were collected to ensure the *separability* of the position and ambiguity parameters in the least-squares solution. As the observation session lengthened, not only did the accuracy of the ambiguities increase but also the coordinate parameters increased. In the limit, there would be no need to resolve the ambiguities because the coordinates would have been determined to centimeter-level accuracy. Only with the drive to shorten observation sessions did it become crucial that AR be undertaken so that there was a *jump* in the coordinate accuracy when using an observation model at Equation 9.20 instead of Equation 9.3. This accuracy jump is illustrated in Figure 9.1, and discussed in Section 9.4.3.

There are several steps in the baseline determination process:

1. Define the a priori values of the ambiguity parameters, generally from the least-squares procedure using model Equation 9.3.
2. Use a *search* algorithm to identify likely integer values for the ambiguities.
3. Employ a *decision making* algorithm to select the "best" set of integer values.
4. Apply some *validation* tests to check whether these integer ambiguity values are indeed correct.
5. If the ambiguity values are reliable then convert the carrier phase data to carrier-range, and use another least-squares procedure with model Equation 9.20.

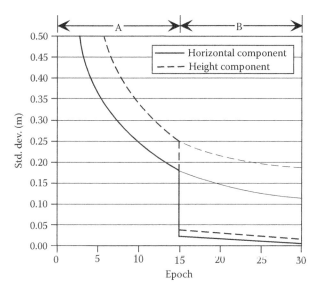

FIGURE 9.1 The evolution in quality (precise) of a baseline solution as ambiguities are resolved.

Considerable R&D has been invested in developing AR procedures for the *high productivity* GPS surveying techniques described in Section 11.1 (e.g., Han and Rizos 1997). The quest has been for techniques that ensure reliable AR even with relatively small amounts of data. AR *reliability* [steps (3) through (5) above] is a function of

- Baseline length (the shorter the better, and typically the receivers are not separated by more than about 20–30 km)
- The number of satellites (the more the better)
- Whether satellites rise or set during the session (the continuous tracking of satellites across the entire observation session is preferred)
- The satellite–receiver geometry, as characterized by DOP values (the lower the better)
- The degree of multipath disturbance to the measurements (the less the better)
- Whether observations are made on both carrier frequencies (it is much easier to resolve ambiguities when dual-frequency observations are available)
- The length of the observation session (the longer the better)

Although the above remarks imply that AR is a process that can only be applied to static GPS carrier phase data, during the last decade and half the AR algorithms have been significantly refined and the receiver hardware improved to such a point that AR can be carried out using just a few tens of seconds of tracking (or even less), *even if the receiver is in motion*. However, such "on-the-fly" (OTF) performance requires (a) dual-frequency carrier phase and precise (P-code level) pseudorange data, (b) high-quality GPS measurements, (c) tracking to six or more satellites, and (d) the baseline length to be less than about 20 km.

9.4.3 COMMENTS TO THE CARRIER PHASE-BASED SOLUTIONS

There are two types of DD carrier phase data solutions:

1. A so-called ambiguity-free solution (also commonly referred to as an "ambiguity-float" or "bias-float" solution), in which both the coordinates (one end of a baseline) and the ambiguity parameters are estimated as real-valued quantities. This solution is based on the observation model defined by Equation 9.3.
2. A so-called ambiguity-fixed solution (also sometimes referred to as a "bias-fixed" solution), in which only the coordinate parameters are estimated. This solution is based on the observation model defined by Equation 9.20, and is therefore a very "strong" (and accurate) solution. However, if the AR is unreliable, and one or more of the ambiguity values is incorrect by one or more integer values (for L1 each integer corresponds to an error of about 19 cm in the derived carrier range), then the subsequent ambiguity-fixed solution is *biased* by one or more decimeters.

The significant impact that AR makes to carrier phase-based GPS positioning is best understood with reference to Figure 9.1, where the solution quality (in terms of the standard deviation of the estimated coordinate components) is plotted against measurement epoch. With increasing epoch number extra carrier phase observations have been included in the least-squares solution, leading to a decrease in the standard deviations (or in other words, an increase in precision, and hopefully also accuracy).

There are several comments that can be made regarding Figure 9.1:

- In "region A" the precision (and accuracy) of the coordinates steadily improves as more data is collected.
- As soon as sufficient data is available to resolve the ambiguities (at epoch 15) a dramatic improvement in the coordinate parameter precision is evident. This is the rationale behind the "rapid-static" GPS technique (Section 11.1.1); that is, collect just enough data to ensure that an ambiguity-fixed solution is obtained.
- In "region B," when the unambiguous carrier-range data are processed, there is no significant improvement in the quality of the coordinate solution, and in effect there is no justification for continuing to collect data past epoch 15. This is the scenario encountered in "stop-and-go" GPS positioning (Section 11.1.2), where just a few epochs of measurements are sufficient to obtain centimeter-level baseline accuracy, or in the case of kinematic positioning (Section 11.1.3) there would be just independent epochs along the receiver trajectory.
- If enough data is collected over an observation session, the precision of the ambiguity-free solution will improve, converging to that obtained from an ambiguity-fixed solution. This means that AR is not required in order to obtain high-accuracy coordinate results. In other words, AR is optional and if there is any indication that the AR is incorrect that solution may be discarded in preference to the ambiguity-free solution.
- In conventional static GPS surveying, the data is post-processed, and it is therefore not known a priori at what point (or even if) sufficient data has been collected to ensure that an ambiguity-fixed solution is obtained. Hence *conservative* observation session lengths are recommended. Furthermore, accuracy is assured because *all* the measurements (including those taken before epoch 15) contribute to the ambiguity-fixed solution.
- The benefit of real-time carrier phase-based techniques (Section 11.2.2) is that only sufficient data need be collected to derive an ambiguity-fixed solution, and then just sufficient additional epochs of carrier-range data to estimate the coordinates. (Of course, in the kinematic mode this would be just one measurement epoch.)

9.5 CONCLUDING REMARKS

The AR is probably the most identifiable characteristic of high-precision GPS positioning. There have been dramatic improvements in the *productivity* of carrier phase-based GPS surveying techniques because the "time-to-AR" has been significantly reduced; from an hour or more during the 1980s, to several tens of minutes in the early 1990s, to the current observation session lengths of a few tens of seconds (and under ideal conditions even less). There are, however, several trends that indicate continued improvement in AR performance can be expected during the coming years (Sections 15.1, 15.3 through 15.5), primarily through the relaxation of some of the constraints to very rapid AR. In particular the distance between receivers will lengthen several fold, and the length of the observation session will decrease to the point where reliable single-epoch AR may be feasible (and reliable). The latter, when implemented in real-time, will make carrier phase-based relative positioning as robust and easy to use as current pseudorange-based differential GPS techniques (Section 11.2.1).

REFERENCES

Han, S. and Rizos, C., 1997, Comparing GPS ambiguity resolution techniques, *GPS World*, 8(10), 54–61.

Hofmann-Wellenhof, B., Lichtenegger, H., and Wasle, E., 2008, *GNSS Global Navigation Satellite Systems: GPS, GLONASS, Galileo, and More*, Springer Verlag, Wien, Austria/New York, ISBN 978-3-211-73012-6, 516pp.

Langley, R.B., 1991, The mathematics of GPS, *GPS World*, 2(7), 45–50.

Langley, R.B., 1993, The GPS observables, *GPS World*, 4(4), 52–59.

Leick, A., 2004, *GPS Satellite Surveying*, 3rd edn., John Wiley & Sons, Hoboken, NJ, ISBN 0-471-05930-7, 435pp.

Strang, G. and Borre, K., 1997, *Linear Algebra, Geodesy, and GPS*, Wellesley-Cambridge Press, Wellesley, MA, 624pp.

10 GPS Instrumentation Issues

Dorota A. Grejner-Brzezinska

CONTENTS

This chapter introduces the important issues related to the Global Positioning System (GPS) instrumentation and addresses questions such as

- What are the components of a GPS receiver?
- What should I consider when purchasing a GPS receiver?
- Why do GPS receivers for surveying applications cost so much more than for Geographical Information System (GIS) mapping?
- Will my receiver suffer from interference or multipath, and can the right choice of antenna help in this regard?
- How is a dual-frequency receiver different from a single-frequency receiver?
- What does real-time positioning imply for the user?
- What are some of the memory considerations for GPS receivers?
- What are the power requirements for GPS receivers?
- Must I use a choke ring antenna for the highest performance?
- Can the same GPS receiver be used for all, or most, types of positioning applications?
- What configurations are now possible for GPS user equipment?

It is beyond the scope of this chapter to provide the reader with a full list of GPS products, their characteristics, and their manufacturers, especially with new receivers being continually released onto the market. The most complete sources of such information are the annual *GPS World Buyers Guide* (usually the June edition) that provides industry-sourced information on GPS/Global Navigation Satellite System (GNSS)-related products for system developers and integrators (it is not intended to be a consumer product guide), and the *GPS World Receiver Survey* (usually the January edition). Now in its 16th year, the annual *GPS World Receiver Survey* provides the longest-running, most comprehensive database of GPS equipment available in one place. The latest editions of the *Buyers Guide, Receiver Survey,* and e-newsletters covering different *industry sectors* are also available on the *GPS World* Website (GPSWorld 2008).

GPS receivers belong to the *user segment* that covers the entire spectrum of applications, equipment, and computational techniques that are available to the systems' users (Section 7.1.3). Over the past two decades, civilian as well as military GPS instrumentation has evolved rapidly, with one of the most dramatic trends being the simultaneous reduction in hardware costs and the increase in positioning performance. Even though military receivers and many civilian application-specific instruments have evolved in different directions, one can ask, *are all GPS receivers essentially equivalent, apart from the functionality and the user software?*

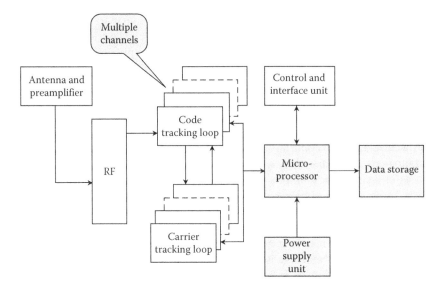

FIGURE 10.1 Conceptual architecture of a GPS receiver.

In general, the answer is *yes*—all GPS receivers consist of essentially the same functional blocks, even if their implementation differs from one *class* of receiver to another. By far, the majority of receivers manufactured today are of the C/A-code single-frequency variety, suitable for most civilian navigation, commercial and general positioning applications. However, for high-accuracy geodesy, and most surveying applications, dual-frequency hardware is standard.

The primary components of a generic GPS receiver are illustrated schematically in Figure 10.1 (Langley 1991). A GPS receiver system must carry out the following tasks:

- Select the satellites to be tracked.
- Search and acquire each of the selected GPS satellite signals.
- Recover the broadcast navigation data for every satellite.
- Make pseudorange, range-rate, and/or carrier phase measurements.
- Compute position, velocity, and time information.
- Optionally record the data for post-processing.
- Optionally transmit the data to another receiver via a radio modem for real-time differential solutions.
- Optionally accept user commands, and display results, via a portable control/display unit or a personal computer (PC).

10.1 ANTENNA AND PREAMPLIFIER

An *antenna* is defined as a device that acts as a transition between a guided wave and a free-space wave, and vice versa (Johnson 1993). On transmission, an antenna receives energy from a transmission line and radiates it into space, and on reception it collects energy from an incident wave and sends it down a transmission line.

The GPS receiving antenna detects an electromagnetic signal arriving from a satellite and, after band-pass filtering, which provides adequate filter selectivity to attenuate adjacent channel interference and initial preamplification, transfers the signal to the radio frequency (RF) section for further processing. A typical GPS antenna is *omnidirectional*, having essentially a nondirectional gain pattern in azimuth, though the pattern does change with the elevation angle.

In general, antennas can be designed as L1-only or dual-frequency, when both L1 and L2 carrier signals are tracked. The most common GPS antenna types currently available are *monopole* (asymmetric antennas) and *dipole* (symmetric antennas) configurations, *quadrifilar helices*, *spiral helices*, and *microstrips*. A helical antenna consists of a single (or multiple) conductor(s) wound into a shape with circular or elliptical cross sections. In general helical antennas are wound with a single conductor; however, a helix can be designed with bifilar, quadrifilar, or multifilar windings (Johnson 1993). A quadrifilar helix antenna is frequently used with handheld GPS receivers as it does not need a *ground plane*. The microstrip antenna, also sometimes called a "patch" antenna, is usually constructed of one or more elements that are photoetched on one side of a double-coated printed-circuit board. The antenna can be circular or rectangular in shape, made up of one or more patches of metal, separated from a ground plane by a dielectric sheet, and may have single- or dual-frequency-receiving capabilities (Langley 1998a,b). Microstrip antennas are the most common, primarily because of relatively easy fabrication and miniaturization, ruggedness, and general suitability for kinematic applications.

GPS antennas are often protected against possible damage by a plastic housing (a "radome") designed to minimize attenuation of the signal (Langley 1995b). For geodetic applications, an additional ground plane or choke ring antenna can be used (Figure 10.2 shows some typical high-precision antennas). The *ground plane* is a horizontally oriented metallic disk, centered at the GPS antenna's base, which shields the antenna from any signals arriving from below the antenna. The *choke*

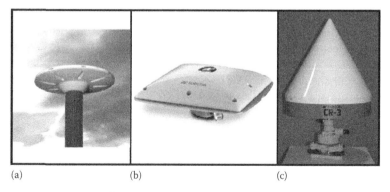

(a)   (b)   (c)

FIGURE 10.2   **(See color insert following page 426.)** (a) Micro-centered L1/L2 antenna with ground plane. (Courtesy of Trimble Navigation Ltd., Sunnyvale, CA) (b) Micro-centered L1/L2 GPS/GLONASS PG-A1 antenna with metal ground plane and rugged housing and (c) Zero-centered L1/L2 CR-3 reference base station antenna mounted on single-depth choke ring, with weatherproof cone environmental cover. (Courtesy of Topcon Positioning Systems, Inc., Livermore, CA.)

ring assembly is essentially a ground plane containing a series of concentric circular troughs one-quarter wavelength deep, designed to eliminate multipath surface waves (Section 8.5.2).

10.1.1 PREAMPLIFIER SECTION

Most GPS antennas are combined with a low-noise preamplifier (sometimes supported by additional filters either before or after the preamplifier). An antenna when combined with a preamplifier (usually housed in the base of the antenna, between the output of the antenna and the feeder line to the receiver) is called an *active* antenna, as opposed to a *passive* antenna that does not have such a preamplifier. The preamplifier boosts the signal level before feeding it to the receiver's RF section. However, caution must be exercised when using an antenna with a preamplifier not supported by the receiver manufacturer. Preamplifier's noise and gain must be within the receiver's acceptable range, and the voltage and current supplied by the receiver to the preamplifier must be compatible with the antenna's characteristics. A similar warning applies when using antenna "splitters" to connect two GPS receivers to a "zero baseline" (i.e., to a single antenna). The total gain of active antennas ranges from 20 to 50 dB (see Section 8.1.1 for definition of *decibel*), while power consumption is between 12 and 32 mA at 5 V DC. An amplified antenna can help prevent such problems as signal loss-of-lock due to antenna dynamics, or partially obstructed skyview due to foliage cover, where a passive antenna may totally fail.

10.1.2 CABLES AND CONNECTORS

An important issue is the proper selection of the antenna cable, as the signal attenuation depends on the type and the length of the cable used. The signal power can be significantly lowered if the coupling between the antenna and the receiver is faulty (e.g., due to different *impedance* of the connecting cable and the antenna, or simply by having dirty connectors), or the cable itself not being of the correct rating (Langley 1998a,b). (Impedance is essentially the ratio of the voltage to the current, measured in Ohms [Ω].) Flexible RG/U coaxial transmission cables with 50 Ω impedance, consisting of a solid or stranded inner conductor, a plastic-dielectric support, and a braided outer conductor, are most commonly used to connect the GPS antenna to the receiver. Since signal loss increases linearly with the cable length (and is slightly higher for higher frequencies), for long cable runs it is recommended that a low-loss cable such as Belden 9913, Belden 8214, RG-58C, RG-62, or RG-62A, be used. Another way to minimize the signal attenuation on long cables is to place an additional preamplifier between the antenna and the cable. For short runs, however, cheaper RG-174 or Belden 9201 cables would be sufficient.

A variety of cable connector types are used for GPS equipment. The most commonly used (in both male and female varieties) are BNC, F, MCX, N-type, lemo, OSX, SMA, SMB, and TNC (Langley 1998a,b). A full list of coaxial cables and frequently used connectors is given in Johnson (1993) and Storm (2009). Antenna cables are dispensed with entirely in cases where the antenna and receiver electronics are integrated within one unit, as in the so-called smart antennas or integrated receivers depicted in Figure 10.3.

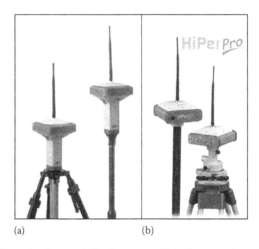

(a)   (b)

FIGURE 10.3   (See color insert following page 426.) Integrated receivers (a) GR-3 and (b) HiPer Pro. (Courtesy of Topcon Positioning Systems, Inc., Livermore, CA.)

10.1.3   ANTENNA PHYSICAL CHARACTERISTICS

In typical GPS antennas, the physical (or geometric) center of the antenna usually does not coincide with the phase (or electrical) center of the antenna where the microwave measurements are made. Moreover, the phase centers for L1 and L2 generally do not coincide, and different types of antennas have different locations of their phase centers. In addition, the location of the phase center can vary with azimuth and elevation of the satellites, and with the intensity of the incoming signal. These phase center variations affect the antenna offsets that are needed to connect GPS measurements to physical monuments. According to the U.S. National Geodetic Survey (NGS), which calibrates geodetic-grade GPS antennas and determine their calibration parameters, ignoring these phase center variations can lead to serious (up to 10 cm) vertical errors (NGS-antcal 2009). Detailed antenna calibration tables that account for the average spatial relationship between the Antenna Reference Point (ARP), such as Bottom of Antenna Mount (BA) and electrical phase centers, are provided by NGS-antcal (2009) and are normally included in any commercial GPS software. It should be mentioned that in high-accuracy static GPS positioning applications it is still a good practice to align the carefully leveled antennas in the same direction (e.g., local magnetic north), which results in the cancellation of both the length and the orientation of the offset between the physical and phase centers when the same type of antenna is used at both ends of short to medium length baselines.

The GPS signal arrives at the phase center (L1 or L2), however for surveying and mapping applications the coordinates of the ground mark are sought, therefore the observations have to be mathematically reduced to the ground mark location using the *antenna height*. The GPS antenna is usually mounted on a tribrach attached to a tripod, pillar, or range pole, and oriented along the local plumb line directly above the ground mark to be surveyed. The antenna height is the vertical distance between the ground mark and the antenna phase center. (Sometimes the slant height is measured to the bottom of the ground plane or the base of the choke rings, and the vertical height is computed using manufacturer-provided information about phase center location above the reference plane.)

10.1.4 Antenna Electrical Characteristics

The measure of antenna gain over a range of elevation and azimuth values is known as the *gain pattern*, where *gain* refers to an antenna's ability to successfully receive a weak signal, or its ability to concentrate in a particular direction the power accepted by the antenna. The gain pattern, similar to the phase center location, changes with the direction of the incoming signal. Since the antenna design must satisfy several (and at times conflicting) requirements, uniform gain in all directions is not necessarily desirable for a GPS antenna. For example, low elevation signals usually have to be filtered out, since too much gain at low elevation will contribute to high multipath and atmospheric signal bias (Section 8.5). At the same time, too much filtering at lower elevations would limit the minimum elevation at which satellites can be observed, resulting in weaker geometry [or higher Dilution of Precision (DOP) values, see Section 9.3.3].

The antenna *bandwidth* is defined as the frequency band over which the antenna's performance (described in terms of gain pattern, polarization, input impedance, etc.) is "sufficiently good" (Langley 1998a,b). Antennas may be *narrowband* or *wideband*. Since GPS is a narrowband system (within a bandwidth of about 20 MHz), most GPS antennas are also of the narrowband type. On the other hand, the bandwidth has to be large enough to assure the antenna's proper functioning over the intended range of Doppler-shifted frequencies. For example, antennas accepting only the central lobe of the C/A-code might have a very narrow bandwidth of ±2 MHz, while antennas accepting more than the central lobe of the C/A-code, or tracking the P(Y)-code, have a wider bandwidth of ±10–20 MHz. Dual-frequency L1/L2 microstrip antennas are usually designed as two-patch antennas, one patch for each L-band frequency, each with a bandwidth of ±10–20 MHz. The bandwidth of microstrip antennas is proportional to the thickness of the substrate used. Interested readers are referred to Johnson (1993), Weill (1997), and Langley (1998a,b), and references cited therein, for more information about antennas.

10.2 RF FRONT-END SECTION AND THE SIGNAL PROCESSING BLOCK

The signal processing functions of a receiver are the "heart" of a GPS receiver, performing the following functions (Parkinson and Spilker 1996):

- Precorrelation sampling and filtering, and Automatic Gain Control (AGC)
- Signal splitting into multiple signal processing channels
- Doppler frequency shift removal
- Generation of the reference Pseudorandom Noise (PRN) codes
- Satellite signal acquisition
- Code and carrier tracking from multiple satellites

- System data demodulation from the satellite signals
- Making pseudorange measurements from the PRN codes
- Making carrier frequency measurements from the satellite signals
- Extracting signal-to-noise information from the satellite signals
- Estimating the relationship to GPS time

The basic components of the RF section are a precision quartz crystal oscillator used to generate the 10.23 MHz reference frequency, multipliers to obtain higher frequencies, filters to eliminate unwanted frequencies, and signal mixers. The RF section receives the signal from the antenna, and translates the arriving (Doppler-shifted) frequency to a lower one, called the *beat* or *intermediate frequency* (*IF*), by mixing the incoming signal with a pure sinusoidal one generated by the receiver's oscillator. As a result, the modulation of the IF remains the same, only the transformed carrier frequency becomes the difference between the original signal and the one generated locally. The IF signal produced by the RF section is subsequently processed by the signal tracking loops.

10.2.1 MULTICHANNEL ARCHITECTURE

An important characteristic of the RF section is the number of tracking channels, and hence the number of satellites that can be tracked simultaneously. Current GPS receivers are based on dedicated channel architecture, where every channel tracks one satellite, either on the L1 or L2 frequency. Since the accuracy, the reliability, and the speed of obtaining position results increases with the number of satellites used in the solution, a good quality receiver should have enough channels to track all visible satellites at any given time. Typically geodetic/survey-grade GPSs are able to track 12 satellites or more (24 or more dual-frequency channels), which is especially important for real-time applications, or when being used as a reference station receiver. If the receiver is capable of tracking other GNSS satellites, such as GLONASS (Section 15.5.1), more channels are used (e.g., 72 channels for dual-frequency GPS, GLONASS, and space-based augmentation system signal support for WAAS/EGNOS/MSAS, see Sections 14.3.1.2 and 15.5.4).

10.2.2 TRACKING LOOPS

The *tracking loop* is a means of tracking a signal that is changing in frequency and in time. Basically, as long as there is signal "lock-on," a tracking loop performs a comparison between the incoming signal and the local one generated within the receiver, and "adjusts" the internal signal to "match" the external one. There are essentially two types of tracking loops used in GPS receivers: *delay-lock* (or code-tracking) loop (DLL) and the *phase-lock* (or carrier) tracking loop (PLL). Dual-frequency receivers have separate channels (and thus separate tracking loops) for both frequencies. A tracking loop must be tunable within a frequency range (i.e., it must have sufficient bandwidth) to accommodate the residual frequency offset of the modulated signal caused by Doppler shifts (maximum ±6 kHz), residual user clock drift, bias frequency offset, and data modulation.

10.2.3 Signal Acquisition with DLL and PLL

The DLL applies a "code-correlation" technique to align a PRN code (either P(Y) or C/A) from the incoming signal with its replica generated within the receiver (Section 8.2.1). The alignment is achieved by shifting the internal signal in time. In principle, the amount of "shift" that is applied corresponds to the time required for the incoming signal to travel from the satellite to the receiver. This time interval multiplied by the speed of light generates the "range" (more correctly the *pseudorange*) between the phase center of the receiver's antenna and the phase center of the GPS satellite transmitting antenna. Once the DLL is locked, the PRN code is removed from the signal, by first mixing it with the locally generated signal, and second, by applying filtering to narrow the resulting bandwidth down to about 100 Hz. Through this process, the GPS receiver reaches the necessary signal-to-noise ratio (SNR) value to compensate for the gain limitations of a physically small antenna (Langley 1991). At this point, the navigation message data can be extracted from the IF signal by the PLL, which performs the alignment of the phase of the beat frequency signal with the phase of the locally generated carrier replica. Once the PLL is locked to the signal, it will continue tracking the variations in the phase of the incoming carrier as the satellite–receiver range changes. The carrier beat phase observable (in cycles) is obtained by counting the elapsed cycles and by measuring the fractional phase of the locked local oscillator signal (Section 8.4.1).

10.2.4 Signal Tracking in the Presence of Anti-Spoofing

The code-correlation tracking procedure described earlier works well when the receiver can generate the replica PRN code. However, under the policy of *anti-spoofing* (AS), which is the current mode of GPS operation, the P-code is replaced by the encrypted Y-code. Hence different tracking techniques must be applied in civilian receivers to make measurements on the L2 frequency (remember, there is no C/A-code on L2 on GPS satellites prior to Block IIR-M, Section 8.7). Thus, AS is the reason why L2 measurements are harder to make than L1 measurements. To overcome the P-code encryption, various codeless squaring, cross-correlation, and quasi-codeless tracking techniques have been developed over the years. Only a small number of GPS receiver manufacturers have patented (or licensed) rights to produce and market dual-frequency instrumentation. This is one of the reasons why dual-frequency instrumentation is much more expensive than single-frequency (L1-only) instruments.

It should be mentioned, however, that all these methods of recovering L2 carrier in the presence of AS show SNR degradation with respect to the code-correlation technique applicable under no-AS conditions (Table 10.1). Consequently, the weaker signal (indicated by the smaller SNR) recovered under AS is more susceptible to jamming (or interference) as well as to ionospheric scintillation effects. As the second (and third) civilian frequencies are broadcast with the progressive launch of new GPS satellites (starting with the first Block IIR-M satellite launched in September 2005), standard code-correlating techniques will be used to make dual- (and triple-) frequency measurements.

Under the *GPS Modernization Program* (see Section 15.4), a new civilian PRN code is modulated on the L2 frequency (Table 8.5), and at the time of writing there were six

TABLE 10.1
L2 Signal Recovery under AS

Component	Squaring Technique	Cross-Correlation	Code-Correlation	Z-Tracking Plus Squaring
L1 C/A code	No	Yes	Yes	Yes
Y2 code	No	Y2–Y1	Yes	Yes
L2 wavelength	Half	Full	Half	Full
SNR (dB)	−16	−13	−3	0
SNR degradation with respect to code correlation (no AS) (dB)	−30	−27	−17	−14

Source: Ashjaee, J. and Lorenz, R., Precision GPS surveying after Y-code, *Proceedings of the Fifth International Technical Meeting of the Satellite Division of the U.S. Institute of Navigation*, Albuquerque, NM, pp. 657–659, 1992.

Block IIR-M GPS satellites in orbit with a civilian "L2C" signal (modulated on the L2 frequency). In 2010, the first of the Block IIF satellites is scheduled to be launched, and will broadcast the third civilian signal at the L5 frequency of 1176.45 MHz.

10.3   OTHER INSTRUMENTAL COMPONENTS

10.3.1   MICROPROCESSOR

GPS receivers perform numerous operations in real time, such as acquiring and tracking of the satellite signals, decoding the broadcast message, timekeeping, range measurement, multipath, and interference mitigation. All of these operations are coordinated and controlled by a *microprocessor*. Other functions that are performed by the microprocessor are data filtering to reduce the noise, position estimation, datum conversion, communication with the user via the control and display unit, and managing the data flow through the receiver's communication ports. With respect to the latter, an industry standard output message format that is used on almost all GPS receivers is known as NMEA 0183 (Langley 1995a). The National Marine Electronics Association, a professional trade association serving all segments of the marine marketplace (NMEA 2009), developed a uniform interface standard for digital data exchange between different marine electronic devices, navigation equipment, and communication equipment. NMEA 0183 data messages use plain text in ASCII (American Standard Code for Information Interchange) format, which makes it comparatively easy to connect a GPS receiver to a PC or other device in order to pass coordinate information to other electronic equipment and users.

10.3.2   CONTROL AND DISPLAY UNIT

The control device, usually designed as a keypad/display unit, will accept commands from the user for selecting different data acquisition options, and other important parameters such as the elevation mask angle, the data recording interval, and the antenna height, as well as for displaying information about the receiver's operations, navigation results, observed GPS satellite constellation, etc. Most receivers intended for surveying and mapping have command/display capabilities with extensive menus and prompting instructions, even offering online help. GPS receivers designed for integration into navigation systems usually do not have their own interface units. In such cases, the communication with the receiver is facilitated through an external PC via messages in the NMEA format (Langley 1995a). Among the currently approved 60 or so NMEA sentence types, nine are specific to GPS receivers: ALM–GPS almanac data, GBS–GPS satellite fault detection, GGA–GPS fix data, GRS–GPS range residuals, GSA–GPS DOP and active satellites, GST–GPS pseudorange noise statistics, GSV–GPS satellite in view, MSK–MSK (minimum-shift keying) receiver/GPS receiver interface, and MSS–MSK receiver signal status. Manufacturers can also define product-specific NMEA messages.

10.3.3 DATA STORAGE

Many GPS receivers, but primarily the ones intended for geodesy, surveying and mapping applications, provide an option for *storing* the measurements and the navigation message for data post-processing. The storage media most frequently used in GPS receivers are removable memory cards or some form of RAM memory. Manufacturers usually offer different memory capacity options, and in the case of extended storage requirements, extra memory can usually be added. The storage capacity needs of a GPS receiver are dependent on the length of the data acquisition session, the type of observable(s) to be recorded, the number of tracking channels, and the data recording interval (which can be a measurement every 120 s for geodesy and static surveying, up to 20–50 times a second for special kinematic applications). A complete set of dual-frequency measurements requires about 100 bytes per epoch; multiplying this by the number of channels (the maximum number of satellites to be observed) and the total number of epochs of observation, gives an estimate of the total memory needed. For example, recording data every 10 s for 10 h with 10 satellites in view will require $360 \times 10 \times 10 \times 100 = 3.6$ MB of data storage.

Receivers storing data internally must have an RS-232 or some other kind of communication port to allow data transfer between the receiver and a PC. For example, transferring data to a PC via a serial port takes about 100 s/MB (thus 16 MB of data takes a little less than half an hour). Many modern receivers use fast parallel ports, which significantly reduce the data transfer time. If a compact flash or a PC card is used for data storage, data can be downloaded to a PC via a specific card reader, a USB port, or a wireless BlueTooth protocol (which utilizes peer-to-peer wireless communications technology facilitating data transmission rates of about 1 MB/s over short distances of the order of a few meters or so). These storage media provide up to many tens of hours of dual-frequency data storage.

10.3.4 POWER SUPPLY

Current GPS instrumentation is comparatively energy efficient and typically uses low-voltage DC power provided by either internal rechargeable batteries, or external batteries (see Table 10.2), allowing for many hours of continuous operation. Typical

TABLE 10.2

Batteries Commonly Used for GPS Equipment

Characteristic	NiCd	Lithium Ion	Sealed Lead Acid (SLA)
Energy density (W/kg)	40–60	100	30
Cycle life	1500	Up to 1000; 2 year life time	200–300
Fast charge time (h)	1–1.5	3–4	8–16
Overcharge tolerance	Moderate	Very low	High
Self-discharge per month (%)	20	10	5
Nominal cell voltage (V)	1.2	3.6	2
Maintenance requirement	30–60 days	Not required	3–6 months
Size/weight	Small for portable equipment	Small	Typically large
Typical cost in U.S.$	50 (7.5 V)	100 (7.2 V)	25 (6 V)
Other characteristics	• Recommended slow charge of a new battery for 24 h • Performs best if fully discharged periodically Sheer endurance • High current capabilities • The only commercial battery that accepts charge at extremely low temperatures	• Does not need periodic full discharges • Lightest of all batteries • May cause thermal runaway • Usually sold in pack, complete with a protection circuit • Non-rechargeable version available • Most expensive	• Good storage life but can never be recharged to its full potential • Should be stored in charged state • Most economic for high-power applications where size and weight is not a consideration

Source: http://www.cadex.com.

power consumption for geodetic/survey-grade GPS receivers ranges from 2 to 9 W. External power can also be supplied to a GPS receiver via an AC–DC converter. The lower the power consumption of the receiver, the more field survey operations can be carried out with a single battery. Fieldwork problems due to the loss of power supply is considered an "ultimate sin," that nevertheless occurs more frequently than it should (Table 13.1)!

10.3.5   RADIO LINKS

Both *Differential GPS* (DGPS) and *Real-Time Kinematic* (RTK) systems require a device to facilitate the reception of messages broadcast by one or more GPS base stations. Each DGPS service provider requires their own specific communications link

(Section 14.3). In general, the appropriate radio/coms equipment can be obtained from third party suppliers, or from the service providers themselves. In the case of RTK systems, the radio link is sold as part of the GPS receiver kit.

10.4 CHOOSING A GPS RECEIVER

The 2009 Receiver Survey in the *GPS World* magazine lists 485 receivers from 72 manufacturers (*GPS World*, January 2009). *Why are there so many GPS receivers on the market?* The answer is *because there are so many different applications of GPS*, and new uses spring up every day. Starting from the lower end of the accuracy requirements, handheld GPS receivers operate at the single-point accuracy level (<10 m), while DGPS receivers that utilize WAAS or EGNOS signals (see Section 14.3.1.2) can provide 1–3 m positioning accuracy, and receivers used for mapping and GIS data collection typically require a positioning accuracy in the range of sub-meter to a few meters. These classes of GPS receivers are single-frequency units, designed to operate in real time. The highest accuracy specifications are naturally associated with geodetic/survey-type applications. In this case, a dual-frequency receiver, collecting data for post-processing, is usually specified. RTK techniques are now increasingly being used when the appropriate reference-to-user receiver data communications link is available (Section 11.2.2 and Table 11.3).

How to define a set of selection criteria? The first task is simply to list the functionalities and characteristics that distinguish one receiver from another. The second task is to assign some scale of values to these, to reflect their "relative level of desirability." Both of these are daunting tasks! An incomplete list of functionalities might include the type of measurements, the number of channels, single- or dual-frequency, the type of data storage, the antenna type, the nature of the instrument "packaging," the functionality of software, pricing, service, and maintenance contract conditions, the range of applications, multipath, and interference rejection qualities.

Those responsible for the selection of GPS equipment for a specific application may find the assignment a challenge if brochures and product specifications provided by the manufacturers are the only information sources consulted. The buyer has to correctly interpret the specifications, and should be able to ask the right questions in order to gain information regarding the conditions (and constraints) under which the specified performance is achieved. Confusion over terminology is also quite common, making it often difficult to compare "apples to oranges." For information on GPS receiver performance, the reader is referred to Van Dierendonck (1995), Langley (1997a,b), Kaplan and Hegarty (2006), and Parkinson and Spilker (1996). (Multi-GNSS receivers as they become available will provide even more choice!) In the following sections, several issues that might aid the selection of the "right GPS receiver" will be discussed.

10.4.1 THE ANTENNA FOR "THE JOB"

The size and physical characteristics, which can be different depending on the application requirements, represent constraints on antenna choice. Hence, in the case of top-of-the-line GPS receivers, the manufacturers offer several different antenna options.

Is there such a thing as an "ideal GPS antenna," or is there always some trade-off to be made? Unfortunately there is no *ideal* antenna suitable for *all* GPS applications. Each user must make a decision based on application-specific criteria, keeping in mind that each antenna type has a character of its own, and even those that appear to be physically similar might have different performance characteristics. This issue becomes of crucial importance when the highest accuracy is required. For example, an antenna that provides a low elevation angle gain cut-off might not be the best to meet the requirements of low phase center variation with zenith angle, and vice versa. Since most multipath signals arrive from directions near the horizon, shaping the pattern to have low gain in these directions may reject such signals. Such a design for a high-profile antenna with a low gain toward the horizon is appropriate for a ground-based reference station. On the other hand, a high profile antenna is not desirable for airborne use where it is necessary to receive signals even while the aircraft undergoes severe pitching and rolling, and where sharp nulls in the gain pattern might cause the loss of signal tracking.

10.4.1.1 Static Positioning

Surveying/geodetic antennas for static positioning are usually much larger, and have a ground plane or choke rings added, compared to compact lightweight antennas intended for kinematic applications. It is worth mentioning here that for some antennas, such as in the case of microstrip antennas, a ground plane not only contributes to multipath mitigation but also increases the antenna's zenith gain (to a limit, the larger the ground plane, the higher the antenna's gain in the zenith direction). Choke ring antennas can almost entirely eliminate the multipath problem for the surface waves and the signals reflected from the ground. (The multipath from objects above the antenna still represents a significant challenge for antenna designers.) The performance of the choke ring antenna is usually better for the L2 frequency than for L1, the reason being that the choke ring can be most easily optimized only for one frequency. If the choke ring antenna is designed for L1, it will have no effect on L2 signals, whereas a choke ring antenna designed for L2 multipath suppression still has some benefits for L1. Designing new antennas capable of tracking the multitude of new frequencies that will be transmitted by new GNSS satellites will be quite an engineering challenge.

10.4.1.2 Kinematic Applications

Airborne antennas, for example, should have the capability of acquiring signals even at the antenna horizon. In addition, bulky antennas are generally inappropriate for aircraft, and hence light-weight, small microstrip antennas are preferred for such applications. The primary criteria for antennas used for kinematic surveying and mapping are small size, light weight, and good multipath rejection (and for high-productivity survey techniques—Section 11.1—dual-frequency antennas must be used). Such antennas can be fitted atop a pole that can be used to position the antenna accurately over a ground mark, or placed on an antenna pole that is carried in a backpack. Hence the small "frisbee" style antenna is favored. For some RTK configurations, the GPS antenna is integrated with the radio link antenna in the one pole-mounted housing. Furthermore, in some instrument models the antenna(s) and the receiver are integrated within the one unit (Section 10.4.4.1).

10.4.1.3   Recreational and DGPS Capable Receivers

The majority of handheld GPS receivers have a very small, single-frequency antenna embedded within the receiver housing. The antennas for handheld receivers usually comprise quadrifilar helices, consisting of two bifilar helical loops, orthogonally oriented on a common axis, which do not require a ground plane (as opposed to the microstrip types used for surveying-type antennas). There is a trend nowadays for antennas intended to receive DGPS messages (e.g., from L-band geostationary satellite transmissions or mobile telephony services) to be integrated within the same housing used for the GPS signal antenna.

10.4.2   GPS Performance

What is meant by GPS performance? It is not simply a question of one receiver brand or model being more "accurate" than another. In general, given a specified accuracy the receiver performance will match that accuracy standard only if a range of criteria are satisfied "in the real world." For example, relative positioning accuracy will typically be a function of the number of satellites being tracked, the measurement type that is being processed, the data processing algorithm used, the length of the observation session, operational mode (such as whether the receiver is moving or stationary), quality of observation conditions, degree of signal obstructions, baseline length, and environmental effects, and consequently the actual performance will vary as these conditions change.

10.4.2.1   Defining Performance Parameters

It should be noted that the commercial data sheets from manufacturers, stating the achievable positioning accuracy, might be misleading because the claimed performance of the receiver is typically achieved under ideal observing conditions. Testing a GPS receiver, on either a preestablished baseline network or using a "zero baseline," is one way to gauge the degree to which the ideal performance degrades when real operating conditions are experienced. In general, the high-productivity carrier phase-based positioning techniques (Section 11.1) will be the most sensitive to deviations of actual observing conditions from ideal ones.

Often more important than the accuracy of the results under certain observing conditions are other measures of "good performance." The following is an incomplete list (and does not distinguish between survey-grade receivers intended for post-processed modes of positioning, and relatively basic navigation receivers): susceptibility to fade or interference, time-to-first-fix, reacquisition time after signal loss, measurement data rate, power consumption, multipath rejection, and so on. Clearly, it is not a simple matter to unambiguously define "good performance."

10.4.2.2   Dual-Frequency or Single-Frequency GPS Receivers

Dual-frequency instruments make twice the number of measurements than single-frequency instruments. Furthermore, the pseudorange and carrier phase measurements on the L1 and L2 frequencies can be linearly combined to generate new observables with specific desired qualities (Section 9.2.1.2). The special combination of $2.546 \times L1 - 1.546 \times L2$ creates an observable that has had the ionospheric delay eliminated, and is known as the "ionosphere-free" observable (Section 9.2.1.1). (L1 and L2 can be either the pseudorange or carrier phase measurements.) Another important dual-frequency carrier phase combination is $\phi_1 - \phi_2$, which is known as the "wide-lane" combination because the effective wavelength is approximately 86 cm (compared with 19 and 24 cm for the L1 and L2 carrier waves, respectively). In summary, dual-frequency instrumentation is insisted upon if

- High-accuracy baseline determination over distances of 20–30 km or more are sought, or
- High-productivity GPS surveying techniques are to be used

Dual-frequency measurements are required for the former because they permit the biases due to the residual ionospheric delay (remaining after differencing the simultaneous measurements from two receivers to the same satellite) to be eliminated. In the case of the latter group of techniques, *rapid* ambiguity resolution (Section 9.4.2) can only be done reliably if combinations of the L1 and L2 measurements are possible. Hence, for these reasons dual-frequency phase-tracking instrumentation represents "top-of-the-line" GPS hardware. As the third frequency (L5) starts transmitting from the new GPS Block IIF satellites, these extra measurements will be welcomed by geodesy and surveying users, as they will improve the accuracy, the speed, and the reliability of carrier phase-based positioning.

10.4.2.3 Radio Interference

GPS receivers can track the satellite signals under favorable conditions, but what increasingly distinguishes one receiver from another is typically how well the tracking can be performed (and the resultant quality of measurements) under less than ideal environmental conditions. Several kinds of interference can disrupt a GPS receiver's operation: in-band emissions, nearby-band emissions, harmonics, and jamming (Parkinson and Spilker 1996). As more and more sources of microwave transmissions appear every day, signal interference has become a major concern to the GPS user community (Johannesse 1997). The radio interference can, at the very least, reduce the GPS signal's apparent strength (the SNR), and consequently the accuracy, or at worse even block the signal entirely. The medium-level interference would cause frequent losses-of-lock or cycle slips, and might render the data virtually useless for high-productivity surveying techniques.

10.4.2.4 Multipath Interference

The multipath is a special form of signal interference. The GPS signal is generated using a spread spectrum technique that inherently rejects interference caused by long-delay multipath (Section 8.5.2). However, short-delay multipath still causes problems, especially for kinematic applications and static positioning using very short observation sessions, thus special techniques have been (and continue to be) developed to reject GPS multipath errors.

10.4.3   HARDWARE COMPONENT QUALITY

Often GPS manufacturers will try to (favorably) differentiate their receiver equipment from that of other manufacturers on the basis of the quality of the hardware components.

10.4.3.1   Receiver Clock

How good is the GPS receiver oscillator (clock)? Is this important to the user? The differential mode of positioning using carrier phase data requires that the measurements undergo "double-differencing" during the baseline estimation process (Sections 8.6.2 and 9.4.1), where the error in range attributable to the receiver oscillator (or clock) is eliminated. There are indeed differences in the quality of the receiver oscillator from one manufacturer to another, however for the vast majority of surveying applications there is no noticeable degradation in the quality of relative positioning results that can be attributed to this source alone.

10.4.3.2   Receiver Noise

Since a GPS receiver, as in the case of any measuring device, is not "perfect," measurements will always be made with a certain level of "noise." The most basic kind of noise is the so-called thermal noise, produced by the random movement of the electrons in any material that has a temperature above $0\,K$ (Langley 1998a,b). The commonly used measure of the received signal strength is the signal-to-noise-ratio, *SNR*. The larger the SNR value, the stronger the signal. In the case of RF and IF, the most commonly used measure of the signal's strength is the carrier-to-noise-power-density ratio, C/N_0, defined as a ratio of the power level of the signal carrier to the noise power in a $1\,Hz$ bandwidth (Van Dierendonck 1995, Langley 1998a,b).

10.4.3.3   Receiver Measurement Precision

C/N_0 is considered a primary parameter describing the GPS receiver performance as its value determines the precision of the pseudorange and carrier phase measurements. Typical values of C/N_0 for modern survey-grade GPS receivers (L1 C/A-code) range between 45 and $50\,dB$-Hz. For example, for C/N_0 equal to $45\,dB$-Hz, and signal bandwidth of $0.8\,Hz$, the RMS tracking error due to receiver thermal noise for the C/A-code is $1.04\,m$. For GPS receivers with narrow correlators (Van Dierendonck et al. 1992) with the same bandwidth and C/N_0, the RMS is only $0.39\,cm$. Hence, there is indeed a difference in the noise level (and therefore the resulting pseudorange measurement precision) between the so-called narrow correlator receivers and standard GPS receivers. The RMS tracking error due to noise for a carrier-tracking loop with a C/N_0 of $45\,dB$-Hz and a signal bandwidth of $2\,Hz$ is only about $0.2\,mm$ for the L1 frequency (Braasch 1994). While similar receiver types from the different manufacturers would be expected to generate L1 carrier phase measurements with very similar precision, the quality of measurements of L2 carrier phase is a function of the signal processing technique used (Table 10.1).

10.4.3.4   Rejection of Radio Interference

The most common interference protection used in GPS receivers are (a) null-steering antennas, also known as a "controlled radiation pattern" antennas (which unfortunately are still rather bulky and expensive, and hence inappropriate for many civilian applications) or (b) narrow front-end filters and narrowed, aided tracking loops (such on-receiver techniques are generally limited in the kinds of interference they can reject). The interested reader is referred to Ward (1994) for a review of interference monitoring, mitigation, and analysis techniques.

10.4.3.5 Multipath Rejection Techniques

The existing multipath rejection technology is intended to improve the C/A-code pseudorange measurement, and can potentially increase the single-point positioning accuracy by about 50%. In general, signal processing techniques can reject the multipath signal only if the multipath distance (the difference between the direct and the indirect paths—Figure 8.11) is roughly more than about 10 m. In a typical geodetic/surveying application, however, the antenna is about 2 m above the ground, thus the multipath distance reaches at most 4 m, and consequently signal processing techniques cannot fully mitigate the effects of reflected signals on pseudorange measurements. In the case of carrier phase measurements, the problem is significantly reduced, being a disturbance of the order of several centimeters. This is, nevertheless, sufficient to impact on the accuracy of techniques that rely on sophisticated "on-the-fly" ambiguity resolution algorithms to achieve high accuracy with very short periods of carrier phase tracking (Section 9.4.2). For more discussion on multipath and its influence on GPS receivers, the reader is referred to Van Dierendonck et al. (1992), Van Nee (1993), Braasch (1994), Weill (1997), Parkinson and Spilker (1996), and Kaplan and Hegarty (2006).

10.4.4 THE RECEIVER "FORM-FACTOR"

GPS receivers come in all "shapes and sizes." In the case of off-the-shelf products (as opposed to *developer's kits* consisting of boardsets or microchips), the GPS "package" is generally designed to suit the application. Hence low-cost recreational receivers consist of a small unit in which all components are integrated into an attractive handheld device. In the case of the top-of-the-line GPS receivers intended for geodesy, surveying, and mapping, several different configurations can be purchased from the manufacturers, or the products are built in a modular fashion so that the various components may be connected together (via cables or direct connectors) in different ways.

10.4.4.1 Smart Antennas/Integrated Receivers

For applications such as tracking weather buoys, the time-tagging of seismic or other events, navigation, or precise timing, and the synchronization of wireless voice and data networks, the so-called smart antennas are often used. These are essentially self-contained, shielded, and sealed units that house a GPS receiver, an antenna, a power supply, a processor, and other supporting circuitry in a single ruggedized enclosure, which mounts like an antenna (Figure 10.3). These units are typically single-frequency instruments, making pseudorange measurements and are intended for automatic, continuous operation. Once power is applied, a smart antenna starts immediately to produce position, course, speed, and time information through a serial interface port (generally in the form of NMEA messages). A second serial port enabling real-time differential GPS operation (Section 14.2.3), or for outputting an accurate timing pulse synchronized to the sub-microsecond level, may also be provided. Perhaps their defining characteristic is that they have a very rudimentary control/display capability consisting of little more than an on–off switch and/or an indicator light (or lights). Even these may be missing in some models.

Smart antennas have also evolved into comprehensive survey packages, for example, the *Locus Survey System* from Magellan Corporation/Ashtech Precision Products, or fully integrated GPS receivers, such as the *GR-3* from Topcon Positioning Systems, or Leica's *SmartStation* (total station with integrated GPS), or Trimble's *R8 GNSS System*. An integrated receiver houses all electrical circuitry, and the data storage medium, and may also include a radio modem, removable battery module, and several serial interfaces. Smart antennas/integrated receivers such as these provide alternative easy-to-use, cable-free, lightweight, low-power options for receivers intended for use as high-productivity GPS surveying tools (Figure 10.3).

10.4.4.2 GIS/Mapping Receivers

In contrast to the "smart antenna"-type instruments, these receivers are distinguished by having both an LCD display/command unit through which instructions and user-entered data is input, and a DGPS signal decoder. In some products, the unit is almost indistinguishable from a basic handheld GPS receiver, with all circuitry integrated into one unit. In contrast, some systems consist of a data logger device that functions as the display/control unit, possibly equipped with a miniature QWERTY keyboard (or other data input devices, such as a barcode reader) to allow feature attribute data to be entered. A laser range finder, with magnetic compass and inclinometer, may be attached to the GPS in order to "shoot" to features in the field-of-view of the GPS point. In this way from one GPS point (which can be surveyed easily because it has clear skyview), many landmarks and features can be coordinated (e.g., for GIS "data capture"), even if they are located where GPS signal shading would have made them difficult to map using GPS alone.

The antenna as well as the receiver box may be in one module, or consist of individual components. For maximum flexibility, the DGPS signal decoder may be a separate interchangeable module (e.g., to allow for the use of a variety of DGPS signal input options—Section 14.3). In such configurations, all components apart from the data-logger device are usually stored in a backpack (which also carries the batteries and the antenna pole), and therefore the system has a similar appearance to a "poleless" RTK instrument.

10.4.4.3 GPS Surveying Receivers

These are the most expensive class of GPS receiver, and are typically able to be used in a variety of positioning modes, such as standard DGPS positioning (e.g., for GIS/mapping), static baseline determination, mounting on aircraft, land or marine platforms for kinematic positioning, and man-portable units for high-productivity GPS surveying.

In cases where the point being coordinated is a ground mark, dual- and single-frequency carrier phase-tracking GPS receivers may be mounted on tripods, control pillars, or special range poles (Figure 10.3). The pole-mounted configuration is generally the most appropriate one for high-productivity survey techniques. The real-time as well as the post-mission data processing operation can be supported. The antenna may be integrated within the same housing as the GPS receiver (Figure 10.3), or separate from it. The latter does allow for more flexibility in that in some circumstances the GPS antenna may be located some distance from where the receiver (and power supply) is housed. Furthermore, such a configuration permits different antennas to be used for certain applications (including choke ring antennas), as well as being more convenient when mounted on a moving platform such as a vehicle or aircraft. The backpack-mounted configuration (Figure 10.4) may be the most convenient for RTK operations, or when the system is used for GIS/mapping, as discussed earlier.

FIGURE 10.4 The backpack-mounted configuration. (Courtesy of Topcon Positioning Systems, Inc., Livermore, CA.)

10.4.4.4 Special High Accuracy Applications

The applications of carrier phase-tracking GPS receivers were traditionally those associated with geodesy and surveying, that is, for static positioning. In such a positioning mode, the receiver was tripod- or pillar-mounted, and was expected to collect data for an observation session that might range from an hour to several days in length. High-productivity techniques have challenged this standard static position mode (Section 11.1). Nevertheless, there is an increasing need for GPS receivers to be set up *permanently* as static reference receivers, at continuously operating base stations, that are part of DGPS service networks (Section 14.4.1), or the International GNSS Service, or a variety of specialized networks addressing such applications as geodynamics, deformation monitoring, and atmospheric remote sensing. In such cases, the form-factor is not critical. Typically, the antenna is of the choke ring variety, the receiver is attached to a PC and/or communications link, and the power supply is assured (and uninterruptible).

Increasingly carrier phase-tracking GPS receivers are being used for kinematic applications such as engineering/farm machine guidance (and even control). The RTK operating mode is used (Section 11.2.2), and a crucial design requirement is that the receiver box must be robust, as it has to operate reliably in extreme environments. The receiver will typically be operating in hot (or very cold) and dusty (or wet) conditions, attached to a vibrating vehicle. Hence, all components (antenna, receiver, radio link, cables, and connectors) must be ruggedized to a significant degree.

10.4.4.5 Software Receivers

The so-called software or software-defined receivers constitute an entirely new category of GPS receivers that has emerged in the last few years. A software GPS receiver mimics the most commonly used GPS receiver design architecture, but it uses minimal electronic hardware in order to acquire and track GPS signals. A minimal RF front-end produces a digital data stream that is input to a general-purpose microprocessor, whose functions are exactly the same as listed in Section 10.3.1. The concept of a software receiver implies that the input signal is digitized as close to the antenna as possible by placing an analog-to-digital converter (ADC) at the earliest possible stage in the receiver (Tsui 2000). By allowing as much digital signal processing on a programmable platform as possible, software receivers feature greater flexibility in functionality, design, and system updates, as compared to the traditional hardware (or "silicon") implementations. Without the traditional hardware design based on GPS chipsets or dedicated application-specific integrated circuits (ASIC), a software receiver can be upgraded inexpensively by uploading new design features and improved algorithms, including new DLL or PLL designs. Flexible software receivers can, in general, offer superior performance in tracking weak signals, on dynamic platforms, and offer better integration capability with other sensor measurements.

In software receiver implementations, the entire receiver processing (signal acquisition and tracking, extracting navigation data, and calculating the user position) is handled by a programmable processor. The programmable processors most commonly used in GPS software receivers are the field programmable gate array (FPGA) and special purpose digital signal processors (DSP), as well as general purpose PCs. The FPGA is a reprogrammable and reconfigurable integrated circuit or a chip composed of an array of configurable logic cells that can be used as building blocks to implement any kind of functionality desired, from low-complexity state machines to complete microprocessors. While a DSP implementation has more flexibility than FPGAs, a PC-based scheme offers the greatest flexibility (Hein et al. 2006).

While their relatively low-development costs, faster adaptability, and short product development cycle are the primary advantages of software receivers in comparison to the traditional hardware implementation, slow throughput, and relatively high power consumption are still major disadvantages. In order to overcome these drawbacks, a software receiver embedded into a single microprocessor or DSP in a software form is emerging in the marketplace. This development has been driven by the mobile application market, such as cellular phones running location-based service applications, and this trend is expected to continue. According to Won et al. (2006) and Hein et al. (2006), it is very likely that software receivers will ultimately penetrate the embedded mass market of multimodal GNSS receivers, and may also succeed as generic top-of-the-line geodetic multi-GNSS receivers.

REFERENCES

Ashjaee, J. and Lorenz, R., 1992, Precision GPS surveying after Y-code, *Proceedings of the Fifth International Technical Meeting of the Satellite Division of the U.S. Institute of Navigation*, Albuquerque, NM, pp. 657–659.

Braasch, M.S., 1994, Isolation of GPS multipath and receiver tracking errors, *Navigation*, 41(4), 415–434.

GPSWorld, 2008, http://www.gpsworld.com/, accessed March 10, 2009.

Hein, G.W., Pany, T., Walner, S., and Won, J.-H., 2006, Platforms for a future GNSS receiver: A discussion of ASIC, FPGA, and DSP technologies, *Inside GNSS*, March, pp. 56–62.

Johannesse, R., 1997, Interference: Sources and symptoms, *GPS World*, 8(11), 44–48.

Johnson, R.C. (ed.), 1993, *Antenna Engineering Handbook* (3rd edn.), McGraw-Hill, New York, 1392pp.

Kaplan, E. and Hegarty, C.J. (eds.), 2006, *Understanding GPS: Principles and Applications* (2nd edn.), Artech House, Norwood, MA, ISBN 1-58053-894-0, 570pp.

Langley, R.B., 1991, The GPS receiver: An introduction, *GPS World*, 2(1), 50–53.

Langley, R.B., 1995a, NMEA 0183: A GPS receiver interface standard, *GPS World*, 6(7), 54–57.

Langley, R.B., 1995b, A GPS glossary, *GPS World*, 6(10), 61–63.

Langley, R.B., 1997a, The GPS error budget, *GPS World*, 8(3), 51–56.

Langley, R.B., 1997b, The GPS receiver system noise, *GPS World*, 8(6), 40–45.

Langley, R.B., 1998a, Propagation of the GPS signals and GPS receivers and the observables, in *GPS for Geodesy* (2nd edn.), Teunissen, P.J. and Kleusberg, A. (eds.), Springer, Berlin, Germany, pp. 112–185.

Langley, R.B., 1998b, A primer on GPS antennas, *GPS World*, 9(7), 50–54.

NGS-antcal, 2009, http://www.ngs.noaa.gov/ANTCAL/, accessed March 10, 2009.

NMEA, 2009, National Maritime Electronics Association, http://www.nmea.org/, accessed March 10, 2009.

Parkinson, B.W. and Spilker, J.J. (eds.), 1996, *Global Positioning System: Theory and Applications*, Vol. I and II, American Institute of Aeronautics and Astronautics, Inc., Reston, VA, ISBN 1-56347-106-X, 793pp.

Storm Products, 2009, http://www.stormproducts.com/, accessed 10 March 2009.

Tsui, J.B.-Y., 2000, *Fundamentals of Global Positioning System Receivers A Software Approach*, John Wiley & Sons, Inc., New York, 238pp.

Van Dierendonck, A.J., 1995, Understanding GPS receiver terminology: A tutorial, *GPS World*, 6(1), 34–44.

Van Dierendonck, A.J., Fenton, A., and Ford, T., 1992, Theory and performance of narrow correlator spacing in a GPS receiver, *Navigation*, 39(3), 265–283.

Van Nee, R., 1993, Spread spectrum code and carrier synchronization errors caused by multipath and interference, *IEEE Transactions on Aerospace & Electronic Systems*, 29(4), 1359–1365.

Ward, P.W., 1994, GPS receiver RF interference monitoring, mitigation, and analysis techniques, *Navigation*, 41(4), 367–391.

Weill, L.R., 1997, Conquering multipath: The GPS accuracy battle, *GPS World*, 8(4), 59–66.

Won, J.-H., Pany, T., and Hein, G.W., 2006, GNSS software defined radio, real receiver or just a tool for experts? *Inside GNSS*, July/August, pp. 48–56.

11 Making Sense of GNSS Techniques

Chris Rizos

CONTENTS

In this chapter, the Global Positioning System (GPS) will be used as the exemplar of a Global Navigation Satellite System (GNSS). All discussions will refer to GPS, its performance, and currently used positioning techniques. Other GNSS as they become fully operational, when used *on their own*, will likely employ the same basic field and data-processing techniques, and deliver similar levels of positioning performance. However, when receivers capable of tracking *multiple* constellations of GNSS satellites are available in the coming 5–10 years, the dramatic increase in the number and the type of observations will have a profound effect on GNSS positioning. This future multi-GNSS capability will be discussed in Chapter 15.

In Section 7.1.3, the user segment was defined as *the entire spectrum of applications, equipment, positioning strategies, and data-processing* techniques that provide the users with the position results. There is a wide variety of GPS applications, which is matched by a similar diversity of user equipment and techniques. Nevertheless, the most fundamental classification system for GPS techniques is based on the type of observable that is tracked (Section 8.2): (a) civilian navigation/positioning receivers using the C/A-code on the L1 frequency, (b) military navigation/positioning receivers using the satellite P(Y)-codes on both L-band frequencies, (c) single-frequency

(L1) carrier phase–tracking receivers, and (d) dual-frequency carrier phase–tracking receivers. When these classes of hardware are used in the appropriate manner for *relative positioning*, the accuracies that are achieved range from a few meters in the case of standard pseudorange-based techniques, to the sub-centimeter level in the case of carrier phase–based techniques. *Although Single-Point Positioning accuracies of 5–10 m are now possible, in this book it will be assumed that for geospatial applications such as geodesy, surveying, and mapping/GIS, only the relative positioning techniques are of relevance.*

The following classes of relative positioning techniques can therefore be identified:

1. *Static and kinematic GPS post-mission surveying techniques*: high-accuracy techniques based on post-processing of carrier phase measurements.
2. *Differential GPS (DGPS)*: instantaneous low-to-moderate accuracy navigation and mapping technique based on pseudorange measurements alone.
3. *Real-time kinematic (RTK) techniques*: versatile centimeter-level accuracy techniques that use unambiguous carrier phase observables in an *instantaneous* positioning mode, that is, there is no delay between making measurements and generating coordinate solutions.

The post-processed GPS surveying techniques are discussed in Section 11.1, with particular emphasis on the so-called high-productivity techniques, which are today widely utilized for many surveying and mapping tasks.

The DGPS and RTK techniques, because they are able to deliver results in real time using a single epoch of measurements, are very powerful GPS positioning technologies. They are discussed in Section 11.2, the communication link issues are identified in Section 11.3, and further discussion given in Section 14.3. In the case of these real-time techniques, there is a clear distinction between the functions of the receiver at the reference station(s) on the one hand and the user equipment on the other. Such a distinction is less obvious in the case of post-processed techniques, where all the survey equipment tends to be owned and/or operated by the surveyor/contractor. In such a case, whether one receiver or the other is designated "reference" or "user" is largely immaterial (especially in the case of static survey techniques). This notion of "ownership" introduces a new element in any consideration of user techniques, namely, some techniques can be seen as requiring the provision of a *service*, while other techniques are completely under the control of the surveyor/contractor.

11.1 HIGH-PRODUCTIVITY GPS SURVEYING TECHNIQUES

There are essentially two types of conventional static GPS surveying techniques:

1. Ultraprecise, long-baseline (hundreds to thousands of kilometers) GPS techniques: relative positioning accuracies of sub-part per million up to several parts per billion, characterized by top-of-the-line GPS receivers and

antennas, many hours (and even days) of observations, and data processing using sophisticated "scientific" software—including GAMIT/GLOBK
(from the MIT/Scripps Institution of Oceanography), BERNESE (from the
University of Bern), GIPSY (from the Jet Propulsion Laboratory [JPL]), and
PAGES (from the U.S. National Geodetic Survey).

2. Medium-to-short baseline GPS survey techniques: accuracies down to a
few parts per million for baselines typically <50 km, to support control network applications, with data processing being carried out by commercial
software packages—examples: Trimble GEOMATICS OFFICE, Topcon's
TOOLS, Leica GEO OFFICE, and Javad's PINNACLE.

The main weaknesses of such procedures are that the observation times are comparatively long, the results are obtained well after the field survey work is completed,
and the field procedures are relatively inflexible. During the late 1980s considerable attention was paid to these issues, as they were considered to be unnecessar-
ily restrictive for the widespread application of precise GPS technology. That is,
if antennas could be *moving during a GPS survey*, then new applications for the
GPS technology could be addressed. If the *length of time required to collect carrier
phase observations* for a reliable solution could be shortened, then the GPS survey
productivity would improve and the technology would be attractive for many more
surveying applications. If the results could be obtained *immediately after the measurements* were made, then GPS could be used for time-critical missions such as
engineering stakeout and machine guidance.

In the last two decades, several GPS surveying methods have been developed with
the two "*liberating*" characteristics of (a) static antenna setups no longer having to be
insisted upon and (b) long observation sessions not being essential in order to achieve
centimeter-level positioning accuracy. These GPS surveying techniques are given
a variety of names by the different GPS manufacturers, but the following generic
terminology will be used in this book:

- *Rapid-static* positioning technique
- *Kinematic* positioning techniques, including what we might refer to as the
  "stop-and-go" technique

*All require the use of specialized hardware and software, as well as new field
procedures*. GPS receivers capable of executing these types of surveys can also be
used for conventional static GPS surveying. Although the field procedures are different from conventional static GPS surveying, the principles of planning, quality
control, network processing, and coordinate transformation are similar (Chapters
12 and 13).

Each of the techniques is a technological solution to the problem of obtaining
high productivity (coordinating as many points in as short a period of time as possible) and/or *versatility* (e.g., the ability to obtain results even while the receiver is
in motion and/or in real time) without sacrificing very much in terms of achieved
accuracy and *reliability*. However, none of these techniques is as accurate or reliable

as conventional static GPS surveying. Each of these techniques has its distinctive characteristics (Section 11.1.4), hence they are offered as positioning options within modern GPS products. The surveyor/contractor, therefore, has a "toolkit" of carrier phase–based GPS techniques to choose from.

11.1.1 RAPID-STATIC GPS SURVEYING

Rapid-static techniques are also sometimes referred to as "fast-static" or "quick-static" techniques. The following characteristics distinguish rapid-static techniques from other methods of static GPS surveying:

- *Observation time requirements*: These are significantly shorter than for conventional static GPS surveying and are a function of baseline length, whether dual-frequency instruments are used, the number of satellites being tracked, and satellite geometry. Typically, the receivers need only occupy a baseline for a period of 10–30 min (the lower value corresponding to baselines <5 km and tracking six or more satellites; the upper value being for longer baselines up to 20 km, where single-frequency receivers are used and/or where tracking is to only four satellites).
- *Specialized software*: The basis of this technique is the ability of the software to *resolve* the ambiguities using a very short observation period. The commercial software has the fundamental requirement of *rapid ambiguity resolution* (AR) capability (Section 9.4.2).
- *Hardware requirements*: In most systems, dual-frequency phase measurements are sufficient, although dual-frequency pseudorange measurements are *also* required for rapid AR.

The field procedures are much like those for conventional static GPS surveying except that (a) the station occupation times (and hence observation session lengths) are shorter, (b) the baselines should be comparatively short, (c) the satellite geometry favorable, and (d) signal disturbances such as multipath should be kept to a minimum. It is not possible to define exactly how much data needs to be collected in order to guarantee quality baseline results every time, that is, "ambiguity-fixed" solutions (Section 9.4.3). *Equipment user manuals typically give guidelines in this regard.* Some receivers also provide an audio and/or visual indication when enough data has been collected in the field (but this cannot be confirmed until the data is downloaded and processing is completed). If the real-time positioning mode is employed (Section 11.2.2), then the "data quantity shortfall risk" for the post-processed mode of rapid-static GPS surveying can be overcome, because in the RTK mode, the operator will be given an indication when an "ambiguity-fixed" solution has been obtained (i.e., enough carrier phase data has been collected by the user receiver, and enough reference station information has been received over a wireless data link—Section 14.3.2.2). See also discussion in Section 12.1.

The rapid-static technique is well suited for short-range applications such as control densification and engineering surveys, or any job where many points need to be

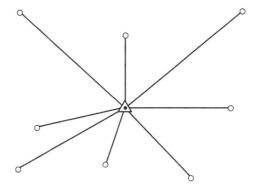

FIGURE 11.1   Field procedure for the rapid-static surveying technique, in a configuration of a reference station at the center while the user receiver moves from point to point generating a radiating baseline pattern.

surveyed (Figure 11.1). Unlike the *kinematic* techniques, there is no need to maintain lock on the satellites when moving from one ground mark to another.

11.1.2 "STOP-AND-GO" GPS SURVEYING

This is a true *kinematic* technique because the user's receiver continues to track satellites while it is in *motion*. The *stop-and-go* technique deserves special consideration because the coordinates of the receiver are only of interest when it is stationary (the "stop" part), but the receiver continues to function while it is being moved (the "go" part) from one stationary setup to the next. (Although GPS receiver manufacturers now typically claim that their equipment operates in only the *static* or *kinematic* mode, it is still useful to highlight the unique aspects of the *stop-and-go* field procedure.) There are in fact three stages to the procedure:

1. *The initial AR*: This is carried out before the stop-and-go survey commences. The determination of the ambiguities by the software can be carried out using any method, but in general it is one of the following:
   * A conventional static (or rapid-static) GPS survey determines the baseline vector from the reference station receiver to the first of the points occupied by the user's receiver. An ambiguity-fixed solution provides an estimate of the integer values of the ambiguities that are then used in subsequent positioning—see Stage 3.
   * Set up the user receiver over a ground mark with known coordinates, usually surveyed previously by GPS, and derive the values of the ambiguities in this way.
   * Employ a procedure known as "antenna swap." The antenna of the reference receiver is set up over the ground mark that has known coordinates (on a tripod or a pillar), the second tripod is set up a few meters away, with the user receiver's antenna placed on it (the exact baseline length need not be known). Each receiver collects data for a few minutes (tracking

the same satellites). The antennas are then carefully lifted from their mounts and swapped, that is, the Receiver 1 antenna is placed where the Receiver 2 antenna had been, and vice versa. After a few more minutes, the antennas are swapped again. The software resolves the ambiguities over this very short baseline. Note that the reference receiver antenna is now occupying the datum station, and the user receiver can proceed with the survey operation.

- The most common technique nowadays is to resolve the ambiguities "on-the-fly" (i.e., while the user receiver is tracking satellites but the receiver/antenna is moving). The impact on precise GPS positioning of on-the-fly AR (OTF-AR) is discussed in Section 11.2.4.

2. *The receiver in motion*: Once the ambiguities have been resolved, the survey can begin. The user's receiver is moved from point to point, collecting just a minute or so of data. *It is vital that the antenna continues to track the satellites.* In this way, the resolved ambiguities are valid for all future phase observations, in effect converting all carrier phase data to unambiguous "carrier-range" data (by applying the integer ambiguities as data corrections—Section 8.4.2). As soon as signal loss-of-lock occurs (causing a cycle slip) then the ambiguities have to be "reinitialized" (or re-estimated). Bringing the receiver back to the last surveyed point, and redetermining the ambiguities can most easily be done using the "known baseline" method (see Stage 1).

3. *The stationary receiver*: The "carrier-range" data is then processed in the double-differenced mode to determine the coordinates of the user receiver relative to the static reference receiver (Section 9.4.1). *The trajectory of the antenna is not of interest, only the stationary points that are visited by the receiver.*

This technique is well suited when many points close together have to be surveyed, and the terrain poses no significant problems in terms of signal disruption (see, e.g., Section 13.2.2). The accuracy attainable is at the few centimeter-level (note that as the user equipment antenna is typically on a range pole or backpack mounted, the verticality and the stability of the pole during the "stop" part will have a significant impact on the final groundpoint positioning accuracy—see Figures 10.3 and 10.4). The software has to sort out the recorded data for the different points and to differentiate the *kinematic* or "go" data (not of interest) from the *static* or "stop" data (which is of interest). (The technique can also be implemented in real time if a wireless communications link is provided to transmit the "carrier-range" data from the reference receiver to the roving receiver, as discussed in Section 14.3.2.2.) The survey is typically carried out in the manner illustrated in Figure 11.2.

It must be emphasized that an important characteristic of this technique is the requirement that the signal lock *must* be maintained on the satellites by the user receiver as it moves from point to point. This may require special antenna mounts on vehicles if the survey is carried out over a large area. One mobile backpack configuration is shown in Figure 10.4.

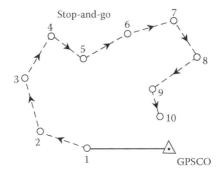

FIGURE 11.2 Field procedure for the "stop-and-go" surveying technique—note the traverse mode of coordination, with the user receiver (typically the antenna is carried on a rangepole) traveling from point to point, and the coordinates derived via baselines radiating from the reference station (as in Figure 11.1).

11.1.3 KINEMATIC GPS SURVEYING

This is a generalization of the *stop-and-go* field procedure. Instead of only coordinating the stationary points and disregarding the trajectory of the roving antenna as it moves from point to point, the objective of *kinematic* surveying is to determine the position of the antenna *while it is in motion* (Figure 11.3). In many other respects, the technique is similar to the stop-and-go positioning technique. That is, the ambiguities must be resolved *before* starting the survey, and the ambiguities must be reinitialized *during* the survey when a cycle slip occurs (as would typically occur in the event of signal blockage by trees or buildings). However, for many applications,

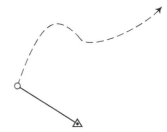

FIGURE 11.3 The kinematic GPS surveying technique—a generalization of the "stop-and-go" positioning technique (Figure 11.2).

such as the positioning of an aircraft or a ship, it would have been impractical to reinitialize the ambiguities if the roving antenna had to return to a stationary control point. Today the kinematic GPS surveying technique routinely uses the OTF-AR procedure, making kinematic surveying techniques ideal for road centerline surveys, topographic surveys, hydrographic surveys, airborne applications, and many more.

11.1.4 CHOOSING THE APPROPRIATE TECHNIQUE

Each of these high-productivity GPS surveying techniques have their strengths and weaknesses; however, all are generally less accurate than the conventional GPS surveying technique. This should not be considered too great a drawback as it is often not required that relative accuracies be higher than 1 ppm. Often a combination of conventional static and GPS techniques such as the ones described earlier are used in a typical survey project. One of the important tasks of the GPS surveyor is to select the best combination of techniques for the terrain, distance, user requirements, and logistical constraints that they face.

The development of OTF-AR algorithms is a dramatic step forward because static ambiguity reinitialization is not necessary for any positioning technique. The ambiguities will be resolved *while the antenna is moving* to the next stationary survey point. If a Point X has been surveyed (i.e., a few minutes of "carrier-range" tracking data have been collected), and as the antenna is moved from Point X to Point Y, an obstruction blocks the signals and causes signal loss-of-lock, then the antenna does not have to go back to Point X. New ambiguities can be resolved *on-the-fly* as the antenna moves from X to Y. However, there must be a sufficient period of uninterrupted tracking for this to take place. During this so-called dead time (top part of Figure 11.4) centimeter-level positioning accuracy is not possible if OTF-AR is implemented in real time, but *is* possible with the post-processing mode because the data before AR has been completed can be "backward" corrected and then used to generate ambiguity-fixed baseline results.

Although the "time-to-AR" varies (and is influenced by factors such as the baseline length, the number of tracked satellites, satellite geometry, multipath disturbance, and several other factors), the required period of continuous carrier phase measurement is *of the order of a few seconds to several minutes*. In an extremely unfavorable scenario, there may be so many signal obstructions that there is insufficient time for the OTF-AR algorithm to work properly during the very short periods of uninterrupted tracking, and hence the survey is not possible using the stop-and-go or kinematic technique. In such circumstances, the rapid-static technique would be

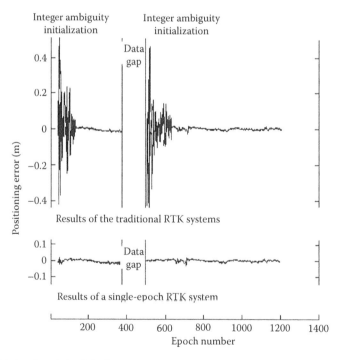

FIGURE 11.4 "Dead time" indicated for OTF-AR after a break in GPS signals, compared with the *ideal* case of single-epoch AR that does not suffer from this problem.

TABLE 11.1

Comparing Conventional Static GPS with Techniques such as Rapid-Static, Stop-and-Go, and Kinematic Techniques

Conventional Static GPS	Rapid-Static, Stop-and-Go, Kinematic GPS
Advantages	Advantages
• Highest accuracy	• Higher accuracy than pseudorange solutions
• Robust technique	• Appropriate for many survey applications
• AR not critical	• High productivity
• Minor impact of multipath	• Similar procedures to modern terrestrial surveying
• Undemanding of hardware and software	
Disadvantages	Disadvantages
• Long observation sessions	• Special hardware and software
• Inappropriate for engineering and cadastral applications	• Susceptible to atmospheric and multipath biases
	• Higher capital costs
	• Ambiguity-fixed or continuous lock required

preferred. The ultimate goal of the GPS manufacturers is to develop systems that reliably resolve ambiguities with a single epoch of data (lower part of Figure 11.4), something that is anticipated as being possible in almost every scenario using triple-frequency and/or multi-GNSS techniques (Chapter 15).

The advantages and disadvantages of the static GPS technique vis-à-vis high-productivity techniques are summarized in Table 11.1.

11.2   REAL-TIME GPS SURVEYING AND MAPPING TECHNIQUES

Traditionally, in geodesy and surveying the measurements are collected using an antenna set up over a (temporarily or permanently) monumented point, and stored for data post-mission processing as discussed in Section 11.1. In many cases, however, such as navigation, mobile mapping, engineering stakeout, and machine guidance, the position of the user receiver antenna has to be determined in *real time*. Naturally, all real-time applications involve some type of wireless data transmission system, such as very high frequency (VHF) or ultrahigh frequency (UHF) systems over short ranges, low-frequency transmitters for medium range distances, geostationary satellites for coverage of entire continents, and increasingly via Internet-enabled devices (e.g., through cellular phone networks). Depending on the accuracy requirements, two basic modes of real-time operation can be used: (a) DGPS or (b) RTK.

DGPS requires a reference station receiver transmitting *pseudorange corrections* to the users, whose receivers use this information together with their measured pseudoranges for positioning at the meter to few meter accuracy level. RTK, on the other hand, is based on transmitting reference station *carrier phase data or corrections* (there are different approaches, see Section 14.2) to the users' receivers. These two techniques are briefly described in the following sections and in Section 14.3. The communications link and other system configuration issues are discussed in Section 11.3.

11.2.1 DGPS TECHNIQUE

In its simplest implementation, a DGPS reference receiver is set up at a site with known coordinates. After it has been configured to operate as the "base station," the reference receiver tracks continuously all the visible satellites and determines the corrections necessary for its pseudorange data to be able to compute the single-point positioning result that is identical to the known coordinates of the site. This correction data is then transmitted to the mobile users (via a wireless data link) whose receivers then apply the corrections to their pseudorange data before computing their SPP solutions (Figure 11.5). The following comments may be made:

- In order to facilitate easy and uniform data transfer to appropriately equipped users, the Radio Technical Commission for Maritime Services (RTCM) in 1985 established a standard format for DGPS correction data transmission (Langley 1994, RTCM 2009). Although other proprietary and industry-specific formats have been developed, the RTCM format remains the most widely used. Furthermore, it is continually being upgraded to handle new data types and to facilitate new real-time services (Sections 11.3.2 and 14.2.3).
- There are two basic implementations of the DGPS technique: one based on using a *single* reference station to generate the correction data, and the other makes use of a *network* of reference stations (Section 11.3.3).
- There are a number of *services* that have been established in order to allow DGPS positioning and navigation to be carried out in many parts of the world. These may be fee based or free to air (see Sections 14.1 and 14.3).

All GPS receivers possessing an I/O port or Bluetooth connection are "DGPS ready." That is, if a data communications link is enabled from a DGPS *service provider*, the GPS receiver is able to decode the received RTCM correction messages and use this information to determine its (relative) position to an accuracy of the order of a

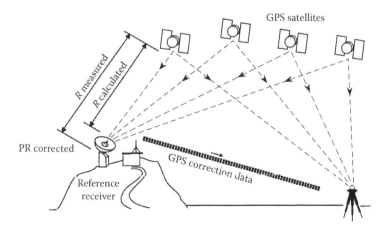

FIGURE 11.5   Range correction implementation of real-time pseudorange-based DGPS.

few meters or less. If the necessary software functionality is present, DGPS-enabled receivers are powerful tools for GIS data capture (Section 10.4.4.2).

Top-of-the-line carrier phase-tracking GPS receivers can be purchased with the option of being able to operate as a DGPS reference receiver. Several receivers may be set up at each DGPS base station, to provide backup or for independent checking of RTCM messages.

11.2.2   RTK-GPS Technique

"RTK" GPS is an attractive technique for many survey applications as there is no post-processing of the carrier phase data (Langley 1998b). The standard scenario requires the surveyor/contractor to use two GPS receivers (one reference receiver and one user's "roving" receiver) connected by a wireless data link. Once set up, the reference receiver will continuously transmit its carrier phase measurements (or some suitably preprocessed set of data) to the roving receiver. The software within the roving receiver then resolves the ambiguities in the shortest time possible (using OTF-AR algorithms), and the resulting "carrier-range" data is used to derive centimeter-level accuracy positioning results. These results may be stored (for later downloading), displayed, and used for field-surveying applications, or processed by a computer to guide construction, mining, or agricultural machinery.

The successful operation of RTK systems using radio modem data links is typically limited to baseline lengths of 5–10 km due to radio range constraints. However, the inter-receiver distance over which *rapid* OTF-AR algorithms work reliably using the current dual-frequency GPS constellation (with good sky visibility so that more than six satellites can be tracked) may be up to 20–30 km. As with the post-processed modes of carrier phase–based positioning, when signals are obstructed the OTF-AR algorithm has to be restarted in order to resolve the (new) ambiguities. As this may take several tens of seconds, and if signal interruptions occur frequently, then this "dead time" (Figure 11.4) can result in RTK being a comparatively inefficient positioning technique. Note that RTK need *not* only be used for purely kinematic applications. RTK equipment can be used in the *stop-and-go* or *rapid-static* modes of surveying as well, the crucial difference (and perhaps important advantage) being that the results are available immediately after data collection is complete.

Some real-time commercial GNSS products claim to operate over longer distances, typically several tens of kilometers, especially when a cellular phone data link is used. In addition, new services by commercial DGPS service providers now boast carrier phase–based solutions with accuracies at the several decimeters, even for inter-receiver distances of several hundreds of kilometers. Increasingly users wishing to obtain centimeter-level accuracies in real time are subscribing to services rather than running their own base stations (Section 14.3.2.2). Real-Time Networks of Continuously Operating Reference Stations (CORS) have been established over the last decade and the trend seems to be accelerating (Section 15.3). New modes of RTK based on a small network of reference stations (rather than a single reference station) are now commonly used in many countries. This RTK technique is also referred to as "network-RTK" (Section 14.3.2.2), and its primary advantage is that the maximum inter-receiver distance (user receiver from the reference stations) can be of the order of 50–70 km or more.

RTK is especially vulnerable to poor satellite visibility, multipath, and an unreliable wireless data link from the reference station(s). This link may be a VHF or UHF wireless connection between the reference and rover receivers (Section 11.3.4), although increasingly cellphone-based services are being used. At the same time, RTK service delivery over the Internet is becoming more popular. Although proprietary data transmission formats are still commonly used for RTK operations, the RTCM (2009) format does provide message types that can be used for single-base RTK and network-RTK (Section 11.3.2). Furthermore, the use of industry standard RTCM formats ensures *interoperability* between reference and user receivers.

11.3 SYSTEM CONFIGURATION AND DATA LINK ISSUES

There are a variety of different implementations of DGPS and RTK (Section 14.3). Some systems are deployed and operated for a specific task by the contractor (this is often the case with RTK systems), or are offered by companies as services to all (as is typically the case for DGPS—see Section 14.1). The following considerations must be addressed by DGPS communication links:

- *Coverage*: This is generally dependent on the frequency of the radio transmission that is used, the distribution and the spacing of transmitters, the transmission power, susceptibility to fade and interference, etc.
- *Type of service*: For example, whether the real-time DGPS service is a "closed" one available only to selected users, such as in the case of a subscriber service, or an "open" broadcast service.
- *Functionality*: This includes such link characteristics as whether it is a one-way or two-way communications link, whether it is continuous or intermittent, whether other data is also transmitted, and so on.
- *Reliability*: Does the communications link provide a reasonable level of service? For example, what are the temporal coverage characteristics? Is there gradual degradation of the link? What about short-term interruptions?
- *Integrity*: This is an important consideration for critical applications, hence any errors in transmitted messages need to be detected with a high probability, and users alerted accordingly.
- *Cost*: This includes the capital as well as ongoing expenses, for both the DGPS service provider as well as users.
- *Data rate*: In general, the faster the data rate, the higher the update rate for range corrections, and hence better the positioning accuracy.
- *Latency*: Refers to the time lag between the computation of correction messages and the reception of message at the rover receiver. Obviously, this should be kept as short as possible.

The RTK link must be at a higher data rate than standard DGPS and is typically required at a 1 s update rate (or higher in the case of machine-guidance applications).

11.3.1 Communications Link Options

In this age of communication technology and information transfer, there are a number of communication options available for DGPS/RTK operation including

1. General systems for two-way communications include:
   - HF/VHF/UHF radio systems: dedicated frequencies, as well as open citizen bands
   - Satellite communications: via geostationary or low-earth-orbiting satellites
   - Cellular phone network: a growing number of options including digital systems, packet-based systems using Internet protocols, etc.
2. Broadcast options are one-way systems suitable for carrying DGPS/RTK messages to the user and include:
   - FM radio ancillary (sub-carrier) channels using RDS or some other protocol
   - LF/MF frequency transmission via marine navigation beacons, AM radios, etc.
   - Television blanking interval transmissions
   - IEEE802.11, and other standards, for WiFi, WiMAX, etc.

Some of the two-way communication systems may already be in place, or can easily be established in virtually any location. The infrastructure for their operation may therefore provide immediate and effective coverage. Satellite communications is a particularly attractive option because of its wide coverage, and hence it is very commonly used for offshore positioning applications (Section 14.3.1). It is, however, a relatively expensive option and is generally used only if there are no other alternatives. Some of the broadcast options are in fact "piggyback" systems that take advantage of a communications infrastructure that is already in place. Direct or dedicated radio systems are the alternatives to piggyback systems. However, these normally would require the establishment of an independent licensed radio transmitting station. One thing is certain, the variety of satellite and ground-based communication systems is likely to grow rapidly over the next decade.

11.3.2 RTCM Data Transmissions

RTCM Special Committee 104 was formed to draft a standard format for the correction messages necessary to ensure an open real-time DGPS system (Langley 1994, RTCM 2009). The format has become generally known as "*RTCM 104.*" According to RTCM recommendations, the pseudorange correction message transmission consists of a selection from a large number of different message types (Table 11.2 lists some of the RTCM Version 3.1 message types). Not all message types are required to be broadcast in each transmission; some of the messages require a high update rate while others require only occasional transmission. (Typically, a DGPS service will broadcast selected messages, depending on its mode of operation.) The RTCM

TABLE 11.2

RTCM SC-104 Recommended Message Types, Version 3.1

Type	Status	Title
1001	Fixed	L1-only GPS RTK observables
1002	Fixed	Extended L1-only GPS RTK observables
1003	Fixed	L1&L2 GPS RTK observables
1004	Fixed	Extended L1&L2 GPS RTK observables
1005	Fixed	Stationary RTK reference station ARP
1006	Fixed	Stationary RTK reference station ARP with antenna height
1007	Fixed	Antenna descriptor
1008	Fixed	Antenna descriptor and serial number
1009	Fixed	L1-only GLONASS RTK observables
1010	Fixed	Extended L1-only GLONASS RTK observables
1011	Fixed	L1&L2 GLONASS RTK observables
1012	Fixed	Extended L1&L2 GLONASS RTK observables
1013	Fixed	System parameters
1014	Fixed	Network auxiliary station data
1015	Fixed	GPS ionospheric correction differences
1016	Fixed	GPS geometric correction differences
1017	Fixed	GPS combined geometric and ionospheric correction differences
1018	Reserved	Reserved for alternative ionospheric correction difference message
1019	Fixed	GPS ephemerides
1020	Fixed	GLONASS ephemerides
1021	Fixed	Helmert/Abridged Molodenski transformation parameters
1022	Fixed	Molodenskir–Badekas transformation parameters
1023	Fixed	Residuals, ellipsoidal grid representation
1024	Fixed	Residuals, plane grid representation
1025	Fixed	Projection parameters, projection types other than Lambert Conic Conformal (2 SP) and Oblique Mercator
1026	Fixed	Projection parameters, projection type LCC2SP (Lambert Conic Conformal [2 SP])
1027	Fixed	Projection parameters, projection type OM (Oblique Mercator)
1028	Reserved	Reserved for global to plate-fixed transformation
1029	Fixed	Unicode text string
1030	Fixed	GPS network RTK residual message
1031	Fixed	GLONASS network RTK residual message
1032	Fixed	Physical reference station position message
1033	Fixed	Receiver and antenna descriptors
4001–4095	Tentative	Message types assigned to specific companies for broadcast of proprietary information

Source: RTCM, Radio Technical Committee for Maritime Services, Web page http://www. rtcm.org, accessed March 10, 2009.

message format is also used to support RTK positioning. As other GNSSs come online, the RTCM format will be updated to accommodate the new measurement types.

One of the most important considerations with respect to the DGPS data link is the rate of update of the range corrections to account for biases due to satellite clock errors and orbit errors. The correction to the pseudorange, and the rate-of-change of this correction, are determined and transmitted for each satellite. If the message *latency* (or age) is too great then temporal decorrelation occurs, and the benefit of the DGPS corrections is diminished. Prior to Selective Availability (SA) being turned off, messages older than about 10 s were typically ignored. With SA no longer implemented, this latency can be quite high, up to 30 s or more.

Note that RTCM is an industry standard, hence "Brand X" rover receiver can apply the corrections even though they were generated by a "Brand Y" reference receiver. RTCM can be implemented in either Local Area DGPS or Wide Area DGPS modes.

11.3.3 LADGPS VERSUS WADGPS

It is possible to differentiate between *Local Area DGPS* (LADGPS) and *Wide Area DGPS* (WADGPS). The assumption made when a pseudorange correction message is generated is that it is a valid calibration quantity representing the "lumped" satellite-dependent and propagation link biases as monitored at the reference receiver—the satellite clock error, satellite orbit bias, troposphere, and ionosphere refraction. (This so-called observation-space representation of DGPS errors is discussed in Section 14.2.2.) However, apart from the first bias quantity, this assumption breaks down as the separation between the reference receiver and the rover receiver increases. Typically, a range of 100 km or so is considered the limit beyond which it is unreasonable to assume that biases will always almost completely cancel when the pseudorange corrections are applied. (In the case of carrier phase data, this assumption generally becomes invalid at distances of 20–30 km and beyond.) *Hence, LADGPS refers to real-time differential positioning typically over distances up to at most a few hundred kilometers using the DGPS corrections generated by a single reference station.* The DGPS correction is generally transmitted to users by some form of terrestrial-based communications system. The free-to-air service provided for maritime users is described in Section 14.3.2.1.

Wide Area DGPS, as the name implies, is a DGPS technique that distributes the accuracy advantages of DGPS across a very wide region. This may be over a continental extent or, in the extreme case, represents a global service. Although there are a number of different implementations of WADGPS (Mueller 1994), all rely on a *network of reference stations distributed across the region of interest* and a communications system with the requisite coverage and availability characteristics. In its simplest form, WADGPS can be considered as a means by which multiple RTCM messages are received at the user receiver (from each of the base stations within the WADGPS network) and the corrections are decoded and, in effect, "averaged" and input through the receivers I/O port as a "synthetic" RTCM message. This implementation is sometimes referred to as *Network DGPS*. More sophisticated implementations model (in real time) the spatial variation of errors due to the atmospheric refraction, etc., in the so-called state-space representation of errors (Section 14.2.2), so that in addition to the WADGPS message being encrypted, a special algorithm computes the rover receiver corrections on the basis of geographic location. Some of these services are described in Section 14.3.1.

11.3.4 RTK Link Options

In RTK-GPS positioning, the speed of the radio link (or data transmission rate that it can handle) is a limiting factor for sending all the required data from the reference receiver to the rover receiver, while the amplification power available at the reference station determines the effective range of RTK-GPS (Langley 1998b). Hence, the quality and the reliability of the radio link impacts directly on the viability of RTK-GPS. In general, the radio link equipment is purchased at the same time as the GPS receivers. This has several advantages, one of which is that radio licensing issues are generally handled by the GPS selling agent. In addition, the frequency has been selected so as not to interfere with the operation of the GPS receiver itself. Some options for the data link are listed in Table 11.3.

In order to extend the range of RTK-GPS, one approach would be to select a surveyed rover location as the site for the next reference station, and move the reference station as the survey progresses. The user should also keep in mind that the radio transmitting antenna choice and its proper location significantly impact the performance of the radio modem, and hence the RTK-GPS operation. Typically, the longer the antenna, the better transmission and reception characteristics it has. For the Spread Spectrum Radio (SSR) modems, 2–6 dB gain antennas are normally used, while for UHF the typical antenna has 6 dB gain. Often, the transmitting antenna has a higher gain than the receiving antenna, although it is common that both are omnidirectional, whip antennas (as can be seen in the backpack of the instrument shown in Figure 10.4). For low-power operations, helical, quarter-wavelength antennas are commonly used (Langley 1998a). Antennas for the SSR modems are sometimes integrated with GPS antennas into a single unit (see Figure 10.3).

The Internet User Datagram Protocol (UDP) and Transport Control Protocol (TCP) are increasingly used to stream GPS data via the Internet. One popular GPS transport protocol, the Network Transport of RTCM via Internet Protocol (NTRIP), is used to facilitate real-time Internet streaming of GPS data (NTRIP 2009). These protocols can be used over digital (packet-based) cellular phone systems.

11.3.5 DGPS/RTK Services

Requiring users to set up a GPS reference station and hook up radios for real-time operations, *all before making a single measurement*, is too cumbersome a procedure to follow for many standard positioning tasks. In fact, for many navigation applications, this was not even feasible. The closest analogy to this situation is to consider mobile communications services. Large organizations such as taxi fleets typically use (or have used) a voice radio communication system consisting of mobile transceivers (and dedicated frequencies) and their associated repeater/base stations.

TABLE 11.3

Summary of the Radio Modem Characteristics Typically Used in RTK-GPS

Component	Frequency Allocation in the United States	Range (km)	Amplifiers	Special Characteristics
SSR[a]	902–928 MHz; 2.4–2.4835 GHz; 5.725–5.850 GHz	1.5 km with embedded; 50 mW amplifier	1 W (max. allowable) extends the range to 20 km in open areas	Considered the most advanced and most reliable radio modem for RTK and DGPS; rover can receive data six times faster than in the case of UHF and VHF
UHF	328.6– 450 MHz; 450–470 MHz; 470–806 MHz; 806–960 MHz; 0.96–2.3 GHz; 2.3–2.9 GHz	Up to 30 km (with amplification of 2–35 W and proper antenna set up)	Typically 2–35 W	Usually applied baud rate[b] allows for the transmission of RTK correction messages one or two times per second
VHF[c]	30–328.6 MHz	Exceeds 20 km	Similar to UHF, or higher	Usually applied baud rate allows to transmit RTK correction one or two times per second

[a] SSR was almost exclusively used by the military until 1985, when the FCC allowed spread spectrum's unlicensed commercial use in three frequency bands: 902–928 MHz, 2.4–2.4835 GHz, and 5.725–5.850 GHz. SSR differs from other commercial radio technologies because it spreads, rather than concentrates, its signal over a wide frequency range within its assigned bands. This is similar to the GPS signals themselves, and helps make SSR less susceptible to intentional or unintentional interference.

[b] The minimum baud rate for UHF in RTK is 2.4 K bps (bits per second). The baud rate used for RTK generally depends on the frequency that is used. SSR has a much higher baud rate than VHF and UHF, and thus can transfer data much faster.

[c] The choice between UHF and VHF depends on the licensing requirements in the area, and what frequencies are already being used.

Yet almost all functions can nowadays be satisfactorily carried out using the cellular phone services provided by telecommunication companies. The former typically requires the purchase and operation of all the mobile user equipment (as well as the repeater/base station infrastructure), while the latter is more convenient as it implies contracting a *service provider* to address the user's communications needs. In fact the most significant characteristic of real-time GPS techniques is the dominant role played by service providers. Almost all real-time DGPS is now undertaken using a single GPS instrument that receives RTCM transmissions from one of a number of possible services (Section 14.3). In the case of carrier phase–based RTK-GPS, the provision of commercial services is growing rapidly as more GPS CORS networks incorporate the appropriate user data link options (Section 15.3.2).

REFERENCES

Langley, R.B., 1994, RTCM SC-104 DGPS standards, *GPS World*, 5(5), 48–53.

Langley, R.B., 1998a, A primer on GPS antennas, *GPS World*, 9(7), 50–54.

Langley, R.B., 1998b, RTK GPS, *GPS World*, 9(9), 70–74.

Mueller, T., 1994, Wide area differential GPS, *GPS World*, 5(6), 36–44.

NTRIP, 2009, Networked Transport of RTCM via Internet Protocol, Web page http://www. ntrip.org/, accessed March 10, 2009.

RTCM, 2009, Radio Technical Committee for Maritime Services, Web page http://www.rtcm. org, accessed March 10, 2009.

12 GPS Projects: Some Planning Issues

Chris Rizos, Dru Smith, Stephen Hilla,
Joe Evjen, and William (Bill) Henning

CONTENTS

Project planning is an important task, as careful planning maximizes the chances of the global positioning system (GPS) survey achieving the desired accuracy, within a reasonable time, and according to budget. Before commencing the planning, the following must first be established (*Note*: this is a partial list only):

- What is the purpose of the survey?
- What are the accuracy and reliability requirements?
- What resources and budget are available?
- What previous surveys have been carried out?
- Are there any continuously operating GPS reference stations nearby that could supply GPS data or provide access to the reference frame?
- Are there any special (or unusual) characteristics of the project?
- Is the contractor suitably equipped to carry out the GPS survey for the client?

Planning therefore encompasses a wide range of tasks including ensuring that the appropriate hardware is available and operational, providing adequate power supplies, collecting local knowledge of the terrain and site conditions, ensuring that adequate maps and information about existing control marks (both their availability as well as the reliability of their previously surveyed coordinates) and GPS reference stations are available, and ensuring that access to the land has been authorized by the relevant authorities. Some of these issues are not only common to GPS but are also relevant to all types of surveying or mapping tasks. However, before proceeding, it is necessary to identify the different GPS techniques that can be used and discuss the characteristics of each technique in relation to its impact on the execution of the surveying task. Although this was already attempted in Section 11.1 (for the post-processed techniques) and Section 11.2 (for the real-time techniques); in this chapter the factors influencing the *selection* of the GPS technique to be used will be examined. It must, however, be emphasized that carrying out a GPS survey task is nowadays more convenient and efficient (and hence lower in cost) than in the past. This is primarily because of the proliferation in many countries of permanent, continuously operating reference stations (CORS), which means that the surveyor may not need to establish his or her own base station (Section 14.4.1). Furthermore, productivity is significantly improved using real-time differential techniques (Section 11.2).

In Section 7.3, the issue of GPS accuracy as a function of position mode and type of observable was first introduced. Chapter 9 developed the mathematical models and processing procedures associated with the different levels of positioning accuracy. In this chapter, it is assumed that current single point–positioning accuracies are insufficient for the applications discussed herein, and therefore, this positioning technique will not be considered further in the context of surveying and mapping applications. Although precise point positioning (PPP) using International GNSS Service (IGS) precise orbits and satellite clocks can attain 1–2 cm accuracy using 12–24 h data sets (or maybe less), this technique will not be discussed further as it is still not a widely used techniques for GPS surveying and mapping.

Relative GPS accuracy and precision range from the several meter level to the sub-centimeter level: There are a wide spectra of hardware, observation, and processing techniques that can be used to address these accuracy requirements. It is this choice of *survey technique* that is perhaps the most critical decision in the planning process, as the selected GPS technique will strongly influence subsequent planning decisions and field/office operations.

12.1   GPS TECHNIQUES

Tables 12.1 and 12.2 summarize the main choices available to a geospatial professional using the full spectrum of today's differential GPS technology. Note that the ranking of techniques in Table 12.1 is according to distance to the nearest reference receivers (in order to highlight the distance versus accuracy relationship inherent in many precise GPS positioning techniques), while Table 12.2 ranks the techniques in order of increasing coordinate accuracy. Coordinate accuracy in the case of relative positioning techniques, such as the differential GPS (DGPS) and baseline or network carrier phase–based GPS positioning, may be expressed in either the *absolute* sense (i.e., with respect to the reference frame or datum) or in a *relative* sense (i.e., with respect to the nearest GPS reference station or stations). In the former case, the authors suggest the use of the expression "accuracy measure." In the latter case, the expression "relative precision measure" is often used in North America, and this convention will be followed in this book. In the case of the carrier phase–based techniques, the principal differentiator between accuracy at the centimeter level and at the several decimeter level is whether the solution is an "ambiguity-fixed" solution or not (Section 9.4.3). In general, resolving ambiguities becomes harder as the baseline length increases (especially beyond 20 km) unless the observation session length also increases, or network-based techniques are used. Hence, if baselines are kept as short as possible (and especially if <10 km), then there are techniques, using dual-frequency GPS receivers, which have reduced the observation session length to below 1 min (see also Sections 11.1 and 11.2). If surveyors are to take advantage of the extensive CORS networks now found in many countries, some additional planning issues need to be addressed. For example, the distance from the nearest CORS, and whether the CORS operators are able to provide real-time services [real-time kinematic (RTK)-GPS or network-RTK], and whether these reference stations are tied in with the national CORS network, will be important.

Each of these techniques is briefly described in the following text, and the criteria for selecting the *appropriate* GPS survey technique for a project are discussed in Section 12.2. (The baseline length versus observation length versus accuracy relationship is much more complex than described here, with many other factors influencing accuracy—including the number of visible satellites and their geometry, the degree of multipath error, the level of ionospheric activity—however, Tables 12.1 and 12.2 are useful for the following discussions.)

12.1.1   STATIC (DUAL-FREQUENCY) GPS

Static GPS conducted with dual-frequency receivers and post-processing software remains the most powerful (read "accurate") positioning tool available to the GPS surveyor. Originally developed to be used for geodetic control surveys, the technique

TABLE 12.1

Primary GPS Survey Options (Ordered by Increasing Inter-Receiver Distance)

Range (km)[a]	Accuracy (m)[b]	Survey Type[c]	Occupation Time per Point	Operational Mode[d]	Vulnerability[e]	Infrastructure[f]
>100	0.5–2	WADGPS (RTCM)	Instantaneous	RT, PR	***	SP
>100	0.1–0.5	STATIC (DF)	<30 min	PP, CPH+	**	CORS
>100	<0.02	STATIC (DF) (geodesy)	Several hours	PP, CPH	*	CORS
<100	0.5–5	DGPS (RTCM)	Instantaneous	RT, PR	***	SP
<100	<1	DGPS (DD)	Instantaneous	PP, PR	****	CORS
<100	<0.2	STATIC (DF)	>30 min	PP, CPH	**	User
<50–70	<0.02–0.2	NETWORK-RTK	Instantaneous[g]	RT, CPH	*****	SP
<50	<0.1	STATIC (DF)	<30 min	PP, CPH	**	User
<50	0.2–0.5	STATIC (SF)	>30 min	PP, CPH	***	User
<20	<0.02–0.2	RTK	Instantaneous[g]	RT, CPH	*****	SP or user
<20	<0.02	RAPID-STATIC	<30 min	PP/RT, CPH+	****	CORS or user
<10	<0.02	RTK	Instantaneous[g]	RT, CPH+	****	User

[a] Indicates distance from the nearest GPS reference station (limits are approximate only).

[b] 95% confidence interval, relative precision or accuracy within the geodetic or reference station network.

[c] DF, dual-frequency instrumentation; SF, single-frequency instrumentation; RTCM, differential PR corrections; DD, double-differenced data; RTK, real-time kinematic; geodesy, scientific SW, precise orbits, network processing.

[d] RT, real time with data link; PP, post-processed; PR, pseudorange; CPH, carrier phase; CPH+, CPH plus PR.

[e] The susceptibility of the technique to degradation due to poor sky visibility (* little or no problem; ***** very vulnerable).

[f] Includes ambiguity initialization (and re-initialization) time using OTF-AR algorithm (at best a few seconds).

[g] SP, service provider; CORS, continuously operating reference station; User, assumes user operates own reference station.

is ideal for coordinating ground points, where the highest accuracy is required and the inter-receiver distance is many tens of kilometers. Under such circumstances, the observation session length may be many hours in order to ensure the relative precision of less than one part per million (ppm)—equivalent to a 1 cm error in a baseline of 10 km in length.

Another distinguishing characteristic of *static GPS* is that to achieve such accuracy the ambiguities need *not* be resolved, making this technique relatively robust

TABLE 12.2
Primary GPS Survey Options (Ordered by Increasing Accuracy)

Range (km)[a]	Accuracy (m)[b]	Survey Type[c]	Occupation Time per Point	Operational Mode[d]	Vulnerability[e]	Infrastructure[f]
<100	0.5–5	DGPS (RTCM)	Instantaneous	RT, PR	***	SP
>100	0.5–2	WADGPS (RTCM)	Instantaneous	RT, PR	***	SP
<100	<1	DGPS (DD)	Instantaneous	PP, PR	****	CORS
<50	0.2–0.5	STATIC (SF)	>30 min	PP, CPH	***	User
>100	0.1–0.5	STATIC (DF)	<30 min	PP, CPH+	**	CORS
<100	<0.2	STATIC (DF)	>30 min	PP, CPH	**	User
<50	<0.1	STATIC (DF)	<30 min	PP, CPH	**	User
>100	<0.02	STATIC (DF) (geodesy)	Several hours	PP, CPH	*	CORS
<50–70	<0.02–0.2	NETWORK-RTK	Instantaneous[g]	RT, CPH	*****	SP
<20	<0.02–0.2	RTK	Instantaneous[g]	RT, CPH	*****	SP or user
<20	<0.02	RAPID-STATIC	<30 min	PP/RT, CPH+	****	CORS or user
<10	<0.02	RTK	Instantaneous[g]	RT, CPH+	****	User

[a] Indicates distance from the nearest GPS reference station (limits are approximate only).

[b] 95% confidence interval, relative precision or accuracy within the geodetic or reference station network.

[c] DF, dual-frequency instrumentation; SF, single-frequency instrumentation; RTCM, differential PR corrections; DD, double-differenced data; RTK, real-time kinematic; geodesy, scientific SW, precise orbits, network processing.

[d] RT, real time with data link; PP, post-processed; PR, pseudorange; CPH, carrier phase; CPH+, CPH plus PR.

[e] The susceptibility of the technique to degradation due to poor sky visibility (* little or no problem; ***** very vulnerable).

[f] Includes ambiguity initialization (and re-initialization) time using OTF-AR algorithm (at best a few seconds).

[g] SP, service provider; CORS, continuously operating reference station; user, assumes user operates own reference station.

and not as vulnerable to multipath interference and sky visibility problems as in the case of other *high-productivity* techniques. However, modern geodetic techniques can achieve even higher relative accuracies, of the order of 10 parts per billion or less, for baselines thousands of kilometers in length, although it must be emphasized that such results require the use of sophisticated data processing software, precise GPS satellite orbit information (e.g., from the IGS—Section 15.3.1), and several days of observation. However, if the baseline length is greater than about 30–50 km, and the observation session is comparatively short (say, less than about an hour), it may *not* be possible to obtain an "ambiguity-fixed" solution, and hence the positioning accuracy will suffer, and likely be at the several decimeter level. Due to the relative precision being a function of distance to the nearest one (or more) permanently operating reference station(s), distance-to-CORS will be one of the critical factors in defining the observation session length that should be used to satisfy the survey accuracy requirement (see Table 12.1). In other words, for a certain specified survey accuracy, the longer the distance-to-CORS, the longer the observation session length should be. However, for most GPS surveying applications, the recommendations of GPS user manuals in this regard may be taken as a guide. Only in the case of the highest-order geodetic control surveys will GPS specifications likely to be the authorative guide.

12.1.2 RAPID-STATIC GPS

For distances less than about 20 km, *rapid-static GPS* (Section 11.1.1) is a widely used, high-accuracy GPS surveying technique. *Rapid-static GPS* is a post-processed or real-time technique that uses the same dual-frequency receivers as for *static GPS*, and achieves similar accuracies. The main difference is that an "ambiguity-fixed" solution must be obtained (Section 9.4.3) even as comparatively short observation sessions are used (of the order of a few tens of minutes—the length being a function of the number of satellites and the geometry), meaning that the *rapid-static GPS* technique is less robust, and more vulnerable to poor sky visibility, than the conventional *static GPS* technique. CORS networks are rarely dense enough to *always* satisfy the 20 km distance constraint for *rapid-static GPS* (an exception is, for example, the Hong Kong CORS network—Figure 14.8). Sometimes the GPS surveyor is lucky to be within 10–20 km of a CORS, in which case *rapid-static GPS* is an attractive technique. A network-based RTK technique can be used for sparser CORS networks—typically 50–70 km inter-receiver spacing—but only if the CORS operator offers such a service. Online data processing services (Section 14.4.3) use scientific GPS data processing software, and therefore the distance-to-CORS constraint is relaxed significantly, even by a factor of 10 or more. For example, the National Geodetic Survey (NGS) service Online Positioning User Service (OPUS) can deliver centimeter-level relative precision with 15 min observation files even if the CORS stations are over 100 km away (NGS 2009a).

12.1.3 STATIC (SINGLE-FREQUENCY) GPS

Static single-frequency GPS surveying is a *cheaper* alternative to *rapid-static GPS* and *RTK-GPS* techniques for baselines of 20 km or less in length. This is because an "ambiguity-fixed" solution is generally possible for observation sessions of the order of 30 min. Relatively low-cost GPS hardware is used, and the data can be post-processed or the coordinate results obtained in real time. However, it may be a little less productive than techniques such as *rapid-static GPS* because of the increased length of the average observation session. Again, as for *rapid-static GPS*, the availability of CORS data (post-processed or real time) benefits the GPS surveyor only if the distance constraints are satisfied, that is, <20 km for a single-base CORS mode, or <50–70 km if a network-RTK service is offered in the survey area.

12.1.4   KINEMATIC AND STATIC GPS SURVEYS USING ON-THE-FLY ALGORITHMS

As already mentioned several times, in order to obtain centimeter-level position-ing accuracy an "ambiguity-fixed" solution is necessary. The "time-to-ambiguity-resolution" (time-to-AR) is therefore the crucial factor in *high-productivity* static GPS surveying. The *rapid-static GPS* technique is relatively inefficient because of the necessity to collect sufficient data (over tens of minutes) to ensure AR. However, *after* AR the time required to determine a high-accuracy coordinate solution for a ground mark is of the order of several seconds (marked as "instantaneous" in Tables 12.1 and 12.2). If the resolved ambiguities can be *preserved*, that is, there is a contin-ued lock on the satellite signals, then after the initial period of static initialization all subsequent ground points can be surveyed very quickly. This is the basis of the *stop-and-go GPS* technique for static point coordination (Section 11.1.2), as well as the *kinematic GPS* survey mode (Section 11.1.3). However, if there is loss-of-lock on the signals, then the AR process must be *re-initialized* or *restarted*. This is a nuisance for static surveys, but can be disastrous for kinematic surveys if this re-initialization *must* be carried out by the antenna remaining stationary. To overcome this con-straint, the so-called on-the-fly (OTF) AR algorithm was developed so that AR can be performed while the receiver antenna is moving. When the baseline is compara-tively short (say <10 km), using dual-frequency GPS receivers, it is generally possible to perform OTF-AR with less than 1 min of (kinematic) data. Hence, even if the signals are lost on a regular basis as the receiver moves from point to point (due to buildings and foliage obstructing the signals), fast OTF-AR can keep the positioning productivity high. As this technique is also implemented in post-processed mode, data can be processed *backward* and the coordinates of points surveyed *before* AR can also be determined. As with the previous two GPS techniques, CORS data for OTF-AR is only useful if the distance constraints are satisfied. However, if AR is not possible sub-meter accuracy is nevertheless achievable using precise pseudorange data and/or "ambiguity-free" carrier-phase solutions (Section 9.4.3).

12.1.5   REAL-TIME KINEMATIC GPS TECHNIQUES

RTK-GPS is in many ways the most attractive technique for high-accuracy and high-productivity GPS surveying (Section 11.2.2). It is versatile because the coordinate solutions are obtained in the field and can be used immediately for time-critical static and kinematic applications such as precise navigation, engineering stake-out, machine guidance, and control, etc. The basis of *RTK-GPS* is the implementation of the OTF-AR algorithm in *real time*. However, the time-to-AR is essentially "dead time" (Figure 11.4) because during this initialization period the accuracy is degraded (to the sub-meter level) until sufficient data is processed to resolve the ambiguities. There are a number of constraints to the use of *RTK-GPS*, for example, it is gen-erally used for baseline lengths <30 km (if using wireless data links—Table 11.3), although the results of OTF-AR using a single-base (reference) station can be unreli-able for baselines longer than about 10 km. Network-based RTK techniques these days can operate over reference receiver distances of greater than 50 km (Section 14.3.2.2) using cellular phone–based transmission of Radio Technical Committee for Maritime Services (RTCM) messages. As the number of Global Navigation Satellite System (GNSS) satellites broadcasting signals increases over the coming decade (Sections 15.4 and 15.5), simulation studies indicate that the robustness of *RTK-GNSS* will *increase* so that centimeter-level relative coordinate precision will be easily achievable with just a few seconds of data over baseline lengths up to many tens of kilometers. It must be emphasized that the expression "accuracy" in the context of relative positioning techniques, such as those discussed here, should be reserved to measures of coordinate quality relative to the reference frame or datum. As this measure includes the reference station coordinate errors—assumed to be error free for these GPS positioning techniques—it is preferable to refer here (and in Tables 12.1 and 12.2) only to the inherent accuracy of the baseline or relative coordinate components of the unknown GPS-surveyed ground mark as "relative precision" or "accuracy within the network," as these are directly attributable to the GPS technique and operational conditions employed in the survey.

12.1.6 KINEMATIC AND "STOP-AND-GO" GPS SURVEYS

As discussed earlier, high-accuracy positioning over short baselines, using very short observation sessions (for positioning in the static mode) or single-epoch positioning (as in the case of kinematic surveys) requires the use of "carrier-range" data. *Carrier range* is what the ambiguous carrier-phase observations are transformed into once the ambiguities have been resolved (Section 9.4.2). Hence, the survey operation consists of two parts: (a) the initial AR and (b) the subsequent carrier range–based positioning. The initial AR can be carried out in OTF mode, using a short period of data, if the baseline satisfies the distance constraint for single-base or network-RTK operations, six or more GPS satellites are tracked, and dual-frequency instrumentation is used. Sub-meter accuracy can be obtained if the ambiguities are *not* resolved. The time-to-AR does not impact on the efficiency of the subsequent "carrier-range" positioning, which can still be carried out (nearly) instantaneously. However, what is crucial is that ambiguity *re-initialization* should not be necessary very often, otherwise the time-to-AR each occasion this happens could adversely impact on survey productivity. Of course, in the limit, this technique degenerates to the standard *static GPS* and *rapid-static GPS* techniques.

12.1.7 DGPS TECHNIQUES

The DGPS is the pseudorange-based relative positioning technique that delivers sub-meter to a few meter accuracy (Section 11.2.1). There are, in general, two classes of DGPS techniques. One is the real-time DGPS technique based on the transmission of RTCM (Section 11.3.2) data corrections (using the so-called observation-state representation—Section 14.2.2). A less-common DGPS technique is based on *double-differenced* pseudorange data (analogous to how carrier phase data is processed for baseline solutions—Section 9.4.1). In both cases, the accuracy is mainly a function of distance to the reference station. The former is the dominant real-time implementation, not requiring the establishment of a reference station by the user (Section 14.1) as a DGPS *service* broadcasts the RTCM messages. In general, if the distance from the reference station is less than 100 km or so (the distance over which it is assumed that there is a cancellation of the atmospheric refraction effects), the major impact on coordinate accuracy is satellite geometry and the quality of the pseudorange measurements. *Wide area* DGPS (WADGPS) is an enhancement of the basic real-time DGPS technique in that a network of reference stations located across a region (typically set up many hundreds of kilometers apart) is used to generate the DGPS data corrections that are then transmitted to the user (Section 11.3.3). Such an implementation permits the modeling of distance-dependent biases such as atmospheric refraction and satellite orbit errors, delivering a meter-to-sub-meter-level accuracy service even if the nearest WADGPS reference station is located hundreds of kilometers away. Wide Area Augmentation System (WAAS) is one WADGPS method that has an advantage over traditional DGPS in that an extra DGPS receiver/antenna is not needed, the corrections are transmitted by the WAAS satellites on the L1 frequency (Section 14.3.1.2).

12.2 STANDARDS AND SPECIFICATIONS

The accuracy requirements for a GPS survey project are usually defined by the client, either explicitly (in the form of a contractual requirement to deliver coordinates with a certain accuracy) or implicitly (defined by the task or survey application having to meet a certain accuracy standard). For the United States *accuracy standards* refer to the Federal Geographic Data Committee documents (FGDC 2009). *Specifications*, on the other hand, set out how the GPS survey is to be carried out so as to satisfy the specified accuracy, and generally consist of a set of strict requirements, or "looser" recommendations or suggestions with regard to survey design, instrumentation, check procedures, monumentation, data processing, etc. There are no international specifications for GPS surveys. Many survey authorities at the national and state level have developed *specifications* or recommended practices or guidelines, often at the application-specific level (e.g., for cadastral or engineering surveys). For example, Schinkle (1998) describes a set of guidelines on how the New York State Department of Transport undertakes GPS engineering control surveys. Such specifications or "standard operating procedures" (SOPs) may be used to assist in the design of guidelines if none are available for a particular survey task. However, the latest techniques based on RTK and network-RTK, or CORS networks, may not be mentioned.

Some so-called *standards and specifications* (S&S) combine the accuracy standards and GPS specifications into a single document, others keep them separate. Accuracy *standards* do not change very often and hence it would be appropriate for them to be in a separate document (or documents). On the other hand, *specifications* are technology specific. Furthermore, the pace at which the GPS technology has changed over the past decade or so means that many specification documents are somewhat dated. Some documents, such as the Australian S&S for geodetic control (ICSM 2006), have undergone regular revision in an effort to keep them up to date as far as the survey technologies is concerned. Other GPS specifications, such as the Canadian (GSD 1992) and the U.S. (FGCC 1989) specifications, are essentially pre-RTK and pre-CORS documents that have lost much of their relevance in the face of rapid technological developments.

The published guidelines for GPS surveys (some of which are referred to earlier) may be the only *official* S&S documents available to guide the surveyor. (The GPS manufacturers' user manuals do not address national or state survey practice, and are therefore of limited use in this regard.) Nevertheless, the surveyor/contractor should be familiar with relevant S&S as the client requirements may be couched in the language of such documents. In addition, *quality assurance practice* demands that GPS field and office procedures are consistent, and adherence to a GPS specification ensures such consistency. Of course, one danger with GPS positioning is that even if good specifications are followed for this field work, an analyst could specify the wrong options when using post-processing software and adversely affect the results. This issue is much harder to address within specifications.

Different national geodetic control authorities typically use different terminology for the *classes* of GPS (and other) surveys, adopt different accuracy measures, and have varying numerical accuracy limits used to define them. In general, these categories of survey are distinguished by relative accuracy measures. Furthermore, some national standards have different accuracy standards for the horizontal position components compared to those for the vertical component, with the added complication that *ellipsoidal heights* may be treated differently to the *orthometric heights* or height above a national datum (NGS 1997). In general, however, preexisting accuracy standards are maintained, but augmented by several higher-accuracy standards that are only applicable for GPS surveys. Many countries are recognizing that there are now essentially *two* types of accuracy/precision measures:

- An *internal* one based on an adjustment of the GPS-only survey
- An *external* one based on a constrained adjustment, where the coordinates of the GPS network define the geodetic framework (either "active" CORS stations or "passive" ground marks occupied by reference receivers)

In Australia and New Zealand, the former corresponds to a survey's *class*, while the latter defines its *order* (ICSM 2006, LINZ 2003). In the case of the United States, the former refers to the classification of the survey according to a *local accuracy* measure, while the latter is the classification according to a *network accuracy* measure (FGDC 2009). For each category of survey accuracy (or standard class), there may be recommended practices for

- Network design
- Instrumentation and survey technique
- Field procedures and monumentation
- Office reduction procedures
- Calibration and result validation procedures

Different countries may not only have a different classification system for accuracy, but also have different recommended practices for each class of GPS survey. Sections 13.1 and 13.2 describe the U.S. accuracy standards and some typical applications.

12.3   SELECTING THE APPROPRIATE GPS TECHNIQUE

One of the most important planning issues is selecting the *appropriate* GPS survey technique from the range listed in Table 12.1 (or 12.2). Factors such as the cost of the equipment, CORS service charges, the occupation time-per-point and the mode of processing, and the level of redundancy (checking) required will affect the budget of a GPS survey project. The final factor, vulnerability to poor sky visibility, may rule out certain observation techniques in environments where GPS signals are obstructed. Note that many of these considerations are in fact addressed in S&S documents and in the GPS manufacturers' user manuals.

12.3.1   Distance from Reference Station(s)

The distance between the survey and reference receiver can be a primary concern because it does impact on the survey technique that can be used if factors such as the accuracy and the observation session length are predefined (Table 12.1). For example, as the inter-receiver distance increases, more sophisticated (and expensive) techniques must be adopted if the accuracy standard in length units (centimeters, decimeters, or meters) is to remain the same. For example, the DGPS is the primary tool for mapping/geographical information system (GIS)-type surveys out to distances of 100 km, but for longer distances the WADGPS technique may be used. The most popular of the DGPS/WADGPS techniques operate as real-time RTCM-based services. Achievable accuracy is typically at the few meter level (though under favorable conditions of good geometry and low residual systematic atmospheric biases the accuracy can be at the sub-meter level), with very short (almost "instantaneous") observation sessions. Other implementations of DGPS/WADGPS, such as the use of double-differenced and/or carrier phase-smoothed pseudorange measurements, are not offered as real-time services. Positioning accuracy may be comparatively high (sub-meter), but the increased logistical complexity of operating one's own reference station is a drawback to their use.

In the case of carrier phase–based positioning techniques, to satisfy a 1–2 cm accuracy requirement, RTK-GPS and rapid-static GPS techniques can be used if inter-receiver distances are less than 10–20 km, or 50–70 km if a network-based RTK service is available. The critical operation is AR, within a "reasonable" time-to-AR. (Reasonable time-to-AR in the context of RTK-GPS, and any OTF-AR GPS survey technique, is tens of seconds to perhaps a few minutes.) If AR is unsuccessful, the degraded decimeter-level accuracy may not meet the project accuracy requirements. Given a certain distance constraint, and accuracy standard to be met, the only option is therefore to extend the observation session length. Perhaps 30 min observation sessions may be sufficient to achieve centimeter-level accuracy for inter-receiver distances of the order of several tens of kilometers. The greater the inter-receiver distance is, the longer the observation session length will typically be. In the limit, static GPS survey techniques are the most robust of the high-accuracy techniques, because the long observation session lengths (several hours to days) overcomes the challenge of relative positioning over very long inter-receiver distances (several tens to hundreds of kilometers) if the data is processed using the appropriate software.

12.3.2 ACCURACY VERSUS COST

Table 12.1 lists the viable techniques to achieve requisite accuracies, although it must be emphasized that care must be taken when using a GPS technique close to the limits of its specifications—though specifications tend to be conservative, and give guidelines for use of particular GPS techniques that have relatively wide safety margins. For example, while most DGPS systems can easily deliver 5 m accuracy, obtaining 0.5 m accuracy or better will require the appropriate software and/or carrier phase–tracking GPS hardware and "ideal" observing conditions. It is therefore important to keep in mind that the stated accuracies assume good sky visibility, low multipath disturbance, and favorable satellite geometry.

GPS technology offers obvious choices in the 0.5–5 m accuracy range (DGPS, WADGPS) and less than 0.05 m accuracy range (RTK, static, rapid-static). However, the centimeter-to-half-meter accuracy "gap" poses special challenges. This is partly due to the vast difference in raw measurement precision of the pseudorange (Section 8.2) vis-à-vis carrier phase (Section 8.4), and the data processing procedures. Carrier phase–based techniques give relative precision at the few-decimeter-to-sub-meter level only when the ambiguities cannot be resolved using the data that has been collected. In such circumstances, an "ambiguity-free" solution (Section 9.4.3) will *fail* to deliver centimeter-level accuracy, but could still be acceptable for mapping/GIS surveys seeking only sub-meter accuracy. An appropriate strategy may be to consider only those GPS techniques that can be relied on to supply results that are *more* accurate than the client's requirements, and to charge for such services. For example, use carrier phase–based techniques for any specification calling for an accuracy higher than a few decimeters. However, if centimeter-level accuracy must be *assured*, the GPS survey will require special care (and will be significantly more expensive), see comments in Section 12.3.1.

12.3.3 OCCUPATION TIME-PER-POINT

In general, the quality of a computed coordinate increases with longer *occupation time*. However, the length of time a GPS receiver must physically occupy a single point will significantly impact *productivity*, and therefore the cost of the survey. *WADGPS* and *DGPS* techniques offer robust, continuous point coordination, but at comparatively low accuracies (here considered to be sub-meter or worse). *Single- and dual-frequency kinematic and static GPS* surveys also offer continuous point coordination so long as the receiver maintains lock on most of the tracked satellite signals while moving from point to point. Initialization and re-initialization, to resolve the carrier phase ambiguities, can significantly increase the time taken to complete these types of surveys, even with OTF-AR algorithms. *RTK-GPS* and *network-RTK* are centimeter-level accuracy, real-time, rapid point coordination techniques, typically requiring just a few seconds at each point (not including time for OTF-AR if it is required due to loss-of-lock).

The rapid-static GPS is another commonly used technique for point coordination if the inter-receiver distances are less than about 20 km (though the U.S. National Geodetic Survey offers a rapid-static service for points surrounded by three CORS stations up to 250 km away—NGS 2009a). However, the *rapid-static GPS* technique is considerably less productive than RTK techniques as the occupation time at a point may be 10 min or more. For high-accuracy surveys beyond about 20 km from a reference station (the distance beyond which rapid AR becomes very difficult), conventional *static GPS* techniques still are preferred, with station occupation times varying from 30 min to several hours depending on the accuracy required and inter-receiver distances. This distance to reference station(s) is difficult to define precisely as it is a function of many factors, including the magnitude of residual biases, multipath errors, the number of visible satellites and their geometry, and so on. Recommended observation session lengths are given in the GPS manufacturers' manuals.

12.3.4 PROCESSING MODE

Techniques that rely on real-time data processing require that the receiver in the field computes the position and the associated quality information. Post-processing techniques rely on receivers logging the raw measurements, which are then downloaded into data reduction software. Both modes have their advantages and disadvantages.

Post-processing allows more flexibility in terms of control over the processing options, hence the results tend to be more reliable. Sophisticated data (post-processing) software can, in principle, provide for greater quality control as there is the opportunity to scan the data for outliers and edit out low-quality data. Many CORS networks will permit free download of RINEX (Gurtner 1994) data files. Hence, the surveyor can use the nearest CORS station to define one end of the baseline, and is saved the trouble of establishing reference stations (Sections 14.4.1 and 14.4.2). The computation step may be simplified further by submitting the survey data (not the CORS data) to online data processing services (Section 14.4.3), and receiving in return the point coordinate. Most of these services currently only support *static GPS* positioning, typically requiring RINEX data files several hours in length. NGS's OPUS (NGS 2009a) is a service that can process files as short as 15 min, and hence is suitable for *rapid-static GPS* surveys. All such online services require the use of dual-frequency instrumentation.

Real-time systems, on the other hand, have access only to prior data (they have no knowledge of future data), and internal real-time QC indicators must be estimated based on relatively scant data (Section 12.4.1). However, the main advantage of real-time processing is the speed at which points can be surveyed and the coordinates determined (some applications such as engineering stake-out and machine guidance systems require the coordinate solutions *immediately*). In addition, coordinate data and feature attributes can be logged in real time and later downloaded directly into GIS or CAD software. Undertaking real-time GPS surveys is more complex if the surveyor has to also operate the reference receiver(s), and the wireless data link between reference and survey receivers. Increasingly CORS networks support real-time *RTK-GPS*, and sometimes *network-RTK* as well, through RTCM-enabled centimeter-level accuracy services (Section 14.3.2.2). However, the distance constraints must be satisfied if the accuracy requirements are to be met, as discussed in Section 12.3.1.

12.3.5 DILUTION OF PRECISION

As a rule-of-thumb, the shorter the occupation time-per-point, the more critical the number of tracked satellites will be on the quality of the resulting coordinate solution. In other words, the *sensitivity* of a coordinate solution to low satellite numbers and poor geometry (defined as a high DOP value—Section 9.3.3) decreases as more data are collected, that is, as the observation session becomes longer. Therefore, conventional *static GPS* surveys are the least sensitive to low numbers of visible satellites and/or high DOP, while real-time GPS techniques such as DGPS, WADGPS, and RTK are more sensitive. RTK-GPS surveys, requiring a minimum of five or six satellites for ambiguity initialization (or re-initialization if signal lock is lost), are the most vulnerable to poor geometry. DOP values are almost always greater than unity, and a "low" value is typically one that is less than 3, an "acceptable" one may be in the range 3–6, a "high" value greater than 6, and an "unacceptable" one greater than 10. S&S may specify maximum DOP values and the minimum number of satellites for particular GPS techniques or applications. Note that DOP is only a measure of geometry, not a predictor of coordinate accuracy. That is, the relative magnitude of two DOP values (corresponding to two different satellite constellations) does not necessarily translate into a linear measure of positioning accuracy. In other words, a DOP value of 2 does not imply a positioning result that is twice as accurate as that when the DOP value is 4. In fact, if the systematic errors are significant, the DOP value has little relation to accuracy at all, and a scenario of high-DOP-low-systematic-errors could result in a higher accuracy than a scenario of low-DOP-high-systematic-errors. Nevertheless, the DOP can be viewed simplistically as a way of amplifying the effect of random errors on position solutions (Milbert 2008).

For a baseline 24 satellite constellation, the standard positioning service document (DoD 2008) states that, with an elevation masking angle of 5°, there will be a position dilution of precision (PDOP) of 6 or less for 98% (or more) of the time. The current constellation consists of more than 30 satellites, and hence the percentage of the time the PDOP value is greater than 6 will be even less. However, these are "open sky" standards, and the "true" PDOP at a certain location may be worse due to signal obstructions and the use of a higher elevation mask angle, for example, going from 5° to 15°. It is advised that DOP values not be used as the only QC measure of a GPS survey.

12.3.6 POOR SKY VISIBILITY

The sky visibility may influence the selection of the GPS survey technique. Table 12.1 (and 12.2) gives an indication of how *vulnerable* each measurement technique is to poor observing conditions. As a general comment, all *high-productivity* GPS positioning techniques (rapid-static, stop-and-go, kinematic) are less reliable in suboptimal observing conditions, such as encountered in urban and wooded areas, than static GPS survey techniques. The reason is that visibility (and multipath interference) problems can, to a significant extent, be overcome by occupying the points for longer time periods.

Is a "clear sky" scenario realistic? The planning phase of a GPS survey may include an assessment of the environment in which the survey is to be undertaken through site reconnaissance (Section 12.5.7). If many points are to be surveyed, obviously the degree of sky visibility can vary greatly from point to point. It should always be kept in mind that the stark differences between a constellation viewed in clear sky, and that which can be tracked in an urban area, can greatly reduce the efficiency of GPS in such environments. While it may, in theory, be possible to optimize a GPS survey observation schedule so that sites with obstructed sky view are surveyed when the visible satellite constellation is optimum as far as the observable sky window is concerned, in reality it is impractical. Each site is visited generally according to logistical considerations, and if the sky visibility is too poor—this can be most easily ascertained for real-time techniques—then select an alternate (eccentric) ground mark to survey, and if the site with obstructed sky view *must* be positioned then this can be done later using an optoelectrical instrument to measure the distance and angles from one or more GPS-surveyed eccentric ground marks (Section 13.3.3).

12.3.7 MULTIPATH DISTURBANCE

The multipath can degrade the accuracy specifications indicated in Table 12.1, especially for the pseudorange-based techniques (Section 8.5.2). Multipath is often associated with poor sky visibility (tall buildings and other obstructions typically are also sources of multipath), but can be severe if an antenna is mounted on an aircraft, on vessels, or on vehicles. Rapid positioning techniques are the most sensitive to multipath, hence to some extent multipath can be averaged out by using static observation sessions longer than about 15 min. However, the multipath can affect *all* types of GPS positioning, and if the ground mark to be surveyed is a critical one, then some effort should be made in the planning phase of the project to find a relatively multipath-free site.

12.4 THE ISSUE OF QUALITY CONTROL IN PROJECT DESIGN

Because many QC practices require extra work, the final project design is usually a *compromise* between technical requirements and economics, worked out within the framework of explicit recommended practices for GPS surveys (or at the very least, prudent practices) that ensure the job gets done to the appropriate standards of accuracy and reliability (Section 12.2.2). In other words, sufficient redundant measurements should be made to ensure adequate QC, *but* not such an increased survey effort that would significantly raise the cost of the project.

12.4.1 INTERNAL QUALITY CONTROL INDICATORS

Internal QC indicators are those provided to users by the GPS receivers themselves (when operated in real-time mode), or by GPS data reduction software (in the case of post-processing). Examples of internal quality indicators are a receiver stating that ambiguities have been fixed in a RTK-GPS survey; a receiver observing for a set amount of time in a rapid-static GPS survey and then indicating to the user that a sufficient amount of data have been collected to estimate a reliable solution; the standard deviations of the final baseline vector from a post-processed static survey; and so on.

But there are limits to the usefulness of such internal QC indicators. For example, if an incorrect height of the antenna is measured, the processing software will not be able to identify such a problem, yet the computed coordinates of the affected ground mark will be incorrect. Internal quality indicators may not detect whether GPS surveys are producing spurious results, as, for example, when validating whether the resolved ambiguities are in fact correct. A range of tests are carried out to validate ambiguities (either within the receiver in the case of a RTK-GPS survey or the PC software during post-processing), and it is very rare that false ambiguities are resolved to integer values using well-known AR search methods based on the LAMBDA (least squares ambiguity decorrelation adjustment) algorithm (Teunissen 1995). Furthermore, final coordinate standard deviation values generated by the data reduction software are an internal precision measure (how well the GPS data fits the prescribed mathematical model), rather than an accuracy measure (how well the final solution matches the true coordinates). Therefore, such internal QC indicators cannot provide an indication of the true accuracy of a position solution, without additional external information, such as the true accuracies of the reference station coordinates themselves. Furthermore, the mathematical models do not include all the complex effects of atmospheric refraction, unknown instrument biases, etc. However, these quality indicators may be useful for indicating problems with the data.

12.4.2 EXTERNAL QUALITY CONTROL INDICATORS

The external QC indicators may be considered a form of *independent* check on GPS solutions. Many S&S documents will set out guidelines for external QC indicators, and define the tolerances that must be met by the GPS survey for a specified accuracy standard. Some GPS baseline solution test procedures are (a) repeatability, (b) polygon "misclosure" of independent baseline vectors, and (c) comparison with "known" control station coordinates. In practice, the applicability of each of these procedures depends mainly on the type of GPS technique used for a particular project and how the receivers are deployed in the field to carry out the survey.

12.4.2.1 Quality Control for Monumented Surveys

GPS control surveys invariably involve setting up GPS antennas on *monumented* or pre-marked points. The standard survey practice of observing a *network* of baselines (defined by a pair of simultaneously observing, carrier phase–tracking GPS receivers) connecting points of interest (whose coordinates are to be determined) to one or more known control stations (defining the network datum) is frequently employed. Increasingly CORS will form part of the network, providing the tie to the geodetic datum (e.g., in the United States this is the realization of the geometric component of the National Spatial Reference System known as "NAD 83"—Section 13.1).

The survey techniques likely to be used for such a project are *static GPS* (single- or dual-frequency instrumentation) or *rapid-static GPS*.

Let us assume that all measurements are made and recorded, and then processed post-survey in the baseline mode. External QC indicators can then be derived from (a) the comparison of individual baseline results that have been measured more than once (repeat baselines), (b) linking several *independent* baseline vectors into a polygon and determining the misclosure (3D coordinate components), or (c) a three-dimensional least-squares adjustment of the GPS-derived baselines or coordinates. The latter may be carried out using commercial GPS data reduction software that rigorously accumulates multi-session baseline results into one solution, or less rigorously combines independently computed baseline vectors in a secondary geometric network adjustment (Craymer and Beck 1992). Additional advantages of the 3D adjustment are that the adjusted coordinates of some points can be compared to their "known" values, and the horizontal and vertical error ellipses can be computed. The difficulty with this method is the all-too-often assumption that a monument's coordinates are "fixed" and "known" forever. This assumption ignores the dynamics of the planet as well as the refinement of the datum itself. Because of this, some countries (such as the United States) are moving toward a purely "active control" realization of their datums, with passive control being reduced in its role. See, for example, the NGS 10 year plan (NGS 2008). On the other hand, Australia has adopted the approach that datum coordinates (for active CORS as well passive ground marks) do *not* change with time, hence ignoring the influence of global tectonics on the national datum. Often S&S documents will specify the maximum allowable size of error ellipses for a certain *class* or *category* of survey (corresponding to various accuracy standards).

When planning a static control survey project, a *reasonable* number of "redundant baselines" (extra baselines linking stations previously surveyed but with different pairs of stations) and "repeat baselines" (resurveying pairs of stations defining the same baselines) should be observed. This requirement is usually specified in S&S documentation, and may be in the form of a guideline such as, for example, that 10% of baselines are resurveyed (repeat baselines) and that each point be connected to a minimum of three *independent* baselines (via redundant baselines). An independent baseline is one that is computed using a set of simultaneous measurements made at the two receivers, and at least one set of measurements has not been used in any other baseline computation. All baselines computed using measurements made during different sessions are considered independent, as, for example, in the case of a single reference station (such as a CORS) and a "roving" receiver moving from point to point (stop-and-go, rapid-static, or static GPS techniques). However, splitting a single observation session into two sessions is not considered to have generated two independent baselines, as many of the observing conditions remain unchanged (e.g., same antenna height, very similar satellite geometry and atmospheric conditions, and same instrument operator), and hence the data is not sufficiently uncorrelated. Confusion arises when more than two receivers are operated simultaneously. In such cases, the number of independent baselines is $(n - 1)$, where n is the number of receivers deployed during the observing session. The choice of data from which pair of receivers are to be processed is arbitrary. For example, assume three receivers at sites A, B, and C, then the possible independent baselines are A-B and B-C, A-B and A-C, and A-C and B-C. The notion of independent baselines only makes sense if the data from the pairs of receivers are indeed processed separately (i.e., independently). Polygons can only be constructed, and misclosures computed, from independent baselines. Furthermore, if the data reduction software is capable of multi-receiver and multi-session processing, then there are effectively no "baselines," just network coordinates.

12.4.2.2   Quality Control for Fast Coordination Surveys

Fast coordination (or "detail") surveys tend to be somewhat different to the monumented surveys referred to earlier. Although some of these survey points may be pre-marked (or, at the very least, located at revisitable locations), the majority of points are likely to be *unmonumented*. If a point is unmonumented, it is not possible to return to the precise location of the initial survey in order to make additional (redundant or check) observations. Typically, these surveys are conducted using *RTK-GPS* (or *network-RTK*) or post-processed techniques such as *stop-and-go*, with a "roving" receiver deployment mode known as "star" or "radial" (Figure 12.1). Each surveyed point is in fact an independent baseline radiating out from a single reference station (which could of course be a CORS). Hence, there are no independent baselines *between* the surveyed points, and the network adjustment-based QC procedure mentioned in Section 12.4.2.1 cannot be used.

Some possible QC techniques for fast coordination surveys include detecting outlier results by using a priori knowledge of the geometry of the surface or feature being surveyed; using a second, independent positioning technique (e.g., a second reference station); ensuring that the observation conditions are good (e.g., maintaining PDOP <6 and tracking at least five satellites); maintaining the same antenna height during the entire survey. However, it must be conceded that many of the coordinated points from such surveys are "temporary," surveyed in order to collect data for as-built surveys, topographic or engineering detail surveys, or rapid mapping, and hence isolated coordinate errors are unlikely to be critical with respect to the project objectives. A simple rule-of-thumb for QC is *if the point to be positioned is important, then do a repeat occupation of that point.* If mixed

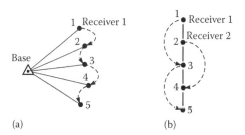

FIGURE 12.1   Receiver deployment strategies: (a) "star" or radial mode and (b) "traverse" mode.

antennas are being used during the project, occupy the point using different GPS equipment (antennas).

12.5   PROJECT PLANNING TASKS

Planning typically comprises the following tasks:

1. *Project design*: involves project layout and network design, and is driven primarily by accuracy and station location/density requirements (defined by the client), productivity/economic considerations (of concern to all parties), and standards and specifications (promoted by the state or national survey authority).
2. *Observation schedule*: giving consideration to such factors as the number of GPS receivers, the GPS technique to be used, occupation time per point, the number of points to be coordinated per day, requirements for multiple point occupancy, etc.
3. *Instrumentation and personnel*: instrumentation appropriate for the GPS technique to be used (which, if not available in-house, could be rented), trained personnel hired, etc.
4. *Logistical considerations*: issues such as transportation and receiver deployment strategies (appropriate to ensure observation schedule can be adhered to), special site requirements (e.g., power availability and station inter-visibility for subsequent non-GPS surveys), and factors related to network design and QC such as multiple point occupancy, etc.
5. *Reconnaissance*: which may or may not be necessary, depending upon how critical the surveyed GPS points are to the overall project, whether permanent marks will be established, the GPS technique that will be used, etc.

Note that it is not always necessary to have *elaborate* project plans. For example, if the intention is merely to coordinate a large number of points (or mapping a continuous trajectory such as a road centerline) from a single reference station, then a high-productivity technique such as *RTK-GPS* using the nearest real-time CORS service does not require extensive project planning. Such surveys could in fact be undertaken on the spur of the moment, by simply driving to the survey site and deploying "roving" RTK-GPS equipment "out of the trunk of a car."

The earlier discussion in this chapter drew attention to the complex relationship between the GPS survey technique to be used, standards and specifications to be considered (and crucial issues that arise), and quality control procedures. Additional planning issues that should be addressed (if appropriate) include:

- The receiver deployment strategy to be used
- The placement of survey marks
- The datum for the survey and the occupation of existing control stations

- The transformation of the height information
- Site reconnaissance
- Sky visibility

12.5.1   RECEIVER DEPLOYMENT

There are a number of possible receiver deployment schemes that can be used. Each has its advantages and disadvantages with respect to logistical considerations such as cost, time, and manpower. In general, some reference station (or stations) is (are) occupied by the surveyor for all or some of the project, or use is made of CORS operated by a government agency or private company (capable of supporting either real-time or post-processed modes of GPS survey). The latter is a very attractive option that is contributing to the increased popularity of GPS among geospatial professionals, as it means they do not need to deploy and operate their own reference station—hence saving on costs and project complexity. The survey requires field receivers that either move between static sessions to predefined points, or ground marks may be occupied on a random basis during *RTK-GPS*, *kinematic*, or *stop-and-go GPS* surveys. A combination of the "star" or "radial mode" and the "leap-frog" or "traverse" mode, as shown in Figure 12.1, is usually used. *Static GPS* or *rapid-static GPS* surveys in which pairs of receivers are operated typically use the "traverse" mode of deployment if no CORS are available in the survey area. *RTK-GPS*, *kinematic*, and *stop-and-go* techniques, or those that use the nearest CORS as the base, generally favor the "star" mode of deployment as only the "roving" receiver moves during the survey.

Care should be taken when using the "star" mode of survey because, unless all the points are visited more than once, all *radial* baselines are in reality "nonredundant" or "no check" baselines. *It must be emphasized that single occupations of surveyed points are all nonredundant.* It could be argued that a two-reference station configuration has natural *built-in* redundancy (there are two independent baselines coming into each point, one from each of the reference stations). However, although the point is no longer fixed by a single "nonredundant" radial baseline, the additional baseline from the second reference station *may* not be considered truly independent as, for example, the same error in the height of antenna or station misidentification can still be made.

12.5.2   PLACEMENT OF GROUND MARKS

One of the advantages of the GPS survey technique over conventional surveying techniques is that surveyed points may be placed *where they are desired* (*subject to sky visibility*), irrespective of whether intervisibility between ground marks is preserved. Generally, the GPS points would be expected to be clustered around the project *focus*, for example, a road, dam, powerline corridor, construction site, etc. This is in contrast to a traditional (pre-GPS) geodetic control network that was geographically evenly spaced, and the control stations were at prominent locations such as at the tops of hills to ensure that they were visible from afar. Nevertheless, some intervisibility of stations may be necessary to permit the use of conventional

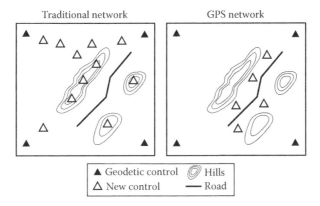

FIGURE 12.2   Terrain need not influence point selection.

terrestrial survey techniques as well. In addition, extra survey stations that "carry in" the *control* or *datum* from the nearest geodetic control stations to the project area are not necessary for GPS work. *Hence, insisting on evenly spaced stations or the selection of stations on the basis of terrain are no longer important considerations* (Figure 12.2). Monumentation guidelines are usually provided by the national or state geodetic agency.

12.5.3   DATUM ISSUES

Of critical importance at the planning stage is to ensure that the *datum* (Chapter 3) to which the results of the survey are related is the one required by the client. It may be necessary to execute a final transformation to change the surveyed coordinates to the required datum and coordinate type (e.g., E, N, U or lat, long, height) using the appropriate transformation parameters (Chapter 4). The number and distribution of "known" control stations, and the accuracy with which the coordinates of the known stations are required, is strongly dependent on the use to which these control stations will be put. Guidelines can usually be found in the relevant GPS S&S. For most purposes, three or four control stations located around the perimeter of the project area is sufficient. Increasingly CORS networks play that function, as they provide a direct connection to the geospatial reference frame or geodetic datum, as well as being actively monitored (and thus have coordinates more "assured" than an unmonitored passive ground control point).

12.5.4   SURVEYING EXISTING CONTROL STATIONS

Some reasons for surveying both existing (i.e., "known") control stations and new points are the following:

- Required by the relevant GPS survey standards and specifications
- For the determination of local transformation parameters between datums
- For quality control purposes

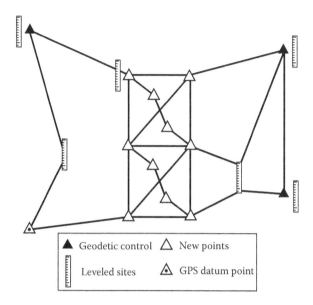

FIGURE 12.3   A GPS network linking existing control stations, new sites, and a GPS datum station.

- To determine how the coordinates of the "known" stations have changed since last being surveyed
- In order to determine the transformation between a leveling-based (orthometric) and GPS-based (ellipsoidal) height datum
- In order to connect new GPS points into the surrounding control network, however, a minimum of one known station must be used as the datum station in the GPS survey.

The control station coordinates may be in the GPS satellite datum, the national geospatial reference frame (e.g., in the United States, the NAD 83), or a local system. Furthermore, perhaps only the horizontal coordinates are known, or their leveled height, or both, as indicated in Figure 12.3.

12.5.5   HEIGHT REDUCTION

It may be necessary to convert the GPS ellipsoidal heights to orthometric heights (Section 13.2.4). This reduction requires that *geoid* information for the local area be available, or at the very least there are some stations that are surveyed in the project area, which also have orthometric heights (likely from leveling within the vertical datum of the region, which may or may not agree with the actual geoid). If several stations have leveled heights, and these points are surveyed using GPS to determine their ellipsoidal heights as well (relative to the datum station), then the difference in the two heights at these stations is a local estimate of the geoid–ellipsoid separation, or geoid height if the vertical datum's "zero surface" is actually the geoid. In the United States, the *GEOID09* model is currently recommended for this purpose (NGS 2009b), but this model is what is known as a "hybrid geoid," where the actual model is not the geoid–ellipsoid separation but is instead the separation between the ellipsoid of the national geometric datum, NAD 83, and the zero height surface of the national vertical datum, NAVD 88. This distinction is critical, as very few national vertical datums actually use the geoid as their zero height surface.

12.5.6 SITE RECONNAISSANCE

Reconnaissance issues relevant to GPS surveying include:

- Satellite availability: satellite selection, satellite health, observation window, etc.
- Satellite visibility: checking on-site obstructions, foliage, and trees
- Clearly identifying the ground mark over which the GPS antenna is to be set up
- Identifying, if necessary, eccentric stations to be occupied if the primary ground mark cannot be used, and other azimuth stations if required
- Station access: critical for minimizing travel times and unscheduled delays in getting on-site
- Site conditions: on-site power? multipath environment?

The information on station access should be clear and unambiguous. This is critical for minimizing down-time due to difficulties in finding stations, or if access involves the assistance of caretakers (e.g., to visit the roofs of buildings), etc. Depending on the objective of the project, and its accuracy requirements, this task may be quite elaborate and include:

- Investigating the provision of on-site power
- Testing soil stability and defining the appropriate antenna mount (tripod, pillar, etc.)
- Selecting sites that avoid potential multipath causing structures
- Selecting sites that are not too close to UHF, TV, radio, microwave, cellular phone, or radar transmitters, as they could affect a GPS receiver's operation (perhaps >50 m distant)
- Establishing permanent monumentation—using previous marks helps avoid this
- Establishing nearby azimuth marks
- Clearing the area of possible obstructions caused by trees or shrubbery
- Consulting satellite image catalogues or maps, including using Google Earth, Google Maps, and similar Web sites
- Taking photographs of the surrounding area, including any tree cover

In some countries, account may have to be taken of the season. For example, tree cover may be much thicker during summer than at other times of the year.

12.5.7 ASSESSING SKY VISIBILITY

To schedule a GPS survey, the following factors need to be taken into account:

- Satellites are not normally tracked below an elevation of 10° due to the larger atmospheric refraction errors at low elevation angles.
- The GPS is always available, and hence surveys can even be conducted at night.
- The length of observation session depends on the GPS technique being used (Tables 12.1 and 12.2).
- The satellites' positions in the sky are predictable. They can be computed and output in a convenient graphical form, and taken out into the field during reconnaissance.
- Satellite availability is best visualized in the form of a *skyplot* a graph of satellite tracks on a zenithal projection centered at the GPS ground station (the satellite azimuths and elevations are shown as a function of time)— which can be computed using standard GPS mission planning software or using online web tools (e.g., Trimble 2009).

This chapter was not intended to be a "cookbook" for executing GPS surveys. The aim was to draw attention to the various factors that should be considered when planning and executing GPS projects. The next chapter will illustrate how many of the issues raised in this chapter are addressed by U.S. surveyors in typical GPS survey and mapping applications.

REFERENCES

Craymer, M.R. and Beck, N., 1992, Session versus baseline GPS processing, *Proceedings of the ION GPS-92, Fifth International Technical Meeting of the Satellite Division of the U.S. Institute of Navigation*, Albuquerque, NM, September 16–18, pp. 995–1004, reprint available at: ftp://geod.nrcan.gc.ca/pub/GSD/craymer/pubs/gps_ion1992.pdf

DoD, 2008, *Global Positioning System Standard Positioning Service Performance Standard*, 4th edn., U.S. Department of Defense, Washington, DC, September 2008, 160pp, available at: http://www.navcen.uscg.gov/gps/geninfo/2008SPSPerformanceStandardFINAL.pdf

FGCC, 1989, *Geometric Geodetic Accuracy Standards and Specifications for Using GPS Relative Positioning Techniques*, Federal Geodetic Control Committee (now the Federal Geodetic Control Subcommittee of the Federal Geographic Data Committee), National Ocean & Atmospheric Administration (NOAA), version 5, dated May 11, 1988 and reprinted with corrs. August 1, 1989, 48pp.

FGDC, 2009, *Geospatial Positioning Accuracy Standards, Parts 1–5*, Federal Geographic Data Committee, Washington, DC, available at: http://www.fgdc.gov/standards/projects/FGDC-standards-projects/accuracy/

GSD, 1992, *Guidelines and Specifications for GPS Surveys*, Geodetic Survey Division, Geomatics Canada, Natural Resources Canada, Ottawa, Canada, release 2.1, 63pp, available at: http://www.geod.nrcan.gc.ca/publications/pdf/guidelinesspecifications.pdf

Gurtner, W., 1994, RINEX: The receiver-independent exchange format, *GPS World*, 5(7), 48–52.

ICSM, 2006, *Standards and Practices for Control Surveys*, Australian Inter-Governmental Committee on Surveying and Mapping, version 1.6, special publication no. 1, available at: http://www.icsm.gov.au/icsm/publications/sp1/sp1v1–6.pdf

LINZ, 2003, *Accuracy Standards for Geodetic Surveys*, Land Information New Zealand, New Zealand, version 1.1, 27pp., available at: http://www.linz.govt.nz/geodetic/standards-publications/standards/index.aspx

Milbert, D., 2008, Dilution of precision revisited, *Navigation*, Spring 2008, 55(1), 67–81.

NGS, 1997, *Guidelines for Establishing GPS-Derived Ellipsoid Heights (Standards: 2 cm and 5 cm)*, National Geodetic Survey Tech. Manual NGS-58, Silver Spring, MD, available at: http://www.ngs.noaa.gov/PUBS_LIB/NGS-58.html

NGS, 2008, The NGS 10 Year Plan, http://www.ngs.noaa.gov/INFO/NGS10yearplan.pdf

NGS, 2009a, National Geodetic Survey Online Positioning User Service, http://www.ngs.noaa.gov/OPUS/index.html, accessed April 5, 2009.

NGS, 2009b, The NGS Geoid, http://www.ngs.noaa.gov/GEOID/, accessed April 5, 2009.

Schinkle, K., 1998, A GPS how-to: Conducting highway surveys the NYSDOT way, *GPS World*, 9(2), 34–40.

Teunissen, P.J.G., 1995, The least-squares ambiguity decorrelation adjustment: A method for fast GPS integer ambiguity estimation, *Journal of Geodesy*, 70(1–2), 65–82.

Trimble, 2009, Planning software. http://www.trimble.com/planningsoftware.shtml, accessed April 5, 2009.

13 Carrying Out a GPS Surveying/Mapping Task

Chris Rizos, Dru Smith, Stephen Hilla,
Joe Evjen, and William (Bill) Henning

CONTENTS

GPS works differently from the techniques based on optical–electrical instrumentation that surveyors have used for decades. Chapters 11 and 12 have dealt at length with the GPS techniques available to the surveyor/mapper, with particular emphasis given to issues that influence the *quality* of outcomes. Hence, in addition to discussing the characteristics and limitations of the various GPS techniques, the authors have also dealt with such topics as GPS standards and specifications, and best practice guidelines for planning and executing successful GPS surveys. In this chapter, the lessons of earlier chapters are applied to some typical GPS surveys. Although the examples are drawn from the United States, many of the principles have general relevance to the practice in other countries.

A GPS surveying project usually begins with a client approaching a geospatial professional to generate a set of coordinates or to prepare a map using the GPS technology. In the United States, most states require that GPS measurements that are used to provide geographic coordinates for surveys, mapping, or for GIS databases must be performed under the direct supervision of a licensed land surveyor in the state where the work is being performed. Four GPS surveying/mapping tasks are examined in this chapter:

- Using GPS to establish passive geodetic control networks
- Using GPS for ground mapping or engineering projects
- Using GPS for GIS data capture
- Using GPS to survey heights in the local vertical datum

Before proceeding further, it is necessary to consider in a little more detail the GPS standards and specifications (S&S) for surveys in the United States.

13.1 U.S. NATIONAL GPS STANDARDS AND SPECIFICATIONS

In the United States, there is a clear distinction between *standards* and *specifications*. Standards are tied to accuracy. Specifications refer to specific survey methodology. There is therefore no true "GPS standard," although the Federal Geodetic Control Commission (FGCC) (now the Federal Geodetic Control Subcommittee of the Federal Geographic Data Committee) did produce a document that defined *both* the geodetic accuracy standards and recommended GPS specifications (FGCC 1989). The geospatial positioning accuracy standards are now described in a series of documents, available on the Federal Geographic Data Committee Web site (FGDC 2009), which define positional accuracy in relative terms, either in relation to neighboring ("directly connected") control points ("local accuracy" or "relative precision") or with respect to the *datum* ("network accuracy") realized by, for example, permanently operating GPS reference stations or other geodetic control points (see below). Despite their rigor, these standards are vague in their definitions (such as stating that CORS are effectively errorless or stating that "directly connected" points are "local," without considering that GPS allows extraordinarily long baseline lengths to "directly connect" points). The National Geodetic Survey (NGS) is the federal agency responsible for establishing and maintaining the National Spatial Reference System (NSRS)—including the national database of geodetic control stations and other geospatial data sets (NGS 2009a)—a realization of which is the coordinates of both CORS stations and passive geodetic control stations determined in a nationwide network adjustment—the latest being "NAD 83 (NSRS2007)." The *core* group of passive stations that were included in the NSRS2007 adjustment were part of several statewide High-Accuracy Reference Networks (HARNs), which were surveyed by NGS and other state, local, and private agencies, using GPS techniques from the late 1980s through to the early 2000s. Another realization of the NSRS will come from periodic updates to the CORS coordinates through multiyear solutions that will provide the position and velocity of the CORS.

National GPS specifications have been developed for geodetic class surveys, but are over 20 years old (FGCC 1989). In many states, the state's Department of Transportation (DOT) has adopted its own GPS specifications. Many of these include specifications for GPS positioning (see, e.g., Schinkle 1998). Some DOTs, for example, do not accept elevations produced by GPS and require that heights be determined using conventional differential leveling techniques. While many municipalities, states, and even federal organizations in the United States have established surveying standards for cadastral surveys, most have not defined GPS specifications for such surveys. On the other hand, the U.S. Forest Service and the Bureau of Land Management have put together a set of guidelines for using GPS to carry out cadastral surveys related to the Public Land Survey System of the United States (FS/BLM 2001). Therefore, it is the surveyor's responsibility to use GPS specifications that meet the positional requirements of a project. Generally, by default, the national specifications for GPS positioning techniques are applicable.

However, as GPS has matured as a tool of choice for geodetic surveys, the need for specifications on establishing a network of passive control stations has fallen away. Surveyors who need to position a mark within the NAD 83 datum have tools such as OPUS (the Online Positioning User Service) offered by the National Geodetic Survey (NGS 2009b), which yield coordinates that are tied to the NSRS, by using absolutely no passive ground control points. This break from the traditional need for passive control to "bring in the datum" (Section 12.5.2 and Figure 12.2) is more fully outlined in the NGS 10 Year Plan (NGS 2008a), where tools such as "OPUS-GNSS" and CORS will mitigate much of the need to establish a local geodetic control network of passive stations around areas of interest.

This should not be taken as a statement that passive control itself is not needed. The NGS 10 Year Plan clearly articulates that passive control will continue to be used as a way to monitor changes to the environment, as well as provide stepping-off points for surveys that require traditional methods of traversing. But because passive control will be "secondary access" to the NSRS in the future, while CORS will be "primary access," the message is clear—the specifications that lay out how to establish a network of passive control around an area of interest will soon be, if not already are, out of date.

In other words, what this means is that in the United States, like many other countries, the trend is, wherever practicable, to get coordinates by setting a GPS receiver over a point of interest and tying to a perpetually monitored CORS network—not by surveying of a passive mark and using precomputed (and potentially out of date) coordinates of that mark.

National geodetic authorities typically use a variety of terminology for the *classes* of GPS survey and have different numerical accuracy limits to define these categories (Section 12.2). In the United States, the term "order" (rather than "class") was historically used to describe the procedures used, and therefore accuracy expected, in a geodetic survey to establish passive control points. For the most part, this terminology is growing in obsolescence in the United States as the need for campaign-style GPS surveys to establish networks of passive marks are becoming less and less needed for accessing the NSRS. As a brief primer, below is a description of how the highest orders of surveys were performed in order to establish what were called (at different dates) the High-Accuracy Reference Networks (HARNs) or Federal Base Networks (FBNs).

13.1.1 HARNs/FBNs

HARNs/FBNs generally correspond to the 1 cm and 2 cm accuracy standards (FGDC 2009). Such surveys are intended mostly for establishing and monitoring passive marks to measure crustal movement (due to tectonics, subsidence, glacial isostatic adjustment, or other factors), and for various scientific purposes. In the past, such surveys were used to coordinate the HARN stations at the federal and state levels. Surveyors established these networks by following the specifications originally published by the FGCC (1989). Those specifications were very strict, requiring, for example, dual-frequency receivers, long static observation sessions, the measurement of redundant baselines, and strict guidelines for monitoring weather conditions and other factors.

13.1.2 ORTHOMETRIC HEIGHTS

NGS S&S for obtaining orthometric heights exists for optical leveling techniques using either conventional geodetic levels or the digital electronic barcode level:

* Geodetic Leveling: NOAA Manual NOS NGS 3 (August 1981)
* Interim FGCS Specifications and Procedures to Incorporate Electronic Digital/Bar–Code Leveling Systems, Version 4.0 (July 1994)

In 2008, NGS issued guidelines on obtaining orthometric heights using GPS techniques (NGS 2008b), but not specifications. Those guidelines, however, were built around the concept of propagating differential orthometric heights in a region, from known benchmarks ("GPS replacing leveling") but they also strongly encourage a different method of simply removing a "hybrid geoid" from ellipsoid heights, which then yields orthometric heights in the national vertical datum, the NAVD 88. The latest hybrid geoid model in the United States is GEOID09 (NGS 2009c). Though this geoid model is claimed to be accurate to a few centimeters, it is important to confirm geoid model performance by surveying several benchmarks that have orthometric height values, following the procedure described in Section 13.2.4. An alternative to using a geoid model is to determine the offset between the GPS height datum and the leveled height datum using the *geometric* technique described in Section 13.2.4.

13.2 SURVEYING/MAPPING APPLICATIONS: GUIDELINES FOR GPS

13.2.1 APPLICATION 1: USING GPS TO ESTABLISH PASSIVE GEODETIC CONTROL NETWORKS

In 1987, the NGS began establishing state-by-state HARNs for the U.S. states and territories, with a station spacing of about 75–125 km (the spacing varies from state to state). At present there are over 3700 HARN stations, all having horizontal and vertical (orthometric and ellipsoidal) values to the highest accuracy standard (FGCC 1989). However, while these stations are (currently) important for the nation's geodetic framework, they are impractical for everyday survey applications because of the long distances between stations.

In many states, the NGS, in coordination with the state DOT or the state geodetic agency, *densified* the HARN to produce station spacing of the order of 25–30 km. Many of the densification projects were performed to either the old Order A or Order B specifications (or 1–2 cm according to the new standards). The coordinates generated by such surveys were submitted to the NGS for incorporation in the NSRS, a process generally referred to as *"bluebooking."* (As a reminder, the NGS is moving toward a realization of the NSRS, which will not have passive marks "in the NSRS" but rather "tied to the NSRS.") When a GPS project is "bluebooked" (NGS 2009d), the stations and data associated with them are maintained by the NGS, for possible inclusion in any future readjustment of the passive national network (see Section 13.4.3). However, considering the trends toward purely virtual networks, as laid out in the NGS 10 Year Plan, the need for submitting and adjusting decades of GPS data on passive control is becoming less and less important in the realization of the NSRS.

Over the past decade, many county and municipal governments have implemented Geographic Information Systems (GISs) (see Part IV) to manage their geospatial information. In most cases, a passive GPS control network has been established to provide the coordinate foundation. Although creating a passive geodetic control framework is a significant cost, it is relatively insignificant compared to the cost of creating a complete GIS database. However, the GPS survey techniques used to establish this framework should be accurate enough to adequately support current and future applications of GIS. If the GIS will be a long-term geospatial information management system, then the passive geodetic control network should be designed to support such an effort, but care should be taken not to rely entirely on passive ground control points that can move, be disturbed, or destroyed. Considering the trends toward RTK and network-RTK, the need for a dense passive control network may ultimately be alleviated through purely virtual means, provided adequate sky coverage is available. Also, with the advent of further GNSS constellations (GALILEO, GLONASS, COMPASS, and others—Sections 15.4 and 15.5), it may be that the "urban canyon" argument against GPS is partially alleviated in the future by the number of GNSS satellites in the sky.

Static GPS surveying is currently the primary method used by the National Geodetic Survey for establishing passive geodetic (ellipsoidal) control networks—currently through "bluebooking" but in the future through tools such as OPUS. Typically, what is defined is

1. The GPS receiver and antenna hardware to be used
2. Guidelines concerning maximum and minimum interreceiver distances, ancillary measurements, type of monumentation, length of observation sessions, testing and validation procedures to be followed, etc.
3. The GPS network design principles to be followed, including the number of repeat baselines to be measured, the extra (redundant) baselines to be measured, the number and type of control stations to be connected to, and so on
4. The data processing software and options that must be used
5. The documentation and reporting guidelines to be followed

Figure 12.3 represents the design of a static GPS survey for the establishment of a passive control network, where it can be seen that there are several *redundant* baselines that have been measured. The final result will be a set of coordinates for the stations generated either by a least-squares adjustment of the individual baselines, or by applying a session-by-session processing strategy, or even by scientific GPS software that can process the data from the entire campaign in a single step adjustment. The processing of individual baselines, and their combination into geometric network adjustments, is the standard methodology employed by commercial software (Section 13.4.2) (although it is unknown to what extent this is done rigorously).

13.2.2 Application 2: GPS for Mapping or Engineering Projects

Mapping projects are often performed to support a city- or county-wide GIS, and may be carried out using airborne or terrestrial techniques. As digital photogrammetry and airborne laser scanning nowadays rely on direct "georeferencing," there is minimal use of ground control points. However, the onboard GPS is used in the high-accuracy *kinematic positioning* mode (Section 11.1.3), and hence the local GPS reference stations (set up on passive ground marks, or based on active CORS) must be used to control the horizontal and vertical accuracy of the imaging sensors. A similar technique is used for the determination of the trajectory of a vehicle-mounted mobile mapping system (MMS), which travels along roads in order to map their centerlines and/or to coordinate points of interest that can be observed from the road vehicle (such as traffic signs, road assets, street frontages, and so on). The MMS consists of one or more digital cameras (to permit the photogrammetric operation to be carried out on the terrestrial images), and the navigation subsystem for georeferencing of images, consisting of carrier phase-tracking GPS receiver (for sub-decimeter accuracy positioning) and an Inertial Navigation System (INS) to determine the attitude (or tilt angles) of the imaging subsystem. In both cases, increasingly use is made of CORS (Section 14.4.1), and all coordinate results are obtained from post-processing. The accuracy requirements in turn are mainly dependent on the mapping scale of the project and the national mapping accuracy standards (FGDC 2009, Part 3).

Planimetric mapping with contours for engineering applications (often referred to as "detail surveys") may require the coordination of many points. This is most economically done by using pre-established control network stations established using static or rapid-static techniques or CORS (if they are nearby), and to then densify this network using the post-processed kinematic or real-time GPS techniques (Section 11.1.2). If dual-frequency GPS receivers are used, ambiguity resolution is possible "on-the-fly" (Section 11.1.4), so that, for example, ambiguities can be resolved while the surveyor is moving from point to point, even if the satellite lock is lost or interrupted during travel. In this mode, the surveyor travels from point-to-point collecting enough data at the ground marks to achieve the results desired. The following comments may be made with regard to these surveys:

- The NSRS horizontal and vertical control should be used as the basis of the control project. This relates the local control network to the national reference system.
- Control stations (or benchmarks in the case of vertical surveys) should be evenly distributed across the region. A good practice is to strive to have control points in a minimum of three of the four compass quadrants.
- Static and rapid-static networks should be used to establish the local passive control network upon which the kinematic survey (post-processed or real-time) is based. Where GPS specifications are not clear on such matters as observation session length (as a function of baseline length and number of observed satellites), the GPS manufacturer's recommendations are usually sufficient.
- It is a good practice when performing rapid-static surveys for a minimum of 50% of the points to be observed two times. In the case of kinematic surveys, perhaps two base stations should be used. The use of fixed height tripods, or bipods, is recommended in order to minimize the possibility of antenna height errors.
- In the case of post-processed kinematic surveys, it is recommended that several reference stations be used, and to compute the baselines independently, as indicated in Figure 13.1, then compare the coordinate values calculated from the different reference stations to check the repeatability of these measurements.
- RTK-GPS is best suited for surveying a large number of points in a relatively small area (Section 11.2.2). It allows the surveyor to obtain coordinate

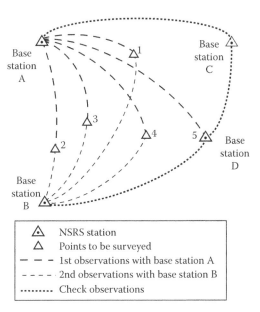

FIGURE 13.1 Performing a post-processed kinematic GPS survey for establishing ground control for engineering applications using two reference stations (for checking).

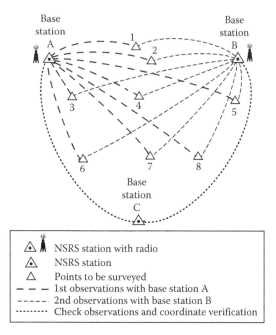

FIGURE 13.2 Performing an RTK-GPS survey for establishing ground control for engineering applications using two reference stations (for checking).

values immediately in the field. When linked to a pen-based computer or PDA, and a mobile CAD package, RTK-GPS allows the surveyor to create a finished mapping product while in the field.

- In order to quality check RTK-GPS surveys, coordinates should be established using one reference station and then be observed a second time using a second RTK reference, as illustrated in Figure 13.2. Although using two simultaneous RTK reference stations is an acceptable check (if the RTK-GPS equipment has the facility to switch between the two transmitting stations), if possible the surveyor should observe the same station several hours later to minimize the chances of introducing systematic errors.
- NGS has produced draft guidelines for RTK-GPS (NGS 2008c).

13.2.3 APPLICATION 3: USING GPS FOR GIS DATA CAPTURE

The differential GPS (DGPS) positioning technique is frequently used for GIS data capture. DGPS is ideally suited for determining coordinates of points of interest, with accuracy in the range of sub-meter to a few meters, using very little data—typically, the station occupation time is less than 30 s. The technique is robust, with no need to solve for any ambiguities. This means that the satellite lock can be lost frequently without the need to reinitialize, making it easier to coordinate points near tall buildings and under heavy tree canopies. Either the Local Area DGPS or the Wide Area DGPS technique may be used (Section 11.3.3), and there are several commercial and free-to-air services available to users (Section 14.3). Although it is possible to use DGPS techniques in the post-processed mode, the majority of DGPS applications are addressed in real time using an appropriate communications data link to carry the DGPS correction messages to users. WAAS is a free-to-air service that does not require a special communications device to access the DGPS correction messages (Section 14.3.1.2). While DGPS can yield sub-meter coordinate results, the height information obtained is typically of lower accuracy (perhaps only of the order of a few meters, or worse). The data collected typically includes point, polygon, or line attribute information suitable for populating GIS database fields.

One common GIS-type application is to support *asset management* and *utility maintenance*. DGPS is very cost-effective when compared with rapid-static GPS, or RTK-GPS. Besides saving time and being less logistically challenging, DGPS systems are less expensive than carrier phase-based equipment. In addition, the equipment is smaller and the GPS receiver can be linked to pen-based computers to create a system ideally suited for rapid GIS-type mapping. Other devices can be added to this mobile mapping system, including laser distance and compass direction measuring devices, barcode scanners, digital cameras, and video. The same system may be used later to perform maintenance and work order management. This can be conducted either at a central docking station or through wireless technology. Instructions on location and work to be performed are sent directly from the central office to maintenance crews in the field. The maintenance crew, using DGPS linked to a mobile computer, can navigate to the location, access the work order, and perform the necessary work. Once the work has been completed, the maintenance crew can then transmit the information back to the central office, where the GIS database is updated, and the crew can then navigate to the next assignment.

13.2.4   Application 4: Using GPS to Determine Heights

Static and rapid-static GPS techniques are capable of establishing the relative *ellipsoidal height* between two simultaneously tracking GPS receivers to a few centimeters accuracy. The accuracy of height difference determination using kinematic or stop-and-go techniques is, however, a little worse. For many engineering and mapping applications, the height quantity required is the *orthometric height*. The relationship between orthometric height and ellipsoidal height is represented by the simple formula (Hofmann-Wellenhof et al. 2008): $H = h - N$, where H is the orthometric height, h is the ellipsoidal height, and N is the geoid height (or the geoid-ellipsoid separation). In the GPS case, the relative height quantities are of interest, that is, $\Delta H = \Delta h - \Delta N$, where Δ represents the *difference* in the respective height quantities between the reference station and the survey point.

The quantity ΔN can be derived from two sources:

1. Using a geoid model such as *GEOID09* in the United States (NGS 2009c). It should be emphasized that GEOID09 allows for the transformation between the ellipsoid height in the official geometric datum, the NAD 83, and the official geopotential datum, the NAVD 88, *even though both datums are known to have systematic errors*. It is common in many countries that the

errors in official datums gradually become well defined, but the process of updating a datum can be very slow and arduous. (A more accurate geoid-ellipsoid separation model would be USGG2009 that makes no attempt to be "beholden" to the official datums of the United States.) The NGS, recognizing that decimeter-to-meter level errors in both NAD 83 and NAVD 88 should not be tolerated, has proposed a plan to update the official datums of the United States, obviating the need for two different sets of geoid models (NGS 2008a).

2. If there is no official geoid model, estimating the difference from surveys of several points that have heights in both systems, and then interpolating to other surveyed points.

Typically, the accuracy of the geoid height difference is a few centimeters over baseline lengths of the order of a few tens of kilometers (perhaps a little worse in mountainous areas). Hence, the accuracy of the GPS-derived orthometric height difference may only be 2–10 cm. However, as new gravity missions (CHAMP, GRACE, and GOCE) are launched, and new geoid theory is developed, the differential geoid height accuracy is expected to reduce.

The latter method may be applicable in countries that do not have high-quality geoid model information. It has several advantages in that it does not require an explicit knowledge of the geoid-ellipsoid separation, but derives that information— that is, the *difference* between the ellipsoid height datum of the GPS network and official leveled (orthometric) height datum (usually based on Mean Sea Level at tide gauges)—from those GPS surveyed stations that have heights in both systems. That is, the GPS heights are related to the adopted ellipsoidal height of the datum station (or stations) held fixed in the baseline survey (Section 9.4.1), but the orthometric heights have been independently determined using standard leveling techniques. If there is a systematic offset between these two datums, this *geometric* technique will be able to accommodate this.

The implementation of the above geometric technique for correcting GPS-derived ellipsoidal height differences to heights on the local orthometric height datum is as follows (this technique is similar to that called "Height Modernization Surveys" in the United States—NGS 2008b):

1. A number of benchmarks with leveled (orthometric) heights are identified and included in the set of points that are surveyed using GPS. Ideally, these points should be well distributed across the area, as well as having some of them around the perimeter of the survey area.

2. As a result of the GPS survey, the ellipsoidal heights of the abovementioned benchmarks are determined. *The two sets of heights permit the offset between the orthometric and the ellipsoidal height datums to be determined.* If the orthometric/leveling datum is coincident with the geoid, then the offset is identical to the geoid height. On the other hand, if the leveling datum is not the geoid, then the offset cannot be interpreted as the geoid height but fulfills the function of a correction—from the GPS height datum to the local leveling datum.

3. The offset will typically *vary* from benchmark-to-benchmark, but the pattern of offsets can be used to predict the height correction at other points where only the GPS height has been determined. This spatial offset pattern can be realized in the form of a simple model, for example as a plane of best-fit through the benchmark corrections. *The height correction is then interpolated at any other point in the survey area using this model.*

This geometric technique is a *pragmatic* solution to the problem of how to convert heights on the GPS datum to heights consistent with the local leveling datum. As geoid models and GPS processing improve, it may become less desirable to disseminate the old benchmark based datum through GPS, but rather to establish new heights directly without relying on benchmarks with questionable height values.

13.3   ON-SITE PROCEDURES

The objective of well-designed procedures is to ensure that the survey is carried out according to plan, in order that good quality data is collected, from which coordinates results are generated to the required level of accuracy and with the appropriate QC measures. Obviously, the higher the survey accuracy requirements, the more elaborate are the field and office procedures. In this section, brief comments are made about issues that impact the efficient execution of GPS fieldwork.

13.3.1   ANTENNA SETUP AND HEIGHT MEASUREMENT

Antenna setup and height measurement errors are, unfortunately, very common. The following are some procedures that minimize the chances of making such errors:

- The antenna normally bears a *direction indicator* that should be oriented in the same direction at all sites using a compass. This ensures that any antenna center offsets (as measured from the mechanical center to the electrical phase center—Section 10.1.3 and NGS 1998) will propagate into the baseline solution (mark-to-mark) in a systematic manner (Schmid et al. 2007).
- The same antenna, receiver, and cabling should be maintained together in a "kit."
- Because of the high precision of GPS surveys, the *centering of the antennas is important*. Optical plummets should be regularly calibrated, as should bulls-eye (or circular) bubbles on fixed-height poles.
- The antenna assembly should be mounted on a standard survey tribrach with an optical plummet, on a good quality survey tripod. Fixed-height poles (usually 2 m) eliminate the need for slant-height measurements.
- Setting up on a geodetic pillar is, of course, reasonably effortless and to be preferred.
- Care must be taken with *antenna height measurement*.

As the latter is probably the most critical of all antenna setup operations, some further comments can be made. *The height of the antenna above the station marker, measured to the standard reference point on the antenna housing, should be measured to the nearest millimeter,* and the measurement checked (e.g., by another person, or by measuring the height in both feet and meters). Although different antenna types have different recommendations for height measurement, all antenna height measurements must be carefully noted, preferably with a diagram. In the case of kinematic or stop-and-go surveys, the antenna is usually mounted on a pole or bipod with constant height. Hence, the chances of making a height measurement error are lowered.

13.3.2 FIELD LOG SHEETS

In the case of RTK-GPS and DGPS, all data is electronically recorded for subsequent download. The following is therefore mostly relevant for post-processed techniques. Two approaches may be used: paper based or via the receiver's control and display unit (CDU) (Section 10.3.2). In the former approach, field log sheets are used, on which pertinent information concerning the site being occupied and the data collection process itself is entered. These sheets may be very comprehensive (running to many pages) in the case of the highest accuracy GPS surveys, or simply a few lines on a standardized booking sheet for rapid mapping techniques such as stop-and-go, RTK-GPS, or DGPS. Increasingly such information is entered directly into the appropriate menu on the CDU. However, a log sheet for a *static* GPS survey would typically contain some or all of the following information:

- Date and time, field crew details, etc.
- Station name and number (including aliases, site codes, etc.)
- Session number, or other campaign indicator
- Serial numbers of receiver, antenna, data logger, memory card, etc.
- Start and end time of observations (actual and planned)
- Satellites observed during session (actual and planned)
- Antenna height (several measurements), and eccentric station offsets (if used)
- Weather (general remarks about temperature, relative humidity, wind speed, storm fronts passing, etc.), and meteorological observations (if measured)
- Receiver operation parameters such as data recording rate, type of observations being made, elevation mask angle used, data format used, etc.
- Any receiver, battery, operator, or tracking problems that were noticed

NGS has developed some draft guidelines (NGS 2009e).

13.3.3 ECCENTRIC STATION SURVEY

Unlike conventional ground surveys, GPS techniques require a comparatively clear skyview above the elevation mask angle (typically set at 10°). Sometimes the ground mark that is to be surveyed does not satisfy this condition, perhaps because it is a previously monumented station. In such circumstances, an *eccentric station* may be surveyed by GPS, and the necessary site measurements made using classical surveying techniques in order to determine the coordinates of the required ground mark.

13.3.4 CHECKLISTS

From the *Quality Management* point of view (Section 12.2.2) it is a good practice to develop a checklist of onsite procedures—referred to as Standard Operating Procedures (SOPs)—that must be followed (some of which would require appropriate entries in the field log sheets):

- GPS receiver initialization procedures
- Setup and orientation of the antenna
- Correct cable connection of antenna-to-receiver, receiver-to-battery, receiver-to-radio, etc.
- Double-checking of centering and antenna height measurement
- Receiver startup procedure (e.g., entry of site number, height of antenna)
- Survey of eccentric station
- Temperature, pressure, and humidity measurements (if required)
- Monitoring receiver operation and data recording (see below)
- Field log sheet entries (see above)
- Photographs or sketches of point occupancy (e.g., obstruction diagrams)
- Procedures at completion of session (e.g., communications, data transfer to PC)
- Instructions in the event of receiver problems (i.e., contingency plans)

A GPS receiver can display a lot of information; hence, a checklist concerning the monitoring of receiver operation is useful. This checklist could include monitoring: battery status; memory capacity remaining; satellites being tracked; real-time navigation position solution; satellite health messages; date and time (UTC or local); elevation and azimuth of satellites; signal-to-noise ratios; antenna connection indicator; tracking channel status; and amount of data being recorded. The "ultimate GPS fieldwork sins" are summarized in Table 13.1. The development of SOPs is one way to guard against survey errors.

13.3.5 FIELD OFFICE PROCEDURES

It cannot be overemphasized that data should be processed as soon as possible after the observation session in order to assure the quality of the survey at an early stage. As a prerequisite therefore, all data should be systematically catalogued and archived between observation sessions (if there is time), or at the end of the working day at the very latest. Many problems can be identified at this stage. The following are some typical field office procedures:

- Data handling tasks—*transfer of data from receiver to PC*
- Data verification, backup, and archiving in field office

TABLE 13.1
GPS Fieldwork Blunders

Common Problems	Remedies/Advice
Power loss causing termination of survey	Always have backup power supplies
Cable problems affecting operations	Keep them in good condition
Incorrect receiver operation	Field staff must be trained, and follow appropriate checklists
Data collection not coordinated	Good teamwork, and have well-designed logistical procedures and checklists and reliable communications devices with each observation team
Loss of data after survey	Follow systematic data management procedures, including backups
Setup on wrong station	Reconnaissance and good onsite procedures
Antenna height measurement error	Check, and recheck readings

- Preliminary computation of baseline(s) (in the United States this can also be done using online services such as OPUS–NGS 2009b)
- Preliminary QC checks, such as the inspection of repeated baselines, loop closures, and evaluation of (incomplete) minimally constrained network
- Management of field crews—*develop contingency plans for repeated observation sessions*
- Preparation of campaign report, and maintain reporting to head office and/ or the client

Without data safely downloaded from the GPS receiver and successfully processed by the data reduction software, the survey work should never be considered complete. Therefore: (1) download data from the receiver "ASAP" (GPS receivers have many hours of internal or removable memory, so daily download is a reasonable routine); (2) download to PC hard disk, then to CDs, and then make backup copies; (3) store backup disks separately; (4) label CDs (be ruthlessly systematic in following a disk labeling convention); (5) cross-reference field log sheets to data files; (6) verify data download (e.g., check number and size of files); (7) delete files from receiver memory when data download procedure has been verified; and (8) process data through reduction software. Of course, if using real-time procedures, the above guidelines refer to the computed coordinate results rather than the raw data itself.

13.4   TYING UP "LOOSE ENDS"

There are additional tasks that might be considered to be part and parcel of GPS surveying. Some of them are mentioned here. The reader is referred to GPS textbooks (such as Hofmann-Wellenhof et al. 2008), and to GPS manufacturers' user manuals for further details.

13.4.1 RINEX FILES

RINEX (*R*eceiver *IN*dependent *EX*change) is a data file format devised in 1989 for geodetic applications requiring the international exchange of GPS data sets, gathered during global campaigns, by different brands of receivers (Gurtner 1994). It is now used as the standard exchange and archive format for GPS surveying and precise navigation applications. RINEX Version 2.3, which is able to handle kinematic GPS data as well as GLONASS data, is now in common use, and RINEX 3.0 is able to handle other GNSS data including the new L5 observations (Gurtner 2007). The RINEX format has the following characteristics:

- ASCII format, with a maximum of 80 characters per record
- Phase data recorded in cycles of the L1 or L2 carrier frequency, pseudorange data in meters
- All receiver-dependent calibrations are assumed to have been applied to the data
- Time-tag is the time of the observation in the receiver clock time frame
- Separate measurement, navigation message, and meteorological data file formats

Many GPS receivers will *output* RINEX formatted data files—or at the very least, the binary data files can be converted to RINEX files using utilities such as TEQC (Estey and Meertens 1999). The GPS data processing web services listed in Section 14.4.3 require RINEX data files. Nowadays, commercial GPS software will also permit the *input* of RINEX data files for processing. This has the following implications: (a) data from one brand of receiver may be processed within another commercial software package, and (b) data from surveys using a mixed set of GPS receivers can be processed within one software package.

13.4.2 NETWORK ADJUSTMENT

In general, a GPS survey project involves the use of a small number of receivers to coordinate a large number of points. The area of operations may span distances of merely a few kilometers (as on an engineering site), to several hundred kilometers, or even thousands of kilometers in the case of scientific geodetic surveys. A typical GPS survey, such as for mapping or control densification, involves distances of the order of several tens of kilometers. The survey may be carried out using conventional static GPS survey techniques, or high-productivity techniques such as stop-and-go, RTK-GPS, etc. For moderate to low accuracy GPS surveys (e.g., as for the applications described in Sections 13.2.2 and 13.2.3):

- Points are either surveyed as "no check" baselines from a single reference station (Section 12.5.1), or are surveyed from two or more reference stations.
- The datum for the surveyed points is the reference station(s)'s coordinate(s) that are assumed to be known in the "*GPS framework*," in the United States it is NAD83 (CORS96), or one of the ITRS reference frames (Section 7.2).

- The coordinates of the surveyed points are easily derived from one or more baseline solutions *without* the need for a network adjustment. (Where there are two coordinates derived from two independent baselines, using the GPS survey mode depicted in Figure 13.1 or 13.2, the mean is taken if the difference—which is a QC measure—is not larger than some specified threshold value.)

In the case of surveys for setting up a passive control network *linked* to the national geodetic framework (i.e., the CORS in the United States), or which is a *densification* of the national geodetic framework (Section 13.2.1), a network adjustment is generally necessary, unless rigorous multisession (even multiyear!) GPS data processing using scientific software is carried out. The following comments may be made with respect to this *secondary* operation:

- There must be a sufficient number of *redundant* baselines to warrant a network adjustment.
- Redundant baseline surveys are usually mandated in national S&S.
- The various baselines may be connected to each other and to the surrounding passive geodetic control network in a complex manner (e.g., see Figure 12.3).
- Several existing passive geodetic control stations are usually surveyed, and hence are linked to the GPS survey (Section 12.5.4).
- The network adjustment may require only one control station's coordinates to be held fixed (in a so-called *minimally constrained adjustment*), or several of them are held fixed, or nearly so (which is usually the case for densification projects). The handling of the existing errors in pre-surveyed passive control is critical. The propagation of errors should be well documented, including statistical assumptions made (e.g., fixed vs. stochastic control).
- The network adjustment requires all the baseline or session (multi-baseline) solutions, and their associated variance–covariance matrices, as *input*—the output is an adjusted set of coordinates for the network, in the datum defined by the fixed control station(s), and the resulting quality information from the final variance–covariance matrix.
- Network adjustment capability is usually provided as part of the commercial GPS data processing software package.

13.4.3 NGS Bluebook Submission

Currently, the National Spatial Reference System (NSRS) in the United States contains both active and passive geodetic control. The future of geodetic control in the United States is developing along the lines of the NGS 10 Year Plan (NGS 2008a). In that plan, local users will establish passive geodetic control with significantly less involvement by NGS, due to its secondary reliability when compared to active control such as CORS. Historically, passive geodetic control was "part of" the NSRS. In the future, much of that passive control will be considered "tied to" the NSRS and the NSRS itself will be accessed primarily through active control. Until that paradigm shift arrives, however, passive geodetic control does continue to be submitted to NGS for inclusion in the NSRS.

Much of the new passive control is submitted from state agencies wishing to densify their local passive control. In the United States, a GPS survey established as a base layer for a state, county, or municipal GIS, or for general control purposes that users wish to have included in the NSRS, must be submitted to the National Geodetic Survey through a process known as *"bluebooking."* The NGS will only "bluebook" projects that meet or exceed the appropriate Federal Geodetic Control Subcommittee (FGCS) specifications, and that were *preapproved* by the NGS for inclusion. The following steps must be followed before any GPS data is acquired (NGS 2009d):

1. Contact the NGS adviser in the state where the GPS work is being performed (if no state adviser exists, contact the NGS headquarters)
2. Inform the NGS of the project's purpose, and ask for any recommendations or suggestions
3. Submit a preliminary control network station map showing all existing NGS controls
4. Submit a station observation plan for approval
5. Perform GPS observations
6. Submit the final report (computed coordinates, etc.)

After the final report has been submitted, the data is checked by NGS for completeness, whether it adheres to FGCS specifications, and for the quality of the computed coordinates. Once the coordinates have been accepted, the NGS will include the station and coordinate information in the NSRS database (referred to internally as the "NGS IDB"), whose data are then made available to the public in a format known as "datasheets." In addition, these GPS solutions can be included in any future spatial readjustment of the geometric portion of the NSRS.

REFERENCES

Craymer, M.R. and Beck, N., 1992, Session versus baseline GPS processing, *Proceedings of the ION GPS-92, Fifth International Technology Meeting of the Satellite Division of the U.S. Institute of Navigation*, Albuquerque, NM, September 16–18, pp. 995–1004, reprint available at: ftp://geod.nrcan.gc.ca/pub/GSD/craymer/pubs/gps_ion1992.pdf.

Estey, L.H. and Meertens, C.M., 1999, TEQC: The multi-purpose toolkit for GPS/GLONASS data, *GPS Solutions*, 3(1), 42–49, available at: http://facility.unavco.org/software/teqc/teqc.html, accessed April 5, 2009.

FGCC, 1989, *Geometric Geodetic Accuracy Standards and Specifications for Using GPS Relative Positioning Techniques*, Federal Geodetic Control Committee (now the Federal Geodetic Control Subcommittee of the Federal Geographic Data Committee), NOAA, Silver Spring, MD, version 5, dated May 11, 1988 and reprinted with corrs. August 1, 1989, 48pp., available from http://www.ngs.noaa.gov/FGCS/tech_pub/GeomGeod.pdf.

FGDC, 2009, *Geospatial Positioning Accuracy Standards, Parts 1–5*, Federal Geographic Data Committee, available from http://www.fgdc.gov/standards/projects/FGDC-standards-projects/accuracy/.

FS/BLM, 2001, *Standards and Guidelines for Cadastral Surveys Using Global Positioning System Methods*, version 1.0, May 9, 2001, U.S. Forest Service, Bureau of Land Management, Washington, DC, available at: http://www.fs.fed.us/database/gps/documents/GPS4CAD_Stds.pdf.

Gurtner, W., 1994, RINEX: The receiver-independent exchange format, *GPS World*, 5(7), 48–52.

Gurtner, W., 2007, *RINEX: The Receiver-Independent Exchange Format*—version 3.0, ftp://ftp.unibe.ch/aiub/rinex/rinex300.pdf.

Hofmann-Wellenhof, B., Lichtenegger, H., and Wasle, E., 2008, *GNSS Global Navigation Satellite Systems: GPS, GLONASS, Galileo, and More*, Springer Verlag, Wien/New York, ISBN 978-3-211-73012-6, 516pp.

NGS, 1997, *Guidelines for Establishing GPS-Derived Ellipsoid Heights* (*Standards: 2 cm and 5 cm*), National Geodetic Survey Tech. Manual NGS-58, Silver Spring, MD, available at: http://www.ngs.noaa.gov/PUBS_LIB/NGS-58.html.

NGS, 1998, *GPS Antenna Calibration at the National Geodetic Survey*, available at: http://www.ngs.noaa.gov/ANTCAL/images/summary.pdf.

NGS, 2008a, *The NGS 10 Year Plan*, http://www.ngs.noaa.gov/INFO/NGS10yearplan.pdf.

NGS, 2008b, *Guidelines for Establishing GPS-Derived Orthometric Heights*, https://www.ngs.noaa.gov/PUBS_LIB/NGS59%20-%202008%2006%209-FINAL-2.pdf.

NGS, 2008c, *National Geodetic Survey User Guidelines for Classical Real Time GNSS Positioning*, version 2.0.3, September 2008, http://www.ngs.noaa.gov/PUBS_LIB/NGSRealTimeUserGuidelines.v2.0.4.pdf.

NGS, 2009a, National spatial reference system, *and Its Relation to the National Geodetic Adjustment NAD 83*, see for example http://www.ngs.noaa.gov/NationalReadjustment/about.html, accessed April 5, 2009.

NGS, 2009b, *National Geodetic Survey Online Positioning User Service*, http://www.ngs.noaa.gov/OPUS/index.html, accessed April 5, 2009.

NGS, 2009c, *The NGS Geoid*, http://www.ngs.noaa.gov/GEOID/, accessed April 5, 2009.

NGS, 2009d, *National Geodetic Survey "Bluebook"*, http://www.ngs.noaa.gov/FGCS/BlueBook/, accessed April 5, 2009.

NGS, 2009e, *Draft GPS Survey Observation Guidelines*, http://www.ngs.noaa.gov/PROJECTS/GPSmanual/observations.htm, accessed April 5, 2009.

Schinkle, K., 1998, A GPS how-to: Conducting highway surveys the NYSDOT way, *GPS World*, 9(2), 34–40.

Schmid, R., Steigenberger, P., Gendt, G., Ge, M., and Rothacher, M, 2007, Generation of a consistent absolute phase center correction model for GPS receiver and satellite antennas, *Journal of Geodesy*, 81(12), 781–798, doi: 10.1007/s00190-007-0148-y.

14 Servicing the GPS/ GNSS User

Gérard Lachapelle, Pierre Héroux, and Sam Ryan

CONTENTS

Servicing the GPS/GNSS user means adding value to positioning capabilities, principally in terms of accuracy, either in real-time or in post-mission mode. Accuracy can be improved by providing supplementary information that *augments* what is available from the standard GPS (and progressively from GLONASS and other GNSS) broadcast signals and navigation message content. This additional information can be made available in different forms, accessed or disseminated over various channels, and applied by the user in different ways, depending on his or her requirements.

Servicing the GPS/GNSS user also means raising awareness of the technical issues he or she may face, and identifying sources of information and knowledge they may tap into to better meet positioning and navigation needs. This chapter will address these two basic service concerns, using GPS as the exemplar of a GNSS. Reference to GLONASS augmentation services will be made where applicable; however, by and large almost all augmentation services support GPS users only.

14.1 GPS AUGMENTATION

14.1.1 RATIONALE FOR AUGMENTATION

While it is often said that "precision is addictive," the reasons for investing in GPS augmentation services are not limited to accuracy improvements alone. Augmentation services may also address integrity, coverage, reliability, and availability requirements. They will impact service cost and affect, one way or another, positioning and navigation performances. For example, an augmentation service developed for an application where the safety of navigation is a prime concern will pay particular attention to its integrity component, while another service addressing the needs of machine guidance will typically demand more accuracy.

Once the necessary infrastructure for an augmented GPS service is in place (reference receivers, communications links, data exchange protocols, etc.), the critical service parameters to consider are

- *Availability*: The percentage of time that the services of the system are usable.
- *Coverage*: The surface area over which correction messages are accessible and applicable to meet the service's specified level of accuracy.
- *Reliability*: The probability that the system will perform to specifications over a defined time period under prescribed operating conditions.
- *Integrity*: The ability of the system to provide timely warnings to users when the service should not be used.

Accuracy, often perceived as the most desirable feature of a positioning service, can itself be subdivided into two categories, namely, predictable and repeatable accuracy:

- *Predictable accuracy* is a measure of the level of agreement between a position solution and a coordinated or mapped point. Predictability assumes that both the estimated position and the known coordinate value are referenced to the same geodetic datum. This is also often referred to simply as "accuracy."
- *Repeatable accuracy* is a measure of the consistency with which a user can return to a position whose coordinates have been measured at a previous time with the same system. This is also often referred to as "precision."

14.1.2 Characteristics of GPS Augmentation

The needs and requirements for augmented GPS services for various air, marine, and land positioning or navigation applications are defined, for example, in the U.S. Federal Radionavigation Plan (FRP 2008). In general, GPS augmentation services can be classified by the following characteristics:

- Satellite tracking infrastructure (single-station and network)
- Communication channels (ground, space, or both)
- Coverage area (local, regional, and global)
- Targeted application or market (navigation, geomatics, and machine automation)
- The service business model (private, public, and partnership)
- The content and protocol for information dissemination
- The data storage and archival specifications
- The accuracy and robustness of the solution (availability, reliability, and integrity)

With augmentation services combining GPS and communications technologies and responding to markets from local to global scales, they may involve a number of agencies and industries. Some are commercial services, while others broadcast messages as free-to-air services to users. Several services are mature, while others are expanding as user markets develop. Stakeholders in these enterprises include GPS/GNSS manufacturers and communication service providers, agents of their products, government agencies, national professional associations, academic institutions, standards organizations, and many others.

As this book focuses on supporting geospatial applications, this chapter will pursue two main objectives: explain the different approaches used for GPS augmentation and provide the reader with information on where to access additional resources.

14.2 AUGMENTATION OPTIONS

14.2.1 Access Channels

GPS users may access various information sources that can assist them in planning, executing, and analyzing GPS surveys. These sources can provide GPS status forecasts, post-mission or real-time data or products, publicly or privately, and in open or proprietary protocols and formats. The information sources can be generated from GPS tracking stations operated and maintained in the public or private domain, depending on the interests being served. The information may support applications at the local, regional, national, or global scale, depending on system specifications and user requirements. From a practical perspective, the usefulness of a particular augmentation service will be determined primarily by whether or not the information it provides is accessible, applicable at the user's location, and compatible with his or her GPS/GNSS equipment or processing software.

Access to the augmentation sources can be achieved using different communication channels, depending on whether the information is required in real-time or post-mission. Internet connectivity facilitates the acquisition and dissemination of the GPS tracking station data and derived products, both locally and globally. Local and regional wireless data services are also becoming a popular choice in urban areas, while dedicated beacons and radio modems are likely to continue broadcasting corrections to marine users along the coast and waterways, and in rural and remote areas on land. Finally, geostationary communication satellites are well suited to seamlessly distribute corrections over large areas and reach remote regions, both on land and at sea.

While real-time GPS augmentation services have traditionally been categorized as land or space based, reflecting the communication channel used for distribution, this distinction is becoming more difficult to maintain. As real-time Internet use becomes more popular, the land- versus space-based distinction between augmentation systems may vanish, as the Internet does not differentiate between carrier technologies. Nevertheless, the distinction will continue to be used in the following as a convenient means to differentiate between different augmentation services.

14.2.2 SPATIAL REPRESENTATION

The majority of GPS augmentation services provide information to users in one of two spatial representations (Figure 14.1). An *observation-space representation* (OSR) is said to be used when code or carrier phase observations or corrections are computed from a single base or reference station, by "lumping" all biases into a single correction for each observation. The information in OSR is useful to remote users who apply the differential processing methodology to perform relative positioning.

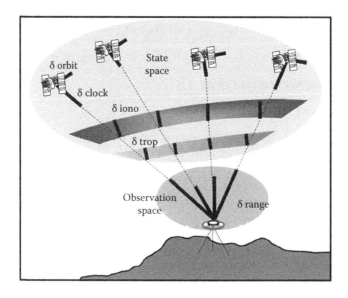

FIGURE 14.1   Spatial representation of corrections.

The alternative to an OSR is a *state-space representation* (SSR) that provides corrections to individual error sources. More specifically, these are satellite orbit and clock errors and estimated atmospheric delays (ionospheric and/or tropospheric). The information in SSR can then be converted into the observation space for a specified user location or applied directly in the receiver's navigation software.

Given the current evolving context with increased Internet usage, it is useful to point out the practical differences between the two spatial representations. Regarding accuracy, the SSR usually offers more uniform accuracy over a larger coverage area as correction accuracy does not deteriorate as rapidly as a function of distance from the reference stations, as OSR does. The SSR approach is used in so-called Wide Area DGPS implementations. On the other hand, an OSR service can often provide more accurate relative positioning over a local area where spatial correlation among error sources is high (e.g., as in so-called Local Area DGPS). In SSR, update rates may vary for the different correction components as a function of their respective correlation time, hence reducing bandwidth requirements. In contrast, OSR updates must be at the frequency of the error source that varies the most rapidly. The SSR network-based approach greatly reduces station-dependent errors and mitigates the impact of missing observations, but calls for more complex standards and end-user algorithms, which are usually slow to be adopted across an industry.

Traditionally, one particular spatial representation has been assumed to be better than the other based on the size of the service area and its point of distribution. OSR has been common in local-area services delivered with land-based communication links, while SSR has been preferred for wide-area services delivered with space-based links. These assumptions linking the spatial representation to the point of distribution are becoming obsolete as Internet services that are not concerned with the underlying communications technology become more commonly used. Additionally, some network-based local-area services are also adopting an SSR to maintain high accuracy over extended areas of coverage.

14.2.3 STANDARD PROTOCOLS AND MESSAGING FORMATS

Since augmentation services consist of GPS information sources being disseminated to end-users over various communications channels, they require that interface protocols and messaging formats be clearly defined for a broad-based user access. Given that the most common spatial representations of the information are OSR or SSR, two standard messaging formats for real-time broadcasts were developed by the marine and aviation sectors. They are usually referred to as "RTCM-104" and "RTCA-159," respectively. The contents of these standard messages are embedded into open protocols for reliable over-the-air transmission that ensures message integrity and facilitates user access (see Table 11.2 for a list of the RTCM v3.1 messages). The Internet User Datagram (UDP) and Transport Control Protocols (TCP) are also used increasingly to stream GPS (and GLONASS) data and products. This fact has led the RTCM committee to recently adopt a GNSS specific transport protocol, the *Network Transport of RTCM via Internet Protocol* (NTRIP), to facilitate real-time Internet exchange of GNSS data and products (NTRIP 2009).

For post-mission access to the information sources, the Internet and its File Transfer Protocol (FTP) have been used extensively. The RINEX file format (Gurtner 1994) has become the standard to exchange station-based observation data and broadcast ephemerides for differential processing using the OSR model. The SP3, CLX, and INX formats are also commonly used to disseminate and archive precise orbits, clocks, and ionospheric maps, respectively, for post-mission processing.

While the correction standards mentioned above are most common, a number of enhancements have been proposed and are in various stages of adoption. In addition, a number of proprietary formats and protocols have been developed and coexist in the real-time kinematic (RTK) GPS service industry (Section 11.2.2), where different GPS manufacturers offer off-the-shelf network solutions as they compete for market share.

14.2.4 EMERGING TRENDS

The OSR approach has been used extensively in local and regional Differential GPS (DGPS) services providing pseudorange corrections for land and marine navigation requiring meter-level positioning accuracy (Section 11.2.1). It has also been popular for carrier phase–based single-base station RTK-GPS services offering centimeter-level relative accuracy over local areas (typically within 10–20 km separation from the base station receiver). The SSR concept has also been recently introduced in regional "network-RTK" systems (Section 14.3.2.2), where the state of the troposphere is estimated at the network level and transformed to observation space for dissemination in RTCM standard messages.

While network-RTK systems adopt an SSR to increase spatial coverage, the availability of precise real-time global orbit products is also raising interest in the "RT-PPP" (Real-time Precise Point Positioning) approach for wide-area applications. PPP uses precise satellite orbit and clock estimates computed from global networks along with atmospheric models for undifferenced single-point positioning (SPP). Centimeter-level accuracies in relation to the origin of the global network can now be achieved with dual-frequency receivers. Unlike RTK-GPS with "on-the-fly ambiguity resolution," PPP still requires a lengthy initialization period (of the order of several minutes or more) to reach the centimeter accuracy level. Nevertheless, recent advances in user algorithms and improvements in global tropospheric models indicate that PPP fast ambiguity resolution may become possible in the future. Synergies between the RTK-GPS and PPP approaches to precise positioning may be exploited in the coming years, offering users optimal precision regardless of the spatial extent of their application.

As new signals become available from modernized and emerging GNSS constellations (Sections 15.4 and 15.5), additional codes modulated on more signals will be broadcast. Providers of augmentation services may not be able to track all the available signals from their reference networks. Maintaining compatibility between all types of measurements at both the provider and user ends will become more complex using OSR. It may become simpler to apply SSR corrections to all observation types, instead of relying on the assumption that there will always be compatible observation types between the user and reference receivers. Recognizing this trend and its future potential, RTCM SSR messages are currently under development and are expected to support all levels of positioning accuracies, including centimeter accuracy RTK-GPS.

14.3 REAL-TIME GNSS AUGMENTATION SERVICES

With the removal of Selective Availability (SA) in May 2000, SPP now satisfies the majority of real-time GPS users. Nevertheless, DGPS is still a requirement for many applications. This is generally the case for any application where accuracy consistently better than a few meters is required, and where reliability and integrity are of concern. This is because the total effect of atmospheric biases on SPP is still larger than 5 m in many cases (Section 8.5.1). Since these errors are spatially (and temporally) correlated, DGPS is effective in reducing them (Section 11.2.1). In real-time, users operating a wide range of GPS receiver equipment must be able to receive and process transmitted GPS data or corrections. As mentioned previously, the Radio Technical Committee for Maritime Services (RTCM 2009) Special Committee 104 supports an open standard format for real-time dissemination of OSR pseudorange or carrier phase corrections. Similarly, the Radio Technical Commission for Aeronautics (RTCA 2009) Special Committee 159 supports an open standard format for real-time dissemination of SSR corrections for satellite orbits, clocks, and ionospheric delays.

14.3.1 SATELLITE-BASED REAL-TIME AUGMENTATION SYSTEMS

The ground segment consists of a network of reference stations that are geographically distributed to ensure continuous satellite tracking over the area of interest with sufficient redundancy to meet correction availability and reliability targets. It should also include geographically separated redundant processing centers, eliminating potential single-site failures that may result from loss of power, computing, or communications. Processing centers host the software to calculate differential corrections and perform various integrity functions. Preferably, redundant Ground Earth Stations (GES) should uplink the GPS corrections and timing information to geostationary satellites.

In addition to the GPS constellation, the space segment may consist of several communication satellites serving one or two major functions, namely, the transmission of DGPS corrections and the provision of a carrier PRN modulation (Section 8.2.1) supporting a ranging capability to improve service availability, reliability, and continuity over the targeted coverage area. By transmitting correction messages in a frequency band close to the GPS signal, the corrections, subject to the line-of-sight constraint, can be received with a single antenna—a practical advantage in terms of user hardware requirements.

Providers of commercial wide-area GPS augmentation services purchase bandwidth on communication satellites to provide privileged access to their users globally, using proprietary over-the-air protocols and formats. Public satellite-based augmentation services, primarily for the aviation industry, also broadcast correction messages for *en route* navigation and final approach in various regions of the world. Both of these classes of services are briefly reviewed below, with their main characteristics summarized in Table 14.1.

TABLE 14.1

Summary of Major GNSS Augmentation Services

System	Network	Broadcast (Format)	Coverage	Horizontal, Accuracy 95%	User Equipment	Sector (Sponsor)	Industry/Access
Satellite based							
WAAS	NSTB	GEO (RTCA)	North America	L1: 1–2 m	WAAS Rx (most makes)	Public (FAA)	Aviation, no fee
EGNOS	ESTB	GEO, Internet (RTCA/Sisnet)	Europe	L1: 1–2 m	WAAS-Rx (most makes)	Public (ESA)	Aviation, no fee
MSAS		GEO (RTCA)	Japan	L1: 1–2 m	WAAS-Rx (most makes)	Public?	Aviation, no fee
CDGPS	CACS/IGS	GEO (MRTCA)	North America	L1: 1–2 m; Dual: 0.5 m	CDGPS-Rx, NovAtel™	Public (CGRSC)	Geomatics, no fee
OMNIStar	NASA/JPL	GEO (proprietary)	Global	VBS: 1 m; XP: 15 cm; HP < 10 cm	Omnistar Rx	Private (Fugro)	Precise Ag; Geomatics; $800–$2500/month
StarFire	NASA/JPL	GEO (proprietary)	Global	SF1: 30cm; SF2: 10cm	Starfire Rx	Private (NavCom)	Precise Ag; Geomatics; $15,000
CNAV GPS	NASA/JPL	GEO (proprietary)	Global	20cm	Cnav Rx	Private (C&C Tech)	Marine $1200/month
Ground based							
Maritime DGPS	DGPS	Beacons (RTCM)	Coasts and waterways	Single: 1–5 m; 99.7%/month	DGPS Rx	Public Coast Guard	Marine; no fees
Local DGPS/RTK	DGPS/RTK	GSM/GPRS (RTCM)	Local area	<5 cm	Wireless	Public, private, and partnerships	Engineering Cadastral $100–300/month

Notes: GEO broadcasts do not cover above 72° latitude. The vertical accuracy is on average 2–3 times worse than that of the horizontal.

14.3.1.1 Commercial Satellite-Based Real-Time Augmentation

There are currently only a few commercial providers of global DGPS corrections catering to users mainly in the precision agriculture and precise offshore positioning sectors. These services rely on a number of ground-based reference stations distributed globally that permit continuous tracking of all satellites of the GPS constellation (no other GNSSs are tracked at this time) throughout their orbits. Reference station data are forwarded to one or more central processing centers where corrections are generated and passed to a GES that uplinks them to a geostationary communications satellite. The satellite acts as a "bent pipe," and broadcasts these corrections to users. The procedure is identical across the world except that different geostationary satellites may be used. In the early days, the Inmarsat satellites were used (for which users required a gimbal-mounted directional antenna). Today, L-band mobile satellite communication systems are generally preferred, enabling in-band reception via small antennas that track signals originating from both the GPS and communication satellite broadcasts.

The commercial space-based services provide either OSR or SSR corrections and usually offer different levels of positioning accuracy at different costs, to address the varying needs of their users. Because of the proprietary nature of their over-the-air broadcast, users require a special decoder or must use a particular receiver (purchased or leased from the DGPS service provider) to access the message content. As it has become possible to access data from global tracking networks, and real-time precise satellite orbits and clock estimates are becoming available, the high precision of the carrier phase measurements will be able to be exploited to compute corrections approaching centimeter accuracy. These high-accuracy systems require dual-frequency observations as input into advanced algorithms to generate the user's navigation solution. The advantage of these services is that their corrections have global coverage, available anywhere in the world (onshore and offshore), within the footprint of the communication satellite. Given that most communication satellites orbit above the equator, they are still weak at servicing users at the higher northern or southern latitudes. The cost of operating the ground and space infrastructure supporting these services is usually recovered by the sale of specific equipment and recurring subscription costs for access to the broadcast messages. With user charges up to 10 times higher than those charged for local ground-based augmentation, the advantages provided by broad coverage and consistent accuracy across the service area appeal mainly to users operating over remote wide-areas requiring a high-reliability service.

14.3.1.2 Public Satellite-Based Real-Time Augmentation

Public satellite-based real-time augmentation services have been developed mainly to support *en route* air navigation. They are usually referred to generically as *wide-area augmentation systems* (WAAS), and their main components are shown in Figure 14.2. Because they ensure the safety of air passengers, the reliability and integrity functions of these augmentation services are extremely important. While intended for air navigation, WAAS-type corrections are publicly available and anyone equipped with a suitably adapted receiver can access them. This means that for

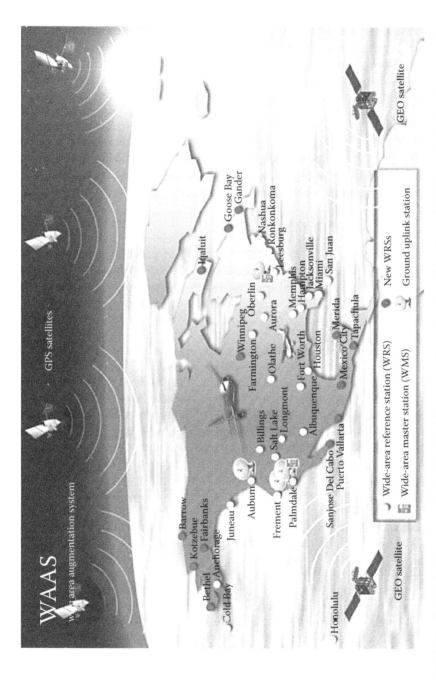

FIGURE 14.2 U.S. Federal Aviation Authority's WAAS concept. (From http://en.wikipedia.org/wiki/Image:FAA_WAAS_System_Overview.jpg. With permission.)

geospatial applications requiring sub-meter-level positioning accuracy, WAAS is a viable alternative to commercial services in those areas of the world where WAAS coverage is provided.

The U.S. Federal Aviation Administration (FAA 2009) was the first to propose and implement the WAAS model and has extended its continental U.S. coverage into Alaska, Canada, and Mexico. Other countries or continents are also deploying similar augmentation systems. The European Geostationary Navigation Overlay System (EGNOS) and the Japanese Multi-Function Satellite-Based Satellite Augmentation System (MSAS) are examples of public wide-area augmentation systems in different regions of the world. All such services are interoperable, transmitting standard format messages so that aviation users can seamlessly travel across the WAAS, EGNOS, or MSAS service areas. Many single-frequency handheld GPS receivers also have WAAS/EGNOS/MSAS tracking capability, and users benefit from higher (differential) positioning accuracy.

Public wide-area augmentation systems with space-based correction delivery, mainly for geospatial and land-based precision navigation applications, also exist. One example is the nationwide Canadian Differential GPS Service (CDGPS), a service developed under a partnership between public survey and land agencies to provide enhanced positioning capabilities within Canada's adopted reference frame. This model was seen as an effective way to deliver enhanced positioning precision uniformly across the country. It is a unique solution tailored to the Canadian context, where a sparse population distributed over a large landmass is provided seamless coverage from mobile satellite broadcasts at no user cost.

While global real-time corrections are currently available commercially for redistribution, public Internet access to real-time data from a global tracking network is becoming a reality under the leadership of the *International GNSS Service* (IGS) with contributions from many national agencies (see Section 15.3.1). There is little doubt that public access to tracking data from a global real-time network will eventually lead to the computation of real-time orbit, clock, and ionospheric products for distribution. Ultimately, many regional to national tracking networks will evolve into a multifunctional infrastructure to support the full range of positioning and scientific needs for the earth and atmospheric sciences.

14.3.2   Ground-Based DGPS Services

Ground-based DGPS services are mainly of the local-area variety and use a range of delivery models, communications infrastructure, and technologies. While many of these services are commercial operations, mainly in populated urban areas, a public service for the maritime industry, recognized internationally, continues to transmit correction messages free-to-air along the coast and inland waters of North America (and elsewhere around the world).

14.3.2.1   Public Marine DGPS Services

Public marine DGPS services are unique because they are intended for maritime applications, yet are compatible with international systems being established around the world. This international standard ensures that marine users are able to acquire and use transmitted RTCM messages wherever they go. This service utilizes the marine beacon frequencies of 285–325 kHz for data transmission. This dedicated band is advantageous as low-frequency ground waves propagate long distances, several hundreds of kilometers over water or land, which is well beyond the radio horizon. Many of the international stations are shown in Figure 14.3, with the stations shown in dark gray and their coverage contours in light gray.

The marine radio beacon DGPS systems are a worldwide endeavor with national services being set up and maintained by the maritime authority of each country. Both the International Marine Organization (IMO) and the International Association of Lighthouse Authorities (IALA) have produced specifications and guidelines for marine DGPS systems (IALA 2009). In the United States and Canada, the U.S. Coast Guard (USCG) and Canadian Coast Guard (CCG) are responsible for these services. The USCG and the CCG declared their DGPS systems to be fully operational on March 12, 1999, and May 25, 2000, respectively. The initial objectives of maritime DGPS services were to support the harbor entrance and approach phase of navigation, vessel traffic control, as an aid to navigation, and for mapping the Exclusive Economic Zone. Navigation for inland waterways was a later application. The service available in the coverage areas has proven reliable, and has received acceptance by users well beyond the maritime community, particularly by farmers in coastal areas and along inland waterways for precision agriculture applications. Note that it is an L1-only service, as is WAAS/EGNOS/MSAS, for single-frequency users.

A plan to extend DGPS coverage to the entire continental United States under the Nationwide DGNSS (NDGNSS) Service project (USCG 2009) has been pursued over the past decade but limited support and conflicting interests has not permitted its full realization. Figure 14.4 shows the current coverage—note the complete coverage along the coast and inland waters, but less complete coverage elsewhere.

14.3.2.2 Single Reference Receiver and Network-RTK Services

In many developed countries, private companies are offering terrestrial local-area DGPS via cellular phones, mainly in populated areas where the wireless communication infrastructure can be exploited. The services usually support RTK-GPS applications (Section 11.2.2) by disseminating corrections enabling relative positioning with centimeter accuracy using double-differencing algorithms with fast "on-the-fly" integer ambiguity resolution (Section 9.4.2). Initially, these corrections were generated from a single reference receiver at a fixed base station, but ambiguity resolution could only be achieved with confidence over baselines up to about 20 km or so in length, particularly during periods of high ionospheric activity. The drive to maintain positioning accuracy while extending the correction "area of applicability" has led to the development of multi-station reference receiver networks, which are now becoming the norm for reliable RTK operations.

As markets evolved, the multi-reference RTK, or "network-RTK," functionality has been implemented in software solutions offered by major GPS manufacturers, facilitating the distribution of RTK-GPS corrections, while providing the opportunity to offer benefits to end-users selecting compatible equipment. Given this capability, various models for distribution of RTK-GPS corrections have evolved in different

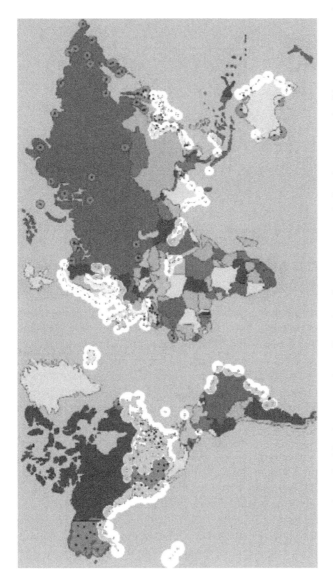

FIGURE 14.3 Worldwide marine DGNSS reference stations. (From http://www.puertos.es/en/ayudas_navegacion/sistemas_navegacion_por_satelite/Sistemas_GPS_y_DGPS.html. With permission.)

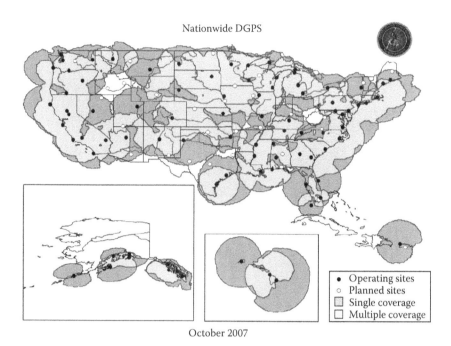

Nationwide DGPS

- • Operating sites
- ○ Planned sites
- ☐ Single coverage
- ☐ Multiple coverage

October 2007

FIGURE 14.4   The U.S. Nationwide DGNSS reference station network. (From http://www. navcen.uscg.gov/dgps/coverage/CurrentCoverage.htm. With permission.)

regions of the world depending on user requirements, communications infrastructure, and business model pursued. In some instances, partnerships between public survey/ land agencies and private providers of GPS solutions have resulted in the deployment of RTK infrastructure at the local and regional level, often extending across a state or country. This type of partnership usually involves some effort to implement (partial or full) cost-recovery measures through user-fees. One example is the Ordnance Survey Smartnet network (Figure 14.5) spanning most of the United Kingdom. Other initiatives are entirely funded in the public sector and justified by the cost-savings realized by "active control." Under the terms of an agreement, access to these correction services is usually provided on a best-effort basis at no cost to the user. The Departments of Transportation of some U.S. states have justified investment in multi-reference RTK networks based on their own requirements, while providing public access to their infrastructure for use in other sectors. Attempts are being made by the U.S. National Geodetic Survey (NGS 2009) to coordinate the development of disparate Real-Time Networks (RTNs) of continuously operating GPS receivers that can provide centimeter-level accuracy positioning to suitably equipped users.

   With the expansion of free-to-air public DGPS services and improved stand-alone SPP in the post-SA era, the market for augmentation services for land vehicle applications has contracted. Nevertheless, RTK-GPS services continue to gain in popularity as cellular networks for wireless communications expand, and the demand for higher accuracy grows. RTK-GPS service delivery over the Internet is also becoming more popular with the expansion of wireless communication channels and their connection

FIGURE 14.5 Regional RTK network in Great Britain. (From http://smartnet.leica-geosystems. co.uk/spiderweb/weblink/information2.htm. With permission.)

through Internet protocols. As already mentioned, a transport protocol known as NTRIP has been adopted by RTCM, and the availability of open source software supporting this protocol is leading GPS manufacturers to embed NTRIP clients in their products (NTRIP 2009). This development will facilitate access to RTK-GPS correction streams wherever Internet connectivity is available.

14.3.2.3 Local Area Augmentation System

The U.S. Federal Aviation Authority is deploying Local Area Augmentation Systems to support precision approach and landing Category II/III aviation applications (FAA 2009). Systems will be deployed at selected airports and will service a limited area around these airports. A local ground system will generate differential L1-only corrections from the GPS, WAAS geostationary satellites, and airport pseudolites (when used in the future). A VHF transmitter will then be used to broadcast the correction and integrity messages to the local aviation users. However, this system, due to its localized application and other restrictions, is of little interest to users outside the aviation community.

14.4   POST-MISSION GPS AUGMENTATION SERVICES

Modern high-productivity GPS surveying, in rapid-static and kinematic modes, is performed relative to a base station with known coordinates, at which a reference receiver is operated (Section 11.1). For the purpose of GPS augmentation, the base station must operate continuously for users to reliably match the times of observations made at single or multiple remote receivers. Since base-station coordinates implicitly carry the underlying reference frame to the remote users, making base-station data available is equivalent to providing access to the positioning datum (Section 7.2). This fact has long been recognized by public agencies who promote the use of a consistent geospatial reference frame for positioning across their jurisdiction. For that reason, public geodetic agencies have played a significant role in establishing continuous local, regional, and national GPS tracking networks to enable integration of surveys and preserve their spatial compatibility in time. In doing so, they have also reduced the need for users to establish their own base station for each particular project.

Many countries have established networks of continuously operating reference stations (CORS), also referred to as "active control" or "fiducial" networks. These networks are generally operated by government organizations, with station tracking data and coordinates typically made publicly available at no user cost for post-mission positioning via RINEX data files. Since these networks generally offer only sparse spatial coverage, denser networks operated by the private sector have appeared to provide tracking data for a fee to more demanding user groups, especially for real-time services (Section 14.3.2.2).

By accessing data from CORS, users can perform high-accuracy positioning using a single GPS receiver, thus halving their equipment costs. While carrier phase–based positioning is always possible, mitigating systematic errors and resolving ambiguities rapidly over long baselines is complex. This is resulting in the establishment of denser active networks for centimeter-level positioning. While CORS networks deployed by national agencies could be enhanced to offer RTK-GPS services, their tracking stations are not always located in populated areas where such services are in the greatest demand. Therefore, the majority of publicly operated CORS networks have to date only collected measurements and archived data files for post-processing, although a real-time data streaming capability is becoming more desirable as the generation of global real-time orbit products will become possible through the efforts of the Real-Time IGS Pilot Project (RT-IGS 2009).

14.4.1 CONTINUOUSLY OPERATING GPS/GNSS TRACKING NETWORKS

Many CORS networks were originally established in support of local geodynamic applications. In terms of tracking station density and coverage, one notable network is Japan's Geographical Survey Institute's (GSI) GEONET, established in the aftermath of the catastrophic 1995 Kobe earthquake. GEONET consists of over 1200 GPS receivers deployed across the country, linked to a central archive at GSI's headquarters in Tsukuba (GSI 2009). Another significant North American example is the Plate Boundary Observatory (PBO 2009) component of EarthScope, comprising over 800 GPS tracking stations (Figure 14.6) in the western United States to study deformation across the active boundary zone between the Pacific and North American tectonic plates. A global network of over 350 tracking stations has also been created with contributions from many countries and agencies participating in the IGS. This network contributes, along with other space geodetic techniques, to the definition and maintenance of the International Terrestrial Reference Frame (ITRF) (Section 7.2.2).

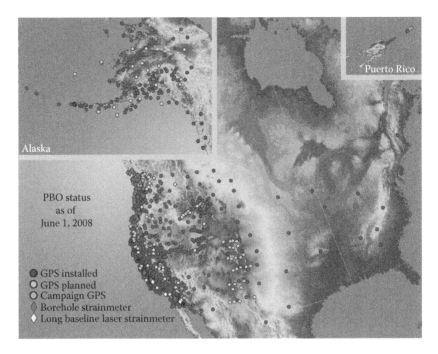

FIGURE 14.6 **(See color insert following page 426.)** PBO Network Design showing proposed GPS site locations. (From http://pbo.unavco.org/~kyleb/blm.htm. With permission.)

The CORS networks have been established by geodetic organizations in many countries to maintain and provide access to their geospatial reference frame and hence to reduce reliance on their traditional monumented networks. It is not possible to list all operational or planned national networks but a few examples will be used to highlight some related issues. In North America, the Canadian Active Control System (CACS) and the U.S. CORS Network are two examples of cooperative programs coordinated by national geodetic agencies. The CACS (Figure 14.7) is a multipurpose tracking network of about 50 CORS with varying equipment configurations depending on the purpose they serve (CACS 2009). While a sparse subset of sites report real-time data for wide-area correction computation in support of CDGPS, others are clustered in areas where geophysical or atmospheric processes are of interest. The U.S. CORS Network consists of over 1000 reference stations (including PBO sites), operated by a variety of government, academic, commercial, and private organizations (NGS 2009), including some reference stations of the U.S. Nationwide DGNSS network (Figure 14.4).

In contrast to these national scale networks, the Hong Kong Lands Department's CORS network (Rizos 2000) (Figure 14.8) serves an area of approximately $1000 \, \mathrm{km}^2$.

Such networks typically log dual-frequency GNSS data and make it available as RINEX files. They may also make these data sets freely available over the Internet in support of automated data processing services (Section 14.4.3).

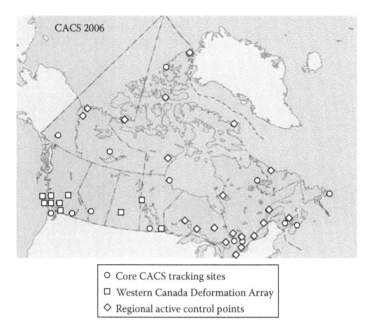

FIGURE 14.7 Continuously operating stations of the Canadian Active Control System. (From *Canadian Active Control System*, Web page, http://www.geod.rncan.gc.ca/acp/cacs_e. php, accessed March 15, 2009. With permission.)

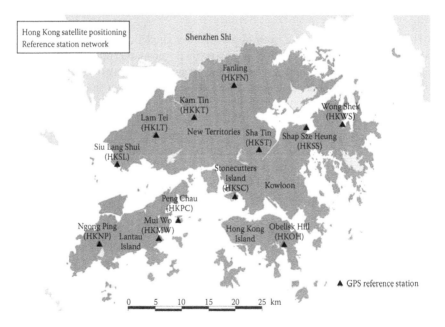

FIGURE 14.8 The Hong Kong GNSS permanent station network. (Courtesy of Survey and Mapping Office, Lands Department of Hong Kong.)

14.4.2 IMPACT OF CORS ON GPS SURVEYING IN THE POST-MISSION MODE

As CORS networks continue to be deployed, the question whether or not they can support centimeter-level carrier phase–based static and kinematic positioning in post-mission remains. While they are being used to support the maintenance of global and national reference frames, further densification is often required for more demanding GPS positioning applications. By allowing users to download reference station data and coordinates, service providers simultaneously give access to a specific geospatial reference frame and enable users to achieve higher relative accuracy. While this method of accessing the reference frame, and at the same time improving positioning accuracy, is efficient, it is also limited by the validity of the models implemented within the user receiver software and their performance over long baselines.

Currently, most software implementations require that data from the nearest GPS base station be downloaded to the user's computer for processing. In practice, they often require that a reference station be within 20–30 km of the user's location for integer ambiguities to be resolved rapidly with a high level of confidence. (The further away the base station, the longer the observation time for a baseline determination, and hence the less productive GPS surveying techniques become— Section 7.4.3.) On the other hand, the establishment of public networks of CORS at a density to support GPS surveys at the highest accuracy level is generally not economically justifiable as a "public good." This has led to the development of private RTK-GPS services in local areas.

Even with the station density of GEONET in Japan, the average receiver spacing remains 20–30 km, which may be insufficient for the most demanding users to perform OTF ambiguity resolution. Most Canadian CACS and many U.S. CORS are more than 100 km apart and therefore there is little likelihood of a user operating in close enough proximity to a CORS station that their commercial GPS processing software can resolve ambiguities to deliver centimeter-level relative accuracy. Scientific software packages may be capable of this, but they require longer observing periods. In the case of networks intended to service local areas, such as in Hong Kong, users can depend on having at least one reference station within a radius of 10 km, and there are no constraints to high-accuracy GPS surveys. In fact, the Hong Kong network is so designed that the vast majority of users can use data from two reference stations (Figure 14.8), and hence are able to compute two independent baselines connecting the user receiver to the CORS network. In this way, the extra baseline can be used as a *quality control* measure (Section 12.4.2.1). One possible solution to this "distance to base station" problem over large regions is to increase the density of reference stations in urban areas, where user demand is highest. The increasing use of techniques that take advantage of network-based processing strategies (in place of single baseline processing) does overcome, to a large extent, the "distance to base station" constraint.

14.4.3 GPS DATA PROCESSING SERVICES

In the client-server world in which we live, another innovative means of serving users is to offer online tools that process their GPS data. This can be accomplished in two ways. A differential processing approach can be applied where the user data is combined with the data from one or several surrounding CORS. Another possibility is to perform undifferenced PPP using precise orbits and clocks. In both approaches, the user records their GPS data and then submits (or uploads) the data files to a central server via email or by accessing a web page (this can even be done from the field via "smart" cell phones and Internet-ready PDAs). Such services effectively eliminate the requirement for users to either install and maintain their own data processing software, or to operate a reference receiver, but usually offer limited flexibility in selecting specific processing options.

In services using differential processing, the central server selects and accesses the data from the nearest GPS reference stations and performs the computation of the user receiver coordinates. After the processing is complete, an email is sent to the user with the results or the Internet FTP location (URL) from which the result file can be downloaded. Such services automate the selection of the most suitable station data and reference coordinates. They attempt to optimize the use of more sophisticated scientific processing software and ensure that the highest possible accuracy is attained, especially over long baselines.

In services using the undifferenced PPP approach, no reference station data or coordinates are required; precise satellite clocks and orbits are used instead. These replace the orbit and clock information provided in the broadcast navigation message (Section 8.3).

There are several online processing services currently available, and more are expected to be established in the coming years. While some services use a PPP approach, most of them apply a double-differencing (DD) method and accept only dual-frequency data sets observed in static mode. While additional functionality supporting kinematic mode and single-frequency processing is possible, it is currently available only from the Canadian CSRS-PPP service. All services allow submission of compressed files, require only a few minutes of computing time, and provide results on anonymous FTP sites. While computing is fast, it can only be initiated once the station data files (for DD) or precise orbit and clock products (for PPP) become available, which may vary from an hour to a day, depending on data type and source. While all these services can generate user coordinates with centimeter-level accuracy, a significant amount of observation time is required for ambiguities to converge. This generally varies from 15 min to 1 h or more depending on the geometry of the observed satellites and user dynamics. While these services have proven very useful for global reference frame access and densification, they do not yet provide the instant solution with centimeter-level accuracy available from local-area OTF ambiguity resolution. Table 14.2 lists some of the most popular public services of this type.

AUSPOS: http://www.ga.gov.au/bin/gps.pl
Auto-GIPSY: http://milhouse.jpl.nasa.gov/ag/
CSRS-PPP: http://www.geod.nrcan.gc.ca/products-produits/ppp_e.php
SCOUT: http://sopac.ucsd.edu/cgi-bin/SCOUT.cgi
OPUS: http://www.ngs.noaa.gov/OPUS/index.html

TABLE 14.2
Online GNSS Processing Services

Online Service	AUSPOS	CSRS-PPP	OPUS	Auto-Gipsy	SCOUT
Agency	AUSLIG	NRCAN	USNGS	JPL	SCRIPPS
Software	MicroCosm	GPSPPP	PAGES	GIPSY	GAMIT
Model	DD	PPP	DD	PPP	DD
# of files	Single	Multiple	Single	Single	Single
Submission	ftp, upload	Upload	Upload	Email	upload
Min. session	2 h	1 epoch	2 h	5 min	1 h
Max. session	7 days	7 days	1 day	1 day	1 day
Frequency	Dual	Dual and single	Dual	Dual	Dual
Kinematic mode	No	Yes	No	No	No
Processing delay	Next-day	Same-day	Next-day	Next-day	Next-day
Reference frame	ITRF/GDA94	ITRF/NAD83	ITRF/NAD83	ITRF	ITRF

14.5   SERVICE PROVIDERS FOR GEOSPATIAL PROFESSIONALS

There is a broad range of information sources concerning GPS services, data, and products, as well as surveying/mapping in general. The Internet is now the primary means of accessing such information.

14.5.1   NATIONAL LAND, MAPPING, OR GEODETIC AGENCIES

These provide a wealth of information, data and services concerning the GPS status, mapping products, datum information, geoid height models, transformation parameters, GPS (and sometimes also GLONASS) data and ephemeris information, standards and specifications, and useful links to government and nongovernment WWW sites. Examples include:

U.S. National Geodetic Survey, http://www.ngs.noaa.gov
Geodetic Survey Division, NRCan, http://www.geocan.nrcan.gc.ca/org/indexe.html
Geoscience Australia, http://www.ga.gov.au
Land Information New Zealand, http://www.linz.govt.nz

14.5.2   ACADEMIC INSTITUTIONS

Many geography, geomatics, geospatial, and surveying departments in North America and elsewhere offer short courses in pertinent subjects. Lists of departments and their specializations can be found on the Internet. Good starting points are the web pages of the national geodetic agencies (Section 14.5.1).

14.5.3   INTERNATIONAL SCIENTIFIC AND PROFESSIONAL ASSOCIATIONS

There are many international associations, some scientific, others discipline based, that are sources of information on geospatial matters. Examples include:

International Association of Geodesy, http://www.iag-aig.org
International Geographic Union, http://www.igu-net.org/
International Cartographic Association, http://cartography.tuwien.ac.at/ica/
International Hydrographic Organization, http://www.iho-ohi.net/english/home/
International Federation of Surveyors, http://www.fig.net/figtree/
International Society for Photogrammetry and Remote Sensing, http://www.isprs.org/
International GNSS Service, http://www.igs.org/

14.5.4   NATIONAL GEOMATICS/GEOSPATIAL ASSOCIATIONS AND INSTITUTIONS

National geomatics organizations generally deal with professional matters such as
training, ethical standards, newsletters, and so on, but also host conferences, publish
journals, and convene working groups on a variety of topics. There are associations
or groups organized at the state or province level, or for a particular class of geospatial professional or industry (e.g., hydrographic surveyors, government employees),
or common interest groups (e.g., GIS-based resource mapping). The main national
bodies in North America are

American Congress on Surveying and Mapping, http://www.acsm.net/
Canadian Institute of Geomatics, http://www.cig-acsg.ca/

14.5.5   PRIVATE ORGANIZATIONS

As far as GPS is concerned, we generally refer to GNSS equipment manufacturers
when we speak of private organizations, but these organizations also include specialist consulting companies. It is not possible to list such organizations in this book,
but the reader will find many such lists under "Web links" at various sites. Some
comprehensive and up to date sites include:

http://www.navtechGNSS.com/links.asp
http://gauss.gge.unb.ca/GNSS.INTERNET.SERVICES.HTML
http://www.gmat.unsw.edu.au/snap/gps/gps_links.htm

REFERENCES

CACS, 2009, *Canadian Active Control System*, Web page, http://www.geod.rncan.gc.ca/acp/
cacs_e.php, accessed March 15, 2009.
FAA, 2009, *Federal Aviation Authority—Navigation Services*, Web page, http://www.faa.gov/
about/office_org/headquarters_offices/ato/service_units/techops/navservices, accessed
March 15, 2009.
Federal Radionavigation Plan, 2008, 184pp., published by Department of Defense, Department
of Homeland Security & Department of Transportation, DOT-VNTSC-RITA-08–02/
DoD-4650.5, available from http://www.navcen.uscg.gov/pubs/frp2008/2008_Federal_
Radionavigation_Plan.pdf, accessed March 15, 2009.
GSI, 2009, *Geographical Survey Institute GEONET*, Web page, http://www.gsi.go.jp/
ENGLISH/index.html, accessed March 15, 2009.
Gurtner, W., 1994, RINEX: The receiver-independent exchange format, *GPS World*, 5(7), 48–52.

IALA, 2009, *International Association of Lighthouse Authorities*, Web page, Table of DGNSS stations, http://www.iala-aism.org, accessed March 15, 2009.

NGS, 2009, *National Geodetic Survey*, Web page, http://www.ngs.noaa.gov/, accessed March 15, 2009.

NTRIP, 2009, *Networked Transport of RTCM via Internet Protocol*, Web page, http://www.ntrip.org/, accessed March 15, 2009.

PBO, 2009, *Plate Boundary Observatory*, Web page, http://pboweb.unavco.org/, accessed March 15, 2009.

Rizos, C., 2000, GNSS surveying technology—Why doesn't every surveyor own a kit? *Surveying World*, 8(4), 26–29.

RTCA, 2009, *Radio Technical Commission for Aeronautics*, Web page, http://www.rtca.org/default.asp, accessed March 15, 2009.

RTCM, 2009, *Radio Technical Committee for Maritime Services*, Web page, http://www.rtcm.org, accessed March 15, 2009.

RT-IGS, 2009, *Real-time IGS Pilot Project*, Web page, http://www.rtigs.net/, accessed March 15, 2009.

USCG, 2009, *U.S. Coast Guard National Differential GNSS Project*, Web page, http://www.navcen.uscg.gov/ndgps/default.htm, accessed March 15, 2009.

15 GPS, GNSS, and the Future

Chris Rizos

CONTENTS

Since the launch of the first satellite in 1978, GPS has evolved from a U.S. military system to a *global utility* that benefits users around the world for many different applications. GPS has ushered in a new era in which any person, no matter where they are, has access to a low-cost positioning technology that does not require special skills to operate. Many new applications are being supported by GPS technology, and associated products and services developed by the private, government, or academic sectors, and others are launched each year. GPS is the first technology that has brought the concept of *ubiquitous* positioning closer to reality.

Surveying and geodesy were among the very first civilian applications of GPS. Over the last three decades, high-accuracy GPS positioning instrumentation, software, and techniques have been progressively refined to the point that *GPS surveying* nowadays challenges traditional terrestrial surveying procedures for most geospatial applications. In Part II of this book, the authors have attempted to draw attention to those developments that are of particular significance for geospatial applications. *Is there room for further improvement in GPS? Will the GPS techniques ultimately supersede all other technologies?* Predicting the future is always a risky enterprise; however, as stated in Section 7.4.3, GPS is likely to *complement* the traditional EDM-theodolite techniques for routine surveying activities for many years to come. Nevertheless, there are many applications, including some forms of mapping and low-to-moderate accuracy positioning, high-accuracy machine automation applications, and geodetic positioning, for which GPS (and other GNSS) is the *preferred* technology. In the following sections, summary remarks will be made with regard to:

1. Trends in user technology
2. Institutional issues such as datum definition, new applications, regulation, etc.
3. Improvements to the GPS space segment
4. Ground-based infrastructure to support high-accuracy positioning
5. Plans over the next 5–10 years for the development and deployment of new *global navigation satellite systems* and *regional* satellite-based augmentation systems

It must be emphasized that no discussion concerning the future of GPS can ignore GNSS developments in general. *Hence, many of the statements made in this chapter should be viewed as being relevant to GNSS in general, and not just in relation to GPS.* Sometimes that will be obvious when the acronym "GNSS" replaces "GPS," but the acronym "GPS" may still be used if the comment is explicitly made in relation to *current* trends (even though they will also be valid for *future* GNSS).

15.1 GPS/GNSS INSTRUMENTATION AND TECHNIQUES

It is possible to identify several user equipment/technique trends in GPS/GNSS that are based on historical developments:

- Reduction in size, power consumption, and cost of basic single-frequency GPS receivers, especially for navigation-type receivers intended for general use. *This is primarily the result of new receiver chip designs and the development of a mass market for car navigation system, GPS-enabled cell phones, and personal navigation devices (PNDs).*
- Continuing improvements in GPS/GNSS receiver firmware that increase reliability, increase sampling rates, lower noise, and make more multipath-resistant observations.
- Increasing interest in GPS/GNSS receivers based on low-cost navigation instruments, but that have carrier-phase data-processing capabilities added (and can be used to smooth pseudorange measurements as well). These are the single-frequency, L1-tracking receivers, but in the coming decade this category will also likely include dual-frequency receivers tracking the L1 and L5 signals.

- Increasing use of development kits to permit the customization of systems for specific applications, including software-based receiver designs (Section 10.4.4.5). *GNSS will become increasingly integrated within complex mobile positioning, telecommunications, and control systems.*
- Continuing deployment of permanent, continuously operating GNSS receivers for base station operations to support a range of navigation, surveying, and geodetic applications. *This will be coupled with improvements in network communications and ubiquitous provision of wireless access to the Internet.*
- Real-time operation will be the norm for all GNSS surveying applications. *Increasingly the user will not operate a base station, but will be relying on real-time augmentation services based on continuously operating reference receiver networks.*
- Increasing variety of positioning techniques, often optimized for specific applications. For example, real-time GNSS-based deformation monitoring systems (including raw data processing, trend analysis, etc.) are best implemented at a central server, and not on the user receiver hardware itself.
- Increasing efficiency in GNSS carrier phase–based positioning as a result of a range of improvements in ambiguity resolution algorithms. *The "Holy Grail" is a reliable single-epoch ambiguity resolution enabling instantaneous centimeter-level accuracy positioning.*
- Increasing integration of GPS/GNSS with other navigation technologies such as gyroscopes, magnetic compasses, tiltmeters, accelerometers, vision systems, laser scanners, and microwave signals from terrestrial transmitters and non-GNSS satellites. *The development of multisensor systems able to operate in environments even when there is little or no GNSS signal availability.*

Many of these trends arise from developments in several crucial areas. These are briefly discussed in the following sections.

15.1.1 Some GPS/GNSS Hardware Trends

There are two basic developments: (a) the GPS receivers (and ancillary equipment) will continue to improve, and (b) integrated sensors (in which GPS is but one component) will become more common. Both developments will not only assist traditional surveying and mapping applications, but will also mean that the technology will increasingly be able to address many other applications that require high-accuracy positioning information.

The dramatic improvements in the *productivity* of carrier phase–based techniques have been primarily the result of several innovations: improved GPS receivers, implementation of real-time algorithms, and the development of sophisticated "on-the-fly" ambiguity resolution (OTF-AR) algorithms. The top-of-the-line dual-frequency GPS receivers are capable not only of making L1 and L2 carrier phase measurements, but also precise pseudorange measurements. Improvements in antenna technology, signal processing, and tracking electronics have resulted in better quality measurements and lower multipath disturbance, both of which have resulted in a shortening of the "time-to-AR" to just a few seconds if the geometric and environmental conditions are right. Only with a significant increase in the number of GNSS satellites/signals will OTF-AR performance improve further.

The key to the development of *next generation* receivers is new chip designs that include the fabrication of a complete GPS receiver on a single chip, with tracking channels also able to track other GNSS signals such as from the GLONASS constellation. This has already resulted in a further reduction in size, power consumption, and cost of basic receiver components that has led to an increased integration of the GPS technology within consumer electronic devices. Examples include the integration of GPS/GNSS into PDAs, PNDs, cellular phones, laptop and palmtop computers, and so on. Although GPS receivers will be an *enabling* technology within more complex integrated systems, even if our attention is restricted to "geospatial sensors" these developments are indeed significant. For example, in the near future *all* surveying/mapping instruments will likely have a GPS/GNSS receiver embedded within them. This would permit all geospatial measurements to be ultimately linked via GNSS to the global datum.

Perhaps the most significant advance in precise GPS in the last decade and a half has been the development and widespread adoption of the *real-time kinematic* (RTK) technique (Section 11.2.2). The continued development of RTK-GPS techniques, including the use of new wireless communication technologies, ensures that cm-level accuracy GPS/GNSS will be the preferred technology for time-critical, outdoor positioning applications such as engineering stake-out, machine guidance and control, and online deformation monitoring systems. The trend is for RTK-GPS techniques to become *almost* as easy and reliable as the standalone GPS positioning capability enjoyed by casual users. This is discussed further, in the context of the provision of the appropriate *GPS/GNSS infrastructure* to support precise positioning, in Section 15.3.

15.1.2 COMMENTS ON GPS/GNSS SOFTWARE TRENDS

While it is fair to say that software has always been a crucial component of GPS receivers, nowhere has software been as important as in the case of carrier phase–based techniques. From the beginning of the use of GPS for geodetic surveying, sophisticated data processing software has been required to convert the raw GPS measurements to coordinates. In particular, the OTF-AR algorithm has been most responsible for the increased productivity and flexibility of today's GPS surveying techniques. However, most GPS researchers admit that the development of data processing algorithms has nearly reached its limit in terms of ensuring cm-level positioning accuracy with the minimum amount of tracking time. It is likely that further improvements in carrier phase–based techniques will therefore be restricted to:

- New strategies of combining data from mobile GPS users with data from permanent continuously operating GPS reference stations.
- The transmission of additional signals by GPS.
- The transmission of other GNSS and navigation satellite signals.

- Augmenting GPS/GNSS measurements with other data, such as from a variety of ground-based or space-based systems.
- Special applications in which additional constraints can be imposed (e.g., fixed distance between moving antennas as in the case of GNSS-based attitude determination systems).
- The combined processing of GPS/GNSS with inertial navigation systems measurements in modern positioning-and-attitude determination systems.

The first three of these issues are discussed further in Sections 15.3 through 15.5.

15.2 INSTITUTIONAL ISSUES

A range of *institutional issues* impact on the adoption and use of GPS technology for geospatial applications. One of these institutional issues, the installation of increasing numbers of permanent GNSS receivers around the world, deserves special mention and will be discussed in Section 15.3.

15.2.1 GPS System Control

For many years, there has been a "tug-of-war" between civilian and military users. The former wishing to have a greater *say* in the running of GPS, while the latter have sought to maintain a tight control on what is considered a strategic asset in modern warfare. The dilemma is that GPS is a "dual use" technology, investing significant advantages to both classes of users. The debate concerning GPS *control* has been played out mainly in the United States, between industry and several user communities on the one hand, and the Department of Defense on the other. However, while the rest of the world appreciates the enormous benefits of GPS, there has also been increasing unease with the degree of reliance on GPS for many crucial applications. The following comments may be made in this regard:

- *Civilian users far outnumber military users.* Yet, there are few means by which civilian users can influence the Department of Defense. *GPS modernization* plans (Section 15.4.1) do, however, go a significant way toward addressing the needs of civilian users for increased accuracy and signal availability.
- *GPS is a global utility.* Many users outside the United States resent the fact that they are relying on the use of a system controlled by the military of one country. This is therefore one of the reasons why several countries have plans to develop new GNSS, as well as *regional navigation satellite systems* (RNSS) and augmentation systems (Section 15.5).
- *The control of access to the highest single-receiver performance by the U.S. military* has meant the imposition of policies of selective availability (SA) on March 25, 1990, and anti-spoofing (AS) on January 31, 1994. Both policies have resulted in extra cost and a reduced level of service for civilian users. (SA was abandoned on May 1, 2000—OoP 2000; and the impact of AS on civilian users will diminish as progress is made in the implementation of the *GPS modernization program*—Section 15.4.)

In late 2004, the National Space-Based Positioning, Navigation, and Timing (PNT) Executive Committee was established (PNTexec 2009), and among its responsibilities, the Executive Committee will:

1. Advise and coordinate on matters pertaining to the U.S. space-based PNT policy with member departments and agencies
2. Resolve interdepartmental issues regarding space-based PNT systems and services
3. Establish and oversee a process to conduct studies and assessments pertaining to space-based PNT that have national benefit beyond the scope of a single department or agency
4. Establish and implement an interagency process for the full and appropriate consideration of national security, homeland security, and civil requirements … to ensure that the United States remains the preeminent military space-based PNT service and that U.S. civil space-based PNT services exceed, or are at least equivalent to, those routinely provided by foreign space-based PNT services

The Executive Committee has also established an Advisory Board with representation from many industry sectors, including international organizations.

15.2.2 GPS DATUM ISSUES

One of the first major impacts of GPS on many countries has been on *national geodesy*. GPS is unchallenged as the geodetic tool for establishing, maintaining, or monitoring national control networks. Under pressure from users to produce maps on the "GPS datum" (understood by most users to be WGS 84), many countries have redefined their national geodetic datums. Although there is some confusion in the general community between WGS84 and the International Terrestrial Reference Frame (ITRF), many (if not all) redefined datums are in fact some realization of the International Terrestrial Reference System (Section 7.2). Most international maps and charts are produced on the WGS84 datum, which essentially is equivalent to ITRF datums at the centimeter level. Other impacts of GPS on datum and control networks include:

- In addition to the traditional levels of national control network (first order/ class, second order/class, etc.), additional high-accuracy control points established using ultra-precise GPS geodesy techniques now form the "backbone" of new datums.
- There is increasing acceptance of international standards and associated services such as those provided by the International GNSS Service (IGS 2009). Many geocentric datums that were established in the last decade or so are therefore examples of the *densification* of the ITRF.
- Although GPS is used to establish 3-D coordinate networks, many applications still distinguish horizontal positioning/mapping from the determination of heights above sea level. *Hence, GPS, geoids, and leveling are inextricably linked.*

- With the increased efficiency of GPS for establishing control-point coordinates, there are proposals in some countries to reduce the number of monumented ground control marks.
- Where monumented control marks are established, they will be increasingly located where they are most needed and not on inaccessible hilltops, as in the past.

15.2.3 GPS ACCEPTANCE AND REGULATION

There are a range of issues that must be dealt with at a national, and even state/province level if GPS is to be accepted for certain applications. Some of these are

- The blurring of responsibilities for the provision of GPS infrastructure such as, for example, permanent GPS reference stations to support real-time DGPS, geodesy, integrity monitoring, etc. Greater efficiencies could be won, for example, if such infrastructure were *multifunctional*, rather than having many stations being operated by different organizations and agencies.
- How to regulate GPS services provided by real-time service providers? Should there be independent *quality assurance*? It is not clear, who is liable for damages due to faults in the GPS system, or its local augmentation, including reference station networks?
- Issues such as GPS system testing, accreditation of GPS surveyors, procedures for GPS cadastral surveys, legal traceability of GPS results, etc., are increasingly being raised.
- Increasingly, high-accuracy GPS positioning is not only the prerogative of the geospatial professionals, with many non-surveyors implementing and using cm-level accuracy technology. This raises a host of legal and administrative issues concerned with education, legal liability, community safeguards, etc.
- GPS and GIS are crucial technologies for the development of the so-called *Spatial Data Infrastructure*. While there is increasing recognition of SDI at the international level, most of the groundwork is being done at the national level. Hence, the challenge is to develop national guidelines and procedures for the SDI that are also compatible with international initiatives and standards.
- The challenge for government is to promote the adoption of GPS and the development of associated services (all of which contribute to a nation's productivity) without undermining customary practices that have well served the community in the past. This is generally addressed at the application or industry-specific level.

15.3 CONTINUOUSLY OPERATING GPS/GNSS NETWORKS

Continuously operating reference station (CORS) networks have been established in many countries to initially support geodesy-type applications. However, increasingly, such GPS *infrastructure* is used to support a range of other high-accuracy applications, including surveying.

15.3.1 CORS Networks and Geodesy

GPS in the 1980s was almost exclusively used for geodetic control surveys, and the inter-receiver distances were at first several tens of kilometers, being the average distance between first-order geodetic control marks. However, at about this time, GPS was also proving itself to be an effective tool of *space geodesy* for measuring crustal motion and establishing the global reference frame (Seeber 2003). As a consequence, the distances between receivers increased progressively to hundreds and then thousands of kilometers, while the relative accuracies simultaneously increased. These developments ensured cm-level accuracy within GPS receiver networks even as inter-receiver distances grew significantly. These GPS geodetic stations inevitably became permanent reference stations for (a) the monitoring of the station motion itself (due to horizontal and vertical crustal motion), (b) defining modern geocentric geodetic datums at the national level, and (c) the extension and increasing density of the geodetic control (ground mark) networks using GPS techniques.

It is important to recognize the significant contribution of the "super-network" of reference stations of the *International GNSS Service* (IGS) to geodesy and to the GNSS community in general. The IGS was established in January 1994 as a service of the International Association of Geodesy (IAG 2009). Since June 1992, the IGS—originally known as the "International GPS Service for Geodynamics," from 1999 simply as the "International GPS Service," and finally since March 2005 as the "International GNSS Service"—has been making freely available to all users, raw GNSS tracking data from its global network (Figure 15.1), and high-accuracy satellite ephemerides and other derived products. Several hundred globally distributed GPS receivers (many now with GLONASS tracking capability) operate on a continuous basis, contributing data to a variety of IGS analysis centers and other users (Dow et al. 2007, IGS 2009).

The mission of the IGS is "to provide the highest-quality GNSS data and products in support of the terrestrial reference frame, Earth rotation, Earth observation and research, positioning, navigation, and timing, and other applications that benefit

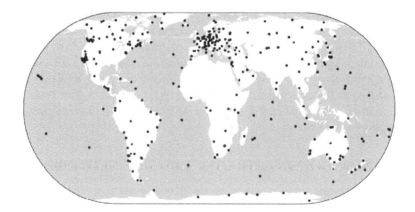

FIGURE 15.1 Global distribution of IGS stations. (From http://igscb.jpl.nasa.gov/network/netindex.html.)

society" (Dow et al. 2007). The IGS activities are fundamental to scientific disciplines concerned with climate, surface weather, sea level change, gravity, space weather research, and more. However, the IGS also supports many other applications including precise navigation, machine automation, time transfer, and surveying and mapping.

15.3.2 CORS Networks and GPS Surveying

One of the defining characteristics of modern carrier phase–based GPS systems is the gradual abandonment of the requirement for user ownership and/or operation of GPS base station receivers. CORS networks offer the possibility of users being able to perform differential positioning using a single GPS user receiver, hence *halving their equipment costs*. Post-processed implementations require that data from the *nearest* GPS base station is downloaded to the user's computer before processing can be carried out (Section 14.4.2). Alternatively, user GPS receiver data can be uploaded to online web services (Section 14.4.3), where the coordinate computations are performed automatically and made available to the user (see Table 14.2 for a summary of some of these services).

There has been a very strong trend in recent years for CORS networks to provide RTK-GPS services. (These networks are sometimes referred to as "real-time networks" (RTN), to distinguish them from standard CORS networks, but with the majority of recently established private and government CORS networks supporting surveying and mapping users via real-time services, it is not necessary to explicitly make this distinction, and hence, the acronym "CORS" will continue to be used in this book in place of "RTN.") Current carrier phase–based RTK-GPS techniques require the reference receiver to be within 20–30 km of the user receiver to ensure reliable OTF-AR. However, the establishment of a network of GPS reference receivers at a density to support all (or most) GPS surveys may not be feasible (Section 14.4.1). In order to overcome the baseline length limitation, the development of network-based methods to model and predict differential errors commenced in the mid-1990s (e.g., Wübbena et al. 1996, Raquet and Lachapelle 2000). In essence, the carrier-phase observations from all reference stations are streamed to a data center where the differential errors are modeled for the entire region covered by the network of reference stations. Corrections to the carrier-phase observations for the reference stations and users are calculated. These corrections can be broadcast to users in an analogous manner to pseudorange-based wide area DGPS (Section 11.3.3) and space-based augmentation systems (SBAS–Sections 14.3.1.2 and 15.5.4). Once they are applied to the carrier-phase observations, the OTF-AR procedure is effectively equivalent to using a single reference station.

Several commercial real-time network-based services operate which, *unlike* conventional RTK-GPS techniques, do not require L1 and L2 AR (Section 14.3.1.1). Differential corrections are transmitted to users via geostationary satellites (as with WADGPS and SBAS services) and although carrier-phase observations are indeed processed, only the "widelane" ambiguities (formed by differencing the L1 and L2 measurements, Section 9.2.1.2) are resolved to their integer values. As a consequence, positioning accuracy is only at the decimeter level. However, the advantage is that distances between reference and user receivers may be many hundreds of kilometers. Such services therefore address the accuracy "gap" between conventional WADGPS/SBAS techniques on the one hand and true centimeter-accuracy network-based RTK-GPS systems on the other.

15.3.3 NETWORK-BASED GPS POSITIONING

Commercial network-based RTK-GPS services capable of cm-level accuracy have been established over the past decade in many countries (Section 14.3.2.2). As described above, in such scenarios, as far as the user is concerned, the GPS hardware and field procedures will be no different to single reference station RTK-GPS systems, except that the distance to the CORS network stations is of the order of many tens of kilometers as opposed to 20–30 km or less. However, there are a number of other advantages of *network-based processing strategies* apart from permitting an increase in inter-receiver distances:

How can network-based techniques improve GPS surveying performance? This question can be answered in a number of ways. If the baseline length constraint is the crucial factor that affects the static or kinematic GPS, then overcoming this constraint using network-based processing techniques can result in significant performance improvements. If improving performance were associated with lowering receiver costs, then a multi-reference receiver network would allow the use of single-frequency GPS receivers instead of expensive dual-frequency receivers. If "time-to-AR" were the critical performance indicator, then network-based processing strategies would contribute to faster AR as well as enhancing the reliability of AR, even for baselines many tens of kilometers in length. However, the implementation of a GPS-positioning technique based on data from a network of GPS base stations is much more complicated than the standard single-base station scenario.

Rarely would the benefits of improved positioning performance be great enough for a single user that they would be compelled to establish and operate their own CORS network. Instead, a *service* could be provided by a government agency, or by private companies on a fee-for-service basis. Given that CORS networks can address many applications (including non-positioning ones), there is increasing interest in providing enhanced user services based on the combined processing of *all* reference receiver data. Such CORS-based surveying services are being established to address the GPS survey needs of large metropolitan areas or even states and countries, and increasingly these services are offered to support the "network-RTK."

15.3.4 NEXT GENERATION CORS INFRASTRUCTURE

The expression "GNSS system of systems" is now increasingly used for receiver hardware, or data-processing software, or CORS infrastructure, that can track signals, process measurements, or provide user services from the different GNSS/RNSS constellations, which will be operational during the next 5–10 years (Sections 15.4 and 15.5). It is worth speculating on what a future "GNSS system of systems" CORS infrastructure may consist of (Rizos 2008):

1. True "GNSS system of systems" CORS receivers able to track *all* GNSS signals—across four distinct L-band frequencies—will likely be established with relatively large inter-receiver separations, perhaps of the order of several hundred kilometers or more. These may be the future backbone of the IGS and other regional scientific GNSS networks.
2. Lower cost multi-constellation GNSS CORS receivers, probably with dual-frequency tracking capability, and possibly established at closer receiver spacings, from just several kilometers apart (to support structural deformation monitoring and single-base RTK applications) and up to several tens of kilometers (to support most DGNSS and network-RTK users). The interoperable frequencies are most probably the L1 (1575.42 MHz) and L5 (1176.45 MHz) frequencies that all GNSS are likely to be transmitting.

It should be noted that all, or even most, of the signals from a "GNSS system of systems" will not be transmitted before 2013 at the earliest, hence, the current investment in CORS infrastructure for the IGS (and other government and private CORS networks) will continue to be in GPS + GLONASS capable receivers. However, the upgrade of the CORS infrastructure after 2013 will have to incorporate multi-constellation GNSS tracking capability.

Next-generation GNSS will also have important implications for the development of public and private CORS networks in other respects. For example, with the benefits of extra satellite visibility (or availability), and carrier phase–tracking of three or more different transmitted L-band frequencies, the inter-receiver separations within GNSS CORS networks could be relaxed even further. "Single-base RTK" could be possible over baseline lengths that are over a hundred kilometers with cm-level accuracy, albeit with lowered reliability *vis-à-vis* "network-RTK" techniques. Furthermore, no discussion on future CORS infrastructure would be complete without addressing the question: *Who will establish and maintain future CORS networks, government or private service providers?* Government agencies and organizations typically justify the costs of implementing CORS networks by citing the principle of "preventable costs." This is similar to the strategy used to finance the establishment of classical geodetic control networks decades earlier. The return on the original investment is not measured in terms of the revenue earned, but justified as a means of keeping the costs borne by the community lower than the alternative (i.e., having no geodetic control infrastructure). This approach also encourages network standardization and avoids the establishment of a patchwork of private, ad hoc networks for project-specific purposes. Rizos (2008) identified the following models for CORS networks, and services derived from them:

1. Institutional CORS infrastructure with no commercial services
2. CORS infrastructure supporting government commercial services
3. Government CORS infrastructure but data licensed to the private sector
4. Cooperative privately owned CORS infrastructure that operate commercial services
5. Privately owned CORS infrastructure that operate commercial services

Which of these models—or variations of the above—will prevail in different countries is impossible to answer at this point in time. There are already several different *business models* for CORS services (real-time positioning, or raw GNSS measurements) in the United Kingdom, Germany, United States, Japan, and elsewhere (Section 14.4.1). However, in many countries the trend is increasingly toward option (3), where the CORS operator *wholesales* GNSS data to one or more *service providers*, who then market RTK-GNSS services to different user communities.

15.4 GPS SATELLITE AND MODERNIZATION PLANS

In 1996, former U.S. President Bill Clinton gave an assurance that GPS would be freely available well into the first decades of the new millennium (OST 1996). Since then, a number of initiatives have been launched which promise to improve the performance of GPS. Before discussing these let us summarize the current (March 2009) situation as far as satellite-based PNT is concerned:

- The current GPS constellation consists of 31 active satellites, with 32 being the maximum number that can be accommodated using the current PRN codes loaded within the GPS receiver firmware (although several others are defined for SBAS satellites).
- In 1998, the first of the Block IIR *replenishment* satellites became operational, replacing the older Block IIA satellites. The current 31-satellite constellation consists of 13 Block IIA, 12 Block IIR, and 6 Block IIR-M satellites. The oldest satellite is PRN24 (of the Block IIA family), launched on July 4, 1991. (Note, GPS nominally comprises only a 24-satellite constellation, hence the extra satellites are "doubling-up" on old satellites, and do not fly in their own orbital "slots" and hence do not necessarily improve availability in urban or forested environments, where there is restricted sky view.)
- The Block IIR-M satellite PRN17 was launched on September 26, 2005, and is the first of the GPS satellites that broadcast a new civilian signal, in addition to the full set of "legacy" signals transmitted by the Block IIA and IIR satellites (Table 8.5).
- The design life of GPS satellites is between 7.5 and 10 years, and typically these satellites have exceeded their design life—the current average age of GPS satellites is 8.9 years. However, the GPS operators have been reluctant to turn off old satellites, and this has meant that next-generation replenishment satellites have not been able to be launched as fast as many in the civilian user communities would prefer.
- The first of the Block IIF *follow-on* satellites is scheduled for launch in 2010, and will broadcast a *third* civilian signal at the L5 frequency, as well as the full set of "legacy" signals transmitted by the Block IIA, IIR, and IIR-M satellites.
- The Russian Federation's GLONASS satellite constellation will be fully deployed by 2010, becoming the second operational GNSS (Section 15.5.1).

- In 1999, the European Union committed itself to the development of a global navigation system known as "GALILEO" (Section 15.5.2).
- In the last few years, several countries have announced plans to develop their own global navigation system (China), or their own *regional naviga-tion system* (India, Japan, and South Korea) (Sections 15.5.3 and 15.5.4).
- The U.S. Federal Aviation Administration's Wide Area Augmentation System (WAAS) is operational, and over the next few years, the European and Japanese SBASs will also be brought into service for the aviation com-munity, as well as other users (Section 14.3.1.2 and 15.5.4).

The GPS improvements identified for implementation out to the year 2020 are (see NavCen 2009, PNTexec 2009):

- New signals and changed signal codes, including the separation of civilian and military codes and frequencies
- Abandonment of the *selective availability* policy, and mitigation of the effect of the *anti-spoofing* policy on civilian users
- Improvements to the GPS control segment
- Increased number of satellites and the power levels of the transmitted signals
- Increased use of SBASs

The Presidential decision document on GPS of March 29, 1996 (OST 1996), stated that both the L1 and L2 frequency bands would be available for civilian use, and that a third civil frequency signal would be transmitted (Van Dierendonck and Hegarty 2000). The former would alleviate the impact of the anti-spoofing policy on civilian users through the broadcast of a new "open" code on the L2 frequency that would more easily permit dual-frequency operation (Fontana et al. 2001), while the latter set off an intensive search for new frequencies that would satisfy a range of require-ments, including the military and the aviation users.

An improved PRN code family (instead of the current L1 C/A-codes) on the L2 frequency of GPS (the so–called *L2C-code*) is being implemented on the last 8 Block IIR satellites—referred to as the "Block IIR-M" satellites, of which at the time of writing six had been launched—to enable civilian receivers to better account for ionospheric error, as well as to be more immune to interference and multipath. The L2C-code is an "open" signal with a published PRN code algo-rithm that can be easily implemented in civilian receivers (L2Csignal 2004). The former U.S. vice president Al Gore announced on January 25, 1999, that the third frequency had been identified at 1176.45 MHz (just below the GPS L2 frequency)—in the aeronautical radio navigation service band—that satis-fies aviation *safety-of-life* requirements (Zeltser 2000, L5signal 2003). The new GPS frequency plan is illustrated in Figures 8.14 and 15.2, and summarized in Table 8.5.

The launch schedule to replace existing satellites is difficult to predict but full operational capability for the L2C will not be declared until 24 satellites in the con-stellation are broadcasting the new signal (8 GPS Block IIR-M plus 16 GPS Block

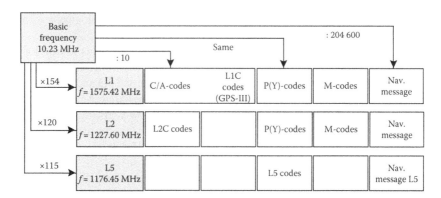

FIGURE 15.2　Proposed structure of the new GPS L1, L2, and L5 satellite signals (cf. Figure 8.1).

IIF satellites), which is not expected to occur until about 2016. True L1 and L2 dual-frequency civilian tracking capability, using the C/A-codes and L2C-codes, has incrementally become available since September 26, 2005, with the launch of the first Block IIR-M satellites. There are not yet enough satellites broadcasting L1 and L2 open signals for there to have been a dramatic impact on survey/geodetic-grade user equipment design, capability, or cost. The older GPS Block IIA (the last one launched on November 6, 1997) and GPS Block IIR (the last one launched on November 6, 2004) satellites will continue to broadcast signals until they are decommissioned. It is possible that by 2013, there may still be several functioning GPS Block IIR satellites, and *new-generation* dual-frequency receivers may *not* be able to make measurements on the L2 frequency because Block IIR satellites cannot broadcast the open L2C-code signal (only the encrypted Y-code on the L2 frequency). Furthermore, the U.S. government has announced that it proposes to discontinue the broadcast of the encrypted Y-code on L2 from 2020 by the modernized GPS satellites, and hence old dual-frequency receivers will not be able to make measurements on the L2 frequency.

The L5 frequency will be available for the first time on the Block IIF satellites, which have their first scheduled launch in 2010. Fully operational *triple-frequency* L1–L2–L5 tracking will commence when 24 satellites are launched—a combination of 16 GPS Block IIF and 8 GPS Block III satellites. This is unlikely until 2018, although suitably equipped civilian users will be able to take advantage of the new codes on (some of) the signals well before then. The dual- and triple-frequency carrier phase measurements will significantly benefit "on-the-fly" ambiguity resolution, and reliable single-epoch ambiguity resolution will be a reality when at least four Block IIF/III satellites (with low PDOP) can be tracked simultaneously.

The next generation Block III family of satellites will incorporate the extra L2 and L5 signals of the Block IIR-M and Block IIF satellites, as well as a new PRN code on the L1 frequency (the so–called *L1C-code*), which will be interoperable with GALILEO's E1 signal (L1Csignal 2006). Note, to preserve backward compatibility with legacy user equipment, all current and planned Block II signals will also be broadcast (apart from the announced plan to cease broadcasting the Y-code on L2 in 2020). The 30 Block III satellites are planned for launch from the year

2014 onward. What remains unchanged throughout the modernization process will be (a) GPS will be controlled and operated by the U.S. Department of Defense, and (b) there will be no user charges introduced.

The military dependence on the C/A-codes to provide the timing information necessary to acquire the P(Y)-codes is progressively being wound back with the transmission of new military codes, called the *M-codes* (these codes are encrypted) on the Block IIR-M (and following) satellites. These M-codes are transmitted on a "split spectrum" scheme, with their carrier frequencies offset 12 MHz above and below the centers of the GPS L1 and L2 frequencies (Figure 8.14). Such a scheme will at last ensure that the civilian and military users will have separate signals as well as codes, albeit transmitted by the same constellation of GPS satellites.

These plans are welcomed, and should cement the dominant role of GPS as the *de facto* standard for satellite system performance, and a crucial component of any international *interoperable* global navigation satellite system.

15.5   OTHER SATELLITE-BASED POSITIONING SYSTEMS

GPS is currently the only fully operational GNSS. Next-generation GNSSs will include not only the United State's modernized GPS and planned GPS-III (discussed earlier), but also Russia's revitalized GLONASS, Europe's GALILEO system, and China's planned COMPASS system. Furthermore, a number of space based augmentation systems (SBASs) and regional navigation satellite systems (RNSSs) will add extra satellites and signals to the multi-constellation GNSS/RNSS "mix."

15.5.1   THE GLONASS SYSTEM

The abbreviation GLONASS is derived from the Russian "Global'naya Navigatsionnaya Sputnikovaya Sistema" (Hofmann-Wellenhof et al. 2008). The first four operational GLONASS satellites were launched by the then Soviet Union in January 1984, with deployment of a full 24-satellite constellation completed in the mid-1990s shortly after GPS's full operational capability was achieved. The design of Russia's GLONASS is similar to GPS except that each satellite broadcasts its own particular frequency, all modulated with the same PRN code. This is known as a frequency division multiple access (FDMA) scheme, as opposed to the code division multiple access (CDMA) scheme used by GPS (Section 8.1.1). Unlike GPS, the GLONASS ground control segment is currently entirely located in Russia and some of the former territories of the Soviet Union, although there are indications that Russia plans to expand the GLONASS control station network beyond its borders, including stations in the southern hemisphere. For a detailed description of GLONASS see GLONASS (2002), Prasad and Ruggieri (2005), and Hofmann-Wellenhof et al. (2008). Information on the status is available from the Roscosmos Information Analytical Center Web site at GLONASS (2009).

Following the dissolution of the Soviet Union, the Russian Federation initially struggled to find sufficient funds to maintain GLONASS. Renewed national confidence, improved economic fortunes of the Russian Federation, and the recognition that GNSS is a critical military technology has resulted in the Russian government commencing a program to *revitalize* GLONASS, with a planned initial 24-satellite constellation by 2010. The basic characteristics of GLONASS as a fully operational GNSS are

- 24 satellites, distributed in three orbital planes (at the time of writing there were 19 functioning satellites).
- 64.8° inclination, 19,100 km altitude (hence an 11 h 15 min period).
- Current GLONASS-M satellites support dual-frequency signal transmission: 1598–1606 MHz for the GLONASS L1 band and 1242–1249 MHz for the GLONASS L2 band.
- Each satellite transmits a different frequency on "G1" (=1602 + K × 0.5625 MHz; K in range [−7, 6]) and "G2" (=1246 + K × 0.4375 MHz; K in range [−7, 6]).
- From 2010 it is planned to launch GLONASS-K satellites with improved performance, and these will also transmit a third civil signal known as "G3" in the aeronautical radio navigation services band near (but not identical) to GPS's L5 frequency. Current information suggests that the "G3" frequency band will be 1198–1208 MHz. However, a full constellation broadcasting three sets of civil signals is unlikely before near the end of the next decade.
- Spread-spectrum PRN code signal structure based on an FDMA scheme. Recently there has been an announcement that CDMA signals on the L1 and L5 frequencies will be transmitted by GLONASS-K satellites in order to make these signals *interoperable* with other GNSS.
- Global coverage for navigation based on simultaneous pseudoranges, with an autonomous positioning accuracy of better than 20 m horizontal, 95% of the time.
- The GLONASS datum "PZ90.02" is not WGS/ITRF. However, PZ90.02 is "close" to ITRF—at the 20–30 cm level—and plans are to progress to an ITRF-equivalent datum.
- Commitment to provide comparable single-point positioning performance to GPS and GALILEO by 2011.
- Can also provide a different level of service to military users compared to civilian users, as in the case of GPS.

Although the frequencies of GPS and GLONASS are different, a single antenna can track the transmitted signals. The data-modeling challenges for integrated GPS+GLONASS processing have already been addressed, and survey-grade receivers capable of making GPS and GLONASS measurements have been available for many years. These combined receivers have demonstrated a marked improvement in reliability and availability in areas where satellite signals can be obstructed, such as in urban areas, steep terrain, or in open-cut mines. All major manufacturers of survey-grade GNSS receivers now supply integrated GPS+GLONASS receivers.

Efforts are being made to increase the degree of *interoperability* of GPS and GLONASS (and other GNSS/RNSS as they come online) by having frequency overlaps at the L1/G1 and L5/G3 bands, at the very least. (This would certainly satisfy the International Civil Aviation Organization—ICAO.) In addition, the recent announcements by the Russian Federation of plans to adopt an ITRF-type datum, and the transmission of CDMA signals on future generations of GLONASS satellites are significant developments. Interoperability, for the benefit of all users, is one of the goals of the United Nation's Office of Outer Space Affairs *International Committee on GNSS* (ICG 2009).

15.5.2 THE GALILEO SYSTEM

Perhaps the most exciting development on the future of GNSSs was the decision by the European Union to develop "GALILEO." For a detailed description of GALILEO, see GALILEO (2009), Prasad and Ruggieri (2005) and Hofmann-Wellenhof et al. (2008). The following is a summary of the characteristics of the proposed GNSS (Rizos 2008):

- The GALILEO design calls for a constellation of 30 satellites (including three in-orbit spares) in a similar *medium earth orbit* (MEO) configuration to GPS—but at an increased altitude (approximately 3000 km higher than GPS) which will enable better signal availability at high latitudes—in three orbital planes inclined at about 56°.
- GALILEO satellites will broadcast signals compatible with the L1/G1 and L5/G3 GPS and GLONASS frequency bands. Those GALILEO frequencies are designated as "E1" and "E5a/E5b." GALILEO will also broadcast in a third frequency band at "E6," which is not at the same frequency as the GPS L2 or GLONASS G2.
- There will be up to *ten* trackable signals—though not all will be "open" (Hein et al. 2007a,b).
- GALILEO proposes to offer *five* levels of service, using different combinations of "open" and "authorized" trackable signals (GALILEO 2009):

1. The *open service* (OS) uses the basic L1/L5 frequency band signals, "free-to-air" to civilians, offering PNT performance similar to single- or dual-frequency GPS and GLONASS.
2. The *safety of life* (SoL) Service allows similar accuracy as the OS but with increased guarantees of the service, including improved integrity monitoring to warn users of any problems. This is intended for mission-critical users such as aviation and maritime navigation, machine guidance, etc.
3. The *public regulated service* (PRS) is to be available to E.U. public authorities providing civil protection and security (e.g., police, border patrol, and quasi-military), with encrypted access for users requiring a high level of performance and protection against interference or jamming.
4. The *search and rescue service* (SAR) is designed to enhance current space-based services (such as COSPAS/SARSAT) by improving the time taken to respond to alert messages from distress beacons, as well as by providing position information to SAR organizations.
5. The *commercial service* (CS) allows for tailored solutions for specific applications based on providing better accuracy (via DGNSS correction messages), improved service guarantees and higher data rates. This is proposed to be a fee-based service.

- The design of the GALILEO open service signal at the E1 frequency is intended to be largely identical to the GPS Block III L1C signal (L1Csignal 2006), a significant concession to GNSS *interoperability.*
- The GALILEO ground control segment has elements similar to the GPS or GLONASS networks of tracking stations and master control stations. Forty ground stations will be deployed globally, and the GALILEO datum will be (for all intents and purposes) the ITRF.
- In the case of GPS and GLONASS, *augmentation systems* to improve accuracy or reliability are operated completely external to the system architectures. Such services are available from third parties such as FUGRO's Omnistar and NAVCOM's Starfire (Section 14.3.1.1), or maritime DGPS beacons operated by national maritime safety authorities (Section 14.3.2.1), or RTK-capable CORS networks (Section 14.3.2.2). GALILEO plans to have a more open architecture whereby subsystems to improve service may be brought "inside" the system through a provision for *regional elements* and *local elements.* The GALILEO system architecture allows for regional *uplink stations* to facilitate those improved services tailored to local applications in different regions of the world.

The current development status is (Rizos 2008):

- It was originally planned that the majority of GALILEO's capital cost would be borne by a public private partnership (PPP)—with the necessary funds jointly raised by the European Union and a consortium of private companies (mainly from the aerospace and telecommunications sectors). However, negotiations progressed so slowly that by 2007 effectively the whole GALILEO project had stalled because of the reluctance of the private sector to invest in the satellite and ground infrastructure due to the high technical and economic risks. In late 2007, it was decided that 100% of the cost of designing, building, and deploying the satellite constellation and ground control segment would be borne by the European Union's taxpayers.
- Final operational capability (FOC) is planned for 2013, perhaps only a few years before GPS-III's FOC.
- GALILEO has moved out of its development phase and into the *in orbit validation* (IOV) phase. The first satellite (the Galileo IOV Experiment—GIOVE-A) was launched on December 28, 2005, and commenced broadcasting signals two weeks later. The second IOV satellite (GIOVE-B) was launched on April 27, 2008.
- GALILEO will be operated by a *civilian agency* under a PPP model whereby the European Commission owns the system (satellites, ground stations, etc.) as a public asset on behalf of the European Union, but another agency or company (or consortium of companies) will be responsible for day-to-day operations.

15.5.3   China's **COMPASS**

For a number of years China has had its own *regional navigation satellite system*, known as "BEIDOU" (2009). In 2006, China announced it would develop its own GNSS, commonly referred to as "COMPASS" (though sometimes BEIDOU/COMPASS—here we will distinguish between the original RNSS and the planned GNSS by referring to the latter simply as "COMPASS"). Few details are available (COMPASS 2009), but it appears that COMPASS will be modeled on GPS and GLONASS in as far as it will be a dual use system, fully funded by the government, with an open service (available to all users globally) as well as restricted, military/security signals. The COMPASS constellation will consist of 30 satellites in MEO, plus 5 in geostationary earth orbit (GEO) (Hein et al. 2007a,b). The first MEO satellite was launched on April 17, 2007. It appears that the ground control segment will be located only within the Chinese land territory. It was announced that FOC would be by the year 2015, although regional capability is expected several years before this date.

There is little evidence so far that COMPASS will have the same level of interoperability as other GNSSs have committed to. The United States and the European Union have separately commenced bilateral discussions with China; the former due to possible incompatibility with the GPS M-code, and the latter because of the concern that some of the COMPASS signal frequencies filed with the ITU overlay those of GALILEO's PRS. More recently, China has espoused the principles of "compatibility" and "interoperability," and hence, it is hoped that their signals will be interoperable and compatible with other GNSSs and RNSSs.

15.5.4   SBAS and **RNSS**

While GPS, GLONASS, GALILEO, and COMPASS are the generally recognized GNSSs (Hofmann-Wellenhof et al. 2008), a discussion on future navigation satellite systems would not be complete without mentioning space-based augmentation systems (SBASs) and regional navigation satellite systems (RNSSs).

SBASs are essentially extra satellites (in addition to the GNSS constellations) transmitting signals intended to address shortcomings of GNSSs for enhanced accuracy, availability, and integrity for civil aviation users. An SBAS constellation usually numbers two or three satellite in GEO. They are supported by a network of ground stations (similar to GNSS control segments) that collect data that is used to generate the differential GPS/GNSS and integrity messages. There are currently three distinct SBASs in orbit covering different parts of the world (Prasad and Ruggieri 2005): the United States's wide area augmentation system (WAAS); the European Union's European geostationary navigation overlay system (EGNOS); and Japan's MTSAT satellite augmentation system (MSAS).

It must be emphasized that all SBASs transmit the same set of signals so that, for example, a WAAS-capable receiver can also track EGNOS and MSAS signals. Although it was intended that only GPS receivers designed for aviation applications could track these signals, the reality is that many low-cost single-frequency GPS receivers today are SBAS-capable. The primary advantage of SBAS is the higher single-point positioning accuracy through the implementation of DGPS-capability, through the user receiver RF front-end. (The DGPS corrections are modulated on the broadcast L1 signal using special message types.) Such enhanced (meter-level) accuracy can benefit some geospatial applications. Other countries may also launch one or more SBAS satellites if they feel strongly enough about civil aviation in their airspace being a sovereignty issue. It is known that India, and possibly South Korea, will follow this path. (Only GNSSs that transmit signals in the L1 and L5 frequency bands will be certifiable by ICAO for use in aircraft, however because to date no satellites transmit L5/G3 signals, only the GPS L1 and GLONASS G1 signals have ICAO certification and hence current SBAS satellites only transmit in the L1/G1 frequency band.)

RNSSs, on the other hand, are intended to provide signal coverage over a nation or region. Some RNSSs are intended to provide full satellite navigation capability *independently* of any GNSS. Other RNSSs may just augment other GNSS signals (in which case they must be 100% interoperable with those GNSS signals). RNSS constellations are much smaller than GNSS constellations, perhaps consisting of only 5–7 satellites (depending upon the chosen orbit configuration). In general, an RNSS need not transmit at the standard GNSS L-band frequencies associated with GPS, GLONASS, and GALILEO. In fact, both China and India have filed applications with the ITU to use downlink frequencies other than those at L1/G1/E1, L2/G2, E6, or L5/G3/E5. The following additional comments may be made with respect to announced RNSSs (Rizos 2008):

1. The *quasi-zenith satellite system* (QZSS) is a multi-satellite *augmentation* system based on three satellites broadcasting GPS-like (plus SBAS and GALILEO-like) signals in a highly inclined (approximately 43° inclination) elliptical orbit (HEO) configuration (with "figure-8" ground track) that increases the number of navigation satellites available at high elevation angles over Japan (hence the term "quasi-zenith" to describe this satellite augmentation system). This would benefit users equipped with modified GNSS receivers (different PRN codes to those used by GPS would be needed) operating in areas with significant signal obstructions such as downtown and mountainous areas. The basic services will be offered for free, although there may be some fee-based services (e.g., transmission of DGNSS correction messages). It is expected that a demonstration QZSS satellite will be launched in mid-2010, with the remaining two satellites launched after the assessment of the results of the first satellite (QZSS 2009).
2. However, Japan has also made reference to a seven satellite RNSS known as JRNSS (*Japanese regional navigation satellite system*), which will consist of the three QZSS satellites plus the MTSAT satellite (part of MSAS) in a GEO, and three additional HEO satellites. This would permit full navigational capability without relying on another nation's GNSS. However, given its QZSS and MSAS "heritage," it is very likely that the JRNSS would have a very high degree of interoperability with other GNSSs. Hence, ultimately all GNSS users in the Asia-Australia region of the world will benefit from these extra signals.
3. The *GPS and geo augmented navigation* (GAGAN 2009) system is India's SBAS (and hence will be compatible with GPS, and perhaps

GLONASS/GALILEO as well). It is intended to support aviation in the Indian subcontinent/oceanic region.

4. The *Indian regional navigational satellite system* (IRNSS 2009) is also intended for use only in the Indian area of interest. Little is known about the IRNSS except for what appears in documents filed for frequency allocation with the ITU. It is a multifrequency system (transmitting two frequencies, one at L5 and the other in the S-band), and will comprise a seven satellite constellation in a combination of GEO and HEO (orbit plane inclination 29°). The first satellite launch will be in 2010, with an announced FOC in 2012.

5. For a number of years, China has had its own RNSS, known as BEIDOU (2009). It currently consists of five satellites in GEO. It is a two-way ranging system, used principally by the military, and is no doubt a system from which China has learned much about satellite-based navigation. Due to its design, it cannot be interoperable with other GNSSs/RNSSs. According to the Chinese government, BEIDOU will be an integral part of COMPASS. (There is some confusion regarding the names, as "BEIDOU" is the Chinese name for "COMPASS".)

6. South Korea has announced studies into the feasibility of developing both an SBAS and an RNSS.

These SBASs and RNSSs plans are part of a trend to the "regionalization" of navigation satellite systems. What is remarkable is that all these plans originate from countries in the Asia-Australia hemisphere. Therefore, suitably equipped users in Australasia and South/East Asia will "see" all these extra signals. However, in the case of high accuracy users, the next generation CORS ground infrastructure must be capable of supporting multi-constellation GNSS/RNSS tracking.

15.6   CONCLUDING REMARKS

The future of GNSS is bright. More satellites and more signals will be welcome by many user communities, and especially the geospatial industry. There are many benefits of the expected extra satellites and their signals in terms of continuity, accuracy, efficiency, availability, and integrity. However, the challenge will be to integrate all these satellite systems into one *global navigation satellite system of systems*. Although RNSSs may be justifiable in the nearer term, and at the narrow national level in terms of providing independent PNT capability, *compatibility*, and *interoperability* of all GNSSs, SBASs, and RNSSs would ensure the benefits of satellite-based PNT are far greater than what any individual system can provide. Let us hope that the compatibility and interoperability of GNSSs and RNSSs will be achieved.

REFERENCES

BEIDOU, 2009, http://en.wikipedia.org/wiki/Beidou_navigation_system (accessed March 15, 2009).
COMPASS, 2009, http://en.wikipedia.org/wiki/COMPASS_navigation_system (accessed March 15, 2009).

Dow, J., Neilan, R.E., and Rizos, C., 2007, The international GNSS service (IGS): Preparations for the coming decade, *20th International Technical Meeting of the Satellite Division of the U.S. Institute of Navigation*, Fort Worth, TX, September 25–28, pp. 2136–2144.

Fontana, R.D., Cheung, W., and Stansell, T., 2001, The modernized L2 civil signal leaping forward in the 21st century, *GPS World*, 12(9), 28–34.

GAGAN, 2009, *GPS and Geo Augmented Navigation*, http://en.wikipedia.org/wiki/GAGAN (accessed March 15, 2009).

GALILEO, 2009, *GALILEO—European Satellite Navigation System*, Web page, http://ec.europa.eu/comm/dgs/energy_transport/galileo/ (accessed March 15, 2009).

GLONASS Interface Control Document (version 5), 2002, downloadable from http://www.glonass-ianc.rsa.ru/i/glonass/ICD02_e.pdf (accessed March 15, 2009).

GLONASS, 2009, *Russian Space Agency Information Analytical Center GLONASS*, Web page, 2009, http://www.glonass-ianc.rsa.ru/ (accessed March 15, 2009).

Hein, G.W., Avila-Rodriguez, J.A., Wallner, S., Pany, T., Eissfeller, B., and Hartl, P., 2007a, Envisioning a future: GNSS system of systems, part 1, *Inside GNSS*, Jan/Feb 2007, pp. 58–67.

Hein, G.W., Avila-Rodriguez, J.A., Wallner, S., Pany, T., Eissfeller, B., and Hartl, P., 2007b, Envisioning a future: GNSS system of systems, part 2, *Inside GNSS*, Mar/Apr 2007, pp. 64–73.

Hofmann-Wellenhof, B., Lichtenegger, H., and Wasle, E., 2008, *GNSS Global Navigation Satellite Systems: GPS, GLONASS, Galileo, and More*, Springer-Verlag, Wien/New York, ISBN 978-3-211-73012-6, 516pp.

IAG, 2009, *International Association of Geodesy*, Web page, http://www.iag-aig.org/ (accessed March 15, 2009.

ICG, 2009, *International Committee on Global Navigation Satellite Systems*, Web page, http://www.unoosa.org/oosa/en/SAP/gnss/icg.html (accessed March 15, 2009).

IGS, 2009, *International GNSS Service*, Web page, http://www.igs.org (accessed March 15, 2009).

IRNSS, 2009, *Indian Regional Navigation Satellite System*, http://en.wikipedia.org/wiki/Indian_Regional_Navigational_Satellite_System (accessed March 15, 2009).

L1Csignal, 2006, *Navstar GPS Space Segment/User Segment L1C Interfaces*, downloadable from http://www.navcen.uscg.gov/gps/modernization/L1/IS-GPS-800_19_DRAFT_Apr06.pdf (accessed March 15, 2009).

L2Csignal, 2004, *Navstar GPS Space Segment/Navigation User Interfaces, IS-GPS-200D Updated for L2C Signals*, downloadable from http://www.navcen.uscg.gov/gen-info/IS-GPS-200D.pdf (accessed March 15, 2009).

L5signal, 2003, *Navstar GPS Space Segment/User Segment L5 Interfaces, IS-GPS-705*, downloadable from http://www.navcen.uscg.gov/gps/modernization/Number.pdf (accessed March 15, 2009).

NavCen, 2009, *U.S. Coast Guard Navigation Center GPS Modernization Information*, Web page, http://www.navcen.uscg.gov/gps/modernization/default.htm (accessed March 15, 2009).

Office of Science and Technology USA, 1996, *U.S. Global Positioning System Policy, U.S. Presidential Decision Directive*, March 29, see Web page http://www.navcen.uscg.mil/pubs/gps/white.htm (accessed March 15, 2009).

Office of the US President's Press Secretary OoP, 2000, *Statement by the President Regarding the United States' Decision to Stop Degrading Global Positioning System Accuracy*, May 1, see Web page http://www.ngs.noaa.gov/FGCS/info/sans_SA (accessed March 15, 2009).

PNTexec, 2009, *Space-Based Positioning, Navigation & Timing National Executive Committee*, http://pnt.gov (accessed March 15, 2009).

Prasad, R. and Ruggieri, M., 2005, *Applied Satellite Navigation: Using GPS, GALILEO, and Augmentation Systems*, Artech House Mobile Communications Series, Boston, MA, ISBN 1-58053-814-2, 290pp.

QZSS, 2009, *Quasi Zenith Satellite System Interface Specification*, Web page, http://qzss.jaxa.jp/is-qzss/index_e.html (accessed March 15, 2009).

Raquet, J. and Lachapelle, G., 2000, Development and testing of a kinematic carrier-phase ambiguity resolution method using a reference receiver network, *Navigation*, 46(4), 283–295.

Rizos, C., 2008, Multi-constellation GNSS/RNSS from the perspective of high accuracy users in Australia, *Journal of Spatial Science*, 53(2), 29–63.

Seeber, G., 2003, *Satellite Geodesy*, 2nd edn., Walter de Gruyter, Berlin/New York.

Van Dierendonck, A.J. and Hegarty, C., 2000, The new L5 civil GPS signal, *GPS World*, 11(9), 64–71.

Wübbena, G., Bagge, A., Seeber, G., Bäder, V., and Hankemeier, P., 1996, Reducing distance dependent errors for real-time precise DGPS applications by establishing reference station networks, *9th International Technical Meeting of the Satellite Division of the U.S. Institute of Navigation*, Kansas City, MO, September 17–20, pp. 1845–1852.

Zeltser, M.J., 2000, The status of aviation-related improvements, *U.S. Institute of Navigation Newsletter*, 10(1), Spring, also available at http://www.ion.org/vol101/vol10.htm.

Part III

Remote Sensing

16 Photogrammetry for Remote Sensing

Rongxing (Ron) Li and Chun Liu

CONTENTS

16.1 OVERVIEW

Photogrammetry is the science or art of obtaining reliable measurements by means of photographs (McGlone et al., 2004). Although the practice of photogrammetry is most commonly applied to aerial photography, its principles apply to other photographs and, by extension, to images collected by other instruments, including the remote-sensing systems described in Chapter 18. Chapter 4 provides an introduction to the concepts and algorithms used in photogrammetry. The practice of photogrammetry yields positional information extracted from the aerial imagery. The elevation information is sometimes used to compile large-scale topographic maps.

Traditionally, photogrammetry has been applied to aerial imagery collected by frame cameras, principally data with fine resolutions and a high overlap rate, whereas remotely sensed imagery has been collected by scanning or pushbroom sensors onboard satellites, collecting data at coarser resolutions and multiple spectral channels. As stated above, the former often forms the basis for producing 3D digital terrain models (DTM) and large-scale maps with accurate representations of the three-dimensional (3D) positions of objects. The latter often forms the basis for small-scale maps using digital image classification (Chapter 19).

However, recent developments in sensor technology and advances in sensor platforms have blurred these distinctions. This trend has extended the spectral sensing capability from a few channels (multispectral) to a high number of channels (hyperspectral) to improve object detection through refined spectral signals. Many pushbroom sensors (Chapter 18) have been used on airborne platforms to perform accurate 3D stereo mapping due to increasingly improved sensor modeling and platform control. Furthermore, very high resolution satellite imaging systems have been declassified by the military and made available for use in civilian applications. These systems have stereo capabilities, are often multispectral, with spatial resolutions comparable to those of airborne sensors. This chapter introduces the reader to applications of photogrammetric principles to these remotely sensed images, formerly considered as distinct from conventional photogrammetric practice. This chapter also highlights recent developments in geometric processing of new high-resolution satellite images. These developments have minimized the distinctions between photogrammetric applications in the realms of satellite-borne and airborne imagery.

The launch of a new generation of high-resolution (submeter) commercial earth-imaging satellites in the late 1990s and thereafter, including QuickBird and IKONOS, marked the start of a new era of space imaging for earth observations (Fritz, 1996; Li, 1998), providing the public with satellite imagery with qualities approaching that which previously could be associated only with aircraft imagery (Figure 16.1). With such detailed ground resolution, for example, products of most national digital mapping agencies, such as digital elevation models (DEMs), digital orthophoto quadrangles (DOQs), digital line graphs (DLGs), and digital shorelines (DSLs), can now be constructed (Ellis, 1978; Lockwood, 1997; U.S. Geological Survey, 1997). Revisit rates of 1–4 days, depending upon the particular satellite system and upon latitude, make it possible to map an area frequently without the special flight planning and scheduling that was required for aerial photogrammetric data acquisition. Stereo models thus formed are valuable for mapping product updating and accurate change detection.

Accurate elevation information is crucial to many spatial information products derived from satellite data. The SPOT imaging system has an enhanced resolution of 2.5 m in the panchromatic channel with a cross-track stereo capability. Elevation accuracies of around 10 m have been achieved with the aid of ground control points (GCPs) (Al-Rousan et al., 1997; Cuartero et al., 2005). The pushbroom imaging technique using one or more linear charge-coupled device (CCD) arrays has been adopted by the higher-resolution systems that can provide a so-called along-track stereo mode where stereo pairs necessary for deriving the horizontal and elevation information of objects can be acquired in quasi real time or real time during a single orbital pass. In contrast, cross-track stereo (which acquires stereo pairs from separate orbital tracks) requires additional time to allow the satellite to point to the same ground area from a neighboring track. Stereo mapping capabilities of similar airborne systems mounted onboard the space shuttle orbiter (which provides along-track stereo imagery with relatively high pointing accuracy) have been demonstrated (Heipke et al., 1996; Fraser et al., 1997). Thus, 3D geopositioning accuracy is closely related to stereo-imaging geometry and the quality of orbital pointing data, which in turn depend upon the design specifications of sensors and the satellite system. One of the ways for acquiring accurate pointing data is through the use of a star tracker, to be discussed in the following text. Table 16.1 presents technical specifications of the IKONOS system.

FIGURE 16.1   GeoEye image depicting Central London, representing properties of fine-resolution commercial satellite imagery. Photogrammetric analysis of stereo satellite imagery of this form permits the extraction of accurate positional and elevation data. Satellite image by GeoEye.

TABLE 16.1
Selected Technical Specifications
for the IKONOS System

Item	Technical Data
Pixel size	12 μm (panchromatic)
Ground resolution	0.82 at nadir (panchromatic)
Multispectral bands	
Blue (4 m)	0.45–0.52 μm
Green (4 m)	0.52–0.60 μm
Red (4 m)	0.63–0.69 μm
Near-infrared (4 m)	0.76–0.90 μm
Focal length	10 m
Image recording	CCD linear arrays (one for IKONOS I)
Orbit height	680 km
Scene coverage	11 km by 11 km
Stereo mode	Along-track and/or cross-track stereo
Revisit rate	1–4 days

16.2 THREE-DIMENSIONAL IMAGING AND STEREO MODEL FORMATION

It is important that a strong photogrammetric geometry is achieved by means of stereo model formation, including along-track and cross-track stereo models. They are the basis of high-precision measurements of horizontal and vertical positions of objects in the ground coordinate system.

16.2.1 ELEVATION INFORMATION IN IMAGERY

The photo scale of a vertical aerial photograph (horizontal image plane) is defined as f/H (where f is the focal length and H is the flying height of the sensor, both expressed in the same linear units). For example, aerial photographs for national mapping taken by a camera with a focal length f of 152 mm at a flying height H of 6000 m have a photo scale of approximately 1:40,000. IKONOS imagery has a photo scale of 1:68,000 ($f/H = 10$ m/680 km). (The unusually long focal lengths of such instruments are feasible because of the reduction in volume and mass of cameras built by Kodak, which folded the light path of its 10-m focal length using a Korsch three-mirror anastigmat [TMA] telescope [Lampton and Sholl, 2007; Bethel, 2008]). In order to derive the elevation, the relevant information must be present in the images in some form. One of these is terrain relief displacement (Figure 16.2).

Assuming that an object on the ground has a height of Δh and a distance R from the nadir point of a vertical aerial photograph with a flying height of H, the focal length of the aerial camera is f (Figure 16.2). The relief displacement d of the object caused by the elevation difference Δh in the aerial photograph can be calculated as (Moffitt and Mikhail, 1980; Wolf, 1983)

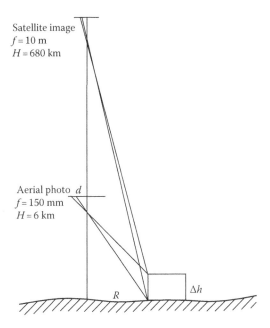

FIGURE 16.2   Schematic diagram representing relief displacement in an aerial photograph and a satellite image. (From Li, R., et al., Integration of IKONOS and QuickBird imagery for geopositioning accuracy analysis, *ASPERS 2007 PE&RS*, 73(9), 1067, 2007. With permission.)

$$d = \Delta h \frac{f}{H} \frac{R}{(H - \Delta h)} \tag{16.1}$$

Based on Equation 16.1, it is obvious that relief displacement is proportional to the elevation difference Δh. Here, f/H is the photo scale. Small-scale photographs are thus less capable of representing relief displacement information. Furthermore, objects close to the nadir point have small distances R and, therefore, will have small relief displacements. The same elevation difference Δh would produce a larger relief displacement as the object is situated farther away from the nadir point. The terms of Equation 16.1 are examined here for both aerial photography and satellite imagery (Figure 16.2) where the object height Δh remains the same.

- The flying height increases greatly from aerial photography to satellite imaging. However, a high-resolution camera (IKONOS, for example) has a focal length of 10 m, which makes its photo scale f/H comparable to that of an aerial photograph.
- Because H is much greater than Δh, the value of the term $R/(H - \Delta h)$ decreases rapidly from aerial photography to satellite imaging. Note that the satellite image covers a much larger ground area. In its nadir area, R (the distance from the nadir point to the object) is small, and, therefore, the corresponding terrain relief displacement will be small. However, in the most frequently imaged area, R is sufficiently large in comparison to H that the ratio $R/(H - \Delta h)$ becomes close to that for aerial photography.

Based on the earlier analysis, the terrain relief displacement information existing in the high-resolution satellite imagery reaches the similar level of that in airborne imagery, which ensures that accurate 3D geopositioning can be performed and fine topographic variations can be determined.

In addition to the relief displacement information in single images, parallax (a positional change in the flying direction of an object point in a vertical stereo image pair caused by motion of the imaging platform) can be used to derive the elevation or the elevation difference (Wolf, 1983):

$$p = f \frac{B}{(H - \Delta h)} \tag{16.2}$$

where
   p is the parallax of the top point of the object in Figure 16.2
   B is the baseline distance between the two perspective centers of the camera

In the high-resolution satellite imaging case, the orbit height H is large and makes parallax p small. However, B is large in both the cross-track stereo mode and the along-track stereo mode. The very long focal length f makes a further contribution to increasing parallax, making it sufficiently large to be measured and used to derive elevation information.

16.2.2 STEREO MODEL FORMATION

Figure 16.3 illustrates three CCD linear arrays (fore, nadir, and aft along the track) "looking" at the same ground profile across the track. Note that the three looking angles can be characterized as $\alpha°$, $0°$, and $-\alpha°$, respectively. Each CCD linear array produces one strip along the track. An object on the ground is usually covered by two or more image strips for stereo mapping. The specific image lines of the strips containing the object are acquired at different times. Thus, the along-track stereo mode may have different combinations of stereo pairs, namely F–N (fore–nadir), N–A (nadir–aft), and F–A (fore–aft). Baselines for IKONOS images (e.g., $\alpha = 15°$) may be about 182 km for F–N and N–A stereo pairs, and 364 km for F–A pairs. Figure 16.4 shows a situation where QuickBird and IKONOS stereo images taken at different times and from different orbits for the same ground area are combined to form various stereo models (Li et al., 2007).

The base–height ratio B/H is a critical factor for accurate 3D mapping (also see Equation 16.2). Aerial photography for national mapping usually has a base–height ratio of 0.6. In the case of IKONOS imagery, for instance, the orbit height is 680 km. Considering $\alpha = 15°$ and the possible combinations, the base–height ratio is 0.27 for F–N and N–A pairs and is 0.54 for an F–A pair. This means that satellite images with $\alpha = 15°$ are able to provide a base–height ratio close to that of aerial photographs for national mapping. Larger looking angles were also used in IKONOS imaging. This partly ensures the quality of the 3D spatial data derived from the imagery. The pointing capability across track makes it possible to flexibly form cross-track stereo strips.

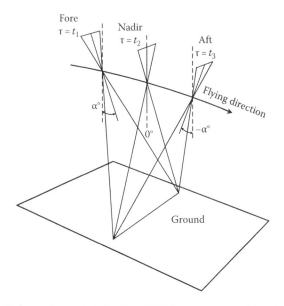

FIGURE 16.3   Fore-, nadir-, and aft-looking CCD linear arrays and their looking angles.

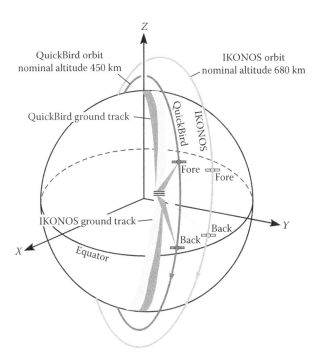

FIGURE 16.4   Illustration of the orbital geometry of QuickBird and IKONOS satellites for stereo model formation. The Cartesian coordinates (X, Y, Z) in this figure are earth centered, earth fixed Cartesian coordinates (ECEF). The inclination angles of the satellites are QuickBird 97.2°, IKONOS 98.1°.

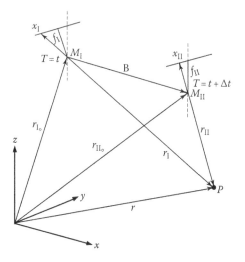

FIGURE 16.5 Photogrammetric intersection in along-track stereo mode.

The accuracies of object positions derived from the cross-track stereo strips are close to those from in-track strips, with some small variations (Li et al., 2007, 2008).

Exterior orientation parameters (see Chapter 4) are used to relate the platform to the ground coordinate system. At any time, the exterior orientation parameters of the sensor are only associated with one image line, including the perspective center and three rotational angles (ω, ϕ, k). They are estimated by measurements of onboard kinematic DGPS (differential global positioning system) to an accuracy of about 3 m and star trackers to an accuracy of about 2 arc seconds. In fact, the sampling rates of DGPS and star trackers are usually lower than that of imaging. As DGPS and star tracker data are not available for every image line, there are polynomial interpolations of the exterior orientation parameters between "orientation lines," which are image lines with actual DGPS and star tracker navigation data. Each exterior orientation parameter is modeled by a second- or third-order polynomial (Li et al., 1998, 2002). Such interpolations also can be performed using orbital parameters (Slama et al., 1980).

Suppose that a stereo line pair consists of an image line of Array I (fore-looking) at time t and an image line of Array II (nadir-looking) at time $t + \Delta t$ (Figure 16.5). Image coordinates x_I and x_{II} of a ground point P in both image lines are measured. The ground coordinates of P in the ground coordinate system (X–Y–Z), or vector r, can be computed (Li et al., 1998).

16.3 PHOTOGRAMMETRIC PROCESSING OF SATELLITE IMAGERY

16.3.1 RATIONAL FUNCTIONS FOR CAMERA MODELING

Photogrammetric processing of stereo images usually applies a rigorous sensor model that reconstructs the physical imaging setting and creates transformations between the 3D object space and the image space. Rational functions (RFs) (Di et al., 2003) have been applied in photogrammetry and remote sensing to represent the transformation between the image space and the object space whenever the rigorous model is unavailable intentionally or unintentionally. The topic has special significance now because some high-resolution commercial images are released to users with only RF coefficients.

The usual rigorous sensor model includes physical parameters characterizing the camera such as focal length, principal point location, pixel size, and lens distortions, together with orientation parameters of the image, such as position of the perspective center and attitude of the camera. Such rigorous models are conventionally applied in photogrammetric processing because of the clear separation between various parameters representing different physical settings. Consequently, these parameters can be modeled and calibrated for the high level of accuracy required in many mapping and other applications (Mikhail et al., 2001).

RFs have recently drawn considerable interest in the civilian remote-sensing community, especially in light of the trend that some commercial high-resolution satellite imaging data (such as IKONOS) is supplied with an RF model instead of a rigorous sensor model. An RF model, which does not reveal the sensor parameters, is generally the ratio of two polynomials derived from the rigorous sensor model and the corresponding terrain information. RFs are used for different reasons, such as to supply data without disclosing the sensor model or to achieve generality of a sensor model that can work across different types of sensors. Research on and applications of RF have been conducted using different satellite imaging systems (SPOT, IKONOS, and QuickBird), airborne systems, and both linear CCD scanners and frame cameras (Whiteside, 1997; Madani, 1999; Dowman and Dolloff, 2000; Tao and Hu, 2001; Di et al., 2001, 2003; Grodecki, 2001; Fraser et al., 2001; Open Geospatial Consortium, 2004).

RFs perform transformations between the image and the object spaces through a ratio of two polynomials. The image coordinates (x, y) and the ground coordinates (X, Y, Z) of a point are normalized to a range of -1.0 to 1.0 by their image size and geometric extent, respectively, for computational stability and to minimize computational errors. The RF can be expressed as

$$x = \frac{P_1(X,Y,Z)}{P_2(X,Y,Z)} \tag{16.3a}$$

$$y = \frac{P_3(X,Y,Z)}{P_4(X,Y,Z)} \tag{16.3b}$$

Polynomials P_i ($i = 1, 2, 3,$ and 4) have the general form:

$$P(X,Y,Z) = \sum_{i=0}^{m_1} \sum_{j=0}^{m_2} \sum_{k=0}^{m_3} a_{ijk} X^i Y^j Z^k \tag{16.4}$$

Usually, the order of the polynomials is limited by $0 \le m_1 \le 3$, $0 \le m_2 \le 3$, $0 \le m_3 \le 3$ and $m_1 + m_2 + m_3 \le 3$. Each $P(X, Y, Z)$ is then a third-order, 20-term polynomial:

$$P(X,Y,Z) = a_0 + a_1X + a_2Y + a_3Z + a_4X^2 + a_5XY + a_6XZ + a_7Y^2 + a_8YZ + a_9Z^2$$

$$+ a_{10}X^3 + a_{11}X^2Y + a_{12}X^2Z + a_{13}XY^2 + a_{14}XYZ + a_{15}XZ^2 + a_{16}Y^3$$

$$+ a_{17}Y^2Z + a_{18}YZ^2 + a_{19}Z^3 \tag{16.5}$$

Replacing the polynomials in Equation 16.3 by Equation 16.5 and eliminating the first coefficient in the denominator, the RFs become

$$x = \frac{\begin{pmatrix} 1 & X & Y & Z & \cdots & YZ^2 & Z^3 \end{pmatrix}\begin{pmatrix} a_0 & a_1 & a_2 & a_3 & \cdots & a_{18} & a_{19} \end{pmatrix}^{\mathrm{T}}}{\begin{pmatrix} 1 & X & Y & Z & \cdots & YZ^2 & Z^3 \end{pmatrix}\begin{pmatrix} 1 & b_1 & b_2 & b_3 & \cdots & b_{18} & b_{19} \end{pmatrix}^{\mathrm{T}}} \tag{16.6a}$$

$$y = \frac{\begin{pmatrix} 1 & X & Y & Z & \cdots & YZ^2 & Z^3 \end{pmatrix}\begin{pmatrix} c_0 & c_1 & c_2 & c_3 & \cdots & c_{18} & c_{19} \end{pmatrix}^{\mathrm{T}}}{\begin{pmatrix} 1 & X & Y & Z & \cdots & YZ^2 & Z^3 \end{pmatrix}\begin{pmatrix} 1 & d_1 & d_2 & d_3 & \cdots & d_{18} & d_{19} \end{pmatrix}^{\mathrm{T}}} \tag{16.6b}$$

where there are 39 coefficients: 20 in the numerator and 19 and the constant 1 in the denominator. In order to solve the RF coefficients, at least 39 control points are required.

16.3.2 GROUND CONTROL

Once RF coefficients are estimated, a photogrammetric intersection can be used to compute the 3D ground coordinates of a point from the measured image coordinates of conjugate points in multiple images (Yang, 2000; Di et al., 2001, 2003).

RF coefficients are usually provided with the images, known as *vendor-provided RF coefficients* (Li et al., 2001). These coefficients are generated from onboard navigational data without ground control. Therefore, some biases may not be corrected and, thus, are inherent in vendor-provided RFs and may be reflected in the achieved geopositioning accuracy. Systematic errors of 16–25 m have been found between RF-derived coordinates and the ground reference (Di et al., 2003; Li et al., 2003; Li et al., 2007). Similar results were reported in Fraser and Hanley (2003). Consequently, vendor-provided RF coefficients may not be sufficiently accurate to perform many photogrammetric/remote-sensing applications.

For the available IKONOS and QuickBird imagery, the impact of these errors are presented as a dominant shift in either image space or object space, which can be corrected by using a few ground points (Di et al., 2003; Fraser and Hanley, 2003; Wang et al., 2005). The correction of this error can improve the accuracy of ground coordinates derived from IKONOS stereo images to about 1 m in planimetry and 2 m in height. A further study by Li et al. (2007) on the potential of an integration of IKONOS and QuickBird images revealed a general trend of increased coordinate accuracy as the convergent angle formed at the ground point from the two stereo images increases (Table 16.2).

TABLE 16.2

Geopositioning Accuracy of Stereo Pairs Compared to Convergent Angle

Image Combination	Combination ID	3D Geopositioning Accuracy (RMSE: Meter)			Convergent Angle (δ_i)
		σ_x	σ_y	σ_z	
QuickBird (F)—QuickBird (A)	1	0.546	0.339	0.623	61.64337°
IKONOS (F)—QuickBird (A)	2	0.421	0.476	0.648	56.76096°
IKONOS (A)—QuickBird (F)	3	0.846	0.661	1.100	37.67997°
IKONOS (F)—IKONOS (A)	4	0.877	0.791	1.091	30.22133°
IKONOS (A)—QuickBird (A)	5	0.437	1.002	1.308	27.53404°
IKONOS (F)—QuickBird (F)	6	1.143	2.011	3.339	11.76017°

Note: F, fore; A, aft. The RMSEs are derived from the check points.

16.4 APPLICATIONS FOR NATIONAL MAPPING

16.4.1 TOPOGRAPHIC MAPPING

It is important to find an effective way to update the massive amount of existing products of middle-scale to some large-scale mapping produced by using airborne images. High-resolution satellite imagery has demonstrated its potential and characteristics to meet the demand. U.S. national topographic products, including 1:24,000-scale DEMs, DOQs, and DLGs, can be generated and updated by using the high-resolution satellite images. Before the satellite era, aerial photographs were the primary data used to derive information necessary for these products. These national products in their analog formats were scanned or digitized to form the basis of the national digital mapping products that have been used by uncountable national and international users for various applications. Now, with the availability of high-resolution satellite imagery and its associated stereo mapping capability, use of such data for updating the national mapping products is of great interest. To meet requirements for producing national mapping products at a scale of 1:24,000, aerial photographs usually have a base–height ratio of 0.63. This requirement can be met by forming stereo pairs of IKONOS or QuickBird imagery as discussed in the previous sections. To meet this standard, other requirements must also be met. The elevation error σ_h in a DEM is required to be less than 15 m. Further, the elevation error of a DEM needed to produce a DOQ is required to be less than 7 m. DLG requires the horizontal error to be less than 12 m and the vertical error to be less than 15 m. DOQs and DOQQs (digital orthophoto quarter-quadrangles) have resolutions of 2 m and 1 m, respectively. To meet these combined requirements for national mapping products at a 1:24,000 scale, based on the discussion in this chapter and relevant literature, it is recommended that at least two GCPs be used. GCPs are distinct points, clearly identifiable both on an image and on a positionally correct map or image, that permit an analyst to bring an image into a planimetrically correct representation of the area it represents.

In addition to geometric accuracy, the representation of topographic objects and the extraction of information from satellite imagery are of great interest. Coupled with high resolution, the panchromatic and multispectral channels available in satellite systems are capable of supplying a sufficient number of image features for interpretation and digital image matching to generate DEM grid points spaced every 30 m. Additional 4-m resolution multispectral DOQs may be produced as byproducts using the multispectral channels and the same DEMs. Attributes of 1:24,000 scale DLG quadrants are derived from maps of the same scale (U.S. Geological Survey, 1997). Their base categories include (1) political boundaries, (2) hydrography, (3) public land survey system (PLSS) data, (4) transportation data, (5) data on other significant manmade structures, (6) hypsography, (7) surface cover (e.g., vegetative surface cover), (8) nonvegetative surface features (e.g., lava and sand), and (9) surface control and marks. Categories (3) and (9) are usually well documented by relevant federal agencies. Their update in the DLG quadrants can be performed either manually or automatically. An update of the rest of the categories, in principle, can be carried out by using high-resolution satellite imagery. Positive results of attribute information extraction from high-resolution satellite imagery have been reported (Lee et al., 2003). However, the current methods still cannot achieve a high percentage of classification accuracy and manual work is necessary. Therefore, improved stability and capability of handling data sets covering large areas need to be developed.

16.4.2 SHORELINE MAPPING

Shorelines are one of the most important features in many national and local spatial databases. They are recognized as unique features on earth and are 1 of the 27 global "Geo-Indicators" referred to by the International Union of Geological Science (Lockwood, 1997). Shoreline mapping and shoreline change detection are critical for safe navigation, coastal resource management, coastal environmental protection, and sustainable coastal development and planning. However, there exist various definitions of shorelines, and they change both spatially and temporally. Deriving correct and updated shorelines are of national and economic importance. With the high-resolution satellite imagery, there will be repeated coverages that capture various positions of the shoreline and can be used to support digital modeling of the coastal environment.

Theoretically, an instantaneous shoreline is the line of intersection between a coastal terrain model (CTM), which is a digital surface model of an area along a shoreline with onshore elevation and nearshore bathymetry, and a water surface (Figure 16.6). The shoreline changes as the water surface increases or decreases. Therefore, the definition of a tide-coordinated shoreline should be based on a particular water datum. The National Geodetic Survey (NGS) of the U.S. agency

FIGURE 16.6 (See color insert following page 426.) Instantaneous shoreline superimposed on a 4-m resolution IKONOS image, Lake Erie, Ohio.

NOAA (National Oceanic and Atmospheric Administration) produces the nation's tide-coordinated mean lower-low water (MLLW) and mean high water (MHW) shorelines on U.S. nautical charts. These datums refer to the synoptic averages over a 19.2-year lunar/solar cycle. NGS derives tide-coordinated shorelines using periodic aerial photographs taken at the time of the desired water level using coordinated quasi real-time hydrographic observations (Slama et al., 1980). Because satellites have a prescribed orbit, there is no control over imaging the coastal area at the time of the desired water level. Also, satellites have a revisit rate of 1–4 days. The CTM may change with time because of erosion, land use, and other natural or human activities in the coastal zone. Thus, instantaneous shorelines derived from high-resolution satellite imagery vary according to both the CTM and the water level.

Stereo satellite image strips can be used to determine the CTM. Onshore land elevations of grid points of the CTM are computed by either image matching or interactive measurements on panchromatic images. Offshore grid points within the area of shallows may be determined in the same way from multispectral band images (4-m resolution), which penetrate water better than visible band images (Slama et al., 1980; Lillesand et al., 2008). Within a set time period, a number of low-water satellite images can be accumulated and used to generate a set of reliable terrain points. Water-penetrating LiDAR and historical bathymetric data can be employed to provide information on the depth of deeper areas. A combination of the land topographic data and bathymetric data produces the required CTM. On the other hand, water surface levels can be modeled by a computer modeling system using tide-gauge observations and other hydrographic data along the shoreline (Bedford and Schwab, 1994). Because historical water level data have been archived, both MLLW

FIGURE 16.7   (**See color insert following page 426.**) Digital shoreline (red) produced from a coastal terrain model and a water surface model versus a shoreline from a map (blue).

and MHW can be determined. As illustrated in Figure 16.7, a spatial operation of the CTM and a desired water surface model can be used to produce a DSL.

16.5   SUMMARY

This chapter provides an analytical discussion of important aspects of the photogrammetric processing of commercial high-resolution satellite imagery. The estimation and analysis methods presented here show the great potential of such newly available imaging technology for national mapping products. Conclusions and recommendations based on the analysis are made for applications of the imagery in national mapping production, including topographic and 3D shoreline mapping. The latter presents a great challenge because of its complexity, spatiotemporal variation, and socioeconomic importance.

These issues have broad significance throughout the practice of remote sensing. Multispectral images that are formed from several spectral bands, including color or color infrared imagery, permit the extraction of very specific classes of information such as water depth, vegetation density, or soil moisture. In the analysis of such imagery, our attention often focuses upon the details of the complex strategies that can be applied to extract thematic data from such imagery. However, the validity of the thematic information depends not only upon our ability to extract reliable information, but upon our ability to associate thematic information with its correct location on the earth's surface. It is the practice of photogrammetry that permits such information to be placed in its correct positional context, so locations, distances, and areas can be accurately assessed.

REFERENCES

Al-Rousan, N., Cheng, P., Petrie, G., Toutin, T., and Zoej, M.J.V., Automated DEM extraction and orthoimage generation from SPOT level 1B imagery, *Photogrammetric Engineering & Remote Sensing*, 63(8):965–974, 1997.

Bedford, K. and Schwab, D., The Great Lakes forecasting system—An overview. In *Hydraulic Engineering (Proceedings of the 1994 National Conference on Hydraulic Engineering)*, eds. Cotroneo, G.V. and Rumer, R.R., pp. 197–201. New York: ASCE, 1994.

Bethel, J.C., http://cobweb.ecn.purdue.edu/~bethel/cams2.pdf (accessed December 9, 2008).

Cuartero, A., Felicisimo, A.M., and Ariza, F.J., Accuracy, reliability, and depuration of SPOT HRV and Terra ASTER digital elevation models, *IEEE Transactions on Geoscience and Remote Sensing*, 43(2):404–407, 2005.

Di, K., Ma, R., and Li, R., Deriving 3-D shorelines from high-resolution IKONOS satellite images with rational functions. In *Proceedings of ASPRS Annual Convention*, April 25–27, 2001, St. Louis, MO, CD-ROM. Bethesda, MD: American Society for Photogrammetry and Remote Sensing, 2001.

Di, K., Ma, R., and Li, R., Rational functions and potential for rigorous sensor model recovery, *Photogrammetric Engineering & Remote Sensing*, 69(1):33–41, 2003.

Dowman, I. and Dolloff, T., An evaluation of rational functions for photogrammetric restitution, *International Archives of Photogrammetry and Remote Sensing*, 33(B3):254–266, 2000.

Ellis, M.Y. ed., *Coastal Mapping Handbook*, Cooperating organizations: U.S. Department of the Interior, Geological Survey [and] U.S. Department of Commerce, National Ocean Survey [and] Office of Coastal Zone Management. Washington, DC: U.S. Government Print Office, 1978.

Federal Geographic Data Committee (FGDC) FGDC Standards Working Group, Public comment on the proposal to develop the National Shoreline Data Standard as a Federal Geographic Data Committee Standard, Washington, DC, *Federal Register*, 62(156):43,342–43,344, 1997.

Fraser, C., Collier, P., Shao, J., and Fritsch, D., Ground point determination using MOMS-02 earth observation imagery, *Geomatica*, 51(1):60–67, 1997.

Fraser, C. and Hanley, H., Bias compensation in rational functions for IKONOS satellite imagery, *Photogrammetric Engineering & Remote Sensing*, 69(1):53–57, 2003.

Fraser, C.S., Hanley, H.B., and Yamakawa, T., Sub-meter geopositioning with IKONOS GEO imagery. In *Proceedings of ISPRS Joint Workshop "High Resolution Mapping from Space,"* September 19–21, 2001, Hanover, Germany, 2001.

Fritz, L.W., Commercial earth observation satellites. In *International Archives of Photogrammetry and Remote Sensing, ISPRS Com. IV*, Vienna, Austria, pp. 273–282, 1996.

Grodecki, J., IKONOS stereo feature extraction—RPC approach. In *Proceedings of ASPRS Annual Convention*, April 25–27, 2001, St. Louis, MO, CD-ROM. Bethesda, MD: American Society for Photogrammetry and Remote Sensing, 2001.

Heipke, C., Kornus, W., and Pfannenstein, A., The evaluation of MEOSS airborne three-line scanner imagery: Processing chain and results, *Photogrammetric Engineering & Remote Sensing*, 62(3):293–299, 1996.

Lampton, M. and Sholl, M., Comparison of on-axis three-mirror-anastigmat telescopes, *Proceedings of the SPIE*, 6687:66870S–66870S-8, 2007.

Lee, D.S., Shan, J., and Bethel, J.S., Class-guided building extraction from IKONOS imagery, *Photogrammetric Engineering & Remote Sensing*, 69(2):143–150, 2003.

Li, R., Potential of high-resolution satellite imagery for national mapping products, *Photogrammetric Engineering & Remote Sensing*, 64(2):1165–1169, 1998.

Li, R., Zhou, G., Gonzalez, A., Liu, J.-K., Ma, F., and Felus, Y., Coastline mapping and change detection using one-meter resolution satellite imagery, Project report submitted to Sea Grant/NOAA. Columbus, OH: The Ohio State University, 1998.

Li, R., Di, K., Zhou, G., Ma, R., Ali, T., and Felus, Y., Coastline mapping and change detection using one-meter resolution satellite imagery. Project report submitted to Sea Grant/NOAA. Columbus, OH: The Ohio State University, 2001.

Li, R., Zhou, G., Schmidt, N.J., Fowler, C., and Tuell, G., Photogrammetric processing of high-resolution airborne and satellite linear array stereo images for mapping applications, *International Journal of Remote Sensing*, 23(20):4451–4473, 2002.

Li, R., Di, K., and Ma, R., 3-D shoreline extraction from IKONOS satellite imagery, Special Issue on Marine & Coastal GIS, *Journal of Marine Geodesy*, 26(1/2):107–115, 2003.

Li, R., Zhou, F., Niu, X., and Di, K., Integration of IKONOS and QuickBird imagery for geopositioning accuracy analysis, *Photogrammetric Engineering & Remote Sensing*, 73(9):1067–1074, 2007.

Li, R., Deshpande, S., Niu, X., Zhou, F., Di, K., and Wu, B., Assessment of geopositioning accuracy in integration of aerial and high-resolution satellite imagery and application in shoreline mapping, *Marine Geodesy*, 31(3):143–159, 2008.

Lillesand, T.M., Kiefer, R.W., and Chipman, J.W., *Remote Sensing and Image Interpretation*. New York: John Wiley & Sons, Inc., 2008.

Lockwood, M., NSDI shoreline briefing to the FGDC Coordination Group, Project report to NOAA/NOS, Washington, DC, 1997.

Madani, M., Real-time sensor-independent positioning by rational functions, In *Proceedings of ISPRS Workshop on Direct Versus Indirect Methods of Sensor Orientation*, pp. 64–75, November 25–26, Barcelona, Spain, 1999.

McGlone, J.C., Mikhail, E.M., and Bethel, J. eds., *Manual of Photogrammetry* (5th edn.). Bethesda, MD: American Society for Photogrammetry and Remote Sensing, 2004.

Mikhail, E.M., Bethel, J.S., and McGlone, J.D., *Introduction to Modern Photogrammetry*. New York: John Wiley & Sons, Inc., 2001.

Moffitt, F.H. and Mikhail, E.M., *Photogrammetry*. New York: Harper & Row Publishers, Inc., 1980.

Open Geospatial Consortium (OGC), 2004. The OpenGIS® abstract specification, Topic 7: The earth imagery case, version 5. OpenGIS® Project Document Number, pp. 04–107. http://www.opengeospatial.org/standards/as (accessed Jan. 09).

Slama, C.C., Theurer, C., and Henriksen, S.W. eds., *Manual of Photogrammetry*. Falls Church, VA: American Society of Photogrammetry, 1980.

Tao, C.V. and Hu, Y., A comprehensive study of the rational function model for photogrammetric processing, *Photogrammetric Engineering & Remote Sensing*, 67(12):1347–1357, 2001.

U.S. Geological Survey, Fact sheets and WWW pages of GeoData: Digital line graphs, digital elevation models, and digital orthophotos. Reston, VA: U.S. Geological Survey, 1997.

Wang, J., Di, K., and Li, R., Evaluation and improvement of geopositioning accuracy of IKONOS stereo imagery, *ASCE Journal of Surveying Engineering*, 131(2):35–42, 2005.

Whiteside, A., Some image geometry models, http://www.opengeospatial.org/standards/dpd, 04–071 Arliss Whiteside, 2004–10–04 (accessed Jan. 09).

Wolf, P.R., *Elements of Photogrammetry, with Air Photo Interpretation and Remote Sensing*. New York: McGraw-Hill, Inc., 1983.

Yang, X., Accuracy of rational function approximation in photogrammetry. In *Proceedings of ASPRS Annual Convention*, May 22–26, 2000, Washington, DC, CD-ROM. Bethesda, MD: American Society for Photogrammetry and Remote Sensing, 2000.

17 Remote-Sensing Analysis: From Project Design to Implementation

Timothy A. Warner

CONTENTS

Remote sensing is a powerful method for acquiring geospatial information, and has been found to be useful for a wide range of applications (Jensen 2005, Campbell 2007). As an area of active research, much has been written on the topic of remote sensing. However, the scientific literature, although very effective in documenting advanced remote-sensing algorithms and analysis methods, is not necessarily helpful in planning new remote-sensing projects. In general, information on how to plan and implement a remote-sensing project is relatively sparse, making the undertaking of a new remote-sensing project a potentially daunting task. This chapter is an attempt to address this deficit by providing a broad and systematic discussion of issues involved in designing and implementing a remote-sensing project. The focus of this chapter is on terrestrial remote sensing, although many of the same issues may be involved in other remote-sensing applications.

This chapter is divided into two main sections, followed by a short conclusion. The first section (Section 17.1) is about project design. This section starts with identifying objectives and proceeds with defining analysis methods and data needs. As with any major endeavor, time spent in planning is important in ensuring the success of the entire project. The second section (Section 17.2) is about project implementation. After finding data, a pilot study provides a useful initial check on the research design, and may prompt further refinements of the planned approach. Projects that require field work pose special challenges. Communication of project results may involve maps, tables, graphs, and summary descriptions. An important component of the communication step is a report of the methods and outcomes of the validation of the results.

17.1 PROJECT DESIGN

17.1.1 DEFINE OBJECTIVES

The process of defining project objectives may seem a relatively simple task. However, in practice, identifying the objectives is made complex because of the need to be comprehensive. Thus, it is usually necessary at least to specify

- The thematic feature(s) of interest
- The level of thematic detail required
- The level of spatial detail required, including the minimum contiguous area required to depict a thematic feature (i.e., the minimum mapping unit) (Saura 2002)
- The geographical extent of the study area
- The time period, including the optimal season and whether current or historic data are required

It is also important to consider the ultimate use of the project, including the environment that the resulting information will be used in. For example, the mismatch between the pixel-based raster data model of remote sensing and the traditional vector model typical of geographic information systems (GIS) (Merchant and Narumalani 2009) poses challenges for using remotely sensed information in a GIS.

The definition of the objectives is an important step because any confusion at this stage will necessarily cause cascading problems throughout the entire project. Because projects are generally constrained by limited data and resources, it is worthwhile to specify both a desired set of objectives and a minimum set of objectives. Specifying the minimum objectives will also ensure that if project methods are revised at a later stage, the original project aims will not be lost.

17.1.2 Literature Search

Remote sensing is a very rapidly changing field, and therefore, literature search is an important first step. Fortunately, with the Internet, a literature search today is much easier than it used to be. Online search engines, such as Google Scholar, help in finding articles, and most journal Web sites allow free searching of abstracts. However, the dense language and sometimes rather arcane jargon of remote-sensing technical literature can pose challenges for understanding the material found. In addition, the fact that negative research results are only rarely published makes it hard to find cautionary information.

17.1.3 Define Analytical Methods

Defining analytical methods is a major component of project planning. The analysis planning starts with a consideration of the products required to meet the project objectives. Methods and data necessary to generate those products are then identified. Special consideration should be given to planning the evaluation of the uncertainty associated with the information generated.

17.1.3.1 Type of Output Required

Information from a project may be either *categorical*, such as a land cover class (Anderson et al. 1976), or *continuous*, such as biomass. In practice, this conceptual division, though useful, is a bit simplistic. Continuous variables can be quantized by dividing the output into discrete classes. On the other hand, if the proportion of the classes in each pixel is the focus of the analysis, continuous measures are required for mapping the otherwise categorical classes. For example, with moderate-resolution imagery, impervious surfaces are often mapped as a proportion of cover within each pixel because the spatial resolution is generally not sufficient to resolve the individual areas of impervious cover (Phinn et al. 2002). Thus, the scale of the remotely sensed data relative to the scale of the phenomenon of interest (Strahler et al. 1986, Woodcock and Strahler 1987) forms a key component of any remote-sensing project, as discussed in Section 17.1.3.3.1.

Another way in which the categorical class is made continuous is through the concept of *fuzzy* classes (Zadeh 1965, Warner and Shank 1997a). Although often used to deal with the type of mixed pixel problem described earlier, fuzzy classification is also a powerful tool for addressing real-world variability in the concept of each class (van de Vlag and Stein 2007). This conceptual fuzziness is distinguished from the mixed pixel issue in that increasing the spatial resolution does not change the nature of the problem. For example, the discrete map classes of wetlands and uplands may in reality represent a continuum of landscape drainage properties. Although many sophisticated applications of fuzzy analysis have been developed, in practice, many users of remote-sensing products seem to struggle to deal with the complexity of fuzzy analysis. Consequently, the final output from fuzzy analysis is often "hardened" to produce traditional categorical classes, particularly if the results need to be incorporated in vector-based GIS analysis, which does not easily represent continuous data.

17.1.3.2 Defining Analytical Methods

A vast array of analysis methods is available to the remote-sensing analyst. However, three broad types of analyses can be identified: (1) enhancement and visualization, (2) classification, and (3) biophysical modeling. The division into these topics is somewhat arbitrary, and many methods can be placed in more than one of these categories. Any one remote-sensing project is likely to include methods from at least two, and typically all three, types of analyses.

In recent years, with improved data quality, increased computer power, and general availability of more sophisticated remote-sensing software, the emphasis in remote-sensing analysis has changed from ad hoc qualitative approaches to more quantitative and model-driven approaches (Liang 2004), often emphasizing biophysical transformations. Nevertheless, much remote sensing still relies on empirical techniques. In empirical approaches, areas where the phenomenon of interest is well characterized are identified in an image and then used to develop an empirical association between image digital numbers (DNs) and that phenomenon. The observed association is then applied to the rest of the image to infer biophysical properties of each pixel. Empirical models have the advantage that they potentially can provide an analysis method that is optimal for the particular data sets used in the study. On the other hand, biophysical model-driven approaches are ideal for operational remote sensing and for large projects, where it is impractical to acquire empirical data to redevelop the empirical relationships each time a new data set is acquired. Biophysical models also allow the application of a more complete understanding of relationship between the physical scene and the remotely sensed image (Moghaddam 2009).

17.1.3.2.1 Enhancement and Visualization

The combination of the human eye and brain is extraordinarily effective at identifying spatial patterns in data. Therefore, visualization is an important component of most analysis strategies, especially for exploratory investigation. However, visual comparisons of the relative value of band combinations or processing steps should be undertaken with caution, as differences in preprocessing and contrast stretches can potentially result in large variations in the apparent relative information content of images. The most basic visualization of remote-sensed imagery is the display of a single band image as a black-and-white image (i.e., in tones of gray), or with a color look-up table. Another common visualization is the three-band, false-color composite.

The two spatial dimensions of the image can be extended by draping the image over a surface, for example, topographic elevation, creating what is termed a *2.5D representation*. In a true three-dimensional (3D) representation, pixels represent volumes, known as *voxels*. Most remote-sensing software packages offer only 2D or 2.5D visualizations. Nevertheless, there are interesting potential applications for 3D representation, for example, viewing and analyzing spatial patterns in multi-temporal data, where the third spatial dimension represents time.

Another area of potential application of visualization to remote sensing is in the area of virtual reality. Virtual reality is a type of human–computer interface (Brodlie et al. 2002) for generating 3D visualizations that allow multiple perspectives on the data (Fisher and Unwin 2002, Warner et al. 2003). Virtual reality has found many applications for simulating the natural environment for scientific research purposes (Grunwald et al. 2006, Hodza, in press). Although less commonly used in remote sensing, virtual reality has great potential for representing complex 3D data sets, such as individual LIDAR returns (Warner et al. 2003). Unfortunately, current visualization software typically does not provide the interactivity necessary to realize the benefits of this technology for remote sensing.

Because human color perception is limited to the three additive primaries (loosely termed blue, green, and red), much emphasis has been placed on research into data transformations and feature selection for visualization. Feature selection is the capturing of most of the information in a multiband data set in just a small number of bands, or features (Warner et al. 1999, Serpico and Moser 2007), and can be useful for predicting the best bands for a false-color composite (Chavez et al. 1982). Transformations are reprojections of the data from the original spectral bands to new bands that are combinations of some or all of the spectral bands. Biophysical transformations, designed to quantify physical properties of the scene, will be covered in Section 17.1.3.2.3. A variety of statistically based transformations have been proposed. One of the most successful is principal component analysis (PCA) (Jensen 2005). In using PCA, it is important to consider that the underlying assumption that data variance, especially variance correlated between bands, represents information. As an alternative, a spatial version of PCA (Wartenberg 1985) may have value for enhancing different spatially varying spectral patterns in an image (Warner 1999).

17.1.3.2.2 Classification

Classification is the categorization of each image pixel in one of a finite number of classes (Jensen et al. 2009). Classification tends to be empirical, using within-scene knowledge to determine class spectral characteristics. In supervised classification, training data is used to represent the typical or central DN values of each class in order to train the classifier. Generally, the aim is to find examples of the phenomenon of interest that are as pure and as representative as possible. An interesting exception to this rule is the classification technique called support vector machines (SVM), for which samples near the decision boundaries between classes (i.e., samples that are potentially confused, such as mixed pixels) are the most useful (Foody and Mathur 2006). Maximum likelihood classification is one of the most powerful statistical classifiers and, despite assumptions of multivariate normality, is relatively robust. Neural networks and SVM are nonstatistical classifiers that do not make assumption about the data distribution, and yet may not require as many training samples as the more standard statistical methods. Expert systems typically also do not assume a specific data distribution. Expert systems facilitate the exploitation of human expert knowledge, but can also draw on automated methods to develop rules for categorizing the data (Jensen et al. 2009).

In unsupervised classification, inherent clusters are identified in the data, and the human analyst makes the decision regarding the class label to apply to each cluster. Unsupervised classification is particularly valuable as an exploratory tool, and when used in combination with iterative methods, such as "cluster-busting," can provide precise refinement of a classification (Jensen et al. 2009).

One of the challenges with unsupervised methods is that typically the user is asked to provide a large number of parameters, and the values that are chosen can have a notable effect on the outcome. This issue is also true for some supervised classification methods, particularly neural networks. On the other hand, SVM classification has been noted for its comparatively robust behavior with respect to initialization parameters (Melgani and Bruzzone 2004).

An area of particular current interest is object-oriented classification (Jensen et al. 2009). Traditional classification is aspatial; each pixel is classified independent of its neighbors. In object-orientated classification, however, the image is segmented in groups of contiguous pixels, which are classified as a single object. This approach allows the use of a much richer range of attributes in the classification, including the object spectral mean, variance, covariance (Lee and Warner 2006), and texture (Kim et al. 2009), as well shape and size. The challenge with object-oriented classification lies in selecting the optimal spatial scale for the segmentation, which is normally chosen subjectively, using trial-and-error methods. However, Kim et al. (2008) have shown that autocorrelation analysis may offer a way to identify an optimum scale associated with maximum relative classification accuracy.

17.1.3.2.3 Biophysical Modeling

Biophysical methods attempt to estimate physical properties, such as water depth, soil moisture, clay percentage, vegetation biomass, and canopy chemistry, from remotely sensed measurements. Although most biophysical variables are continuous, not all are. For example, techniques for identifying minerals in rocks in hyperspectral data (Chen et al. 2007) are usually included in the category of biophysical methods, especially if those methods draw on library spectra.

One of the earliest, and still commonly applied, biophysical models is the tasseled cap (TC) transformation (Kauth and Thomas 1976). The TC transformation is a linear combination of the four Landsat MultiSpectral Sensor (MSS) spectral bands to generate four new bands, including brightness and greenness, from the original multispectral data. With Landsat thematic mapper (TM) data, an additional biophysical dimension, Wetness, is identified (Crist and Cicone 1984). The TC transformation was originally proposed as a method of summarizing spectral and temporal trends in agricultural fields, but has found broad application, to include, for example, urban and natural environments (Dymond et al. 2002).

A wide range of biophysical variables have also been proposed based on image ratios, which typically use two or three bands (Campbell 2007). The most common ratio index is the vegetation index, especially the normalized difference vegetation index (NDVI). A wide range of vegetation indices have been proposed to overcome drawbacks with NDVI, such as sensitivity to soil color (Jensen 2005). Recently, the enhanced vegetation index, which has improved sensitivity in high biomass regions and a reduced influence from atmospheric noise, has been adopted as one of the moderate-resolution imaging spectroradiometer (MODIS) products (Huete et al. 2002). Ratios have also been used to estimate other biophysical parameters, including rock and soil minerals (Jensen 2005, Campbell 2007, Nield et al. 2007).

17.1.3.3 Defining Remotely Sensed Data Needs

Characteristics of the required remotely sensed data are developed from the project aims. The four main image attributes to consider are the spatial, spectral, radiometric, and temporal properties (Warner et al. 2009). Scale is a useful concept for describing these image attributes. Scale can be defined as having two components: grain, or the finest unit of measurement, and extent, or the range over which the measurements are made (Turner et al. 2001). Thus, for the spatial scale, the grain is the image ground resolution element, which is often approximated by the linear dimension of the pixel size. The extent is the area covered by the entire image.

Considerable research has been invested in efforts to estimate optimal scales for particular projects, particularly with respect to spatial (Woodcock and Strahler 1987, Hengl 2006) and spectral (Serpico and Moser 2007) properties, although radiometric (Narayanan et al. 2000) and temporal properties (Key et al. 2001) have also been studied. For example, these studies have shown that too coarse a pixel size will not allow differentiation of the features of interest, but also that a pixel size that is too fine has its own penalties, including resolving extraneous internal variation that causes unnecessary data costs and processing expenses (Latty et al. 1985).

17.1.3.3.1 Spatial Properties

In considering the optimal spatial grain, the first step is to decide whether the desired output is a map of the boundaries of a feature of interest, or merely the proportions of the feature of interest within each pixel. The first of these two options requires H-resolution imagery, defined as having a pixel size that is much smaller than the size of the objects to be mapped, whereas the second implies L-resolution imagery, in which the pixel size is much larger than the individual objects (Strahler et al. 1986).

Between the H- and L-resolution scales is the situation where mapping the boundaries of features is not important, but merely the presence or absence of a feature. For example, in a project to count the number of houses, it might appear that one could specify a pixel size no bigger than the smallest house. However, this conflation of spatial resolution and pixel size is misleading, because, first, an object would typically have to be at least twice the linear dimension of the pixel in order to result in even just one pure pixel of that object. Furthermore, the contrast between the object and the background against which it is imaged as also an important component of whether an object can be discriminated. Therefore, more comprehensive measures of image resolution quantify how contrast or sharp changes in the scene are represented in the image, for example, the modulation transfer function and the point spread function (Huang et al. 2002). However, because these measures are rather complex, the pixel remains a common summary measure of spatial grain.

The geographic extent of the area of interest defines the formal spatial boundary of a project. For planning purposes, the analyst usually defines a compact polygon that encompasses the actual project area. Often, political or natural boundaries may define a project's spatial extent. Further, limitations in the footprints of individual images, and the consequent requirement in many cases to mosaic images of different acquisition dates, adds complexity to image analysis because of the necessity of normalizing between images. Therefore, for satellite images, if it is feasible to limit the study area to a single image, it is advisable to do so.

A related spatial issue is image georeferencing. Consideration of the quality of the georeferencing required is particularly important if the analysis includes data from multiple sources, or the output results are to be used as a map or in a GIS. To ensure a high-quality georeferencing, it is important to have sufficient numbers of ground control points (GCPs), and this may require expanding the geographic extent to include a larger area than might otherwise be considered.

17.1.3.3.2 Spectral-Radiometric Properties

Knowledge of the spectral properties of the objects of interest, as well as the background or any other potentially similar targets, is important in defining the required spectral and radiometric properties of the remotely sensed data needed for the project. The spectral properties define the wavelength regions of the image bands, and the radiometric properties define the relationship between radiance and the pixel values.

The required spectral-radiometric information for project design can potentially come from a general understanding of the spectral properties of common earth materials (van Leeuwen 2009) found in the scene, or from the acquisition of custom spectra. The ability to capture the spectral characteristics of the objects of interest depends on the width, number, location, and radiometric sensitivities of the sensor's spectral bands. For example, a small number of broad spectral bands in the visible, near-infrared, and shortwave infrared can capture basic information about vegetation pigments, leaf structure, and moisture content (van Leeuwen 2009). In comparison, many narrow spectral bands in the shortwave infrared are required to differentiate common clay minerals, and a small number of narrow bands in the thermal infrared may be used to differentiate the silicate minerals that make up many common rocks (Chen et al. 2007).

Generally speaking, each spectral band potentially adds information; therefore it may seem best to specify hyperspectral data wherever possible. However, as with excessive spatial detail, excessive spectral detail imposes its own penalties, and may even result in lower accuracies (Hughes 1968). Feature selection, already mentioned in the context of choosing three bands for a false composite, is often also used to identify a subset of bands for image processing. Feature selections methods typically focus on uncorrelated variance as a measure of information (Chavez et al. 1982). Alternatively, multispectral autocorrelation has been proposed as a measure of information for feature selection (Warner and Shank 1997b, Warner et al. 1999).

17.1.3.3.3 Temporal Properties

Determination of the required temporal properties includes specifying the optimal date, whether recent or historic, season, and for aerial imagery, even time of day for the image acquisition. Vegetation phenology, at least in relatively moist regions, can be important in determining the optimal season for image acquisition (Key et al. 2001). Phenology can influence image usefulness, even for studies that do not focus on vegetation, because vegetation may be correlated with features of interest, or may simply be the background against which objects are viewed. Another seasonal issue to consider is local climate, including seasons of wet and dry periods, cloudiness, and snow.

Most single images capture an essentially instantaneous view of the earth. Remote sensing data archives, especially those with multiple image acquisitions of the same geographic location, offer a series of historical and time-specific records of past conditions on earth's surface. In particular, the long history of satellite acquisitions by the Landsat program is an unrivaled resource for moderate-resolution global change studies (Goward and Masek 2001).

There is generally an inverse relationship between image-acquisition frequency and pixel size. Coarse-resolution sensors can potentially acquire daily coverage of the globe; there is as yet not a single comprehensive data set of satellite-based fine-resolution imagery of even one continent. High temporal resolution of coarse-resolution imagery offers the possibility of combining multiple images over short periods, such as a week or a month, to produce cloud-free composite images (Holben 1986). These cloud-minimized multi-temporal composites are particularly useful for tracking subtle environmental changes, particularly natural and agricultural vegetation phenological patterns (Myneni et al. 1997, Delbart et al. 2006).

17.1.3.4 Incorporating Ancillary Data

Remote sensing is a powerful method for providing geospatial information, and often is the only source for which particular information can be generated. Nevertheless, it is important to recognize the limitations of remote sensing. In particular, the phenomenon may not have a sufficiently distinct spectral characteristic to be mapped reliably. To give two examples, many tree species tend to have rather similar spectral properties, and surface mining can be confused with other disturbances, including natural events such as the aftermath of flooding. Another limitation of remote sensing is that imagery generally provides information only about the uppermost surface of the earth and, if there is vegetation present, the upper surface of the vegetation.

Non-remote–sensing data can potentially help overcome some of the limitations of remotely sensed data. A commonly used ancillary data set is a digital elevation model (DEM) (Campbell 2007). (A broad definition of remotely sensed data would include DEMs, and therefore not place DEMs in a separate category.) DEMs provide a 3D context for image analysis, including raw elevations and derived macro- and micro-topographic variables (Deng 2009), which potentially can be related to climate, predicted solar insolation, moisture runoff, and even type of anticipated land use or disturbance. Another useful ancillary data set is cadastral information for land use studies, because land use often changes at parcel boundaries.

Ancillary data may require special processing. In particular, ancillary data should not be combined directly with other remote-sensing data in image analysis routines without checking whether inclusion of the data violates any assumptions in the analysis methods, for example, multivariate normality. However, some remote-sensing methods, such as decision trees and neural networks, are particularly attractive for the very reason that they can accommodate categorical data. The Dempster–Shafer theory provides a formal framework for combining disparate data with varying degrees of uncertainty, and is especially useful where the ancillary data is partial or incomplete (Dempster 1967, Shafer 1976, van de Vlag and Stein 2007).

17.1.3.5 Software Needs

A wide range of sophisticated commercial remote sensing and photogrammetric software is now available (Chapter 20). However, for advanced analysis, it may be necessary to develop custom programs. Some remote-sensing software packages have libraries of basic and advanced functions that can be used in custom programs, thus facilitating the development of computer programs from scratch. Most remote-sensing programs employ a wide range of important and export functions, but in choosing the software for the image analysis, it is important to verify that the package supports the format desired for the final product output.

17.1.3.6 Accuracy Evaluation

It is common practice in remote sensing to specify the uncertainty in the derived information (Chapter 21). For categorical map data, the convention is to use a random sample, not used in developing the processing method, to provide a comprehensive confusion matrix, or error matrix, as well as an overall accuracy estimate (Stehman and Foody 2009). A key issue with the error matrix is that it too has uncertainty, which is a function of the sample size and the number of classes present.

Collecting data for the accuracy assessment can be expensive, and may comprise a significant portion of the total project cost. It is important that the sample be selected randomly, although modifications such as stratified random sampling should be employed to overcome challenges posed by sampling classes with very different proportions of the total map coverage. To save on the cost of field work, manual interpretation of imagery with a higher spatial resolution than that used for the analysis is often used.

Continuous variables, for example, estimates of leaf area index (LAI), are usually evaluated in terms of metrics such as root mean square error. With the development of new classification approaches, such as object-oriented classification and fuzzy classification, adaptations to the traditional approach to the error matrix are required. For example, Jäger and Benz (2000) present summary accuracy measures based on fuzzy similarity, and Brandtberg et al. (2003) propose a fuzzy method of handling agreement between object-based classifications.

17.1.3.7 Field Work Requirements

As discussed earlier, many remote-sensing methods are empirical, and thus require detailed information from within the scene. Data for the evaluation also requires independent information from within the scene. This information is ideally obtained from field observations made contemporaneously with the image acquisition (Johannsen and Daughtry 2009). Field data collection can be very expensive, and thus planning is important. Base maps, including topographic maps, aerial photography, and other imagery, can assist this process. Clearly, however, contemporaneous information cannot be acquired for historic imagery, and even for current imagery may be prohibitively expensive. Therefore, a common alternative is to use the visual interpretation of higher-spatial–resolution aerial or satellite imagery as the reference data set.

17.2 PROJECT IMPLEMENTATION

This chapter so far has focused on project design. As will be seen, separating design and implementation is in reality a somewhat artificial distinction, because project design is often iteratively refined during the early stages of implementation.

17.2.1 Finding and Obtaining Data

17.2.1.1 Matching Data Availability to Data Requirements

The availability of both aerial and satellite image data has increased greatly over the last decades. Nevertheless, data availability still remains one of the major constraints for remote-sensing project implementation. Practical constraints in instrument design and data collection normally cause grain and extent attributes of scale to be inversely related. In addition, although there has been a general trend toward increasing the availability of finer grained data with respect to spatial, spectral, radiometric, and temporal dimensions of imagery, usually such improvements are at the expense of one or more of the other image qualities. For example, fine spatial resolution data generally have fewer bands, and are acquired less frequently than coarser-resolution data.

17.2.1.1.1 Spatial Properties

Imagery with a wide range of pixel sizes is available. Fine-resolution images may include government aerial imagery and commercial satellite data. Landsat TM remains the workhorse for moderate-resolution imagery, and Landsat remains as the one moderate-resolution sensor with effective global seasonal repeat coverage (Goward and Masek 2001). However, the failure of the scan line corrector on Landsat 7, the general aging of Landsat 5, and delays in the Landsat Data Continuity Mission (Wulder et al. 2008), call into question the future role of Landsat. In response to these concerns, the U.S. Department of Agriculture Satellite Imagery Archive has already transitioned to the India Remote Sensing Advanced Wide Field Sensor (AWIFS) (Dave et al. 2006) as a replacement for their Landsat TM acquisitions. AWIFS has fewer spectral bands, but a much wider field of view than Landsat TM. The longest time series of coarse-resolution data is provided by the National Oceanic and Atmospheric Administration (NOAA) advanced very high-resolution radiometer (AVHRR) instruments. A wide range of very high-quality MODIS data is also available now. The French SPOT Vegetation sensor also provides useful coarse-resolution data (Campbell 2007).

17.2.1.1.2 Spectro-Radiometric Properties

Most satellite-borne sensors acquire data in a small number of spectral bands. An exception is the hyperspectral sensor, Hyperion, which is flown on the earth observing 1 (EO1) satellite (Campbell 2007). Hyperion is an experimental sensor, with a limited duty cycle, and therefore acquires only a small number of hyperspectral images, each of which has a relatively narrow swath.

Most multispectral sensors have bands in the same general regions: usually two to three in the visible, one in the near-infrared, and sometimes one in the mid-infrared, and only very occasionally one in the shortwave infrared. ASTER is a notable exception, since it has one band in the mid-infrared and five in the shortwave infrared. ASTER also has five thermal bands, making it the only moderate spatial resolution, multispectral thermal instrument. However, ASTER also has a limited duty cycle.

Radiometric properties of sensors have improved greatly over the years, from 6–7 bit for Landsat MSS, to 10–12 for commercial fine-resolution sensors and MODIS. In addition, modern data tends to be calibrated better, making conversion to at-sensor radiance or estimated ground reflectance possible.

17.2.1.1.3 Temporal Properties

Nadir-viewing satellites have a revisit cycle that is generally set and which depends on the orbital characteristics. Such sensors are ideal for acquiring systematic global coverage. Nevertheless, in the past, image acquisition has been limited by constraints on available local ground receiving stations, limited on board storage, or limited satellite power. Pointable satellites (Toutin 2009), which can observe regions at relatively high angles, have much more frequent potential revisit frequency, which can be important for time-crucial events, for example, disasters.

Disasters often occur with little advance warning, and the affected areas may not have the experience or resources to respond rapidly. Therefore, to expedite acquisition, processing, and distribution of imagery in support of disaster relief, the International Charter on Space and Major Disasters (International Charter 2008, Harris 2009) was established to provide an international protocol and infrastructure for emergency support.

17.2.1.1.4 Other Data Issues

Data costs can vary greatly for a project from nothing, if free data are available, to requiring significant financial investment, if a custom aerial or satellite data set is required. In the United States, federal agencies generally sell image data at the cost of filling the user request (COFUR), and do not attempt to recover the original acquisition cost (Harris 2009). Another important U.S. policy is that federal image data are placed in the public domain, and as a result a large amount of both U.S. aerial and global satellite data is available free over the Internet. Notably, the entire U.S. Geological Survey archive of Landsat imagery is available for instant download at no charge. In other countries, the availability of aerial imagery varies greatly, and may even be treated as a state secret.

Commercial satellite imagery generally is more expensive, and since data are usually only collected in response to user requests, archived imagery may be limited. Normally a charge is made for tasking the satellite to acquire requested images. Commercial satellite imagery also usually has strict licensing limits that may not allow the distribution of the data to others, even within the same organization.

17.2.2 PILOT STUDY

A pilot study is an excellent way to determine the feasibility of a planned project. A relatively quick initial evaluation of data quality and exploratory analysis may show limitations in the planned approach, as well as unexpected opportunities. The human eye is very effective at identifying patterns, and therefore a qualitative visual analysis can be very helpful. Following the pilot study, it may be necessary to revise the project methods, and possibly even the project goals.

17.2.3 FIELD DATA COLLECTION

Modern global positioning system (GPS) equipment has made the collection of spatial attributes for fieldwork much easier. Indeed a GPS unit can become a basic field-recording device, using forms for systematizing data-collection protocols.

Some units can run GIS software, thus allowing the direct input of location and other attributes into a fieldwork database. Digital photographs provide excellent documentation of field samples and general ground conditions.

A particular challenge with field work is scaling from the field observation to the pixel. Part of the problem is that even for an orthorectified image, the specific ground area represented by each pixel has considerable uncertainty. First, the boundaries on the ground representing the pixel area are in reality not sharp, and any pixel will include some radiance from adjacent pixels. Second, georeferencing has error, typically one-quarter to a half a pixel or more. The net result is that if a direct correlation of a field sample to specific pixels is required, the field sample should ideally be acquired over an area representing multiple pixels.

If spectral reflectance measurements are made, for example, with a field spectrometer, the question of scaling from the field observation to the aerial or satellite pixel should be considered carefully. It may be difficult to gain sufficient height to view the scene from above, and to integrate the spectrometer field of view over a large enough area. For vegetation, an alternative is to acquire separate spectra of leaves and woody material, but shadow may also be an important component of the image pixel. For geological investigations, characterizing the effect of weathering may be important, and the variability of the surface, unconsolidated material.

17.2.4 ANALYSIS

The importance of documentation of analysis procedures cannot be overstated. Most remote-sensing analysis is done using software with a graphical user interface. The sequence of steps involved may be complex, requiring multiple iterations of analysis. Modern computers are so fast that the analyst usually does not have time to keep detailed handwritten notes. A partial substitute may be to organize data very systematically, and to use a simple naming convention for files that indicates the processing steps and important parameters involved in producing an image, although using this approach the file names may become cumbersomely long.

Some software packages also address this problem by providing automated metadata tracking for images, or by generating a log file documenting each step taken in the analysis. Software such as Imagine and IDRISI that allow the graphical creation of processing scripts facilitate not only the speed up of repeat analyses but also provide comprehensive documentation of methods (Warner and Campagna 2004).

17.2.5 COMMUNICATING RESULTS

Despite the importance of communicating results, the topic is often given short shrift. Communication to nontechnical audiences poses particular challenges. For example, simply the use of the name "near-infrared," and the typical depiction of near-infrared in images in red tones, may incorrectly suggest to the general public that heat information is displayed. Thus, for nontechnical audiences, careful consideration should be given to not just the technical correctness of the information, but also how the information may be perceived.

Usually remote-sensing map results are supplemented by graphs, tables, and of course text. However, maps are generally the primary method of presenting the spatial component of remote-sensing results. The types of remote-sensing maps include traditional thematic maps, false-color composites, and various image visualizations and enhancements. Most remote-sensing software and GIS packages have map functionality for developing the basic components of maps. At a minimum, a map should have an indication of scale, orientation, location, and source data, including the date of acquisition. For false-color composites, the bands used and the color assignments for those bands, should be indicated. Remote-sensing data sets are often large; for illustrative purposes, it can be useful to depict a small example region. Consistency in scale and area depicted in a series of maps facilitate comparisons.

Traditional paper maps have a long heritage in remote sensing. A relatively recent innovation is the ability to provide dynamic maps through the Internet. Internet map services provide a bridge between maps and the underlying data, and break down the limitations of traditional static maps.

An important component of most remote-sensing analyses is summary information on the uncertainty in the results. Challenges posed by the confusion matrix include obtaining a sample sufficiently large to provide a meaningful estimate of the true map accuracy, the communication of the results to nontechnical audiences, and capturing any spatial variation that may exist in map accuracy.

17.3 CONCLUSIONS

This chapter has emphasized the significance of the planning component in practical applications of remote sensing. However, an ability to apply flexibility may be equally significant, as the analyst must often overcome unforeseen problems. Therefore, even as remote sensing moves to a more quantitative and model-driven approach (Liang 2004), an empirical structure still underpins many remote-sensing projects.

Remote sensing as an analytical method poses many challenges. Nevertheless, remote sensing offers the potential for providing geospatial information that cannot be acquired in a similarly comprehensive and systematic manner through any other method. Much of the past emphasis of automated image analysis has been at regional to global scales, where remote sensing alone can provide rapid and objective information. However, in recent years, especially since the advent of fine-resolution commercial satellites (Toutin 2009), digital analysis methods have increasingly been applied to remote sensing of local regions. With the growing interest and availability of Internet-based image visualization tools, such as Google Earth (Landenberger et al. 2006), remote sensing is likely to grow in importance.

ACKNOWLEDGMENT

The contributions of Jim Campbell to this chapter are gratefully acknowledged.

REFERENCES

Anderson, J.R., Hardy, E.E., Roach, J.T., and Witmer, R.E., *A Land Use and Land Cover Classification System for Use with Remote Sensor Data*. United States Geological Survey Professional Paper 964. Washington, DC: U.S. Government Printing Office, 1976.

Brandtberg, T., Warner, T.A., Landenberger, R., and McGraw, J., Detection and analysis of individual leaf-off tree crowns in small footprint, high sampling density LIDAR data from the eastern deciduous forest in North America. *Remote Sensing of Environment* 85:290–303, 2003.

Brodlie, K., Dykes, J., Gillings, M., Haklay, M.E., Kitchin, R., and Kraak, M., Geography in VR: Context. In: P. Fisher and D. Unwin (Eds.), *Virtual Reality in Geography*. London, U.K.: Taylor & Francis, Chapter 2, pp. 7–16, 2002

Campbell, J.B., *Introduction to Remote Sensing*. New York: Guilford Press, 2007.

Chavez, P.S., Berlin, G.L., and Sowers, L.B., 1982. Statistical methods for selection Landsat MSS ratios. *Journal of Applied Photographic Engineering* 8(1):23–30, 1982.

Chen, X., Warner, T.A., and Campagna, D.J., Integrating visible, near-infrared and short-wave infrared hyperspectral and multispectral thermal imagery for geological mapping at Cuprite, Nevada. *Remote Sensing of Environment* 110:344–356. DOI:10.1016/j.rse.2007.03.015, 2007.

Crist, E.P. and Cicone, R.C., A physically-based transformation of Thematic-Mapper data—The TM Tasseled Cap. *IEEE Transactions on Geoscience and Remote Sensing* GE-22(3):256–263, 1984.

Dave, H., Dewan, C., Paul, S., Sarkar, S.S., Pandya, H., Joshi, S.R., Mishra, A., and Detroja, M., AWiFS camera for Resourcesat. *Proceedings of the SPIE* 6405, 64050X. DOI:10.1117/12.693971, 2006.

Delbart, N., Le Toan, T., Kergoat, L., and Fedotova, V., Remote sensing of spring phenology in boreal regions: A free of snow-effect method using NOAA-AVHRR and SPOT-VGT data (1982–2004). *Remote Sensing of Environment* 101:52–62, 2006.

Dempster, A.P., Upper and lower probabilities induced by a multivalued mapping. *Annual Mathematical Statistics* 38(2):325–339, 1967.

Deng, Y., Making sense of the third dimension through topographic analysis. In: T.A. Warner, M.D. Nellis, and G.M. Foody (Eds.), *Handbook of Remote Sensing*. London, U.K.: Sage, Chapter 22, 2009.

Dymond, C.C., Mladenoff, D.J., and Radeloff, V.C., Phenological differences in Tasseled Cap indices improve deciduous classification. *Remote Sensing of Environment* 80:460–472, 2002.

Fisher, P. and Unwin, D., Virtual reality in geography: An introduction. In: P. Fisher and D. Unwin (Eds.), *Virtual Reality in Geography*. London, U.K.: Taylor & Francis, Chapter 1, pp. 1–4, 2002.

Foody, G.M. and Mathur, A., The use of small training sets containing mixed pixels for accurate hard image classification: Training on mixed spectral responses for classification by a SVM. *Remote Sensing of Environment* 103:179–189, 2006.

Goward, S.N. and Masek, J.G., Landsat—30 years and counting. *Remote Sensing of Environment* 78(1):1–2, 2001.

Grunwald, S., Ramasundaram, V., Comerford, N.B., and Bliss, C.M., Are current scientific visualization and virtual reality techniques capable to represent real soil-landscapes? In: P. Lagacherie, A.B. McBratney, and M. Voltz (Eds.), *Digital Soil Mapping: An Introductory Perspective*. Berlin, Germany: Elsevier, pp. 571–580, 2006.

Harris, R., Remote sensing policy. In: T.A. Warner, M.D. Nellis, and G.M. Foody (Eds.), *Handbook of Remote Sensing*. London, U.K.: Sage, Chapter 2, 2009.

Hodza, P. in press. Evaluating user experience of Experiential GIS, Transactions in GIS.

Hengl, T., Finding the right pixel size. *Computers & Geosciences* 32:1283–1298. DOI:10.1016/j. cageo.2005.11.008, 2006.

Holben, B., Characteristics of maximum-value composite images from temporal AVHRR data. *International Journal of Remote Sensing* 7(11):1417–1434, 1986.

Huang, C., Townshend, J.R.G., Liang, S., Kalluri, N.V., and DeFries, R.S., Impact of sensor's point spread function on land cover characterization: Assessment and deconvolution. *Remote Sensing of Environment* 80(2):203–212, 2002.

Huete, A., Didan, K., Miura, T., Rodriguez, E.P., Gao, X., and Ferreira, L.G., Overview of the radiometric and biophysical performance of the MODIS vegetation indices. *Remote Sensing of Environment* 83:195–213, 2002.

Hughes, G.F., On the mean accuracy of statistical pattern recognizers. *IEEE Transactions on Informational Theory* IT-14:55–63, 1968.

International Charter, 2008. International charter: Space and major disasters. http://www.disasterscharter.org (last accessed July 29, 2008).

Jäger, G. and Benz, U., Measures of classification accuracy based on fuzzy similarity. *IEEE Transactions on Geoscience and Remote Sensing* 38(3):1462–1467. DOI:10.1109/36.843043, 2000.

Jensen, J.R., *Introductory Digital Image Processing: A Remote Sensing Perspective.* Upper Saddle River, NJ: Prentice Hall, 2005.

Jensen, J.R., Im, J., Hardin, P., and Jensen, R.R., Image classification. In: T.A. Warner, M.D. Nellis, and G.M. Foody (Eds.), *Handbook of Remote Sensing.* London, U.K.: Sage, Chapter 19, 2009.

Johannsen, C.J. and Daughtry, C.S.T., Surface reference data collection. In: T.A. Warner, M.D. Nellis, and G.M Foody (Eds.), *Handbook of Remote Sensing.* London, U.K.: Sage, Chapter 17, 2009.

Kauth, R.J. and Thomas, G.S., The tasselled cap—A graphic description of the spectral-temporal development of agricultural crops as seen by Landsat. *Proceedings of the Second Annual Symposium on Machine Processing of Remotely Sensed Data*, Purdue University, West Lafayette, IN, pp. 4B-41–4B-51, 1976.

Key, T., Warner, T.A., McGraw, J., and Fajvan, M.A., A comparison of multispectral and multitemporal imagery for tree species classification. *Remote Sensing of Environment* 75: 100–112, 2001.

Kim, M., Madden, M., and Warner, T.A., Estimation of the optimal image object size for the segmentation of forest stands with multispectral IKONOS imagery. In: T. Blaschke, S. Lang, and G. J. Hay (Eds.), *Object-Based Image Analysis—Spatial Concepts for Knowledge-Driven Remote Sensing Applications.* Berlin, Germany: Springer-Verlag, Chapter 3.2, 2008.

Kim, M., Madden, M., and Warner, T.A., Forest type mapping using object-specific texture measures from multispectral IKONOS imagery: Segmentation quality and image classification issues. *Photogrammetric Engineering and Remote Sensing* (in press), 2009.

Landeberger, R.E., Warner, T.A., Ensign, T., and Nellis, M.D., Using remote sensing and GIS to teach inquiry-based spatial thinking skills: An example using the GLOBE program's integrated earth systems science. *Geocarto International* 21(3):61–71, 2006.

Latty, R.S., Nelson, R., Markham, B., Williams, D., Toll, D., and Irons, J., Performance comparison between information extraction techniques using variable spatial resolution data. *Photogrammetric Engineering and Remote Sensing* 51(9):1459–1470, 1985.

Lee, J.Y. and Warner, T.A., Segment based image classification. *International Journal of Remote Sensing* 27(16):3403–3412, 2006.

Liang, S., *Quantitative Remote Sensing of Land Surfaces.* New York: John Wiley & Sons, Inc., 2004.

Melgani, F. and Bruzzone, L., Classification of hyperspectral remote sensing images with support vector machines. *IEEE Transactions on Geoscience and Remote Sensing* 42:1778–1790, 2004.

Merchant, J. and Narumalani, S., Integrating remote sensing and geographic information systems. In: T.A. Warner, M.D. Nellis, and G.M. Foody (Eds.), *Handbook of Remote Sensing*. London, U.K.: Sage, Chapter 18, 2009.

Moghaddam, M., Polarimetric SAR phenomenology and inversion techniques for vegetated terrain. In: T.A. Warner, M.D. Nellis, and G.M. Foody (Eds.), *Handbook of Remote Sensing*. London, U.K.: Sage, Chapter 6, 2009.

Myneni, R.B., Keeling, C.D., Tucker, C.J. Asrar, G., and Nemani, R.R., Increased plant growth in the northern high latitudes from 1981 to 1991. *Nature* 386:698–702, 1997.

Narayanan, R.M., Sankaravadivelu, T.S., and Reichenbach, S.E., Dependence of information content on gray-scale resolution. *Geocarto International* 15(4):15–27, 2000.

Nield, S.J., Boettinger, J.L., and Ramsey, R.D., Digitally mapping gypsic and nitric soil areas using Landsat ETM data. *Soil Science Society of America Journal* 71:245–252. DOI:10.2136/sssaj2006–0049, 2007.

Phinn, S., Stanford, M., Scarth, P., Murray, A.T., and Shyy, P.T., Monitoring the composition of urban environments based on the vegetation-impervious surface-soil (VIS) model by subpixel analysis techniques. *International Journal of Remote Sensing* 23(20):4131–4153, 2002.

Saura, S., Effects of minimum mapping unit on land cover data spatial configuration and composition. *International Journal of Remote Sensing* 23(22):4853–4880, 2002.

Serpico, S.B. and Moser, G., Extraction of spectral channels from hyperspectral images for classification purposes. *IEEE Transactions on Geoscience and Remote Sensing* 45(2):484–495, 2007.

Shafer, G., *A Mathematical Theory of Evidence*. Princeton, NJ: Princeton University Press, 1976.

Stehman, S.V. and Foody, G.M., Accuracy assessment. In: T.A. Warner, M.D. Nellis, and G.M. Foody (Eds.), *Handbook of Remote Sensing*. London, U.K.: Sage, Chapter 21, 2009.

Strahler, A.H., Woodcock, C.E., and Smith, J.A., On the nature of models in remote sensing. *Remote Sensing of Environment* 20:121–139, 1986.

Toutin, T., Fine spatial resolution optical sensors. In: T.A. Warner, M.D. Nellis, and G.M. Foody (Eds.), *Handbook of Remote Sensing*. London, U.K.: Sage, Chapter 8, 2009.

Turner, M.G., Gardner, R.H., and O'Neill, R.V., *Landscape Ecology in Theory and Practice: Pattern and Process*. New York: Springer, 2001.

van de Vlag, D.E. and A. Stein, Incorporating uncertainty via hierarchical classification using fuzzy decision trees. *IEEE Transactions on Geoscience and Remote Sensing* 45(1):237–245. DOI:10.1109/TGRS.2006.885403, 2007.

van Leeeuwen, W.J.D., Visible, near-IR & shortwave IR spectral characteristics of terrestrial surfaces. In: T.A. Warner, M.D. Nellis, and G.M. Foody (Eds.), *Handbook of Remote Sensing*. London, U.K.: Sage, Chapter 3, 2009.

Warner, T.A., Analysis of spatial patterns in remotely sensed data using multivariate spatial correlation. *Geocarto International* 14(1): 59–65, 1999.

Warner, T.A. and Campagna, D., IDRISI Kilimanjaro review. *Photogrammetric Engineering and Remote Sensing* 70(6):669–673, 684, 2004.

Warner, T.A. and Shank, M., An evaluation of the potential for fuzzy classification of multispectral data using artificial neural networks. *Photogrammetric Engineering and Remote Sensing* 63(11):1285–1294, 1997a.

Warner, T.A. and Shank, M., Spatial autocorrelation analysis of hyperspectral imagery for feature selection. *Remote Sensing of Environment* 60:58–70, 1997b.

Warner, T.A., Steinmaus, K., and Foote, H., An evaluation of spatial autocorrelation-based feature selection. *International Journal of Remote Sensing* 20(8):1601–1616, 1999.

Warner, T.A., Nellis, M.D., Brandtberg, T., McGraw, J., and Gardner, J., 2003. The potential of virtual reality technology for analysis of remotely sensed data: A LIDAR case study. *Geocarto International* 18(1):25–32, 2003.

Warner, T.A., Nellis, M.D., and Foody, G.M., Remote sensing data selection issues. In: T.A. Warner, M.D. Nellis, and G.M. Foody (Eds.), *Handbook of Remote Sensing*. London, U.K.: Sage, Chapter 1, 2009.

Wartenberg, D., Multivariate spatial correlation: A method for exploratory geographical analysis. *Geographical Analysis* 17(4):263–283, 1985.

Woodcock, C.E. and Strahler, A.H., The factor of scale in remote sensing. *Remote Sensing of Environment* (21):311–332, 1987.

Wulder, M.A., White, J.C., Goward, S.N., Masek, J.C., Irons, J.R., Herold, M., Cohen, W.B., Loveland, T.R., and Woodcock, C.E., Landsat continuity: Issues and opportunities for land cover monitoring. *Remote Sensing of Environment* 112(3):955–969, DOI:10.1016/j.rse.2007.07.004, 2008.

Zadeh, L.A., Fuzzy sets. *Information and Control* 8:338–353, 1965.

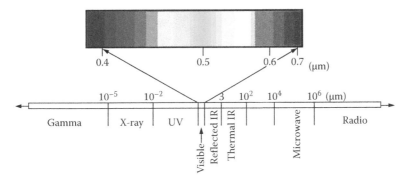

FIGURE 5.2 Spectral regions of the electromagnetic spectrum.

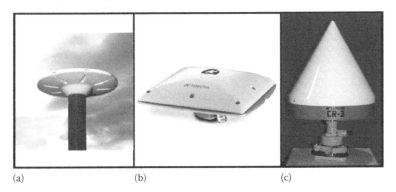

(a) (b) (c)

FIGURE 10.2 (a) Micro-centered L1/L2 antenna with ground plane. (Courtesy of Trimble Navigation Ltd., Sunnyvale, CA) (b) Micro-centered L1/L2 GPS/GLONASS PG-A1 antenna with metal ground plane and rugged housing and (c) Zero-centered L1/L2 CR-3 reference base station antenna mounted on single-depth choke ring, with weatherproof cone environmental cover. (Courtesy of Topcon Positioning Systems, Inc., Livermore, CA.)

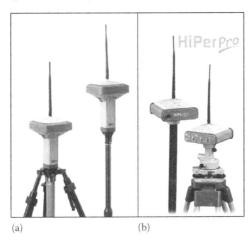

(a) (b)

FIGURE 10.3 Integrated receivers (a) GR-3 and (b) HiPer Pro. (Courtesy of Topcon Positioning Systems, Inc., Livermore, CA.)

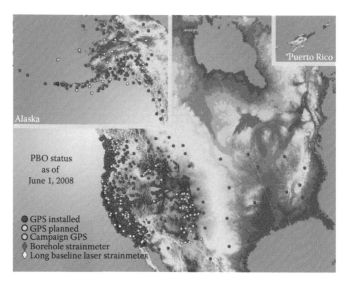

FIGURE 14.6 PBO Network Design showing proposed GPS site locations. (From http://pbo.unavco.org/~kyleb/blm.htm. With permission.)

FIGURE 16.6 Instantaneous shoreline superimposed on a 4-m resolution IKONOS image, Lake Erie, Ohio.

FIGURE 16.7 Digital shoreline (red) produced from a coastal terrain model and a water surface model versus a shoreline from a map (blue).

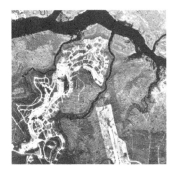

FIGURE 18.2 Digital orthophoto prepared from color infrared aerial photography. The southeast portion of the Cainhoy SC digital orthophoto quarter quadrangle (DOQQ) created from NAPP color infrared aerial photography acquired on February 9, 1995 (1 × 1 m spatial resolution). A golf course community and a small airport lie south of the Wando River. (Courtesy of U.S. Geological Survey, Reston, VA.)

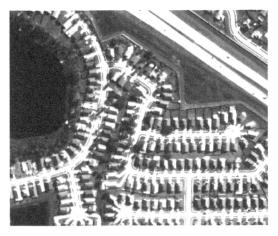

FIGURE 18.3 ADAR 5500 digital photograph. A natural color ADAR 5500 digital photography (RGB = bands 3, 2, 1) of a suburban community east of Orlando, FL acquired on November 5, 1995 (1 × 1 m spatial resolution). (Courtesy of Positive Systems, Inc., Whitefish, MT.)

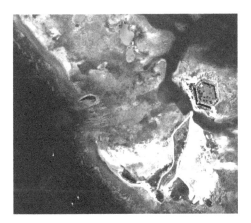

FIGURE 18.4 Landsat Enhanced Thematic Mapper Plus (ETM+). Landsat Enhanced Thematic Mapper Plus (ETM+) image representing Abu Dhabi, United Arab Emirates, acquired on July 11, 1999 (RGB = 7,4,3) at 30 × 30 m spatial resolution. (Courtesy of U.S. Geological Survey, Reston, VA.)

FIGURE 18.5 The NASA HyspIRI mission. The NASA HyspIRI mission (to be confirmed) will be based on space-proven components used in the Moon Mineralogy Mapper (M3) platform. (Courtery of Rob Green, NASA JPL.)

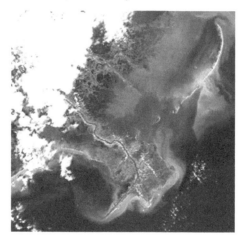

FIGURE 18.6 An image of the Mississippi acquired by the MODIS system onboard NASA's Terra satellite. This scene, acquired on February 24, 2000, depicts the coastal features that surround the delta, including sediment plumes, shallow water bays, and barrier islands. (Courtesy of National Aeronautics and Space Administration, Washington, DC.)

(a)   (b)

FIGURE 19.1 (a) U.S. Geological Survey Geologist uses a pocket stereoscope to examine aerial photography for geological information, 1957. (Courtesy of USGS Photographic Library; photography by E.F. Patterson.) (b) Analysts examine digital imagery displayed on a computer screen using specialized software, 2008. (Courtesy of Virginia Tech Center for Geospatial Information and Technologies.)

FIGURE 19.8 Landsat thematic mapper (TM) false-color composite (RGB—bands 4, 3, 2) Dawson County, Nebraska, August 8, 2007 (see Figure 19.9). (Courtesy of U.S. Geological Survey).

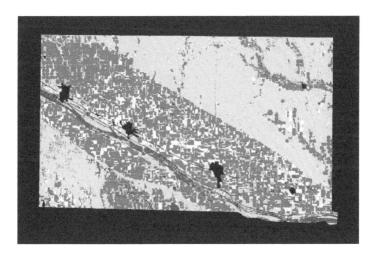

FIGURE 19.9 Results of supervised agricultural crop classification of 1997 multi-date Landsat TM data for Dawson County, Nebraska. Legend: dark green = corn; light green = soybeans; orange = sorghum; yellow = alfalfa; pink = small grains; tan = rangeland, pasture, grass, black = urban land and roads; blue = open water; red = riparian forest and woodlands; cyan = wetlands and subirrigated grasslands; gray = other agricultural land; white = summer fallow (see Figure 19.8). (From Narumalani, S. et al., Information extraction from remotely sensed data, in *Manual of Geospatial Science and Technology*, Bossler, J. (ed.), Taylor & Francis, London, U.K., 2002, pp. 298–234. With permission.)

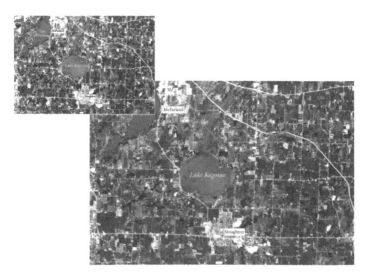

FIGURE 19.13 Change detection, tornado damage, Stoughton, WI, August 18, 2005. ASTER and TM imagery (Chapter 18) provided before (inset) and after (larger image) imagery to document the track of the tornado, as indicated by the debris and damage represented in red by the change detection analysis. (Courtesy of Environmental Remote Sensing at the Space Science & Engineering Center, University of Wisconsin-Madison, http://ersc.ssec.wisc.edu/tornadoes.php.)

FIGURE 24.2 Field photographs of airborne LIDAR site taken in May 2008. (a) A side-view of the houses during low tide, (b) a front-view of one house during the low tide, and (c) a ground-view of one house during high tide. The large sand bags cover the septic tank of the house. (Photography by Y. Wang.)

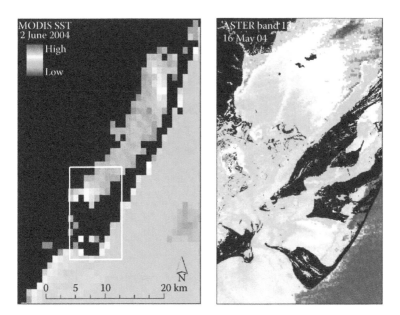

FIGURE 24.4 Comparison of thermal sea-surface temperature from MODIS and ASTER in Chincoteague Bay and nearshore Atlantic. Imagery acquired from the USGS land processes distributed active archive data pool. (From http://lpdaac.usgs.gov/datapool/datapool.asp. With permission.)

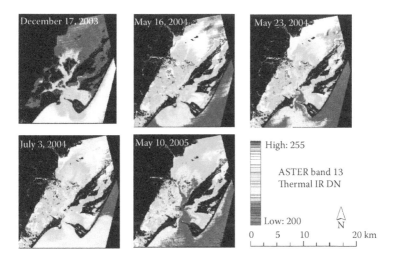

FIGURE 24.5 Time series of nighttime ASTER thermal band 13 images for Chincoteague Inlet, Virginia. Colorized thermal DN values contrast seasonal temperature differences between oceanic coastal waters and estuarine water with high resolution, 90 m. Land pixels are masked using daytime multispectral reflectance.

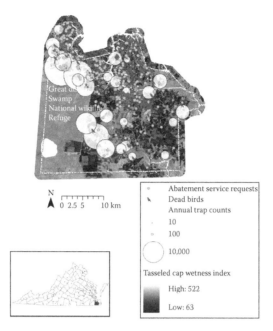

FIGURE 24.6 Integrated remote sensing and field mosquito data for Chesapeake, Virginia, showing mosquito trap counts, dead bird counts, resident abatement service requests, and Landsat ETM+ Tasseled Cap wetness index. Imagery acquired from the University of Maryland global land cover facility. (Courtesy of http://www.landcover.org/index.shtml.)

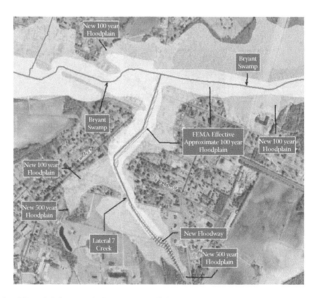

FIGURE 25.1 Floodplains as defined by traditional methods (blue) and using more detailed LIDAR data (pink). The higher detail and finer resolution of the LIDAR data provide a broader definition of the floodplain. (From North Carolina Floodplain Mapping Program. With permission.)

FIGURE 25.2 Top: Downtown Blacksburg, Virginia as imaged by a conventional aerial photograph. (Courtesy of Aerial Imagery.) Bottom: Nighttime aerial imagery of the same region, June 8, 2007, approximately 12:30 a.m. (Courtesy of Peter Sforza.) Center: Composite photography showing nighttime lights against the background of the daytime photograph of the same area as a locational reference. (Courtesy of Peter Sforza.)

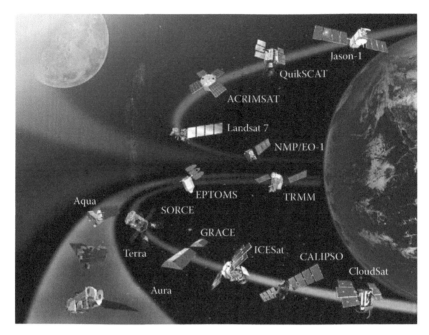

FIGURE 25.3 NASA Earth Science Missions as of 2008.

EOSDIS context

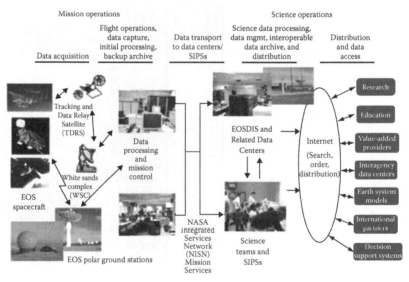

FIGURE 25.4 NASA Earth Observing System (EOS) Data Information System (DIS).

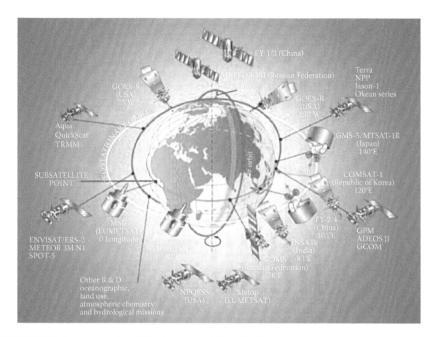

FIGURE 25.5 International Environmental Satellite Systems.

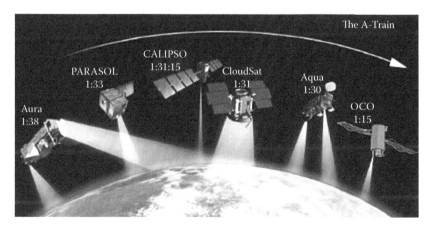

FIGURE 25.6 NASA Earth Observing System (EOS) "A-train."

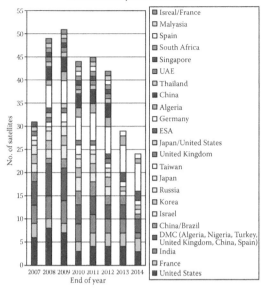

FIGURE 25.7 Optical imaging satellites up through 2014. (Modified from Stoney, W.E., *ASPRS Guide to Land Imaging Satellites*, American Society for Photogrammetry and Remote Sensing, Bethesda, MD, 2008.)

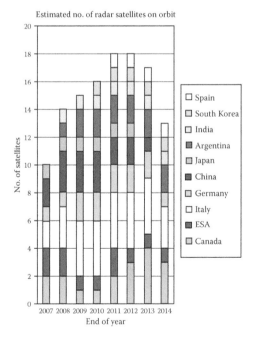

FIGURE 25.8 Radar imaging satellites through 2014. (Modified from Stoney, W.E., *ASPRS Guide to Land Imaging Satellites*, American Society for Photogrammetry and Remote Sensing, Bethesda, MD, 2008.)

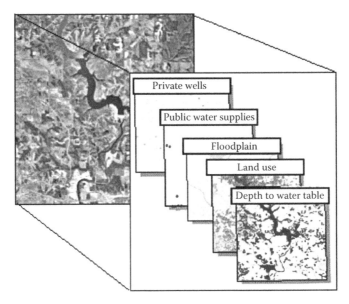

FIGURE 27.5 Entity types, for example, landfill location problem, include land use, public water supplies, private wells, water table, and floodplain.

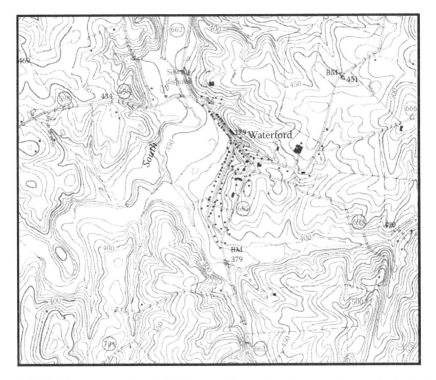

FIGURE 28.1 An example of USGS topographic quadrangle maps showing both objects (roads, buildings, etc.) and fields (contours).

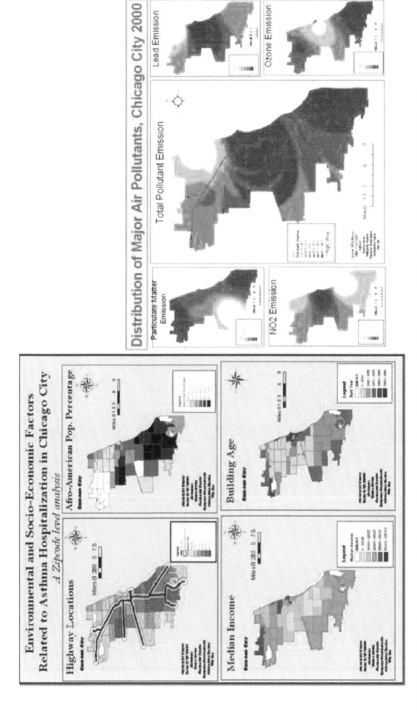

FIGURE 35.2 Occurrence of asthma and its causes, City of Chicago, IL. (Courtesy of U.S. Census Bureau; UIUC, UP418 GIS class project.)

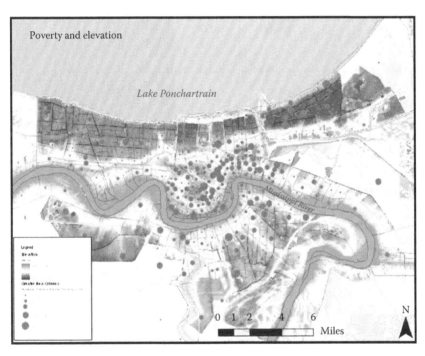

FIGURE 35.3 City of New Orleans–Presence of population in poverty. (Courtesy of USGS; U.S. Census Bureau; UIUC, UP418 GIS class project.)

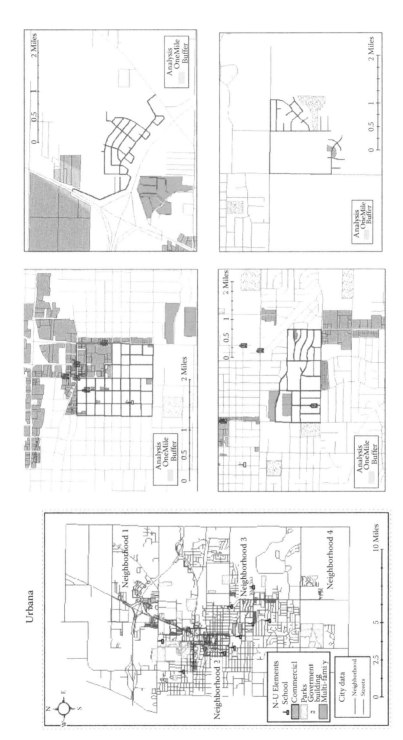

FIGURE 35.4 Street pattern analysis. City of Urbana, IL. (Courtesy of U.S. Census Bureau TIGER files and city of Urbana; UIUL, UP418. GIS class project.)

FIGURE 35.5 Tax increment district (TIF) performance analysis, City of Champaign, IL. (Courtesy of City of Champaign, IL; UIUC, UP418, GIS class project.)

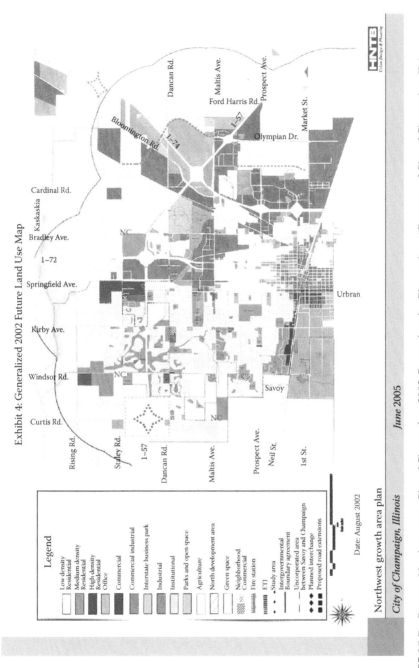

FIGURE 35.7 Proposed land use map, City of Champaign, 2002 Comprehensive Plan update. (Courtesy of City of Champaign, IL.)

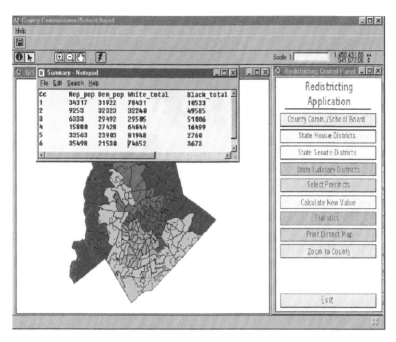

FIGURE 35.10 Election districts generated by GIS districting function. (Courtesy of Macklenburg County, North Carolina, GIS Department.)

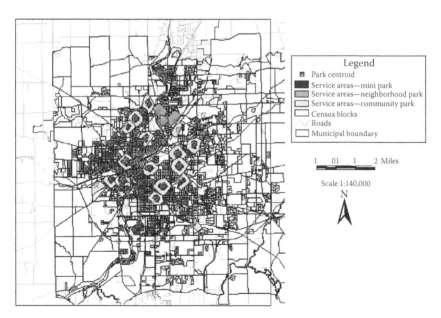

FIGURE 35.11 Delineation of service areas for mini, neighborhood, and community parks. (Courtesy of Illinois Department of Natural Resources; UIUC, GIS-based Illinois Recreational Facilities Inventory Project (G-IRFI).)

Urbana Park District Service Areas

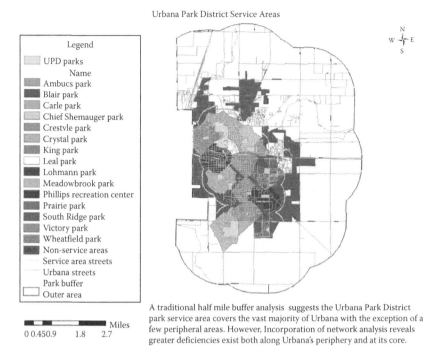

Legend
UPD parks
Name
Ambucs park
Blair park
Carle park
Chief Shemauger park
Crestvle park
Crystal park
King park
Leal park
Lohmann park
Meadowbrook park
Phillips recreation center
Prairie park
South Ridge park
Victory park
Wheatfield park
Non-service areas
Service area streets
Urbana streets
Park buffer
Outer area

Miles
0 0.45 0.9   1.8   2.7

A traditional half mile buffer analysis suggests the Urbana Park District park service area covers the vast majority of Urbana with the exception of a few peripheral areas. However, Incorporation of network analysis reveals greater deficiencies exist both along Urbana's periphery and at its core.

FIGURE 35.12 Park service analysis, City of Urbana, Illinois. (Courtesy of UP418, UIUC GIS class project, City of Urbana Park District.)

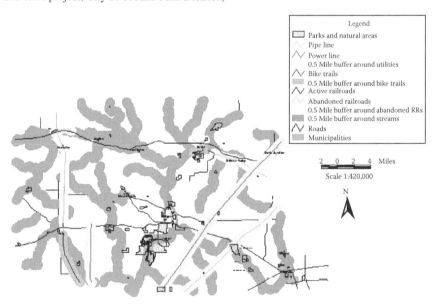

Legend
Parks and natural areas
Pipe line
Power line
0.5 Mile buffer around utilities
Bike trails
0.5 Mile buffer around bike trails
Active railroads
Abandoned railroads
0.5 Mile buffer around abandoned RRs
0.5 Mile buffer around streams
Roads
Municipalities

2   0   2   4   Miles
Scale 1:420,000

N

FIGURE 35.13 Connectivity analysis for regional recreational/open space, City of Rockford, Illinois. (Courtesy of Illinois Department of Natural Resources; UIUC, GIS-based Illinois Recreational Facilities Inventory Project (G-IRFI).)

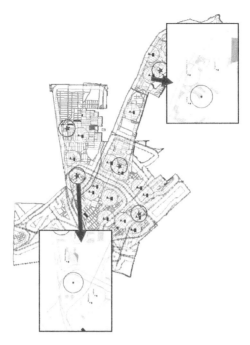

FIGURE 35.14 Suitability analysis of elementary schools' accessibility to walking students, East St. Louis, IL. (Courtesy of U.S. Census Bureau, City of East St. Louis; UIUC, ESLARP project and UP418, GIS class project.)

FIGURE 35.15 Parcel mapping, City and County of Honolulu, Hawaii. (Courtesy of City of Honolulu, HI, GIS, http://gis.hicentral.com/.)

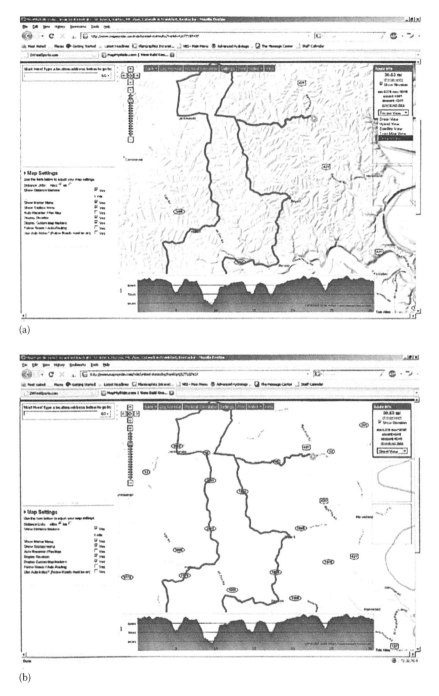

FIGURE 38.1 Screenshots courtesy of www.mapmyride.com: One route shown against three different base maps—(a) shaded digital relief map. (b) state highway map.

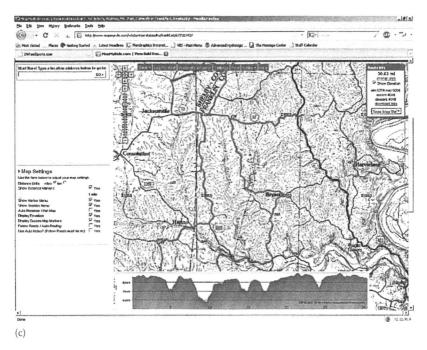

(c)

FIGURE 38.1 (continued) Screenshots courtesy of www.mapmyride.com: One route shown against three different base maps—(c) USGS topo map. (Screenshots from www.mapmyride.com)

FIGURE 38.2 Index map to lineament structures and radon detector readings with embedded image metadata. (From Bobbitt, R. and Johnson, C., Boosting fieldwork productivity with photomaps, *ArcUser Online by ESRI*, July–September 2007. With permission.)

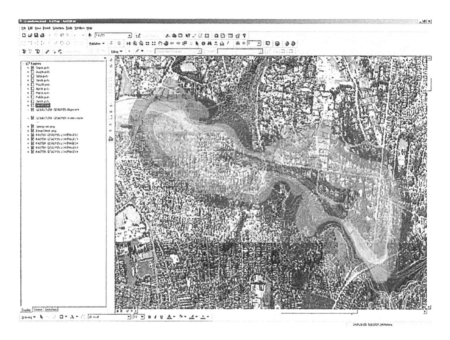

FIGURE 38.3 Water table impact zones resulting from tunneling for the East Link Project, Melbourne, Australia. (From Developing Australia's Largest Road Project; ArcNews online Summer 2007. With permission.)

18 Remote-Sensing Systems for Operational and Research Use

Jan A. N. van Aardt, Melanie Lück-Vogel,
Wolfgang Lück, and John D. Althausen

CONTENTS

18.1   INTRODUCTION

This chapter introduces the geospatial analyst and researcher to a variety of remote-sensing platforms that are commonly used for commercial and research purposes. Our use of the word "operational" in this chapter should be highlighted. "Operational" is meant to imply remote-sensing data that are readily available, either through tasking of systems, for example, airborne campaigns, or through data dissemination services, for example, the U.S. Geological Survey (USGS); it excludes data acquired using experimental or research platforms. Even so, a host of options are available to users of remote-sensing data today, leading to a variety of sensor types and associated characteristics that must be carefully considered when selecting a remote-sensing system. This chapter summarizes these considerations, presents opportunities that currently exist for the remote-sensing analyst and researcher, and provides a brief overview of future or planned missions. These considerations are especially important not only to prepare for today's needs but also to anticipate capabilities that will be available for future geospatial-related activities.

When selecting a candidate remote-sensing system to carry out a study, the following should be considered:

- Photo/imagery type (analog and digital)
- Platform type
- Ordering specifications
- Data structure and format
- Sensor type and characteristics
- Spatial, spectral, temporal, and radiometric resolutions
- Pricing
- Continuity and longevity of the supplying organization

18.2 PHOTO/IMAGERY TYPE

One of the first concerns that must be addressed is the selection of imagery type. Imagery type, as used in this context, differentiates between hard copy and digital (soft copy) formats. The project to be carried out by the remote-sensing analyst normally dictates the imagery type needed. The selection is often hard-copy imagery. If image maps or photographs are required for manual (visual) interpretation, If digital processing is necessary then digital imagery is preferred. (These alternatives are also discussed in Chapter 19.) However, the general trend increasingly favors the use of digital system formats, since advanced visualization technology allows for on-screen interpretation, and soft copy data and products can be converted into hard-copy derivatives as required.

18.2.1 ANALOG PHOTOGRAPHIC PRINTS

Other than traditional topographic maps, analog hard-copy aerial photographs have been the most popular format for data used to visualize the earth's landscape and their continued use, especially for historic analysis, warrants a discussion of past and current analog data products. However, with the advent of digital, high spatial resolution, multispectral cameras, an increasing number of providers are migrating to the routine use of the digital format throughout their operations. Traditional analog aerial cameras typically expose black-and-white film through a Kodak Wratten 12 minus-blue (yellow) filter that records only green and red wavelength energy and minimizes haze effects in the blue spectral region (Lillesand et al., 2008). Cameras may also expose natural color film that records blue, green, and red wavelength energy or false color-infrared film that records green, red, and near-infrared energy through a Wratten minus-blue (yellow) filter. The image analyst may request that the exposed film be processed to yield positive prints or diapositive film (i.e., the translucent slide format that once was common for personal photography). Diapositive film is practical for professional or scientific applications, when the photograph is to be backlit to facilitate visual photo interpretation or, in the past, for conversion to digital remote sensor data via scanning processes. For photographic processing, direct contact printing is preferred, usually using a negative size of 23×23 cm.

One of the most common remote-sensing applications is the preparation of hard-copy image maps (sometimes referred to as photomaps, characterized by planimetrically correct presentation of aerial images). Image maps take advantage of (a) the bird's eye view or vertical perspective of the terrain and (b) the broad geographic coverage that remote sensing can provide. Image maps are very useful when topographic maps are unavailable, outdated, or in cases where imagery and products are required on an operational basis and cannot rely on satellite overpasses or weather-dependent spaceborne imagery. Image maps can be customized for the analyst by size, scale, color (wavelength) selection, and view perspective. Under most conditions, the ancillary information on an image map mimics the traditional topographic map. Additional information that may be present includes platform/sensor name, date of imagery acquisition, and color-coded legend (indicating typical land covers and land uses).

Current remote-sensing analysis and mapping software, for example, Erdas Imagine™, ENVI™, and ESRI ArcGIS™ (Chapter 20), provide the templates to customize image maps to look and feel like standard USGS and National Imagery and Mapping Agency (NIMA) products, with reference grids and elevation contours draped over imagery (ERDAS, 1999). The requirements to enhance features and select specific sensor spectral sensitivities typically dictate the type of information conveyed on the image map and whether it should be a true or false-color presentation.

Typical interpretation tasks for both hard-copy photographs and image maps include feature detection, identification, description, and technical analysis. Imagery interpretability rating systems are commonly developed by agencies to assess the usefulness of photography and image maps. Rating systems are subjective and depend on the ability of the remote-sensing analyst and the smallest element that can be resolved (Jensen and Cowen, 1999).

18.2.2 DIGITAL IMAGERY

Aerial photography and other types of hard-copy imagery such as thermal infrared (TIR) or radar images can be converted into a digital format through the process of image digitization. Such data are generally labeled *digital imagery* when supplied or acquired and analyzed in the raster (matrix) environment. Typical image-processing techniques that are carried out on digital imagery include radiometric, geometric, and atmospheric corrections, followed by application-specific analyses, for example, land cover/land use classification, change detection, modeling, and biophysical information extraction. Applications of visual and digital image-processing techniques are discussed in Chapter 19.

18.2.2.1 Data Structure and Format

Spatial modeling usually incorporates two data structures: raster and vector. Raster-based structures are defined as pixel elements in a grid-based system that are effective for spatially continuous analyses and are amenable to advanced software and processing approaches, whereas the vector-based structures in remote sensing are considered point-line-polygon systems that are utilized for discrete, localized mapping. Raster-based analysis often generate inputs to vector-based systems, for example, a land-cover classification based on satellite imagery, followed by import and spatial analysis of the resultant thematic layer in a vector-based system. Most remote sensor data are collected in raster format (e.g., Landsat ETM+ or SPOT XS), although remote-sensing systems such as light detection and ranging (LIDAR) sensors and the compact airborne spectral imager (CASI) can collect discrete point data when in a profiling mode. There are benefits and cost trade-offs associated with using the two different approaches to structuring data files. In general, raster data require considerably more disk space for storage, and more processing time than do vector data.

18.2.2.2 Image Format

When handling raster files, the remote-sensing analyst must be aware of the particular file format the imagery is stored in. The most common generic binary formats are band-sequential (BSQ), band-interleaved-by-pixel (BIP), and band-interleaved-by-line (BIL).

BSQ stores each band as a separate file, which allows for each band to be analyzed as a separate and individual entity. Remotely sensed data are normally stored in this format so that each spectral band can be accessed separately. BIP stores all data layers of one pixel as adjacent members of the data set. It is a useful format if the remote-sensing analyst is looking at information at specific locations within numerous data bands or if there are hundreds of bands present, as is the case for hyperspectral remote sensor data. The BIL format is a compromise between BSQ and BIP. Rows from each band are stored adjacent to each other and thus different features of a row that correspond to one another are located near each other in the data set. BSQ, BIP, and BIL formats do not compress the raw data in their native formats.

Generic binary file formats such as BIP, BIL, and BSQ are often encapsulated within a wrapper that represents a proprietary format such as the tagged image file format (TIFF). TIFF is the most common format for storing digital photographs and images. This file format is a standard in the field of graphics and was developed in 1986 by the Aldus Corporation. The purpose of TIFF is to describe and store raster images. The TIFF can handle both grayscale and full multiband color images and thus is useful for many applications of remotely sensed data, for example, land-cover mapping and applications that require only common wavelength combinations. The raw TIFF is also amenable to inclusion of georeferencing information, using a newer TIFF format, called GeoTIFF. The GeoTIFF imbeds georeferenced metadata, a property that has popularized GeoTIFF as a useful format for transferring geo-rectified imagery and has become the standard within the USGS for storing and transferring their digital orthophoto quadrangles (DOQs).

Companies that develop digital image–processing software often have their own specialized file formats. Such file formats maintain the raw data and include header files that contain information about the raw data (number of bands, rows, and columns), image statistics, georeferencing information, and sensor parameters. Most company-specific formats can be converted and exchanged through import and export modules that are supplied with digital image–processing software. Many of the most important digital image–processing systems are described in Chapter 20.

18.3 PLATFORM TYPE

In this chapter, "*platform type*" is defined as either an aircraft-based or satellite-based data collection system. The advantages and disadvantages that the remote-sensing analyst must take into account when selecting either platform are described in the following text.

18.3.1 ADVANTAGES OF AN AIRCRAFT PLATFORM

A wide variety of airborne remote-sensing systems is available to image analysts. Aircraft platforms offer the possibility of low flying altitudes, acquiring high-resolution imagery (down to 0.1 m pixels), frequent revisits that can be tailored to acquire data during optimal weather conditions, and variable flight paths. Overall, collection parameters for aircraft are more flexible than satellite platforms, with the aircraft platform being more amenable to changing mission plans and relocation. With most aircraft systems, there is potential for simultaneous collection of differential global positioning system (DGPS) (see Chapter 11) and inertial navigation information along with the imagery, thus providing the possibility for highly accurate georeferencing.

18.3.2   DISADVANTAGES OF AN AIRCRAFT PLATFORM

The most obvious disadvantage of selecting an aircraft platform is that the lower flying altitude will limit the area of coverage. Unlike a spaceborne sensor system that can have areal coverage on the order of hundreds to thousands of square kilometers, an aircraft system must acquire data for multiple flight lines to cover an area equivalent to that of a typical single satellite image. The mosaicing of multiple photographs or flight lines thus becomes an important factor and processing consideration when selecting an aircraft platform. Although imaging systems onboard aircraft are continuously improving, the stability of these systems in terms of geometric distortions is often a concern. The analyst must always take into consideration the effects of aircraft roll (rise–dip of aircraft wings), pitch (up–down aircraft nose movement), and yaw (left–right aircraft movement) on the resulting remote-sensing data collection.

18.3.3   ADVANTAGES OF A SATELLITE PLATFORM

Typical earth observation satellites orbit in repetitive geosynchronous or sun-synchronous orbits, while geostationary satellites remain stationary above a predefined location. The orbits are relatively stable and result in remote sensor data with limited geometric distortion. A satellite's altitude, orbit, and path are generally fixed with each mission and monitored throughout its lifespan by ground stations located across the globe. Different data sets from a single sensor thus have similar spatial resolution and swath (or fields-of-view [FOVs] for geosynchronous satellite) characteristics. Many new satellite-borne sensor systems have the capability to acquire images of the earth not just directly below the satellite at nadir, but at locations off-nadir. This off-nadir pointing capability is especially useful when performing disaster assessment, given more frequent revisit times (higher temporal resolution), and for applications that require stereo-imagery, for example, the extraction of digital elevation models (DEMs) from overlapping, stereo-image pairs (Chapter 16).

18.3.4   DISADVANTAGES OF A SATELLITE PLATFORM

In general, there are relatively few limitations to consider when selecting a satellite platform. A lack of available high-resolution data, due to high, fixed orbits, was a concern in the past; however, current new-generation satellite-based sensors offer high spatial-resolution data (ground resolution often <1 m) at relatively high, fixed orbits. Timeliness may also be perceived as a limitation, but currently there are several satellite constellations that provide near real-time data. Unfortunately, when a satellite remote-sensing system encounters a platform/sensor/engineering problem, it is not possible to retrieve, fix, and recalibrate it as is the case with aircraft-based sensor systems.

18.4 SENSOR TYPE AND RESOLUTION CHARACTERISTICS

It could be argued that, for the analysts' choice of imaging system, the aspects that require the most careful attention are those that concern the resolution characteristics of a sensor. Broadly stated, resolution refers to the levels of detail conveyed by an image, defined by four interrelated characteristics:

- *Spatial resolution* is a measure of the finest spacing of objects on the ground that can be differentiated by a remote-sensing sensor. In practical terms, one can generalize that the minimum acceptable spatial resolution is half the diameter of the smallest dimension of the object of interest. As an approximation, spatial resolution corresponds to pixel size for digital imagery, while the concept is roughly analogous to "grain" in photographic images. Typical differentiation is made among very high (<2.5 m), high (<10 m), medium (10–100 m), low (100–500 m), and very low (>500 m) resolution digital imagery, although these designations are subjective (Lillesand et al., 2008).
- *Spectral resolution* is regarded as the number and width (i.e., spectral bandwidth) of wavelength intervals that a sensor can collect for any given pixel or location. Sensors are typically divided into panchromatic or monospectral (1 spectral band), multispectral (2–8 bands), and hyperspectral (>20 contiguous spectral bands) systems.
- *Temporal resolution* is a measure of sensor revisit time for a given location. In other words, it describes the periodicity (repetitive character) of the sensor's data acquisition capabilities over a fixed target. Note that temporal resolution is usually defined for at-nadir sensing, that is, perpendicular to the target surface; many vendors claim high temporal resolutions, but fail to specify that these are only achievable at off-nadir viewing angles. Revisit time is a function of a satellite's orbit, altitude, sensor optics, and sensor pointability. For nadir-only viewing sensors, the revisit time is primarily a function of the size of the imaging swath of the system and orbital path. For a pointable optics sensor, the revisit time is a function of how far off-nadir the sensor can point and the altitude at which the data are being collected, as well as imaging swath and orbital path. Temporal resolution is a critical consideration in applications related to dynamic phenomena, such as forest and range fires, floods, and other natural disasters, and when changeable weather conditions prevail. It should be noted that off-nadir imagery, taken at an inclination of >30° from nadir, is of limited use for mapping applications and quantitative image analysis techniques due to induced radiometric

and geometric distortions. These distortions can no longer be corrected for at extreme off-nadir angles.

* *Radiometric resolution* describes the sensor's ability to distinguish between objects of similar reflectance and hence represents the sensitivity of a sensor to fine or subtle differences in captured electromagnetic radiation. Sensor bit-range is used to define these characteristics, for example, a 2 bit sensor can differentiate between $2^2 = 4$ gray levels in a specific band, while an 8 bit sensor can capture $2^8 = 256$ gray levels in that same band. Many modern sensors capture imagery in 12 bit (4096 gray levels) radiometric resolution, which theoretically allows the discrimination of features with only subtle radiation differences.

Generally speaking, there exists a trade-off between spectral, spatial, and temporal characteristics, where high spatial resolution typically results in low spectral (limited measurable energy on a per-pixel basis) and low temporal (small pixels result in narrow swath-widths, which equals long revisit times) resolutions. Although the opposite is typically true of low spatial resolution sensors, advances in sensor technology have invalidated many such stereotypical assumptions. Radiometric resolution, on the other hand, is very much a function of the sensor quality and materials. Figure 18.1 provides a rough overview of the various trade-offs between resolution characteristics, all of which should be carefully weighed when the analyst has to choose a sensor for a specific project, for example, does he or she want to differentiate between various tree species at the tree level, between crops at the farm level, or perform a regional land cover classification on a regular basis?

The remainder of this section identifies many of the available remote-sensing systems that are operating, or are expected to be operating in the near future, while also elucidating resolution specifications. It also identifies some of the historical data sets (e.g., National High Altitude Photography (NHAP) Program and Landsat multispectral scanner [MSS]) that are available in archives and that may be of use to the remote-sensing analyst.

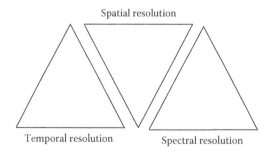

FIGURE 18.1 A generalized overview of the trade-offs among spatial, spectral, and temporal resolutions, with triangle width being indicative of resolution magnitude. It is obvious that these sensor characteristics, along with radiometric resolution, are critical considerations when a system is selected for a specific earth observation application.

18.5   SENSORS PROVIDING SPECTRAL INFORMATION

The sensors that relate most directly to our everyday experience provide information concerning the spectral characteristics, or "colors," of objects considered in a multispectral context. Such sensors are designed to acquire information about the brightness of objects and features in carefully defined regions of the electromagnetic spectrum. Each sensor is designed for a typical field of application. It is thus recommended that the user, who wants to apply remote-sensing data to a specific task, carefully determines if the data are applicable to that task: for example, is the temporal resolution adequate for repetitive monitoring purposes, or is the spatial resolution sufficient to assess the ground features of interest?

18.5.1   AIRCRAFT REMOTE-SENSING SYSTEMS

The discussion begins with a description of aircraft mapping programs and collection systems and concludes with an overview of many of the satellite-based sensor systems. Although satellite sensors typically can be divided into very high, high, medium, low, and very low spatial resolutions, the two latter categories do not apply to airborne systems.

18.5.1.1   High Spatial Resolution Systems

These systems collect imagery at relatively fine spatial resolutions (<10 m), offering the ability to examine detail in urban regions, including individual structures, transportation systems, and land use patterns. By necessity, the finer detail conveyed by such images means that individual scenes represent rather small geographic regions, meaning that coverage must be carefully targeted areas of significance to customers.

NHAP Program: The Soil Conservation Service of the U.S. Department of Agriculture carried out the NHAP program between 1980 and 1989. This was a cooperative effort by federal agencies to acquire cartographic quality 23 × 23 cm false-color infrared and panchromatic aerial photographs. The flight lines ran north–south through the center of USGS 7.5 min topographic quadrangles and each false-color infrared photo at 1:58,000 covered the geographic area of 1 quad. One full frame of false-color infrared imagery covered approximately 69 sq mi (176 km). The panchromatic photography was acquired at 1:80,000. Archived NHAP data are available and may be utilized by the remote-sensing analyst for a historical perspective of a region. The National Cartographic Information Centre (NCIC) in each state can be contacted to determine coverage or a search may be conducted via the Internet (see http://edc.usgs.gov/guides/nhap.html).

National Aerial Photography Program (NAPP): The requirement for higher spatial resolution aerial photography by many federal agencies led to the replacement of the NHAP by the NAPP in 1987. This false-color infrared coverage is flown at an altitude of 20,000 ft above ground level, and the frames are cantered on quarter sections of USGS quadrangles. The photography has a nominal scale of 1:40,000 and one full frame covers approximately 32 sq mi (83 km²). The intent is to rephotograph areas every 5–7 years for the entire conterminous United States. Many of the recent NAPP scenes, especially from the 1993 to 1995 aerial campaigns, are available as DOQs through the USGS and numerous state agencies (see http://edc.usgs.gov/guides/napp.html). An example of NAPP color-infrared photography is shown in Figure 18.2.

Both the NHAP and NAPP missions obtained leaf-off (winter) and leaf-on (summer) aerial photography depending upon mission requirement. If the use of the photography was primarily for contour mapping, then the requirements of the photography were better met if the ground surface can be seen. This necessitates data collection in the early spring for optimal leaf-off conditions. If the need is to map the different types of vegetation based on their spectral characteristics, then of course aerial photographs that capture leaf-on trees are more useful.

18.5.1.2 Aircraft Videography

A relatively recent development in the field of remote sensing is real-time true-color and false-color infrared video. This technology allows an interpreter to view color or false-color infrared imagery as it is being recorded, or in "real time." The impact of this development is of major significance to many fields that focus on the earth's surface condition. The main difficulties with videography are rectifying and registering information gathered from such systems in a geographic information system (GIS) format and carrying out radiometric corrections across the entire series of frames. Also, the individual frames of videography contain relatively few

FIGURE 18.2 (**See color insert following page 426.**) Digital orthophoto prepared from color infrared aerial photography. The southeast portion of the Cainhoy SC digital orthophoto quarter quadrangle (DOQQ) created from NAPP color infrared aerial photography acquired on February 9, 1995 (1 × 1 m spatial resolution). A golf course community and a small airport lie south of the Wando River. (Courtesy of U.S. Geological Survey, Reston, VA.)

picture elements. This means that it is necessary to fly many flight lines at low altitudes in order to acquire high spatial resolution imagery. Videography imagery is notoriously difficult to edge-map and to convert to a controlled mosaic format. However, it is relatively inexpensive with the major mission expense being the mobilization of the aircraft.

18.5.1.3 Aircraft Digital Photography/Multispectral Imagery

A variety of commercial companies have sprung up to fill the niche of providing high spatial resolution digital airborne imagery. Many users have made the transition from hard-copy film cameras to digital data products, given that digital airborne sensors have developed to where high quality, a large degree of automation, and more integrated workflows are commonplace even at high spatial resolutions. The digital airborne sensing domain has evolved from services provided by companies such as Positive Systems, Inc., who supported Airborne Data Acquisition and Registration Systems (ADAR 3000 and ADAR 5500), to companies such as Fugro Earthdata, capable of providing a range of airborne remote-sensing data and products through their Leica ADS40-SH52 and LIDAR systems.

The ADAR 3000 is a single camera system that is capable of acquiring digital panchromatic, true color, or color-infrared photographs. The ADAR 5500 system (Table 18.1) is a four-camera system that provides a four-band multispectral image. Data collected by the ADAR 3000 system are ideally suited for GIS and computer-aided design (CAD) applications, while data collected by the ADAR 5500 systems can be digitally processed and utilized for environmental and wetlands monitoring, forestry applications, and precision agriculture. Data from both systems are spatially and spectrally of high fidelity and can provide the remote-sensing analyst with a seamless mosaic of an area at a spatial resolution of 0.5×0.5 m with spectral sensitivity in the visible and near-infrared regions. An example of ADAR 5500 imagery is shown in Figure 18.3.

The Leica ADS40 airborne digital system, on the other hand, represents the more recent spate of airborne image acquisition systems and can acquire digital aerial imagery at spatial resolutions ranging from 0.07 to 2 m in panchromatic, red, green, blue, and near-infrared spectral bands. Derived products include ortho-imagery at 1:500–1:24,000 scales, DEMs, and a variety of classified products, for example, land use/land cover, impervious surfaces, etc.

TABLE 18.1
Airborne Data Acquisition and Registration System 5500 Characteristics

Sensor Parameter	ADAR 5500 System Specifications
Imaging frame	1500×1000 pixels
Instantaneous field of view	0.44 mrad
Spatial resolution	Depends on flight mission
Spectral filters	Four programmable bands between 0.400 and 1.000 µm (blue to near-infrared)

FIGURE 18.3   (**See color insert following page 426.**) ADAR 5500 digital photograph. A natural color ADAR 5500 digital photography (RGB = bands 3, 2, 1) of a suburban community east of Orlando, FL acquired on November 5, 1995 (1 × 1 m spatial resolution). (Courtesy of Positive Systems, Inc., Whitefish, MT.)

18.5.1.4   Aircraft Hyperspectral Imagery

18.5.1.4.1   Airborne Imaging Spectroradiometer for Applications

Spectral Imaging Ltd. and 3Di, LLC (Easton, MD) have developed and marketed the Airborne Imaging Spectroradiometer for Applications (AISA). It offers the remote-sensing analyst a hyperspectral data set of 10–70 programmable bands extending from the visible into the near- and mid-infrared spectral regions (Table 18.2). AISA has been used by 3Di and its customers in a number of applications including precision agriculture, wetlands delineation, and coral reef mapping.

18.5.1.4.2   Compact Spectrographic Imager-2

The Compact Airborne Spectrographic Imager-2 (CASI-2) is a hyperspectral imager that utilizes a push-broom spectrograph. It is a programmable system that offers

---

TABLE 18.2

Airborne Imaging Spectroradiometer for Application Characteristics

Sensor Parameter	AISA Specifications
Instrumentation	Pushbroom linear array
Imaging swath	364 pixels
Instantaneous field of view	1.0 mrad
Spatial resolution	Depends on flight mission
Spectral bands	10–70 programmable bands between 0.430 and 0.900 μm (blue to near-infrared) (also now mid-infrared)

TABLE 18.3
CASI-2 Characteristics

Sensor Parameter	CASI-2 Specifications
Instrumentation	Pushbroom linear array
Imaging swath	512 pixels
Instantaneous field of view	1.34 mrad
Spatial resolution	Depends on flight mission
Spectral bands	19–288 programmable bands between 0.400 and 1.00 µm (blue to near-infrared)

the user a choice of 288 different visible and near-infrared wavelengths. CASI can supply hyperspectral data in spatial resolutions from 0.5 to 3 m. This type of remote-sensing system has been utilized in a number of different environmental applications including forestry, agriculture, and wetlands mapping. With the ability to select specific wavelengths, CASI-2 offers the remote-sensing analyst a potential to effectively target and study specific environmental features, for example, spectral regions and absorption features that are associated with vegetation health and stress. Table 18.3 summarizes the characteristics of CASI-2.

18.5.1.4.3   HyMap

HyMap represents a series of sensors developed by Hyvista, Australia. The first HyMap sensor had 96 spectral channels in the 550–2500 nm wavelength range and was optimized for mineral exploration, with high signal to noise ratio (SNR) in the 2000–2500 nm spectral range. More recent sensors have been able to collect 128 bands in the 440–2500 nm spectral range, as well as two thermal bands (3–5 and 8–10 µm). These types of spectral characteristics have expanded the utility of HyMap to applications such as pollution monitoring, agriculture and forestry (production systems), soil mapping, and the assessment of invasive species (Cocks et al., 1998).

The internals are characterized by an opto-mechanical scanning system that incorporates spectrographic/detector array modules, an onboard reference lamp, and a shutter for dark current monitoring. Spectral calibration serves to determine the band center wavelengths and bandwidths. The HyMap sensor is spectrally calibrated in the laboratory for at-nadir viewing. Detailed specifications of the HyMap sensor are provided in Table 18.4 (Cocks et al., 1998).

18.5.1.4.4   Airborne Visible-Infrared Imaging Spectrometer

The Airborne Visible/Infrared Imaging Spectrometer (AVIRIS; Table 18.5), developed by the Jet Propulsion Laboratory (JPL) in 1983, arguably is the benchmark for airborne imaging spectroscopy. AVIRIS first started operating in 1987 and has the distinction of being the first imaging spectrometer to measure the electromagnetic spectrum from 400 to 2500 nm. AVIRIS measures up-welling radiance through 224 contiguous spectral channels at 10 nm intervals across the spectrum. Image dimensions typically are 11 km width by up to 800 km length at a 20 m spatial resolution

TABLE 18.4

HyMap Imaging Spectrometer Sensor Specifications

Sensor Parameter	HyMap Specifications
Instrumentation	Whiskbroom scanner
Imaging swath	60°–70°
Instantaneous field of view	1.34 mrad
Spatial resolution	2–10 m
Spectral bands	100–200 bands; 10–20 nm bandwidths VIS (0.45–0.89 μm), NIR (0.89–1.35 μm), SWIR (1.40–1.80 μm and 1.95–2.48 μm), MWIR (3–5 μm), TIR (8–10 μm)
SNR (30° > SZA, 50% reflectance)	500:1

Source: Cocks, T. et al., The HyMap Airborne Hyperspectral sensor: The system, calibration, and performance, *Proceedings of 1st EarSeL Workshop on Imaging Spectroscopy*, Zurich, Switzerland, 1998, 6.

TABLE 18.5

Airborne Visible-Infrared Imaging Spectrometer (AVIRIS) Characteristics

Sensor Parameter	AVIRIS Specifications
Instrumentation	Whiskbroom linear array
Imaging swath	614 pixels
Instantaneous field of view	1.0 mrad
Spatial resolution	Depends on flight mission (ranges between 3.4 and 20 m)
Spectral bands	224 bands at 10 nm intervals between 0.400 and 2.500 μm (blue to mid-infrared)
Radiometric resolution	12 bit

(20 km flying height). AVIRIS imagery is acquired from the Q-bay of a NASA ER-2 aircraft at 20,000 m. Low-altitude collections also are possible, with a flying height of approximately 13.2 km resulting in 3.4 m spatial resolution, for instance. The spectral, radiometric, and spatial calibration of AVIRIS are measured preflight in laboratory and monitored in-flight each year (Green et al., 1998). Applications involving AVIRIS data are numerous and include ecological assessment, coastal/marine mapping, geology/mineralogy, and atmospheric profiling. Appendix 18.A provides Web sites for data from AVIRIS and other sensors.

18.5.1.5 Aircraft Laser Systems

There is a variety of airborne laser, or LIDAR, systems currently under operation. An explanation of LIDAR systems is provided in Chapter 23. We will not attempt a detailed description of all available systems, although most of these are produced by either Optech, Inc., or Leica Geosystems. These LIDAR systems are not imagers per se, since they acquire range/distance data on a per-point basis, which effectively results in text files with multiple x, y, and z entries. Discrete return (point based, individual return) LIDAR systems currently operate at more than 100 kHz (100,000 pulses/s), can register more than one return per pulse, which leads to the extraction of vegetation structural information, and typically have a small ground footprint (<1 m). Variations in the specifications can lead to very advanced systems, such as those operated by the Carnegie Airborne Observatory (CAO; http://cao.stanford.edu/), which also includes a waveform LIDAR, capable of digitizing the waveform energy response of a pulse as it interacts with a target.

We will limit ourselves to the description of a couple of established airborne laser systems; one is used for fluorescence mapping, while the other is used for topographic modeling.

18.5.1.5.1 Airborne Oceanographic LIDAR-3

The first Airborne Oceanographic LIDAR system (AOL) was developed by National Aeronautics and Space Administration (NASA) in the late 1970s. The current system, AOL-3, is flown for validation data collection by the moderate-resolution imaging spectroradiometer (MODIS) (see Section 18.5.2.3.2). Another application to which the AOL-3 is well suited is imaging of fluorescence on the ocean surface of dissolved organic material and chlorophyll. Imaging spectrometers are also flown by NASA to collect ocean color information simultaneously with the AOL-3. Table 18.6 summarizes the basic characteristics of the AOL-3 system.

18.5.1.5.2 Airborne Topographic Mapper

Developed and marketed by NASA, the Airborne Topographic Mapper (ATM) is a LIDAR instrument designed to conduct high-resolution surface topographic mapping. As part of a joint effort with National Oceanic and Atmospheric Administration (NOAA), NASA has used the ATM to map beach profiles from Delaware to South Carolina. When collected with DGPS, ATM instruments (Table 18.7) have collected elevation information at accuracies approaching 10 cm. As mentioned earlier, numerous commercial firms now offer DEM information derived from LIDAR systems.

TABLE 18.6
AOL-3 Characteristics

Sensor Parameter	AOL-3 Specifications
Instrumentation	Dual wavelength laser fluorospectrometer
Pulse width	12 ns
Instantaneous field of view	1.0 mrad
Spectral bands	0.355 μm (ultraviolet)
	0.532 μm (green)

TABLE 18.7
ATM Characteristics

Sensor Parameter	ATM Specifications
Instrumentation	Laser transmitter system
Pulse width	7 ns
Instantaneous field of view	2.1 mrad
Spectral bands	0.532 μm (green)

18.5.1.6 Medium Spatial Resolution Systems

Airborne imaging systems are typically of the high spatial resolution type, given the typical acquisition's altitude above-ground-level (AGL) and sensor FOV. However, the spatial resolution characteristics of such systems are largely a function of just that, the AGL and FOV specifications. It therefore follows that, if a lower spatial resolution and subsequent increased coverage are desired, one could merely specify a higher acquisition altitude for a given FOV. This would effectively result in medium spatial resolution imagery, with the stated benefit of larger area coverage than would have been the case for a low altitude acquisition. One additional benefit that often goes unnoticed is that of increased SNR in such imagery. By integrating the spectral signal across a larger per-pixel area, more upwelling electromagnetic radiation reaches the sensor, which improves the signal. This is rarely required for multispectral sensors, which gain signal strength by increasing the sensor bandwidths, i.e., we are now integrating not across space, but across spectral response. Hyperspectral sensors, however, stand to gain from lower spatial resolutions or increased pixel size, since these sensors typically have narrow, contiguous bands. These narrow bandwidths benefit from larger pixels that allow more upwelling radiation reaching the sensor on a per-pixel basis, which results in more signal/energy for sensing purposes. Hence, this category could apply to most airborne systems discussed earlier in terms of higher AGL flights and increased spatial coverage, while such higher altitudes are especially useful in the case of hyperspectral sensors or imaging spectrometers, specifically AVIRIS, discussed in Section 18.5.1.4.

18.5.2 SATELLITE REMOTE-SENSING SYSTEMS

Satellite observation systems offer the characteristics already outlined in Sections 18.3.3 and 18.3.4. They have greatly expanded the reach of overhead imaging systems, especially by increasing our ability to acquire repetitive coverage of a single region, to image remote regions, and to collect imagery of regions affected by natural disasters.

18.5.2.1 Very High and High Spatial Resolution Systems

As was the case with aircraft-/airborne sensors, these satellite systems collect imagery at relatively fine spatial resolutions (<10 m), and offer the same types of applications and restrictions in terms of pixel size.

18.5.2.1.1 Corona

Although IKONOS (see the following paragraphs) can rightfully be regarded as the first civilian, high-resolution imaging system, such systems have been in use for military purposes since the late 1950s. The U.S. military space imaging reconnaissance program, Corona, delivered high-resolution data from 1959 to 1972. Images were acquired using an analog photo system and the films were returned to earth in a sealed capsule. The panchromatic imagery from satellites launched after 1963 had spatial resolutions between 1.8 and 2.75 m. Data used to be classified, but can be requested from USGS at relatively low cost since 1995. Today these data deliver useful information for environmental and urban monitoring and change analysis, covering time periods before the Landsat era began.

18.5.2.1.2 GeoEye: Operating IKONOS and GeoEye 1

IKONOS was the civilian world's first high spatial resolution satellite remote-sensing system and is operated by GeoEye (formerly owned by Space Imaging, Inc.). After IKONOS-1 failed to achieve orbit on April 27, 1999, IKONOS-2 (then renamed to IKONOS) was successfully launched and deployed on September 24, 1999, and jump-started the high-resolution, commercial satellite imaging industry (Fritz, 1996). The Kodak camera system onboard IKONOS has one panchromatic band at 0.82 m spatial resolution and four multispectral bands at 3.2 m resolution. IKONOS has a small swath width of 11.3 km (the size of a U.S. township), while revisit times are 3–4 days. The system can view both along-track and cross-track, and has fore-and-aft imaging capabilities for stereo-viewing. Data from IKONOS are useful for application in urban planning, natural resource investigations, and disaster management. Table 18.8 summarizes the basic characteristics of the IKONOS system. Several other high-resolution satellite sensor systems have since been launched or are scheduled for launch.

The GeoEye-1 satellite was launched on September 6, 2008, and boasts the highest commercially available spatial resolution to date at 0.41 m panchromatic spectral

TABLE 18.8

IKONOS Characteristics

Sensor Parameter	HRV/HRBIR (IKONOS) Specifications
Instrumentation	Kodak linear array digital camera
Imaging swath	11 km
Instantaneous field of view	1.0 mrad
Spatial resolution	3.2 m (multispectral); 0.8 m (panchromatic)
Spectral bands	0.45–0.52 μm (blue)
	0.52–0.60 μm (green)
	0.63–0.69 μm (red)
	0.76–0.90 μm (near-infrared)
	0.45–0.90 μm (panchromatic)
Repeat coverage	3–4 days

capability and 1.65 m for its four-band multispectral imagery. It has a revisit time of less than three days at off-nadir image acquisition. The system was undergoing its 45 day engineering evaluation phase at the time of this publication, after which GeoEye will start to provide imagery.

18.5.2.1.3 DigitalGlobe: Quickbird and Worldview-1 and -2

DigitalGlobe systems that are currently operating include the Quickbird and Worldview-1 satellites (http://www.digitalglobe.com/). Quickbird has a spatial resolution of 0. 61 (panchromatic) and 2.4 m (four multispectral bands: 450–520, 520–600, 630–690, and 760–900 nm). Typical applications include local to regional environmental analyses of land use, agriculture, and forestry, as well as urban applications. Worldview-1, on the other hand, has only a panchromatic band at 0.5 m spatial resolution. This sensor platform, launched on September 18, 2007, has a ground swath width of 17.6 km and revisit times of 4.6 days and 1.7 days at nadir-approximating and off-nadir angles, respectively. Worldview-2, slated for launch in 2009, will again push the envelope in terms of spatial resolution (0.46 m panchromatic), but will also have multispectral capabilities (2.4 m; RGB and NIR, as well as red-edge, coastal, yellow, and a second NIR band).

18.5.2.1.4 RapidEye: RapidEye Constellation

RapidEye is a German company that specializes in providing fast end-to-end geo-information products derived from remote-sensing imagery. They operate a constellation of five microsatellites, the RapidEye 1 constellation, launched in 2008 (http://www.rapideye.de/). These satellites have an identical payload, provide imagery with five spectral bands (blue, green, red, red-edge, and NIR) at 5 m spatial resolution and 12 bit radiometric resolution, and have swath widths of 70 km each. The constellation is able to cover in excess of 4.5 million km^2 daily and provide a seamless coverage of large countries within five days. The red-edge band was selected for applications in precision agriculture, which is the target market for RapidEye.

18.5.2.1.5 Other

Other very high-resolution (VHR) satellites worth mentioning are the Israeli EROS A and B satellites, which boast panchromatic bands with resolutions of 1.8 and 0.7 m, respectively. India, on the other hand, launched Cartosat 1 in 2005, which generates 2 m panchromatic stereo-imagery, while Cartosat 2 (2007) provides imagery at a spatial resolution of 0.8 m. Other international initiatives include

- South Korea, who is operating Kompsat 2 that provides imagery with a spatial resolution of 1 m in the panchromatic and 4 m in the VNIR spectral range (four multispectral bands).
- Taiwan owns Formosat 2 with a spatial resolution of 2 m in the panchromatic and 8 m in the VNIR spectrum. Formosat 2 has a fixed sun-synchronous orbit, which implies that it covers certain parts of the globe on a daily basis, while never imaging other areas.

- Finally, the French–Italian Pleiades satellites, to be launched in 2010, are agile, small satellites that will capture 0.50 m panchromatic and 2 m multispectral imagery.

18.5.2.2   Medium Spatial Resolution Systems

Medium resolution satellite systems offer a compromise between synoptic coverage and fine detail; they sacrifice the ability to see individual trees to gain the advantage of seeing the entire forest. Such systems enable analysts to acquire imagery of entire countries and to collect the repetitive coverage that permits the monitoring of landscape changes over years and decades—capabilities not offered by aerial photography and fine-resolution satellite systems.

18.5.2.2.1   Landsat Multispectral Scanner

The MSS was an optical–mechanical system that had a ground resolution of 79 × 56 m and recorded energy in four broad spectral bands (green, red, and two near-infrared). The MSS was flown onboard Landsat-1 through Landsat-5 (Table 18.9). Geographic coverage was approximately 185 × 170 km. Since this sensor collected data between 1972 and 1992, the remote-sensing analyst could utilize this data set for historical analysis to supplement current data collection.

18.5.2.2.2   Landsat Thematic Mapper

Landsat-4 and -5 included the thematic mapper (TM) sensor payload in addition to the MSS (http://edc.usgs.gov/guides/landsat_tm.html). The TM had three visible bands (30 m spatial resolution), one near-infrared and two short-wave infrared (SWIR) bands (30 m spatial resolution), and a thermal band (120 m spatial resolution) (Engel and Weinstein, 1983). The TM infrared bands allowed for vegetation discrimination, geological interpretation, and soil moisture differentiation. The utilization of TM data for various interpretive purposes is facilitated by the system's unique combination of bandwidths, spatial resolution, and geometric fidelity. Table 18.10 summarizes the basic characteristic of the Landsat TM system.

TABLE 18.9
Landsat MSS Characteristics

Sensor Parameter	Landsat MSS Specifications
Instrumentation	Cross-track scanner
Imaging swath	185 km
Spatial resolution	79 × 56 m
Spectral bands	0.5–0.6 μm (green)
	0.6–0.7 μm (red)
	0.7–0.8 μm (near-infrared)
	0.8–1.1 μm (near-infrared)
Repeat coverage	18 days (Landsat—1, 2, 3); 16 days (Landsat—4 and 5)

TABLE 18.10

Landsat Thematic Mapper Characteristics

Sensor Parameter	Landsat TM Specifications
Instrumentation	Cross-track scanner
Imaging swath	185 km
Spatial resolution	30 × 30 m (bands 1–5, 7); 120 × 120 m (band 6)
Spectral bands	1: 0.45–0.52 μm (blue)
	2: 0.52–0.60 μm (green)
	3: 0.63–0.69 μm (red)
	4: 0.76–0.90 μm (near-infrared)
	5: 1.55–1.75 μm (mid-infrared)
	6: 10.4–12.5 μm (thermal-infrared)
	7: 2.08–2.35 μm (mid-infrared)
Repeat coverage	16 days

18.5.2.2.3   Landsat Enhanced TM Plus

Landsat-7 Enhanced TM Plus (ETM+) is an environmental satellite launched on April 15, 1999. Landsat-7 is a continuation of the Landsat satellite series, which had a major setback when Landsat-6 was lost on October 5, 1993, when it did not achieve orbit (Table 18.11).

The ETM+ sensor collects eight bands of data from the visible to thermal region (Table 18.12). The data are quantized to 8 bit with a resultant range of values from 0 to 255. Like its predecessor, Landsat TM, the spatial resolution for the visible

TABLE 18.11

Timetable of Landsat Launches Dating Back to July 23, 1972

System	Launch	End of Service	Sensor	Resolution (m)	Original Communication	Altitude (km)	Revisit Time (Days)
Landsat-1	7/23/72	1/6/78	RBV	80	Direct downlink with recorders	918	18
Landsat-2	1/22/75	2/25/82	MSS	80			
			RBV	80			
Landsat-3	3/5/78	3/31/83	MSS	80			
			RBV	80			
Landsat-4	7/16/82	1993	TM	80	Direct downlink	705	16
			MSS	80			
Landsat-5	3/1/84	MSS. 1995	TM	30	TDRSS		
			MSS	80			
Landsat-6	10/5/93	10/5/93	TM	30	Direct downlink with recorders		
			MSS	80			
Landsat-7	4/15/99	Still active	ETM+	30/15			

TABLE 18.12

Landsat Enhanced TM Plus Characteristics

Sensor Parameter	Landsat ETM+ Specifications
Instrumentation	Cross-track scanner
Imaging swath	185 km
Spatial resolution	30×30 m (bands 1–5, 7); 60×60 m (band 6); 15×15 m (band 8)
Spectral bands	1: 0.45–0.515 μm (blue)
	2: 0.525–0.605 μm (green)
	3: 0.63–0.69 μm (red)
	4: 0.75–0.90 μm (near-infrared)
	5: 1.55–1.75 μm (mid-infrared)
	6: 10.4–12.5 μm (thermal-infrared)
	7: 2.09–2.35 μm (mid-infrared)
	8: 0.53–0.90 μm (panchromatic)
Repeat coverage	16 days

and SWIR bands (1–5, 7) for ETM+ is 30×30 m. For historical continuation of the Landsat band numbering sequence, ETM+ band 6 is designated as the thermal band. The spatial resolution of the thermal band was an improvement over the previous TM thermal band to 60×60 m, while band 8 was designated as the new panchromatic band. This latter spectral region has a spatial resolution of 15×15 m and offers improved mapping in urban areas for Landsat data users. However, since May 2003 the sensor has been affected by a scan line correction error (SLC error) leading to severe artifacts at the image periphery. Even if Landsat continues to acquire in the SLC-off mode and although methods have been developed to address the missing data within the image, the use of the data remains limited to mono-temporal applications or change analysis over long time-intervals. Landsat-7 imagery is thus not recommended for the analysis of short-term interval changes. Figure 18.4 illustrates the high-quality imagery from the Landsat-7 ETM+, available for the period from end of 1999 until May 2003.

The next generation Landsat, "Landsat-8" or "operational land imager (OLI)," is currently in the system development phase and is expected to be launched in 2013. It will have the same set of non-thermal spectral bands (30 m) found in Landsat-7, but will also include two narrow-bands focused on coastal/aerosol applications (433–453 nm) and cirrus cloud assessment (1360–1390 nm), while also maintaining the 15 m panchromatic band (500–680 nm). At the time of writing this chapter, the inclusion of a thermal band was under discussion by mission scientists.

18.5.2.2.4 SPOT—Systeme Pour L'Observation de la Terre

Systeme Pour I' Observation de la Terre (SPOT) was conceived and designed by the French Centre National d'Etudes Spatiales (CNES). Five satellites have been launched since 1986 and are operated by SPOT Image (Table 18.13) (http://www.spot.com/).

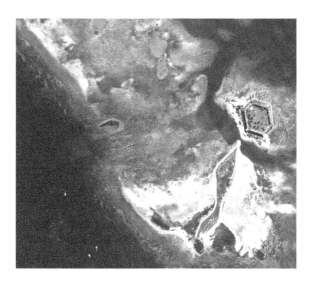

FIGURE 18.4 (**See color insert following page 426.**) Landsat Enhanced Thematic Mapper Plus (ETM+). Landsat Enhanced Thematic Mapper Plus (ETM+) image representing Abu Dhabi, United Arab Emirates, acquired on July 11, 1999 (RGB = 7,4,3) at 30 × 30 m spatial resolution. (Courtesy of U.S. Geological Survey, Reston, VA.)

TABLE 18.13
Timetable of SPOT Launches Since February 22, 1986

System	Launch	End of Service	Sensor	Resolution (m)	Original Communication	Altitude (km)	Revisit Time (Days)
SPOT 1	2/22/86	11/03	PAN	10	Direct downlink with recorders	822	26: nadir
SPOT 2	1/22/90	Still active	XS PAN	20 10			1–5: off-nadir
SPOT 3	9/26/93	11/4/96	XS PAN	20 10			26: nadir
SPOT 4	3/24/98	Still active	XS PAN VEGETATION	20 10 1000			
SPOT 5	5/03/2002	Still active	HRG HRS VEGETATION	5(2.5)/10/20 10 1000		822	2–3: off-nadir 26: nadir

SPOT was the first civilian satellite to include a linear array sensor, employing a push-broom scanning technique, and to have pointable optics for increased revisit time and acquisition of stereoscopic imagery. The payload for SPOT 1, SPOT 2, and SPOT 3 (launched in 1986, 1990, and 1993, respectively) consisted of two identical high-resolution-visible (HRV) imaging systems: One multispectral with a green, red, and near-infrared band (20 m spatial resolution), and another with a panchromatic band (10 m spatial resolution). SPOT 4 added a mid-infrared band with 20 m spatial resolution (HRVIR: HRV infrared). The red band on SPOT 4 (1998) furthermore was improved to 10 m spatial resolution in order to replace the panchromatic band. SPOT 5 (HRG—high-resolution geometric) was launched in 2002 and boasts enhanced spatial resolutions in the green, red, and near-infrared bands of 10 m; the SWIR band remained at 20 m resolution. SPOT 5 is equipped with two panchromatic sensors at 5 m spatial resolution each, which allows for the calculation of a 2.5 m resolution panchromatic product through the combination of the two sensors (Table 18.13).

SPOT 1 retired from operational service at the end of 1990, SPOT 2 is to be decommissioned in 2009, and SPOT 3 became inoperable in 1996. This left only SPOT 4 and 5 for continued data acquisition. The low spatial resolution SPOT Vegetation sensors onboard SPOT 4 and 5 will be described in Section 18.5.2.3.3. Spot 5 also carries the HRS instrument, which captures high-resolution along-track stereo-imagery with a resolution of 5 × 10 m. These images are suitable for the extraction of high-resolution DEMs.

Due to pointability and several satellite constellations, the SPOT sensor can obtain repetitive coverage of the globe every 1–5 days. The SPOT sensor suite can provide the remote-sensing analyst with increased frequency of coverage for areas with high cloud coverage, stereoscopic viewing, and DEM generation. Typical applications for SPOT multispectral imagery include precision agriculture, natural resource management, wetlands inventory, and urban planning and topographic mapping at 1:50,000 scales. Table 18.14 summarizes the basics characteristics of the SPOT system.

18.5.2.2.5 Hyperion

NASA developed and launched the first spaceborne imaging spectrometer (Hyperion) in November 2000. Hyperion is part of the payload of the Earth Observing 1 (EO-1) platform, inserted into an orbit in formation with Landsat-7 (705 km sun-synchronous, 98° inclination, 10:00 h descending node orbit). This sensor boasts continuous spectral coverage over 220 bands, with a spatial resolution of 30 m (http://eo1.gsfc.nasa.gov/Technology/Hyperion.html). A Hyperion scene is comprised of a narrow strip of approximately 7.7 km in the across-track direction and 42 or 185 km in the along-track direction. The telescope consists of two separate grating image spectrometers to improve SNR and collects a 10 nm (average) sampling interval over the contiguous reflected spectrum from 356 to 2577 nm. The first spectrometer covers the visible-near-infrared (VNIR) portion of the spectrum, while the second spectrometer collects SWIR energy. Spectral overlap between the two spectrometers occurs between 852 to 1058 nm (Pearlman et al., 2001; Ungar, 2001).

Although Hyperion samples the VNIR–SWIR electromagnetic spectrum in 220 channels, not all of the acquired bands are calibrated. This is due to decreasing detector response at suboptimal wavelengths, as well to the area of spectrometer overlap. The non-calibrated sections are bands 1–7 (356–417 nm) and bands 225–242 (2406–2578 nm), as well as bands 58–70 (VNIR instrument) and bands 71–76

TABLE 18.14

Systeme Pour l'Observation de la Terre Characteristics

Sensor Parameter	SPOT HRV/HRVIR Specifications
Instrumentation	Pushbroom linear array
Imaging swath	60 km
Spatial resolution	20 × 20 m (multispectral); 10 × 10 m (panchromatic); 15 ×15 m (band 8)
Spectral bands	B1: 0.50–0.59 µm (green)
	B2: 0.61–0.68 µm (red)
	B3: 0.79–0.89 µm (near-infrared)
	B4: 1.58–1.75 µm (mid-infrared)—on SPOT 4 only
	P: 0.51–0.73 µm (panchromatic)—on SPOT 1, 2, and 3
	P: 0.61–0.68 µm (panchromatic)—on SPOT 4
Repeat coverage	1–5 days
Sensor parameter	SPOT HRG specifications
Instrumentation	Pushbroom linear array
Imaging swath	60 km
Spatial resolution	20 × 20 m (band 4); 10 × 10 m (band 1–3); 5 × 5 m (panchromatic)
Spectral bands	B1: 0.50–0.59 µm (green)
	B2: 0.61–0.68 µm (red)
	B3: 0.78–0.89 µm (near-infrared)
	B4: 1.58–1.75 µm (mid-infrared)
	P:0.48–0.71 µm (panchromatic)
Repeat coverage	2–3 days off-nadir (26 days nadir)

(SWIR instrument). The final data product therefore consists of 198 bands for continuous spectra from 427 to 2395 nm. The actual bandwidths for Hyperion vary by band, even though the average bandwidth is indicated as 10 nm. Bandwidths are derived from the full-width half-maximum (FWHM) values that result from instrument spectral response curves (NASA, 2007).

Various international space agencies have identified the need for developing the next generation spaceborne imaging spectrometers. Two planned missions will be discussed briefly in this context, namely, the German Space Agency's (DLR) Environmental Mapper (EnMAP) and NASA's HyspIRI or "Plant Physiology and Functional Types (PPFT)" mission.

The DLR (German space agency) initiated a selection process in 2003 for a future earth observation mission whereby two proposals were selected for phase A (concept) studies. These were the EnMAP (hyperspectral mission) and TanDEM-X (SAR interferometry mission) platforms. The Phase A study was finalized at the end of 2005, after which both the EnMAP and TanDEM-X missions were selected for implementation. EnMAP now has to undergo phases B through D (construction, integration, testing), with an envisaged launch date of end-2011.

TABLE 18.15
EnMAP (ESA) Mission Specifications

Spectral Range (nm)	Spectral Sampling Interval (nm)	Spectral Bandwidth (FWHM)(nm)	Number of Bands (Total of 236)
VNIR-range			
420–1000	5	5 ± 1	114
SWIR-range			
900–1390	10	10 ± 1.5	44
1480–1760	10	10 ± 1.5	28
1950–2450	10	10 ± 1.5	50

The EnMAP (Table 18.15) mission goals are

- To provide high-spectral resolution observations of biogeochemical and geophysical variables
- To observe and develop a wide range of ecosystem parameters encompassing agriculture, forestry, soil/geological environments, and coastal zones/inland waters
- To enable the retrieval of presently undetectable, quantitative diagnostic parameters needed by the user community
- To provide high-quality calibrated data and data products to be used as inputs for improved modeling and understanding of biosphere/geosphere processes

NASA also is investigating the need for and capability of launching a high-fidelity spaceborne imaging spectrometer in the near future. Lessons learned from the Hyperion and NASA Moon Mineralogy Mapper (M3; Figure 18.5), launched October 22, 2008, on India's Chandrayaan-1 platform, will undoubtedly come in handy in the design, construction, and management of such a mission. The HyspIRI mission has as objective to help answer the following questions:

- Ecosystem function and diversity
  - What are the spatial distributions of different plant functional groups, diagnostic species, and ecosystems?
  - How do their locations and function change seasonally and from year to year?
- Biogeochemical cycles
  - How do changes in the physical, chemical, and biotic environment affect the productivity, carbon storage, and biogeochemical cycling processes of ecosystems?
  - How do changes in biogeochemical processes feed back to other components of the earth system?
- Ecosystem response to disturbance
  - How do human-caused and natural disturbances affect the distribution, biodiversity, and functioning of ecosystems?

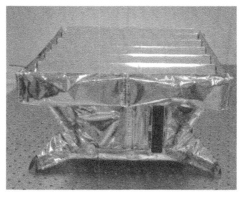

FIGURE 18.5 **(See color insert following page 426.)** The NASA HyspIRI mission. The NASA HyspIRI mission (to be confirmed) will be based on space-proven components used in the Moon Mineralogy Mapper (M3) platform. (Courtesy of Rob Green, NASA JPL.)

- Ecosystems and human well-being
  - How do changes in ecosystem composition and function affect human health, resource use, and resource management?

Initial specifications indicate a spectral range of 380–2500 nm in 10 nm bands, a 60 m spatial resolution, 19 day revisit time, and specific focus on terrestrial and shallow water areas. The platform likely will consist of a visible shortwave infrared (VSWIR) imaging spectrometer and a multispectral TIR scanner.

18.5.2.2.6   ASTER

ASTER (Advanced Spaceborne Thermal Emission and Reflection radiometer) was launched onboard NASA's TERRA satellite in 1999 in cooperation with Japan's Ministry of International Trade and Industry (http://asterweb.jpl.nasa.gov/). ASTER provides multispectral imagery at spatial resolutions of 15 m (visible and near-infrared; bands 1–4), 30 m (shortwave infrared; bands 5–10), and 90 m (thermal infrared; bands 11–15). The spatial and spectral properties allow for mapping of land cover, land surface temperature, and emissivity. Bands 3 and 4 have an identical spectral range (0.760–0.860 μm), but differ in their viewing angle. The generation of stereo-images and derivation of DEMs in 30 m resolution are thus facilitated with band 3 viewing in the nadir direction and band 4 viewing backward. ASTER also serves to provide a "zoom function" for the other instruments onboard TERRA, for example, MODIS (Section 18.5.2.3.2).

18.5.2.2.7   Resourcesat-1 (P6)

Resourcesat-1, also known as P6, is an Indian remote-sensing satellite launched in 2003 to continue the services provided by the successful IRS-1C and IRS-1D missions (http://www.isro.org/pslvc5/index.html). P6 has three sensors onboard—(1) LISS-III with green, red, NIR, and SWIR bands at 23 m spatial resolution and a swath width of 141 km, (2) LISS-IV with green, red, and NIR bands at 5.8 m spatial resolution and a 26.8–30 km swath width, or a panchromatic band at 5.8 m spatial resolution and 70 km swath width, and (3) AWiFS, which provides four-band imagery at 56–70 m spatial resolution with a swath width of 737 km. While LISS-II and LISS-IV supplies 7 bit imagery, AWIFS has a radiometric resolution of 10 bits. Resourcesat-1 has become a popular alternative to Landsat 5 and 7, mainly due to end-of-life complications onboard the two U.S. satellites. Resourcesat-2, an exact replica of Resourcesat-1, will be launched in 2009 to provide even more imaging capabilities.

18.5.2.2.8   Other

Other medium resolution remote-sensing satellites include CBERS (Chinese Brazilian Earth Remote-Sensing satellite) 1, 2, and 2B, which are built and operated by Brazil and China (http://www.cbers.inpe.br/en/index_en.htm). Only CBERS 2B, which was launched in 2007, is currently operational. The five band (blue, green, red, near-infrared, and panchromatic) imagery collected by CBERS is of particular interest to developing countries as it is made available free for Africa and South America. Imagery from the CCD sensor has a spatial resolution of 20 m and a swath width of 113 km.

The European Union has decided to fund a constellation of earth observation satellites, dubbed "Sentinels," under the Global Monitoring of Environment and Security (GMES, now known as Kopernikus) program. Sentinel-1 is intended to provide continuity of ERS 1 and 2 and ASAR data and Sentinel-2 for SPOT and Landsat data, while Sentinel-3, -4, and -5 will be ocean and atmosphere-monitoring systems that build on the Envisat heritage. Sentinel-2, of specific relevance to this section, is designed to capture 13 spectral bands at 10, 20, and 60 m spatial resolutions with a 290 km swath width and a revisit time of 10 days. The first Sentinel-2 satellite is slated for a 2012 launch, with identical platforms to follow toward establishing a Sentinel-2 constellation.

18.5.2.3   Coarse Spatial Resolution Systems

Although the advantages of collecting imagery at spatial resolutions of 0.5–1 km or more may not be immediately obvious, it is the coarse resolution of such imagery that permits the observation of entire continents for observation of broad-scale patterns of vegetation and climate. Further, the coarse spatial resolution permits the acquisition of imagery at frequent intervals. This capability enables the accumulation of cloud-free scenes over a week or so to allow for the compilation of composites that can produce snapshots of the earth's surface at intervals necessary to reveal seasonal changes in both agricultural and natural systems. In addition, such systems are often engineered to high standards to permit the acquisition of very high-quality image data.

18.5.2.3.1 NOAA Advanced VHR Radiometer Very High Resolution (AVHRR)

The NOAA's AVHRR is one of the oldest and most widely utilized satellite systems for global studies (Table 18.16) (http://noaasis.noaa.gov/NOAASIS/ml/avhrr.html). It was developed for global and repetitive (daily) monitoring of cloud coverage, sea surface temperature, oceanic currents, terrestrial vegetation, and polar ice (Cracknell, 1997). Nowadays, AVHRR data are frequently used for long-term environmental monitoring and extraction of phenological vegetation information in the context of global change assessment.

The AVHRR has one visible band (red), one near-infrared band, and three thermal bands, all with a ground resolution of 1.1 km (Table 18.17). Data are ideally suited for input into global models and databases, given global mapping at 1 and 4 km grids.

The AVHRR constellation consists of 14 satellites dating back to 1978, with the latest, NOAA-18, being launched in June 2005. NOAA-14 to NOAA-18 are still operational. The sensors consist of identical or comparable spectral bands. Unfortunately, onboard radiometric calibration is not possible, which has the disadvantage that an intersensor comparison and correction for ageing effects for single sensors are not

---

TABLE 18.16
Timetable of AVHRR Launches Since October 13, 1978

System	Launch	End of Service	Sensor	Resolution (km)	Original Communication	Altitude (km)
TIROS-N	10/13/78	1/30/80	AVHRR	1.1	Direct downlink	833
NOAA-6	6/27/79	11/16/86			with recorders	(NOAA-
NOAA-B	5/29/80	5/29/80				6,8,10,12,
NOAA-7	6/23/81	6/7/86				15,17)
NOAA-8	3/28/83	10/13/85				870
NOAA-9	12/12/84	5/11/94				(NOAA-
NOAA-10	9/17/86	Stand-by				7,9,11,14,
NOAA-11	9/24/88	9/13/97				16,18)
NOAA-12	5/14/91	12/15/94				
NOAA-13	8/9/93	8/9/93				
NOAA-14	12/30/94	Still active				
NOAA-15	5/13/98	Still active				
NOAA-16	9/21/2000	Still active				
NOAA-17	6/24/2002	Still active				
NOAA-18	6/20/2005	Still active				

TABLE 18.17
Advanced Very High Resolution Radiometer Characteristics

Sensor Parameter	AVHRR Specifications
Instrumentation	Cross-track scanner
Imaging swath	2700 km
Spatial resolution	1.1 km
Spectral bands	1: 0.58–0.68 µm (red)
	2: 0.725–1.10 µm (near-infrared)
	3A: 1.58–1.64 µm (only NOAA 15–17)
	3B: 3.55–3.93 µm
	4: 10.50–11.5 µm (NOAA-6,8,10,12,15,17)
	4: 10.3–11.3 µm (NOAA-7,9,11,14,16,18)
	5: 11.50–12.50 µm (not on NOAA-6,8,10,12, 15,17)
Repeat coverage	Daily

possible. Long-term monitoring and change detection analyses have to take this into consideration. In addition to the spectral data listed earlier, the USGS also produces a series of derived AVHRR normalized difference vegetation index (NDVI) composites and global land cover characterization (GLCC) data. The NOAA system has a daily repetition rate, which is nowadays referred to as "hyper-temporal." Daily data are available, but composites of 7 or 10 days, which reduces typical noise in the daily data (e.g., cloud coverage), are also used often.

18.5.2.3.2 *Moderate-Resolution Imaging Spectroradiometer (MODIS)*

Two MODIS sensors were launched onboard NASA's TERRA (1999) and AQUA (2002) satellites. These missions were designed to contribute to the AVHRR mission with enhanced technology. The TERRA and AQUA satellites have a temporal resolution of 1–2 days, which allows for real-time daily monitoring and early warning systems. As TERRA passes the Equator in the morning and AQUA in the afternoon, two images per day can be obtained. The MODIS sensors, with its 36 spectral bands, are designed to comprehensively provide data on ocean, land, and atmospheric processes. Applications range from assessment of clouds/aerosols, ocean color, vegetation/land surface, surface/atmospheric temperature, and include system calibration efforts. The spatial resolution is 250 m for the red and NIR bands (bands 1 and 2), 500 m for the blue, green, and three SWIR bands (bands 3–7), and 1000 m for the remaining 29 bands, covering the visible, infrared, and thermal range. The swath width of MODIS is 2330 km, but images are usually released in tiles of approximately 1200 × 1200 km (Table 18.18).

In contrast to AVHRR, MODIS has an onboard calibration system that allows for the conversion of raw digital numbers to reflectance values (Salomonson et al., 1995; Barnes et al., 1998). Figure 18.6 shows an example of coastal images acquired by the MODIS sensor. Quality flags for each pixel indicating false or missing pixel data, for example, those caused by clouds, are provided for some of the products. MODIS

TABLE 18.18
MODIS Characteristics

Sensor Parameter	MODIS Specifications
Instrumentation	Whiskbroom linear array
Imaging swath	2700 km
Spatial resolution	250 m (bands 1–2)
	500 m (band 3–7)
	1 km (bands 8–36)
Spectral bands	20 bands: 0.4–3.0 μm (blue to mid-infrared)
	16 bands: 3.0–15 μm (thermal-infrared)
Repeat coverage	1–2 days

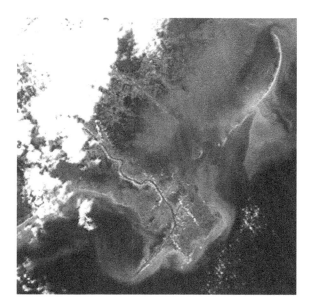

FIGURE 18.6 **(See color insert following page 426.)** An image of the Mississippi acquired by the MODIS system onboard NASA's Terra satellite. This scene, acquired on February 24, 2000, depicts the coastal features that surround the delta, including sediment plumes, shallow water bays, and barrier islands. (Courtesy of National Aeronautics and Space Administration, Washington, DC.)

data are available at low cost. They are frequently used for large-scale (regional, continental, global) analysis, for example, assessing land cover change and vegetation phenology, early warning systems for wild fires, and global change studies.

18.5.2.3.3   SPOT 4 and 5 VEGETATION

VEGETATION is a wide angle (2250 km swath) multispectral earth observation instrument onboard the SPOT 4 (VEGETATION-1) and SPOT 5 (VEGETATION-2) satellites, with a spatial resolution of 1 km. It provides four spectral bands (blue, red,

TABLE 18.19
SPOT 5 VEGETATION Characteristics

Sensor Parameter	VEGETATION Specifications
Instrumentation	Pushbroom linear array
Imaging swath	2250 km
Spatial resolution	1 km
Spectral bands	B0: 0.45–0.52 μm (blue)
	B2: 0.61–0.68 μm (red)
	B3: 0.79–0.89 μm (near-infrared)
	B4: 1.58–1.75 μm (mid-infrared)
Repeat coverage	Daily

NIR, SWIR) and is primarily intended for vegetation monitoring and oceanographic applications (blue band). The VEGETATION instrument (Table 18.19) flying on SPOT 4 and SPOT 5 provides global coverage on an almost daily basis, thus making it an ideal tool for observing long-term environmental changes on a regional- and global scale, comparable to MODIS and AVHRR. In contrast to the other SPOT products, VEGETATION is operated by the Flemish Institute for Technological Research (VITO) in Belgium.

18.5.2.3.4   Sea-Viewing Wide Field-of-View Sensor (SeaWiFS)
SeaWiFS was launched onboard the Seastar satellite (OrbView-2) on August 1, 1997 (http://oceancolor.gsfc.nasa.gov/SeaWiFS/). It was developed and is operated by GeoEye (previously Orbimage, Inc.) and NASA. SeaWiFS constitutes the long-awaited follow up to NASA's coastal zone color scanner. SeaWiFS is optimized for use over water, even though terrestrial sensing also was given some consideration in its design (Table 18.20). SeaWiFS applications include phytoplankton mapping, modeling of nitrogen/carbon cycling, and delineation of major ocean features.

TABLE 18.20
SeaWiFS Characteristics

Sensor Parameter	SeaWiFS Specifications
Instrumentation	Cross-track scanner
Imaging swath	2800 km
Spatial resolution	1.1 km
Spectral bands	0.402–0.422 μm/0.433–0.453 μm/0.480–0.500 μm (blue)
	0.500–0.520 μm/0.545–0.565 μm (green)
	0.660–0.680 μm (red)
	0.745–0.785 μm/0.845–0.885 μm (near-infrared)
Repeat coverage	Daily

SeaWiFS data are available in three formats. High-resolution picture transmission (HRPT) data are full resolution image data transmitted to a ground station, as data are collected (1.1 km spatial resolution). Local area coverage (LAC) scenes are also full resolution data, but recorded with an onboard tape recorder for subsequent transmission during a station overpass (1.1 km resolution). Finally, global area coverage (GAC) data provide daily, subsampled global coverage recorded on the tape recorders and then transmitted to a ground station (4 km resolution).

18.5.2.3.5   ENVISAT

The principal objective of the ENVISAT program is defined as "... to endow Europe with an enhanced capability for remote-sensing observation of earth from space, with the aim of further increasing the capacity of participating states to take part in the studying and monitoring of the earth and its environment" (ESA, 2007) (http://envisat.esa.int/). The primary objectives of the program are (1) to ensure the continuity of observations initiated with the European Remote-sensing Satellites (ERS) (including radar-based sensing), (2) to expand the ERS mission, especially the ocean and ice missions, (3) to bolster the range of parameters observed toward improved knowledge of the factors that define environments, and (4) to contribute significantly to environmental studies, especially in atmospheric chemistry and oceanographic sciences. The mission thus aims to continue and improve upon measurements initiated by ERS-1 and ERS-2, and to take into account the requirements related to the global study and monitoring of the environment. ENVISAT is an essential element in providing long-term continuous data sets that are crucial for addressing environmental and climatological issues (ESA, 2007).

ENVISAT occupies a sun-synchronous polar orbit of 800 km altitude with a 10:00 h mean local solar time (MLST) descending node and a 98.55° inclination. The temporal resolution of 35 days allows the platform the ability to provide a complete coverage of the earth in 1–3 days for most sensors; profiling instruments MWR and RA-2 are exceptions. The sensor of interest for our purposes is MERIS, a programmable, medium-spectral resolution, imaging spectrometer that spans the visible and near-infrared spectral range. Fifteen spectral bands can be selected from ground stations, while each band has a programmable width and spectral location in the 390–1040 nm spectral range (ESA, 2007). Although MERIS really is a super-spectral (spectral coverage between multispectral and imaging spectroscopy/hyperspectral) and not an imaging spectrometer, this programmability arguably lends itself to coverage in contiguous wavelengths, albeit at different time intervals.

MERIS scans the target surface using a pushbroom scanner. A charge-coupled device (CCD) setup allows for spatial sampling in the across-track direction, while the satellite forward motion provides scanning in the along-track direction. The instrument FOV is 68.5° at nadir, which results in a swath width of 1150 km. The spatial resolution is 300 m (at nadir), but is effectively reduced to 1200 m when the onboard combination of four neighboring samples (across-track) over four successive lines is implemented (ESA, 2007).

18.6   SENSORS PROVIDING STRUCTURAL
AND OTHER INFORMATION

Whereas previous sections dealt with sensors that record the spectral properties of features at the earth's surface, this section describes systems that detect physical characteristics of these features. In urban regions, these physical properties pertain to the nature of buildings, highways, transmission lines, and similar structures. In rural regions, such imagery will be sensitive to the sizes and arrangement of trees, brush, vegetation in general, and the details of the landscape, including slopes, drainage, and other surface irregularities. Because this information is completely independent of the spectral properties discussed previously, such imagery offers valuable information that aids our ability to understand the nature of the landscape.

18.6.1   AIRBORNE STRUCTURAL SYSTEMS

A variety of airborne systems that are aimed at the improved structural characterization of natural and man-made environments exist today. These systems range from radio detection and ranging (radar) to LIDAR platforms, although LIDAR has become increasingly prevalent in natural resource research and applications.

LIDAR involves the emission of a laser pulse from an airborne sensor, measurement of the pulse's return-travel time from sensor to target, and calculation of the distance traveled by the laser beam. The distance from the sensor to the target is often converted to target height above sea level, given that the sensor (flying) height is known (Baltsavias, 1999a). The main purpose of all laser-scanning systems therefore is to accurately measure the distance between the target and the sensor. The ranging unit of such a system includes both the emitting laser and the electro-optical receiver. The apertures of these two components are mounted such that the transmitting and receiving paths share the same optical path, thereby ensuring that laser-illuminated objects are in the FOV of the optical receiver. The distance to the target is calculated by halving the time elapsed between emission and arrival of the reflected laser pulse at the receiver (Ackermann, 1999; Wehr and Lohr, 1999). Some benefits of LIDAR technology versus traditional photogrammetric techniques include LIDAR's ability to operate during night time, the ease with which interpretation can be automated, inherent structural information, canopy penetration capabilities, and high point density, while the lack of full area coverage and cost can be listed as drawbacks (Ackermann, 1999; Baltsavias, 1999b).

Laser-ranging systems can be divided into pulse and continuous wavelength (CW) lasers, with pulsed lasers more prevalent in current systems than CW systems. Pulsed laser altimeters can further be subdivided into discrete return or waveform sampling sensors.

Discrete return sensors measure single- or multiple return distances by evaluating the returned energy signal to find a peak or peaks that define discrete objects in the laser's path. Either the distance to peak-edge or the maximum power of a peak is recorded (Wehr and Lohr, 1999). Multiple return systems are often used to extract both canopy and ground returns, assumed to be represented by first and last returns, respectively. Vegetation canopy height is then defined as the difference between the first and last returns (Lefsky et al., 1999).

Return waveform systems operate on the assumption that the shape of the waveform from the returned signal represents a vertical distribution of the intercepted surfaces within a given laser footprint. Such a waveform accounts for the spatial distribution of laser beam intensity along and across the laser beam's path. Discrete systems generally utilize a small footprint (<1 m) sensor, as opposed to waveform systems with large footprint (5–15 m) sensors (Weishampel et al., 1996; Blair and Hofton, 1999; Lefsky et al., 1999). Large footprint systems are useful for waveform sampling devices because recovery from tops of crowns and the ground are possible in the same waveform, while the resolution remains small enough to detect individual crown contribution to such a waveform (Lefsky et al., 1999).

A number of commercial companies perform routine discrete return LIDAR survey these days, but the more prevalent systems are often developed by Optech Inc. and Leica. These systems typically feature high-frequency, multiple returns per LIDAR pulse, and high ground point density, although the latter is very much dependent on flying altitude and aircraft speed. Examples include Optech Inc.'s Gemini system (167 kHz; 4 returns per pulse; up to 0.05 m accuracy) and Leica's ALS60 Airborne Laser scanner (200 kHz; 4 returns per pulse). Waveform LIDAR is still very much a research tool, with the CAO (cao.stanford.edu) being a good example of its research application toward the improved understanding of natural systems.

18.6.2 SPACEBORNE STRUCTURAL SYSTEMS

In contrast to passive sensors discussed in Section 18.5, active sensors emit signals that interact with a surface, after which the backscattered signal is detected by the sensor. Spaceborne active systems currently consist of low spatial resolution LIDAR and radar at variable resolutions. This section will only focus on spaceborne radar systems. Radar sensors emit a signal at a determined frequency, intensity, and with a given polarization (horizontal [H] or vertical [V]). The sensor subsequently measures the frequency, polarization, phase, and time delay of the returning signal. These information components allow for accurate determination of the distance between the sensor and the target. Most radar sensors of relevance to earth observation are of the synthetic aperture radar (SAR) type. A SAR uses the change in signal frequency, caused by the Doppler effect and induced by the moving spacecraft, to locate the position of the ground feature with which the signal has interacted. SAR will only work when signals interact with an off-nadir target, with the range resolution increasing with larger incidence angles. Radar geometry is thus affected by foreshortening, layover, and shadow in rugged terrain. Wavelengths used by radar systems have been divided into K (1.11–1.67 cm), X (2.5–3.75 cm), C (3.75–7.5 cm), S (7.5–15 cm), L (15–30 cm), and P (>1 m) bands. Although signal penetration is largely determined by the relationship between surface element size and wavelength, longer wavelengths typically provide superior penetration, while shorter wavelengths allow image assembly at higher spatial resolutions.

18.6.2.1   ERS1 and ERS2 and ASAR

ERS1 was the first satellite owned and operated by ESA (http://www.esa.int/esaCP/index.html) and was launched in 1991 (ERS2 was launched in 1995). Both satellites had identical SAR instruments on board and captured imagery in the C-band at single VV polarization and a fixed incidence angle, subsequently producing imagery with a 30 m spatial resolution for a 100 km swath width. ESA operated ERS 1 and 2 in tandem, with a 24 h spacing between the two satellites, for a period of nine months. Such a short time delay between two image acquisitions with the same radar geometry allowed for interferometric analysis. However, due to a decorrelation of SAR signals over time, caused by slight surface movements, the revisit time of 35 days for a single satellite proved insufficient. The systems did enable the imaging of large land masses, followed by generation of accurate DEMs and assessment of woody biomass in the boreal regions through interferometric analysis. A gyro on board ERS1 malfunctioned in 2000, thereby leaving only ERS2 in operation today. ESA launched Envisat in 2002 with ASAR (advanced SAR) on board to ensure continuity of the ERS program. This instrument captures imagery in C-band with the choice of five polarization modes, namely, VV, HH, VV/HH, HV/HH, or VH/VV. VV implies the sending and receiving of data in vertical polarization and VH for sending in horizontal and receiving in vertical polarization. A wide range of acquisition modes are supported, ranging from global monitoring to wave mode, with spatial resolutions varying from 1000 m to less than 10 m, respectively.

18.6.2.2   RADARSAT-1 and -2

RADARSAT is an advanced earth observation satellite system developed to monitor environmental change and to support resource sustainability. With the launch of RADARSAT-1 on November 4, 1995, RADARSAT provided access to a fully operational civilian radar satellite system capable of large-scale production and timely delivery of data (Table 18.21). Applications include ice reconnaissance, coastal surveillance (Figure 18.7), and soil/vegetation moisture studies. RADARSAT was developed by the Canadian Space Agency and Canada Centre for Remote Sensing. Radarsat International commercialized the system, but was eventually acquired by MDA. Radarsat-1 has varying imaging modes in C-band with HH polarization. RADARSAT-2 is owned and was built by MDA and was launched in December 2007. Radarsat-2 is the first commercial, fully polarimetric C-band satellite

TABLE 18.21
RADASAT-1 Characteristics

Sensor Parameter	RADARSAT-1 Specifications
Instrumentation	Active microwave C band
Imaging swath	50–500 km
Spatial resolution	8–100 m (multi-beam)
Polarization	HH
Repeat coverage	24 days
Depression angles	40°–70°

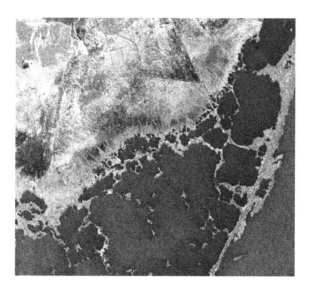

FIGURE 18.7 RADARSAT-1 image of the southern end of the Florida Everglades and the northeastern corner of Florida Bay. The scene was acquired on September 27, 1997, in a Standard 1 Beam mode (12.5 × 2.5 spatial resolution). (Courtesy of RADARSAT International, Richmond, Canada.)

(http://www.radarsat2.info/). It provides imagery that ranges between 3 and 100 m spatial resolutions at varying swath widths.

18.6.2.3　TerraSAR-X and TanDEM-X

TerraSAR-X is the world's first commercial high-resolution radar satellite and was launched in June 2007. TerraSAR-X is owned and operated by the German Space Agency (DLR) and commercialized by Infoterra GmbH. TerraSAR-X has spotlight, stripmap, and scansar imaging modes and can generate scenes with spatial resolutions of 1, 3, and 16 m, respectively. TanDEM-X, an exact replica of TerraSAR-X, will be launched in 2009 and fly adjacent to TerraSAR-X in a helical orbit. The satellites will acquire three single pass interferometric image pairs, covering the world's landmasses, over a period of three years. These image pairs will be used to produce a global DEM of HRTI 3 standard at 10 m spatial resolution and with an accuracy of 3 m, while DEMs at HRTI 4 standard with a 3 m spatial resolution and a vertical accuracy below 1 m will be generated for selected areas only. TerraSAR-X 2 will be owned by Infoterra and is currently in a design phase.

18.6.2.4　JERS and ALOS

The Japanese Earth Resources Satellite, JERS-1, was launched in 1992 and was decommissioned in 1998 (http://www.eorc.jaxa.jp/JERS-1/). It had an L-band SAR on board and captured imagery at 18 m spatial resolution. The Advanced Land Observation Satellite (ALOS, or Daichi in Japanese) was launched in 2006 to ensure data continuity of the JERS mission. The platform's PALSAR instrument images operationally in the L-band in single and dual-polarimetric modes and experimentally in full polarimetric mode. Spatial resolutions vary from 7 to 100 m, depending on the mode of acquisition. ALOS is primarily a monitoring system with acquisition schedules designed to cover all landmasses at regular intervals. PALSAR is used extensively for biomass estimations and the generation of DEMs under moderate vegetation cover conditions, given that L-band generally results in superior penetration ability when compared to either X- and C-bands.

18.7   ORDERING SPECIFICATIONS

Now that many of the most widely used aircraft- and satellite-based remote-sensing systems have been identified, the next step is to determine the requirement of the specific imagery needs of a project.

18.7.1   Aircraft Film/Digital Photography

When recording photography, whether hard copy or digital, it is important to understand how the data were or will be collected by the aircraft system. For example, before any photography is processed or interpreted the user should have predetermined specifications on collection date and time, flight parameters, and camera and/or film specifications.

18.7.1.1   Collection Date and Time

For aerial imagery to be properly interpreted and analyzed, the data should be collected at a very specific time of day and year. Under normal conditions, remotely sensed data are collected during mid-morning hours. This minimizes the effect of cloud cover and other meteorological effects, allows the interpreter to take advantage of a sun angle that will supply sufficient scene brightness, and allows for the casting of sufficient shadows that can be utilized as an element toward improved image interpretation. The time of year will vary depending on geographical region and the information that is needed. For example, when studying wetlands productivity, it is important to collect data during peak biomass. If aerial photography of the Florida wetlands were needed, it would be best to collect the data in April or May, which also coincides with the end of the dry season and expected lowest water levels.

18.7.1.2   Flight Parameters

The scale of the photography is determined by the flying altitude of the aircraft and optics (focal length) of the camera system (refer to Section 16.2.1 for the scale equation). A prior knowledge of the focal length that will be used by the aerial mapping firm will enable the user to determine what flying altitude above ground level is necessary for collecting a scale of photography that is suitable for the project. Once this has been determined, the flight line must be generated that indicate where exposures should be made, along with the amount of overlap and side-lap between individual frames of aerial photography.

After collection, the aerial mapping firm should provide the user with a detailed map showing the actual flight path of the aircraft. Maps should include topographic ground information and supply geo-referenced principal points for each photograph, as well as altitude above sea level and percent overlap between successive photographs, while percent side-lap between adjacent flight lines should also be reported. If available, aircraft altitude information should be provided for each exposure frame.

It is recommended that aerial mapping firms report standard flight mission parameters as specified by the American Society for Photogrammetry and Remote Sensing (http://www.asprs.org/), including a map illustrating plotted locations of each frame of coverage from the film roll, which in turn includes frame center coordinates from an onboard GPS system.

18.7.1.3   Camera and Film Specification

To understand and properly interpret aerial photography, several types of information need to be provided to the user. The variables shown in Table 18.22 represent the minimal amount of information that should be provided in terms of camera and film specifications. Having this information becomes even more critical. If the photography is going to be ortho-rectified.

18.7.2   SATELLITE SENSOR IMAGERY

Satellites and the sensors they carry have certain features that influence how they operate, and you might select one over another (Jensen, 1996; Kramer, 1996; Rees, 1999). Some parameters that might influence your decision to select one satellite-based sensor over another are given in the following paragraphs. Again, as with aircraft-based systems, your decision to choose any one specific sensor should be based on project requirements.

18.7.2.1   Orbit Parameters

All civilian remote-sensing satellites map the earth in near-polar sun-synchronous or geosynchronous orbits. Once in orbit, satellites generally remain at a constant speed and fixed in their paths. Therefore, most satellites cannot be moved or tasked to

---

TABLE 18.22

Film and Camera Specifications That Need to Be Identified to Carry Out Analysis and Product Generation

Film Parameter	Camera Parameter
Film type (panchromatic, true, false color)	Lens type
Film frame (e.g., 23 × 23 cm)	Lens angular field of view
Film production (e.g., print or diapositive)	Lens focal length
	Lens F-stop/aperture
	Camera type
	Camera shutter speed setting
	Filter type
	Filter "T" (transmittance) value

other areas not covered by their orbital paths. (This generalization does not apply to military and strategic reconnaissance satellites, which are not discussed in this chapter.) There are times when earth-resource satellites are tasked to collect nighttime imagery, e.g., thermal sensors. It is important to remember that microwave satellite systems can collect data both during the day and the night (Way and Smith, 1991; Jensen, 2007; Lillesand et al., 2008).

18.7.2.2 Viewing Geometry

Satellite observation systems can either be fixed at a certain viewing angle or have pointable optics. Fixed sensors are most often constrained to at-nadir viewing (i.e., aimed perpendicular to earth's surface). It is obvious that sensors designed for nadir viewing can only acquire imagery that is directly below the path of the satellite. In contrast, pointable sensors can be aimed as desired to view side-to-side or aft-to-fore relative to the orbital path, thereby acquiring imagery that is not directly beneath the satellite's track. Adjustable viewing is extremely important because it can help increase temporal resolution (repeatable coverage times) and introduce the opportunity for stereo-imaging. However, the remote-sensing analyst should keep in mind that off-nadir pointing can introduce geometric/dimensional problems related to the pixel sizes across the imagery (Westing, 1990; Seto, 1991), and increase atmospheric effects at edges of images due to the longer length of the atmospheric path traveled by reflected solar radiation. Off-nadir viewing also introduces bidirectional reflectance distribution function (BRDF) characteristics in the imagery that are beyond the scope of this discussion, but can be very important when monitoring vegetation obtained at different viewing angles (Jensen and Schill, 2000).

18.7.3 COSTS

The price of aircraft and satellite photography and imagery varies in relation to spatial and spectral resolution, areal coverage, processing level, and (historical) acquisition date. One way to look at the relative cost of remote sensor data is to determine the cost of collecting imagery for a specific size of geographic area (e.g., cost/km^2). Table 18.23 provides pricing for imagery collected by several of the current cameras or sensors. It is frequently less expensive to purchase archived data. Unfortunately, it is difficult to set prices on older photography and imagery products, given that prices fluctuate with age of data, associated products, and changing pricing policies of the data providers.

18.8 SUMMARY

As outlined in Chapter 17, the practitioner of remote sensing must master the many components that characterize each project, anticipating the interrelationships between the multifaceted operational, disciplinary, and technical dimensions that prevail in any specific situation. This chapter has addressed the central role of the analyst's choice of sensor system, which in itself requires consideration of several

TABLE 18.23

Price Comparison of Selected Recently Collected Imagery Using Remote-Sensing Systems[a]

System	Spectral Range	Spatial Resolution	Cost/km² ($)
Aerial photography	PAN or VIS/NIR	1:12,000	200[a] (free products from USGS)
Quickbird	PAN or VIS/NIR	0.61/2.4 m	22[b]
IKONOS	PAN or VIS/NIR	0.8 or 3.2 m	27.50[c]
RapidEye	VIS/NIR	5 m	0.45–0.50
Resourcesat	G/R/NIR/SWIR	5/60 m	0.005 (AWiFS)—4.80 (LISS4)
Landsat-7 ETM+	VIS/NIR/MIR/TIR/PAN	30/60/15 m	Free[d]
SPOT 5 imagery	PAN or VIS/NIR/MIR	10 or 20 m	0.98[e]
RADARSAT	Fine-beam	10 m	1.30[c]
	Standard-beam	30 m	0.3[c]
LIDAR	Typically 532 or 1064 nm	Depends on FOV/ altitude	144[f]

[a] Cost based on a study area of 250 km².
[b] http://www.eurimage.com/products/docs/eurimage_price_list.pdf.
[c] Cost based on one full scene with no advanced post-processing.
[d] Landsat archive data are now free at USGS; media cost.
[e] Cost based on 10 m color and 5 m panchromatic.
[f] Cost based on a large data collect (430,000 acres or 174,000 ha).

interrelated decisions. The choice of sensor impacts scientific concerns, such as the abilities to acquire appropriate areal coverage, scale and resolution, and spectral coverage. The analyst must also consider the sensor system's impact upon practical concerns such as mission planning, timing, accurate cost estimates, and data quality. Because no single decision can be considered in isolation from the others, the analyst must be prepared with the technical information required to reach sound decisions, but even more significantly, be equipped with a broad perspective that allows identification of latent interconnections between ostensibly distinct aspects of a project. Because few individuals can master all of the required expertise, the ability to communicate across disciplinary, organizational, and (sometimes) international differences forms an invaluable asset.

APPENDIX 18.A: SENSORS DESCRIBED IN THIS CHAPTER

WWW addresses (or postal addresses) are provided to give the reader an additional resource for information and specifications on selected remote-sensing systems.

Sensor/System/Platform	Address
Aircraft film photography	
NHAP	http://edc.usgs.gov/guides/nhap.html
NAPP	http://edc.usgs.gov/guides/napp.html
Aircraft videography	
V-STARS—Videography Systems by Geodetic Services, Inc.	http://www.geodetic.com
Aircraft digital photography	
Pictometry	http://www.pictometry.com
Fugro Earthdata—Leica ADS40-SH52	http://earthdata.com/
Aircraft hyperspectral imagery	
AISA	http://www.specim.fi/
CASI-2	http://www.itres.com/
AVIRIS	http://aviris.jpl.nasa.gov/
Satellite high-resolution multispectral imagery	
IKONOS and GeoEye	http://www.geoeye.com/
Quickbird and WorldView-1	http://www.digitalglobe.com/
RapidEye	http://www.rapideye.de/
Satellite moderate-resolution multispectral imagery	
Landsat MSS/TM/ETM+	http://landsat.gsfc.nasa.gov/
SPOT	http://www.spot.com/
ASTER	http://asterweb.jpl.nasa.gov/
EARTH EXPLORER	http://earthexplorer.usgs.gov
Satellite moderate-resolution microwave imagery	
RADARSAT	http://radarsat.space.gc.ca/
TerraSAR-X	http://www.infoterra.de
Satellite coarse-resolution multispectral imagery	
AVHRR	http://noaasis.noaa.gov/NOAASIS/ml/avhrr.html
MODIS	http://modis.gsfc.nasa.gov/
SeaWiFS	http://oceancolor.gsfc.nasa.gov/SeaWiFS/

REFERENCES

Ackermann, F., 1999. Airborne laser scanning—Present status and future expectations, *ISPRS Journal of Photogrammetry and Remote Sensing*, 54, 64–67.

Baltsavias, E.P., 1999a. Airborne laser scanning: Basic relations and formulas, *ISPRS Journal of Photogrammetry and Remote Sensing*, 54, 199–214.

Baltsavias, E.P., 1999b. A comparison between photogrammetry and laser scanning, *ISPRS Journal of Photogrammetry and Remote Sensing*, 54, 83–94.

Barnes, W.L., Pagano, T.S., and Salomonson, V.V., 1998. Prelaunch characteristics of the Moderate Resolution Imaging Spectroradiometer (MODIS) on EOS-AM1, *IEE Transactions on Geoscience and Remote Sensing*, 36(4), 1088–1100.

Blair, J.B. and Hofton, M.A., 1999. Modeling laser altimeter return waveforms over complex vegetation using high-resolution elevation data, *Geophysical Research Letters*, 26(16), 2509–2512.

Cocks, T., Jensen, J.R., Stewart, A., Wilson, I., and Shields, T., 1998. The HyMap Airborne Hyperspectral sensor: The system, calibration, and performance, *Proceedings of 1st EarSeL Workshop on Imaging Spectroscopy*, Zurich, Switzerland, 6 pp.

Cracknell, A.P., 1997. *The Advanced Very High Resolution Radiometer*, Taylor & Francis, London, U.K.

Engel, J.L. and Weinstein, O., 1983. The Thematic Mapper: An overview, *IEEE Transactions on Geoscience and Remote Sensing,* 21(3), 258–265.

ERDAS, 1999. *ERDAS Imagine Field Guide*, ERDAS Inc., Atlanta, GA.

ESA, 2007. ESA Missions, http://envisat.esa.int

Fritz, L.W., 1996. The era of commercial earth observation satellite, *Photogrammetric Engineering and Remote Sensing,* 62(1), 39–45.

Green, R.O., Eastwood, M.L., Sarture, C.M., Chrien, T.G., Aronsson, M., Chippendale, B.J., Faust, J.A., Pavri, B.E., Chovit, C.J., Solis, M., Olah, M.R., and Williams, O., 1998. Imaging spectroscopy and the Airborne Visible/Infrared Imaging Spectrometer (AVIRIS), *Remote Sensing of Environment*, 65, 227–248.

Jensen, J.R., 1996. *Introductory Digital Image Processing: A Remote Sensing Perspective*, Prentice Hall, Upper Saddle River, NJ, 318 pp.

Jensen, J.R., 2007. *Remote Sensing of the Environment: An Earth Resource Perspective,* Prentice Hall, Upper Saddle River, NJ, 592 pp.

Jensen, J.R. and Cowen, D.C., 1999. Remote sensing of urban/suburban infrastructure and socio-economic attributes, *Photogrammetric Engineering and Remote Sensing*, 65(5), 611–622.

Jensen, J.R. and Schill, S.R., 2000. Bidirectional reflectance distribution function (BRDF) characteristics of smooth cordgrass (*Spartina alterniflora*) obtained using a SandMeier field goniometer, *Geocarta International,* 15(2), 21–28.

Kramer, H.J., 1996. *Observation of the Earth and Its Environmental: Survey of Mission and Sensors* (3rd edn.), Springer-Verlag, Berlin, Germany, 960 pp.

Lefsky, M.A., Harding, D., Cohen, W.B., Parker, G., and Shugart, H.H., 1999. Surface LIDAR remote sensing of basal area and biomass in deciduous forests of eastern Maryland, USA, *Remote Sensing of Environment*, 67, 83–98.

Lillesand, T.M., Kiefer, R.W., and Chipman, J.W., 2008. *Remote Sensing and Image Interpretation*, John Wiley & Sons, New York, 756 pp.

NASA, 2007. Earth Observing-1 (EO-1) Baseline Mission Overview, http://eo1.usgs.gov/userGuide/ index.php? page=hyp_prop

Pearlman, J., Carman, S., Segal, C., Jarecke, P., Clancy, P., and Browne, W., 2001. Overview of the Hyperion Imaging Spectrometer for the NASA EO-1 mission, *IEEE International*, Vol. 7, Sydney, Australia, July 9–12, 2001, pp. 3036–3038.

Rees, G., 1999. *The Remote Sensing Data Book*, Cambridge University Press, Cambridge, U.K.

Salomonson, V.V., Barker, J., and Knight, E., 1995. Spectral characteristics of the Earth Observing System (EOS) Moderate-Resolution Imaging Spectroradiometer, *Imaging Spectrometry*, 2480, 142–152.

Seto, Y., 1991. Geometric correction algorithms for satellite imagery using a bi-directional scanning sensor, *IEEE Transactions on Geoscience and Remote Sensing,* 29(2), 292–299.

Ungar, S.G., 2001. Overview of EO-1: The first 120 days, *Proceedings of the Geoscience and Remote Sensing Symposium*, Sydney, Australia, July 9–13, 2001, pp. 43–45.

Way, J. and Smith, E.A., 1991. The evolution of synthetic aperture radar systems and their progression to the EOS SAR, *Transactions on Geoscience and Remote Sensing*, 29(6), 962–985.

Wehr, A. and Lohr, U., 1999. Airborne laser scanning—An introduction and overview, *ISPRS Journal of Photogrammetry and Remote Sensing*, 54, 68–82.

Weishampel, J.F., Ranson, K.J., and Harding, D.J., 1996. Remote sensing of forest canopies, *Selbyana*, 17, 6–14.

Westing, T., 1990. Precision rectification of SPOT imagery, *Photogrammetric Engineering and Remote Sensing*, 56(2), 147–253.

19 Information Extraction from Remotely Sensed Data

James B. Campbell

CONTENTS

The practice of remote sensing relies not only upon acquisition of imagery but also upon our ability to extract useful information from image data, so that users can receive reliable information relevant to their needs. Even a novice can easily realize that a remotely sensed image conveys a wealth of information. Yet, one of the most obvious advantages of remotely sensed imagery—that it conveys this wealth of multifaceted information—forms one of the principal obstacles for those who use the imagery, as this information is interwoven with many other forms of information that may not be relevant for the purpose at hand.

Analysts therefore must be able to identify and separate information of interest from noise, error, and other extraneous information recorded on the image. Thus, the analyst's function can be compared to that of a filter that sifts and isolates the useful components of a complex image that pertain to a specific purpose—for example, identifying a specific kind of forest, or delimiting areas within a forest that have been affected by disease or insect infestations. Although in some instances extraction of useful information is straightforward, more typically it requires application of established procedures shown to be reliable, economical, accurate, and applicable in diverse circumstances.

This chapter outlines some of the principal strategies used to derive information from remotely sensed imagery. Although it may seem that extraction of such information is intuitive, the process is subject to hidden difficulties and errors not immediately obvious to the novice. In everyday experience, we can routinely see the meaning in images presented in newspapers and magazines without specialized training. However, in the context of examining aerial images, extraction of information requires an astute understanding of the imagery at hand, the topic of the interpretation, and the geographic region in question. Because this process can be so complex, the reader should recognize that this chapter can offer only an outline of relevant topics and should be prepared to refer to more complete presentations such as those of Jensen (1996, 2007), Lillesand et al. (2008), or Campbell (2007), as well as more specialized references as may be needed.

Most observers distinguish between *visual interpretation*, in which a skilled analyst visually examines imagery to extract information, chiefly through qualitative analysis, and *image processing*, in which digital image data are analyzed quantitatively to extract statistically defined features. Even though the actual practice of remote sensing is too complex and too varied to cleanly fit this dichotomy, it does form a useful framework to start this discussion.

19.1 VISUAL INTERPRETATION

This section describes techniques that an analyst uses to extract information from an image by close visual examination and analysis. In past decades, this process would be applied by *photointerpreters* (now referred to as *image interpreters*, because the images today examined are seldom, strictly speaking, really photographs) who used magnifying glasses or stereoscopes to examine paper prints (Figure 19.1a), usually marking the photograph or a translucent overlay. Today, image interpreters examine images using a variety of computer displays and specialized software (Figure 19.1b).

For this discussion, it is useful to distinguish between two kinds of information based upon their spatial characteristics—*point* and *thematic* information. Point data consist of individual objects or features defined by their individual sizes, shapes identities, and placement—objects, buildings, vehicles, and highways, for example. In contrast, thematic information conveys information about the nature and distribution of areal features, such as forests, agriculture landscapes, or urban land as classes formed by assemblages of individual objects and features.

(a)   (b)

FIGURE 19.1   **(See color insert following page 426.)** (a) U.S. Geological Survey Geologist uses a pocket stereoscope to examine aerial photography for geological information, 1957. (Courtesy of USGS Photographic Library; photography by E.F. Patterson.) (b) Analysts examine digital imagery displayed on a computer screen using specialized software, 2008. (Courtesy of Virginia Tech Center for Geospatial Information and Technologies.)

19.2   INFORMATION EXTRACTION BY IMAGE INTERPRETATION

Image interpretation is the practice of visual examination of imagery to identify objects and features, and to assess their significance. In its essence, image interpretation applies the same innate skills that we use every day to understand the photographs we see in newspapers and magazines. However, in the context of aerial imagery, image interpretation requires development of specialized interpretive skills that define and focus our normal abilities in the context of unfamiliar perspectives. The practice of image interpretation originated as aerial photography became a reliable source of information, initially for military information in World War I, then formalized by the time of World War II, and still later applied to tasks in the civil sector. Although formal instruction in photointerpretation declined in the 1960s and 1970s as multi-spectral imagery and, later, digital image analysis, matured, the practice has revived recently as fine-resolution commercial satellite imagery has provided increased detail that requires the skills of image interpretation—techniques that were not as relevant in the context of coarser-resolution imagery that previously prevailed.

19.2.1   ELEMENTS OF IMAGE INTERPRETATION

Image interpreters apply some combination of the eight elements of image inter-pretation, which characterize objects and features as they appear on aerial imagery. Although in practice image interpreters integrate their use of these eight elements, it is convenient to list them separately, to highlight their significance, and provide a com-mon vocabulary for communication among analysts. Figure 19.2 offers examples.

- *Image tone* denotes the lightness or darkness of a specified section of an image (Figure 19.2). For black-and-white images, tone may be character-ized as "light," "medium gray," "dark gray," "dark," and so on, as the image assumes varied shades of white, gray, or black. As applied to color or false-color

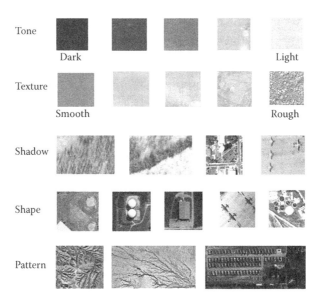

FIGURE 19.2 Illustrations of selected dimensions of the elements of image interpretation: tone, texture, shadow, shape, and pattern. (Images courtesy of U.S. Geological Survey and USDA.)

imagery, tone designates color, described as "light brown," "dark blue," or "pale green," for example. It is significant that image tone can be influenced not only by the nature of the features represented on the image but also by the nature of the solar illumination, and upon idiosyncrasies of imaging systems that may cause the variations in brightness unrelated to the features on the image. Thus, the interpreter must employ caution in relying solely on image tone for an interpretation, as it can be influenced by factors other than the absolute brightness of the target. Experiments have shown that interpreters tend to be consistent in interpretation of tones on black-and-white imagery, but less so in interpretation of color imagery (Cihlar and Protz, 1972). Their results imply that human interpreters can provide reliable estimates of relative differences in tone, but are not capable of accurate description of absolute image brightness.

• *Image texture* refers to the apparent roughness or smoothness of a region within an image. Image texture is usually caused by microshadows, the highlighted and shadowed patterns formed when an irregular surface is illuminated at an oblique angle. Contrasting examples (Figure 19.2) include the rough textures of a mature forest and the smooth textures of a wheat field. The character of the texture depends not only upon the nature of the surface itself but also upon the angle and direction of illumination, so will vary as illumination varies. Human interpreters are very good at distinguishing subtle differences in image texture, so it forms a valuable aid to interpretation.

• *Shadow.* Features illuminated at an angle by solar radiation cast shadows that often reveal their size or shape in ways that may not be obvious from an overhead perspective (Figure 19.2). The role of shadow is obvious in the instance of structures, and less obvious, but equally significant in natural settings. Shadows can enhance the boundaries between land cover parcels, and can contribute to the distinctive appearances of land cover parcels.

- *Pattern* identifies the placement of objects into distinctive, repetitive, arrangements (Figure 19.2). Pattern on an image usually originate from functional relationships between the individual features that compose the pattern. Thus, the buildings in an industrial plant may have a distinctive pattern originating from their organization to permit economical flow of materials through the plant from receiving raw material to shipping of the finished product. The distinctive spacing of trees in an orchard arises from careful planting of trees at intervals that minimize competition between individual trees and permit convenient movement of equipment through the orchard.
- *Association* specifies the mutual occurrence of certain objects, usually without the strict spatial arrangement implied by pattern. In the context of military photointerpretation, the association of specific items has great significance, as, for example, when the identification of a specific class of equipment implies that other, more significant, items are likely to be found nearby.
- *Shapes* of features are obvious clues to their identities (Figure 19.2). Individual objects have distinctive shapes, which provide the basis for identification. Features in nature often have such distinctive shapes that shape alone might be sufficient to provide clear identification. For example, ponds, lakes, and rivers occur in specific shapes unlike others found in nature. Often specific agricultural crops tend to be planted in fields that have characteristic shapes (perhaps related to the constraints of equipment used or the kind of irrigation that the farmer employs).
- *Size* has significance in two ways. First, relative sizes of objects in relation to other objects offer interpreters an intuitive notion of scale and resolution, even without measurements or calculations. This use of size is ultimately based upon recognition of familiar objects (structures, forests, highways, etc.) and extrapolation to relate the sizes of these known features to estimate sizes of other objects that might not be as easily identified. Second, use of the scale of an image (Chapter 16) to extrapolate a measured image distance to estimate an actual length on the ground can confirm a tentative identification, especially if an object's dimensions are distinctive. Furthermore, absolute measurements permit the calculation of lengths, volumes, or (sometimes) rates of movement (e.g., of vehicles or ocean waves as they are represented by sequential images).
- *Site* refers to topographic position. For example, water towers are positioned at elevated topographic sites to provide gravity drainage to lower regions. Orchards may be positioned at characteristic topographic sites—often on hillsides (to avoid cold air drainage to low-lying areas) or near large water bodies (to exploit cooler spring temperatures near large lakes to prevent early blossoming).

In practice, the analyst applies such elements in combination as needed for specific tasks, including

- Identification—specifically identify objects with respect to specific classification
- Delineation—outlining extents of specific facilities, land cover classes, or distinct regions
- Change detection—comparing images of the same region as observed at differing times to summarize the changes that have occurred (Section 19.4.1)
- Enumeration—counting numbers of specified items
- Mensuration—measurement of lengths, widths, or volumes.

A critical dimension of this approach to extraction of information is the recognition that the goal of image interpretation is not limited to simple description or inventory of objects, but that it also offers the opportunity to assess the significance of the information in the context of previous interpretations of the same site, related sites, or surrounding features.

19.2.2 Interpretation of Digital Imagery

Prior to the routine availability of computer systems for display and analysis aerial imagery (Chapter 20), photointerpretation was practiced within the realm of photographic prints and transparencies, using magnifiers and stereoscopes. As digital analyses have increased in significance, so has the importance of interpretation of imagery presented on computer displays. Although such interpretations apply the same principles outlined earlier for conventional imagery, digital data have their own characteristics that require special considerations in the context of visual interpretation.

19.2.2.1 Image Enhancement

Image enhancement adjusts the distribution of brightness values to improve the visual appearance of digital images, as judged for a specific application. Therefore, it is important to emphasize that enhancement is an arbitrary exercise—what is successful for one purpose may be unsuitable for another image or for another purpose. In addition, image enhancement is conducted without regard for the integrity of the original data. The original brightness values will be altered as their visual qualities are improved; so they lose their relationships to the original brightnesses as measured by the sensor. Therefore, enhanced images should not be used as input for additional analytical techniques; rather, further analysis should be based upon the original (unenhanced) values.

Contrast refers to the range of brightness values present on an image; *contrast enhancement* refers to the practice of manipulating the range of brightnesses to improve the visual appearance of an image. Contrast enhancement is often required because digital imagery may have been acquired using brightness ranges that do not match the capabilities of the human visual system, nor those of photographic films and computer displays. Therefore, for analysts to optimize their ability to

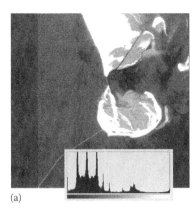
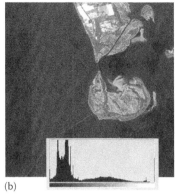

(a) (b)

FIGURE 19.3 SPOT HRV panchromatic image of Fisherman's Island, VA, before and after image enhancement. The insets show frequency histograms, with the horizontal axis representing brightness, from dark on the left, to bright on the right, and the vertical axis representing numbers of pixels at each brightness level. (a) Image prior to enhancement; the distribution of brightness values represents details within the land areas as very bright pixel values. (b) The same image after application of image enhancement, which has redistributed brightness values to reveal details within the land surface. The histogram for the enhanced image shows the reallocation of brightness values to the central region of the brightness range, so the medium gray tones are available to represent the detail of the land regions.

extract information from a specific image, the range of brightnesses displayed must be adjusted for the purpose at hand. Figure 19.3a illustrates the practical effect of image enhancement—before enhancement, significant detail is lost in the darker regions of the image; after enhancement has stretched the histogram of brightness values to take advantage of the capabilities of the display system, the lost detail is clearly visible.

Several alternative strategies have been developed for image enhancement, including

- *Linear stretch* converts the original digital values into a new distribution, using new minimum and maximum values as specified by the analyst. The algorithm then matches the old minimum to the new minimum and the old maximum to the new maximum. All of the old intermediate values are then scaled proportionately between the new minimum and maximum values.
- *Histogram equalization* reassigns digital values in the original image such that brightnesses in the output image are equally distributed among the range of output values (Figure 19.4). Unlike contrast stretching, histogram equalization is achieved by applying a nonlinear function to reassign the brightnesses in the input image such that the output image approximates a uniform distribution of intensities. The histogram's peaks are broadened, and the valleys are made shallower. Histogram equalization has been widely used for image comparison processes (because it is effective in enhancing image detail) and for the adjustment of artifacts introduced by digitizers or other instruments.

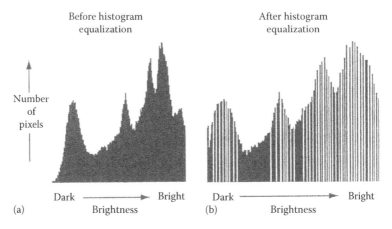

FIGURE 19.4 Schematic illustration of histogram equalization image enhancement technique. (a) Image brightness values are compressed, perhaps concealing image detail that requires a broader range of brightnesses to be visible. (b) Spreading the brightnesses over a broader range reveals features previously concealed. See Figure 19.3 for an example. (Reprinted from Campbell, J.B., *Introduction to Remote Sensing*, 4th edn., Guilford Press, New York, 2007, pp. 155, 454. With permission.)

- *Density slicing* is accomplished by arbitrarily dividing the range of brightnesses in a single band into intervals, then assigning each interval to a color. Density slicing may have the effect of emphasizing certain features that may be represented in vivid colors, but, of course, does not convey any more information than the single image used as the source.

19.2.2.2 Spatial Filtering

Another strategy for image enhancement focuses upon the application of spatial filters to highlight or suppress specific features in an image based on the spatial frequencies within the patterns represented on the image. The spatial frequency component of an image can be visualized as the image texture, determined by the scale of the variations in tone within a small region of an image. "Rough"-textured areas of an image, where the changes in tone are abrupt over a small area, have high spatial frequencies, while "smooth" areas with little variation in tone over several pixels, have low spatial frequencies. A common filtering procedure involves moving a "window" of a few pixels in dimension (e.g., 3×3, 5×5) over each pixel in the image, applying an averaging algorithm to either enhance or suppress brightness differences within the window, and replacing the central pixel with the new value. The window is moved along in both the row and column dimensions one pixel at a time, repeating the calculation until the entire image has been filtered and a new version of the image has been generated. By varying the calculation, and the weightings of the individual pixels in the window, filters can be designed to enhance or suppress different types of features, as discussed in Section 29.4.2.

Edge enhancement is a specialized version of spatial filtering that reinforces the visual boundaries between regions of contrasting brightness. Human interpreters usually perform best when they can detect sharp edges between adjacent regions.

Without sharpening   With sharpening

FIGURE 19.5 Example of edge enhancement. The right-hand image illustrates that the application of an image sharpening algorithm to the original image (left) enhances edges between regions of contrasting brightness.

In practice, the presence of noise, coarse resolution, and other factors often tend to blur or weaken the distinctiveness of these edges. Edge enhancement strengthens the local contrast of tonal differences within a local region (Figure 19.5). A typical edge-enhancement algorithm is implemented using a moving window that is systematically repositioned across the image—at each position, it is then possible to calculate a local average of values within the window; the central value can be compared to the averages of the adjacent pixels. If that value exceeds a specified difference from this average, the value can be altered to accentuate the difference in brightness between the two regions.

19.2.3 Who Can Implement Manual Interpretation?

In some situations, such as identification of locations of manhole covers for a study of urban infrastructure, an employee familiar with the location can be trained in fundamentals of image interpretation to extract the necessary information. In other situations, such as interpretation of geologic or hydraulic information, more complex skills, more subtle distinctions, and a wider range of experience, are required, so a seasoned professional experienced in a specific region, topic, and imagery is required.

Thus, an effort to develop an in-house capability for image interpretation depends first upon the availability of a well-prepared and experienced staff. Usually image interpretation skills require a depth of knowledge and experience not easily acquired, so it may be difficult to assign existing staff to image interpretation duties. Because effective image interpretation requires topic-specific skills, image-specific skills, and knowledge tailored to certain locales, it is often believed that it is more effective to train someone already prepared in a subject area as an image analyst rather than to train someone prepared as an image analyst in the subject matter in question.

For example, an experienced forester is likely to master the practice of image interpretation as applied for forestry more effectively than an image interpretation specialist can acquire the depth of knowledge already mastered by an experienced forester.

Second, visual extraction of information from imagery may require specialized equipment or software. Visual extraction of information from imagery on an *ad hoc* basis does not necessarily require specialized equipment. In practice, however, visual information must be extracted such that it can be presented in its spatial context, so systematic image interpretation requires the use of geographic information system (GIS) or image-processing systems to compile and maintain metadata in a rigorous manner that can be used by the organization in question.

It is unlikely that most organizations not already focused on image analysis will have the capabilities and resources to develop their own in-house capabilities. Therefore, organizations with a need to extract information from remotely sensed data will likely seek the services of organizations with established capabilities, rather than to attempt to develop internal capabilities. Most often, such organizations will provide image interpretation support as an ancillary to their principal activities, perhaps listed as photogrammetry, GIS services, cartography, or environmental analysis. The skills and experiences of such organizations are diverse, and clients' needs are so specialized that it is essential that the client carefully assess in advance the capabilities of a prospective service provider in relation to the very specific requirements of the project.

19.3   INFORMATION EXTRACTION BY SPECTRAL ANALYSIS

Whereas previous sections of this chapter are devoted to the extraction of information related to objects or distinct features based upon their distinctive physical features, this section focuses on the extraction of information based upon spectral properties—that is, the varied patterns of brightneess reflected from the surfaces in question. By analogy to our everyday experience, this process compares to the use of color to extract information from an image. However, the practice of remote sensing employs instruments that observe the landscape across many spectral channels rather than the three primary colors available to us in the visible region of the spectrum. Such instruments collect data across dozens, or hundreds, of spectral channels, as discussed in Chapter 18. These spectral properties can be regarded, loosely, as the "colors" of objects, although in remote sensing, these wavelengths extend far outside the visible spectrum, so they are not visible to the human eye, and we lack the vocabulary to discuss these ideas in everyday conversation. These analyses are applied most commonly to extraction of thematic data, seldom for point data.

19.3.1   Vegetation Indices

The field of remote sensing has developed several strategies for extracting the spectral information. The use of *band ratios* illustrates a simple example of how the practice of remote sensing can exploit the spectral information in an image. A remotely sensed image can be considered as an array of values, known as *pixels*, which record the brightness over a broad range of spectral values. A pixel, therefore, consists of a set of values, with each value representing its brightness within a specific region of the spectrum. A band ratio is formed by dividing the digital value of a pixel as represented in one region of the spectrum by the value of another region of the spectrum with respect to the same pixel. This operation has value because division tends to cancel out the brightness information in each pixel, leaving only a pure measure of the differences between the spectral values of the pixels in the two spectral regions. One benefit of this operation is that it tends to reduce effects of shadows, leaving the spectral information intact.

A special case of band ratios, *vegetation indexes*, are valuable as tools to enhance the record of vegetative surfaces. For example, the spectral reflectance of leaves' living vegetation is characterized by comparatively low values in the blue and red regions because photosynthetic processes absorb blue and red radiation. Within the visible region of the spectrum, the same vegetation is bright in the green region because of the reflectance of green radiation by the chlorophyll present in living leaves. Thus, healthy vegetation appears green to the eye because its brightest visible reflectance is within the green region (Figure 19.6). However, if a broader region of the spectrum is considered, it is clear that other tissues within the living leaf reflect near-infrared (NIR) radiation much more than any of the regions within the visible. Thus, sensors that can view densely vegetated areas in the NIR region record these surfaces as intensely reflective surfaces.

A band ratio formed using the red and NIR regions of the spectrum will further emphasize the presence of vegetated surfaces. In its most basic form, such a ratio takes the form of the NIR:R ratio, in which NIR signifies the brightness of a pixel in the NIR region of the spectrum, and R signifies the brightness of the same pixel in the red region of the spectrum. Because living vegetation reflects strongly in the NIR, but tends to absorb red radiation during photosynthesis, this ratio will be large for pixels that represent vegetated surfaces, and small for pixels characterized by sparse vegetation cover, dead or stressed vegetation, or by barren or nonvegetated surfaces.

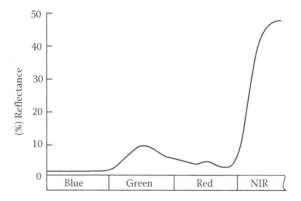

FIGURE 19.6 Typical spectral reflectance from a living leaf. The leaf is brightest in the NIR region, although in the visible region, the maximum brightness lies within the green region. (Reprinted from Campbell, J.B., *Introduction to Remote Sensing*, 4th edn., Guilford Press, New York, 2007, pp. 155, 454. With permission.)

Such ratios are referred to as *vegetation indices*. Although a multitude of varied strategies for calculating vegetation indices have been proposed (Curran, 1980), comparative evaluations tend to indicate that they tend to have similar effectiveness (Perry and Lautenschlager, 1984). Much of the remote-sensing community has accepted the effectiveness of a specific form of the basic ratio described earlier, known as the normalized difference vegetation index (NDVI):

$$NDVI = \frac{IR - R}{IR + R} \tag{19.1}$$

This variation of the basic ratio records the same relationships defined by Equation 19.1, but has been defined to optimize its statistical properties.

Price (1987) and others have found that efforts to compare ratios or indices over time, or from different instruments, should present brightnesses as radiances before calculating ratios, to account for differences between sensors and their calibration. Note also that, because atmospheric scattering is wavelength-dependent, the red (R) portion of the equation will be altered more than the NIR portion. As a result, atmospheric scattering distorts the observed value of the ratio relative to its "true" value as observed at ground level, or under ideal conditions of a clear atmosphere. Further, even under ideal atmospheric conditions, the atmospheric path from a pixel at ground level to a satellite sensor differs greatly between a pixel at nadir and a similar pixel positioned at the edge of the image, contributing also to differences in observed values of the ratio. Under some circumstances, analysts have the ability to compensate for such effects, so they do not eliminate the value of such ratios, although they do constitute cautions for their use.

Figure 19.7 illustrates the application of such a ratio in viewing a vegetated surface; the bright white areas in the NDVI image depict pixels that are dominated by dense, healthy vegetation. For this region, these areas are mainly irrigated agricultural fields (the circular fields irrigated by center-pivot irrigation systems), and riverine vegetation within the floodplain of the stream visible at the upper left region of the image. In contrast, the black areas are plowed fields totally devoid of vegetative cover, and the various shades of gray indicate unirrigated cropland and rangeland, both characterized by sparse, less-vigorous, vegetative cover at this time of year. It should be noted that although much of this information can be extracted from other channels of a multispectral image, use of the NDVI and similar measures provides a clear and unambiguous signal for the presence of vegetative cover, with clarity not always present in the original, unratioed, channels.

19.3.2 DIGITAL IMAGE CLASSIFICATION

Digital image classification is the process of assigning pixels to classes. Usually each pixel is treated as an individual unit composed of values in several spectral bands. By comparing pixels to one another, and to pixels of known identity, it is possible to assemble groups of similar pixels into classes that are associated with the informational categories of interest to users of remotely sensed data. These classes form regions on a map or an image, so that after classification the digital image

FIGURE 19.7   The NDVI applied to view an image of a vegetated landscape. The bright white areas in the large NDVI image depict pixels that are dominated by dense, healthy vegetation. It should be noted that although much of this information can be extracted from the several channels of a multispectral image (as shown in the inset), the use of the NDVI and similar measures provides a clear and unambiguous signal for the presence of vegetative cover, with a clarity not always present in the original, unratioed, channels.

is presented as a mosaic of uniform parcels, each identified by a color or symbol. These classes are, in theory, homogeneous: pixels within classes that are spectrally more similar to one another than they are to pixels in other classes. In practice, of course, each class will display some diversity, as each scene will exhibit some variability within classes. Figures 19.8 and 19.9 illustrate image classification, showing an original image (Figure 19.8) as a three-band composite, and the classified image (Figure 19.9), symbolizing each class as a separate color keyed to the menu of classes required for the analysis.

The term *classifier* refers loosely to a computer program that implements a specific procedure for image classification. Over the years, scientists have devised many classification strategies, as reviewed in Chapter 20; the analyst must select a classifier that will best accomplish a specific task. It is not possible to state that a given classifier is "best" for all situations because the characteristics of each image and the circumstances for each study vary so greatly. Therefore, each analyst must understand the alternative strategies for image classification to be prepared to select the most appropriate classifier for the task at hand.

The simplest form of digital image classification is to consider each pixel individually, assigning it to a class based upon its several values measured in separate spectral bands. Sometimes such classifiers are referred to as *spectral* or *point* classifiers because they consider each pixel as a "point" observation (i.e., as discrete values isolated from their neighbors). Although point classifiers offer the benefits of simplicity and economy, they are not capable of exploiting the information contained in

FIGURE 19.8   **(See color insert following page 426.)** Landsat thematic mapper (TM) false-color composite (RGB—bands 4, 3, 2) Dawson County, Nebraska, August 8, 2007 (see Figure 19.9). (Courtesy of U.S. Geological Survey).

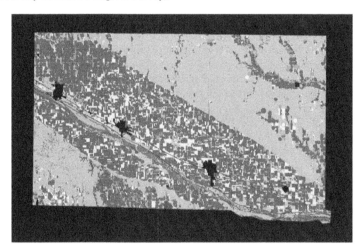

FIGURE 19.9   **(See color insert following page 426.)** Results of supervised agricultural crop classification of 1997 multi-date Landsat TM data for Dawson County, Nebraska. Legend: dark green = corn; light green = soybeans; orange = sorghum; yellow = alfalfa; pink = small grains; tan = rangeland, pasture, grass, black = urban land and roads; blue = open water; red = riparian forest and woodlands; cyan = wetlands and subirrigated grasslands; gray = other agricultural land; white = summer fallow (see Figure 19.8). (From Narumalani, S. et al., Information extraction from remotely sensed data, in *Manual of Geospatial Science and Technology*, Bossler, J. (ed.), Taylor & Francis, London, U.K., 2002, pp. 298–234. With permission.)

relationships between each pixel and those that neighbor it. Human interpreters, for example, could derive little information using the point-by-point approach, because humans derive less information from the brightnesses of individual pixels than they do from the context and patterns of brightnesses of groups of pixels. To exploit information conveyed by pixels with their spatial context, analysts can employ *spatial*, or *neighborhood*, classifiers, which examine small areas within the image using both spectral and textural information to classify the image. Spatial classifiers are necessarily more complex to design, and more difficult to apply than are point classifiers. In some situations, spatial classifiers have demonstrated improved accuracy, but few have found their way into routine use for remote-sensing image classification. *Object-oriented classifiers* apply a specialized approach to image classification, using techniques for identifying homogenous patches in pixels ("objects"), which are then classified as discrete features (Chapter 20).

19.3.2.1 Informational Classes and Spectral Classes

Informational classes are the categories of interest to the users of the data, such as, for example, the different kinds of forests, or land uses that convey information to the planners, managers, and scientists who will use the information in their routine activities. Such classes convey the information that we extract for the imagery. However, remotely sensed imagery cannot record such information directly—we extract them indirectly, using the brightnesses recorded by each image. These brightnesses form *spectral classes*—groups of pixels that are uniform with respect to their brightnesses in several spectral channels. If an analyst can match spectral classes within remotely sensed data to informational classes, then the image forms a valuable source of information. Thus, in essence, the practice of image classification proceeds by matching spectral categories to informational categories.

Seldom can we expect to find exact one-to-one matches between informational and spectral classes. Any informational class includes spectral variations arising from natural variations within the class. For example, a region of the informational class "forest" may display variations in age, species composition, density, and vigor, which all lead to differences in its spectral appearance. Furthermore, variations in illumination and shadowing may create additional variations within otherwise spectrally uniform classes. Thus, informational classes are typically composed of numerous *spectral subclasses*—spectrally distinct groups of pixels that together may be assembled to form an informational class. To conduct digital image classification, the analyst must treat spectral subclasses as distinct units during classification, but then display several spectral classes under a single symbol for the final image or map to be used by planners or administrators.

19.3.2.2 Supervised Classification and Unsupervised Classification

Another important distinction in image classification separates supervised classification from unsupervised classification. *Supervised classification* procedures require considerable interaction with the analyst, who must guide the classification by identifying areas on the image that are known to belong to each category. In contrast, *unsupervised classification* requires only minimal interaction with the analyst, in a search for natural groups of pixels present within the image. Although the distinction between supervised and unsupervised classification is useful, the two strategies are not as distinct as these definitions suggest—some methods, known as *hybrid classifiers*, share characteristics of both supervised and unsupervised methods.

Unsupervised classification is the identification of natural groups, and their interrelationships, within multispectral imagery. Although the existence of natural, ordered, groups of similar spectral values within an image may not be obvious to the novice, it can be demonstrated that digital remote-sensing imagery are composed of spectral classes that are reasonably uniform internally with respect to brightnesses in several spectral channels. Unsupervised classification is the definition, identification, labeling, and mapping of these natural classes.

Unsupervised classification offers advantages (relative to supervised classification) of (1) minimal interaction with the analyst, (2) ability to identify unique or special classes that may occupy only very small areas, and (3) extensive prior knowledge of the region is not required. Although this third characteristic is often cited as an advantage of unsupervised classification, but it is more accurate to state that the nature of knowledge required for unsupervised classification differs from that required for supervised classification. To conduct supervised classification, detailed knowledge of the area to be examined is required to select representative examples of each class to be mapped. To conduct unsupervised classification, no detailed prior knowledge is required, but knowledge of the region is required to interpret the meaning of the results produced by the classification process. Thus, any effort to apply image classification requires an astute knowledge of the region classified.

The limitations of unsupervised classification arise primarily from reliance upon the naturally occurring groups, and difficulties in matching these groups to the informational categories that are of interest to the analyst. Thus, the analyst has limited control over the menu of classes, the spectral properties of specific informational classes, and may experience difficulty in matching spectral classes to informational classes.

Supervised classification can be defined informally as the process of using samples of known identity (i.e., pixels already matched to informational classes) to classify pixels of unknown identity (i.e., to assign unclassified pixels to one of several informational classes). Samples of known identity are those pixels located within *training areas*, or *training fields*, defined by the analyst to encompass pixels that can be clearly matched to areas of known identity on the image. Such areas should typify spectral properties of the categories they represent, and, of course, must be homogeneous with respect to the informational category to be classified. Pixels located within these areas form the training samples used to guide the classification algorithm to assign spectral values to the correct informational classes. Clearly, the careful selection of these training data is a key step in supervised classification.

Advantages of supervised classification (relative to unsupervised classification), include (1) clearer control over the menu of informational categories matched to a specific purpose and geographic region; (2) the correspondence of the classification to areas of known identity; and (3) ability to detect serious errors in classification by thorough examination of training data. Limitations of supervised classification include (1) imposition of a classification structure, which may not match the natural classes that exist within the data; (2) definition of training data with reference to informational categories and only secondarily with reference to spectral properties, and (3) selection of training data that may not represent conditions encountered throughout the image.

19.3.2.3   Classification Algorithms

To illustrate some of the specifics of the practice of image classification, the following paragraphs introduce some of the basic strategies employed, including *minimum distance to means*, *parallelepiped classification*, and *maximum likelihood classification*. Software for implementing these techniques is reviewed in Chapter 20, and in the material referenced by that chapter. Further, the classification process must be planned together with a plan for *accuracy assessment* (Chapter 21) necessary to evaluate the success of the classification effort.

Minimum distance to means is a classification strategy that assesses distances between clusters, and between cluster centers (sometimes called "centroids") and individual pixels, then using distances to define clusters, and allocate pixels to clusters. This procedure relies upon successive iterations to optimize definitions of clusters, and the assignment of pixels to clusters. In Figure 19.10, the two axes represent brightnesses in two spectral channels selected from a larger number of multispectral channels, increasing in brightness from dark near the origin to bright at the left and upper extremes. In this multispectral data space, pixels from a remotely sensed image are plotted as black dots, representing pixels already assigned to spectral classes; the larger black dots represent the centroids (centers) of these clusters of pixels. The crosses represent unassigned pixels, not yet allocated to any of the clusters.

The minimum distance classification strategy allocates these unassigned pixels by assessing the multispectral distance from each pixel to the center of each cluster (represented in the diagram by the lines connecting each pixel to the center of the nearest cluster). As new pixels join each cluster, its center shifts slightly as its

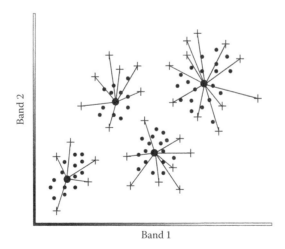

FIGURE 19.10   Minimum distance classification. Here, the two axes represent brightnesses in two spectral channels selected from a larger number of multispectral channels, increasing in brightness from dark near the origin to bright at the left and upper extremes. In this multispectral data space, pixels from a remotely sensed image are plotted as black dots, representing pixels already assigned to spectral classes; the larger black dots represent the centroids (centers) of these clusters of pixels. The crosses represent unassigned pixels, not yet allocated to any of the clusters.

statistical properties change with the addition of new pixels. The allocation process is iterated repeatedly until the shift in the position of the centroid is minimal, and the classification process terminates. Normally the analyst can specify a minimum threshold for assessing the change in centroid position, and/or set minimum or maximum numbers of iterations. For unsupervised classification, the cluster centers can be assigned at random; they migrate toward the values that characterize a particular image. Alternatively, if the procedure is applied for supervised classification, the starting points can be set by training data, and the movement of their centroids constrained within limits set by the analyst.

Such procedures can identify natural structures within the image data, producing accurate identification of spectral classes. Astute application can be very effective in identifying spectral classes within an image; analysts may be able to match spectral classes to informational uses.

Parallelepiped classification defines polygons in multispectral data space, as illustrated in Figure 19.11. Here, the illustration shows that the sizes of the polygons are set by the extreme values of each training data set. The two axes represent brightnesses in two spectral channels selected from a larger number of multispectral channels, increasing in brightness from dark near the origin to bright at the left and upper extremes. The ranges in brightness along each axis define the sizes of the polygons (known as parallelepipeds), as shown in Figure 19.11. Here the dots represent the training data used to define the parallelepipeds, and the crosses represent other

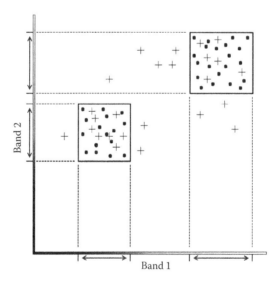

FIGURE 19.11 Parallelepiped classification. Here, the illustration shows that the sizes of the polygons are set by the extreme values of each training data set. The two axes represent brightnesses in two spectral channels selected from a larger number of multispectral channels, increasing in brightness from dark near the origin to bright at the left and upper extremes. The ranges in brightness along each axis define the sizes of the polygons (known as parallelepipeds). The dots represent the training data used to define the parallelepipeds, and the crosses represent other pixels, yet to be classified. Those that fall within the parallelepipeds are assigned to the categories defined by the training data.

pixels, yet to be classified. Those that fall within the parallelepipeds are assigned to the category defined by the training data. Parallelepiped classification is one of the oldest and best-established classification strategies, but in practice, implementation can be difficult. In Figure 19.11, those crosses that fall outside the parallelepipeds remain unclassified—a practical problem for this strategy. Further, sometimes the parallelepipeds can overlap in multispectral data space, creating the potential of a single pixel to belong to two informational classes. Often these difficulties are addressed by procedures that use a minimum distance rule to allocate the pixels affected by these problems.

Maximum likelihood classification is based upon the premise that training data form typical and normally distributed samples of the values within the spectral classes within an image. Then it uses characteristics of the training data to assess probabilities that each of the unclassified pixels belong to each of the training data, and uses the probabilities to assign pixels to most likely classes. Figure 19.12 shows a schematic view of training data plotted in multispectral data space, with the ellipses representing the varying numbers of pixels within the training data. In a three-dimensional view, these training data could be visualized as three-dimensional topographic surfaces (the third dimension would represent numbers of pixels), with peaks at the central values for each training data set, declining in height near the edges. Such surfaces can be regarded as probability functions (*probability density functions*) (see Chapter 6), with the lines representing contours of equal probability for a pixel to belong to that class, with probabilities increasing near the center of the training data, and decreasing near

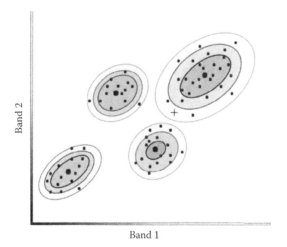

Band 1

FIGURE 19.12 Maximum likelihood classification. A schematic view of training data plotted in multispectral data space, with the ellipses representing the varying numbers of pixels within the training data. Such surfaces can be regarded as probability functions (*probability density functions*), with the lines representing contours of equal probability for a pixel to belong to that class, with probabilities increasing near the center of the training data, and decreasing near its edges. Pixels falling near the center of a training data set, at the peak, will have a high probability of belonging to the class, whereas those near the periphery, where the number of pixels within the training data for that class are lower, will have a much lower probability of membership in that class.

its edges. Pixels falling near the center of a training data set, at the peak, will have a high probability of belonging to the class, whereas those near the periphery, where the numbers pixels within the training data for that class are lower, will have a much lower probability of membership in that class.

This strategy, known as *maximum likelihood classification*, is extremely accurate and powerful, provided that the training data are representative of the image data, and follow normal frequency distributions. Because remotely sensed data do not usually meet these requirements, in practice, maximum likelihood classification cannot always meet its potential. However, if applied with care, it is one of the most flexible and useful classification strategies for remotely sensed data. If training data are not selected with care, the maximum likelihood strategy will assign pixels, but not necessarily to the correct classes.

Accuracy assessment. Image classification process concludes with the application of accuracy assessment (Chapter 21), characterized by category-by-category evaluation of classification against a set of reference data. Pixel-by-pixel assessment, class-by-class evaluation permits user to assess the quality of the classification, but also permits the analyst to assess the weaknesses of the classification and to refine it to address weaknesses. As discussed in Chapter 21, accuracy assessment must be planned an integral component of the project, so that the collection of reference data necessary for the accuracy assessment are coordinated with the classes to be used in the classification process.

19.3.2.4 Who Can Execute Digital Extraction of Image Information?

An organization that requires analysis to extract information from remote-sensed imagery faces the same choice posed by any newly defined need—developing an in-house capability or contracting for the services from an external provider. Developing an in-house capability is not beyond the capabilities of many organizations of modest size. But such an effort will require computer hardware, including workstations with ample storage, displays, and special-purpose software (Chapter 20), and internet connectivity. These investments are likely to be worth their costs if there is a recurring need over the long term. If the services are to be contracted, it is essential to confirm that they can provide the expertise that matches the client's needs. Given appropriate imagery and software, a competent novice can complete the steps required to conduct an image processing project. However, production of quality work requires the expertise of staff prepared with combined experience applying image-processing techniques, knowledge of local setting, and needs of clients.

19.4 EXTRACTING TEMPORAL INFORMATION

Because remote-sensing imagery records information associated not only with a specific place but also with a specific time, it forms a record that can be analyzed to extract temporal data that reveals changes, or the absence of changes, for phenomena represented by the imagery. Changes significant in many contexts, including, for example, changes in landcover, growth and losses of forests, landscape differences related to climatic change, or identification of areas damaged by fire, flood, or tornadoes. Business enterprises use change-detection procedures to identify changes in the road networks used for personal navigation systems (Chapter 18). The significance of reliable change detection algorithms is not often recognized, but they underlie many of the routine applications that drive the commercial applications outlined in Chapter 22.

19.4.1 DATA FOR TEMPORAL ANALYSIS

Although there are many sources of data providing temporal sequences of imagery, this section introduces the reader to a few of the collections that are easily accessible to the public at minimal cost. Many other specialized local or private collections are often available for the determined investigator.

During the late 1930s, in the United States, the federal government began systematic programs to acquire aerial photography to support environmental remediation, crop monitoring to the agricultural economy, and forestry programs. As a result, there is now an archive, in the public domain, that represents a very large portion of the nation over a span of 50 years or more. These images are now available for studies to examine (for example) environmental, land use, and coastal changes of intervals of many decades. The U.S. Department of Agriculture's (USDA) Aerial Photography Field Office (APFO) (http://www.fsa.usda.gov/FSA/). In the 1980s, the U.S. federal government began systematic photography of the entire United States. National High Altitude Photography Program (NHAP) (1980–1989), and its successor, the National Aerial Photography Program (NAPP) acquired high-altitude aerial photography of the United States on a systematic basis.

These programs supported by numerous federal agencies with requirements for aerial photography, coordinated by the U.S. Geological Survey, which provides access through the GloVis (Global Visualization, which provides use with the ability to search for imagery and download digital versions of the imagery without cost). In other nation, similar services are often available, but with details that differ greatly with respect to specifics. In other nations, corresponding national mapping agencies, such as France's *Institut Géographique National* (*IGN*) (http://www.ign.fr/) and Britain's Ordnance Survey (OS) (http://www.ordnancesurvey.co.uk/) provide comparable services. Further, readers should examine TerraServer (http://www.terraserver.com/) and related Web sites.

The U.S. Landsat archive includes imagery representing most of the earth's land areas since 1972. These data are accessible using GLOVIS (global visualization viewer) (http://glovis.usgs.gov/), an online search utility that provides convenient access to the complete U.S. Landsat archive. The U.S. Geological Survey has initiated a policy of providing no-cost access to the entire archive through internet downloads (at the time of writing this chapter, this service is available to selected portions of the archive, which will soon extend to the entire collection). These data provide an extensive resource for those who wish to examine temporal changes.

19.4.2 SEQUENTIAL AERIAL PHOTOGRAPHY

The simplest application of change detection, known as *sequential aerial photography*, applies multiple aerial photographs of a specific region to identify changes that have occurred during the interval between pairs of photographs. This relatively simple approach has been used in many contexts to extract useful information (e.g., Erb et al., 1981; Walsh and LaFleur, 2005). In the ideal, photographs selected for sequential aerial photography should be compatible with respect to their key characteristics, including scale, emulsion, season of acquisition, and time of day to assure that differences in the imagery reflect genuine differences relating to the topics at hand, rather than differences in images unrelated to the topic of the interpretation. Usually change detection requires the application of image registration (Chapter 17) to assure that photographs share same scales and geometries, so that image differences indicate real differences in the subject to be analyzed.

19.4.3 DIGITAL CHANGE ANALYSIS

In the realm of digital remote sensing, comparisons of imagery over time are conducted by applying *change detection* algorithms—procedures designed to search for changes between images acquired at different dates. Such images, like the aerial photographs mentioned previously, must be compatible with respect to basic image qualities, such as resolution, spectral channels, and geo-referencing. Such procedures compare registered digital images pixel by pixel, to assess the degree of differences between the brightnesses of the pixels that form a pair at each location. The algorithm records a change when these differences exceed a certain threshold, which can be adjusted by the analyst to apply the sensitivity necessary to attain the requirements for a specific project.

Figure 19.13 illustrates an application of change detection to represent the track of a tornado that struck near Stoughton, WI, August 18, 2005. This tornado, assigned a F3 rating on the Fujita scale (winds estimated at 200 mph or higher), left a track about a half-mile wide and 10 mi in length, causing one death, and profound damage in both residential and agricultural landscapes. In this application, change detection is based upon the spectral differences between the undamaged landscape recorded in the "before" image and the field of debris and damage within the tornado track recorded in the "after" image.

The issue of compatibility retains its significance in the digital domain. Images must register and have comparable levels of spatial, spectral, and radiometric resolution. Unless the objective is to compare images from one season to another, the two images typically represent the same season but different years. Jensen (1996) lists a selection of preprocessing operations effective in preparing data for use in change detection comparisons. In any specific situation, the analyst must employ an intimate knowledge of the area to be studied, then apply a selection of methods tailored to the specifics of the study area. It seems unlikely that there is any single procedure that can be equally effective in all situations. Jensen's (1996) list of digital change detection procedures includes the following:

1. *Image algebra* apples arithmetic operations to pixels in each image, then forms the change image from the resulting values. *Image differencing*, for example, simply subtracts one digital image from another digital image of the same area acquired at a different date. After registration, the two

FIGURE 19.13 **(See color insert following page 426.)** Change detection, tornado damage, Stoughton, WI, August 18, 2005. ASTER and TM imagery (Chapter 18) provided before (inset) and after (larger image) imagery to document the track of the tornado, as indicated by the debris and damage represented in red by the change detection analysis. (Courtesy of Environmental Remote Sensing at the Space Science & Engineering Center, University of Wisconsin-Madison, http://ersc.ssec.wisc.edu/tornadoes.php.)

images are compared pixel by pixel to generate a third image composed of numerical differences between paired pixels from the two images. Values at or near zero identify pixels that have similar spectral values, and therefore presumably have experienced no change between the two dates. This procedure is typically applied to a single band of a multispectral data set; usually a constant value is added to eliminate negative values. The analyst must select (sometimes by trial and error) a threshold level to separate those pixels that have experienced change from those that have not changed land cover, but may exhibit small spectral variations caused by other factors. Jensen (1996, 2007) reports that image differencing is among the most accurate change detection algorithms, but that it is not equally effective in detecting all forms of land-use change.

2. *Postclassification comparison* requires two or more independent classifications of each individual scene, using comparable classification strategies. The two classifications are then compared, pixel by pixel, to generate an image that shows pixels placed in different classes on the two scenes. Successful application of this method requires accurate classifications of both scenes so that differences between the two scenes portray true differences in land use rather than differences in classification accuracy. In urban and suburban landscapes, the high percentages of mixed pixels (at Landsat

MSS resolution) have tended to decrease classification accuracy, and therefore to generate inflated estimates of the numbers of pixels that have experienced change. Because postclassification comparison permits compilation of a matrix of *from–to changes*, it provides more useful results than some other methods (e.g., image algebra that simply identifies pixels that have changed, without identifying the classes involved).

3. *Multidate composites* are formed by assembling all image bands from two or more dates together into a single data set. The composite data set can then be examined using multivariate analyses described in earlier chapters, including principal components analysis and image classification.

Finally, it should be emphasized that the practice of change detection, regardless of the technique applied, depends upon the use of imagery collected at different dates that is compatible with respect to its basic properties (such as resolution, illumination, and spectral channels) so that the analysis addresses genuine changes in the land surface, rather than differences in the imaging technique.

19.4.4 Land Surface Phenology

Change analysis can be conducted at finer temporal resolution using broad-scale sensors that collect data at coarse spatial resolution but at frequent temporal intervals. An excellent example of the value of this capability is offered by the field of *phenology*, the study of seasonal patterns of the emergence, growth, and decline of vegetative tissue, as observed both in agricultural and natural landscapes. *Land surface phenology* is the use of broad-scale instruments such as Advanced Very High Resolution Radiometer (AVHRR) and Moderate Resolution Imaging Spectroradiometer (MODIS) (Chapter 18) to observe very large areas at frequent intervals to examine broad-scale phenological patterns. These instruments permit the collection of NDVI (Section 19.3.1) to measure the presence and the vigor of vegetation cover, to track seasonal rises and declines of vegetative growth within very large regions over extended time intervals.

Observed over time, values of the NDVI depict a seasonal pattern of increase and decrease. Values rise as spring approaches at the onset of longer days, warmer temperatures, and spring rains. The seasonal biological response in NDVI records the emergence of buds, and then leaves, known as *green-up*. As the full summer season arrives, the green-up process peaks; mature plants invest energy in growing the seeds and fruits necessary to prepare the plant for success in the next year's growing season—green-up slows, stops, and then declines as autumn begins. The decline in greenness, as observed by NDVI decline from mid- to late summer is known as *brown-down*. Brown-down records the decline in vigor as senescence, and then harvest time, approaches. The interval between the onset of green-up and the end of brown-down defines a region's growing season. Thus, these data can provide a means of assessing place-to-place variation in the length of growing seasons. The result produces the typical seasonal patterns depicted in Figure 19.14.

Reed et al. (1994) and Loveland et al. (1995) were among the first to systematically use remote-sensed imagery to track seasonal vegetation patterns for

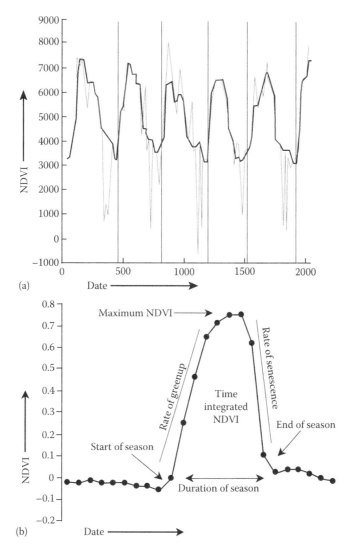

FIGURE 19.14 Phenological data. (a) NDVI data collected by MODIS illustrating phenological patterns over five seasons. The thin, jagged line represents the individual NDVI values as observed by the MODIS sensor. The heavy line shows the values as smoothed by a moving average to eliminate the noise in the individual observations, and thereby better represent the seasonal trajectory of vegetation changes. (Courtesy of Aaron Dalton and Peter Sforza.) (b) An idealized annual cycle as captured by the changes in NDVI, annotated to indicate the key components of a phenological cycle, as discussed in the text. (Courtesy of Kirsten deBeurs.)

continental areas. They accumulated cloud-free AVHRR composites over 14 day intervals, covering, altogether, a four-year interval. They calculated NDVI values for each pixel, so, in the aggregate, they recoded the response of local vegetation to seasonal changes, to variations in weather and climate, and to differences in human influences upon vegetation of agricultural and rangeland landscapes. Examination of NDVI over such long periods and such large areas permits the inspection of phenological patterns of varied land cover classes, responses to climatic and meteorological events, including drought, floods, and freezes. They used AVHRR 14 day composites, basically collected twice a month during the March–October growing season and once a month during the winter season. Such products are useful for monitoring agricultural regions, assessing climatic impact upon ecosystems, production of food, and cover for wildlife. Agricultural systems can be observed on a broad scale to assess crops' response to meteorological and climatic variations throughout the growing season, and forecast yields. Subsequent applications of the same approach, some using MODIS data to provide additional spatial detail within the framework of broad coverage (Kang et al., 2003; Zhang et al., 2003; Stöckli and Vidale, 2004). de Beurs and Henebry (2008) examined land surface phenology of Afghanistan recorded by MODIS imagery over several decades to observe changes in land cover patterns related to environmental, social, and governmental changes during that interval. Other applications have examined departures from normal phenological cycles to aid in locations of insect infestations upon forest cover (de Beurs and Townsend, 2008), and to assess patterns of surface mining in Appalachia (Townsend et al., 2008).

There are some precautions necessary in making detailed interpretations of such data. NDVI is, of course, subject to atmospheric effects, and the coarse resolution of AVHRR data encompasses varied land cover classes with varied phonological responses, especially in regions where landscapes are locally complex and diverse. The data accumulate noise created by the atmospheric contamination of the brightness values, so it is necessary to smooth the observations to filter out the noisy artifacts of atmospheric effects, snow cover, and the like upon NDVI values.

19.5 SUMMARY

Extraction of information from the imagery collected by remote-sensing instruments is an essential component of any remote-sensing system, equal in significance to the components that are often more visible. Analysts apply alternative strategies to extract useful remote-sensed data, selecting methods that match the nature of the task, information required, and the imagery at hand. For analyses that are based upon large scale and fine detail, practitioners often apply the image interpretation techniques introduced in Section 19.2.1 to outline areas, identify and mark features, enumerate items, and to measure volumes, areas, and distances. Vegetation indices provide the ability to extract information concerning the nature and the status of the vegetative cover of a landscape; the field of landscape phenology applies broad-scale satellite sensors to record and analyze seasonal variations in vegetative cover, revealing information concerning dynamics of both natural and human processes. Image classification can be employed to examine spectral qualities of data to extract classes that identify spectral classes within the data, match them to the informational classes of significance for the client. Together, these strategies provide the ability to extract a wide range of data from remotely sensed imagery.

REFERENCES

Campbell, J.B., *Introduction to Remote Sensing* (4th edn.), New York: Guilford, 2007.

Cihlar, J. and Protz, R., The perception of tone differences from film transparencies, *Photogrammetria*, 8, 131–140, 1972.

Curran, P., Multispectral remote sensing of vegetation type and amount, *Progress in Physical Geography*, 4, 315–341, 1980.

de Beurs, K.M. and Henebry, G.M., War, drought, and phenology: Changes in land surface phenology in Afghanistan since 1982, *Journal of Land Use Science*, 3, 95–111, 2008.

de Beurs, K.M. and Townsend, P.A., Estimating the effect of gypsy moth defoliation using MODIS, *Remote Sensing of Environment*, 112, 3983–3990, 2008.

Erb, T.L., Philipson, W.R., Teng, W.L., and Liang, T., Analysis of landfills with historic air photos, *Photogrammetric Engineering & Remote Sensing*, 47, 1363–1369, 1981.

Jensen, J.R., *Introductory Digital Image Processing: A Remote Sensing Perspective* (2nd edn.), Saddle River, NJ: Prentice Hall, 1996.

Jensen, J.R., *Remote Sensing of the Environment: An Earth Resource Perspective*, Saddle River, NJ: Prentice Hall, 2007.

Kang, S., Running, S.W., Lim, J.-H., Zhao, M., Park, C.-R., and Loehman, R., A regional phenology model for detecting onset of greenness in temperate mixed forests, Korea: An application of MODIS leaf area index, *Remote Sensing of Environment*, 86, 232–242, 2003.

Lillesand, T.M., Kiefer, R.W., and Chipman, J.W., *Remote Sensing and Image Interpretation* (6th edn.), New York: John Wiley & Sons, 2008.

Loveland, T.R., Merchant, J.W., Brown, J.F., Ohlen, D.O., Read, B., and Olsen, P., Seasonal land cover regions of the United States, *Annals of the Association of American Geographers*, 85, 339–355, 1995.

Narumalani, S., Hlady, J.T., and Jensen, J.R., Information extraction from remotely sensed data, Chapter 19 in *Manual of Geospatial Science and Technology*, (J. Bossler, ed.), London, U.K.: Taylor & Francis, pp. 298–234, 2002.

Perry, C.R. and Lautenschlager, L.F., Functional equivalence of spectral vegetation indices, *Remote Sensing of Environment*, 18, 35–48, 1984.

Price, J.C., Calibration of satellite radiometers and comparison of vegetation indices, *Remote Sensing of Environment*, 18, 35–48, 1987.

Reed, B.C., Brown, J., VanderZee, T., Merchant, J.W., and Ohlen, D., Measuring phenological variability from satellite imagery, *Journal of Vegetation Science*, 3, 703–714, 1994.

Stöckli, R. and Vidale, P.L., European plant phenology and climate as seen in a 20 year AVHRR land-surface parameter dataset, *International Journal of Remote Sensing*, 25, 3303–3330, 2004.

Townsend, P.A., Helmers, D.P., McNeil, B.E., de Beurs, K.M., and Eshleman, K.N., Changes in the extent of surface mining and reclamation in the Central Appalachians: 1976–2006, *Remote Sensing of Environment*, 113, 62–72, 2008.

Walsh, D.C. and LaFleur, R.G., Landfills in New York, 1844–1994, *Ground Water*, 33, 556–560, 2005.

Zhang, X., Friedl, M.A., Schaaf, C.B., Strahler, A.H., Hodges, J.C.F., Gao, F., Reed, B.C., and Huete, A., Monitoring vegetation phenology using MODIS, *Remote Sensing of Environment*, 84, 471–475, 2003.

20 Image Processing Software for Remote Sensing

Matthew Voss, Ramanathan Sugumaran, and Dmitry Ershov

CONTENTS

This chapter provides an overview of commercial and open source remote sensing softwares (RSSs) and their functionalities. It outlines the major features that RSS provides and illustrates how some of these programs implement those features.

Currently, none of the available image processing systems dominates the RSS world. As such, there are many competing programs that can perform the basic operations required for the analysis of remotely sensed data. An additional complicating factor is that there are a wide variety of uses for remote sensing imagery, ranging from employing imagery as a background to integration with other data for advanced classification and analysis. Costs of RSS range from free to tens of thousands of dollars. These extreme ranges in capabilities and costs make it difficult for both novice and advanced users to select a software that meets their needs. One of the goals in developing this chapter was to create a Web-based software database that would assist users in choosing software that works with their hardware and performs the analyses they require.

20.1   DATABASE DESIGN AND DEVELOPMENT

In order to summarize the existing RSS, we designed and developed a database by reviewing existing publications (e.g., Campbell, 2007; Jensen, 1995; Jensen and Jensen, 2002; and GIM International, 2004), examining Web sites, and reviewing product specifications from software developers. A full list of the software developers and their Web sites is available at http://itt240-04.geog.uni.edu/rs-software/index.php.

Users should access this site using their usual browsers; the page opens to show several small search windows, each representing varied characteristics of software systems, including operating system (OS), classification technique, the form of the software license, and file format. For each characteristic, the system lists available options. The user highlights the key characteristics of interest by left-clicking with the mouse to highlight (in the gray tone) the desired item on the list, using the scroll bars at the right of each window to search for the items of interest. Then, right-clicking with the mouse in the appropriate "search" icon launches a search of the database to identify the systems with the selected characteristics. In the "results" list, the names are linked to the web pages for each provider, so the user can easily view further details of each system by clicking on the provider's name. The "search by all attributes" capability provided by the link at the bottom of the page permits the user to search by all the attributes highlighted in the windows on the page. Thus, the use of this utility can assist analysts in finding those systems that have those characteristics most relevant to their needs. An obvious benefit of such a Web site is that it can be easily edited as new software is introduced or features are added to existing software; the Web site can easily be updated to reflect such changes. Additionally, the Web site is freely accessible to anyone with access to the Internet.

The data was compiled into an ACCESS database that could be used to populate data fields in the Web site, focusing principally upon data formats and functionalities of the software. The information collected included the name of the software, the company, the company Web site, the data formats, image classification options, and so on. A simple Web site was developed using ASP.net to query this database and return the software that meets the criteria specified by the user. Examples of potential queries might include

- What are the open-source RSS available on the market?
- Which software provides an artificial neural network classifier?
- What are the radiometric functionalities available in a particular RSS?
- What is the hardware requirement for a particular software system?

These capabilities permit users to easily search for desired capabilities, to effectively compare qualities of alternative systems, and to quickly see the range of capabilities and qualities offered by the suite of systems available to practitioners of remote sensing.

20.2 ASSESSMENT

There are many factors to consider when selecting an RSS package, including the hardware the software will run on, the proposed budget for the software, the types of data the software will be required to work with, and the functions the software is expected to perform. There are dozens of remote sensing packages and hundreds of different file formats. Some of the most basic factors to be considered in selecting software for a particular facility or application include

- The OS the software supports
- Whether it is free/open source, or a commercial product
- Whether it can display multiple band combinations and perform histogram stretches
- What type of image correction the software offers
- Whether it can perform preprocessing operations, such as principal component analysis
- The ability of the software to perform unsupervised and supervised image classifications
- The ability to perform advanced classification such as object-based, neural networks and decision trees
- Its ability to perform basic geographic information system (GIS) functions

Table 20.1 provides a brief summary of the software and the functions they provide.

20.2.1 COMPUTER CONSIDERATIONS

Computer capabilities have grown by leaps and bounds since digital image analysis first became common in the early 1970s (Botkin et al, 1984; Jensen, 1995). During this interval, processing power has grown from 8 mHz 16-bit processors to 3.4 GHz 64-bit processors. Processes that took 20 min in the late 1970s and early 1980s take less than a minute using current computer technology (Landgrebe, 1980). The amount of processing power required for a given remote sensing task can vary greatly. Simply displaying remote sensing data is not particularly computationally intensive. However, performing an object-based classification can be a more time-consuming process, as the computer is required to iteratively perform calculations on each object and its neighboring objects thousands of times.

The two major OSs that support RSS are Unix (and its offshoot Linux) and the Microsoft Windows platforms. Linux and Unix may be installed on nearly any computer and can frequently share hard disk space with a Windows installation. The Macintosh OS is becoming a more popular software development platform. As the Macintosh OS is loosely based on Unix, it is not uncommon to see software available for Unix/Linux also available for Macintosh (e.g., ITT ENVI, http://www.ittvis.com/ProductServices/ENVI.aspx). However, Windows and Unix/Linux are the dominant platforms for remote sensing products. Generally, commercial software is developed for the Microsoft Windows platform. However, ENVI provides support for all three major OSs.

TABLE 20.1

Selected Commercial and Open Source/Free Softwares for Remote Sensing Applications

Commercial Software	Operating System	Image Display	Geometric Correction	Radiometric Correction	Preprocessing	Classification	Advanced Classification	Object-Based Image Classification	Fuzzy/Soft Classification	Cartography Tools	GIS
AGI	XP	+									−
ArcMap	XP	+								+	+
Definiens developer	XP Vista	+			+	+	+	+	+		
ER mapper	XP	+	+	+		+	+			+	−
ERDAS imagine	XP L	+	+	+	+	+	+	+		+	−
IDRISI andes	XP	+	+	+	+	+	+		+		+
Intergraph geomedia image	XP	+	+								+
ITT ENVI	XP L Mac Unix	+	+	+	+	+	+	+			−
MapInfo	XF	+									
PCI geomatica	XF L Solaris	+	+	+		+				+	+

Free/open source

Software	Platform								
Bilko	XP	+	+	−	+				− +
CHIPS	XP	+	+	−	+				+
GRASS	XP L Mac	+	+	−	−				
LAS	Unix / Solaris / AIX	?	+	+	+				
Multispec	XP Mac	+				+	+		
OpenEV	XP L Irix / Solaris	+				+			
Opticks	XP Solaris	+	+	−				−	
OSSIM	XP Mac	+	+	−					
SPRING	XP	+	+	+	−			−	+
U.S.GS	XP	+	+	+	−				−
MIPS									

Notes: + signifies extensive capabilities; − indicates limited capabilities; there are no known capabilities when the space is left blank. ? indicates unknown capabilities. XP refers to Microsoft Windows XP compatibility; L is Linux compatibility; Mac refers to the Macintosh OS X operating system.

File sizes for remote sensing data are steadily increasing. Fortunately, hard drive prices per gigabyte continue to decline as spatial and spectral resolution increases. It is becoming very easy to have a folder rapidly grow to dozens of gigabytes while working on a project. Operations such as resolution merges or principal component analyses can create more and more files. It is easy to be surprised by the number images and the size of the files.

20.2.2 CUSTOMIZATION

Depending on the project, it may be desirable to customize the software to enable it to repeat tasks automatically. For example, a script file could be written to download a given satellite scene daily or weekly from the Internet and create an NDVI layer from it. Alternatively, tools may be written to allow people who are not familiar with the RSS to select options from a drop-down menu to perform a basic task.

Some software companies have developed their own languages for this customization. ITT has the IDL language development environment that allows users to create custom data display applications. Leica has the ERDAS macro-language (EML) that provides the necessary tools to create scripts and customize the user interface. Alternatively, users of open source software may inspect and modify the code as they see fit. For example, Opticks, an open source RSS sponsored by Ball Aerospace, has an online forum that invites users to participate in the development of plug-ins. This is true of many of the open source projects. Tools and questions are shared via the Internet. It may also be necessary to have a compiler or software development environment such as Microsoft Visual Studios in the event that an analyst wishes to customize software. The most common programming languages for software development are C, C++ , Visual Basic, JAVA, and Python.

20.2.3 LICENSE TYPE

All computer softwares available can be divided into two broad groups: open source and commercial (or proprietary) softwares. Open source software is free to all users. In addition to the executable file, users may also download the source code to change as they see fit. This may be desirable to those who are experienced programmers. Often it is distributed under the General Public License (gnu.org-http://www.gnu.org/copyleft/gpl.html). Commercial software is developed by companies that wish to profit from the development and sale of software.

Commercial software, although costly, tends to be more polished and often provides a wider range of features, and this is evident from a cursory glance at Table 20.1. Many of the open source RSS packages available perform a small list of functions such as image display and basic image preprocessing. Softwares such as ENVI or Imagine will provide the tools to take many projects from start to finish including georeferencing, spectral enhancement, many types of preprocessing and image classification.

A further benefit of commercial software is the availability of technical support from the manufacturer. Normally, there are not companies that stand behind an open source product. Advocates of open source software would say that technical support is also available for free software in the form of Web forums where users share technical difficulties and their solutions.

It is important to note that although people frequently equate Linux with open source, there are many open source projects that run on Windows just as there are several commercial products that run on Linux. Overall, approximately 60% of the RSS is commercial.

20.2.4 DATA TYPES

There is no shortage of data types in remote sensing. With RADAR, LIDAR, and more traditional imagery, there are hundreds of different file formats. Generally, the predominant image formats are tagged image file format, GeoTIFFs, Erdas IMAGINE. img files, and ESRI GRID files (Chapter 18). LIDAR (Chapter 23) files can present special difficulties, as the native LIDAR format (the .las file) is not accepted by many remote sensing image processing packages. There are only about a half dozen RSS that can open this file type without any add-on extensions. Additionally, the viewing of digital elevation models is an increasingly important part of the remote sensing world as high-resolution elevation data is becoming more commonplace (Chapter 31). The commercial packages ENVI and ERDAS Imagine provide the ability to open a large number of file types.

20.2.5 DISPLAY

RSS should be able to display individual bands and three-band color composites. It should also allow the user to adjust contrast and provide the ability to perform histogram stretches and algebraic and linear band combinations. Simply displaying an image and viewing it can provide the user with some preliminary analysis of a scene. By changing the band viewing combinations, certain aspects of the imagery can be emphasized or enhanced. This can allow a user to make some preliminary judgments about their study area. This is a simple function that nearly all RSSs provide. However, basic software packages may not provide histogram or linear stretches that can be used to brighten dark imagery (Chapter 19).

20.2.6 PREPROCESSING: GEOMETRIC AND RADIOMETRIC CORRECTIONS

Geometric and radiometric transformations are frequently performed on remote sensing data prior to analysis. Images are subject to random and systemic errors in geometry (Lillesand et al, 2008). Systemic errors, such as skew distortion caused by the rotation of the earth, can be corrected by applying formula derived from previously developed models. Random errors require the use of ground control points that are used to reference locations on the imagery. Most RSS packages provide some means of geometric correction. Softwares capable of performing these transforms typically use nearest neighbor, bilinear interpolation, or cubic convolution to process the image (Lillesand et al, 2008). The accuracy of the transform is typically indicated as root mean squared error (Chapter 6). Most of the RSS (both commercial and open source) offer geometric correction or georeferencing capabilities.

Radiometric errors are typically caused by illumination and atmospheric effects (Franklin and Giles, 1995). More broadly, these errors can be introduced by the sensor itself or by atmospheric influence (Jensen, 1995). Sensor errors tend to involve the failure of the sensor to record a pixel or a line of pixels. Typically, these flaws are corrected by determining the average values of surrounding pixels. The two main sources of atmospheric error are the absorption or the scattering of light as it passes through the atmosphere or topographic attenuation, which can be something as simple as the study area being in an area of shadow (Jensen, 1995). The means for correcting atmospheric error are widely varied, and a proper RSS package should provide a number of different tools to correct for haze or striping. Additionally, some software offer tools have been specifically developed for a particular sensor. ENVI, for example, provides a variety of calibration tools for sensors carried by AVHRR, Landsat and Quickbird, among others (Chapter 18).

Image processing can be considered largely as an extension of the field of statistics and probability, and as a result, statistical operations may be performed on the datasets before final analysis. In particular, hyperspectral data may need to be reduced in dimensionality, to reduce the amount of storage space and processing overhead required to perform image processing. There are several ways in which this is done. Principal components analysis (Eklundh and Singh, 1993; Fung and Ledrew, 1987), minimum noise fraction (Boardman et al, 1995), and, more recently, independent component analysis (Lennon et al, 2001) have been used to reduce correlation between bands and reduce noise. Principal component analysis appears to be a fairly popular method of data preprocessing and is fairly easy to find in open source RSS. Independent component analysis is a relatively recent addition to the field of remote sensing, and it is currently offered in the commercial packages of ENVI and Imagine.

20.2.7 CLASSIFICATION

Classification is one of the more important aspects of remote sensing image analysis (Chapter 19). In classification, the software groups pixels together on the basis of their similarity to other pixels using statistics and probability.

Unsupervised classification is the means of classification that requires the smallest amount of user input prior to classification. Here the pixels are divided into an arbitrary number of classes based on their similarity using methods such as K-means or ISODATA. This method requires more work from the user if a useful product is desired as the classes must be grouped into meaningful associations.

Supervised classification aims to reduce the amount of work done by the user after the classification by establishing definitions of classes prior to classification. Considerable attention must be paid to the data that defines the classes. There is much variation in the means by which the software defines these classes (more so than unsupervised classification), so attention must be paid to this area when selecting RSS. The maximum likelihood classifier (Lillesand et al, 2008) is probably the most common of the supervised classifiers. However, recent developments have focused on machine learning in the form of support vector machines. This classification method attempts to create a hyperplane that exists in multidimensional space that defines the boundaries between two classes. Many commercial and several free/open source RSS offer supervised and unsupervised classification. Open source support for this feature tends to be more limited in scope than commercial offerings. While a free RSS may offer maximum likelihood classification, a commercial offering such as ENVI will offer maximum likelihood, Spectral Angle Mapper, and Binary Encoding supervised classifiers (Lillesand et al, 2008).

20.2.7.1 Advanced Classifiers

Advanced classifiers are more frequently offered in commercial RSS. Advanced classifiers generally require more interaction from the user (such as a decision tree classifier) or use more advanced statistical analyses than maximum likelihood classifiers (support vector machines). Commercial classifiers that offer advanced classification include IDRISI Andes, ERDAS Imagine, Definiens Developer (http://www.definiens.com/), and ITT ENVI.

Knowledge-based classifiers are a type of supervised classification that requires outside knowledge based on the reflectance of the classes. With a knowledge-based classifier, the user defines the parameters for each class. This is also the type of classifier required for implementing a classification tree analysis (Lawrence and Wright, 2001). This type of classifier may also be called a decision tree. Both classifiers work in a similar manner in that the user first separates all the pixels into two classes and continues to split those classes into more narrowly defined groups until all the target classes are established. One of the advantages of this classifier is the ability to incorporate multiple datasets into a single classification. For example, an elevation dataset can be incorporated into a project using multispectral data.

Fuzzy classification is comparatively rare among classifiers. Using fuzzy classification methods, the software does not simply assign a single class to each pixel; rather, it assigns each pixel a set of values that represent degrees of membership in each of the classes under consideration. This allows the user the chance to analyze potential sources of classification errors. These classifiers are generally thought to more readily adapt to variations in reflectance among objects in a single class. Two of the more common RSS that support this sort of classification are IDRISI Andes and Definiens Developer.

Although it is not itself a type of image classification, object-based image segmentation is a powerful tool that many remote sensing scientists have utilized to address classification problems. Object-based segmentation combines traditional pixel-based classification with the ability to define classes with additional rules based on the size, the shape, or the texture of a group of pixels. This presents the user with a wide array of tools to apply in defining a class. In this sense, classifiers using object-based methodologies are knowledge-based as well. More of the commercial software packages are offering the ability to segment pixels, however the German company Definiens is most well known for their effective object-based classifiers. Both ERDAS Imagine and ENVI provide image segmentation capabilities that can be incorporated within image classification; however, with these programs, the segmentation process forms an optional step in the classification process. Definiens software has been built around the segmentation process, so classification cannot proceed without image segmentation when using Definiens Developer—this characteristic follows from the

use of so many image object statistics in the application of the classification parameters offered by the software. Through the past several versions of the software,
multiple segmentation models have been developed that allow the user more flexibility in defining both the segment size and shape. Segmentation is optional with both
Imagine and EVNI, which are not as advanced as the segmentation process used by
Definiens Developer.

20.2.8 POST-CLASSIFICATION

Accuracy assessment is a key aspect of image classification (Chapter 21). A software
should offer the user some automated methods for assessing the accuracy of the classification. Most commonly an error or confusion matrix is used that simply provides
a table with details about how many times a pixel is classified correctly and how
many times one class is confused with another. Accuracy is also stated in terms of
overall accuracy, producer's accuracy, user's accuracy, and the Kappa coefficient of
agreement (Jensen, 1995). Accuracy can be defined using reference samples, specifically assigned for the task of assessing accuracy. Some software packages include
utilities to generate random sample locations to identify the placement of such reference samples for accuracy assessment.

It is important to inform potential users of a classified data set of as many details
as possible about the sources of data and the analytical processes used to create a
specific product. Therefore, it is important to consider the ability of a software system to generate metadata files that can capture the sequence of analytical techniques
applied for a specific project. Some software packages provide templates for the creation of metadata files. This capability might include a summary of the image source
files, information about the geographic projection of the dataset, any preprocessing
steps, information about the classification, and how the accuracy assessment was
performed (Jensen, 1995).

20.2.9 CARTOGRAPHIC OUTPUT/UTILITIES

To present a completed product that can be disseminated to a community of nonspecialists, the image analyst must consider the use of processes that present the
results in formats convenient for customers. Data can either be registered to a specific coordinate system or unregistered (Jensen and Jensen, 2002). Unregistered data
will likely be disseminated as a JPEG or PNG file that will be incorporated into a
presentation, report, or article created using a word processor or some image editing software. Alternatively, to create a map product with the output requires a GIS
system such as ESRI's ArcMap or an RSS system that offers cartographic layout
tools. These packages will include templates that allow for the easy addition of the
registration marks used for U.S. Geological Survey (USGS) quadrangle maps, pre-
made north arrows, automated legends, and other map essentials. Such templates
make creating a professional quality map much easier than the alternatives. As more
imagery is being served over the Internet through web map service servers, the ability to access that data via an RSS package is increasingly important (Percivall and
Plesea, 2003; Vatsavai et al, 2006).

20.2.10   GIS Functionality

GIS functionality (Chapter 27) can be an important addition to a remote sensing package. Many projects have integrated GIS and remote sensing capabilities (Blaschke et al, 2000; Câmara et al, 1996; Goodchild, 1994; Percivall and Plesea, 2003). Therefore, the ability to display GIS data and conveniently incorporate it into analysis can simplify many tasks. For example, training and reference data can be easily developed in a GIS environment. RSS systems that offer the ability to directly import shapefiles and use those files for image classification greatly simplify the overall process.

Additionally, while some remote sensing packages have often been somewhat weak in terms of the capabilities they offer to generate attractive maps, these capabilities have mainly fallen within the domain of GIS software. The ability to easily share data between GIS and RSS could be an important factor for the analyst to consider in the purchase of software. GIS integration is becoming more commonplace in RSS. ITT recently announced a partnership with ESRI that will facilitate better exchanges of data between the two platforms (Business Wire, 2008).

20.3   CONCLUSION

The world of RSS is expanding as the amount of remotely sensed imagery continues to grow. There are approximately 90 different software packages that are capable of displaying remote sensing data, of those, one-third are open source products. There are also a wide range of analytical capabilities offered by software packages, ranging from simple image viewing to complete statistical analysis. Our resource demonstrates the capabilities of different open source and commercial softwares for a particular remote sensing project through simple Web-based approaches. This kind of capability empowers the user community to quickly and effectively evaluate and select the software required to meet their needs.

REFERENCES

Blaschke, T., Lang, S., Lorup, E., Strobl, J., and Zeil, P., Object-oriented image processing in an integrated GIS/remote sensing environment and perspectives for environmental applications, *Environmental Information for Planning, Politics and the Public*, 2, 555–570, 2000.

Boardman, J.W., Kruse, F.A., and Green, R.O., Mapping target signatures via partial unmixing of aviris data, *Summaries of the Fifth Annual JPL Airborne Geoscience Workshop*, Pasadena, CA, Vol. 1, pp. 23–26, 1995.

Business Wire, ITT introduces ENVI integration with ESRI® ArcGIS®; latest release supports the integration of imagery to GIS workflows with seamless data exchange between ENVI and ArcGIS desktop from ESRI, *Business Wire* (http://www.lexisnexis.com/), August 2008.

Botkin, D.B., Estes, J.E., MacDonald, R.M., and Wilson, M.V., Studying the Earth's vegetation from space, *BioScience*, 34, 508–514, 1984.

Câmara, G., Cartaxo, R., Souza, M., Freitas, U.M., and Garrido, J., Spring: Integrating remote sensing and gis by object-oriented data modelling, *Computers & Graphics*, 20, 395–403, 1996.

Campbell, J.B., *Introduction to Remote Sensing*, Guilford Press, New York, 2007.

Eklundh, L. and Singh, A., A comparative analysis of standardized and unstandardised principal components analysis in remote sensing, *International Journal of Remote Sensing*, 14, 1359–1370, 1993.

Franklin, S.E. and Giles, P.T., Radiometric processing of aerial and satellite remote-sensing imagery, *Computers and Geosciences*, 21, 413–423, 1995.

Fung, T. and Ledrew, E., Application of principal components analysis to change detection, *Photogrammetric Engineering and Remote Sensing*, 53, 1649–1658, 1987. gnu.org; http://www.gnu.org/copyleft/gpl.html.

GIM International, Product survey on remote sensing processing software, *GIM International*, 18, 2004, 45–49, 2004.

Goodchild, M.F., Integrating GIS and remote sensing for vegetation analysis and modeling: Methodological issues, *Journal of Vegetation Science*, 5, 615–626, 1994.

Jensen, J.R., *Introductory Digital Image Processing: A Remote Sensing Perspective*, Prentice Hall, Upper Saddle River, NJ, 1995.

Jensen, J.R. and Jensen, R.R., Remote sensing digital image processing system hardware and software considerations, *Manual of Geospatial Science and Technology*, Taylor & Francis, London, U.K., 2002.

Landgrebe, D., The development of a spectral-spatial classifier for earth observational data, *Pattern Recognition*, 12, 165–175, 1980.

Lawrence, R. and Wright, A., Rule-based classification systems using classification and regression tree analysis, *Photogrammetric Engineering and Remote Sensing*, 76, 1137–1142, 2001.

Lennon, M., Mercier, G., Mouchot, M.C., and Hubert-Moy, L., Independent component analysis as a tool for the dimensionality reduction and the representation of hyperspectral images, *Geoscience and Remote Sensing Symposium, 2001. IGARSS'01. IEEE 2001 International*, 6, 2001, 2893–2895, 2001.

Lillesand, T.M., Kiefer, R.W., and Chipman, J.W., *Remote Sensing and Image Interpretation*, John Wiley & Sons, New York, 2008.

Percivall, G. and Plesea, L., Web map services (WMS), *Proceedings, The Third International Symposium on Digital Earth* (Milan Konecny, ed.), Brno, Czech Republic, September 21–25, 2003. *Global Mosaic*, p. 99, 2003.

Vatsavai, R.R., Shekar, S., Burk, T.E., and Lime, S., UMN-Mapserver: A high-performance, interoperable, and open source web mapping and geo-spatial analysis system, *Geographic Information Science, Proceedings*, 4197, 400–417, 2006.

21 How to Assess the Accuracy of Maps Generated from Remotely Sensed Data

Russell G. Congalton

CONTENTS

This chapter is part of a manual on geospatial science and technology. Manuals deal with the "how to" component of a discipline. This chapter is devoted to assessing the accuracy of maps created from remotely sensed data. It would be great to list the five steps that the reader must follow in order to conduct such an accuracy assessment. Unfortunately, map accuracy assessment does not follow such a simple recipe. Instead, there are a number of considerations that must be included from the beginning of the project as well as statistical and practical methodologies that

403

must be balanced to achieve a successful assessment. Therefore, instead of a few simple steps, this chapter presents a flowchart (Figure 21.1) to help the reader begin to see all the components that must be considered and planned out to conduct a valid accuracy assessment. The rest of this chapter deals with each of the parts of this flowchart.

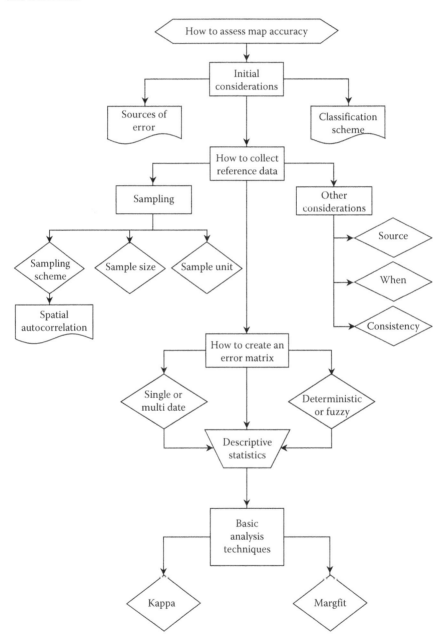

FIGURE 21.1 Flowchart for assessing map accuracy.

21.1 INITIAL CONSIDERATIONS

Assessing the accuracy of maps generated from remotely sensed data is widely accepted as a vital step in most mapping projects. The purpose of quantitative accuracy assessment is the identification and measurement of map errors. There are many reasons to conduct an accuracy assessment including:

- To evaluate just how good the map really is
- To understand the errors in the map (so they can be corrected)
- To provide an overall assessment of the reliability of the map (Gopal and Woodcock, 1994)
- To use the map in a geographic information system (GIS) requires some knowledge of its accuracy
- Because it is a requirement of the project (part of the contract)

The history of accuracy assessment of digital remotely sensed data is relatively short, beginning in the mid-1970s. Before this time, maps derived from analog remotely sensed data (i.e., photointerpretation) were rarely subjected to any kind of quantitative accuracy assessment. Field checking was performed as part of the interpretation process, but no overall map accuracy or other quantitative measure of quality was generally produced. It was only after photointerpretation began to be used as the reference data to evaluate maps generated from digital remotely sensed data that issues concerning the accuracy of the photointerpretation arose, and were systematically debated by those practicing remote sensing.

The history of accuracy assessment can be effectively divided into four developmental stages or epochs. Initially, no real accuracy assessment was performed but rather an "it looks good" mentality prevailed. This approach is typical of any new, emerging technology in which methodology is changing very rapidly and little time is available for much review. Despite the maturing of the technology over the last 30 years, some remote sensing analysts still adhere to the "it looks good" mentality. The second stage is called the epoch of non-site-specific assessment. During this period, overall acreages were compared between ground estimates and the map without regard for location (e.g., Meyer et al., 1975). This second epoch was relatively short-lived and quickly led to the age of site-specific assessments. In a site-specific assessment, actual places on the ground were compared to the same place on the map and a measure of overall accuracy (i.e., percent correct) determined. Finally, the fourth and the current age of accuracy assessment could be called the age of the error matrix. This epoch includes a significant number of analysis techniques, most importantly the kappa analysis. A review of the techniques and considerations of the error matrix age can be found in Congalton (1991) and in great detail in Congalton and Green (2009).

21.1.1 SOURCES OF ERROR

It is critical to realize that errors exist in every component of a mapping project and failure to consider these errors does not minimize their impact. A number of papers have been published in the literature with various representations of possible errors. Congalton and Green (1993) looked specifically at the error or sources of confusion between the map created from the remotely sensed data and the reference data. They include:

1. Registration differences between the reference data and the remotely sensed map classification.
2. Delineation error encountered when the sites chosen for accuracy assessment are digitized.
3. Data entry error when the reference data is entered into the accuracy assessment data base.
4. Error in interpretation and delineation of the reference data (e.g., photointerpretation error).
5. Changes in land cover between the date of the remotely sensed data and the date of the reference data (temporal error). For example, changes due to fires, urban development, or forest harvesting can be misinterpreted as mapping errors.
6. Variation in classification and delineation of the reference data due to inconsistencies in human interpretation of heterogeneous vegetation.
7. Errors in the remotely sensed map classification.
8. Errors in the remotely sensed map delineation.

Lunetta et al. (1991) took a broader approach and looked at a more inclusive list of errors and how they accumulate throughout a remotely sensed mapping project. These errors can best be represented in a diagram (Figure 21.2). It should be noted that errors accumulate throughout a mapping project and can be quite significant.

Given the many potential sources of error and the relative magnitude of each, it is critical that accuracy assessment form an integral component of any mapping project. Failure to incorporate the process into a mapping project dooms the map to misuse and misinterpretation, and the map user to potential trouble.

21.1.2 CLASSIFICATION SCHEME

The classification scheme must be agreed upon early in the project. Failure to have a well-designed and complete scheme determined early in the project dooms the project to significant problems including economic inefficiencies and wasted efforts. Classification schemes or systems are used to categorize remotely sensed map information into a meaningful and useful format. The rules used to label the map must therefore be rigorous and well defined. The best way to assure that the scheme is valid is to define a classification system that is totally exhaustive, mutually exclusive, and hierarchical (Congalton and Green, 1999).

A totally exhaustive classification scheme ensures that *everything* in the image fits into a category; that is, nothing is left unclassified. A mutually exclusive classification scheme further assures that everything in the image fits into one *and only one* category, that is, every object in the image can have only one label. The ability of the classification to meet these important requirements depends on two critical components: (1) a set of labels (e.g., white pine forest, oak forest, nonforest) and (2) a set of rules or definitions (e.g., white pine forest must comprise at least 70% of the stand). Without these components, the image classification would be arbitrary and inconsistent.

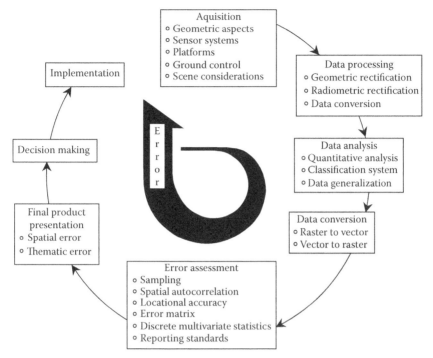

FIGURE 21.2   Sources of error in remotely sensed data. (Reproduced from Lunetta, R. et al., *Photogramm. Eng. Remote Sens.*, 57, 677, 1991. With permission.)

Finally, hierarchical classification schemes—those that can be collapsed from specific categories into more general categories—can be advantageous (Congalton and Green, 1999). For example, if it is discovered that white pine, red pine, and hemlock forests cannot be reliably mapped, these three categories could be collapsed into one general category called coniferous forest.

21.2   HOW TO COLLECT REFERENCE DATA

Reference data collection is the first step in any assessment procedure, and may be the single-most important factor in accuracy assessment, since an assessment will be meaningless if the reference data cannot be trusted. There are a number of important factors that must be considered when collecting reference data. These include the issues related to sampling: (1) sampling scheme, (2) sample size, and (3) sample unit; and related considerations including (1) the source of the reference data, (2) the timing of the collection of reference data, and (3) the instillation of objectivity and consistency within the collection process.

21.2.1   SAMPLING SCHEME

An accuracy assessment very rarely involves a complete census of the map derived from the remotely sensed data (i.e., every pixel in the image), since this is too large a data set to be practical (van Genderen and Lock, 1977; Hay, 1979; Stehman, 1996).

Evaluating the accuracy of a remotely sensed map therefore requires sampling to determine if the mapped categories agree with reference data categories (Rosenfield et al., 1982).

In order to select an appropriate sampling scheme for accuracy assessment, some knowledge of the distribution of the vegetation/land cover classes should be known. For example, if there are cover types that comprise only a small area of the map, simple random sampling and systematic sampling may under-sample these land cover classes, or completely omit them. Stratified random sampling, in which a minimum number of samples are selected from each land cover category, would alleviate this issue.

Ginevan (1979) suggested general guidelines for selecting a sampling scheme for accuracy assessment: (1) the sampling scheme should have a low probability of accepting a map of low accuracy; (2) the sampling scheme should have a high probability of accepting a map of high accuracy; and (3) the sampling scheme should require a minimum number of "ground truth" samples. While the term "ground truth" has often been used to refer to the reference data— this author believes that it is a poor choice. While the reference data should be of higher accuracy than the map it is being used to assess, it is rarely the case that the reference data are perfect or could be considered as a "truth." Therefore, the use of the term "ground truth" is inappropriate and misleading. The use of the term "reference data" is preferred and will be used throughout this chapter.

Stratified random sampling has historically prevailed for assessing the accuracy of remotely sensed maps (van Genderen and Lock, 1977; Jensen, 1996). Stratified sampling has been shown to be useful for adequately sampling important minor categories, whereas simple random sampling or systematic sampling tended to over-sample categories of high frequency and under sample categories of low frequency (van Genderen et al., 1978; Card, 1982). Systematic sampling, however, has been less widely agreed upon. Cochran (1977) found that the properties of the mapped data greatly affected the performance of systematic sampling in relation to that of stratified or simple random sampling. He found that systematic sampling could be extremely precise for some data and much less precise than, say, simple random sampling for other data.

No one method can be recommended globally for accuracy assessment. Rather, the best choice of sampling scheme requires knowledge of the structure and the distribution of characteristics in the map so that cover types will be sampled proportionately and a meaningful assessment produced. One important factor that must be considered when selecting the proper sampling scheme is spatial autocorrelation.

21.2.1.1 Spatial Autocorrelation

Because of sensor resolution, landscape variability, and other factors, remotely sensed data are often spatially autocorrelated (Congalton, 1988). Spatial autocorrelation involves a dependency between neighboring pixels such that a certain quality or characteristic at one location has an effect on that same quality or characteristic at neighboring locations (Cliff and Ord, 1973; Congalton, 1988). Spatial autocorrelation can affect the result of an accuracy assessment if an error in a certain location can be found to positively or negatively influence errors in surrounding locations.

Spatial autocorrelation is closely linked with sampling scheme. For example, if each sample carries some information about its neighborhood, that information may be duplicated in random sampling where some samples are inevitably close, subsequently violating the assumption of sample independence (Curran and Williamson, 1986). For this reason, Curran and Williamson (1986) found that this duplication of information could be minimized by using systematic sampling, "which ensures that neighboring sample points are as far from one another as is possible for a fixed sample size and site area." However, spatial autocorrelation could also be responsible for periodicity in the data, which could dramatically affect the results of a systematic sample (Congalton and Green, 1999). Indeed, Pugh and Congalton's (1997) spatial autocorrelation analysis indicated that systematic sampling should not be used when assessing error in Landsat TM data for New England. Therefore, although we may not be able to completely escape the effects of spatial autocorrelation, its effects must be considered when selecting an appropriate sample size and sampling scheme.

21.2.2 SAMPLE SIZE

An appropriate sample size is essential in order to produce a valid accuracy assessment. In particular, small sample sizes can produce misleading results. That is, small samples that imply that there are no errors can be deceptive since the error-free result may have occurred by chance, when in fact a large portion of the classification was in error (van Genderen and Lock, 1977). For this reason, sample sizes of at least 30 per category have been recommended (van Genderen et al., 1978), while others have concluded that fewer than 50 samples per category are not appropriate (Hay, 1979). Sample sizes can be calculated from equations from the multinomial distribution, ensuring that a sample of appropriate size is obtained (Tortora, 1978). Most recently, sample sizes of 50–100 for each cover type are recommended, so that each category can be assessed individually (Congalton and Green, 1999).

21.2.3 SAMPLE UNIT

In addition to determining effective sample sizes, an appropriate sample unit must be chosen. Historically, the sample unit may be a single pixel, a cluster of pixels, a polygon, or a cluster of polygons. A single pixel has traditionally been a common, but poor choice of sample unit (Congalton and Green, 1999), since it is an arbitrary delineation of the land cover and may have little relation to the actual land cover delineation. Further, it is nearly impossible to align one pixel in an image to the exact same area in the reference data. In other words, the positional accuracy is often insufficient to allow for a single pixel to be used as the sample unit for thematic accuracy assessment. A cluster of pixels (e.g., a 3 × 3 pixel square) is thus often a better choice for the sample unit, since it minimizes registration problems. A good rule of thumb is to choose a sample unit whose area most closely matches the minimum mapping unit of the reference data. For example, if the reference data have been collected in 2 ha minimum mapping units, then an appropriate sample unit may be a 2 ha polygon.

21.2.4 SOURCE OF THE REFERENCE DATA

Reference data have been collected from many sources including photointerpretation; aerial reconnaissance with a helicopter or airplane; video; drive-by surveys; and visiting the area of interest on the ground (Congalton and Biging, 1992). Ground visits, themselves, have ranged from visual calls by walking through the area to detailed measurements of parameters such as species, size class, and crown closure (Congalton and Biging, 1992). Despite these varied approaches, reference data have historically been assumed to be 100% accurate, or at least substantially more accurate than the map (Card, 1982). Congalton and Green (1993), however, have shown that differences between photointerpreted reference data and mapped data are often caused by factors *other* than map/classification error, including photointerpretation error. While photointerpretation remains a widely used means of collecting reference data, Congalton and Biging (1992) demonstrated that at least some ground data should be collected using field measurements. Indeed, they found that field measurements coupled with other visual estimates provided the most efficient ground reference data.

21.2.5 WHEN SHOULD THE REFERENCE DATA BE COLLECTED?

The information used to assess the accuracy of remotely sensed maps should be of the same general vintage as those originally used in map classification. The greater the time period between the media used to label the map and that used in assessing map accuracy, the greater the likelihood that differences will occur due to changes in land cover (from harvesting, urbanization, etc.) rather than misclassification. Therefore, reference data collection should be timed to occur as close as possible to the date of acquisition of the remotely sensed data. In some instances, the reference data collection must be at the exact or near exact time as the imagery is acquired to minimize any change that could occur. In other situations, reference data collected within a year or two of the imagery acquisition would be sufficient.

21.2.6 CONSISTENCY AND OBJECTIVENESS

It is critical that the procedures used for assessing the accuracy of any map derived from remotely sensed data be consistent and objective. Failure to achieve this goal renders the assessment worthless. Certain conditions must be followed to promote consistency and objectiveness. The first of these conditions involves the independence of the reference data from the training data. In addition to using reference data to assess map accuracy, discovering confusion between cover types, and perhaps improving the classification; reference data are also used initially to "train" the classification algorithm. In the past, most assessments of remotely sensed maps were conducted using the same data set used to train the classifier. For example, in a supervised classification, spectral characteristics of known areas were used to train the algorithm that then classified the rest of the image, and these known areas were also used to test the accuracy of the map. This training and testing on the same data set resulted in an invalid accuracy assessment that clearly overestimated classification accuracy (Congalton, 1991). In order for accuracy assessment procedures to be valid and truly representative of the classified map, data used to train the image processing system should not be used for accuracy assessment. These two data sets must be independent of each other.

The second condition that promotes consistency and objectiveness is the development of procedures to ensure that the same methods are used to collect reference data at each sample site. The training of data collection personnel and the use of a field form significantly aid in improving consistency. Field data collectors may have a variety of backgrounds (e.g., forestry, wildlife, planning, ecology, geography) and various skill levels. Training is required to coordinate the effort and make sure that everyone is following the same procedures. A field form or better yet, an automated data logger, can be used to guide the data collection process and guarantee consistency.

Finally, some quality control measures should be implemented for all steps in the collection procedure to minimize errors. For example, some sample sites should be visited by a second data collector and the data between the two collectors compared for consistency. Sample sites that do not agree should be revisited to discover what issues caused the disagreement. A final component to the entire process should be the complete and total documentation of all the methods, procedures, considerations, and decisions made throughout the assessment. Given the complexity of map accuracy assessment, thorough documentation of the entire process is a must.

21.3   HOW TO CREATE AN ERROR MATRIX

Once the reference data has been collected in a valid, well-documented way, then the assessment process continues with the creation of an error matrix (Table 21.1). An error matrix is a square array of numbers, or cells, set out in rows and columns, which expresses the numbers of sample units assigned to each land cover type as compared to what is observed on the ground (see Chapter 6). Columns in the matrix represent the reference data (actual land cover) and rows represent assigned (mapped) land cover types. The major diagonal of the matrix indicates agreement between the reference data and the map. The process of generating an error matrix is summarized in Figure 21.3.

The error matrix is useful for both visualizing image classification results and, perhaps more importantly, for statistically measuring the results. Indeed, an error matrix is the only way to effectively compare two maps *quantitatively*.

21.3.1   Descriptive Statistics

Several descriptive statistics can be computed from the error matrix. A measure of overall accuracy can be calculated by dividing the sum of all the entries in the major diagonal of the matrix by the total number of sample units in the matrix (Story and Congalton, 1986). In the ideal situation, all the nonmajor diagonal elements of the error matrix would be zero, indicating that no area had been misclassified (Congalton et al., 1983). The error matrix also provides accuracies for each land cover category as well as both errors of exclusion (omission errors) and errors of inclusion (commission errors) present in the classification (Card, 1982; Congalton, 1991; Congalton and Green, 1999).

TABLE 21.1

Sample Error Matrix

			Reference data					Land cover categories
		D	C	AG	SB	Row total		
Map	D	63	3	18	22	106		D = deciduous
	C	2	76	3	8	89		C = conifer
	AG	0	6	85	19	110		AG = agriculture
	SB	1	5	2	87	95		SB = shrub
	Column total	66	90	108	136	400		

Overall accuracy = (63 + 76 + 85 + 87)/400 = 311/400 = **78%**

Producer's accuracy	User's accuracy
D = 63/66 = 95%	D = 63/106 = 59%
C = 76/90 = 84%	C = 76/89 = 85%
AG = 85/108 = 79%	AG = 85/110 = 77%
SB = 87/136 = 64%	SB = 87/95 = 92%

Omission errors can be calculated by dividing the total number of correctly classified sample units in a category by the total number of sample units in that category from the reference data (the column total) (Story and Congalton, 1986; Congalton, 1991). This measure of omission error may be referred to as "producer's accuracy," because from this measurement the producer of the classification will know how well a certain land cover category was classified (Congalton, 1991). For example, the producer may be interested in knowing how often "deciduous" was in fact classified as deciduous (and not, say, conifer). To determine this, the 63 correctly classified deciduous samples (see Table 21.1) would be divided by the total 66 units of deciduous from the reference data, for a producer's accuracy of 95%. In other words, deciduous forest was correctly identified as deciduous 95% of the time.

Commission errors, on the other hand, are estimated by dividing the number of correctly classified sample units for a category by the total number of sample units that were classified in that category (Story and Congalton, 1986; Congalton, 1991; Congalton and Green, 1999). This is also called "user's accuracy," indicating for the user of the map the probability that a pixel classified on the map actually represents that category on the ground (Story and Congalton, 1986; Congalton and

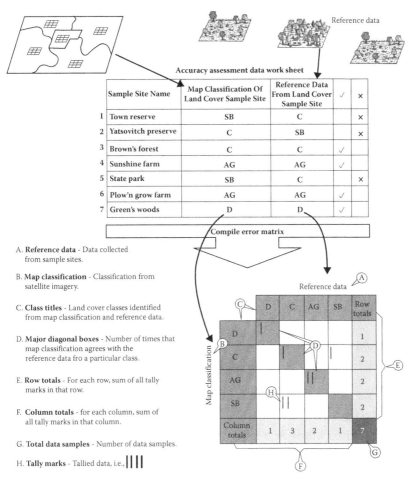

FIGURE 21.3 Diagram of error matrix generation process.

Green, 1999). So, while the producer's accuracy for the deciduous category is 95%, the user's accuracy is only 59%. That is, only 59% of the areas mapped as deciduous are actually deciduous on the ground. However, because each omission from the correct category is a commission to the wrong category, it is critical that both producer's and user's accuracies are considered, since reporting only one value can be misleading (Story and Congalton, 1986; Congalton and Green, 1999).

21.3.2 Single-Date versus Multi-Date Thematic Accuracy

Single-date thematic accuracy refers to the accuracy of a mapped land cover category at a particular time compared to what was actually on the ground at that time. The methods described for an error matrix up to this point apply to a single-date error matrix.

Table 21.2 presents a comparison between a single-date error matrix and the corresponding error matrix for change detection or multi-date assessment (Congalton and Green, 1999). The table shows a single date error matrix for three vegetation/land cover categories (A, B, and C). The matrix is of dimension 3 × 3. As we saw in the previous section, one axis of the matrix represents the three categories as derived from the remotely sensed classification, the other axis shows the three categories identified from the reference data, and the major diagonal of this matrix indicates correct classification.

Table 21.2 also shows a change detection error matrix generated for the same three vegetation/land cover categories (A, B, and C). Note, however, that the matrix is no longer of dimension 3 × 3 but rather 9 × 9. This is because we are no longer looking at a single classification but rather a change between two different classifications generated at different times (i.e., a multi-date assessment). Note that in a typical single-date error matrix there is only one row and column for each map category. However, in a multi-date assessment the error matrix is the size of the number of categories squared. Therefore, the question of interest is, "What category was this area at time 1 and what is it at time 2?" The answer has nine possible outcomes for each dimension of the matrix (A at time 1 and A at time 2, A at time 1 and B at

---

TABLE 21.2

A Comparison between a Single-Date Error Matrix and a Change Detection Error Matrix for the Same Land Cover Categories

time 2, A at time 1 and C at time 2, ..., C at time 1 and C at time 2) all of which are indicated in the error matrix. It is then important to note what the remotely sensed data said about the change and compare it to what the reference data indicate. This comparison uses the exact same logic as for the single classification error matrix, it is just complicated by the two time periods (i.e., the change). Again, the major diagonal indicates correct classification while the off-diagonal elements indicate the errors or confusion.

The change detection error matrix can also be simplified into a no change/change error matrix. The no change/change error matrix can be formulated by summing the cells in the four appropriate sections of the change detection error matrix (Table 21.2). For example, to get the number of areas that both the classification and the reference data correctly determined that no change had occurred between two dates, you would simply add together all the areas in the upper left box (the areas that did not change in either the classification or reference data). You would proceed to the upper right box to find the areas that the classification detected no change and the reference data considered change. From this no change/change error matrix, analysts can easily determine if a low accuracy was due to a poor change detection technique, misclassification, or both.

The change detection thematic accuracy presents even more difficulties and challenges than single-date thematic accuracy. Many questions arise such as how does one obtain information on reference data for images that were taken in the past? Likewise, how can one sample enough areas that will change in the future to have a statistically valid assessment? Until recently, most studies on change detection did not present any quantitative results with their work.

National Oceanic and Atmospheric Administration (NOAA), as part of its Coastal Change Analysis Program (C-CAP), funded a task force that investigated and proposed methods for assessing the accuracy of change detection. This task force wrote a monograph to recommend appropriate techniques and suggest future work (Khorram et al., 1999). The change detection error matrix (Table 21.2) was recommended and sampling techniques using unequal probability sampling were developed. One of the biggest problems with conducting a change detection accuracy assessment is to insure sampling in the areas that have actually changed, since change is typically such a rare event (i.e., most of the mapped area has not changed between the two dates). Another issue is finding accurate reference data for some date in the past so that the map could be assessed for that date. The monograph is recommended to the interested reader for a valuable review of many of these complex issues.

21.3.3 DETERMINISTIC VERSUS FUZZY ERROR MATRIX

All of the discussion regarding the error matrix to this point in the chapter has referred to the traditional or deterministic error matrix. One of the assumptions of the deterministic error matrix is that an accuracy assessment sample site can have only one label. This requirement was part of the characteristics of a good classification scheme (i.e., the mutually exclusive criterion) discussed earlier in this chapter. However, classification scheme rules often impose discrete boundaries on continuous conditions in nature. In many situations, the classification scheme represents artificial breaks along a continuum of land-cover (i.e., measures of forest canopy density or mixtures of coniferous versus deciduous forest). In such cases, observer variability is very difficult to control and, while unavoidable, can have profound effects on the results (Congalton and Green, 1999). One solution to this issue is to use a fuzzy accuracy assessment approach to compensate for differences between reference and map data that are not caused by map error, but rather by variation in interpretation (Gopal and Woodcock, 1994; Green and Congalton, 2004).

In the situation where significant variation exists in our ability to label the reference data, the use of a fuzzy error matrix is warranted. In generating a fuzzy error matrix, one reference label is still determined to be the most appropriate, however other labels can be considered acceptable and the rest considered unacceptable. For example, a forest that is 90% conifer and 10% deciduous should be labeled as conifer in the reference data. However, a forest that is 55% conifer and 45% deciduous should still be called conifer, but may acceptably be called deciduous. In other cases, depending on the source of the reference data, a young stand of trees may acceptably be labeled as shrubs because of their height. Therefore, in the case of the fuzzy error matrix, there can be more than one acceptable answer.

The fuzzy error matrix is generated in much the same way as the deterministic matrix (Green and Congalton, 2004). The reference label deemed to be the correct one is compared with the map. The major diagonal still represents agreement between the map and the reference data. However, now there are two values in each off-diagonal cells of the matrix. The first value in the off-diagonal represents the acceptable reference calls while the second value represents the unacceptable calls. The error matrix then allows for the computation of both the traditional descriptive statistics using only the correct values as indicated by the major diagonal, but also allows for the calculation of fuzzy descriptive statistics by adding the major diagonal value to the first number in the off-diagonal cells to compute overall, producer's, and user's accuracies. Table 21.3 presents an example that shows both the deterministic and fuzzy descriptive statistics being computed from a single matrix. In this matrix, the deterministic overall accuracy is 54.8% while the fuzzy overall accuracy is 77.1%. A quick look at the matrix reveals that some map classes had acceptable confusion between them due to the variation in the classification scheme and the ability of the analyst to interpret these classes (e.g., grassland versus other agriculture, deciduous versus evergreen, and grassland versus barren/ sparse vegetation). The computation of individual class accuracies (producer's and user's accuracy) is performed in the usual way for the deterministic approach and as described earlier for the fuzzy approach. For example, computing the producer's accuracy for the deciduous forest class is performed by dividing the major diagonal value (54) by the sum of the deciduous column (64) and obtaining 84.4%. The fuzzy producer's accuracy is computed by summing the major diagonal with all the values in the first position of the off-diagonal cells in the deciduous column (54 + 4 + 2 = 60) and dividing by the sum of the deciduous column (64) and obtaining 93.8%. In this way, acceptable variation in the accuracy assessment can be accounted for while still maintaining all the power of the error matrix approach to accuracy assessment.

TABLE 21.3
An Example of a Deterministic and Fuzzy Error Matrix

	Reference Data											User's Accuracy			
Map	Deciduous Forest	Evergreen Forest	Shrub/ Scrub	Grassland	Barren/ Sparse Veg	Urban	Ice/Snow	Agri-culture Other	Agri-culture Rice	Water	Cloud/ Shadow	Deter-ministic Totals	Percent Deter-ministic	Fuzzy Totals	Percent Fuzzy
Deciduous forest	54	24.7	0.1	0.3	0.0	0.1	0.0	0.10	0.0	0.18	0.0	54/118	45.8%	78/118	66.1%
Evergreen forest	4.1	22	0.1	0.0	0.0	0.0	0.0	0.1	0.0	0.3	0.0	22/32	68.8%	26/32	81.3%
Shrub/scrub	2.0	0.1	16	7.2	0.0	0.0	0.0	2.2	0.0	0.0	0.0	16/32	50.0%	27/32	84.3%
Grassland	0.1	0.0	5.1	24	0	0.0	0.0	3.0	0.0	0.0	0.0	24/34	70.5%	32/34	94.1%
Barren/sparse veg	0.0	0.0	0.2	0.0	0	0.0	0.0	0.1	0.0	0.0	0.0	0/3	0.0%	0/3	0.0%
Urban	0.0	0.0	0.0	0.0	0.0	21	0.0	2.0	0.0	0.0	0.0	21/33	91.3%	23/33	100.0%
Ice/snow	0.0	0.0	0.0	0.0	0.0	0.0	0	0.0	0.0	0.0	0.0	NA	NA	NA	NA
Agriculture other	0.2	0.1	7.12	18.6	0.0	2.0	0.0	27	0.0	1.2	0.0	27/78	34.6%	55/78	70.5%
Agriculture rice	0.0	0.0	0.0	0.0	0.0	0.0	0.0	0.0	0	0.0	0.0	NA	NA	NA	NA
Water	0.0	0.0	0.0	0.0	0.0	0.0	0.0	0.0	0.0	25	0.0	25/25	100.0%	25/25	100.0%
Cloud/shadow	0.0	0.0	0.0	0.0	0.0	0.0	0.0	0.0	0.0	0.0	0	NA	NA	NA	NA
Producers' accuracy															
Deterministic totals	54/64	22/55	16/45	24/60	NA	21/24	NA	27/48	NA	25/49	NA				
Percent deterministic	84.4%	40.0%	35.5%	40.0%	NA	87.5%	NA	56.3%	NA	51.0%	NA				
Fuzzy totals	60/64	46/55	28/45	49/60	NA	23/24	NA	34/48	NA	26/49	NA				
Percent fuzzy	93.8%	83.6%	62.2%	81.7%	NA	95.8%	NA	70.8%	NA	53.1%	NA				

Overall Accuracies

Deterministic	Fuzzy
189/345 54.8%	266/345 77.1%

21.4 BASIC ANALYSIS TECHNIQUES

Once an error matrix has been properly generated, it can be used as a starting point to calculate various measures of accuracy in addition to overall, producer's, and user's accuracy.

21.4.1 KAPPA

A discrete multivariate technique called Kappa (Bishop et al., 1975) can be used to statistically determine if (1) the remotely sensed classification is better than a random classification and (2) two or more error matrices are significantly different from each other. Kappa (κ) (Cohen, 1960) is estimated by $\hat{\kappa}$ ("κ hat," commonly designated as "KHAT") (Congalton and Mead, 1983) by the following relationship:

$$\text{KHAT} = \frac{p_o - p_c}{1 - p_c} \tag{21.1}$$

where
   p_o is the actual agreement, or the overall proportion of the map correctly classified
   p_c is the "chance" or random agreement

The actual agreement is the total number of correctly mapped samples (i.e., the summation of the major diagonal in the error matrix). The chance agreement is calculated by the summation over all categories of the proportion of samples in each category from the map data multiplied by the proportion of samples in each category from the reference data. In this way, it incorporates all off-diagonal cell values in the error matrix.

KHAT is thus a measure of the *actual* agreement of the cell values minus the *chance* (i.e., random) agreement (Congalton and Mead, 1983; Rosenfield and Fitzpatrick-Lins, 1986). The KHAT value can therefore be used to determine whether the results in the error matrix are significantly better than a random result (Congalton, 1991). The KHAT value inherently includes more information than the overall accuracy measure since it indirectly incorporates the error (off-diagonal elements) from the error matrix (Rosenfield and Fitzpatrick-Lins, 1986; Chuvieco and Congalton, 1988).

In addition, confidence limits can be calculated for the KHAT statistic, which allows for an evaluation of significant differences between KHAT values (Aronoff, 1982; Congalton and Green, 1999). Because the Kappa analysis is based on the standard normal deviate and the fact that although remotely sensed data are discrete, the KHAT statistic is asymptotically normally distributed, confidence intervals can be calculated using the approximate large sample variance (Congalton and Green, 1999). To test if a classification is better than a random classification, the Z test is performed, as follows:

$$Z = \frac{\text{KHAT}}{\sqrt{\hat{\text{var}}(\text{KHAT})}} \tag{21.2}$$

where Z is standardized and normally distributed. To test if two error matrices are significantly different from one another, the following Z test is used:

$$Z = \frac{|KHAT_1 - KHAT_2|}{\sqrt{v\hat{a}r(KHAT_1) + v\hat{a}r(KHAT_2)}} \qquad (21.3)$$

where
  $KHAT_1$ represents the KHAT statistic from one error matrix
  $KHAT_2$ is the statistic from the other matrix

The Kappa analysis thus brings to accuracy assessment the power and efficiency of a parametric model with the ability to detect smaller differences than a nonparametric alternative (Rosenfield, 1982).

Further, Kappa can be used to measure accuracy for individual categories by summing respective numerators and denominators of KHAT separately over all categories (Rosenfield and Fitzpatrick-Lins, 1986; Congalton and Green, 1999). This results in conditional agreement, or conditional kappa.

21.4.2 MARGFIT

The analysis of the error matrix can be taken yet another step further by normalizing the cell values. For example, in an iterative process called *margfit*, the matrix is normalized to one (Table 21.4). Because the cell values in each row and column in the matrix are forced to sum to one, each cell value becomes a proportion of one, which can easily be multiplied by 100 to obtain percentages. Consequently, there is no need for producer's and user's accuracies since the cell values along the major diagonal

---

TABLE 21.4
Example of Normalized Matrix

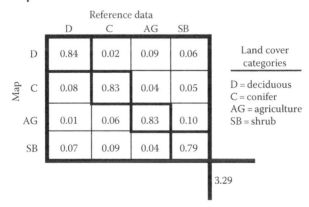

Normalized accuracy = 0.84 + 0.83 + 0.83 + 0.79 = 3.29/4.0 = **82%**

---

represent the proportions correctly mapped. Congalton et al. (1983) argued that the normalized accuracy is a more inclusive measure of accuracy than either KHAT or overall accuracy because it *directly* includes the information in the off-diagonal element of the error matrix. Because each row and column sums to the same value, different cell values (e.g., different forest cover classes) within an error matrix and among different error matrices can be compared despite differences in sampling scheme (Aronoff, 1982). Aronoff (1982) cautioned, however, that normalizing can tend to hide the data on sample size. This can be important, since, for example, an accuracy of four out of five sample points correct has less confidence than 40 out of 50 correct.

21.5   CONCLUSIONS

Assessing the accuracy of maps generated from remotely sensed data is not a simple task. However, these maps are used to make decisions that may have a global impact—decisions regarding land use/tenure, climate change effects, resource treatments, water quality, ecosystem health considerations, wildlife habitat, sustainability, and a multitude of other applications and issues. In order to make any effective, intelligent decisions, however, the data must be accurate and reliable. For this reason, accuracy assessment will continue to be an integral component of any map generated from remotely sensed data. This chapter has outlined the various issues that must be considered and has presented an overview of the methodologies that must be applied when performing any map accuracy assessment.

REFERENCES

Aronoff, S., Classification accuracy: A user approach, *Photogrammetric Engineering and Remote Sensing*, 48(8):1299–1307, 1982.

Bishop, Y., S. Fienberg, and P. Holland, *Discrete Multivariate Analysis—Theory and Practice*, MIT Press, Cambridge, MA, 575p, 1975.

Card, D., Using known map category marginal frequencies to improve estimates of thematic map accuracy, *Photogrammetric Engineering and Remote Sensing*, 48(3):431–439, 1982.

Chuvieco, E. and R. G. Congalton, 1988, Using cluster analysis to improve the selection of training statistics in classifying remotely sensed data, *Photogrammetric Engineering and Remote Sensing*, 54(9):1275–1281, 1988.

Cliff, A. D. and J. K. Ord, *Spatial Autocorrelation*, Pion Limited, London, U.K., 178p, 1973.

Cochran, W. G., *Sampling Techniques* (3rd edn.), John Wiley & Sons, New York, 428p, 1977.

Cohen, J., A Coefficient of agreement for nominal scale, *Educational and Psychological Measurement*, 20:37–46, 1960.

Congalton, R. G., Using spatial autocorrelation analysis to explore errors in maps generated from remotely sensed data, *Photogrammetric Engineering and Remote Sensing*, 54:587–592, 1988.

Congalton, R. G., A review of assessing the accuracy of classifications of remotely sensed data, *Remote Sensing of Environment*, 37:35–46, 1991.

Congalton, R. G. and G. S. Biging, A pilot study evaluating ground reference data collection efforts for use in forest inventory, *Photogrammetric Engineering and Remote Sensing*, 58(12):1669–1671, 1992.

Congalton, R. G. and K. Green, A practical look at the sources of confusion in error matrix generation, *Photogrammetric Engineering and Remote Sensing*, 59:641–644, 1993.

Congalton, R. G. and K. Green, *Assessing the Accuracy of Remotely Sensed Data: Principles and Practices* (2nd edn.), CRC Press, Boca Raton, FL, 183p, 2009.

Congalton, R. G. and R. A. Mead, A quantitative method to test for consistency and correctness in photo-interpretation, *Photogrammetric Engineering and Remote Sensing*, 49:69–74, 1983.

Congalton, R. G., R. G. Oderwald, and R. A. Mead, Assessing landsat classification accuracy using discrete multivariate statistical techniques, *Photogrammetric Engineering and Remote Sensing*, 49:1671–1678, 1983.

Curran, P. J. and H. D. Williamson, Sample size for ground and remotely sensed data, *Remote Sensing of Environment*, 20:31–41, 1986.

van Genderen, J. L. and B. F. Lock, Testing land use map accuracy, *Photogrammetric Engineering and Remote Sensing*, 43:1135–1137, 1977.

van Genderen, J. L, B. F. Lock, and P. A. Vass, Remote sensing: Statistical testing of thematic map accuracy, *Remote Sensing of Environment*, 7:3–14, 1978.

Ginevan, M. E., Testing land-use map accuracy: Another look, *Photogrammetric Engineering and Remote Sensing*, 45:1371–1377, 1979.

Gopal S. and C. Woodcock, Theory and methods for accuracy assessment of thematic maps using fuzzy sets, *Photogrammetric Engineering and Remote Sensing*, 60:181–188, 1994.

Green, K and R. G. Congalton, *An Error Matrix Approach to Fuzzy Accuracy Assessment: The NIMA Geocover Project*. A peer-reviewed chapter in: Lunetta, R. S. and J. G. Lyon (Eds.), *Remote Sensing and GIS Accuracy Assessment*, CRC Press, Boca Raton, FL, 304p, 2004.

Hay, A. M., Sampling designs to test land-use map accuracy, *Photogrammetric Engineering and Remote Sensing*, 45:529–533, 1979.

Jensen, J. R., *Introductory Digital Image Processing: A Remote Sensing Perspective*, Prentice Hall, Upper Saddle River, NJ, 316p, 1996.

Khorram, S., G. Biging, N. Chrisman, D. Colby, R. G. Congalton, J. Dobson, R. Ferguson, M. Goodchild, J. Jensen, and T. Mace, Accuracy assessment of remote sensing-derived change detection, Monograph, American Society for Photogrammetry and Remote Sensing, Bethesda, MD, 64p, 1999.

Lunetta, R., R. G. Congalton, L. Fenstermaker, J. Jensen, K. McGwire, and L. Tinney, Remote sensing and geographic information system data integration: Error sources and research issues, *Photogrammetric Engineering and Remote Sensing*, 57:677–687, 1991.

Meyer, M., J, Brass, B. Gerbig, and F. Batson, ERTS data applications to surface resource surveys of potential coal production lands in southeast Montana, IARSL Research Report 75-1 Final Report, University of Minnesota, Minneapolis, MN, 24p, 1975.

Pugh, S. and R. G. Congalton, Applying spatial autocorrelation analysis to evaluate error in new England forest cover type maps derived from thematic mapper data, *Proceedings of the Sixty-Third ASPRS Annual Meeting*, Seattle, WA. American Society of Photogrammetry and Remote Sensing, Vol. 3, pp. 648–657, 1997.

Rosenfield, G. H., The analysis of areal data in thematic mapping experiments, *Photogrammetric Engineering and Remote Sensing*, 48(9):1455–1462, 1982.

Rosenfield, G. and K. Fitzpatrick-Lins, A coefficient of agreement as a measure of thematic classification accuracy, *Photogrammetric Engineering and Remote Sensing*, 52:223–227, 1986.

Rosenfield, G. H., K. Fitzpatrick-Lins, and H. Ling, sampling for thematic map accuracy testing, *Photogrammetric Engineering and Remote Sensing*, 48:131–137, 1982.

Stehman, S., Estimating the kappa coefficient and its variance under stratified random sampling, *Photogrammetric Engineering and Remote Sensing*, 62(4):401–402, 1996.

Story, M. and R. G. Congalton, Accuracy assessment: A user's perspective, *Photogrammetric Engineering and Remote Sensing*, 52:397–399, 1986.

Tortora, R., A note on sample size estimation for multinomial populations, *The American Statistician*, 32:100–102, 1978.

22 Emerging Markets for Satellite and Aerial Imagery

Joel Campbell, Travis E. Hardy, and Robert Chris Barnard

CONTENTS

As the demand for geographic data continues to grow in all segments of society, the remote sensing industry continues to grow to meet these challenges. This chapter focuses on the driving forces behind the growth of outside the traditional markets for aerial imagery and how the industry has responded. In addition, we look at future trends and their impact on the practice of remote sensing.

The fuel for this growth comes in many different varieties. National governments increasingly rely on commercial remote sensing platforms to provide imagery for large mapping projects, both civil and defense in nature. In many parts of the world, access to high-resolution imagery is the best, if not only, source from which to create national scale maps. These maps support a wide range of applications, including agricultural resource management, urban and regional planning, and infrastructure planning and management. Further, defense organizations use these forms of data to develop common understandings of national geographies in order to defend borders, manage conflict zones, and better understand the physical geographies of national territories of both allies and adversaries. In many cases, these images have been acquired using satellite platforms, which have proven to be the most cost-effective systems for the acquisition of such imagery. Aerial remote sensing, on the other hand, has evolved into a highly specialized domain with a focus on urban areas, engineering project support, and related infrastructure applications.

Other factors effecting growth include the business adoption of geospatial data and analysis, as well as the increase in consumer awareness of geographic data and its usefulness in our everyday lives. Every day that passes reveals additional examples of business adoptions—trucking companies use geographic systems to route and track their vehicle; retail establishments provide locations and directions to store fronts; insurance companies model risk in areas prone to natural disasters; election campaigns segment demographics to target potential voters (to name only a few of many examples).

Likewise, there is a rapid growth in the number of Internet sites that provide consumers driving directions, locations of real estate for sale, or simply locations of friends in a mobile social networking environment. These applications, combined with the use of imagery and maps by media to tell stories, have created an explosion of interest in consumer-ready geographic information. As geography becomes the heart or the centerpiece of our lives and decision-making processes, the need for geospatial information that is accurate, up to date, and complete becomes paramount.

Recent market surveys have quantified the speed of growth for this industry, and actions of companies to meet the demands both in the United States and around the globe. The remote sensing marketplace in total was estimated to be $7.3 billion in 2007 and growing at a compound annual growth rate (CGAR) of 6.3% per year globally. A more narrow focus on very high-resolution imagery (collected at a detail finer than 1 m in resolution) reveals a global market of nearly $2 billion in 2007. It is estimated, based on the variety of platforms under development and the overall growth in demand, that this segment of the market will reach $3.2 billion by 2012 (BCC Report, February, 2007). The following sections examine several dimensions underlying this growth.

22.1 PUBLIC POLICY DRIVERS OF REMOTE SENSING TECHNOLOGY

In the United States, federal policy regarding remote sensing has consistently attempted to promote the development of a commercial capability for satellite remote sensing. Although the government determined that the management of some satellite systems (Landsat) benefited from governmental support, the policy has attempted to keep step with technology by licensing progressively higher-resolution platforms. Here, we briefly summarize those significant policies that form milestones marking key points in the history of the U.S. policy relating to commercial remote sensing.

Presidential Directive 37, issued in June 1978 by President Carter, limited the resolution of U.S. civilian remote sensing satellites to spatial resolutions coarser than 10 m. The Land Remote Sensing Commercialization Act of 1984 (P.L. 98-965) set out terms for transferring the government-owned Landsat satellite program to the private sector. From the onset, this Landsat commercialization policy proved to be problematic. Government studies had pointed to the need for as much as $500 million in subsidies if the program was to remain viable and to continue to develop new satellite platforms. The only company willing to pursue the commercial operations of Landsat at the time was the earth observation satellite company (EOSAT) Corporation, a joint venture between the satellite manufacturer Hughes and the electronics firm RCA. This firm was to operate Landsats 4 and 5, which had reached orbit in 1982 and 1984, respectively; build two new satellites, Landsats 6 and 7; and hold exclusive rights to market photos and other data. It soon became apparent that there was insufficient demand to support the business without considerable subsidy and that, at the time, there was no governmental champion for the program. EOSAT quadrupled the price of its imagery and collected fees from overseas receiving stations but never brought in sufficient funds so, according to Heppenheimer (2008), "it limped from one financial crisis to the next." EOSAT continued to struggle in the face of competition from international ventures that were willing to subsidize their commercial ventures. Faced with the continued resistance by the U.S. government to provide subsidies, a new law, the Land Remote Sensing Policy Act of 1992 (P.L. 102-555), repealed the 1984 law and returned Landsat to the government. Decisions during 1994 sorted out the responsibilities. EOSAT continued with difficulty to operate Landsats 4 and 5 and retained the right to sell their photos. NASA took over the responsibility for building Landsat 7, with national oceanic and atmospheric administration (NOAA) agreeing to operate this spacecraft in orbit. The U.S. Geological Survey (USGS) took over the task of marketing its data, while maintaining an archive of photos for sale to customers. EOSAT was ultimately acquired by Space Imaging in 1996.

The Land Remote Sensing Policy Act of 1992 (P.L. 102-555) also established procedures for licensing remote sensing operators (P.L. 102-555, Sections 2 and 201). The act also amended the 1978 limitation on resolution increasing allowable resolution to 0.82 m. Presidential Decision Directive 23 (PDD-23) issued on March 24, 1994, by President Clinton set forth guidelines for anyone applying for a remote sensing operating license under the 1992 remote sensing act. The guidelines cover any activity that involves foreign access to images, technology, and systems. One feature of this directive established that the federal government would have the right to limit a licensee's observations for national security reasons (a restriction known as "shutter control").

In the Commercial Space Act Public Law 105-303, October 28, 1998, the Congress further declares that free and competitive markets create the most efficient conditions for promoting economic development, and should therefore govern the economic development of earth's orbital space. Though this policy was primarily designed to support the International Space Station, NASA used it to support commercial remote sensing requirements through its Earth Science Enterprise.

An important example of the resistance to change in U.S. space policy is the controversy surrounding the establishment of a National Applications Office (NAO) to examine the use of intelligence, surveillance, and reconnaissance satellites for civilian and law enforcement applications. The concern about the improper use of reconnaissance satellites began in the mid-1970s, due to past efforts of the U.S. Central Intelligence Agency (CIA) and other agencies to monitor U.S. persons. The 1975 Rockefeller Commission (the commission on CIA activities within the United States) was asked to review the issues involved in domestic overhead photography due to concerns about past efforts of the CIA and other agencies to monitor U.S. citizens. The commission found that the CIA, then in charge of most satellite strategic reconnaissance efforts, had provided photography for civil activities, such as mapping, assessing natural disasters, conducting route surveys for the Alaska pipeline, national forest inventories, determining the extent of snow cover in the Sierras to forecast the extent of runoff, and detecting crop blight in the plains states. The commission noted that it was possible that a small percentage of aerial photography was being used for law enforcement and was "outside the scope of proper CIA activity." "The Commission believes, however, that the legislators, when they prohibited the CIA from engaging in law enforcement activities in the 1947 enactment of the National Security Act, could not have contemplated the systems presently in use." In 1976, in response to the Rockefeller Commission's conclusions and other concerns, the Civil Applications Committee (CAC) was established in 1976 to serve as an interface through which the needs of civilian agencies for satellite data could be reviewed and prioritized. The CAC was created by a joint memorandum signed by the Assistant to the President for National Security Affairs, the Director of the Office of Management and Budget, and the Director of Central Intelligence (DCI).

The 9/11 attacks led to a general reconsideration of the relationships between law enforcement and intelligence agencies, and in 2002, led to the establishment of the Department of Homeland Security (DHS), which has both law enforcement and intelligence responsibilities. In December 2004, the Intelligence Reform and Terrorism Prevention Act of 2004 (P.L. 108-458) established the position of Director of National Intelligence (DNI), in part as a replacement for the DCI, to coordinate intelligence activities and their relationship with law enforcement. A further review of the potential contribution of satellite surveillance to the civil sector was undertaken in 2005 by an Independent Study Group (ISG) established by the office of the DNI. The ISG noted the opportunities for the domestic applications of satellite reconnaissance, but argued that not enough was being done to take advantage of them. The drafters of the ISG sought to establish a venue through which information from intelligence satellites could be shared with the DHS and law enforcement agencies. They argued: "The root of the problem is a lack of a clearly articulated comprehensive policy on the use of IC [intelligence community] capabilities for domestic needs." The commission suggested that the DHS, a member of the IC, serve as the intermediary between the IC and the state, local, and tribal law enforcement agencies serving as the executive agent of what it termed a new domestic applications office (DAO). In May 2007, the DNI designated the DHS as the executive agent and the functional manager of what was designated as an NAO. There was, however, no public notice of the establishment of the new office at that time. At the time this chapter was written, the Congress has precluded any funds in the Consolidated Appropriations Act, 2008 (H.R. 2764, P.L. 110-161), from being used to "commence operations of the National Applications Office ... until the Secretary [of the Department of Homeland Security] certifies that these programs comply with all existing laws, including all applicable privacy and civil liberties standards, and that certification is reviewed by the Government Accountability Office" (HR 2764, Section 525).

22.2 PERSONAL PRIVACY AND REMOTE SENSING

The potential for remote sensing activities to threaten personal privacy has long attracted the attention of those knowledgeable of its capabilities to capture detailed images of what the public perceives to be private space. For example, Slonecker et al. (1998) noted "… remote sensing technology could soon develop the capability to generate and deliver a level of information detail that could violate common societal perceptions of individual privacy, and a number of direct legal and ethical consequences could result" (p. 589). Indeed, as time has passed since this statement was made, increases in the spatial detail captured by imagery, increases in areas imaged, and the wider distribution of such imagery to the public (especially through the Internet) escalate the potential for these concerns to materialize.

In the United States, court decisions have, for the most part, upheld the legality of conducting what might be considered to be normal acquisitions of remotely sensed imagery. (However, it should be noted that the collection of imagery for law enforcement purposes may have a more uncertain status.) On May 19, 1986, the Supreme Court determined that the U.S. Environmental Protection Agency (EPA) was within their rights to obtain the aerial photography of a Dow Chemical production facility after requests to gain entrance to the facility were refused by their management. Dow had sued the EPA on the grounds that the aerial survey violated its right against unlawful search. On June 11, 2001, in a decision on *Kyllo v. the U.S.*, the Supreme Court upheld a decision by the Ninth Circuit that the use of a thermal imaging system that was used to detect heat from lamps that were used to grow marijuana constituted a search and thus required a warrant.

This trend was upheld in a civil case brought by the entertainer Barbra Streisand against Ken Adelman, a film archivist who had obtained thousands of aerial photos of the California coastline. Ms. Streisand's estate appears in these photos. On December 3, 2003, the Los Angeles Superior Court issued an opinion holding that Ms. Streisand abused the judicial process by filing a lawsuit against Ken Adelman and his Internet service provider layer42.net, and pictopia.com. The court also firmly rejected Streisand's request for an injunction to force the removal from Adelman's Web site (www.californiacoastline.org) of a panoramic photographic frame that happens to include an image of her estate.

It is counterintuitive that the events of 9/11 have had no observable effect on U.S. policy regarding the resolution of remotely sensed imagery. The highest permissible resolution under current federal policy is 0.5 m. DigitalGlobe's Worldview 1 platform, launched in September of 2007, and along with GeoEye's GeoEye-1 platform, launched in September 2008, are the only commercially available platforms to achieve the 0.5 m resolution limitations imposed by regulation. In fact, GeoEye-1 actually collects imagery at 41 m and processes its data to represent coarser detail and downsample their resulting imagery prior to sale to nongovernmental entities. Clearly, we are reaching a tipping point whereby the technology and the regulation are not in sync. Furthermore, planned launches of foreign imaging platforms, not subject to U.S. regulation, will soon break the half-meter barrier.

At the time of this writing, there is no U.S. policy regulating the resolution or the shutter control of aerial photography. The perfection of inertial navigation systems combined with airborne Global Positioning System (GPS) has greatly increased the ability to fly and map almost any region that can be flown and mapped at high levels of accuracy and precision without the need for little or no ground control. There are flight restrictions imposed by the federal aviation administration (FAA) but these impose restrictions on aerial photographic aircraft to prevent them from interfering with commercial aviation, rather than restrict regions that can be photographed.

The U.S. military services also impose some restrictions on overflights of military bases but the FAA does not issue permits for overflights of military bases. As a result, projects measured in thousands of square miles are routinely completed. Many states have gathered consortia and have contracted for statewide data collections. The majority of these programs produce ortho-corrected image maps at resolutions of 1 ft or 6 in." per pixel. (Orthophotographic photomaps are planimetrically correct aerial imagery presented in digital formats, as described in Section 22.7.1.) Camera manufacturers are now manufacturing the second generation of the high-end digital camera systems. These second-generation systems have multispectral capabilities which capture a natural color (red–green–blue) and/or a false color infrared rendition. The post-processing applications are robust and can rectify and mosaic very large mosaics with a seamless appearance. Much of this imagery is in the public domain and form parts of the image backgrounds for Google Earth and Microsoft Virtual Earth.

While existing federal policy governing remote sensing technology is not numerous and tends to be largely focused around issues of permitted usage of space-based platforms—future public policy around remote sensing technologies may be once again approaching a critical nexus. At the time that this chapter was written, the Secretary for the U.S. Department of Interior directed the USGS to make its 35 year Landsat satellite image archive available over the Internet for free in 2009. In this same speech, he also pledged the Department's commitment in making a new image acquisition program, which is in the planning stages, known as imagery for the nation (IFTN), a reality. Both of these large-scale remote sensing initiatives will require new public policy consideration or perhaps have emergent privacy considerations that may need to be addressed from a public policy perspective.

22.3 ONLINE DIMENSION

The online mapping phenomenon is barely more than a decade old. MapQuest, which started as an experiment, is often cited as the first real online mapping site. Its ability to provide point-to-point driving directions was revolutionary when first unveiled in 1996. MapQuest originated from GeoSystems Global Corporation, a company with roots in conventional cartography and map publishing for customers like the National Geographic, the telephone company Yellow Pages, and airline magazines. The MapQuest application proved that a geographic service offered on the Internet could attract users and therefore create a marketplace for advertisers interested in reaching those visitors. During the first 30 days of existence, MapQuest received over 1 million visitors (http://computer.howstuffworks.com/mapquest1.htm). By 2007, MapQuest was the world's most popular online mapping site with nearly 40 million visitors per month. It is clear that the use of geographic content and services attracts visitors to a given site and therefore increases the value to advertisers. It is fair to say that MapQuest was to online mapping what Yahoo was to text search in the early days of the Internet. In both cases, these industry pioneers were surpassed by new innovations that expanded on their themes to supersede them by providing more complete and varied services to Internet users.

Today, there are many generic as well as highly focused Internet sites that provide mapping data and services. Google Maps and Google Earth, as well as Microsoft Virtual Earth and Yahoo Maps, have all become important players in the broad Internet mapping marketplace. In each case, the goal of reaching consumers to effectively deliver advertising results in a minimalist requirement for up-to-date, accurate spatial data, to include satellite and aerial imagery. As these systems continue to expand their coverage globally, their ability to manage image updates is geared toward regular release cycles, much like the software industry or street map data providers, such as Navteq and TeleAtlas.

This gap in the currency of image data has led to an expansion of the market for accurate, current image data for use in many business decision processes. The exposure to the possibilities created by these services has dramatically increased the interest in geospatial thinking and therefore the need for data and solutions. In addition, their success has ushered in a whole new type of spatial solution, the Web-based geospatial serve, which can be consumed by developers to create meaningful applications. Yet again, the need for more current, accurate imagery is largely still an unmet need that can only fully be satisfied by the commercial imaging companies, aerial and satellite, and their business models of delivering imagery on demand. Simply stated, custom-ordered geographies and products meet specific customer needs.

The online mapping revolution has increased awareness in geospatial data and solutions, thereby allowing the entire remote sensing industry to create new applications and markets.

22.4   PERSONAL NAVIGATION DEVICES

The fastest growing market for geospatial data, to include remotely sensed imagery, is the personal navigation market. What began as an extremely expensive option on luxury automobiles has quickly grown into a portable device market generating billions of dollars in sales each year globally (Berg Insight, 2007). With a growth rate in excess of 20% and annual sales of some 20 million units in 2007, this growing market space is certain to impact the geospatial industry in much the same manner the online mapping phenomenon did. It is expected that more than 100 million units will be in service by 2011, creating an overwhelming demand for data (Fabris, 2007).

Current units are limited in their storage capability, limiting opportunities for image integration. That said, the increase in mobile broadband networks is quickly driving interest in so-called connected devices that no longer must be self-contained for their data requirements, but rather can access specialty data over the wireless networks of the world as it is needed. This combination of mobile wireless computing and customized navigation displays will open many new opportunities for using remotely sensed data to enhance the utility of personal navigation.

The resulting opportunities created have led to large business deals finding Nokia acquiring street map maker NAVTEQ and device manufacturer TomTom of the Netherlands acquiring TeleAtlas, the second dominate street map maker. These two acquisitions in 2008 are a key indicator of the growth expected in this market and the overall need to control the primary data content providers to insure the availability of this data in future years.

The next wave of capabilities will include street-level photos and video fully integrated with overhead imagery to provide consumers with views of the landscape from any level. The complexity of integrating these disparate data sources in a meaningful way is the basis for the next wave of research and development in the remote sensing community. As the growth of available data continues, the need is to understand relationships and fully integrate these data to provide the most timely, accurate, and accessible view of an area. The challenge of remote sensing professionals is to manage this plethora of information and present it to users in an authoritative way.

The personal navigation market is currently the fastest growing spatial market and clearly the best opportunity to provide spatial awareness and analysis to the consumer market, creating an insatiable appetite for more current and accurate data. The opportunities created by this market space will push the fields of remote sensing and GIS to new levels of innovation.

Collectively, the online mapping revolution along with personal navigation devices, location-aware cell phones, and emerging societal trends such as social networking and globalization to name a few, has only served to increase awareness in geospatial data and solutions. This pervasive awareness coupled with faster, cheaper, smaller computing technologies is allowing the entire remote sensing industry to rise with the tide. So if the theory of Moore's Law of computing holds for the next 20 years, the computational performance of computers will be improved 10,000 times by 2026. Looking at this statement through the lens of other sciences such as nanotechnology work already underdevelopment, material trends will likely make possible increasingly smaller and smaller implementations requiring even less and less power (Boehm, 2006). This will no doubt give way to concepts already on the drawing board such as microsatellites and sensor platforms.

22.5   MEDIA BECOMES SPATIALLY LITERATE

The increasing caliber of publically accessible remotely sensed data has opened the opportunity for media to use imagery as a source of information in a manner that was previously only available to governmental agencies. The most notable example may be the use of Landsat and SPOT data to investigate the source of radiation release by the failure of the Chernobyl nuclear plant. At the time of the incident, networks monitoring atmospheric contaminants identified the likely release of radiation in Eastern Europe. Initially, the Soviet government would not acknowledge the existence of an emergency, and Western governments were not releasing conclusions of their intelligence agencies. In this context, Western news organizations used the thermal band of the Landsat TM sensor, and the finer detail offer by SPOT imagery to confirm the incident at Chernobyl. This is just one example of the press using imagery to independently investigate a news story of pressing public interest. Recently, news organizations and nongovernmental organizations have used publically available satellite imagery to report on humanitarian crises in Sudan and refugees in South Ossetia.

News media have also used geospatial data to document and illustrate reporting. One need only review the 2008 U.S. presidential election to fully understand the media's growing fascination and the embrace of geospatial technologies. Anyone who watched a major television network witnessed anchors and correspondents at large Map Walls, or touch screen Map Tables, using geography to tell the story of the election results and campaign strategies. Data as detailed as the precinct level was used to analyze voting patterns and voter turnout and present all Americans with a clear picture of how votes are cast and tabulated to elect a president.

The coverage of the election of 2008 is just one example of how geospatial data has entered the mainstream media. Nearly every night one can see some news outlet using Google Earth or other spatial data to establish the context for a news event. In addition to the planned, long running, narratives covering elections, natural disasters, military conflicts, and economic trends have all used spatial technologies to tell the story. The visual representation of the spatial context for a new event has become a commonplace component of twenty-first century media, if not a required element of a news story.

When Hurricane Katrina slammed the Gulf Coast of the United States, not only did the media use imagery to orient the public and tell "the story" but government agencies and relief organizations also counted on post-event imagery to manage various aspects of the recovery. Satellite and aerial imagery acquired immediately after the event was used extensively to analyze the high water lines and the impact of flooding. Further, imagery formed a key component for the calculation of debris volume and provided critical data to the plan for its subsequent removal.

The major Tsunami that struck Banda Ache, Indonesia on December 26, 2004 likewise was a defining moment for the use of imagery by the media. Because of the region's remote location in an underdeveloped region of the world, newly captured imagery formed the principal source of information for understanding the impact of this catastrophic disaster. The use of imagery for rescue and relief efforts was unprecedented. While access to the region was limited, roads and other civil infrastructure completely destroyed, satellites were able to capture imagery that formed the basis for relief efforts, reconstruction planning, and mobilization of public interest in aid programs assisting the victims.

This intersection of the public and private use of remotely sensed data ushered in a new paradigm of using image content by sharing the latest data and methods to inform, protect, and serve the public at large. These new cooperative efforts have no doubt changed the way spatial data and technologies will be exploited for all natural disasters that follow. No longer will these efforts be seen as new or innovative; rather, they will be expected as essential components of any relief effort. If the media can make sense of a disaster using timely image data, then the pressure is clearly placed on public agencies and relief organizations to do likewise in the pursuit of their respective missions.

As the trend toward using geospatial data and tools continues to enter every aspect of our life, the need for professionals in the field is growing exponentially. It is clear that behind every Map Wall, or featured news story, there is a geospatial professional assuring that the data and analysis are valid.

22.6 EMERGING CAPABILITIES

Imagine sitting at your computer late one evening and checking your personal e-mail account. Now, imagine that you have just received a notice from your local discount home and garden center telling you that a specific lawn fertilizer is on sale and it is exactly what your lawn needs. Sound hard to believe? With the power of emerging remote sensing technologies, this level of personalized information is not only possible but will also become increasingly commonplace.

Unmanned aerial vehicles (UAVs) are now being used to monitor the borders of the United States, to provide situational awareness for public safety organizations, of major events like the Super Bowl or a concert or to track vehicles and people in times of crime or war.

UAVs have played a significant role in the conflict in Iraq, providing analysts and commanders with near-real-time information needed to plan and execute their missions. Known as "persistent surveillance" these units fly for long periods of time, focused on the imaging of a single geographic location, to stream imagery in its geographic context, for which can be geographically understood and used in many forms of advanced analysis. The richness of the data captured and received provides many opportunities for modeling and analysis unheard of a few years ago.

Hyperspectral sensors have become more commonplace as scientific research looks for new ways to use electro-optical imaging to derive unique signatures of features on the earth's surface. These systems that measure the intensity of light in hundreds of color bands allow a greater understanding of the uniqueness of features and their relationships to one another.

Light detecting and ranging (LIDAR) systems (Chapter 23) are becoming increasingly popular for imaging the earth's terrain and creating digital elevation models. The ability to analyze millions of points in a cloud and determine the top, middle, and bottom of features, both natural and man-made, has proved an exceptional method for creating digital elevation models and a 3D representation of man-made features like buildings, bridges, and the like. Synthetic aperture imaging radar (SAR) systems (Chapter 18) are likewise becoming more commonplace as systems are developed for use on space-based and aerial platforms. The ability of these systems to image at night, in any kind of weather, expands the imaging window in ways never before imagined.

Like aerial or satellite imaging before them, these new technologies will take years to realize their full potential. Advances in the science of using the data, developing standards for interpreting the data and libraries of spectral signatures will take years to fully develop and adopt. The level of research in these areas is clearly growing making these technologies the bold new frontier in the remote sensing community. All of these systems employ advanced electronics such as GPS and inertial measurement units (IMU) to provide accurate positioning and geo-referencing of the imagery they acquire.

Combined with software advances in feature identification, change detection, and spectral analysis, the possibilities to derive meaningful information from simple imagery grow exponentially. As the volume and the diversity of data continue to grow, the methods and approaches to analyzing these data becomes the great challenge for the next generation of remote sensing scientists.

Now back to that e-mail. A UAV is performing the persistent surveillance of your neighborhood using a hyper-spectral sensor. The data captured is analyzed using the latest software techniques and spectral signatures for vegetation thus allowing your local home and garden center to identify not only the type of grass growing in your yard but also the health of the grass and therefore what fertilizer is best suited to improve the conditions for you specifically. Not only would a service of this type be beneficial to you, it will also be beneficial to the environment. This target approach to vegetation management, and all of the permeations you can imagine, will allow us to be more targeted in our use of chemicals thereby reducing the overfertilization and resulting toxic pollution of the planet. Not a bad vision for employing these emerging technologies in a way that benefits society. What is needed now is the research and development to make ideas like these not only practical but affordable.

22.7 CASE STUDY: EVOLUTION OF THE USDA NAIP DIGITAL IMAGERY PROGRAM

Implementing current technology and procedures often requires the reexamination of accepted procedures and standards—a process illustrated by the U.S. Department of Agriculture's (USDA's) National Agriculture Imagery Program (NAIP) effort, which illustrates how a fresh look at accepted practices and standards for the acquisition and the analysis of imagery can improve the efficiency and the effectiveness of imagery collection and analysis.

22.7.1 BACKGROUND AND IMPACT

The farm service agency (FSA) of the USDA administers and manages farm commodity, credit, conservation, and disaster and loan programs through a network of 2346 FSA county offices across the continental United States. These programs are designed to improve the economic stability of the agricultural industry and to help farmers adjust production to meet demand. Economically, the desired result of these programs is a steady price range for agricultural commodities for both farmers and consumers. A key role of the field offices is to certify farmers for the various farm programs and to pay out farm subsidies and disaster payments.

Compliance with the terms for the payment of subsidies is critical to the viability of the program, and remote sensing has long been a key tool of the compliance verification process. Because of the large areas that must be monitored, and the narrow temporal window when conditions are suitable for acquisition (monitoring must occur during the key portions of the growing season), the compliance program specifically relies on aerial photography. Since the 1940's, crop compliance relied on aerial enlargements that were rectified for basic tip and tilt; field boundaries were manually transferred drafted onto the photographs.

In order to improve the efficiency of the compliance program and to computer-enable the spatial element of farm records, in 1999, USDA decided to embrace digital orthophotography. Digital orthophotography provides planimetrically correct aerial imagery, in which the usual positional errors inherent to aerial photography have been corrected. These products, known as digital orthophoto quadrangles (DOQs), correspond to the USGS's system of topographic quadrangles. (A variant of the DOQ, the digital orthophoto quarter-quadrangle (DOQQ) is a subdivided version of the DOQ that provides a data unit that is more convenient to store and transmit.) Digital orthophoto image bases support large-scale conversion of the farm's field boundaries into a coherent nationwide GIS-enabled digital land base. A joint USDA/USGS effort required five years to complete the acquisition of the needed aerial photography and the production of approximately 200,000 quarter-quadrangles for nationwide coverage in 2004. USDA used the individual orthophoto tiles produced by USGS to create approximately 2500 seamless mosaics for each county (MDOQ). (MDOQs are "mosaiked" DOQQs—digital mosaics that represent entire counties (Mathews and Davis, 2007). This product exceeded the positional accuracy (the correction of offset roads and streams) and the radiometric consistency of the original USGS product. The MDOQs were compressed to facilitate handling at the field offices. The MDOQs were then used to digitize the boundaries of approximately 20,000,000 farm fields from the rectified enlargements to create a uniform, seamless, attributed coverage across the United States. Imagery was collected annually for all of the agricultural lands of the conterminous United States during the growing season which would provide an up-to-date image to document conditions during each season. The imagery was collected using 35 mm color slides that were scanned and geo-referenced and could be easily overlaid with the MDOQ and the digitized common land unit (CLU) data. This product proved to be superb in terms of speed, quality, and ease of handling.

The decision to replace 35 mm film with the national agricultural imagery program (NAIP) creates an excellent compilation of the issues and directions of remote sensing in the future. Some key elements of this include

1. A joint federal/private partnership driving private sector initiative
2. Redesign of business processes to better utilize the technologies and data sets
3. Expansion of remote sensing data usage and adoption in accordance with federal policy
4. Recognition that a successful outcome depended on a total solution rather than simply a data acquisition/ortho-production program

The original specification for the compliance imagery specified a natural color rendition captured during the growing season (typically June through September) and included two possible resolutions/accuracies—a 2-m resolution that must match within 10 m of the reference MDOQ or a 1 m resolution that must match within 5 m of the reference MDOQ. In 2008, the specification was modified to a uniform 1 m resolution and a four-band delivery so users can produce a natural color or false color infrared rendition of the imagery. A preliminary compressed county mosaic (CCM) is required within 30 days of the date of photography. A final corrected version of the CCMs must be delivered in the year of the acquisition.

22.7.2 SIGNIFICANT PARADIGM SHIFTS

This case study illustrates the significance of several key components that contribute to the success of a broad-scale program of national significance by redefining the key processes and standards necessary to achieve the required results:

- *Total solution.* USDA recognized that the successful adoption of digital geospatial data depended on a redefinition of business processes combined with products that met the fundamental requirements of the field offices rather than simply providing new types of geospatial data. Critical elements of the NAIP also included the conversion of farm field data (CLU) into a GIS format. Field offices also required hardware sufficient to manage and manipulate the digital data, software applications, and training.
- *Balancing quality and timeliness.* USDA recognized that specifications used for compliance imagery had to be adapted in order to reflect the requirements of the program but also to provide vendors with maximum flexibility to ensure success. Accuracy specifications were keyed to the existing MDOQ base rather than an absolute accuracy to the ground, limits for permissible cloud cover were expanded, and specifications for tone and color balance requirements were relaxed. USDA was confronted with the challenge of developing standards to attain consistent quality assurance against these specifications. Vendors have continued to improve production processes to improve the speed and the quality of the program. USDA continues to train staff to interpret their specifications to match the product with their true requirements.
- *Shared risk.* NAIP requires that imagery is acquired and delivered in a specific time window. If a contractor is unable to obtain full coverage of the required area within this time window, USDA stops the collection and must resort to more expensive means for collection of the needed information from field investigation for the remaining areas.
- *Rapid delivery.* The NAIP was the first to demonstrate that digital orthophotography could be captured and delivered in a period of weeks rather than months or years. The NAIP further demonstrated the ability of modern airborne sensor technology to address the practical obstacles to rapidly capturing imagery of very large areas of coverage.

22.7.3 OUTCOMES

The NAIP has been extremely successful achieving an average successful collection of over 90% of the contracted area each year. The resulting product has been widely adopted by other federal agencies, commercial providers, and state governments as an image base map. State governments routinely cost-share with the NAIP, so savings and efficiencies have extended beyond the USDA. The USDA routinely makes the NAIP coverage available to state agencies in anticipation of hurricanes or other disasters, so the NAIP imagery has become a critical part of emergency support.

Although the NAIP program has changed the overall expectation for imagery standards and for the speed of delivery. The NAIP has created a catalyst where private business continues to improve production processes to increase speed and efficiency in the hope of capturing a larger share of the annual allotment of the NAIP imagery. Other programs have started to request the delivery of imagery within months of collection rather than a year or more, as was typical prior to the NAIP.

The USDA recognized that the program should be a partnership with industry, and therefore allows NAIP contractors to distribute or repackage the raw imagery to produce value-added products or services to meet local needs. This provides NAIP contractors with an additional potential revenue stream and allows USDA to make a contribution, paying for the raw materials as a means to incubate other types of products.

22.8 SUMMARY

As with all applications of technology in modern society, improvements in geospatial instrumentation and analysis will generate unexpected technological developments, with equally unanticipated societal effects. During the next decades, we will continue to see innovations in sensors, processing software, leading to further business opportunities, further extension of markets, and continuing current redefinition of the remote sensing industry. As the market continues to develop, it is clear that advanced research in methods and practices will continue to outpace that in many similar geospatial market segments. The yet-undiscovered application areas, the growing proliferation of data, and the variety of sensor types will all lead to greater demand for solutions to a growing myriad of challenges. Further, it is clear as the systems become more advanced with enhanced resolution, the policy and privacy implications of the technology advances will need careful attention to ensure that all equities are represented without stifling the marketplace or innovation. Much as the availability of the MP3 player and the IPod created new challenges in the recording industry, these challenges will be difficult but they are not insurmountable. This chapter has explored some of the issues and possibilities in an effort to introduce the reader to the unfamiliar context for next decade, to begin the dialog and prepare for the exciting future that is on our horizon.

REFERENCES

Boehm, B., Some future trends and implications for systems and software engineering processes, *Systems Engineering*, 9, 1–19, 2006.
BBC Market Research Report, Remote sensing technologies and global markets, February 2007, http://computer.howstuffworks.com/mapquest1.htm, last accessed November 2008.
Berg Insight, Berg Insight says PND shipments will reach 53 million units in 2012, http://www.gisuer.com, accessed December 15, 2008, November 7, 2007.

Fabris, N., Personal navigation devices will surpass 100 million units by 2011, *Directions Magazine*, http://www.directionsmag.com/press.releases/?duty=Show&id=19822&trv=1; accessed December 15, 2008, November 8, 2007.

Florini, A.M. and Dehqanzada, Y.A., No more secrets?: Policy implications of commercial remote sensing satellites. Carnegie: Carnegie Paper No.1, 1999.

Florini, A.M. and Dehqanzada, Y.A., The global politics of commercial observation satellites (Chapter 20), in *Commercial Observation Satellites: At the Leading Edge of Global Transparency*, John C. Baker, Kevin M. O'Connell, and Ray A. Williamson (eds.). Carnegie, Washington, DC: RAND Corporation and ASPRS, 432-000, 2001.

Givri, J.R., Satellite remote sensing data on industrial hazards, *Advances in Space Research*, 15, 87–90, 1995.

Heppenheimer, T.A., http://www.centennialofflight.gov/essay/SPACEFLIGHT/remote_sensing/SP36.htm, accessed December 2008.

Livingston, S., Remote sensing technology and the news media (Chapter 23), in *Commercial Observation Satellites: At the Leading Edge of Global Transparency*, John C. Baker, Kevin M. O'Connell, and Ray A. Williamson (eds.). Santa Monica, CA, RAND Corporation and ASPRS, pp. 485–500, 2001.

Mathews, L. and Davis, D., The national agricultural imagery program: Options and challenges, *2007 ERSI International Users' Conference*, San Diego, CA, 2007. http://gis.esri.com/library/userconf/proc07/papers/papers/pap_1104.pdf, accessed Dec 2008.

Richter, R., Lehmann, F., Haydn, R., and Volk, P., Analysis of LANDSAT TM images of Chernobyl, *International Journal of Remote Sensing*, 7, 1859–1867, 1986.

Slonecker, E.T., Shaw, D.M., and Lillesand, T.M., Emerging legal and ethical issues in advanced remote sensing technology, *Photogrammetric Engineering & Remote Sensing*, 64, 589–595, 1998.

23 Airborne LIDAR Mapping

Ayman Fawzy Habib

CONTENTS

Recently, light detection and ranging (LIDAR) systems have emerged as a fast, accurate, and cost-effective technology for direct acquisition of highly dense 3D positional data from physical surfaces. The widespread adoption of LIDAR systems has been propelled by the improved performance and lower cost of modern direct geo-referencing technology (global navigation satellite systems [GNSS], inertial measurement units [IMU], and GNSS/IMU integration techniques). A LIDAR system is a combination of two main components: the direct geo-referencing and laser ranging components. The direct geo-referencing component provides the position and the attitude of the mapping platform. The laser ranging component, on the other hand, provides the distance between the laser-beam firing point and its footprint. The geo-referencing and ranging information are combined to provide the ground coordinates of the laser-beam footprints leading to a highly dense and accurate point cloud covering the mapped area.

The main objective of this chapter is to provide an overview of LIDAR principles, mathematics, error sources, quality assurance and quality control (QA/QC) procedures, data processing techniques, and applications. Section 23.1 provides a brief explanation of the laser-generation and range-derivation principles. This discussion is followed by LIDAR scanning principles and mathematics. The impact of random and systematic errors in the LIDAR system measurements and parameters on the ground coordinates of the derived point cloud will be discussed next. Then, QA/QC procedures of LIDAR systems and derived data are introduced followed by a brief discussion of some of LIDAR data processing activities. Finally, this chapter will conclude with some applications involving the collected LIDAR data (e.g., topographic mapping, orthophoto generation, 3D city modeling, and flood mapping).

439

23.1 PRINCIPLES

Before discussing laser-generation principles, let us start by explaining the process of stimulating an atom by an external energy source and the resulting photon emission. Atoms in their steady state can be stimulated to emit photons through a two-step procedure (Figure 23.1). Energy injected into the atom is absorbed by electrons in low-energy orbits. As a result, these electrons jump to higher-energy orbits (Figure 23.1a). Eventually, these electrons drop to lower-energy orbits while the energy difference between the orbits is released in the form of light photons (Figure 23.1b).

This concept is utilized for the generation of LASER (light amplification by stimulated emission of radiation) (Figure 23.2). The laser-generation unit consists of a tube, with a mirrored surface at one end and a half-mirrored surface on the other end, filled with atoms and a flash tube that provides the initial energy for exciting the atoms in the laser tube. Figure 23.2a shows the laser tube in its steady state. The flash tube injects light/energy into the laser tube (Figure 23.2b). This light/energy excites the atoms in the laser tube. The excited atoms then return to their ground state by emitting photons (Figure 23.2c). The emitted photons run in various directions, and some of them run in a direction parallel to the tube's axis bouncing back and forth off the mirrors on either end of the laser tube (Figure 23.2d). The back and forth bouncing action stimulates other atoms and leads to further emission of more photons. This process repeats itself resulting in an amplification of the atom stimulation and photon emission. Finally, the emitted photons are released in the form of laser light through the half-mirrored surface. Laser light is monochromatic (i.e., it contains one specific wavelength in contrast to white light, which is a mix of several wavelengths).

The wavelength of emitted laser light is determined by the amount of energy released when the electrons of the excited atoms drop to a lower-energy orbit. Laser light usually belongs to the visible or near infrared portions of the electromagnetic radiation spectrum (500–1500 nm). Typical LIDAR systems emit laser with wavelength in the range from 1040 to 1060 nm. Laser light is also coherent (i.e., laser light is organized in the sense that each photon moves in step with other photons). Another property of laser light is that it is directional with a very tight beam and is very strong and concentrated. This directional property is different from white light, which travels in many directions and is very weak. A laser beam, however, is not perfectly

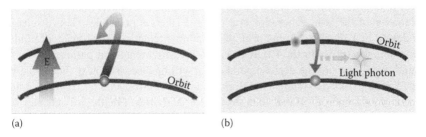

(a)   (b)

FIGURE 23.1 Emission of light photons: (a) excitation of an atom and (b) the release of photons.

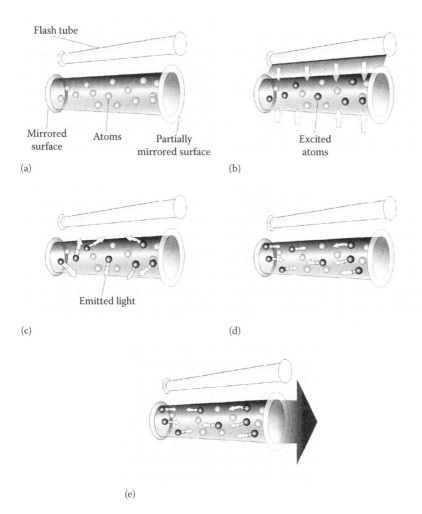

FIGURE 23.2 Generation of laser light: (a) the components of a laser generation system, (b) the injection of light into the laser tube, (c) the emission of photons, (d) the bouncing of photons off either end of the laser tube, and (e) the release of laser light.

FIGURE 23.3 The divergence angle of a laser beam.

cylindrical in shape (Figure 23.3). The waist is the narrowest part of the beam; from the waist, the beam diverges by an angle γ, which is known as the beam divergence angle, and it typically ranges from 0.2 to 1.0 mrad. Therefore, the laser footprint should be thought of as a disc/ellipse rather than a distinct point. The size and the shape of the footprint depends on several factors such as the beam divergence angle,

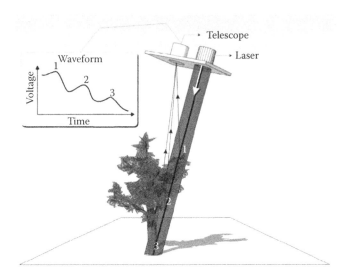

FIGURE 23.4   Laser pulse interaction with objects and range derivation from the digitized waveform.

the distance between the laser-beam firing point and the object, the look angle of the laser beam, and the orientation of the surface of the mapped object.

The emitted light by the laser tube interacts with objects and a portion of this light is reflected back and is detected by the telescope component of the laser system (Figure 23.4). The detected energy is recorded against the time between the signal emission and its reception in a graph, which is known as the waveform. A peak in the waveform signifies a reflected pulse from an object. Depending on the nature of the mapped object, a portion of the laser light might penetrate that object (e.g., a tree canopy). Penetrated energy might interact with other objects leading to several peaks in the waveform (Figure 23.4). The time delay between the laser-pulse emission and the reception of the reflected pulse is used to derive the distance from the laser-beam firing point and the reflecting object. The time delay multiplied with the speed of light is equivalent to double the distance between the laser-beam firing point and the reflecting object. Within the laser unit, the laser beam is rotated to cover a profile in the object space.

In addition, the laser system is usually mounted onboard a platform (such as an aircraft); the forward motion of this platform provides coverage of extended areas on the ground. The orientation of the laser beam is controlled by a steering mirror. The motion of the mapping platform as well as the rotation of the steering mirror controls the shape of the laser scans in the object space. Linear and elliptical laser scanners are the most popular systems. In a linear laser scanner, the steering mirror is rotated in one direction, usually across the flight direction, leading to a zigzag scan pattern in the object space (Figure 23.5a). Elliptical laser scanners, on the other hand, have a steering mirror, which is rotated across two axes (Figure 23.5b). The majority of commercially available laser systems implement linear scanners. A laser system is equipped with encoders that measure the rotation angles of the steering mirror.

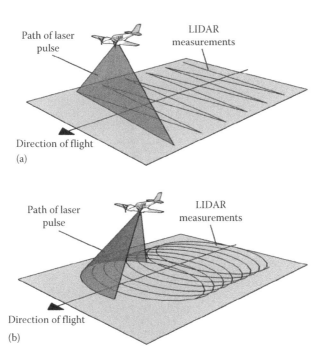

FIGURE 23.5 Scan patterns in (a) linear and (b) elliptical laser systems.

In a LIDAR system, the laser unit is integrated with a direct geo-referencing unit to derive the ground coordinates of the laser-beam footprints. The involved mathematical model for the derivation of the point-cloud coordinates and the impact of systematic and random errors in the system measurements and parameters on these coordinates are discussed in the following sections.

23.2 LIDAR PRINCIPLES AND MATHEMATICS

A typical LIDAR system consists of two main components: a direct geo-referencing unit and a laser-ranging unit. As it can be seen in Figure 23.6, the direct geo-referencing unit is comprised of an integrated GNSS/IMU, which provides the position and the attitude of the LIDAR system. The laser ranging unit provides the range between the laser-beam firing point and its footprint. Figure 23.7 shows a schematic diagram of a LIDAR system, together with the involved coordinate systems. Equation 23.1 is the basic geometric model that incorporates the LIDAR measurements and parameters to derive the ground coordinates of the laser beam footprints (El-Sheimy et al., 2005). This equation relates four coordinate systems: the ground coordinate system, the IMU body frame, the laser-unit coordinate system, and the laser-beam coordinate system.

This equation represents a three-vector summation process. The first vector, $\vec{X}_0$, is the vector from the origin of the ground coordinate system to the IMU body frame, the second vector, $\vec{P}_G$, is the offset vector between the laser unit and the IMU body frame, and the third vector, $\vec{\rho}$, is the range vector between the laser-beam firing

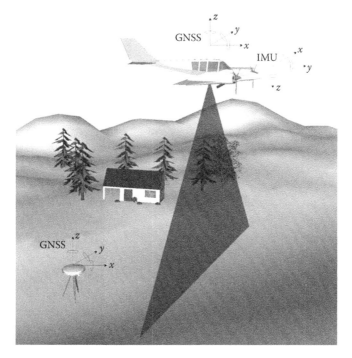

FIGURE 23.6   Basic components of a LIDAR system.

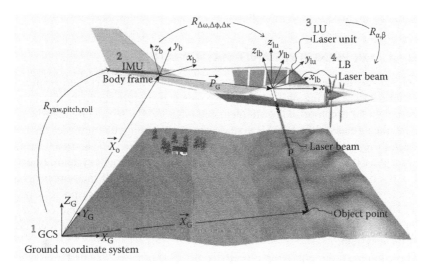

FIGURE 23.7   Coordinates and parameters involved in LIDAR positioning.

point and the object point. The summation of these three vectors, after applying the appropriate rotations ($R_{yaw,pitch,roll}$, $R_{\Delta\omega,\Delta\varphi,\Delta\kappa}$, $R_{\alpha,\beta}$), yields the vector $\vec{X}_G$, which is the ground coordinate of the footprint under consideration. The vector $\vec{X}_o$ and the matrix $R_{yaw,pitch,roll}$ are the position and the attitude of the IMU body frame at the time of a given pulse, and they are determined from the integrated GNSS/IMU measurements. The components of the range vector $\vec{\rho}$ relative to the ground coordinate system is defined by the measured range, the orientation of the laser beam within the laser unit as measured by the steering mirror encoders, and the GNSS/IMU-derived attitude. The spatial offset $\vec{P}_G$ and the rotational offset $R_{\Delta\omega,\Delta\varphi,\Delta\kappa}$ describe the positional and rotational relationships between the IMU body frame and the laser-unit coordinate system. These offsets are commonly known as the bore-sighting parameters and are determined through a calibration procedure.

Based on the discussion so far, it is clear that all the quantities in the right-hand side of Equation 23.1 are either measured or determined from a calibration procedure. Therefore, the ground coordinates of the laser footprint can be derived from a single pulse. The quality of the derived coordinates depends on the accuracy of the measurements from the laser scanner, GNSS, and IMU as well as the accuracy of the bore-sighting parameters relating these subsystems.

$$\vec{X}_G = \vec{X}_o + R_{yaw,pitch,roll}\vec{P}_G + R_{yaw,pitch,roll}R_{\Delta\omega,\Delta\varphi,\Delta\kappa}R_{\alpha,\beta}\begin{bmatrix} 0 \\ 0 \\ -\rho \end{bmatrix} \qquad (23.1)$$

At this stage, it is worth mentioning that LIDAR-point positioning is very similar to object-point positioning using a directly geo-referenced photogrammetric system, Figure 23.8. As it can be seen in this figure, the ground coordinates of an object point,

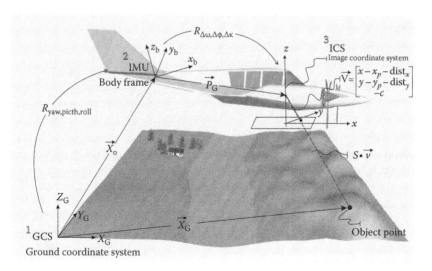

FIGURE 23.8 Coordinates and parameters involved in photogrammetric positioning using a directly geo-referenced imaging system.

whose image has been observed, can be derived according to Equation 23.2. Similar to the LIDAR geometric model, Equation 23.2 is a three-vector summation process. The first two vectors in Equations 23.1 and 23.2 are identical. The main difference between Equations 23.1 and 23.2 resides in the third vector. In the LIDAR geometric model, the range vector between the laser-beam firing point and its footprint is completely defined. In the photogrammetric model, however, the third vector is partially defined. In other words, only the vector from the camera projection center to the image point is defined. The vector from the projection center to the object point is derived by scaling the vector between the projection center and the image point (the scale factor S in Equation 23.2). This scale factor is unknown, varies from point to point in the same image, varies for the same point in overlapping images, and can be only derived by observing the same object point in two or more overlapping images. Therefore, in contrast to the LIDAR positioning, where the ground coordinates of the laser footprint can be derived from a single pulse, the ground coordinates of an object point from photogrammetric processing can be only derived after observing that point in multiple images.

Observing the same object point in overlapping images allows for the estimation of the scale factors associated with the involved images. For example, in a stereo pair, one would have six equations and five unknowns (the ground coordinates of the object point as well as the scale factors associated with this point in the two images). A direct consequence of this difference between LIDAR and photogrammetric positioning is that the photogrammetric reconstruction is based on redundant measurements (e.g., for a stereo-pair, we have six equations in five unknowns, which are determined through a least-squares adjustment procedure). On the other hand, LIDAR positioning is not based on redundant measurements (i.e., a single pulse would lead to three equations in three unknowns). Having nonredundant measurements in LIDAR positioning would have a significant impact on the QC of the derived ground coordinates from the system, which will be addressed in Section 23.4.

$$\vec{X}_G = \vec{X}_o + R_{\text{yaw,pitch,roll}}\vec{P}_G + S\,R_{\text{yaw,pitch,roll}}R_{\Delta\omega,\Delta\varphi,\Delta\kappa}\begin{bmatrix} x - x_p - \text{dist}_x \\ y - y_p - \text{dist}_y \\ -c \end{bmatrix} \quad (23.2)$$

In addition to the ground coordinates of the laser footprints, a LIDAR system generates intensity data, which signifies the reflective properties of the objects covered by the laser footprints and is affected by the orientation of the reflecting surface relative to the laser beam. Due to the scanning nature and variations in the surface elevations, the derived point cloud from a LIDAR system is irregularly distributed. The elevation and intensity data of the irregularly distributed point cloud can be interpolated to produce range and intensity images similar to those in Figure 23.9.

Typical specifications of commercially available LIDAR systems are summarized in Table 23.1. In this table, the pulse repetition rate indicates the number of emitted

(a)　　　　　　　　　　　　　　　　　　(b)

FIGURE 23.9　Sample of interpolated LIDAR imagery: (a) range/shaded relief image and (b) intensity image.

TABLE 23.1
Typical Specifications of Commercially Available LIDAR Systems

Specification	Typical Values
Laser wavelength	900–1550 nm
Pulse repetition rate	25–167 kHz (250 kHz max)
Pulse energy	Up to 100s µJ
Pulse width	<10 ns
Beam divergence	0.25–2 mrad (2.7 mrad max)
Scan angle (full angle)	40° (80° max)
Scan rate	25–90 Hz (415 Hz max)
Scan pattern	Zigzag; parallel; elliptical; sinusoidal
GPS frequency	10 Hz
INS frequency	200 Hz
Operating altitude	80–3500 m (6000 m max)
Footprint	0.25–2 m at 1000 m AGL
Multiple elevation capture	1–4 returns and full waveform
Ground spacing	0.5–2 m
Vertical accuracy	<15 cm at 1000 m AGL
Horizontal accuracy	<50 cm at 1000 m AGL

laser pulses per second. The scan rate, on the other hand, refers to the number of completed full scans per second. These specifications show that current LIDAR systems are fast and accurate data acquisition tools, which are capable of covering tens to hundreds of square kilometers of ground per hour. The achievable accuracy is quite impressive given the pulse repetition rate and possible flying heights of modern LIDAR systems. Data collection is also quite flexible in several ways. LIDAR can penetrate canopy, making ground measurement possible, and may be used in light rain. Data collection may be carried out day or night and is unaffected by sun angle.

In addition, the system can map surfaces with little or no texture (such as ice/snow-covered surfaces, deserts, and wetlands), unlike photogrammetry, which requires the identification of specific object points in overlapping imagery.

Another advantage of LIDAR is that it produces high-resolution 3D surfaces; the collected point cloud can be as dense as millions of points per square kilometer depending on the pulse repetition rate, flying height, and flying speed. Compared to photogrammetric positioning (Chapter 16), derivation of the ground coordinates of the LIDAR point cloud is an automated process. Photogrammetric positioning, on the other hand, requires manual or automated identification of conjugate points in overlapping imagery. Manual measurements of conjugate points in imagery are reliable but time-consuming. Automated matching of conjugate points, however, can be unreliable especially when dealing with large-scale imagery over urban areas. At this stage, one should note that LIDAR and photogrammetric technologies should not be viewed as competing data acquisition systems. Depending on the intended application, photogrammetric or LIDAR systems might be more appropriate. In other applications, such as automated reconstruction and realistic visualization of 3D urban environments, the integration of photogrammetric and LIDAR data is recommended. Tables 23.2 and 23.3 illustrate the pros and cons of photogrammetric and LIDAR systems and how the disadvantages of one system can be compensated for by the advantages of the other.

TABLE 23.2
Photogrammetric Cons as Contrasted by LIDAR Pros

LIDAR Pros	Photogrammetric Cons
Dense information along homogeneous surfaces	Almost no positional information along homogeneous surfaces
Day or night data collection	Day time data collection
Direct acquisition of 3D coordinates	Complicated and sometimes unreliable matching procedures
Vertical accuracy is better than the planimetric accuracy	Vertical accuracy is worse than the planimetric accuracy

TABLE 23.3
LIDAR Cons as Contrasted by Photogrammetric Pros

Photogrammetric Pros	LIDAR Cons
High redundancy	No inherent redundancy
Rich in semantic information	Positional; difficult to derive semantic information
Dense positional information along object space break lines	Almost no information along break lines
Planimetric accuracy is better than the vertical accuracy	Planimetric accuracy is worse than the vertical accuracy

23.3   LIDAR ERROR BUDGET

The quality of the derived point cloud coordinates from LIDAR depends on the random and systematic errors in the system measurements and parameters. A detailed description of LIDAR random and systematic errors can be found in Huising and Pereira (1998), Baltsavias (1999), and Schenk (2001). The magnitude of the random errors depends on the precision of the system's measurements, which include position and orientation measurements from the integrated GNSS/IMU, mirror angles, and ranges. Systematic errors, on the other hand, are mainly caused by biases in the bore-sighting parameters relating the system components as well as biases in the system measurements (e.g., ranges and mirror angles). In the following subsections, the impact of random and systematic errors in the system measurements and parameters on the reconstructed object space will be summarized.

23.3.1   RANDOM ERRORS

The purpose of studying the impact of random errors is to provide sufficient understanding of the nature of the noise in the derived point cloud as well as the achievable precision from a given flight and system configuration. In general, the effect of random errors in the system measurements can be analyzed through two approaches. One approach can be based on a simulation procedure, which starts from a given surface and flight trajectory. The surface and flight trajectory are then used to derive the system measurements (ranges, mirror angles, position and orientation information for each pulse). Then, noise is added to the system measurements, which are later used to reconstruct the surface through the LIDAR equation. The differences between the noise-contaminated and true coordinates of footprints are used to represent the impact of a given noise in the system measurements.

The following list summarizes the effect of noise in the system measurements on the derived coordinates of the LIDAR point cloud.

- Position noise will lead to similar noise in the derived point cloud. Moreover, the effect is independent of the system flying height and scan angle.
- Orientation noise (attitude or mirror angles) will affect the horizontal coordinates more than the vertical coordinates (especially in the nadir region). In addition, the effect is dependent on the system flying height and scan angle.
- Range noise mainly affects the vertical component of the derived coordinates (especially in the nadir region). The effect is independent of the system flying height. The impact, however, is dependent on the system's scan angle.
- Noise in some of the system measurements might affect the relative accuracy of the derived point cloud. As an illustration, Figure 23.10 reveals that a given noise in the GNSS/IMU-derived orientation affects the nadir region of the flight trajectory less significantly than off nadir regions. Such a phenomenon is contrary to derived surfaces from photogrammetric mapping where the measurements' noise does not affect the relative accuracy of the final product.

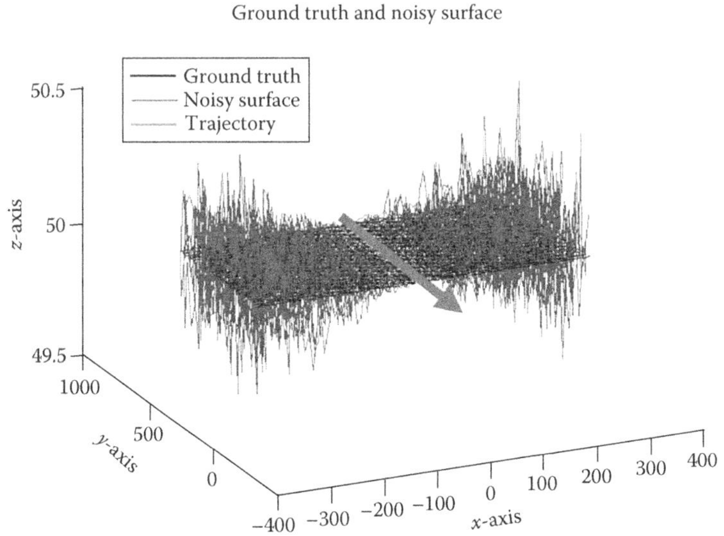

FIGURE 23.10 Effect of noise in the GNSS/IMU-derived orientation on the point cloud coordinates.

The second approach for studying the effect of random errors can be based on the law of error propagation applied to the LIDAR equation. For each of the LIDAR footprints, variance–covariance propagation can be used to estimate the precision of the derived coordinates given the precision of the system measurements. The advantage of such a methodology is allowing for the estimation of the "best achievable precision" from a given system and flight configuration. One should note that the use of the "best achievable precision" expression is based on the fact that error propagation assumes a relatively flat and horizontal solid surface without considering the nature of the interaction of the laser beam with the reflecting object. In other words, the precision of the derived point cloud covering steep and/or forested areas is not considered.

To illustrate such a procedure for an operational system, Table 23.4 shows the expected precision of the measurements from two optech LIDAR systems (ALTM 2050 and ALTM 3100). The manufacturer precision specifications for both systems is as follows: horizontal precision is less than 1/2000 of the flying height in meters while the vertical precision is less than 15 cm at a flying height of 1200 m and less than 25 cm at a flying height of 2000 m. Using error propagation and the specifications in Table 23.4, the expected precision of the derived LIDAR footprints can be computed, refer to Table 23.5. It should be noted that the reported precision in Table 23.5 corresponds to LIDAR footprints at the swath edges (i.e., footprints with the maximum scan angle). In other words, these numbers represent the worst precision within the swath. When comparing the manufacturer precision specifications with the numbers in Table 23.5, one can observe that the manufacturer precision specifications are more conservative than the calculated precision using error propagation, which only considers relatively flat and solid surfaces.

TABLE 23.4

Precision Specifications for the Optech System Components

System Model	GPS (m) Post-Processed	IMU (°) Post-Processed Roll	IMU (°) Post-Processed Pitch	IMU (°) Post-Processed Yaw	Scan Angle (°)	Laser Range (cm)
ALTM 2050	0.05–0.3	0.008	0.008	0.015	0.009	~2
ALTM 3100	0.05–0.3	0.005	0.005	0.008	0.009	~2

Sources: Applanix, POS AV specification, URL: http://www.applanix.com/media/downloads/products/specs/POSAV%20Specs.pdf, last accessed November 20, 2007; Optech, ALTM 3100 specifications, URL: http://www.optech.ca/pdf/Specs/specs_altm_3100.pdf, last accessed November 20, 2007.

TABLE 23.5

Expected Precision of the LIDAR Coordinates Using Error Propagation of the Precision Specifications in Table 23.4

	Estimated Accuracy (at 20.0° Scan Angle) X Flying Height (m) 1200	2000	Y Flying Height (m) 1200	2000	Z Flying Height (m) 1200	2000
System Model						
ALTM 2050	0.27	0.45	0.29	0.48	0.12	0.18
ALTM 3100	0.17	0.27	0.23	0.38	0.10	0.15

23.3.2 SYSTEMATIC ERRORS

Systematic biases in the system measurements (e.g., GNSS/IMU-derived positions and attitudes, mirror angle measurements, and measured ranges) and calibration parameters (e.g., bore-sighting parameters relating the system components) will lead to systematic errors in the derived point cloud. The bias effect can be either derived through mathematical analysis of the LIDAR equation or the use of a simulation process. The simulation process starts with a given surface and trajectory, which are then used to derive the system measurements (ranges, mirror angles, position and orientation information for each pulse). Then, biases are added to the system parameters and measurements, which are used to reconstruct the surface through the LIDAR equation.

The differences between the bias-contaminated and true coordinates of the footprints are used to represent the impact of a given bias in the system parameters or measurements. Table 23.6 provides a summary of the various systematic biases and their impacts on the derived coordinates from a LIDAR system with a linear scanner. As it can be seen in this table, the impact of biases on the derived point cloud coordinates might depend on the flying direction, flying height, and/or the scan angle.

TABLE 23.6

Summary of the Impact of Biases in the Parameters and Measurements of a LIDAR System with a Linear Scanner on the Derived Point Cloud

	Flying Height	Flying Direction	Scan Angle
Bore-sighting offset bias	• Effect is independent of the flying height	• Planimetric effect is dependent on the flying direction • Vertical effect is independent of the flying direction	• Effect is independent of the scan angle
Bore-sighting pitch bias	• Effect is dependent on the flying height	• Planimetric effect along the flight direction is dependent on the flying direction	• Effect is independent of the scan angle
Bore-sighting roll bias	• Planimetric effect across the flight direction is dependent on the flying height • Vertical effect is independent of the flying height	• Planimetric effect across the flight direction and the vertical effect are dependent on the flying direction	• Planimetric effect across the flight direction is independent of the scan angle • Vertical effect is dependent on the scan angle
Bore-sighting yaw bias	• Effect is independent of the flying height	• Planimetric effect along the flight direction is independent of the flying direction	• Planimetric effect along the flight direction is dependent on the scan angle
Range bias	• Effect is independent of the flying height	• Planimetric effect across the flight direction and the vertical effect are independent of the flying direction	• Planimetric effect across the flight direction and the vertical effect are dependent on the scan angle
Mirror angle scale bias	• Effect is dependent on the flying height	• Planimetric effect across the flight direction and the vertical effect are independent of the flying direction	• Planimetric effect across the flight direction and the vertical effect are dependent on the scan angle

Except for corridor mapping, LIDAR flight missions are conducted in several strips with some overlap. Due to the presence of systematic errors in the data acquisition system, derived surfaces from neighboring strips might exhibit systematic incompatibility in the overlap area. The magnitude of this discrepancy, the nature of its variation across the flight line, and the direction of that incompatibility relative to the flight directions can be used to infer the presence of specific systematic errors. Therefore, checking the compatibility among overlapping LIDAR strips can be used as a QC procedure to verify the quality of the system and will be discussed in Section 23.4.

23.4   QA AND QC OF LIDAR SYSTEMS AND DERIVED DATA

Having discussed the impact of random and systematic errors in the LIDAR measurements and parameters on the derived coordinates of the point cloud, we will shift the focus to QA and the QC of LIDAR systems and derived data, respectively. In this context, the term "quality assurance" is used to denote activities focused on ensuring that a process will provide the quality needed by the user. For spatial data acquisition systems, QA mainly deals with creating management controls including the calibration, planning, implementation, and review of data collection activities. QA procedures are usually conducted prior to the surveying/mapping mission.

An example of a LIDAR QA activity is gaining prior knowledge about the area to be surveyed in terms of its extent and terrain coverage (e.g., vegetation and/or buildings) in order to set up the appropriate flight specifications. In forested areas, a slower flying speed, smaller scan angle, higher pulse repetition rate, and/or lower flying height might be necessary to increase the point density and to increase the probability of having more pulses penetrating to the ground. Also, the selection of the appropriate mission time according to the GNSS constellation distribution is another important item. For example, a typical requirement is to have at least four well-distributed satellites with elevation angles above $15°$ throughout the survey. Moreover, it is recommended that the aircraft should stay within a given distance from the GNSS base station.

Other than the previous QA activities, LIDAR system calibration is essential for ensuring the quality of the data to be acquired. LIDAR calibration, which aims at the estimation of system parameters, is usually accomplished in several steps: (1) laboratory calibration, (2) platform calibration, and (3) in-flight calibration. In the laboratory calibration, usually conducted by the system manufacturer, the individual system components are calibrated. In addition, the eccentricity and misalignment between the steering mirror and the IMU unit as well as the eccentricity between the IMU and the sensor reference point are determined. In the platform calibration, the eccentricity between the sensor reference point and the GNSS antenna is determined.

The in-flight calibration utilizes a calibration test field composed of control surfaces for the estimation of the LIDAR system parameters. Observed discrepancies between the LIDAR-derived and control surfaces are used to refine the bore-sighting parameters (misalignment between the mirror and the IMU) and systematic errors in the system measurements (mirror angles and ranges). Current in-flight calibration methods have the following drawbacks: (1) they are time-consuming and expensive; (2) they are generally based on complicated and sequential calibration procedures; (3) they require some effort in surveying the control surfaces; (4) some of the calibration methods involve manual and empirical procedures; (5) some of the calibration methods require the availability of the LIDAR raw measurements such as ranges, mirror angles, as well as position and orientation information for each pulse (Burman, 2000; Filin, 2003; and Skaloud and Lichti, 2006); and (6) there is a lack of a commonly accepted methodology since the calibration techniques are usually based on a manufacturer-provided software package and/or the expertise of the LIDAR data providers. These problems impede the users from having a standard calibration report with the associated measures, which quantify the quality of the calibration procedure. Current research is focusing on the development of a standardized calibration procedure as well as determining the appropriate control and flight configuration for reliable estimation of the system parameters (Skaloud and Lichti, 2006; Habib et al., 2007a). Until such procedures are developed, users of LIDAR data are in great need for practical and effective tools to evaluate the quality of the delivered point cloud. Such tools are usually provided through a QC mechanism.

The term "quality control" is used to denote post-mission procedures, which aim at providing checks to ensure data integrity, correctness, and completeness. In other words, the main objective of QC procedures is to verify whether the desired quality has been achieved or not. To illustrate what is meant by potential QC activities, one can refer to established QC procedures for a photogrammetric reconstruction exercise, where the external/absolute and the internal/relative and accuracy of the final product are checked. The external/absolute quality (accuracy) is evaluated through a checkpoint analysis using independently surveyed targets.

For the evaluation of the internal/relative quality (precision) of the outcome from a photogrammetric reconstruction exercise, we typically use the a posteriori variance factor and the variance–covariance matrix resulting from the bundle adjustment procedure. When deriving similar measures for the LIDAR data, we are faced with two challenges. The first challenge is caused by the irregular nature of the LIDAR footprints. Therefore, the possibility of correlating the LIDAR footprints in overlapping strips with physically identifiable control points in the object space for accuracy evaluation is quite difficult. Second, the ground coordinates of the LIDAR footprints are not based on the manipulation of redundant measurements in an adjustment procedure. Consequently, we do not have measures such as the a posteriori variance factor to evaluate the goodness of fit between the observed quantities and the estimated coordinates as expressed by the sensor model or the variance–covariance matrix, which quantifies the precision of the estimated parameters.

The external/absolute quality of the LIDAR data can be established using independently surveyed check point. As mentioned earlier, the irregularity of the LIDAR footprints makes it difficult to match a specific LIDAR footprint to a distinct object point. To overcome such a limitation, one can use a specially designed LIDAR target. The target design depends on the LIDAR system involved; an example of such a target design is shown in Figure 23.11a (a white circle inside a black ring). The target can be laid out in such a way that it is slightly above the ground surface, as shown in Figure 23.11b. The coordinates of the target centroid is derived from a GNSS survey. The target can be extracted from range and intensity LIDAR imagery using a segmentation procedure leading to an estimate of the LIDAR coordinates of the target centroid. Figure 23.12 illustrates how a target of the form shown in Figure 23.11a would look in each type of image; in the range image, it is a light (protruded) circle on a dark (sunken) background and in the intensity image, one can see the target's black and white pattern. The coordinates of the extracted targets are then compared with the surveyed

(a)   (b)

FIGURE 23.11   Check points for the external QC of LIDAR data: (a) target design and (b) target layout in captured LIDAR data. (From Csanyi, N. and Toth, C., *Photogram. Eng. Remote Sens.*, 73, 385, 2007.)

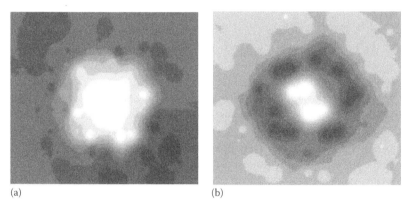

   (a)   (b)

FIGURE 23.12   A control target in LIDAR imagery: (a) range image and (b) intensity image. (From Csanyi, N. and Toth, C., *Photogram. Eng. Remote Sens.*, 73, 385, 2007.)

coordinates using a root mean square error (RMSE) analysis. The resulting RMSE value is a measure of the absolute/external quality (accuracy) of the LIDAR-derived surface. For a meaningful QC, the control target should be at least three times more accurate than the LIDAR surface (ASPRS, 2004). Establishing and surveying LIDAR targets is, however, an expensive and time-consuming procedure, and its implementation depends on the accessibility of the site to be mapped. Moreover, the identification of the targets in the intensity and range data depends on the point density. To ensure the availability of sufficient footprints over the target, we should have a high point density and/or large targets. These requirements limit the practicality of using control targets for the absolute QC of LIDAR data.

The evaluation of the internal/relative quality of LIDAR data, on the other hand, does not require control targets. Since the majority of LIDAR missions are conducted using multiple flight lines with some overlap among them (Figure 23.13), the internal/relative QC can be carried out by checking the relative consistency of the LIDAR data in overlapping strips. In the absence of systematic biases in the system measurements and/or parameters, there should be no systematic discrepancies between

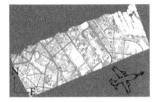

FIGURE 23.13   Three overlapping LIDAR strips.

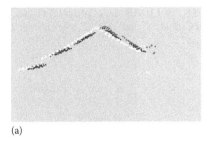

(a)   (b)

FIGURE 23.14   (a) Roof profile in three overlapping strips indicates the presence of systematic biases in the data acquisition system and (b) the quality of fit between conjugate surface elements after bias removal is used to quantify the noise level in the point cloud.

conjugate surface elements in overlapping strips. For non-consistent LIDAR strips, as can be seen in Figure 23.14a, one can use the detected discrepancies to infer the presence of systematic errors in the LIDAR parameters and/or measurements. Due to the irregular-distribution nature of the LIDAR point cloud, one cannot assume point-to-point correspondence in overlapping strips. Therefore, we cannot use conjugate points to test whether there are consistent discrepancies between overlapping strips.

Instead, one can use conjugate linear and areal features, which can be identified in overlapping strips. Such features, however, would require preprocessing of the LIDAR point cloud to extract areal and linear features (e.g., segmentation, plane fitting, and neighboring plane intersection). Moreover, linear and areal features can be reliably extracted only in urban areas. Such a restriction would limit the practicality of the QC procedure. To apply the QC in nonurban areas, one should use other primitives, which can be derived in urban and rural areas with minimal preprocessing of the original LIDAR footprints. To satisfy these objectives, one can represent one strip using the original footprints, while the second strip is represented by triangular patches, which can be derived from a triangular irregular network (TIN) generation procedure. TIN generation from a given set of points is an automated procedure, which is available in several GIS packages.

As an example, Figure 23.15 illustrates the case where the strip denoted by (S_2) is represented by a set of points while the other strip denoted by (S_1) is represented by a set of triangular patches. Due to the high density of the LIDAR data as well as the relatively smooth characteristics of terrain and man-made structures, using TIN patches to describe the physical surface is quite acceptable. It is quite obvious that there are some exceptions where the TIN patches would not represent the physical

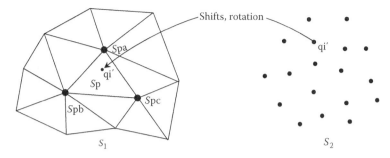

FIGURE 23.15   Point-to-patch correspondence in overlapping LIDAR strips.

surface (e.g., patches in vegetation areas and patches formed by vertices along ground and non-ground objects). Although one cannot assume that there is no point-to-point correspondence in overlapping strips, one can argue that we have point-to-patch correspondence as long as the patch represents the physical surface.

The identification of conjugate point–patch pairs in overlapping strips can be achieved by implementing an iterative closest patch procedure, which is a slight modification of the iterative closest point (ICP) procedure (Zhang, 1994; Cheng and Habib, 2007). Identified conjugate surface elements in overlapping strips can be used to estimate the necessary transformation parameters (e.g., shifts and rotation angles) for the alignment of these strips. Significant deviations from optimum values (i.e., zero shifts and zero rotation angles) can be used to infer the presence of systematic biases in the LIDAR system. After removing the effect of systematic errors (i.e., improving the consistency among overlapping strips by applying the estimated transformation parameters), one can use the quality of fit between conjugate surface elements to quantify the random noise level in the LIDAR data, Figure 23.14b. Such a measure can be evaluated by computing the average normal distance between conjugate point–patch pairs after bias removal. Utilizing such an approach one can quantify the noise level, precision, and biases in the data acquisition system. Therefore, one can argue that checking the consistency among conjugate surface elements in overlapping strips can be used to evaluate the precision and the accuracy of the LIDAR point cloud (i.e., the internal/relative and external/absolute quality). Moreover, the same procedure can be used to check the compatibility of the LIDAR data with an independently collected and more accurate surface model; thus directly quantifying the external/absolute quality of the point cloud.

23.5   LIDAR DATA PROCESSING

A LIDAR system delivers the ground coordinates of the laser beam footprints. Depending on the pulse repetition rate, the flying height, and the flying speed, the collected data set can be extremely large, which might limit the flexibility of dealing with such data. Therefore, following data collection, specialized processing techniques are necessary to abstract the huge number of the collected data to a more manageable size.

LIDAR data classification into terrain and off-terrain points and segmentation are two of the key LIDAR data processing techniques and will be discussed in this section.

Classifying the LIDAR footprints into terrain and off-terrain points is valuable for several applications. For example, the derivation of contour lines, road network planning, and flood-zone delineation require the availability of a terrain model. Building model generation, 3D city modeling, and forest mapping, on the other hand, are based on the manipulation of non-ground points. To satisfy the needs of these applications, the research community has been developing several techniques for the classification of LIDAR data. A substantial portion of existing techniques is based on mathematical morphology and includes the methodologies presented by Kilian et al. (1996), Vosselman (2000), Roggero (2001), and Zhang et al. (2003). In the proposed methodology by Vosselman (2000), the classification of LIDAR data is based on two operations: morphological erosion using a predefined discriminant function and comparison between the eroded and original surfaces. A point is classified as a ground point if its original height does not exceed the height of the eroded surface at the same location. A variation of this methodology was proposed by Roggero (2001), where the terrain is extracted using a local morphological operator. The shape of this operator is designed according to the slope of the bare earth. Since the shape of the bare earth is not known, a local linear regression procedure is used to derive an estimate of its shape. The main limitations of morphological filters are their sensitivity to the chosen discriminant function and the utilized window size as it relates to the size of above-ground objects such as buildings and forest canopy. For example, a small window will wrongly classify large buildings as ground. A larger window, on the other hand, might classify portions of a steep terrain as non-ground. To overcome the window size limitation, multiple window sizes have been proposed by Kilian et al. (1996) and Zhang et al. (2003).

Another group of LIDAR point classification techniques is based on the manipulation of a TIN structure. In Axelsson (2000), ground points are classified through progressive densification of a TIN model. The process starts with a coarse TIN, whose vertices are derived from the lowest points in local areas with a predefined size. Once again, the performance of this procedure depends on the local area size. The algorithm proposed by Sohn and Dowman (2002) fragments a LIDAR DEM, which has been convolved with heterogeneous terrain slopes, into a set of homogeneous subregions. The terrain is defined by subregions, which can be characterized by a single slope. The main problem with this methodology, as reported by Sithole and Vosselman (2004), is the unreliable classification of low and complex objects.

LIDAR classification using linear prediction and hierarchic robust interpolation has been proposed by Kraus and Pfeifer (1998, 2001), Pfeifer et al. (2001), and Briese et al. (2002). In this approach, one starts with a rough approximation of the terrain surface model. The defined surface model is iteratively reduced to the ground surface using linear prediction with the help of a predefined weighting function. The quality of the derived digital terrain model (DTM) depends on the design of the weighting function. Sithole and Vosselman (2004) reported that linear prediction procedures might lead to unreliable classification of low and complex objects. The last group of LIDAR point classification is based on segmentation techniques. Jacobsen and Lohmann (2003) developed a classification procedure, which starts by segmenting the point cloud. The segments are then categorized into ground and non-ground regions using the height difference between neighboring segments. Segmentation techniques are computationally expensive especially when dealing with large areas.

Other filtering algorithms are also introduced in Elmqvist et al. (2001), Haugerud and Harding (2001), Brovelli et al. (2002), Wack and Wimmer (2002), Masaharu and Ohtsubo (2002), and Akel et al. (2007). A detailed comparison of some of the existing LIDAR classification techniques is provided in Sithole and Vosselman (2004).

Figure 23.16 illustrates the LIDAR data classification process. The optical image in Figure 23.16a covers the same area mapped by the LIDAR system. The LIDAR data is represented as a shaded relief image in Figure 23.16b. The classified non-ground points are shown in white in Figure 23.16c. Figure 23.16d shows the generated DTM from the classified LIDAR terrain points. The classified terrain points can be used to derive contour lines. Identified non-ground points, on the other hand, can be further classified into vegetation, mature trees, and man-made structures (e.g., buildings). Such a classification can be carried out with the help of segmentation techniques, which will be discussed in the following paragraph.

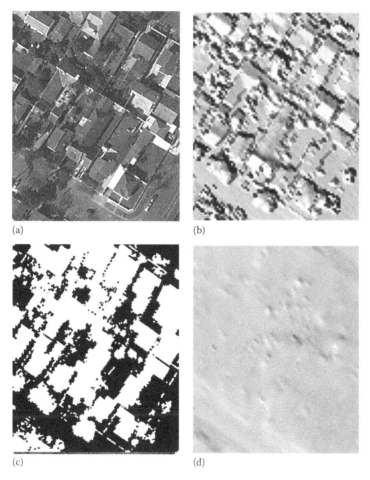

(a)

(b)

(c)

(d)

FIGURE 23.16 (a) Optical image, (b) over the area covered by the LIDAR data, (c) classified off-terrain points in white, and (d) generated DTM from the classified LIDAR terrain points.

In general, the objective of segmentation techniques is to cluster the LIDAR data into groups sharing similar attributes. Segmentation techniques usually start with the definition of the neighborhood for a given point. Neighboring points can be identified according to the proximity among LIDAR points and physical surface characteristics. An example of neighorhood definition is proposed by Filin and Pfeifer (2006), where the neighborhood is defined as neighboring points that belong to the same physical surface. A set of attributes for each point is then computed using its neighboring points. For example, the surface normal to the plane passing through the defined neighborhood can be used as the attribute for the LIDAR point under consideration. Finally, a clustering procedure, which is based on the similarity of the computed attributes of the points and the proximity of these points, can be implemented. In other words, neighboring points that have homogeneous attributes are augmented into one group. Axelsson (2000), Kim et al. (2007), and Filin and Pfeifer (2006) provide alternative procedures for LIDAR data processing and segmentation. As an example of the outcome from such data processing technique, Figure 23.17 shows the result of a segmentation technique applied to LIDAR data covering buildings.

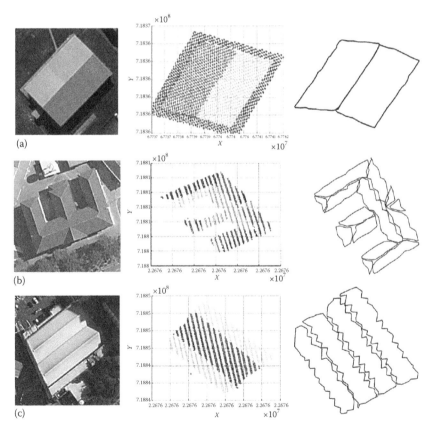

FIGURE 23.17 Aerial photos of the area of interest (left), segmentation results of the LIDAR data (middle), and detected boundaries from the segmentation results (right): (a) building with a gable rooftop, (b) E-shape building rooftop, and (c) saw-tooth building rooftop.

As it can be seen in this figure, the individual planar patches constituting the roof-tops of those buildings are identified (second column in Figure 23.17). The boundaries of these planar patches are also identified (third column in Figure 23.17). Thus, the LIDAR point cloud along the building rooftops can be represented, without any loss of information, by the boundaries of these planar patches. In other words, the LIDAR point cloud is abstracted into fewer points and the abstracted points are correlated with physical objects (these are the key objectives of LIDAR classification and segmentation techniques).

23.6 LIDAR APPLICATIONS

LIDAR data can be used for the delivery of diverse products: full-feature products, bare earth, contour lines, building footprints, land usage maps, transportation network, volumetric computation, power line maps, and utility corridors. The following list provides examples of the utilization of LIDAR data in several application fields:

- *Urban mapping*: 3D city modeling, wireless communications, change detection, emergency route planning, and signal propagation
- *Vegetation and forest mapping*: biomass volume calculations, DTM generation, and change detection
- *Shoreline mapping*: erosion monitoring and flood management
- *Power lines/pipelines/corridor mapping*: tree locations, tower locations, catenary models, and power line-to-vegetation critical distance analysis
- *Integration with photogrammetric data*: orthophoto generation and LIDAR-derived features as control for photogrammetric geo-referencing

Figures 23.18 and 23.19 provide illustrations of some LIDAR applications. Figure 23.18 shows the integration of photogrammetric and LIDAR data for orthophoto generation and the 3D realistic visualization of urban environments. As it can be seen in this figure, the derived visualization from the integrated photogrammetric and LIDAR data—Figure 23.18d—provides a better description of the environment when compared with that derived from either LIDAR or photogrammetric data. The integration of photogrammetric and LIDAR data for better description of the object space requires the registration of these datasets to a common reference frame. The co-registration of photogrammetric and LIDAR data can be achieved by using LIDAR derived features (e.g., linear and areal features) for the geo-referencing of the photogrammetric data (Shin et al., 2007). The next step in the visualization process is relating the positional and spectral attributes in the LIDAR and photogrammetric data, respectively. Establishing such a relationship can be achieved by orthophoto generation techniques, where the spectral attributes in the imagery is projected onto the LIDAR points. Differential rectification is a well-established procedure for orthophoto generation. Such a procedure, however, would lead to serious artifacts, which is commonly known as the double mapping problem, in the generated visualization when dealing with large-scale data over areas with sudden elevation changes (Habib et al., 2007b). These artifacts are caused by the fact that differential rectification does not consider the introduced occlusions by the relief displacement

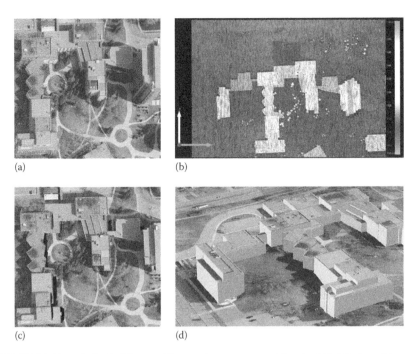

(a)   (b)

(c)   (d)

FIGURE 23.18 (a) Optical image (b) over the area covered by the LIDAR data, (c) generated true orthophoto, and a 3D realistic visualization from the integration of imagery and LIDAR data.

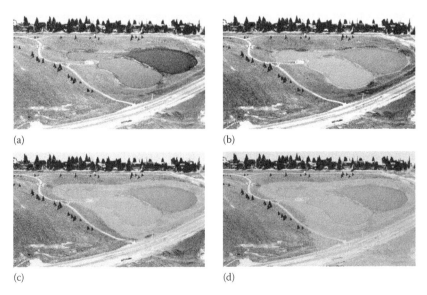

(a)   (b)

(c)   (d)

FIGURE 23.19 (a) Draped orthophoto on top of a LIDAR-derived DTM and (b, c, and d) flooded areas for three different water levels.

in perspective imagery. True orthophoto generation techniques are developed to deal with such a problem (Habib et al., 2007b). Following the generation of a true orthophoto, it can be draped on top of the LIDAR data to produce a 3D visualization, refer to Figure 23.18d. Figure 23.19 shows another illustration of the utilization of LIDAR data for the identification of flood-prone zones. Following the classification of LIDAR data into terrain and off-terrain points, a DTM can be generated using the classified terrain points. The resulting DTM can be used to highlight areas prone to flooding given different water levels. Using draped imagery over the DTM would show the affected areas more clearly.

REFERENCES

Akel, N., Filin, S., and Doytsher, Y., Orthogonal polynomials supported by region growing segmentation for the extracted of terrain from LiDAR data, *Photogrammetric Engineering and Remote Sensing*, 73(11): 1253–1266, 2007.

American Society of Photogrammetry and Remote Sensing LiDAR Committee, *ASPRS Guidelines–Vertical Accuracy Reporting for LiDAR Data and LAS Specifications*, May 2004. Ed. Flood, M. Retrieved April 19, 2007, from http://www.asprs.org/society/divisions/ppd/standards/Lidar%20guidelines.pdf

Applanix, 2007. POS AV specification, URL: http://www.applanix.com/media/downloads/products/specs/POSAV%20Specs.pdf (last accessed November 20, 2007).

Axelsson, P., DEM generation from laser scanner data using adaptive TIN models, *International Archives of the Photogrammetry*, 33(B4/1): 110–117, 2000.

Baltsavias, E., Airborne laser scanning: Existing systems and firms and other resources, *ISPRS Journal of Photogrammetry and Remote Sensing*, 54(2–3): 164–198, 1999.

Burman, H., Calibration and orientation of airborne image and laser scanner data using GPS and INS, PhD dissertation, Royal Institute of Technology, Stockholm, Sweden, 125p, 2000.

Briese, C., Pfeifer, N., and Dorninger, P., Applications of the robust interpolating for DTM determination, *International Archives of Photogrammetry and Remote Sensing*, 34(3A): 55–61, 2002.

Brovelli, M., Cannata, M., and Longoni, U., Managing and processing LiDAR data within GRASS, *Proc. GRASS Users Conference*, September 11–13, University of Trento, Trento, Italy, 2002.

Cheng, R. and Habib, A., Stereo-Photogrammetry for generating and matching facial models, *Optical Engineering Journal*, 46(7): 067203/1–067203/11, June 2007.

Csanyi, N. and Toth, C., Improvement of LiDAR data accuracy using LiDAR-specific ground targets, *Photogrammetric Engineering and Remote Sensing*, 73(4): 385–396, April 2007.

Elmqvist, M., Jungert, E., Lantz, F., Persson, A., and Soderman, U., Terrain modeling and analysis using laser scanner data, *International Archives of the Photogrammetry*, 34(3/W4): 219–227, 2001.

El-Sheimy, N., Valeo, C., and Habib, A., *Digital Terrain Modeling: Acquisition, Manipulation and Applications*, Artech House Remote Sensing Library, Boston, MA, 257p, 2005.

Filin, S., Recovery of systematic biases in laser altimetry data using natural surfaces, *Photogrammetric Engineering and Remote Sensing*, 69(11): 1235–1242, 2003.

Filin, S. and Pfeifer, N., Segmentation of airborne laser scanning data using a slope adaptive neighborhood, *ISPRS Journal of Photogrammetry and Remote Sensing*, 60: 71–80, 2006.

Habib, A., Bang, K., Shin, S., and Mitishita, E., LiDAR system self-calibration using planar patches from photogrammetric data, *The Fifth International Symposium on Mobile Mapping Technology (MMT'07)*, May 28–31, 2007, Padua, Italy, 2007a.

Habib, A., Kim, E., and Kim, C., New methodologies for true orthophoto generation, *Photogrammetric Engineering and Remote Sensing*, 73(1): 25–36, 2007b.

Haugerud, R. and Harding, D., Some algorithms for virtual deforestation (VDF) of LiDAR topographic survey data, *International Archives of the Photogrammetry*, 34(3/W4): 211–218, 2001.

Huising, E. and Pereira, L., Errors and accuracy estimates of laser data acquired by various laser scanning systems for topographic applications, *ISPRS Journal of Photogrammetry and Remote Sensing*, 53(5): 245–261, 1998.

Jacobsen, K. and Lohmann, P., Segmented filtering of laser scanner DSMs, *International Archives of Photogrammetry and Remote Sensing*, 34(3/W13), 2003.

Kilian, J., Haala, N., and Englich, M., Capture and evaluation of airborne laser scanner data, *International Archives of Photogrammetry and Remote Sensing*, 31(B3): 383–388, 1996.

Kim, C., Habib, A., and Mrstik, P., New approach for planar patch segmentation using airborne laser data. Alternative methodologies for the quality control of LiDAR systems, *American Society of Photogrammetry and Remote Sensing (ASPRS) Annual Conference: Identifying Geo-Spatial Solution*, May 7–11, 2007, Tampa, FL, 2007.

Kraus, K. and Pfeifer, N., Determination of terrain models in wooded areas with airborne laser scanner data, *ISPRS Journal of Photogrammetry and Remote Sensing*, (53): 193–203, 1998.

Kraus, K. and Pfeifer, N., Advanced DTM generation from LiDAR data, *International Archives of Photogrammetry and Remote Sensing*, 34(3/W4): 23–30, 2001.

Masaharu, H. and Ohtsubo, K., A Filtering method of airborne laser scanner data for complex terrain, *International Archives of the Photogrammetry, Remote Sensing and Spatial Information Sciences, Commission III*, 09–13, September, Graz, Austria, 34(3B): 165–169, 2002.

Optech, ALTM 3100 specifications, 2007. URL: http://www.optech.ca/pdf/Specs/specs_altm_3100.pdf (last accessed November 20, 2007).

Pfeifer, N., Stadler, P., and Briese, C., Derivation of digital terrain models in the SCOP++ environment, *Proceedings of OEEPE Workshop on Airborne Laserscanning and Interferometric SAR for Detailed Digital Terrain Models*, March 1–3, Stockholm, Sweden, 2001.

Roggero, M., Airborne laser scanning: clustering in raw data, *International Archives of the Photogrammetry*, 34(3/W4): 227–232, 2001.

Schenk, T., Modeling and analyzing systematic errors in airborne laser scanners, *Technical Report in Photogrammetry No. 19*, Ohio State University, Columbus, OH, 2001.

Shin, S., Habib, A., Ghanma, M., Kim, C., and Kim, E., Algorithms for multi-sensor and multi-primitive photogrammetric triangulation, *ETRI Journal*, 29(4): 411–420, August 2007.

Sithole, G. and Vosselman, G., Experimental comparison of filter algorithms for bare-earth extraction from airborne laser scanning point clouds, *ISPRS Journal of Photogrammetry and Remote Sensing*, 59(1–2): 85–101, 2004.

Skaloud, J. and Lichti, D., Rigorous approach to bore-sight self-calibration in airborne laser scanning, *ISPRS Journal of Photogrammetry and Remote Sensing*, 61: 47–59, 2006.

Sohn, G. and Dowman, I., Terrain surface reconstruction by the use of tetrahedron model with the MDL criterion, *International Archives of the Photogrammetry*, 34(3A). 336–344, 2002.

Vosselman, G., Slope based filtering of laser altimetry data, *International Archives of Photogrammetry and Remote Sensing*, 33(B3): 935–942, 2000.

Wack, R. and Wimmer, A., Digital terrain models from airborne laser scanner data—A grid based approach, *International Archives of the Photogrammetry*, 34(3B): 293–296, 2002.

Zhang, Z., Iterative point matching for registration of free-form curves and surfaces, *International Journal of Computer Vision*, 13(2): 119–152, 1994.

Zhang, K., Cheng, S., Whitman, D., Shyu, M., Yan, J., and Zhang, C., A progressive morphological filter for removing non-ground measurements from airborne LiDAR data, *IEEE Transactions on Geoscience and Remote Sensing*, 41(4): 872–882, 2003.

24 Selected Scientific Analyses and Practical Applications of Remote Sensing: Examples from the Coast

Thomas R. Allen and Yong Wang

CONTENTS

Remote sensing contributes profoundly to a wide array of scientific and practical applications. In the context of interdisciplinary research, remote sensing can contribute to inferential and exploratory analyses, deductive or hypothesis-driven research questions, post hoc validation, or continual observation and monitoring. Well-known remote-sensing applications include defense and reconnaissance, resource inventories such as agriculture, forestry, wetlands, land cover and land use patterns, and meteorology. Less well-known but emerging fields of applied remote sensing include health and coastal hazards. Remote sensing of coastal processes, such as sea-level rise, could demonstrate atmospheric and sea-surface temperature analysis for documenting the existence of thermal-induced expansion, apply terrestrial resources satellites (e.g., Landsat) to inventory, classify, and detect changes in shorelines and wetlands, and contribute to the long-term monitoring, erosion and subsidence rates, and hazard mitigation applications using very high resolution aerial photography, hyperspectral sensing, and active sensors including synthetic aperture radar (SAR) and light detection and ranging (LIDAR).

This chapter surveys remote sensing projects encompassing scientific analyses and practical applications. Although the chapter must necessarily compromise with respect to the depth and breadth of the applications presented, it can nonetheless present a wide range of remote-sensing applications. First, we provide an overview of the general approaches and background for the projects we discuss, including:

1. LIDAR data and coastal geomorphology
2. Shoreline delineation and change
3. Coastal floodplain inundation mapping
4. Remote sensing of mosquito breeding habitats
5. Estuary water quality and flushing

These applications are detailed in a series of case studies. In addition, we include a demonstration of a new technique for fusing multiple view-angle satellite data to present higher-resolution imagery. In the second part, we present more detailed project case studies within these larger research topics.

24.1 DATA ACQUISITION FOR REMOTE-SENSING PROJECTS

The design of remote-sensing projects frequently takes the form of a linear process with complexity and feedbacks concomitant to increasingly sophisticated information needs. Since virtually all remote sensing involves the collection, processing, and output of data, projects may begin with an assessment of needs. Very often needs assessments focus on the availability of data, including searches of existing data archives, then retrieval, acquisition, and assessments of the levels of preprocessing (atmospheric, radiometric, or geometric) required. New data collections are intimately involved with business practices and, for public agencies, established procurement processes and policies.

In the examples presented below, nearly every project begins with the assessment of available data and applies the knowledge of data producers, data archives, and geographic and database search engines. Projects relying upon satellite data typically focus on large national or international archives, such as the U.S. Geological Survey (USGS), the U.S. National Aeronautics and Space Administration's (NASA) distributed active archives (DAACs) (http://lpdaac. usgs.gov/), or commercial producers. Examples of data sources in this chapter also include airborne acquisitions of LIDAR data later provided online by the U.S. National Oceanographic and Atmospheric Administration's (NOAA) Coastal Services Center's LIDAR retrieval tool (http://maps.csc.noaa.gov/TCM/), ASTER imagery from the Land Processes DAAC, USGS, interagency multi-resolution land characteristics consortium (MRLC) (http://www.mrlc.gov/), and Alaska Satellite Facility (http://www.asf.alaska.edu/). In addition, sensor-specific databases or projects may be used, or data providers may sell (or resell) public domain data. Embedded in the search and acquisition process are several data- and project-specific decision characteristics, such as (1) spatial and spectral resolution characteristics; (2) extent of coverage, affecting the number of scenes, storage, and acquisition costs; (3) cloud-free conditions and time of day (particularly in cloudy, tropical, or projects conducted within narrow time-spans). Logistical factors are also in need of consideration, including the data format and delivery method. These range from disc-based storage media, portable hard drives, and electronic delivery. Even pervasive electronic delivery has recently morphed, from simple use of the file transfer protocol (FTP) download to vendors using the SKYPE™ voice and data service to ship compressed imagery. Thus, prior to data procurement, a user should have acquired a solid familiarity with remote sensing data providers, such as those presented in textbooks, references throughout this manual, or comprehensive online summaries of data providers. In addition to budgeting monies to cover the image acquisition, time should also be allocated to conduct search and acquisition processes. These issues are included in overviews of a select set of scientific and applied remote sensing.

24.1.1 LIDAR Data and Coastal Geomorphology

When studying morphology and morphologic changes on barrier islands, past researchers have faced the laborious task of surveying beaches, often in limited areal extents, using widely spaced transects and profiles, or by interpreting aerial photography. These time-consuming historic efforts designed to capture coastal topography can now be gathered within hours by an airborne LIDAR mission. Through the combined use of a laser transmitter with high repeating pulse frequency and a high-speed scanning system, as well as GPS and inertial navigation systems onboard, very dense measurements of x, y, and z on the surface by the laser beam are produced. For the purpose of beach, dune, or floodplain topographic mapping, the nominal postspacing of airborne LIDAR points are typically 1.0–1.5 m with a vertical accuracy of ±0.15 m. Thus, the creation of an elevation surface model with fine (submeter) resolution and vertical (z) accuracy less than 20 cm becomes possible.

LIDAR (see Chapter 23) has helped advance several areas of coastal research. Meredith et al. (1999) evaluated hurricane-induced beach erosion between 1997 and 1998 along the entire North Carolina coastline using LIDAR digital elevation models (DEMs) with a spatial resolution of 5 × 5 m. To accurately depict the topography of the dunes, however, Woolard (1999) suggested using DEM of spatial resolution less than 5×5 m. Thus, another volumetric change analysis of deposition or erosion was performed for the barrier islands of the North Carolina coast (White and Wang 2003). They used DEMs at 1.5×1.5 m spatial resolution acquired in 1997, 1998, 1999, and 2000. The high-resolution DEM data allowed comprehensive visual/quantitative investigation into the spatial patterns of morphology and morphologic change that occurred to the barrier islands' oceanfront beaches between 1997 and 2000. For instance, a before and after beach nourishment project was clearly portrayed by the DEMs (Figure 24.1), where the erosional signature is indicated by a drop of surface elevation (darker tone) on the beach and a shoreward shift of the wetted beach. Applying different management practices on barrier islands and continual development along the immediate coastal area will greatly affect the coastline's responses and possibly the outcome of the future coastline. Thus, LIDAR DEMs provide an extraordinary capability for capturing the coastline's ever-changing morphology in a quick, cost-effective manner, and offer enormous possibilities to enhance detailed knowledge of the coastal zone.

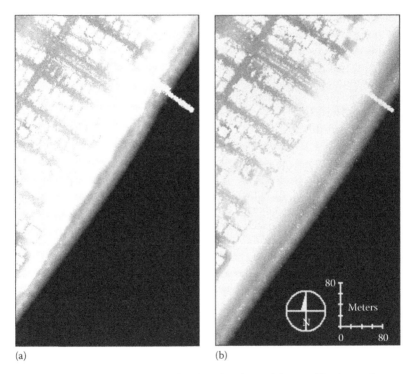

(a)   (b)

FIGURE 24.1 DEMs before (a) and after (b) a beach nourishment. The linear feature near northeast corner is a pier. The beach is roughly oriented about 50° from east toward north, and the ocean is in the east and southeast. LIDAR data for Wrightsville Beach, North Carolina were acquired from the NOAA coastal services center (CSC) topographic change mapping program. (From http://maps.csc.noaa.gov/TCM/.)

24.1.2 Shoreline Delineation and Change

The Climate Change 2007 Fourth Assessment report of Working Group II from the Intergovernmental panel on climate change (IPCC) (IPCC 2007) highlights the potential impacts and management options for policy on global climate change. The value of the coastal zone for fisheries, recreation, housing, coastal ecosystem integrity, water quality, storm damage protection, and other benefits is extremely high. Several technical developments in change detection generally point to the need for applying multifaceted techniques, whether shoreline changes (Allen and Oertel 2005), hurricane damage assessment (Collins 1997), or land use/land cover (Berlanga-Robles and Ruiz-Luna 2002). Some coastal areas are eroding faster than others are, yet extensive shorelines such as those found in major estuaries are difficult to map and monitor routinely. Furthermore, due to landward migration of shoreline, salt marsh, and estuarine ecosystems figure importantly in the study of past and future climate changes (Pizzuto and Rogers 1992, Moorhead and Brinson 1995, Michener et al. 1997, Mendelssohn and Kuhn 2004). However, a critical factor in this research and hazard mitigation is not readily available—the migration rate of estuarine shorelines.

Remote-sensing methods that improve our understanding and estimation of erosion or landward retreat rate (m/year) can be quite beneficial. The use of an active sensor such as synthetic aperture radar (SAR) is even more advantageous over the use of optical sensors because the SAR has its own energy source, and its microwave energy cannot only penetrate cloud but also vegetation canopy. Recently, Wang and Allen (2008) used L-band (24.0 cm wavelength) horizontally transmitted and horizontally received (HH) SAR data to delineate shorelines of the outer Pamlico Peninsula, Dare County, North Carolina in 1994 and 2006. Virtually no discernible shoreline changes on the north and south sides of Pamlico Peninsula were detected. However, significant landward migration in the central east-facing shore was observed. These findings not only support the utility of SAR for further monitoring of shorelines and shoreline changes in estuaries, but also demonstrate the value of the methodology for timely identification of erosional hotspots in coastal zones.

24.1.3 Inundation Mapping of Coastal Floodplains

Mapping inundation extent is a major need following riverine flooding or coastal storm surge events. The analysis includes the delineation of water versus "non-water" areas before and during the flood event of an area of interest, respectively. Once the assessment of water versus non-water areas has been made, one can determine whether an area is flooded or not by using change detection methods. Much attention has been given to mapping the extent of a flood by using optical, radar, DEMs, and river gauge data. The popularity of methods that use these data sets has risen due to their effectiveness, efficiency, availability, and low or zero end-user cost. Success stories can be found (e.g., Brackenridge et al. 1998, Correia et al. 1998, Melack and Wang 1998, Kraus 2000, Colby et al. 2001, Wang et al. 2002, Henry et al. 2006, Martinez and Le Toan 2007, North Carolina Floodplain Mapping Program 2008). In the following sections, the strengths and weaknesses of the individual and combined use of optical, radar, DEM, and river gauge data to map the flood extent will be discussed.

24.1.4 ESTUARY WATER QUALITY AND FLUSHING

Wetlands, estuaries, tidal creeks, and coastal bays often serve as receptacles for pollutants from anthropogenic activities. If pollutant delivery is not buffered, or if pollutants are not flushed from these environments, loading may cause severe damage to natural resources. Pollutant and sediment loading in estuarine environments is related to flushing, particularly tides and basin hypsography (relationship between elevation and surface area), affecting marsh loss, flooding, and vegetation buffers. Stagnant coastal bays will tend to accumulate pollutants from anthropogenic input and are therefore more vulnerable than bays that are characterized by rapid water exchanges. Potential pollutant loading input is related to the watershed land surface area, number of tributaries, and length of shoreline. These important flushing characteristics of basins have not been widely considered in coastal management. Tidal regime, freshwater input, and dynamic shorelines and bathymetry are also factors that affect remote sensing of estuarine features (Jensen et al. 1993).

Remote sensing can provide extensive information on flushing and circulation patterns in estuaries for coastal management using NASA earth observing resources. Potential applications of multi-waveband, thermal, and stereoscopic capable sensors, such as NASA's advanced spaceborne thermal emission and reflectance radiometer (ASTER) satellite appear to offer untapped potential. ASTER collects remote sensing data in three spectral regions, the visible to near-infrared (VNIR), shortwave infrared (SWIR), and thermal infrared (TIR) (Abrams 2000). For estuarine environmental applications, the satellite has five thermal wavelength bands with 90 m spatial resolution, lower than Landsat's single thermal channel (60 m) but with finer spectral resolution in the thermal IR wavelengths and is designed for temperature and emissivity separation (Cjakowski et al. 2004). The thermal IR system has five spectral bands covering 8.125–11.65 μm at 90 m spatial resolution. The scientific product that can be used for surface water temperature requires a temperature-emissivity separation algorithm (Gillespie et al. 1998), leading to a surface kinetic temperature product. An aquatic application of ASTER data was demonstrated by Böhme et al. (2006), who used them to monitor changes in the extent of Lake Urema, Mozambique, that were associated with hydrology and geomorphology. In a study that closely matches the scales of the local bays and estuaries in the United States, Kishino et al. (2005) illustrated the potential for high-resolution ASTER imagery to assess phytoplankton and colored dissolved organic matter (CDOM), using a neural network classifier to separate chlorophyll *a* and suspended matter in optically complex coastal waters, with less success in distinguishing CDOM. Stereoscopic observations taken by ASTER also facilitate measurement of short-term displacement of surface currents and waves (Matthews 2005), and its thermal infrared channels enable high-resolution mapping of variations in sea surface temperatures.

24.1.5 REMOTE SENSING OF MOSQUITO BREEDING HABITATS

The reemergence and diffusion of vector-borne diseases presents a problem of international scope that must often be managed by a frontline of local environmental health and mosquito control specialists. Mosquito-borne diseases such as West

Nile Virus, Dengue Fever, or even malaria, have the potential for reemergence or spread in developed countries. Locating and managing mosquito breeding over large areas is a difficult and labor-intensive task for these personnel. Given the high profile of public health threats, the mosquito control industry and professional organizations are taking advantage of digital satellite data, GIS, and image processing capabilities for surveillance and mosquito control in the United States. Prevention of mosquito-borne disease transmission involves planning for effective surveillance, control, and risk reduction. Surveillance involves the assessment of habitats, seasonal patterns, and ecology of the vectors in a region, as well as the monitoring (spatial and temporal) of traps, complaints, and, when known, cases of transmission. Catch basins, storm water drains, and retention ponds, thus, are critical areas for surveillance and the design of programs to kill mosquito larvae in support of urban mosquito control programs. By virtue of their enclosed character, such basins may require field surveillance or mapping by the local municipality, and most urban centers have a data infrastructure to house these data within their GIS databases. Because mosquitoes retransmit and thus "amplify" the virus in local bird populations, during large mosquito "bloom" events, the presence of rich reservoirs, abundant vectors (insects or animals capable of spreading the disease), and dense human populations in proximity may lead to outbreaks or epidemics. In this context, the ability to map precise locations and conditions of disease transmission is extremely significant to breaking transmission paths and protecting the populace (Meade and Earickson 2005). For instance, vernal forest water pools depend on rainfall and on the morphology of the terrain and vegetation, but once "activated," create breeding habitat that may be monitored or controlled. Prevention is thus accomplished by knowing the spatial domain and temporal cycles of reservoir, vectors, and the populations at risk. Remote sensing can inventory, identify, and monitor trends, to different degrees, for each of these factors in coordination with meteorological and species population cycles.

24.2 PROJECT CASE STUDIES

24.2.1 Mapping Flood Extent Using Remotely Sensed Data Sets

Optical remote sensors measure solar reflectance from objects on the ground but have limitations for mapping the extent of inundated land. Because dry surfaces and wet/open water surfaces have distinctly different reflectance characteristics, optical data can easily distinguish these surfaces. The advantages in using the optical data in flood extent mapping include its high reliability and accuracy, efficiency, cost-effectiveness, and ease of use. A possible drawback is the lack of canopy penetration—flooded areas under the canopy cannot be detected. These limitations are especially troublesome in coastal floodplains, where flooding commonly occurs in summer or fall, when broadleaf forests are observed leaf-on, obscuring flooded regions from optical observation. Applications of optical wavelengths are also limited by limitations imposed by cloud cover and by fixed orbits and predetermined ground tracks that constrain the flexibility of temporal resolution with respect to a specific target or study area. Finally, the relatively coarse spatial resolution of some satellite data may impose constraints. However, due to the current availability of high-resolution data, the resolution issue is much less a problem now (and for the future) than in previous years.

The basic principles of using SAR data to map the flood extent are similar to those used with the optical data discussed above. The advantages of SAR over an optical sensor are the use of its own energy source, its all-weather capability, and its ability to penetrate vegetation canopy. SAR's penetration ability is especially true for systems using long wavelengths, such as an L-band system (relative to radar systems of shorter wavelengths, such as a C-band (5.6 cm wavelength) sensor. Currently, ERS-2 (Earth Resources Satellite-2) SAR and ASAR (Advanced Synthetic Aperture Radar) operated by the European Space Agency, RadarSat-II SAR operated by the Canadian Space Agency, and Advanced Land Observation Satellite (ALOS) PALSAR operated by the Japanese Aerospace Exploration Agency (JAXA) are in orbit. ERS-2 SAR is a C-VV (vertically transmitted and vertically received) system. RadarSat-II SAR and ASAR are C-band and multiple polarization radars. PALSAR is an L-band and multiple-polarization sensor (European Space Agency 2007). Primary concerns include fixed satellite orbits, and difficulty to analyze and interpret the radar data. These sensors may not collect continuously due to high consumption of electric power. In addition, the penetration into tree canopies from the C-band SARs can be limited, and in some cases, the SARs may not even penetrate forested areas with dense canopies.

LIDAR and satellite-derived topography are also useful for flood mapping. The USGS has created comprehensive DEMs for the United States. In the 2000 space shuttle radar topography mission (SRTM), NASA jet propulsion laboratory created global DEMs for land areas between latitudes of 60° N and 60° S. The spatial resolution and vertical accuracies of the DEMs vary. The USGS and the U.S. Environmental Protection Agency (EPA) operate a network of gauge stations on rivers in the United States. At many stations, a stream's water surface height, discharge, and flow velocity are measured daily and are available online. The DEMs and river gauge readings can be combined to study the inundation extent of a flood. The advantages of using the DEM and river gauge data are because they are reliable, accurate, efficient, and allow accessible ground verification. However, one concern is the accuracy of the DEMs. DEM error is the major cause of failure in determining flooded areas away from primary river channels (Wang et al. 2002). However, such inaccuracies can be reduced with the initiation of the national LIDAR topography mapping initiative that aims to create nationwide DEMs with high spatial resolution and high horizontal and vertical accuracies (http://lidar.cr.usgs.gov/). Consequently, the availability of multiple digital data sets and data developments point toward the increased uses of remote sensing data and technology in floodplain mapping. The creation of methods that integrate various types of data sets and take advantage of the strengths of each data set remains the primary challenge.

24.2.2 Topographic LIDAR Analysis of the Outer Banks, North Carolina

A barrier island is an elongated, narrow strip of sand positioned just offshore, and parallel to a mature coastline. Characteristically distinct features include a shoreface, dune fields, overwash fans, grasslands, marshes, shrubs, maritime forest, and tidal flats. The barrier island is a dynamic feature, subject to a complex inter-twining of physical processes, such as the wave action of water and aeolian effects that act to shape and mold their form through time. Weather systems such as a hurricanes, nor'easters, and frontal passages are known to be significant events that can affect the form and topography of the island. Since the mid 1990s, NASA and NOAA Coastal Services Center have conducted airborne LIDAR missions jointly to image North Carolina's Outer Banks. Thus the topography of these barrier island beaches have been scanned multiple times, providing data for the analysis of morphological changes in erosional and depositional landforms, and permitting calculation of volumetric change over large regions of these barrier islands. The Outer Banks, like other developed barrier islands, are experiencing conflicts between development and erosional processes, especially where "hot spots" of erosion that have been difficult to predict. One extreme example is represented in Figure 24.2, depicting nine houses near the town of South Nags Head, Outer Banks, which are literally on the beach or under water during high tide. Figure 24.3 presents a series of LIDAR DEMs near the nine houses and shows the loss of beach and dune, as well as the exposure of the houses to the Atlantic Ocean.

(a)

(b)

(c)

FIGURE 24.2 **(See color insert following page 426.)** Field photographs of airborne LIDAR site taken in May 2008. (a) A side-view of the houses during low tide, (b) a front-view of one house during the low tide, and (c) a ground-view of one house during high tide. The large sand bags cover the septic tank of the house. (Photography by Y. Wang.)

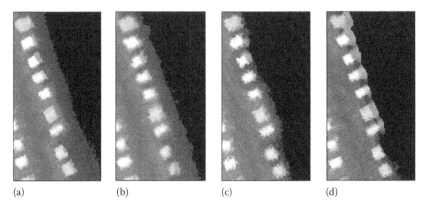

(a)   (b)   (c)   (d)

FIGURE 24.3 LIDAR DEMs in 1997 (a), 1998 (b), 1999 (c), and 2005 (d). The Atlantic is toward the right in dark tone. The nine bright, rectangular shapes are beachfront homes.

24.2.3 THERMAL REMOTE SENSING OF ESTUARINE FLUSHING

This case study evaluates the potential for remote sensing to distinguish relative flushing of coastal bays and estuaries using ASTER thermal band imagery. The application, if shown feasible, would validate locations where water quality monitoring and modeling have indicated problems, as well as opened the door to the operational use of remote sensing for water quality mapping at regional to local scales of observation. Remotely sensed spectral signatures were evaluated for characterizing the exchange of bay water with ocean water (flushing) and in particular, that element of the flushing that floods the lagoon, fills up the tidal prism, and returns to the ocean (repletion). However, daytime multispectral images lacked the ability to discriminate tidal circulation between the lagoon and ocean. These patterns of circulation are also complex, hydrodynamic patterns that require bathymetric, tidal, wind, and other physical factors in order to be mapped and characterized in a GIS (Allen et al. 2007). Thermal distinctions between ocean and bay waters in ebb-tidal, flood-tidal, and seasonal conditions were instead investigated to measure the extent and pattern of flood tide penetration into the lagoon and estimate the volume of water that exchanges during each tidal cycle. First, the moderate resolution imaging spectrometer (MODIS) was evaluated over a long time series of cloud-free, composited imagery (spanning years 2000–2004) in order to assess which season/s provide the strongest thermal differentiation between estuarine and ocean sea surface temperatures (using the MODIS terra sea-surface temperature (SST) products) (Chapter 18). Higher resolution ASTER imagery (Chapter 18) would be analyzed within an estuary to map water masses. Thus, residual waters with longer residence times in the lagoon and potentially higher, undiluted nutrient and pollutant loads would identify "Risk zones" in coastal bays in concert with the distance-to-inlet and signature gradients. The result could be a prototype system for evaluating environmental risks to the numerous coastal bays on the eastern seaboard.

The MODIS time series analyses revealed that stronger contrast in SSTs are found in spring and fall, but seasonal weather variations and the tidal conditions at the

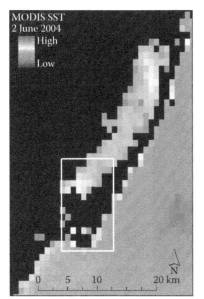
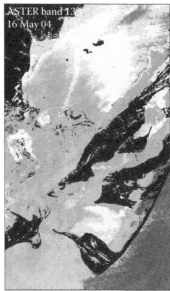

FIGURE 24.4 **(See color insert following page 426.)** Comparison of thermal sea-surface temperature from MODIS and ASTER in Chincoteague Bay and nearshore Atlantic. Imagery acquired from the USGS land processes distributed active archive data pool. (From http://lpdaac.usgs.gov/datapool/datapool.asp. With permission.)

time of acquisition must also be taken into account. An evaluation of ASTER quick look imagery for visual contrasts between the thermal signatures of estuarine and ocean water affirmed the potential to distinguish these features. ASTER thermal bands differentiated distinct temperature gradients and masses of water bodies in the coastal bays, suggesting possible thermal signature extraction and image classification. Figure 24.4 illustrates the contrast in spatial resolution affecting discrimination of thermal features within the lagoon, inlet, and ocean between MODIS and ASTER. Subsequent analyses examined higher spatial resolution ASTER data, but required a time series of imagery to control for diurnal, tidal, seasonal, interannual, and event-driven variations (e.g., upwelling or cyclone storm surges.) The project then proceeded with downloading and further preprocessing of level 1B imagery from the ASTER Data Pool of the US Geological Survey Land Processes DAAC Data Pool (http://lpdaac.usgs.gov/datapool/).

Terrestrial watershed data and a tidal prism model were combined with image analysis to produce information products, including maps showing tidal penetration, spectral and thermal signatures of representative water conditions, potential anoxic water body locations, coastal bathymetry and marsh shorelines, and statistical metrics on the flushing of coastal bays (tidal turnover time, flushing volumes, and summary profiles of bays and watersheds). Figure 24.5 shows a time series of ASTER images taken over winter, spring, and summer months focused on the extent of flood and ebb tidal circulation between Chincoteague Bay and the near shore ocean in Chincoteague Bight.

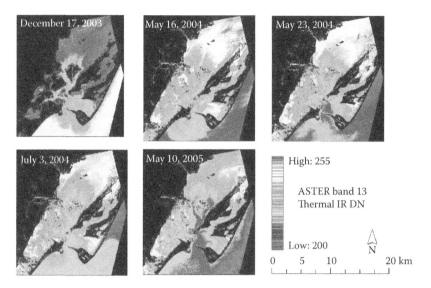

FIGURE 24.5 **(See color insert following page 426.)** Time series of nighttime ASTER thermal band 13 images for Chincoteague Inlet, Virginia. Colorized thermal DN values contrast seasonal temperature differences between oceanic coastal waters and estuarine water with high resolution, 90 m. Land pixels are masked using daytime multispectral reflectance.

The December 2003 image illustrates the typical winter to early spring contrast between ocean and estuarine water temperature as well as the coldest, shallow bay locations. The two, May 2004 images, show a gradual change to a warmer bay versus ocean as well as the degree of mixing indicated by moderate temperatures, notable as these images occurred at mid-ebb tide. The final image of Figure 24.5 shows the strong contrast of a flood tide plume entering Chincoteague Bay via channels constricted by marshes the following May 2005 at the peak of flood tide penetration. Ocean water temperatures appear lower than 2004, but the hydrodynamics persist in creating approximately the same control on tidal flood extent. These inferences were used to affirm the measurement and modeling of tidal flushing and repletion (Allen et al. 2007) and illustrate the potential for mapping and monitoring estuarine processes.

24.2.4 INTEGRATING REMOTE SENSING AND FIELD SURVEILLANCE OF MOSQUITO VECTORS

24.2.4.1 Approach

Remote sensing, coupled with the growing capacity of local-level vector control programs' GIS and GPS infrastructure, could provide the needed technological means to perform their critical functions more efficiently and effectively. Early warning systems, for instance, may combine GIS and remote sensing and include reporting systems, risk mapping, and environmental early warning systems (Myers et al. 2000). The example, detailed in Allen and Shellito (2008), illustrates the integration of remote sensing, field surveillance data, and geostatistics, by a suite of spatial statistics for interpolating values at unsampled locations, gauging uncertainty in the predicted values, and simulating attributes for locations with the limited information available. For the purpose of mosquito surveillance and control, field sampling design and collection are labor-intensive and cost-prohibitive tasks. Geostatistics offers a means to more effectively sample and estimate the distribution of mosquito vectors. Insect populations are typically heterogeneous in their spatial densities, as they are responding to multivariate habitat characteristics and environmental controls. In addition, GIS would provide the integrative analysis and database for combining the remote sensing, field data, and ancillary environmental data (hydrology, runoff, storm water basins, and affected population).

24.2.4.2   Study Area, Mosquito Data, and Imagery Acquisition

Fairfax County, Virginia, is a substantial portion of the Washington, DC, metropolitan area with a population of approximately 970,000 people within an area of 400 square miles (U.S. Census 2000). The City of Chesapeake, Virginia, is a populous suburb of the Norfolk metropolitan area, adjacent to the Great Dismal Swamp National Wildlife Refuge. Both areas are typical of mid-latitude humid subtropical climates, having moderately mild winters and hot, humid summers. Mosquitoes may survive mild winters, and blooms may occur throughout the year but particularly May–October. Data acquisition procedures coupled field-based surveillance of mosquito breeding pools using Centers for Disease Control and Prevention (CDC) light traps and gravid traps with the assessment and acquisition of cloud-free Landsat digital data over northern Virginia. Concurrent to field trapping, Landsat-5 Thematic Mapper (TM) and Landsat-7 Enhanced Thematic Mapper (ETM+) satellite data acquisitions were assessed for cloud-free observation of the study area. The available cloud-free ETM+ imagery and approximately weekly compositing of mosquito abundances combined for a synoptic perspective on the regional landscape while capturing important trends in the developing mosquito population. Landsat digital data were searched using the USGS' Eros Data Center Earth Explorer system (http://edcsns17.cr.usgs.gov/EarthExplorer/) and delivered electronically via FTP.

24.2.4.3   Image Processing and Spatial Analysis

Image enhancements for data analysis and mosquito habitat investigation applied the Normalized Difference Vegetation Index (NDVI) and the Tasseled Cap transformation to Landsat ETM+ data (Kauth and Thomas 1976, Crist and Ciccone 1984, Huang et al. 2002). The NDVI was run within the Erdas IMAGINE software (Chapter 20) using standard normalized difference of bands TM4-TM3 vegetation index for each scene. USGS 1:24,000 DEMs for the study area were used to characterize environmental conditions arising from landform and topography. Elevation, slope, aspect and solar insolation, and topographic soil moisture potential were derived using standard geomorphometric procedures for gridded DEMs (cf., Gallant and Wilson 2000). A suite of kriging analyses (Davis 2002) were performed to assess the utility of remote sensing, GIS, and integrated GIS-RS data for improving the spatial prediction of mosquito abundance. The sampled mosquito pools were aggregated into totals of vector species per site per weekly, monthly, and seasonal abundance. Ordinary kriging (OK) was applied to the mosquito pool data set alone, to produce a prediction of mosquito abundance.

24.2.4.4   Results of Integrated Remote Sensing and Mosquito Surveillance Data Analysis

Co-kriging predictions of primary mosquito abundance were improved by the integration of spatially correlated secondary data, remotely sensed indices and GIS environmental variables. A sample of mosquito abundances predictions are shown in Figure 24.6 for the City of Chesapeake, Virginia, in conjunction with the locations of "found dead birds", "resident requests for mosquito control," and the Tasseled Cap moisture index from Landsat-7 ETM+. The imagery provided rich data, including the best performing sets of co-kriging spatial variables, combined with the biweekly mosquito

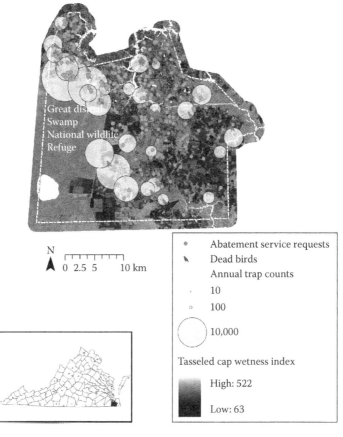

FIGURE 24.6   (See color insert following page 426.) Integrated remote sensing and field mosquito data for Chesapeake, Virginia, showing mosquito trap counts, dead bird counts, resident abatement service requests, and Landsat ETM+ Tasseled Cap wetness index. Imagery acquired from the University of Maryland global land cover facility. (Courtesy of http://www.landcover.org/index.shtml.)

counts occurring simultaneous to the Landsat image acquisition date. The pattern of mosquito abundances exhibit a strong degree of temporal autocorrelation, both in the field observations and the co-kriging predicted surfaces. These results affirm the role of remote sensing imagery for local surveillance and control of mosquito vectors. Given high-resolution cloud-free satellite imagery, geostatistical techniques were demonstrated to successfully interpolate mosquito abundances with limited ground data.

The advantages of remote sensing in this case must be matched to diverse public health monitoring and control operations, rather than to a linear view of what imagery can produce (e.g., land cover/use, environmental parameters, and trends). The scientific application of remote sensing is a useful tool to improve the efficiency and effectiveness of mosquito control operations, advanced as they may be considering the modern use of biomedical techniques to screen for diseases in mosquitoes and human serological samples. While case studies of remote sensing in mosquito-borne disease surveillance are reported in the scientific literature (Imhoff and McCandless 1988, Pope et al. 1992, Washino and Wood 1993), operational and sustained uses of remote sensing in the health industry and public sector offer an opportunity for growth. Bellows (2007) developed and implemented an operational system using mosquito species groups' population cycles to predict mosquito abundance from integrated weather, remote sensing, and GIS data.

24.3 DATA FUSION OF SPATIALLY OVERLAPPED MULTI-VIEW IMAGERY

There is a need for high spatial resolution or super-resolution (SR) data (Park et al. 2003). For instance, to study land cover and land use (LCLU) and their changes in urban settings, the spatial resolution decreases from a pixel of 30–80 × 30–80 m to a pixel of less than 0.5 × 0.5 m in size as the LCLU classification level increases. Investigations of coastal changes, such as storm erosion or sea-level rise, will often present the problem of incipient changes (e.g., widening of tidal creeks), that are not detectable at a resolution of 30 × 30 m (Rogers and Kearney 2004). Spectral unmixing shows promise to resolve these changes, but improving the spatial resolution or merging high spatial resolution and multispectral images could also be explored. Increasing the spatial resolution of an image normally requires the use of an instrument with a narrow instantaneous field of view, usually a prohibitively expensive option. Another alternative is to derive the high-resolution data from a set of low-resolution (LR) data through data fusion or the SR reconstruction. One advantage of the SR reconstruction approach is that an existing LR imaging system and its data sets can still be utilized, effecting cost-reduction.

To map global topography, JAXA scientists developed the panchromatic remote-sensing instrument for stereo mapping (PRISM), carried onboard the ALOS, launched in January 2006. The PRISM sensor consists of three identical sensors (forward, nadir, and backward looking), and operates in nine modes of data acquisition (http://www.eorc.jaxa.jp/ALOS/about/prism.htm). In mode 1, it collects the data in triplets, that is, three nearly fully overlapped stereo images or the same ground area viewed by the forward-, nadir-, and backward- view sensors sequentially. Each triplet covers a ground area of 35 × 35 km. The spatial resolution of the individual view is 2.5 × 2.5 m, which

(a)   (b)

FIGURE 24.7 An agricultural area in North Carolina, November 26, 2006, showing (a) nadir-view zoomed in by 2×, and (b) the 2× high resolution image reconstructed through fusion of stereo Advanced Land Observation Satellite (ALOS) Panchromatic Remote-sensing Instrument for Stereo Mapping (PRISM), from the Japanese Aerospace Agency (JAXA). (Image ordered from the Alaska Satellite Facility.)

is coarser than the spatial resolution of panchromatic data of IKONOS and QuickBird. However, the cost of a triplet (http://www.asf.alaska.edu) is currently US$125, which is much cheaper (per square kilometer) than that of the IKONOS or QuickBird data (http://www.geoeye.com). With the three-looking configurations in geometry, the same ground location is observed at three different viewing points. The observations can have some overlap spatially. Thus, the combined observation of a triplet should have higher sampling density (as compared to each individual image of the triplet) spatially. Therefore, one can create data sets with higher spatial resolution than with individual (forward-, nadir-, or backward-view) images. Figure 24.7 presents an example, in which high spatial resolution (1.25 × 1.25 m) data were created through the reconstruction of the low resolution (2.5 × 2.5 m) triplet using a super resolution reconstruction fusion technique (Park et al. 2003). The technique can be applied to other sensors such as Leica's ADS40 that has the multi-viewing capability. This project demonstrates the utility of emerging remote sensing methods that provide adequate resolution for fine-scale mapping.

24.4 CONCLUSIONS

Drawing from the authors' research programs, in this chapter we provide samples of scientific and applied remote sensing projects. Our emphasis on coastal and satellite-based sensors demonstrates that even within a particular domain, remote sensing can provide a diversity of data sources, image processing techniques, and information products. We have identified challenges and needs assessment requisite for imagery research and acquisition, as well as the general approaches for LIDAR coastal topographic mapping, shoreline mapping and change detection, estuary tidal flushing, and integrated remote sensing in environmental health. Examples included active sensors (SAR, LIDAR) and passive optical and multispectral data (ASTER, MODIS, PRISM, and Landsat.) Image processing techniques ranged from visual image interpretation, multispectral enhancement, image classification, and LIDAR DEM raster analysis. In addition, each case study also describes a scientific contribution, methodological advancement, or environmental resource management application. In the development of the applied projects, especially those devoted to the monitoring of estuarine flushing and mosquito control surveillance, the authors also actively engaged practitioners who were unfamiliar with the potential of remote sensing but who became sufficiently educated to employ it. These experiences parallel many historical phases of user adoption, where an innovation first replicates, automates, and eventually surpasses the analytic capabilities of a traditional technology. Thus, we provide a set of examples that may also exemplify a continuum of "Technology Readiness Levels," a metric developed by NASA to quantitatively gauge the maturity of certain technologies and their suitability for operational use (Mankins 1995). We can conclude that LIDAR coastal morphology, estuarine, and mosquito applications are sufficiently mature for routine implementation. In contrast, the demonstrated projects using multiple view-angle image fusion and SAR analysis of coastal change are still relatively nascent techniques. Collectively, these examples of remote sensing projects, even if extracted from the coastal research of two investigators, are illustrative of the strong continued growth of remote sensing in environmental sciences and natural resource management.

REFERENCES

Abrams, M., The advanced spaceborne thermal emission and reflection radiometer (ASTER): Data products for the high spatial resolution imager on NASA's terra platform, *International Journal of Remote Sensing*, 21(5):847–859, 2000.

Allen, T.R. and Oertel, G.F., Landscape modifications by Hurricane Isabel, Fisherman Island, Virginia, In: Sellner, K.G. Ed., *Hurricane Isabel in Perspective: Proceedings of a Conference*, November 15–17, 2004, Chesapeake Research Consortium Publication 05-160, pp. 73–79, 2005.

Allen, T.R. and Shellito, B.A., Spatial interpolation and image-integrative geostatistical prediction of mosquito vectors for arboviral surveillance, *Geocarto International*, 23(4):311–325, 2008.

Allen, T.R., Tolvanen, H., Oertel, G.F., and McLeod, G.M., Spatial characterization of environmental gradients in a coastal lagoon, *Estuaries and Coasts*, Chincoteague Bay, MD, 30(6):1–19, 2007.

Bellows, A.S., *Modeling* Habitat and environmental factors affecting mosquito abundance, Chesapeake, Virginia, Doctoral dissertation, Ecological Sciences, Old Dominion University, Norfolk, VA, June 2007.

Berlanga-Robles, C. and Ruiz-Luna, A., Land use mapping and change detection in the coastal zone of Northwest Mexico using remote sensing techniques, *Journal of Coastal Research*, 18(3):514–522, 2002.

Böhme, B., Steinbruch, F., Gloaguen, R., Heilmeier, H., and Merkel, B., Geomorphology, hydrology, and ecology of Lake Urema, central Mozambique, with focus on lake extent changes, *Physics and Chemistry of the Earth*, 31:745–752, 2006.

Brackenridge, G.R., Tracy, B.T., and Knox, J.C., Orbital SAR remote sensing of a river flood wave, *International Journal of Remote Sensing*, 19:1439–1445, 1998.

Cjakowski, K., Goward, S.N., Mulhern, T., Goetz, S.J., Walz, A., Shirey, D., Stadler, S., Prince, S.D., and Dubayah, R.O., Estimating environmental variables using thermal remote sensing, in: Quattrochi, D. and Luvall, J.C. Eds., *Thermal Remote Sensing in Land Surface Processes*, CRC Press, Boca Raton, FL, pp. 11–32, 2004.

Colby, J.D., Mulcahy, K., and Wang, Y., Modeling real-time flooding extent due to Hurricane Floyd in the coastal plains, *Global Environmental Change Part B: Environmental Hazard*, 25p, 2001.

Collins, R.F., Assessing the impact of Hurricane Hugo on coastal South Carolina through digital image change detection, *Southeastern Geographer*, 37(1):76–84, 1997.

Correia, F.N., Rego, F.C., Saraiva, M.D.S., and Ramos, I., Coupling GIS with hydrologic and hydraulic flood modeling, *Water Resources Management*, 12:229–249, 1998.

Crist, E.P. and Cicone, R.C., Application of the tasseled cap concept to simulated thematic mapper data, *Photogrammetric Engineering and Remote Sensing*, 50(2):343–352, 1984.

Davis, John C., *Statistics and Data Analysis in Geology*, John Wiley & Sons, New York, 2002.

European Space Agency, Information on ALOS PALSAR products for ADEN users, ALOS-GSEG-EOPG-TN-07-0001, Technical Note, 2007.

Gallant, J.C. and Wilson, J.P., Primary topographic attributes, in Wilson, J.P. and Gallant, J.C. Eds., *Terrain Analysis: Principles and Applications*, John Wiley & Sons, New York, pp. 51–85, 2000.

Gillespie, A., Rokugawa, S., Matsunaga, T., Cothern, J.S., Hook, S., and Kahle, A.B., A temperature and emissivity separation algorithm for advanced spaceborne thermal emission and reflection radiometer (ASTER) images, *IEEE Transactions on Geoscience and Remote Sensing*, 36(4):1113–1126, 1998.

Henry, J.B., Chastanet, P., Fellah, K., and Desnos, Y.L., Envisat multi-polarized ASAR data for flood mapping, *International Journal of Remote Sensing*, 27(10):1921–1929, 2006.

Huang, C., Wylie, B., Homer, C., Yang, L., and Zylstra, G., A tasseled cap transformation for Landsat 7 ETM+ at-satellite reflectance, U.S. Geological Survey http://landcover.usgs.gov/pdf/tasseled.pdf, 2002.

Imhoff, M.L. and McCandless, S.M., Flood boundary delineation through clouds and vegetation using L-band space-borne radar: A potential new tool for disease vector control programs, *Acta Astronautica*, 17(9):1003, 1988.

Intergovernmental Panel on Climate Change (IPCC), Climate change of 2007: Impacts, adaptation and vulnerability, Working Group II Contribution to the Intergovernmental Panel on Climate Change Fourth Assessment Report, 23 p, http://www.ipcc.ch/SPMP13apr07.pdf (last accessed September 2008).

Jensen, J.R., Cowen, D.J., Althausen, J.D., Narumalini, S., and Weatherbee, O., The detection and prediction of sea level changes on coastal wetlands using satellite imagery and a geographic information system, *Geocarto International*, 8(4):87–98, 1993.

Kauth, R.G. and Thomas, G.S., The Tasseled Cap—A graphic description of the spectral-temporal development of agricultural crops as seen by LANDSAT, *Proceedings of the Symposium on Machine Processing of Remotely Sensed Data*, Purdue University, West Lafayette, IN, pp. 4B-41–4B-51, 1976.

Kishino, M., Tanaka, A., and Ishizaka, J., Retrieval of Chlorophyll *a*, suspended solids, and colored dissolved organic matter in Tokyo Bay using ASTER data, *Remote Sensing of Environment*, 99:66–74, 2005.

Kraus, R., Floodplain determination using ArcView GIS and HEC-RAS, In: Maidment, Djokie, D. Eds., *Hydrologic and Hydraulic Modeling Support with Geographic Information Systems*, ESRI Press, New York, 2000.

Mankins, J.C., Technology readiness levels: A white paper, NASA Office of Space Access and Technology, Advanced Concepts Office, http://www.hq.nasa.gov/office/codeq/trl/trl.pdf, (last accessed August 15, 2008), 1995.

Martinez, J.M. and Le Toan, T., Mapping of flood dynamics and spatial distribution of vegetation in the Amazon floodplain using multitemporal SAR data, *Remote Sensing of Environment*, 108:209–223, 2007.

Matthews, J., Stereo observations of lakes and coastal zones using ASTER imagery, *Remote Sensing of Environment*, 99:16–30, 2005.

Meade, M.S. and Earickson, R.J., *Medical Geography*, 2nd edn, Guilford Press, New York, 501pp., 2005.

Melack, J.M. and Wang, Y., Delineation of flooded area and flooded vegetation in Balbina reservoir (Amazonas, Brazil) with synthetic aperture radar, *Verhandlungen des Internationalen Verein Limnologie*, 26:2374–2377, 1998.

Mendelssohn, I.A. and Kuhn, N.L., Sediment subsidy: Effects on soil-plant responses in a rapidly submerging coastal salt marsh, *Ecological Engineering*, 21:115–128, 2004.

Meredith, A.W., Eslinger, D., and Aurin, D., An Evaluation of hurricane-induced erosion along the North Carolina Coast using airborne LIDAR surveys, National Oceanic and Atmospheric Administration Coastal Services Center Technical Report, NOAA/CSC/99031-PUB/001, 1999.

Michener, W.K., Blood, E.R., Bildstein, K.L., Brinson, M.M., and Gardner, L.R., Climate change, hurricanes and tropical storms, and rising sea level in coastal wetlands, *Ecological Applications*, 7:770–801, 1997.

Moorhead, K.K. and Brinson, M.M., Response of wetlands to rising sea level in the lower coastal plain of North Carolina, *Ecological Applications*, 5:261–271, 1995.

Myers, M.F., Rogers, D.J. Cox, J., Flahault, A., and Hay, S.I., Forecasting disease risk for increased epidemic preparedness in public health, in Hay, S.I., Randolph, S.E., and Rogers, D.J. Eds., *Remote Sensing and Geographic Information Systems in Epidemiology*, Academic Press, New York, pp. 309–330, 2000.

North Carolina Floodplain Mapping Program, http://www.ncfloodmaps.com (last accessed August 15, 2008).

Park, S.C., Park, M.K., and Kang, M.G., Super-resolution image reconstruction: A technical overview, *IEEE Signal Processing Magazine*, 20:21–36, 2003.

Pizzuto, J.E. and Rogers, E.W., The Holocene history and stratigraphy of palustrine and estuarine wetland deposits of central Delaware, *Journal of Coastal Research*, 8:854–867, 1992.

Pope, K.O., Sheffner, E.J., and Linthicum, K.J., Identification of central Kenyan Rift Valley fever virus vector habitats with Landsat TM and evaluation of their flooding status with airborne imaging radar, *Remote Sensing of Environment*, 40(3):185–196, 1992.

Rogers, A.S. and Kearney, M.S., Reducing signature variability in unmixing coastal marsh thematic mapper scenes using spectral indices, *International Journal of Remote Sensing*, 25(12):2317–2335, 2004.

U.S. Bureau of the Census. http://www.census.gov (state and county profile), 2000.

Wang, Y. and Allen, T.R., Estuarine shoreline change detection using Japanese ALOS PALSAR HH and JERS-1 L-HH SAR data in the Albemarle-Pamlico sounds, North Carolina, USA, *International Journal of Remote Sensing*, 29(15):4429–4442, 2008.

Wang, Y., Colby, J., and Mulcahy, K., An efficient method for mapping flood extent in a coastal floodplain by integrating Landsat TM and DEM data, *International Journal of Remote Sensing*, 23(18):3681–3696, 2002.

Washino, R.K. and Wood, B.L., Application of remote sensing to vector arthropod surveillance and control, *American Journal of Tropical Medicine and Hygiene*, 50:134–144, 1993.

White, S.A. and Wang, Y., Utilizing DEMs derived from LIDAR data to analyze morphologic change in the North Carolina coastline, *Remote Sensing of Environment*, 85(1):39–47, 2003.

Woolard, J.W., Volumetric change of coastal dunes using airborne LIDAR, Cape Hatteras National Seashore, North Carolina, Master's thesis, Department of Geography, East Carolina University, Greenville, NC, 109pp, 1999.

25 Remote Sensing— A Look to the Future

James B. Campbell and Vincent V. Salomonson

CONTENTS

Any discussion of the future of remote sensing is inherently problematic, in the sense that no one, of course, can presume to know the future—and the alternative—the extrapolation of current trends forward for several years—hardly carries the insight that the activity implies. The result is likely to be less about the future than it is about the present and our understanding of the past. Nonetheless, there is value in thinking about the implications of current developments, as we may be able to envision some of the synergies, unanticipated consequences, or opportunities that may not be immediately obvious in our routine preoccupation with day-to-day activities.

It is relevant to mention some of the previous efforts to see the future of remote sensing. Lee's (1922) review of the capabilities of aerial photography in effect formed a blueprint for the development of the applications of aerial photography. Later, the National Academy of Sciences' (1970) report Remote Sensing with Special Reference to Agriculture and Forestry offered a view forward for the next few decades. In 1973, the Report of the Federal Mapping Task Force on Mapping, Charting, Geodesy, and Surveying, outlined a view of the future of U.S. governmental operations for the subsequent decades; its effects can be seen now in the character of current, established, remote-sensing programs. Such efforts, despite their differing characters, illustrate the value of looking forward to sketch out emerging trends.

More recent perspectives have highlighted current issues that limit successful progress in the advance of the practice of remote sensing. Jensen (2002) identified several key issues necessary for sustaining progress in the field of remote sensing:

- *Data continuity.* Consistency of data collection and calibration necessary for robust analyses of multitemporal data. Systematic collection of data in support. Special concern focuses upon the continuity of the Landsat system, which has been in service long beyond its design life. As the system aged, there is an increasing concern about developing the "data gap"—the loss of continuity that would occur between the failure of the existing system and the beginning of service of the replacement systems. At the time of writing this chapter, a replacement system is under construction as part of the Landsat Continuity Mission (LCM), but because of the timing of the prospective launch of LCM in relation to the deteriorating condition of the TM5 and TM7 instruments, a gap in coverage is inevitable. In fact, a *de facto* data gap had begun well before the initiation of the LCM effort due to the reduction in acquisitions necessary to extend the active service of the instruments already in service.
- *Instrumentation gaps.* Jensen (2002) noted the absence of orbital systems that collect operational data, including LIDAR, hyperspectral, and active microwave. Implicit in his discussion is a concern for operational systems that routinely support enterprises and governmental activities supporting agriculture, forestry, disaster relief, and floodplain mapping.
- *Improved data analysis algorithms.* Advances in information extraction are usually achieved at the cost of increased analytical complexity, tailoring of algorithms to increasing specialized circumstances. Analysts need systems and interfaces that permit users to visualize an entire analytical sequence as a system and to guide the analysis toward the results, rather than forcing the analyst to become immersed, piecemeal, in each separate step. Jensen (2002) also noted the absence of a robust and practical atmospheric correction system, improved strategies, and sampling methods for extracting and evaluating change detection.
- *Data integration, standards, and interoperability.* Advances in these areas are necessary to minimize the effort devoted to importing and exporting data between analytical systems, avoid duplication of data collection efforts, facilitate analysis of multitemporal and multisensor data sets, assure compatibility between analytical systems, and facilitate sharing of data collected by different sensor systems.

None of these issues have been resolved in the interval since Jensen prepared the list, so these topics retain their relevance, and indeed, their significance has increased as time passes.

This chapter extends this perspective by highlighting current developments that underlie trends to identify some issues that seem likely to increase in significance in the near future. Campbell et al. in Chapter 22 reviewed some of the current drivers of the current market for geospatial data, including

- Public policy: Although U.S. governmental policies limit the detail of satellite data released to the public, technological capabilities permit acquisitions at high levels of detail, and practices and policies of systems designed by other nations may control future reality.
- Personal privacy: Court cases have in essence, confirmed the legality of routine collection of remotely sensed data, at least for applications unrelated to law enforcement there are few legal barriers to collection of fine-resolution imagery.
- The growing significance of web delivery of individualized geospatial data to the public though mapping systems such as Google Earth, Microsoft Virtual Earth, and Yahoo Maps.
- The role of personal navigation devices and information as a driver for much of the market for publically distributed imagery.

Campbell et al.'s list differs from that of Jensen's in the sense that it focuses less upon technical and analytical concerns than it does upon the social context for the use of remotely sensed imagery. It highlights the ways by which public interest in remotely sensed data have driven applications, enlarging the role of the public—the nonspecialist—as consumers of remotely sensed imagery, relative to that of the specialist, with highly focused, scientific and technical applications. Such issues have driven recent applications of imagery within the public domain, and help set the context for capabilities in other application spheres. They will likely have a role to play in future development.

25.1 PUBLIC REMOTE SENSING

Public remote sensing refers to applications that present aerial imagery in formats that are immediately accessible to audiences not normally experienced in the use of such imagery. The premier example, of course is Google Earth, which presents georeferenced image composites in a web-based format that permits users to roam the field of view in an intuitive manner. Of course, the public has long had, notionally at least, access to very large archives of remotely sensed imagery. However, this access required at least a rudimentary knowledge of imaging systems, the details of archival organization, searches, and the bureaucracy of ordering and shipping. Such barriers, as minimal as they may seem, are sufficient to screen out large portions of the public from first-hand uses of aerial imagery, which until recently have remained in the domain of the specialist.

Google Earth, as the principal example of public remote-sensing systems, offers composite images in a format that is tailored for use by the nonspecialist, although not in formats that are useful in an analytical context. Whereas in Google Earth, the variations in dates, scales, and qualities of coverages are completely unsatisfactory for conventional uses of aerial imagery, Google Earth's georeferenced format, ease of navigation, and ability to link with related content provide a product that is tailored for use by an audience that can make effective use of such data. Presenting imagery in a format that permits rapid web delivery, and intuitive navigation and changes in scale, have bypassed such barriers to the delivery of imagery to a new audience.

The widespread availability of such systems creates an interface that is common to a broad community of users, thereby forming a large population familiar with its content and its functionality. Such systems are beginning to implement capabilities for providing thematic content individualized for specific communities, including user-specific applications for sharing content across professional and novice users. Examples, current at the time of this writing, include the "National Geographic Layer" and applications to focus on the humanitarian crisis in Sudan, present information documenting tropical deforestation, and tracking the spread of avian influenza. We might expect that continued development of such capabilities could well open privacy issues as noted in Chapter 22, with increases in the number of legal challenges to such uses of high-resolution imagery in the public sphere. Google has been cultivating such applications through the award of "challenge grants" to promote the development of such applications by public charitable organizations.

Although Google Earth forms the most visible of the public access systems, the value of large-scale image display systems developed for cities and counties should be noted. Many local jurisdictions use aerial imagery as backdrops for vector displays that show property boundaries, transportation and utilities, planning projections, and related data. Although such systems typically lack the convenient interfaces that characterize Google Earth and related systems, they have had their own impact upon public discourse. Presentation of imagery in this context provides context for interpretation of vector data, depicting local landmarks familiar within each neighborhood. Use of image data in this context often increases public participation (relative to the display of vector data without the image background), public awareness of local policy issues, and levels of detail entering public discourse.

Although such systems are known for their ability to provide public access to image resources, the same systems have been applied to share information within more restricted professional communities. Google Earth is used as a framework for specialized products, including emergency response systems that identify evacuation routes, damage assessment, and critical infrastructure. For example, the Alabama Department of Homeland Security (ALDHS) uses Google Earth in a professional version tailored for its specific purposes, with access restricted to authorized users, for sharing data across the many local, state, and federal agencies with responsibility for public safety (see Virtual Alabama: http://www.dhs.alabama.gov/virtual_alabama/home.aspx).

Such specialized systems benefit from the same functionality that underlies the public versions—it enables officials across a jurisdiction to simultaneously view the same information and understand the current situation. Because the simple interface, already known to most staff, promotes immediate establishment of collaborative partnerships, it minimizes infrastructure and training costs for participating organizations. Its uniform interface can easily be used by personnel throughout an organization, without the need for prior experience in uses of specialized GIS or image-analysis software.

25.2 SYNERGIES BETWEEN TECHNOLOGIES

Past progress in the field of remote sensing has been characterized by convergences of multiple technologies. It can be said that the combination of the airplane and camera are inherently synergistic in the sense that the two technologies in combination provide a capability far beyond the domain of either. A more current example is the synergistic combinations of GPS, Inertial Measurement Units (IMUs), and laser scanners to provide practical LIDAR systems (Chapter 23). Prior to development of current LIDAR technologies, the fundamentals of laser systems had been used in nonimaging systems to view the Earth, assess atmospheric quality, measure altitudes, and for gross assessment of ground surfaces. Early NASA LIDAR systems were developed in the 1970s and 1980s (Ackermann, 1999); by the early 1980s, second generation LIDAR systems were in use around the world, but were expensive and had limited capabilities. With the enhanced computer capabilities available today, and with the latest positioning and orientation systems, LIDAR systems have become commercially viable alternatives for generation of Digital Elevation Models (DEM) of the Earth's surface (Ackermann, 1999).

These imaging systems became practical not solely because of advances in a single technology, but because of the convergence of improvements in three separate technologies. The Global Positioning System provides precise and reliable positional information that enables precise tracking of position and altitude of the aircraft. Inertial Measurement Units provide accurate information concerning orientation of the aircraft, and high-quality lasers provide the ability to acquire detailed information describing the Earth's surface. The synergistic merging of technologies leads to the formation of true imaging *systems*, rather than use of a single instrument.

In the past, the practice of remote sensing was often characterized by the image acquisition systems. Now, remote-sensing applications depend not only upon image acquisition, but also on the ability to archive, index, transmit, and analyze imagery, and to assemble imagery with related ancillary data, then to analyze and interpret data, and distribute the results too. The current practice is linked to availability of inexpensive computer memory and access availability of a robust Internet. This, of course, includes the ability to transmit data from the archives to users, but also to enable robust searches of archives to allow users to review and assess contents.

When the Internet became a commonly used resource, it was not clear that it could have an impact on remote sensing and geospatial data in general. Yet it has become one of the principal technologies enhancing the value of remote-sensed imagery. It has provided the public at large with direct, first-hand, access to the imagery of their communities and neighborhoods—an achievement that a decade before would have seemed an improbable, if not impossible, prospect. Yet, this synergistic relationship has yet to take its full course. Digital image archives are increasing in size and scope. Internet access to such archives is increasing, enabling a larger population to search, view, and acquire imagery. Prior to this development, knowledge of the existence of an archive itself was restricted principally to those who belonged to a community already initiated to practice remote sensing, and the use of the archive required perhaps not so much a specialized knowledge, but a degree of persistence not possessed by the usual novice. Likewise, the increasing availability of online indexes and archives has greatly broadened the size and diversity of the population of individuals with access to image resources.

25.3   IMAGE ACQUISITIONS BY STATE AND LOCAL GOVERNMENTS

Public access to systematic and high-quality imagery has provided multiple benefits to local jurisdictions. An increasing number of states and local jurisdictions systematically acquire imagery of their territories in support of the numerous governmental functions, including planning, environmental policies, infrastructure management, and maintenance of property records. Hall (2008) reviewed the case for establishing a statewide LIDAR program in Virginia. His survey indicated that 87% of responding businesses and governmental agencies used elevation data in some form, and 74% of existing users reported that existing data did not meet their needs—needed more current, more accurate, and higher precision data. His survey forms another example of increasing awareness among a population of nonspecialists who are aware of the availability, value, and at least some of the technical characteristics of remote sensing imagery in the context of issues within their local communities.

Many jurisdictions provide the public with web-enabled systems for general access to georeferenced versions of such imagery, and often offer the ability to access and view such imagery in the context of other GIS layers, such as property boundaries, soils, and topography. Increases in availability have been accompanied by improvements in quality and detail of imagery. For example, some state governments have acquired statewide LIDAR coverage, which provides topographic detail and accuracy significantly much higher than previously available over wide areas. As the quality and detail of such imagery increases, we can consider the impact upon public awareness of environmental issues that previously could be understood at the local level only at relatively coarse levels of detail.

For example, increased public access to high-quality geospatial data, and products derived from the analysis of such imagery, may well lead to increased, and better-informed, public participation in debates concerning local environmental issues. It seems clear that the statewide LIDAR coverage in North Carolina, for example, has increased the detail and accuracy of flood plain mapping, in some regions noticeably enlarging the defined flood plain (Figure 25.1). The availability of such higher-quality data has the potential to influence public discourse concerning local issues—for example, the improved resolution and accuracy of such imagery changes the nature of floodplain maps by exposing inaccuracies of less detailed and less accurate interpretation based on previous technologies. Especially in urbanized regions, several decades of zoning, planning, and development decisions have been based upon the superseded data. The availability of updated data will in due course have its impact upon insurance evaluations of existing structures, assessment of flood hazards, zoning decisions, and local real estate markets. Thus, local governments may face reevaluation of previous policies and decisions, and participation of a public with a much more informed and critical perspective on issues once seen as residing in the realm of the specialist.

Nighttime photography. Another example of the ability of redefined applications of established imaging capabilities can be found in the growing interest in applications of nighttime aerial imagery within the local community. In earlier decades,

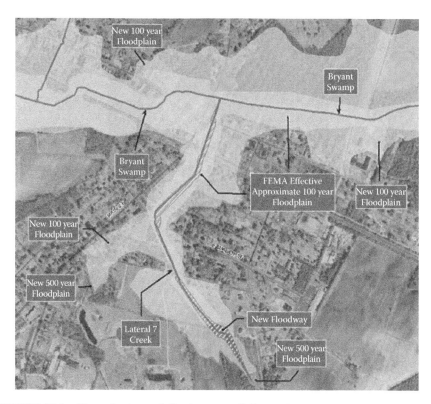

FIGURE 25.1 **(See color insert following page 426.)** Floodplains as defined by traditional methods (blue) and using more detailed LIDAR data (pink). The higher detail and finer resolution of the LIDAR data provide a broader definition of the floodplain. (From North Carolina Floodplain Mapping Program. With permission.)

nighttime photography was implemented using artificial sources of light to illuminate the terrain, usually for military reconnaissance (to detect, for example, rail and bridge repairs, or troop movements conducted under cover of darkness). In contrast, current interest in night photography is targeted at recording the pattern of illumination from artificial light sources, using specialized equipment and techniques tailored to acquire imagery under these unusual conditions. Some imagery offers oblique views; other projects use metric cameras to provide orthographic views that can match to the geometry of traditional daytime photography. A common motivation for such programs focuses upon public safely—especially at universities, where nighttime imagery has been used to evaluate safety issues—not all areas are illuminated equally, and the pattern of illumination does not necessarily correspond to the areas and pathways most frequently used at night (Figure 25.2). Analysis of such imagery can be employed to improve the distribution of lighted pathways, and to designate areas that might require additional safety precautions.

Another concern focuses upon assessing and monitoring the effects of light pollution in cities and neighboring locales. Many readers are familiar with the composite

FIGURE 25.2 **(See color insert following page 426.)** Top: Downtown Blacksburg, Virginia as imaged by a conventional aerial photograph. (Courtesy of Aerial Imagery.) Bottom: Nighttime aerial imagery of the same region, June 8, 2007, approximately 12:30 a.m. (Courtesy of Peter Sforza.) Center: Composite photography showing nighttime lights against the background of the daytime photograph of the same area as a locational reference. (Courtesy of Peter Sforza.)

image compiled from the Defense Satellite Meteorological Satellite (DMSP) imagery to show, worldwide, the pattern of artificial nighttime illumination as viewed from space. Assessed globally, the pattern of illumination indicates the overall pattern of the human population and economic activity. At local scales, the pattern of nighttime illumination also indicates patterns of human activity. Because artificial lighting can be designed to minimize light pollution while retaining its ability to meet safety needs, nighttime imagery can be used to assess the impact of lighting upon human health and upon distributions of wildlife populations, for example.

25.4 EXPLOITING THE TEMPORAL DIMENSION TO REMOTELY SENSED IMAGERY

The conventional analysis of imagery exploits its spectral and spatial content; analysis of its temporal content has usually been restricted to extracting changes between a few images, usually collected at arbitrary times. Thus, the unavailability of imagery collected at frequent intervals has restricted the full exploitation of its temporal dimension.

In the 1980s, remote sensing analysts began to use imagery from meteorological satellites to examine land areas at continental and subcontinental scales (Tucker et al., 1985). The ability of such sensors to image the same regions daily, or more often, allowed analysts to compile composites of large regions, using only the cloud-free images collected within periods of 2 weeks or so. Such techniques permitted the compilation of imagery representing very large regions, and even more significantly, to observe seasonal changes characteristic of specific climatic zones. Further, the development of vegetation indices (Chapter 19) offered the ability to enhance the ability to separate vegetated and non-vegetated surfaces, and to discern differences in the density and vigor of the vegetative cover. A sequence of such images collected over periods of several months allows observation of phenological changes—that is, changes in vegetative cover associated with seasonal changes in weather and climate. By the 1990s, such methods had evolved into techniques that enabled the examination of phenological variability on a pixel-by-pixel basis at continental scales (Loveland et al., 1995). The term *Land Surface Phenology* (Chapter 19) signifies the application of phenological analysis to broad-scale regions to consider both cultivated cropland as well as prevailing vegetation patterns.

Such methods permit analysts to derive information not only about the climate-driven seasonal phenologic variations prevailing within major biomes, but also to examine phenological patterns associated with human occupation of the landscape. Such changes are related to local crop calendars, but also to land abandonment, changes in cropping cycles, introduction of new crops or cultivation practices. For example, de Beurs and Henebry (2008) examined AVHRR and MODIS imagery of Afghanistan acquired since 1982 to identify phenological variations related to warfare and associated disruptions of social systems that characterized Afghanistan during that interval. Such phenological analysis relies upon accurate processing of remotely sensed data, and then the application of time series models characterizing the temporal variation, and derivation of the metrics that characterize the distinguishing features of specific phenological events, such as the beginning and end of growing seasons, onset of droughts, and similar events.

For many, it may be counterintuitive that so much, and such detailed, information can be extracted from imagery with such coarse spatial resolution. A key characteristic relevant for the consideration of future developments is that the analysis that relies not upon fine spatial resolution, but upon frequent observations of the same region—in effect, a substitution of time for space—a use of temporal analysis rather than spatial analysis, to extract information capturing a different dimension of the landscape.

Agnew et al. (2008) examined temporal changes in patterns of nighttime lights of Baghdad as observed by the Defense Meteorological Satellite Program at specific selected periods of dim moonlight during the interval 2003–2007. Their objective was to investigate the relationships between patterns of nighttime lights and the changes in population patterns within and between districts within the city, noting also that the pattern of lighting also indicates zones favored to receive electrical power on a regular basis. In this context, they examined spatial and temporal lighting patterns within neighborhoods of known ethnic identities, and then examined the events associated with sectarian and ethnic violence reported during the same intervals. Their analysis requires careful attention to registration, atmospheric conditions, phase of the moon, and other considerations. Their work illustrates another application of the value of temporal analyses to extract not only biophysical data—a traditional focus of many remote sensing analysis—but also to extract information revealing human behavior. Such information is not immediately associated with the nature of the instrument and its usual applications.

25.5 INSTRUMENTS AND SYSTEMS—A LOOK AHEAD

Ultimately, the practice of remote sensing depends upon the availability of reliable instruments that can collect accurate images of the Earth's surface. The applications outlined above are based upon new or redefined analysis of data collected by established remote sensing systems. The following sections offer a preview of the characteristics of some of the principal systems planned for the future.

25.5.1 BACKGROUND

The last two decades have seen tremendous progress in the use of remote sensing for studying Earth-atmosphere systems and related applications. This has been particularly true regarding land processes and terrestrial ecosystems. In the case of observing systems, there has been a significant set of advances in the operation of aircraft and spaceborne systems using radar, hyperspectral, and LIDAR systems. Airborne LIDAR topography (Chapter 23) has seen remarkable growth. Spaceborne systems operated by government and private industries have also grown in number and capability throughout the world.

In the case of spaceborne systems, a case in point is the successful deployment and operation of the NASA Earth Observing System (EOS) depicted in Figure 25.3. The EOS missions have included, for examples, such sensors and the Landsat-7 Enhanced Thematic Mapper (ETM) that has been operating since 1997, the Advanced Spaceborne Thermal Emission and Reflection Radiometer (ASTER) on the Terra Spacecraft operating since 2000, and the Moderate Resolution Imaging Spectroradiometer (MODIS) operating on the Terra and Aqua missions (Chapter 18).Other systems of relevance for land imaging and studies include the Advanced Land Imager components and the

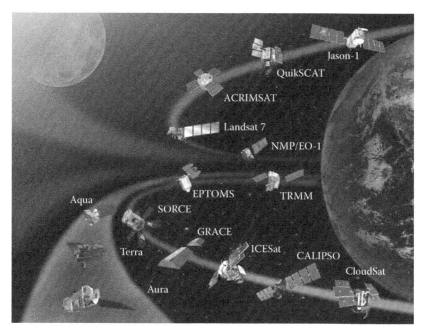

FIGURE 25.3   (See color insert following page 426.) NASA Earth Science Missions as of 2008.

hyperspectral radiometer—Hyperion on the Earth Observer-1 mission; the Advanced Microwave Scanning Radiometer (AMSR-E) that can observe snow and ice properties and soil wetness variations along with many other parameters for studying clouds, precipitation, sea surface temperate, etc. The Icesat mission that includes a LIDAR for mapping the topography of ice sheets such as that in Greenland, vegetation, and land topography. These systems, along with the EOS Data and Information System (EOSDIS; see Figure 25.4) have provided tremendous advances in increasing accessibility of data for a wide variety of purposes (Parkinson et al., 2006).

Internationally there has also been great progress in the development and operation of governmentally sponsored or operated systems by countries such as those associated with the European Space Agency (ESA), Japan, China, India, Canada, and other countries. Figure 25.5 offers an overview of the extent and scope of such systems. Such systems have largely been what can be described as low spatial resolutions systems, having spatial resolutions of 1 km or larger that principally allowed global, daily atmosphere, and ocean processes and trends.

In addition, there has been a tremendous growth of what are termed medium resolution optical systems (e.g., 15–100 m spatial resolution sensors) and high-resolution sensor systems (that is, better than 15 m and now including sub-meter spatial resolutions. These systems are operated by both commercial entities as well as by individual countries. Table 25.1 illustrates the extent of these systems throughout the world (Stoney, 2008). Similarly, there are many relatively medium- to high-resolution radar systems (see Table 25.2).

EOSDIS context

Mission operations

Flight operations,
data capture,
initial processing,   Data transport   Science data processing,
backup archive   to data centers/   data mgmt, interoperable   Distribution
Data acquisition   SIPSs   data archive, and   and data
distribution   access

Science operations

Research

Tracking and
Data Relay
Satellite
(TDRS)   EOSDIS and   Education

Related Data   Value-added
Data   Centers   Internet   providers
processing   (Search,
and   order,   Interagency
mission   distribution)   data centers
White sands   control
complex   Earth system
(WSC)   models
EOS
spacecraft   International
partners

NASA
Integrated   Decision
Services   support systems
Network   Science
EOS polar ground stations   (NISN)   teams and
Mission   SIPSs
Services

FIGURE 25.4 **(See color insert following page 426.)** NASA Earth Observing System (EOS) Data Information System (DIS).

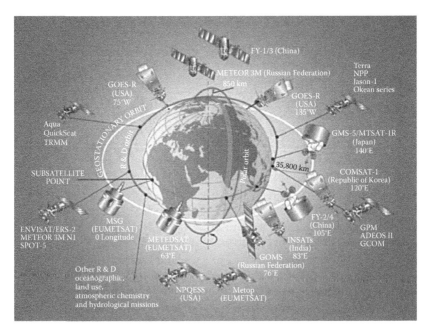

FIGURE 25.5 **(See color insert following page 426.)** International Environmental Satellite Systems.

TABLE 25.1
Land Imaging Optical Satellites

Optical Land Imaging Satellites with 56 m or Better Resolution by Country

Satellite	Country	Launch	Pan Res. (m)	Ms Res. (m)	Swath (km)
Landsat-5	United States	03/01/84		30.0	185
Landsat-7	United States	04/15/99	15.0	30	185
IKONOS-2	*United States*	09/24/99	1.0	4	11
EO-1	United States	11/21/00	10.0	30	37
QuickBird-2	*United States*	10/18/01	0.6	2.5	16
WorldView-1	*United States*	09/18/07	0.5		16
WorldView-2	*United States*	07/01/08	0.5	1.8	16
GeoEye-1	*United States*	08/23/08	0.4	1.64	15
LDCM	United States	07/01/11	10.0	30	177
DMC UK (SSTL)	United Kingdom	09/27/03		32	600
TopSat (SSTL)	United Kingdom	10/27/05	2.5	5	10, 15
DMC UK-2	United Kingdom	11/15/08		22	660
DubaiSat-1	UAE	11/16/08	?	?	?
DMC BilSat (SSTL)	Turkey	09/27/03	12.0	26	24, 52
THOES	Thailand	02/27/08	2.0	15	22, 90
FORMOSAT-2	Taiwan	04/20/04	2.0	8	24
ARGO	Taiwan	07/01/09		6.5	78
DMC Delmos-1	Spain	11/15/08		22	660
SeoSat	Spain	07/01/10	2.5		?
SumbandilaSat	South Africa	04/01/08		7.5	?
X-Sat	Singapore	04/16/08		10	50
MONITOR-E-1	Russia	08/26/05	8.0	20	94, 160
Resurs DK-1 (01-N5)	Russia	06/15/06	1.0	3	28
DMC NigeriaSat-1 (SSTL)	Nigeria	09/27/03		32	600
DMC NigeriaSat	Nigeria	07/01/09	2.5	5, 32	320
RazakSat[a]	Malaysia	03/01/08	2.5	5	?
KOMPSAT-1	Korea	12/20/99	6.6		17
KOMPSAT-2	Korea	07/28/06	1.0	4	15
KOMPSAT-3	Korea	11/01/09	0.7	3.2	?
TERRA (ASTER)	Japan/United States	12/15/99		15, 30, 90	60
ALOS	Japan	01/24/06	2.5	10	35, 70
Venus	Israel/France	08/01/08		10	28
EROS A1	*Israel*	12/05/00	1.8		14
EROS B1	*Israel*	04/25/06	0.7		7
EROS C	*Israel*	04/01/09	0.7	2.8	11
IRS 1D	India	09/29/97	6.0	23	70, 142
IRS ResourceSat-1	India	10/17/03	6.0	6, 23, 56	24, 140, 740
IRS Cartosat 1	India	05/04/05	2.5		30

(*continued*)

TABLE 25.1 (continued)
Land Imaging Optical Satellites

Optical Land Imaging Satellites with 56 m or Better Resolution by Country

Satellite	Country	Launch	Pan Res. (m)	Ms Res. (m)	Swath (km)
IRS Cartosat 2	India	01/10/07	0.8		10
IRS ResourceSat-2	India	12/15/08	6.0	6, 23, 56	24, 140, 740
TWSAT	India	07/01/09		35	140
EnMap	*Germany*	07/01/11		30 Hyp	30
RapidEye-A	*Germany*	04/01/08		6.5	78
RapidEye-B	*Germany*	04/01/08		6.5	78
RapidEye-C	*Germany*	04/01/08		6.5	78
RapidEye-D	*Germany*	04/01/08		6.5	78
RapidEye-E	*Germany*	04/01/08		6.5	78
SPOT-2	France	01/22/90	10.0	20	120
SPOT-4	France	03/24/98	10.0	20	120
SPOT-5	France	05/04/02	2.5	10	120
Pleiades-1	France	03/01/10	0.7	2.8	20
Pleiades-2	France	03/01/11	0.7	2.8	20
SPOT	France	07/01/12	2.0	6	60
Proba	ESA	10/21/97		18 Hyp	14
Sentinel 2 A	ESA	07/01/12		10, 20, 60	285
Sentinel 2 B	ESA	07/01/13		10, 20, 60	285
CBERS-2	China/Brazil	10/21/03	20.0	20	113
CBERS-2B	China/Brazil	09/19/07	20.0	20	113
CBERS-3	China/Brazil	05/01/09	5.0	20	60, 120
CBERS-4	China/Brazil	07/01/10	5.0	20	60, 120
Beijing-1 (SSTL)	China	10/27/05	4.0	32	600
HJ-1-A	China	04/01/08		30, 100 Hyp	720, 50
HJ-1-B	China	04/01/08		30, 150, 300	720
Hi-res Stereo Imaging	China	07/01/08	2.5, 5	10	?
DMC ALSAT-1 (SSTL)	Algeria	11/28/02		32	600
Alsat-2A	Algeria	12/01/08	2.5	10	?
Alsat-2B	Algeria	12/01/09	2.5	10	?

Source: Stoney, W.E., *ASPRS Guide to Land Imaging Satellites*, American Society for Photogrammetry and Remote Sensing, Bethesda, MD, 2008.

Revised on January 21, 2008.

Notes: Countries given in italics indicate satellites used for commercial purposes. Read 4/1 = first quarter; 7/1 = in that year; and 11 and 12s = late in that year.

[a] Near Equatorial orbit.

TABLE 25.2
Radar Imaging Satellites

Radar Land Imaging Satellites

Satellite	Country	Launch	Best Res. (m)	Band
By date				
ERS-2	ESA	04/21/95	30.0	C
RadarSat 1	Canada	11/04/95	8.5	C
ENVISAT	ESA	03/01/02	30.0	C
ALOS	Japan	01/24/06	10.0	L
YaoGan WeiXing 1 (JB-5)	China	04/27/06	5.0	L
YaoGan WeiXing 3 (JB-5–02)	China	11/12/07	5.0	L
COSMO-Skymed-1	Italy	06/08/17	1.0	X
TerraSAR X	Germany	07/15/07	1.0	X
RadarSat2	Canada	09/14/07	3.0	X
COSMO-Skymed-2	Italy	12/08/07	1.0	X
HJ-1C	China	06/15/08	?	S
SAOCOM-1A	Argentina	07/01/08	10.0	L
RISAT	India	07/01/08	3.0	C
COSMO-Skymed-3	Italy	07/01/08	1.0	X
TerraSAR L	Germany	08/15/08	1.0	L
COSMO-Skymed-4	Italy	03/01/09	1.0	X
TanDem-X	Germany	06/30/09	1.0	X
SAOCOM-1B	Argentina	07/01/09	10.0	L
KompSAT 5	South Korea	03/15/10	3.0	X
Radarsat Constellation-1	Canada	07/01/11	?	C
Sentinel 1	ESA	07/01/11	5.0	C
SeoSar[a]	Spain	07/01/11	?	?
Radarsat Constellaiion-2	Canada	07/01/12	?	C
Radarsat Constellation-3	Canada	07/01/13	?	C
By country				
SeoSar[a]	Spain	07/01/11	?	?
KompSat 5	South Korea	03/15/10	3.0	X
ALOS	Japan	01/24/06	10.0	L
COSMO-Skymed-1	Italy	06/08/07	1.0	X
COSMO-Skymed-2	Italy	12/08/07	1.0	X
COSMO-Skymed-3	Italy	07/01/08	1.0	X
COSMO-Skymed-4	Italy	03/01/09	1.0	X
RISAT	India	07/01/08	3.0	C
TerraSAR X	Germany	07/15/07	1.0	X
TerraSAR L	Germany	08/15/08	1.0	L
TanDem-X	Germany	06/30/09	1.0	X
ERS-2	ESA	04/21/95	30.0	C
ENVISAT	ESA	03/01/02	30.0	C
Sentinel 1	ESA	07/01/11	5.0	C

(continued)

TABLE 25.2 (continued)
Radar Imaging Satellites

Radar Land Imaging Satellites				
Satellite	Country	Launch	Best Res. (m)	Band
YaoGan WeiXing 1 (JB-5)	China	04/27/06	5.0	L
YaoGan WeiXing 3 (JB-5–02)	China	11/12/07	5.0	L
HJ-1C	China	06/15/08	?	S
RadarSat 1	Canada	11/04/95	8.5	C
RadarSat 2	Canada	09/14/07	3.0	X
Radarsat Constellation-1	Canada	07/01/11	?	C
Radarsat Constellation-2	Canada	07/01/12	?	C
Radarsat Constellation-3	Canada	07/01/13	?	C
SAOCOM-1A	Argentina	07/01/08	10.0	L
SAOCOM-1B	Argentina	07/01/09	10.0	L

Source: Stoney, W.E., *ASPRS Guide to Land Imaging Satellites,* American Society for Photogrammetry and Remote Sensing, Bethesda, MD, 2008.
Revised on January 21, 2008.
Notes: The Chinese reported the launch of YaoGang WeiXing 2 on May 27, 2007 but have left it out of all later notices. Although their many civil uses were advertised, the YaoGang may be primarily military.
[a] May be operated as part of the German TerraSar-X and TanDem-X constellation.

25.5.2   A Look Ahead

With regard to spaceborne systems, a glimpse of what may be in the future can be derived from the recommendations for the next decade given to NASA by the National Research Council (NRC, 2008). There are several missions covering studies of the land, ocean, and atmosphere. Of those, and others, related to land imaging and processes studies, the following examples and concepts are noted.

25.5.2.1   Optical Hyperspectral and Multispectral Imaging

- *Landsat Data Continuity Mission (LDCM).* http://ldcm.nasa.gov/index. htm would continue Landsat coverage beyond the present capabilities represented by Landsat-7 and Landsat-5. As mentioned previously, the ages of those two missions create the danger of a gap in coverage. The present launch date is 2011 with a design-expected lifetime of 5 years. The LDCM requirements stress data continuity and include a minimum field of view of 185 km cross-track swath width at the equator, spectral bands similar to those of the Landsat-7 mission, a pixel ground sampling distance of between 28 and 30 m for all spectral bands except the panchromatic band (pixel size between 14 and 15 m).
- *HyspIRI: Hyperspectral Infrared Imager.* The Decadal Survey (NRC, 2008) list of missions calls for the HyspIRI mission nominally scheduled

for launch in 2013–2016. It would use a hyperspectral imager and a thermal infrared scanner for advanced ecosystem and deforestation studies and other related phenomena. Both instruments would be pointable to provide global coverage every 30 days at a spatial resolution of 45 m. This capability would represent an advance over the EO-1 mission previously described.

- *NASA National Polar Orbiting Environmental Satellite System (NPOESS) Preparatory (NPP) Mission and NPOESS missions Visible and Infrared Imaging Radiometer Sensor (VIIRS).* http://npoess.noaa.gov/index.php?pg=viirs. The VIIRS sensor has a heritage principally from the MODIS sensor on the EOS Terra and Aqua missions, but also a heritage from the venerable Advanced Very High Resolution Radiometer (AVHRR) that has flown successfully on the NOAA Polar Orbiting Satellite Series (POES) since 1978. The VIIRS with over 20 bands, several of which have considerably better spatial resolution than the AVHRR, represents a considerably better capability and the promise of continuity into the foreseeable future.

25.5.2.2 Active and Passive Microwave Imaging

Missions to study soil moisture, and snow and cold land processes are being considered. A mission focused upon the observation of soil moisture, termed "Soil Moisture Active and Passive" (SMAP) is planned for launch in 2010–2013. The SMAP combines an active radar with a passive radiometer, allowing soil moisture to be measured and analyzed globally at a resolution of 3–10 km every 2–3 days. For the Snow and Cold Land Processes (SLCP) mission, nominally planned for the 2016–2020 timeframe, a pair of synthetic aperture radars will be able to characterize both deep and shallow snowpacks. A dual frequency passive radiometer will provide additional detail, and allow for the comparison with snow- data from similar sensors on other platforms. In both cases, the improvements in microwave monitoring of soil moisture variations and dynamics as well as snow, sea ice, and other cryospheric parameters will be substantial.

25.5.2.3 Active Optical and Microwave Missions

Missions that make use of LIDAR as well as radar are being conceived and planned principally to improve upon capabilities being flown in the EOS and other international systems such as those depicted in Figures 25.4 and 25.5. The radar systems provide all-weather capabilities that are particularly valuable in very cloudy regions in the tropics and for estimating above-ground biomass, for example. Radars with multichannel, multiple polarization, and increased swath and resolutions should be expected for the future. The advancement in the development and operation of the LIDAR systems is also very impressive. Although not indicated explicitly in the examples below, there is an increasing likelihood that LIDARs offering a swath capability will be flown in the future. It is conceivable that LIDAR systems with 5 m spot sizes and as many as a thousand beams side-by-side may be developed to provide highly repetitive coverage over large areas.

- *ICESat-II: Ice, Cloud, and Land Elevation Satellite II.* ICESat-II will seek to improve upon the capabilities of ICESat-I. It is nominally planned for the 2010–2013 timeframe and will continue to use LIDAR capability primarily to measure the height and infer the volume of glaciers, ice sheets, and sea ice. It will also continue to be helpful in gathering observations of vegetation heights leading to data on land carbon storage and better understanding of responses of vegetation to changing climate and land use. The LIDAR will be a single channel LIDAR with GPS navigation and pointing capability to permit repeated high-accuracy sampling of ice elevation.
- *DESDynI: Deformation, Ecosystem Structure, and Dynamics of Ice.* The DESDynI mission planned also for 2010–2013 will complement ICESat-II by providing observations of the effects of changing climate and land use on species habitats, ecosystems, and carbon storage in vegetation, plus providing data on the response of ice sheets to climate change and the resulting impact on sea level. The primary instrument is an interferometric synthetic aperture radar (InSAR). The InSAR's resolution is expected to be better than 35 m and thus will allow depiction of tectonic faults and surface deformation with high precision. The second DES DynI instrument is a multibeam laser altimeter, operating in the infrared range. It will collect data from specific points on the Earth's surface to supplement the InSAR's broader sweep.
- *LIST: LIDAR Surface Topography.* The LIST mission is planned for 2016–2020. The objectives are to use LIDAR to obtain global topography data with a resolution of 5 m and a precision in the tenths of a meter. This level of detail is substantially better than present spaceborne capabilities at 30–90 m horizontal resolution, with a vertical precision of about 10 m. The overall capability should yield mapping of landslides, earthquakes, and flood hazards at scales that are fine to be useful for site-specific land use decisions as well as for research. Scientists should also be able to find active faults, map the global loss of topsoil, and detect signs of potential volcanic activity.
- *TanDEM-X* (Feske et al., 2008). TerraSAR-X is flying since 2007 and providing a radar that permits up to 1 m resolution in a "spotlight" mode that can be used for commercial uses and scientific studies. A follow-on capability called TanDEM-X will operate along with TerraSAR-X in a tandem, formation-flying fashion with the intention of providing a global digital elevation model for all global landmasses that will outdo anything achieved to date.
- *Satellite Formation Flying.* The TerraSAR-X and TanDEM-X examples just described also highlight that there will be an increasing reliance on "formation-flying" to utilize data from multiple instruments on multiple spacecrafts. Already there is utilization of such capabilities as provided by the EOS "A-train." As shown in Figure 25.6 there are presently five satellites in the A-train, not including the Orbiting Carbon Observatory (OCO) presently planned for launch in early 2009. Similar capabilities exist in the EOS morning orbit constellation, where the Terra satellite with its several imaging sensors (MODIS, MISR, and ASTER), flies in formation with the Landsat-7 and the EO-1 Hyperion sensor.

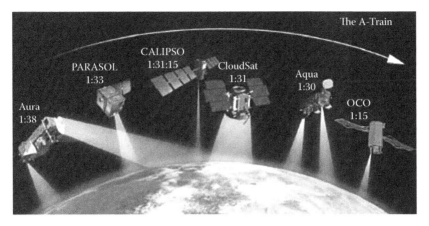

FIGURE 25.6   (See color insert following page 426.) NASA Earth Observing System (EOS) "A-train."

25.5.3   OTHER PROSPECTS, CHALLENGES, AND ISSUES

As noted in the previous edition of this manual (Jensen, 2002), there are many considerations that accompany the advances and prospects previously alluded to above. Certainly, there is distinct growth in, at least, the spaceborne systems for land imaging. Nevertheless, the issue of data continuity remains a challenge. The issue of data continuity certainly exists in relation to Landsat-7 and the LDCM. Similarly, continuity is an issue as EOS instruments such as MODIS evolve to the VIIRS on the NPP missions and the NPOESS operations. It remains to be seen whether gaps in coverage can be avoided. To supplement such gaps calls for a strong consideration of using other similar international systems and working increasingly in an international, cooperative mode with exchange of data and better data standards. To this end, the Global Earth Observing System of Systems (GEOSS) established in recent years offers considerable potential for providing this kind of collaboration and efficient use of observing systems. The GEOSS was organized by the Group on Earth Observations (GEO) (http://www.earthobservations.org/geoss.shtml). GEO comes from the members of the United Nations and the European Council, but is open to all entities involved in Earth observations and related activities. This effort seeks to link together existing and planned observing systems around the world and support the development of new systems where gaps currently exist. It promotes common technical standards so that data from the thousands of different instruments can be combined into coherent data sets. The environmental ministers and representatives of 50 nations including the United States adopted a 10 year implementation plan in April 2004 for the GEOSS.

With the abundance of missions and sensors, many of which provide data at high rates, there will be opportunities and challenges for processing of these data. A fundamental aspect of these efforts will involve what is often termed "data fusion" and involves the use of techniques that combine data from multiple sources in such a way as to achieve accuracies and information that would be achieved from a single source. Clearly fusing or combining data involves optical and microwave

Estimated no. of optical satellites on orbit

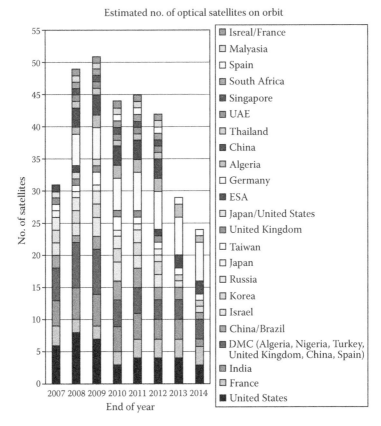

FIGURE 25.7 **(See color insert following page 426.)** Optical imaging satellites up through 2014. (Modified from Stoney, W.E., *ASPRS Guide to Land Imaging Satellites*, American Society for Photogrammetry and Remote Sensing, Bethesda, MD, 2008.)

systems, or LIDAR and radar involve different physics that, if combined, will offer more information than using one source alone. In this case, the challenge is to combine data and information from sensors on multiple platforms including not only spaceborne sensors, but also aircraft and *in situ* sensors. Certainly such an effort would include combining remote sensing sources with geographical information systems (GIS).

In addition, such data fusion efforts as just indicated will necessarily involve the use of advanced technologies and concepts. Such concepts as "cloud computing" may be employed to access computing capabilities, storage resources, and software that is distributed throughout the Internet. Certainly, with the vast extent and array of users, there needs to be creative thinking and research to make the best use of the high volumes of data involved and the depth of information that can be derived. In many cases, remote sensing can view situations that call for the real-time extraction of information and dispersal to the user communities. Examples include the monitoring of air-quality, imminent flooding and disaster response, including forests and

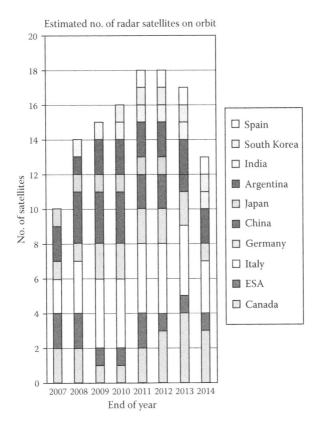

FIGURE 25.8 **(See color insert following page 426.)** Radar imaging satellites through 2014. (Modified from Stoney, W.E., *ASPRS Guide to Land Imaging Satellites*, American Society for Photogrammetry and Remote Sensing, Bethesda, MD, 2008.)

range fires, hurricanes, tsunamis, and the many other broad-scale events that require rapid assessment and response.

Along with data processing, computation, real-time delivery, etc., to make the acquisition, processing, delivery, and analysis of these data effective, there are issues of atmospheric correction and image normalization, the use of expert systems and neural network analysis, change detection, standards, and interoperability that remain as challenges for the future (Jensen, 2002).

In addition to the issues of data continuity and data processing noted above, the appropriate mix of governmental systems and private industry systems will still need to be addressed. As Tables 25.1 and 25.2 indicate, there are so many different satellite systems and countries involved, along with the environmental satellite systems. At a minimum, the remote sensing community needs something like the GEOSS concept to sort out how these resources should be best coordinated and used. Figures 25.7 and 25.8 show the large number of missions involving both relatively high resolution optical and radar imaging sensors.

25.6 SUMMARY

Briefly stated, future prospects for remote sensing reside in the advances in instrumentation and sensor systems. Furthermore, its future lies in the consequences of the expanding community of users, from an established pool of specialists, or those who rely upon the specialized products of the specialists, to a much broader public that will use imagery for a range of basic applications supporting their personal or business needs without specialized processing.

- *Instrumentation*: Advances in instrumentation will contribute to the bridging of gaps in collection capabilities, extending the reach and capabilities of LIDAR, microwave, InSAR, and hyperspectral systems through satellite systems that establish broadly based operational systems for sensors currently employed for much more limited missions. Other efforts will assure continuity in existing capabilities, especially Landsat and AVHRR-class systems that have established global archives, recording decades of observations of the Earth's surface. Such missions will focus on applications areas, including observation of the cryosphere, soil moisture, and topographic elevation that will extend the geographic reach and the level of precision of existing missions.
- *Expanded markets*: Future developments in the practice of remote sensing must consider not only the prospects for new data acquisition capabilities, but also the impact of expanding audiences for imagery from the established specialists into a broader community of nonspecialists. These nonspecialists use imagery for everyday activities (e.g., personal navigation), and for business applications (such as real estate, marketing programs, and specialized locational services) outside the current suite of established applications. Although some of these applications may appear initially to be superficial in character, unrelated to the core concerns of the remote sensing community, it should be recognized that once a capability is successfully established in one applications sphere, it forms the basis for applications in other, ostensibly unrelated application areas. It is through such processes that innovations in one arena spread to influence capabilities in others.
- *International cooperation and collaboration*: Interest in satellite remote sensing extends to a surprising number of nations, as is evident from the inspection of Tables 25.1 and 25.2. The capabilities of such programs are often similar to each other, but differ with respect to critical technical characteristics. Given the expense of operating so many analogous systems, there may be much to be gained by improved coordination of standards and capabilities that could improve and extend the reach of a worldwide archive of remotely sensed data to improve our ability to address common problems.

REFERENCES

Agnew, J., Gillespie, T. W., Gonzalez, J., and Min, B., Baghdad nights: Evaluating the US military "surge" using nighttime light signatures, *Environment and Planning A*, 40, 2285–2295, 2008.

Ackermann, F., Airborne laser scanning present status and future expectations, *ISPRS Journal of Photogrammetry and Remote Sensing*, 54, 148, 1999.

de Beurs, K. M. and Henebry, G. M., War, drought and phenology: Changes in land surface phenology in Afghanistan since 1982, *Journal of Land Use Science*, 3, 95–111, 2008.

Feske, T., Mellot, A., and Drescher, J., TerraSAR-X: The German radar eye in space, *Imaging Notes*, 23, 24–29, 2008.

Hall, S., Building a Case for Lidar in the Commonwealth of Virginia: Issues of Consideration. Virginia Information Technologies Agency. Virginia Geographic Information Network, 2008 (http://www.isp.virginia.gov/geopdf/CBA_Progress-and-Overview_042308.pdf).

Jensen, J. R., Remote sensing—Future considerations, *Manual of Geospatial Sciences and Technology*, Chapter 23 (J.D. Bossler, Ed.), Taylor & Francis, London, U.K., pp. 389–398, 2002.

Lee, W. T., *The Face of the Earth as Seen from the Air*. American Geographical Society Special Publication No. 4, American Geographical Society, New York, 1922.

Loveland, T. R., Merchant, J. W., Brown, J. W., Ohlen, D. O., Read, B., Band, P., and Olsen, P., Seasonal land cover regions of the United States, *Annals of the Association of American Geographers*, 85, 339–355, 1995.

National Academy of Sciences, *Remote Sensing with Special Reference to Agriculture and Forestry*, National Academy of Sciences, Washington, DC, 1970.

National Research Council (NRC), *Satellite Observations to Benefit Science and Society: Recommended Missions for the Next Decade, Committee on Earth Science and Applications from Space: A Community Assessment and Strategy for the Future* (R. Henson, Ed.), National Research Council, Washington, DC, 2008.

Parkinson, C. L., Ward, A., and King, M. D., 2006, *Earth Science Reference Handbook*, National Aeronautics and Space Administration, Washington, DC, 2006.

Stoney, W. E., *ASPRS Guide to Land Imaging Satellites*, American Society for Photogrammetry and Remote Sensing, Bethesda, MD, 2008 (http://www.asprs.org/news/satellites/ASPRS_DATABASE_021208.pdf).

Tucker, C. J., Townshend, J. R. G., and Goff, T. E., African land-cover classification using satellite data, *Science*, 227, 369–374, 1985.

Part IV

Geographic Information Systems

26 Geographic Information Systems and Science

Robert B. McMaster and Steven M. Manson

CONTENTS

26.1   INTRODUCTION

Geographic information science is a relatively new discipline that has emerged from the evolution of digital cartography, and improvements in computer technology over the past 30 years. The term geographic information science, or GISci, applies to the theoretical underpinnings of the technology, including database theory, methods of analysis, visualization techniques, and societal implications of these technologies. The evolution of the field is evident in how the term most commonly used until the mid-1990s was geographic information systems (GISs), which refers to the hardware and software components. Unfortunately, most individuals use the term GIS to refer to the systems themselves, and are unaware of the rapidly evolving science.

From its incipient roots in digital mapping technology, GIS has now become a billion dollar industry in the United States. It is utilized by the private sector in a variety of ways; the public sector across local, state, and the federal governments; and in academia, as witnessed by how nearly all universities offer a series of courses, and often minor/major degrees in GISs.

GIS is becoming a routine analysis and display tool for spatial data, and is used extensively in applications such as

- Land use mapping ranging from urban planning to forest management
- Transportation mapping and analysis, such as determining efficient transportation routes for deliveries and emergency response
- Geodemographic analysis for store location and customer marketing
- Utilities infrastructure mapping, particularly for gas, water, and electric line mapping and maintenance
- Natural resource assessment, including water quality assessment and wildlife habitat studies

GIS allows efficient and flexible storage, display, and exchange of spatial data, as well as use in models of all kinds. More recently, the term geographic information science has emerged as representing "the science of spatial data processing"—including conceptual problems in spatial data acquisition, storage, analysis, and display—while the term geographic information systems is reserved for the actual hardware/software component of the technology.

26.2 WHAT IS A GIS?

The term geographic information system is often applied to any package that involves mapping capability or spatial data. However, as Kraak and Ormeling (1996) point out, there are actually several different types of systems—*spatial* information systems—that may be categorized based on their functionality. Figure 26.1 arrays these systems based

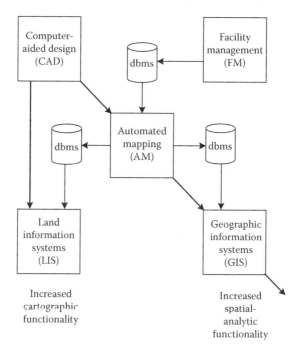

FIGURE 26.1 The relationship between several types of spatial information systems. (Redrawn from Kraak, M.J. and Ormeling, F., *Cartography and the Visualization of Spatial Data*, Longman Scientific, Essex, U.K., 1996, p. 10.)

on both their cartographic and spatial analytic capabilities (Kraak and Ormeling, 1996, p. 10). The simplest systems involve computer-aided design (CAD). CAD systems provide powerful design tools to assist with automated drawing and are used by engineers, architects, and designers. While the architect can design a building in three dimensions, there is no real cartographic or spatial-analytic capability with CAD systems.

Similarly, facilities management software allows for the organization of complex "spatial" databases such as those used by utilities companies for the maintenance of customer accounts, yet rarely does this software allow for analysis or mapping. More sophisticated software involves computer mapping, where spatial databases may be displayed, and complex symbolization types portrayed. Many computer mapping systems allow for limited analyses, such as address matching and non-topological overlay (which does not require a sophisticated underlying mathematical data model), but these systems are still not considered as full GIS. Land information systems (LISs) are designed for the storage and cartographic display of large-scale property or cadastral databases. LISs are typically utilized for maintaining the parcel-level data needed for city management, such as taxation, infrastructure repair, and the mapping of crime. Finally, a true GIS allows for the powerful spatial analysis and cartographic display of spatial databases. A working definition of a GIS is *a computer-based set of methods for the acquisition, storage, analysis, and display of spatially referenced data.*

26.3   CORE KNOWLEDGE IN GEOGRAPHIC INFORMATION SCIENCE

There are a core set of topics that are fundamental to GIS. These include:

- Geometric aspects of coordinate systems and map projections, including construction methods and geometric distortions produced
- Collection, creation, and input of spatial and attribute information
- Basic cartometric techniques including the measurement of distance and area from maps and digital databases
- The notion of map scale as the mathematical relationship between map and earth distance, as well as mathematical transformations of scale in computer-based map display operations
- The derivation of multiple scale databases and automated generalization
- Display and visualization of data, including statistical classification and analysis of attribute data for effective thematic map display
- Geographical data structures including vector and raster-based models
- Database and information retrieval methods to run queries against spatial and attribute information
- Spatial analysis and geographical problem solving using a map algebra along with statistical and mathematical modeling
- Data quality, error propagation through databases, and error assessment
- Principles of cartographic representation, including four-dimensional cartographies and multimedia cartography

26.4 A BRIEF HISTORY OF GIS

A comprehensive history of geographic information systems/science is not possible in this short chapter, but the reader may refer to Foresman's edited volume, *A History of GIS* (1998), for seminal events. Several key events will be reviewed in this chapter, including the State of Minnesota's MLMIS (Niemann and Niemann, 1996), the Harvard Laboratory for spatial analysis and computer graphics, the National Center for Geographic Information and Analysis (NCGIA), and the University Consortium for Geographic Information Science.

26.4.1 THE MINNESOTA LAND MANAGEMENT INFORMATION SYSTEM

The Minnesota land management information system (MLMIS) started in the late 1960s as a joint research project between the State Planning Agency and the Center for Urban and Regional Affairs (CURA) at the University of Minnesota. The seed for MLMIS was planted in 1967, when the Minnesota State Legislature became interested in how Minnesota's lakeshore lands were being developed and perhaps overused. The goal was "to attempt to influence decision making to improve land-use planning along Minnesota's thousands of miles of shoreline, and answer questions such as, 'Display all the government lots with wet soil plus lowland bush vegetation plus flat to gentle shore with a weedy bottom offshore'" (Niemann and Niemann, 1996). The Statewide Lakeshore Development Study led to the creation of the statewide land use map, the basis of MLMIS. This early statewide database utilized the United States Public Land Survey System (PLSS) quarter-quarter section unit (40 acres) as the data storage and graphic link for the study. Land use and land cover were determined from aerial photography, geocoded at the quarter-quarter section, and represented with line-printer characters (3×5 representing a square of "40"). The actual computer production and printing of this first ever statewide land use map represents this country's first true GIS. Additionally, much of environmental planning programming language (EPPL), the first real software designed for mapping and analysis, was developed from these early efforts. Thus, using the maps and analysis, the researchers pressed for better land-use planning. "Environmentally-based land-use planning in Minnesota had become a reality thanks to the computer-based data and analysis" (Niemann and Niemann, 1996).

24.4.2 HARVARD LABORATORY

A second major activity in the history of GIS involved Harvard University's Laboratory for computer graphics and spatial analysis, developed by Howard Fisher. The Harvard Laboratory was responsible for creating the first commercially available computer-mapping package, SYMAP, for introducing a set of new analysis and mapping methods, and for sponsoring a series of computer mapping workshops. SYMAP, which stood for synagraphic mapping, or "acting together graphically" was a computer-mapping program that allowed users to build simple spatial databases and created three types of maps: conformant (choropleth), isarithmic, and proximal, based on Thiessen polygons. As with the MLMIS project, the output from the SYMAP package was grid-based and represented with line-printer characters. SYMAP's structure included a series of modules, or packages, including those for the creation of the database. The major components included

- Outline, to create the bounding polygon of the study area (e.g., U.S. border)
- Conformolines, to create bounding polygons of any subunits (e.g., state borders) used for choropleth mapping
- Data points, creation of centroids of subunits that were used for isarithmic mapping
- Values, centered on statistical values used for mapping
- Map package, which manipulated the size, number of data classes, value range minimum and maximum, and other variables

The Harvard Laboratory, which also developed a vector-based package called Odyssey that allowed for complex topological data structures, was dismantled during the 1970s with many of its key researchers continuing their work at other institutions.

24.4.3 NATIONAL CENTER FOR GEOGRAPHIC INFORMATION AND ANALYSIS

During the mid-1980s, the National Science Foundation quickly realized that some type of coordinated effort in GISs was needed, and a call for proposals to establish a national center was issued in 1987. Eight academic institutions or consortia submitted proposals, and in 1988 the winning group of the University of California, Santa Barbara, State University of New York at Buffalo, and the University of Maine was selected. The NCGIA focused on basic research and held a series of expert meetings on fundamental topics. Below is a listing of these initiatives, along with dates of activity:

- Accuracy of spatial databases (1998–1990)
- Languages of spatial relations (1989–1991)
- Multiple representations (1989–1991)
- Use and value of geographic information (1989–1992)
- Large spatial databases (1989–1992)
- Spatial decision support systems (1990–1992)
- Visualization of spatial data quality (1991–1993)
- Formalizing cartographic knowledge (1993–)
- Institutions sharing geographic information (1992–)
- Spatio-temporal reasoning in GIS (1993–)
- Integration of remote sensing and GIS (1990–1993)
- User interfaces for GIS (1991–1993)
- GIS and spatial analysis (1992–1994)
- Multiple roles for GIS in U.S. global change research (1994–)
- Law, information policy and spatial databases (1994–)
- Collaborative spatial decision-making (1994–)
- The Social implications of how people, space and environment are represented in GIS (1995–)

- Interoperating GISs (1996–)
- Formal models of the common-sense geographic world (1996–).

The NCGIA had a prolific publications program, including software and books, reports of the specialists meetings, and scientific papers. The full listing of NCGIA publications and software may be found at http://www.ncgia.ucsb.edu.

During the mid-1990s, the NCGIA established a new research program, called Project Varenius. Where the initial set of basic research topics attempted to cover all of geographic information science, this new project focused on three specific areas: Cognitive models of geographic space, computational implementations of geographic concepts, and geographies of the information society. As with the previous effort, Varenius was organized around a set of specialist meetings, as listed below. One can see from the topics addressed by the NCGIA the initial set of core research areas considered significant by the GIS research community:

- Discovering geographic knowledge in data-rich environments
- Multiple modalities and multiple frames of reference for spatial knowledge
- Measuring and representing accessibility in the information age
- Cognitive models of dynamic geographic phenomena and representations
- Empowerment, marginalization, and public participation GIS
- Place and identity in an age of technologically regulated movement
- Ontology of fields
- International conference and workshop on interoperating geographic information systems

In addition to forwarding the basic research agenda, the NCGIA also worked on issues of education in the emerging area of geographic information science. In the late 1980s, the Center produced a print version of a model curriculum, which contained modules on introductory material, technical issues, and applications. Later, this project migrated into the virtual domain, where an online version may now be found at http://www.ncgia.ucsb.edu. NCGIA also developed a K-12 school program in order to have the technology move down into the pre-collegiate levels.

26.4.4 University Consortium for Geographic Information Science

During the 1990s, it became clear that the GIS community required a common voice that was broader than the three-member NCGIA. After several years of discussion, the University Consortium for Geographic Information Science (UCGIS) was established in December of 1994. As established at the founding meeting, the UCGIS has several primary purposes:

- To serve as an effective, unified voice for the geographic information science research community
- To foster multidisciplinary research and education
- To promote the informed and responsible use of geographic information science and geographic analysis for the benefit of society

The formal UCGIS Board was elected in the fall of 1995, with the first president being John Bossler of the Center for Mapping at Ohio State University, and the editor of this book.

One of the major contributions of the UCGIS has been the establishment of a series of research, education, and application priorities, or what are now called challenges. The initial series of research challenges established in 1996, and updated in 1998, include:

- Spatial data acquisition and integration
- Distributed computing
- Extensions to geographic representation
- Cognition of geographic information
- Interoperability of geographic information
- Scale
- Spatial analysis in a GIS environment
- The future of the spatial information infrastructure
- Uncertainty in spatial data and GIS-based analyses
- GIS and society

A second set of research challenges was established in 2002, as follows. It is interesting to see how some topics were expanded to reflect the changing face of technology (e.g., the addition of mobile computing to distributed computing) or added due to changing research priorities (e.g., ontological foundations):

- Spatial data acquisition and integration
- Cognition of geographic information
- Scale
- Extensions to geographic representations
- Spatial analysis and modeling in a GIS environment
- Research issues on uncertainty in geographic data and GIS-based analysis
- The future of the spatial information infrastructure
- Distributed and mobile computing
- GIS and society: Interrelation, integration, and transformation
- Geographic visualization
- Ontological foundations for geographic information science
- Remotely-acquired data and information in GIScience
- Geospatial data mining and knowledge discovery

The goal of creating these research challenges was to integrate the various expertises that exist with UCGIS, to identify those true impediments to critical research areas, and to push for increased federal funding for these areas. Readers can find white papers for some of these topics on the UCGIS Web site (www.ucgis.org) and chapters in the volume edited by McMaster and Usery (2004). UCGIS also established a series of education challenges (1997), and application challenges (1999), described at the UCGIS Web site. Finally, the UCGIS has also developed the "Geographic Information Science and Technology Body of Knowledge" to guide educational institutions in their teaching priorities (DiBiase et al., 2006). The body of knowledge identifies ten areas of interest:

- Analytical methods
- Conceptual foundations
- Cartography and visualization
- Design aspects
- Data modeling
- Data manipulation
- Geocomputation
- Geospatial data
- GI S&T and society
- Organizational and institutional aspects

The UCGIS matured into an organization consisting of nearly 90 organizations, with an annual assembly and significant national-level activity. For instance, each January the UCGIS sponsors a Washington-based event, where government agencies learn about the potential for GIS in solving basic societal and environmental problems.

26.5   SOCIETAL IMPLICATIONS OF GIS

An area of increasing research and concern in the discipline of GIS involves work at the interface of geographic information science and the societal implications of these technologies. Without a firm understanding of the consequences of GIS use, much effort may be wasted or lost on technology and good intentions may provide little benefit and possible misunderstanding. Several conferences, and a growing number of publications, represent the growing activity in GIS-society studies. One of the first major efforts was the NCGIA Initiative 19, "The social implications of how people, space, and environment are represented in GIS" that focused attention on the social contexts of GIS production and use and addressed a series of conceptual issues, including:

- In what ways have particular logic and visualization techniques, value systems, forms of reasoning, and ways of understanding the world been incorporated into existing GIS techniques, and in what ways have alternative forms of representation been filtered out?
- How has the proliferation and dissemination of databases associated with GIS, as well as differential access to spatial databases, influenced the ability of different social groups to utilize information for their own empowerment?
- How can the knowledge, needs, desires, and hopes of marginalized social groups be adequately represented in GIS-based decision-making processes?
- What possibilities and limitations are associated with using GIS as a participatory tool for more democratic resolution of social and environmental conflicts?
- What ethical and regulatory issues are raised in the context of GIS and society research and debate?

It is clear that in the discipline of GIS, as with all information technologies, investigators must be both aware of, and concerned with, the potential impact of their analyses and visualizations. As an example, GIS are increasingly being utilized in assessing "environmental justice," or whether toxic waste sites and various other forms of hazards disproportionately affect certain marginalized populations (e.g., the poor, minorities, or children). Although GISs can indeed show relationships between, for instance, toxic sites and low-income populations, might the results of such analyses and the resultant visualizations be used by government and/or insurance companies to argue for altered zoning or higher insurance premiums? As with all such technologies, the best solution is to have well-trained, knowledgeable, and ethical persons working with the technologies, and policy makers who clearly understand the limitations.

26.6 THE LITERATURE OF GEOGRAPHIC INFORMATION SYSTEMS/SCIENCE

The literature in the discipline of GIS has grown rapidly over the past twenty years, and GIS now boasts a number of dedicated journals and trade publications alongside a large number of texts. Some of the resources available for both researcher and practitioners include:

Scholarly journals
Cartography and Geographic Information Systems
International Journal of GIS
Geographical Analysis
Geographical Systems
Transactions in GIS
Geoinformatica
The URISA Journal

Popular magazines
GeoWorld (formerly GIS World)
GeoSpatial Solutions (formerly GeoInfoSystems)
ARC News
ARC User

Proceedings
Auto-Carto-xx, 1972–1997, and 2005–
Spatial Data Handling, 1984 onward
GIS/LIS, 1988–1998
GI Science, 2000 onward

Major textbooks
Burrough, P., 1986.
Principles of GIS for Land Resources Assessment
Aronoff, S., 1989, 1991
GIS: A Management Perspective
Peuquet, D., and D. F. Marble, 1990
Introductory Readings in GIS

Huxhold, W. E., 1991
  An Introduction to Urban Geographic Information Systems
Laurini, R. and D. Thompson, 1992
  Fundamentals of Spatial Information Systems
Chrisman, N., 1997, 2001
  Exploring Geographic Information Systems
Heywood, I., S. Cornelius, and S. Carver, 1998, 2002, 2006
  An Introduction to Geographical Information Systems
Clarke, K. C. 1997, 1999, 2000, 2003
  Getting Started with Geographic Information Systems
DeMers, M. N., 1999, 2002, 2008
  Fundamentals of Geographical Information Systems
Bernhardsen, T., 1992, 1999, 2003
  Geographic Information Systems
Longley, P. A., M. F. Goodchild, D. J. Maguire, and D. W. Rhind, 2005
  Geographic Information Systems and Science
Lo, C. P. and A. K. W. Yeung, 2002, 2006
  Concepts and Techniques of Geographic Information Systems
O'Sullivan, D. and D. J. Unwin, 2003
  Geographic Information Analysis
Harvey, F., 2008
  A Primer of GIS: Fundamental Geographic and Cartographic Concepts

26.7   CONCLUSION

The discipline of geographic information systems/science has witnessed remarkable growth and maturation during its first thirty years. From its roots in rudimentary computer mapping software, GIS now permeates many aspects of society, ranging from being a multi-billion dollar business in the private sector to being used by local, state, and federal government as well as non-governmental organizations. Nearly all universities now have established coursework in GIS and there are many professional master's degrees (Wikle and Finchum, 2003). GIS software is used for simple mapping, especially given the proliferation of web mapping, but is increasingly used for complex spatial analysis and modeling as well. The remainder of this section covers the basics of GIS, including fundamental principles, transformations, geographic data structures, spatial analysis, basics of cartography and visualization, the societal implications of geographic information science and systems, and implementing a GIS. Chapter 27 covers some of the fundamental issues in GIS, including types of data, sources of data, and provides an explanation of GIS modeling. Chapter 28 reviews geographic data structures, and provides examples of both vector and raster approaches, as well as some of the most current approaches. Chapter 29 covers principles of spatial analysis and modeling, while Chapter 30 reviews considerations of spatial data quality. Switching to representation, Chapter 31 addresses basic principles of cartography, symbolization, and visualization while Chapter 32 is on implementing a GIS. The final chapter in this section, Chapter 33, addresses the complex relationship between geographic information science/systems and the society it is embedded in.

REFERENCES

Aronoff, S., 1989. *Geographic Information Systems: A Management Perspective.* WDL Publications, Ottawa, Canada.

Bernhardsen, T., 1999. *Geographic Information Systems,* John Wiley Press, New York.

Burrough, P.A. and R.A. McDonnell, 1986. *Principles of Geographical Information Systems.* Oxford University Press, Oxford, U.K.

Chrisman, N.R., 2001. *Exploring Geographic Information Systems* (2nd edn.). John Wiley & Sons, New York.

Clarke, K.C., 2003. *Getting Started with Geographic Information Systems* (4th edn.). Prentice-Hall, Upper Saddle River, NJ.

DeMers, M.N., 1999. *Fundamentals of Geographic Information Systems.* John Wiley & Sons, New York.

DiBiase, D., M. DeMers, A.T. Luck, A. Johnson, B. Plewe, K. Kemp, and E. Wentz (Eds.), 2006. *Geographic Information Science and Technology Body of Knowledge.* University Consortium for Geographic Information Science, Association of American Geographers, Washington, DC.

Foresman, T. (Ed.), 1998. *The History of Geographic Information Systems.* Prentice Hall, Upper Saddle River, NJ.

Harvey, F., 2008. *A Primer of GIS: Fundamental Geographic and Cartographic Concepts.* Guilford Press, London.

Heywood, I., S. Cornelius, and S. Carver, 2006. *An Introduction to Geographical Information Systems.* Prentice Hall, Upper Saddle River, NJ.

Huxhold, W.E., 1991. *An Introduction to Urban Geographic Information Systems.* Oxford University Press, New York.

Korte, G.B., 1997. *The GIS Book.* Onword Press, Sante Fe, NM.

Kraak, M.J. and F. Ormeling, 1996. *Cartography and the Visualization of Spatial Data.* Longman Scientific, Essex, U.K.

Laurini, R. and D. Thompson, 1992. *Fundamentals of Spatial Information Systems.* Academic Press, London, U.K.

Lo, C.P. and A.K.W. Yeung, 2006. *Concepts and Techniques of Geographic Information Systems.* Prentice Hall, Upper Saddle River, NJ.

Longley, P.A., M.F. Goodchild, D.J. Maquire, and D.W. Rhind (Eds.), 2005. *Geographical Information Systems,* John Wiley & Sons, New York.

McMaster, R.B. and E. Usery (Eds.), 2004. *A Research Agenda for Geographic Information Science.* Taylor & Francis/CRC Press, Boca Raton, FL.

Niemann, B.J. and S.S. Niemann, 1996. LMIC: Pioneers in wilderness protection. *GeoInfo Systems,* 6(9): 16–19.

Obermeyer, N.J. and J.K. Pinto, 1994. *Managing Geographic Information Systems.* The Guilford Press, New York.

O'Sullivan, D. and D. Unwin, 2002. Geographic Information Analysis. John Wiley Press, New York.

Peuquet, D.J. and D.F. Marble (Eds.), 1990. *Introductory Readings in Geographic Information Systems.* Taylor & Francis, London, U.K.

Pickles, J., 1995. *Ground Truth: The Societal Implications of Geographic Information Systems.* The Guilford Press, New York.

Tomlin, C.D., 1990. *Geographic Information Systems and Cartographic Modeling.* Prentice-Hall, Englewood Cliffs, NJ.

Wikle, T.A. and G.A. Finchum, 2003. The emerging GIS degree landscape. *Computers, Environment and Urban Systems,* 27(2): 107–122.

27 GIS Fundamentals

David Bennett and Marc Armstrong

CONTENTS

27.1 INTRODUCTION

Record-keeping systems that help people collect, manage, analyze, and display geographic information can be traced back thousands of years (Dale and McLaughlin, 1988: 46; Thrower, 1972). During the past several decades, these systems have evolved rapidly to keep pace with changes in computer technologies as well as business and government practices. Despite these advances, and the attention that such systems have received in recent years, the concept of a "geographic information system" (GIS) remains elusive since it is subject to a broad range of context-dependent interpretations. In this chapter, we will define commonly used terms, sketch out a framework designed to help practitioners conceptualize, organize, and implement GIS applications, and demonstrate, using a prototype application, how GIS software can be used to address an environmental problem.

The previous chapter discussed the concepts of geographic information systems and science. To help place this discussion in context, we will draw a further distinction between a "geographic information system" and "geographic information system software." A GIS refers to technology embedded in a particular social structure (e.g., the city governance of Iowa City, Iowa) in which information about places and their characteristics is recorded and transformed. The combination of a particular technology and a particular social setting creates a unique set of constraints and opportunities for the design, implementation, and maintenance of a GIS (Chrisman, 1997). The success or failure of a specific GIS implementation often depends on an appropriate match between technology and the institutional culture of a particular agency. Geographic information system software, on the other hand, is a set of stored computer instructions that transforms geographic information. Though the basic concepts that underlie the specification of these instructions may have long histories, the functions available in GIS software, and their applications have evolved rapidly during the past several decades. In the following sections, we first describe what we mean by GIS, and then describe the kinds of functional transformations that are embedded in GIS software.

27.2   BACKGROUND

A GIS must provide some measure of value (e.g., economic, social, political) to ensure its survival. Some researchers have attempted to attach an economic value to implemented systems while others have described methods for conducting cost–benefit analyses prior to system implementation (Obermeyer, 1999; Gillespie, 2000). Value can accrue to businesses, for example, through increased efficiency in the execution of routine business functions (e.g., the delivery of goods and services), or through more effective methods for selecting from among competing plans for infrastructure development (e.g., retail-site selection). Within the government, the assessment of value is less straightforward but equally important. Many of the early failures of GIS can be traced to the use of data that lacked appropriate resolution and to the lack of application-specific functions that were needed to support the activities routinely performed by government agencies (Dueker, 1979; Tomlinson et al., 1976; see Chapter 33 for a summary of applications and Part V for numerous examples). Consequently, continued maintenance and support for the flawed systems could not be easily rationalized within existing bureaucratic structures, funding withered, and the implemented, but not particularly useful, systems were doomed.

Lessons were learned from early failures. Commercial firms engaged in the development and sales of GIS software have paid careful attention to the needs of their users and this has, in turn, shaped the form of available software functions and the rate at which new functionality is developed. Initial concerns with the efficient management of geographic information, while still critical given the explosion in volume of available data (Armstrong, 2000), have begun to be displaced by an increased focus on analytical capabilities. This has led to a compartmentalization of software functions that are tailored to specific application domains. This trend has allowed software engineers to develop products to support specific work tasks (e.g., an individual transformation or entire toolboxes designed explicitly to support location decisions or watershed management activities). Such products are being distributed as open-source software and as component or service-based software modules that connect to the public domain and proprietary GIS software. Improvements to methods of analysis, either as fundamental shifts in routine practices, or in improvements to the performance of existing approaches, can have demonstrable payoffs in time saving and in the quality of analyses provided. Greater access to digital geographical data and advances in data-handling technologies, however, do not guarantee a successful GIS implementation. The development of such systems requires a sound understanding of GIS fundamentals.

27.2.1 A TRANSFORMATIONAL VIEW OF GIS SOFTWARE

The purpose of GIS software is to transform geographically referenced data using a set of software tools that facilitate the capture, storage, manipulation, analysis, and display of geographical data. Each of these processes can be expressed as a transformation of format, attributes, or geometry. Consequently, we can view GIS software as a kind of "transformational engine" that uses data inputs and produces information that has some value-added component (e.g., decision support). This transformational view has a lineage traced back to early work by Tobler (1979) that has persisted through cartography, analytical cartography (Clarke, 1995), and GIS (Chrisman, 1999). Clarke (1995) identifies four types of cartographic transformations that can be placed directly into the context of GIS projects:

1. Geometric (e.g., changes in coordinate systems)
2. Dimension (e.g., from a two-dimensional polygon to a zero-dimensional point)
3. Scale (e.g., from a 1:24,000 scale map to a 1:250,000 scale map)
4. Map symbolization (e.g., the selection, classification and presentation of elements on a map)

While it is true that the functional gap between digital cartographic software and GIS software has essentially disappeared, users of GIS software are, in general, more interested in the analysis and management of geographical data and the phenomena that these data represent (e.g., streets networks, land use, watersheds). GIS software, therefore, typically includes transformational functions that

1. Change the form of geographical data (e.g., capture spatial data by transforming paper maps into digital form, or change digital data (bits) into scientific visualizations (pixels))

2. Change the digital representation of geographical data (e.g., a raster-to-vector data transformation (Flanagan et al., 1994), transforming data stored in spatial data transfer standard (SDTS) format (USGS, 2000) into a proprietary format)

3. Change the content of geographical data (e.g., eliminate unneeded data, add information content to raw data through analysis)

These three basic categories can be further decomposed into more specific functional classes. For example, changes in content can be classified into transformations that assist in the:

1. Modification of data (e.g., update cadastral maps to reflect new subdivisions or new owners). The transformations traditionally associated with automated cartography belong in this category.

2. Extraction of data (e.g., create a new data set as a subset of a larger data set, proximity operators, the extraction of topographic features from digital elevation models, the extraction of the shortest path between two points in a street network).

3. Analysis of data (e.g., synthesize or abstract data to make it compatible with other data sets, provide analytical value to raw data, and integrate multiple data sets into a single data set).

The above lists are only illustrative of what can be accomplished using GIS software. For a more formal review of proposed taxonomies for GIS transformations, the reader is referred to Chrisman (1999).

GIS applications can be described as a sequenced series of transformations that begins with the "capture" of one or more data sets and concludes with the synthesis of these data into a form that helps users to answer questions (Figure 27.1). It is important to note that the way in which geographical data are stored, places real limits on the kinds of transformations that can be performed and, thus, the kinds of analyses that a GIS user can conduct. Furthermore, though each GIS software package implements a particular set of transformational tools, no single software package implements all available tools. The selection of a particular software package might, therefore, affect the kinds of transformations that can be accomplished. It is, therefore, a good practice to begin each GIS project with a careful design process.

During the GIS design process, the application developer must carefully define the problem. This definition can begin by seeking to delimit the scope of inquiry:

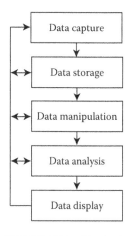

FIGURE 27.1 The basic five steps associated with GIS applications. Note that as the process proceeds, it is often necessary to return to a previous step to, for example, gather new data or expand the analysis.

- What are the key questions that must be answered to solve the problem under consideration?
- What are the essential geographical features that must be captured in the digital domain to answer these questions?
- How do these features vary across space or through time?
- What tools are required to quantify and analyze these spatio-temporal patterns?

The insight gained from this initial problem definition stage will guide key decisions regarding software, geographical representation, database content, scale, resolution, spatial extent, and geographical reference systems. For further information on GIS implementation, see Chapter 32.

27.3 THE LANGUAGE OF GIS

As with many technological fields, GIS has a unique language that must be mastered before one can take full advantage of the capabilities of the associated software. Historically, there were many different "dialects" of this language as the various loci of GIS development generated software to meet the needs of specific application domains. For example, many of the concepts that led Tomlin to develop map algebra (Tomlin; 1990; Tomlin and Tomlin, 1981) were apparently independently developed in other places by researchers with different disciplinary backgrounds (Alsberg et al., 1975; Fabos et al., 1978). While multiple definitions of technical terms still cause some ambiguity, for example, alternative usages of the term *arc*, much progress has been made in the standardization of a GIS language. Here we introduce the reader to terminology that has become somewhat standard within the GIS community. Many of the issues introduced here are discussed in detail in subsequent chapters.

27.3.1 SPATIAL REPRESENTATION

One of the first challenges that a user of digital geographic data must confront is how to best represent the geographical systems of interest within the digital domain (Peuquet, 1988). Real world geographical systems are complex, dynamic, and interrelated. While seemingly complex to the new user, geographical data sets are normally simplified, static models of reality (c.f., Peuquet and Duan, 1995). The choice of a particular model, or representation, will affect almost every aspect of a GIS project, from the cost of data acquisition to the types of conclusions that can be drawn from associated analyses.

In 1982, the U.S. National Committee for Digital Cartographic Data Standards set out to standardize the terminology associated with the digital representation of geographical data. This work evolved into the spatial data transfer standard (SDTS), which was first ratified as U.S. Federal Information Processing Standard #173 in 1994 (FIPS, 1994). The current version of this standard is known as ANSI NCITS 320-1998, which was ratified by the American National Standards Institute in 1998 (ANSI, 2000) and articulates with a series of standards maintained by the International Standards Organization (ISO). These standards provide a conceptual foundation on which application developers can build digital representations of geographical phenomena and a language that allows them to communicate this representation, either verbally or in digital form, to others in an unambiguous manner.

27.3.1.1 The Conceptual Model

GIS data sets store a digital representation of real world geographical phenomena (Figure 27.2a). Such phenomena can be tangible features (e.g., the segment of Main Street between 1st and 2nd Avenue) or they can be more abstract, like a measurement of elevation at a particular point on the Earth's surface. GIS data sets typically represent a specific class of geographical phenomena (e.g., the phenomenon *Main Street between 1st and 2nd Avenue is a kind of road*, Figure 27.2b). These classes are referred to as entity types. While the representation of entity types as independent data sets (often referred to as a theme) is useful from a data management perspective, the choice of a particular data classification scheme can have a significant effect on the outcome of any subsequent analyses (Anderson, 1980: 104; Bowker and Star, 1999).

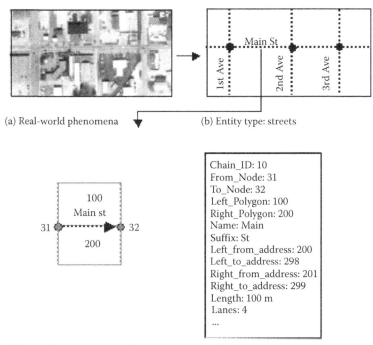

(a) Real-world phenomena

(b) Entity type: streets

(c) Entity object: Main street (between 1st and 2nd)

FIGURE 27.2 Geographical data sets store (a) real-world phenomena (a street network) as (b) entity types (a street data set) that, in turn, are comprised of one or more entity objects (a street segment). Attribute information uniquely identifies (c) each object (e.g., Chain_ID) and provides information useful for the management of the entity type.

Each entity type is associated with a set of attributes that are used to uniquely identify specific phenomena (key attributes) and provide information of value to users (e.g., street name, length, and ownership). A geographical phenomenon (e.g., the segment of Main Street between 1st and 2nd Avenue) is known as an *entity instance* and the digital representation of this instance is an entity object (Figure 27.2c). Entity objects often contain data on location, attribute, and topology (relative location).

27.3.1.2 Dimensionality and Entity Objects

Entity objects are often classified by their spatial dimension. This carryover from traditional cartographic symbolization has been translated into the conceptual conventions of GIS and serves partly as a convenience for programmers and their data structures. Entity objects that represent geographical phenomena are formed from 0, 1, 2, and 3 dimensional spatial objects (these same objects are referred to as 0, 1, 2, and 3 cells in some of the technical literature). For each dimension a set of well-defined spatial objects have been defined (see examples in Figure 27.3). For example, a *point* is defined as "a zero-dimensional object that specifies geometric location", while a *node* is "a zero-dimensional object that is a topological junction of two or more links or chains, or an end point of a link or chain." (USGS, 2000). As this definition suggests, complex objects are often constructed from a collection of more simple objects. For example, a *complete chain* is "a directed nonbranching sequence of nonintersecting line segments and (or) arcs..." that "...references left and right polygons and start and end nodes" (USGS, 2000). The construction of complex objects from simple objects and the explicit representation of topology (e.g., the polygon to the left of Main Street from 1st to 2nd Street represents census block y) provide advantages for data management and spatial analysis (Peucker and Chrisman, 1975).

Two-dimensional entities can be placed into three categories, polygons, grids, and images. As suggested above, polygons may be defined as a set of 1-dimensional spatial objects or as a set of points. Entity types built from points, lines, or polygons are collectively referred to as vector-based representations. Topological vector-based representations explicitly store relative spatial location (Figures 27.2

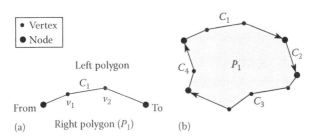

FIGURE 27.3   ANSI NCITS 320–1998 provides an established set of definitions for a variety of spatial objects. For example, C1 in 25.3a is a representation of a *complete chain* that bounds the *GT-polygon* P_1 in 25.3b (a polygon defined by geometry and topology).

and 27.3, Peucker and Chrisman, 1975). Depending on the underlying geographic data structure and associated software, the explicit representation of topological relations may be absent (thus calculated in real-time), predefined by the software vendor, or user defined. Grids are formed from a tessellation of the landscape into cells (typically square). Images are comprised of pixels (picture elements) and are often captured from some form of remote sensing platform (e.g., aerial photographs or satellite images). Collectively, grids and images are classed as raster-based representations.

Though the advantages and apparent flexibility associated with the use of these techniques for the representation of geographical phenomena is real, this approach also can prove to be vexing to novice and expert alike (Buttenfield and McMaster, 1991). For example, a city might be encoded and represented as a point at one scale and as a polygon at another. The way in which such dimensional shifts take place with respect to scale and purpose has a long and active research history (Leitner and Buttenfield, 1995) that has taken on renewed importance with the proliferation of ubiquitous cartography (Nivala and Sarjakoski, 2007). The link between scale and representation also can have important implications when selecting applicable analytical methods.

27.3.1.3 From Entity Objects to Data Structures

The ANSI NCITS 320-1998 standard provides a common language that can be used to communicate information about the structure and content of geographical data sets. It, however, does not specify how these data sets are to be implemented. The specific implementation of an entity object, referred to as a data structure, is left to software producers and, thus, is often viewed as proprietary knowledge. Various tools are being developed to facilitate access to such propriety data structures. For example: (1) the SDTS profile formats in the United States (http://mcmcweb.er.usgs.gov/sdts/) and the National Transfer Format (NTF) in Britain (http://www.bsi-global.com/group.html) provide well defined data structures that vendors can export to and import from; and (2) the Open GIS Consortium promotes access to proprietary data through the specification of a common communication protocol (McKee, 1999; http://www.opengis.org).

27.3.1.4 The Role of Metadata

To assess the utility of a geographic data set for a particular application the GIS analyst must know such things as: (1) its spatial extent; (2) how it was developed (e.g., its lineage); (3) the set of stored attributes; (4) the resolution of the stored data; (5) the coordinate system used to locate the data; and (6) the accuracy of the stored data. Resolution is a statement about the level of detail recorded in the data set (e.g., how many land use classes are used, what is the size of the smallest mapping unit recorded on the map). Accuracy, on the other hand, is a measure of how closely the data reflects a real location or attribute value. Such issues are critical to the assessment of utility. For example, the data must cover the geographical extent of an area of interest, and the spatial and attribute data must be sufficiently detailed, current, and accurate. Other issues are largely a matter of convenience and consistency. For example, coordinate systems and data structures can be transformed and databases can be joined.

The information required to assess the fitness of a data set in the context of a particular problem is stored as *metadata*. The development and maintenance of metadata has become especially important as large volumes of data are being distributed across computer networks and, as a consequence, there is no "institutional" knowledge about the conditions under which the data were produced. From a thorough reading of the metadata, an individual should be able to determine what types of transformations are needed to prepare the data for analysis.

27.3.2 FINDING THE NECESSARY DATA

Historically, the cost of data acquisition has been a significant impediment to the development of GIS applications. Publicly accessible repositories of digital geographical data were rare and when such data were available, they often were stored in proprietary formats or lacked the accuracy and resolution needed to be widely applicable to geographical problem solving. In recent years, the volume of data that can be downloaded from network accessible data repositories has increased rapidly. In the United States, the Federal Geographic Data Committee (FGDC) has acted as a catalyst for the development of online geospatial data repositories through its efforts to coordinate the national spatial data infrastructure (NSDI).

The NSDI is being developed to promote the distribution of geographical data produced by governments, industry, non-profit organizations, and the academic community (http://www.fgdc.gov). At the heart of the NSDI effort is the geospatial data clearinghouse. Each node on the clearinghouse network of data repositories is managed by a sponsoring agency that is responsible for the maintenance of the geographical databases within its jurisdiction. While the clearinghouse network is focused on data repositories within the United States, there are several international sites linked into this network. There are currently over 250 clearinghouse nodes and this number continues to grow. Data can be accessed through regional repositories (e.g., for Iowa, http://www.igsb.uiowa.edu/nrgislibx/gishome. htm), agency specific data distribution centers (e.g., the USGS Earth Resources and Observation and Science Data Center, http://edc.usgs.gov/), and the national clearinghouse (e.g., http://www.fgdc.gov/dataandservices, http://gos2.geodata.gov/wps/portal/gos).

Much of the data that is stored on nodes within the United States can trace their lineage back to data collection efforts initiated by different federal agencies. The following discussion provides a brief introduction to a sample of these data sets. Please keep in mind that this is not meant to be an exhaustive list of data sets or data providers.

27.3.2.1 USGS Data Sets

The United States Geological Survey (USGS) provides four commonly used data sets: digital line graphs (DLG), digital raster graphics (DRG), digital elevation models (DEM), and digital orthophoto quarter-quadrangles (DOQ). A DLG file can (but does not always) contain vector representations of the following entity types: hypsography, hydrology, vegetative surface cover, non-vegetative features, jurisdictional boundaries, survey control markers, transportation, man-made features, and the public land survey systems (for details see http://rockyweb.cr.usgs. gov/nmpstds/acrodocs/dlg-3/1dlg0798.pdf). The USGS has produced DLG files from their 1:24,000, 1:100,000, and 250,000 scale topographic maps. Digital raster graphics files are raster representations of a USGS topographic quadrangle maps. These images have been geographically referenced and rectified to facilitate their use within GIS software packages. Digital elevation models provide topographic data in raster form (USGS, 1993). These files are available at multiple resolutions (from 10 × 10 m to 3 × 3 arc s). DOQs are digital, geographically referenced, and geometrically corrected, aerial photographs with a 1 m resolution (sometimes provided in resampled form to reduce storage requirements).

The USGS distributes these data in SDTS format. However, it is often possible to find these data in proprietary data formats (e.g., ESRI's vector or raster formats) at geospatial data clearinghouse nodes. Note that not all USGS data sets are available for all areas in the United States.

27.3.2.2 U.S. Census Bureau

The U.S. Census Bureau maintains geographically referenced data sets derived from a variety of data collection activities. Among the most commonly used data sets are the decennial census of population and the associated TIGER (topologically integrated geographic encoding and referencing) files. A key to the utility of TIGER files is their rich topological and attribute content. Features (i.e., chains) in a TIGER line file contain information about multiple levels of census geography, and, if appropriate, the street name and associated address range for each side of the street (Broome and Meixler, 1990). Using TIGER files and street address information, census surveys returned by the general population, can be easily linked to census geography. Through this link, spatially aggregated data can be computed and used to produce a wide variety of socio-demographic maps and analyses (Figure 27.4). This same functionality can also be used to geographically locate

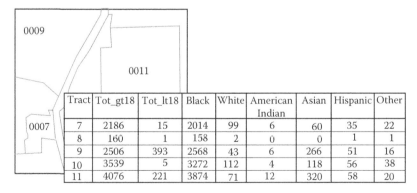

Tract	Tot_gt18	Tot_lt18	Black	White	American Indian	Asian	Hispanic	Other
7	2186	15	2014	99	6	60	35	22
8	160	1	158	2	0	0	1	1
9	2506	393	2568	43	6	266	51	16
10	3539	5	3272	112	4	118	56	38
11	4076	221	3874	71	12	320	58	20

FIGURE 27.4 Census geography can be linked to a wealth of socio-demographic data generated as part of the U.S. decennial census.

any observation that has been assigned a street address (Rushton et al., 2007). For example, retail establishments can quickly map customers based on their home address.

27.3.2.3   Natural Resource Conservation Service

The USDA provides soils data in digital form. These vector data sets represent soil series polygons and can be linked to a wealth of descriptive data that are stored in the soil survey geographic database (SSURGO). These data sets include information on the physical and chemical properties of the soil (e.g., salinity and depth to water table) and interpretive data that suggests appropriate uses for the land. These data can be found at the USDA geospatial data gateway (http://datagateway.nrcs.usda. gov/) and soils mart (http://soildatamart.nrcs.usda.gov/).

27.4   PUTTING IT ALL TOGETHER THROUGH DATA TRANSFORMATIONS

In this section, we illustrate how geographical problem solving can be represented as a well-defined set of transformations. A set of transformations that is designed to address a particular geographical problem is often referred to as a cartographic model and is expressed in a dialect of map algebra. Several alternative taxonomies for GIS-based transformations have been put forth in the literature (e.g., Tomlin, 1990; Maguire and Dangermond, 1991; Chrisman, 1997 and 1999). In Section 27.2, we enumerated seven generic functions that are recurrent themes in most of these taxonomies. These transformations change the geometry, dimension, scale, symbolization, form, structure, or content of geographical data sets. To motivate this discussion we will develop a simple cartographic model designed to facilitate the selection of a sanitary landfill within Johnson County, Iowa. The following is a subset of the locational requirements for a landfill site set forth in the Iowa Administrative Code (http://www.legis.state.ia.us/IAC.html), though these requirements have been slightly modified for the purpose of this illustration:

1. The proposed site must be so situated that the base of the proposed site is at least 5 ft above the high water table.
2. The proposed site must be outside a floodplain.
3. The proposed site must be so situated to ensure no adverse effect on any well within 1000 ft of the site.
4. The proposed site must be so situated to ensure no adverse effect on the source of any community water system within one mile of the site or at least one mile from the source of any community water system in existence at the time of application for the original permit.
5. The proposed site must be at least 50 ft from the adjacent property.
6. The proposed site must be beyond 500 ft from any existing habitable residence.

Some of these regulatory statements provide hard-and-fast rules. These *binary* variables are unambiguous and, in theory, are easy to document. For example, it is a straightforward GIS operation to determine the spatial relation between a proposed landfill site and a well if both are positioned accurately. Other rules are less well defined. When, for example, does the impact of a landfill become "adverse" and how do we evaluate *a priori* what the likely impact of a proposed landfill will be on a water supply? Such an evaluation would require advanced ground water models. While interfaces are being constructed that link such models to GIS software, the representation of dynamic spatial processes is largely beyond the capability of "out-of-the-box" GIS software.

Still other regulations are well defined but require data that are not generally available at the resolution needed to address this particular problem. For example, surveyors can say with some certainty that a point is within or beyond 50 ft of an adjacent property line. Digital data sets that accurately document parcel location, however, are not always available.

Nevertheless, each of these site requirements refers to a geographical phenomenon of interest that can, in theory, be rendered as a separate theme in a GIS database. Not surprisingly, the entity types of interest in this application include land use, public water supplies, private wells, water table, and floodplains (Figure 27.5). To instantiate an entity type as digital entity objects, the GIS analyst must select a spatial object that captures the form and spatial pattern of the associated phenomenon. For example, a GIS analyst may choose to represent land use, depth to water table, and floodplains as polygons (see Figure 27.3), and water supplies and well locations as point objects.

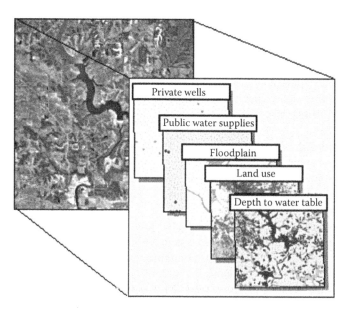

FIGURE 27.5 (**See color insert following page 426.**) Entity types, for example, landfill location problem, include land use, public water supplies, private wells, water table, and floodplain.

27.4.1 Preparing the Data

Table 25.1 presents an abbreviated set of metadata for the geographical data that are needed to address this problem. Note that not all of these data are available in a digital, geographically referenced form. For example, data for land parcels are in hard copy form and private wells developed after 1995 are referenced only by street address. These gaps in the availability of digital data raise three additional important issues.

27.4.1.1 Digitizing and Scale

In the example developed here, parcel data stored as analog maps must be transformed to digital form (i.e., digitized). Several strategies can be explored to convert paper maps to digital form, but one commonly employed approach is referred to as manual digitizing (Marble et al., 1984). The process works by placing a map on a digitizing tablet and tracing selected geographical features with an electronic device called a puck. When the user activates the puck, perhaps to signify a street intersection or a point along a curve in a stream, its location is sensed using the time delay between an emitted electromagnetic impulse and a fine mesh embedded in the tablet. The resolution of a digitizer can be very high (e.g., 0.01 mm). Despite this apparent accuracy, however, the user must be aware of the impact of error on the resulting database.

Consider, for example, the process of digitizing features from a 1:24,000 USGS topographic map. First, the original map document will contain error. Assuming that the map meets National Map Accuracy Standards, the best that we can assume is that roughly 90% of all "well-defined" points are within 12.2 m of their actual location. Second, a digitizing mistake of, for example, a millimeter produces a positional shift of 24 m at a scale of 1:24,000. Third, users must approximate continuous features (e.g., a sinuous stream or the curve of a *cul de sac*) as a series of discrete line segments (for a review of error attributable to the digitizing process see Jenks, 1981; Bolstad et al., 1990).

A review of the metadata associated with the data sets used in our prototypical project (Table 27.1) illustrates why we must be concerned about error resulting from the digitizing or cartographic process even when we download digital data sets from network accessible repositories. All of the data sets used here can trace their lineage to a hard copy map and each has an error component. Positional errors resulting from scale, map generalization, policies and practices, human error, and the discretization of continuous features are all faithfully copied into a digital database by the software and such errors will propagate through any analyses performed using these data (Lanter and Veregin, 1992; Hunter and Goodchild, 1996; Heuvelink, 1998). Thus, while it is true that digital geographic information has no real scale and it can be enlarged or reduced to the limits of precision of a computer system, the raw material for a GIS data set is often taken from analog maps. The scale of the source material is, therefore, integral to the data set and can influence the quality of solutions obtained from GIS-based analyses. This is now widely recognized and consequently scale is an important part of most metadata schemes that have been devised.

TABLE 27.1

An Abbreviated Set of Metadata for Data Sets Used in the Landfill Example

Soils

Projection	UTM, Zone 15, NAD83
Format	Polygon
Attributes	Mapping unit ID
Original source	USDA soil conservation service, 1:15,840 scale Mylar map sheets

Floodplain

Projection	UTM, Zone 15, NAD83
Format	Polygon
Attributes	Flood frequency (FREQ)
Original source	Federal emergency management agency, 1:12000–1:24000 scale paper maps

Private water wells

Projection	UTM, Zone 15, NAD27
Format	Point
Attributes	Permit number, owner, date, address,…
Original source	Iowa Department Of Natural Resources, well reports

Private water wells

Format	Comma separated text file
Attributes	Permit number, owner, date, address,…
Original source	Iowa Department Of Natural Resources, well reports

Community water sources

Projection	UTM, Zone 15, NAD83
Format	Comma separated text file
Attributes	Permit number, public water supply name, depth,…
Original source	Iowa Department Of Natural Resources, 1:100,000 scale maps

Parcel maps

Projection	UTM, Zone 15, NAD83
Format	Mylar maps
Attributes	
Original source	Johnson county assessor's office

Land use

Projection	UTM, Zone 15, NAD83
Format	Grid
Attributes	Land cover (LC)…
Original source	Iowa Department of Natural Resources, supervised classification of Landsat imagery

Soils database

Format	Comma separated text file
Attributes	Mapping unit id, average depth to water table
Original source	National Resource Conservation Service, Map unit interpretation database (MUIR)

Note: While many of these data sets exist, the metadata presented were modified from their original form for the purpose of illustration.

27.4.1.2   The Impact of Map Projections

The co-registration of the various geographic themes requires a consistent coordinate system (see Chapter 2). Here all themes will be *projected* to UTM zone 15 using the North American Datum of 1983 (NAD83). A map projection is the transformation of a three-dimension surface (the Earth) to a plane (a flat map) (Snyder, 1987, see also Chapter 4 of this manual). The projection of a data set to a planar surface simplifies geometrical analyses but introduces distortion into the data set. There are, of course, many ways to perform such transformations and each projection will have its own unique pattern of error and distortion. GIS users should understand the impact of this error on their projects.

27.4.1.3   The Need to Address Match

Note that not all of the data needed for our landfill location problem possess geographical coordinates. For example, our only knowledge about wells drilled after 1995 is through permit references to street addresses. The question becomes: How are addresses converted to plane coordinates? The answer lies in address matching (Dueker, 1974; Drummond, 1995; Rushton et al. 2007). The process of address matching relies on an up-to-date street centerline file that contains attribute data for street names and address ranges, for the left and right side of the road, for each chain in the database. In the United States, TIGER files provide a common source of such information (Broome and Meixler, 1990). However, steps must be taken to ensure that these data sets are sufficiently current and accurate to support a particular application. A number of private firms provide centerline data sets that are continually updated.

Given an appropriate street file, address matching the street addresses associated with the well records becomes, in part, a data extraction (query) procedure. For example, to place a well at 250 Main St onto a map we would find an instance in the *streets* database where street name = "Main", suffix = "St", left_from_address >= 200 and left_to_address < 300, see Figure 27.2c). In the address matching process, it is often assumed that location can be inferred through linear interpolation between the coordinates of the points of the associated street segment. In the absence of additional information, the best estimate for the location of 250 Main Street, for example, is halfway along the selected chain.

27.4.2   Data Extraction and Integration

The transformations presented in Figure 27.6a will generate input data that share a common data structure and coordinate system and, thus, provide raw material for data extraction and analysis (Figure 27.6b). Data extraction, in this situation, is used to subset the data set and, thus, focus attention on the area of interest. Here we perform two common data extraction procedures: (1) a *select* operator is used to extract relevant polygons from the *soils* and *land use* themes; and (2) a *proximity* operator (in this case a buffer function) is used to find all land that may not be appropriate for the development of a landfill because of its proximity to features identified by the state's administrative code (e.g., within one mile of public water supply, see Figure 27.7).

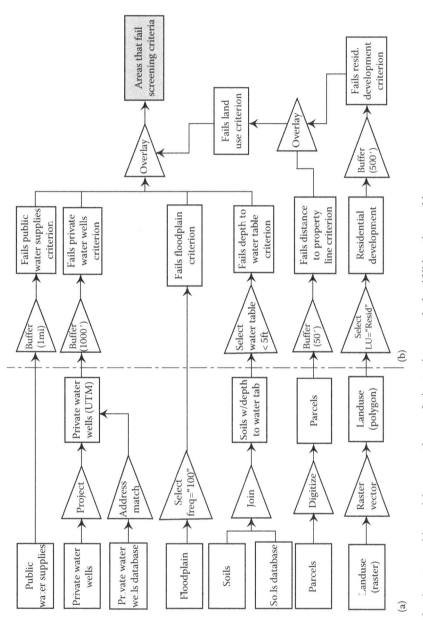

FIGURE 27.6 A cartographic model representing a solution to the example, landfill location problem.

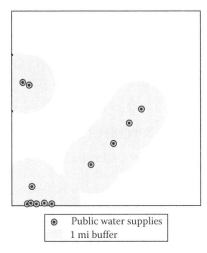

FIGURE 27.7   The results of a proximity operator (buffer) applied to public water-supply locations.

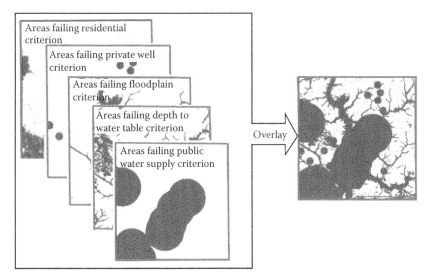

FIGURE 27.8   The results of an overlay operator (union) applied to the example, landfill location problem.

Finally, the various themes produced through the above transformations must be merged into a single theme that represents those areas that are unsuitable for the development of a landfill. This transformation can be implemented using an over-lay function (Figure 27.8). Care must be taken, however, in the interpretation of this analysis. As suggested above each of the individual data sets contains error and the error is propagated into this composite data set. The accumulated error can be significant (MacDougall, 1975; Chrisman, 1987; Lanter and Veregin, 1992) and decisions derived from overlay analysis must take into consideration the uncertainty generated by less than perfect data. The analysis presented here, for example, is best viewed as an inexpensive screening exercise that allows decision makers to focus more detailed analyses on promising sites. Users must also be cognizant that GIS software will allow them to perform analyses that are conceptually flawed (Hopkins, 1977; Malczewski, 2000).

27.5   LIMITATIONS AND OPPORTUNITIES

The conceptual advancements made by GIS researchers over the past 40 years have been impressive and the utility of the resulting software is undeniable. However, it is equally safe to say that much work remains to be done in the field of geographic information science (GISci) before these software packages reach their full potential. Here we select three issues that, on the one hand, limit the utility of current GIS software packages and, on the other, offer research opportunities for those interested in GISci. For a more thorough accounting of current GISci research themes in the United States the reader is directed to the University Consortium for Geographic Information Science (http://www.ncgis.org) and The National Center for Geographic Information and Analysis (http://www.ncgia.ucsb.edu).

27.5.1   Geographic Representation

Geographical systems are dynamic (change through time), interconnected (energy and material flows through space), and three-dimensional. However, the lineage of geographic information system software has its roots in automated cartography and image processing. This heritage has left what has seemed to be an indelible mark on GIS software (although recent developments are promising). This is both good and bad. The ability to communicate the results of geographical analyses in cartographic form has obvious advantages and image-processing techniques provide an invaluable source of derived data for geographical analysis. Thus, the tools of cartography and imaging have found a natural home in the GIS toolbox. Unfortunately, the forms of geographic representation that are well suited to cartographic representation and digital image processing do not necessarily transfer well when one is interested in the dynamic nature of geographical processes (Langran, 1993; Peuquet and Duan, 1995; Yuan, 1999; Yuan and Stewart Hornsby, 2008). Research into the digital representation of dynamic geographical systems has been active over the past several years (Wesseling et al., 1996; Bennett, 1997; Westervelt and Hopkins, 1999). Much of this work has been experimental and is not readily portable to mainstream GIS software packages. However, as software development companies adopt more open and extensible architectures, 4-dimensional GIS software (3 dimensions in space, one in time) becomes more feasible.

27.5.2   Multifaceted Nature of Geographical Problems

Geographical problem solving is often complex and multifaceted. The search for solutions to such problems often requires users to synthesize data from disparate sources and to find common ground among multiple competing stakeholders (Armstrong, 1994; Armstrong and Densham, 2008). One of the most powerful features of GIS software is its ability to integrate information. The GIS application presented is driven by simple exclusionary rules derived, for the most part, from mandatory state regulations. The problem, however, becomes more complicated when multiple stakeholders are considered. For example, landfill operators might be concerned about finding an economical solution, such as one that minimizes land acquisition and transportation costs. The public may be concerned about such issues as the impact of the landfill on property values, agricultural land, local aesthetics, or environmental justice. Balancing issues related to acquisition costs, transportation costs, the politics of the "Not In My Back Yard" problem, and environmental justice is a complicated problem. How does one weigh the relative importance of property values and transportation costs? Can you define and calculate an appropriate index for "local aesthetics" or "environmental justice"?

The integration of multicriteria evaluation techniques with GIS software has become increasingly common (Carver, 1991; Jankowski, 1995; Malczewski, 1999; Xaio et al., 2007). However, there are serious theoretical issues that must be addressed when using any technique that integrates data that are based on different metrics (Chrisman, 1997). Spatial decision support systems are developed to acknowledge this complexity and provide a means for developing alternatives that can be used to foster discussion about facets of a problem that may not be included in the digital domain (Armstrong et al., 1986; Densham, 1991).

27.5.3 Ubiquitous GIS

Individuals with mobile computing technology can produce, consume, and process geographic information at any time and location. This near ubiquitous access to real-time geospatial technologies is causing researchers to fundamentally rethink traditional GIScience concepts. Gartner et al. (2007), for example, outlined a series of research themes ripe for (re)consideration in the context of ubiquitous GIS. These themes include (for a complete list and discussion see Gartner et al., 2007):

1. Representation and use of context. The way a user interacts with spatial technology and data will vary dramatically, depending on who the user is, the location that he or she is at, the time they are at that location, and their immediate objectives (e.g., science or dinner).
2. Location in non-traditional spaces. Ubiquitous geospatial technologies will follow users into spaces that are not traditionally considered by GIS (e.g., through buildings and subway systems). New 4D location technologies are needed to support these kinds of movements.
3. Visualization on mobile devices. A mobile device requires information to be communicated using a small visual display (e.g., a cell phone LCD) in a rapid, concise, and unambiguous manner—a communication paradigm distinctly different from traditional forms of cartography.

4. Representation of mobile entities. The spatio-temporal representation of mobile entities within the context of GIS remains an active area of research.
5. Ethics, privacy, and digital surveillance. Ubiquitous GIS presents important ethical challenges for the data user and producer alike. Ubiquitous GIS facilitates the collection of individual-level data with precise location/time identifiers. While such data might have significant utility in, for example, studies of urban dynamics, the potential for abuse is very real.

27.6   CONCLUSIONS

GIS software provides an engine that, when in the hands of a trained user, can transform spatial data into usable information. The successful application of GIS software and geographical data to a highly diversified set of problems provides solid evidence that the transformational framework on which this software is built is both robust and general. Yet it is equally true that much work remains to done before GIS reaches its full potential. GIS software, for example, is limited by its inability to fully represent the spatiotemporal dynamics of geographical systems. Paradoxically, when and if the potential of GIS is fully realized, it is possible that many of the resulting products will not be recognizable as GIS software. Rather the capabilities of GIS software may become deeply embedded into the functionality of a wide variety of application specific software and, thus, geoprocessing software will become both transparent and ubiquitous (Armstrong, 1997).

However, regardless of the form taken by the next generation of spatially enabled software, the proper use of such technology will require an understanding of the conceptual framework on which geoprocessing tools are built. Our intent here was to introduce the reader to such a framework. Subsequent chapters will flesh out this framework through discussions that are both theoretical and practical in nature.

REFERENCES

ANSI 2000. http://webstore.ansi.org/ansidocstore/product.asp?sku=ANSI+NCITS+320%2D1998.
Alsberg, P.A., McTeer, W.D., and Schuster, S.A. 1975. *IRIS/NARIS: A Geographic Information System for Planners*. CAC Document No. 188, Center for Advanced Computing, Urbana, IL.
Anderson, P.F. 1980. *Regional Landscape Analysis*, Environmental Design Press, Reston, VA.
Armstrong, M.P. 1994. Requirements for the development of GIS-based group decision support systems. *Journal of the American Society for Information Science*, 45(9): 669–677.
Armstrong, M.P. 1997. Emerging technologies and the changing nature of work in GIS. *Proceedings of GIS/LIS'97*, American Congress on Surveying and Mapping, Bethesda, MD, unpaged CD-ROM.
Armstrong, M.P. 2000. Geography and computational science. *Annals of the Association of American Geographers*, 90(1): 146–156.
Armstrong, M.P. and Densham, P.J. 2008. Cartographic support for locational problem-solving by groups. *International Journal of Geographical Information Science*, 22(7): 721–749.
Armstrong, M.P., Densham, P.J., and Rushton, G. 1986. Architecture for a microcomputer based spatial decision support system. In *Proceedings of the Second International Symposium on Spatial Data Handling*, IGU Commission on Geographical Data Sensing and Processing, Williamsville, NY, 120–131.

Bennett, D.A. 1997. A framework for the integration of geographic information systems and modelbase management. *International Journal of Geographical Information Science*, 11(4): 337–357.

Bolstad, P.V., Gessler, P., and Lillesand, T.M. 1990. Positional uncertainty in manually digitized map data. *International Journal of Geographical Information Systems*, 4: 399–412.

Bowker, G.C. and Star, S.L. 1999. *Sorting Things Out: Classification and Its Consequences*, MIT Press, Cambridge, MA.

Broome, F.R. and Meixler, D.B. 1990. The TIGER data base structure. *Cartography and Geographic Information Systems*, 17(1): 39–47.

Buttenfield, B.P. and McMaster, R.B. 1991. *Map Generalization: Making Rules for Knowledge Representation*, John Wiley & Sons, New York.

Carver, S.J. 1991. Integrating multicriteria evaluation with geographical information systems, *International Journal of Geographic Information Systems*, 5(3): 321–339.

Chrisman, N.R. 1987. The accuracy of map overlays: a reassessment. *Landscape and Urban Planning*, 14: 427–439.

Chrisman, N.R. 1997. *Exploring Geographic Information Systems*, John Wiley & Sons, New York.

Chrisman, N.R. 1999. A transformational approach to GIS operations. *International Journal of Geographic Information Science*, 13(7): 617–638.

Clarke, K.C. 1995. *Analytical and Computer Cartography*, Prentice Hall, Englewood Cliffs, NJ.

Dale, P.F. and McLaughlin, J.D. 1988. *Land Information Management*, Oxford University Press, New York.

Densham, P.J. 1991. Spatial decision support systems. In *Geographical Information Systems: Principles and Applications*, eds. D.J. Maguire, M.F. Goodchild, and D.W. Rhind, John Wiley & Sons, New York, pp. 403–412.

Drummond, W.J. 1995. Address matching: GIS technology for mapping human activity patterns. *APA Journal*, Spring: 240–251.

Dueker, K.J. 1974. Urban Geocoding. *Annals of the Association of American Geographers*, 64(2): 318–325.

Dueker, K.J. 1979. Land resource information systems: A review of fifteen years experience. *Geo-Processing*, 1: 105–128.

Fabos, J.G., Greene, C.M., and Joyer, S.A. 1978. *The Metland Landscape Planning Process: Composite Landscape Assessment, Alternative Plans Formulation and Plan Evaluation; Part 3 of the Metropolitan Landscape Planning Model*. Research Bulletin 653, Mass AG Experimental Station, Amherst, MA.

FIPS. 1994. *Spatial Data Transfer Standard (SDTS) Federal Information Processing Standards*, National Institute of Standards and Technology, Washington, DC.

Flanagan, N., Jennings, C., and Flanagan, C. 1994. Automatic GIS data capture and conversion, in *Innovations in GIS 1*, ed. Worboys, M.F., Taylor & Francis, Bristol, PA, pp. 25–38.

Gartner, G., Bennett, D.A., and Morita, T. 2007. Toward Ubiquitous Cartography *Cartography and Geographic Information Science*, 34(4): 247–257.

Gillespie, S.R. 2000. An empirical approach to estimating GIS benefits. *Journal of the Urban and Regional Information Systems Association*, 12(1): 7–14.

Heuvelink, G.B.M. 1998. *Error Propagation in Environmental Modelling with GIS*, Taylor & Francis, Bristol, PA.

Hopkins L.D. 1977. Methods for generating land suitability maps: A comparative evaluation. *Journal for American Institute of Planners*, 34(1): 19–29.

Hunter, G.J. and Goodchild, M.F. 1996. A new model for handling vector data uncertainty in geographic information systems. *URISA Journal*, 8: 51–57.

Jankowski, P. 1995. Integrating geographical information systems and multiple criteria decision-making methods. *International Journal of Geographical Information Systems*, 9(3): 251–273.

Jenks, G.F. 1981, Lines, computers and human frailties, *Annals of the Association of American Geographers*, 71(1): 1–10.

Langran, G. 1993. *Time in Geographic Information Systems*, Taylor & Francis, Washington, DC.

Lanter, D.P. and Veregin, H. 1992. A research paradigm for propagating error in layer-based GIS. *Photogrammetric Engineering and Remote Sensing*, 58: 825–833.

Leitner, M. and Buttenfield, B.P. 1995. Multi-scale knowledge acquisition: Inventory of European topographic maps. *Cartography and GIS*, 22(3): 232–241.

MacDougall, E.B. 1975. The accuracy of map overlays. *Landscape Planning*, 2: 23–30.

Maguire, D.J. and Dangermond, J. 1991. The functionality of GIS. In *Geographical information systems: Principles and applications*, eds. D.J. Maguire, M.F. Goodchild, and D.W. Rhind, John Wiley & Sons, New York, pp. 319–335.

Malczewski, J. 1999. *GIS and Multicriteria Decision Analysis*, John Wiley & Sons, New York.

Malczewski, J. 2000. On the use of weighted linear combination method in GIS: Common and best practice approaches. *Transactions in GIS*, 4(1): 5–22.

Marble, D.F., Lauzon, J.P., and McGranaghan, M. 1984. Development of a conceptual model of the manual digitizing process. In *Proceedings of the International Symposium on Spatial Data Handling* (Vol. 1), Geographisches Institut, Zürich, Switzerland, pp. 146–171.

McKee, L. 1999. The impact of interoperable geoprocessing. *Photogrammetric Engineering and Remote Sensing*, 65: 564–566.

Nivala, A. and Sarjakoski, T.L. 2007. User aspects of adaptive visualization for mobile maps. *Cartography and Geographic Information Science*, 34(4): 275–284.

Obermeyer, N.J. 1999. Measuring the benefits and costs of GIS. In *Geographical Information Systems: Principles and Applications*, 2nd edn, eds. P.A. Longley, M.F. Goodchild, D.J. Maguire, and D.W. Rhind, John Wiley & Sons, New York, pp. 601–610.

Peucker, T.K. and Chrisman, N.R. 1975. Cartographic data structures. *American Cartographer*, 2: 55–69.

Peuquet, D.J. 1988. Representation of geographic space: Toward a conceptual synthesis. *Annals of the Association of American Geographers*, 78(3): 375–394.

Peuquet, D.J. and Duan, N. 1995. An event-based spatiotemporal data model (ESTDM) for temporal analysis of geographical data. *International Journal of Geographical Information Systems*, 9(1): 7–24.

Rushton, G., Armstrong, M.P., Gittler, J., Greene, B.R., Pavlik, C.E., West, M.M., and Zimmerman, D.L. (eds.) 2007. *Geocoding Health Data*, CRC Press, Boca Raton, FL.

Snyder, J.P. 1987. *Map Projections—A Working Manual*. USGS Professional Paper 1395, U.S. Government Printing Office, Washington, DC.

Thrower, N.J.W. 1972. *Maps & Man*, Prentice-Hall, Englewood Cliffs, NJ.

Tobler, W.R. 1979. A transformational view of cartography. *American Cartographer*, 6(2): 101–106.

Tomlinson, R.F., Calkins, H.W., and Marble, D.F. 1976. *Computer Handling of Geographical Data*, The UNESCO Press, Paris, France.

Tomlin, C.D. and Tomlin, S.M. 1981. *An Overlay Mapping Language, Regional Landscape Planning*, American Society of Landscape Architects, Washington, DC.

Tomlin, C.D. 1990. *Geographic Information Systems and Cartographic Modeling*, Prentice Hall, Englewood Cliffs, NJ.

USGS 1993. *Data Users Guide 5: Digital Elevation Model*, United States Department of Interior, Reston, VA.

USGS 2000. http://mcmcweb.er.usgs.gov/sdts/standard.html.

Wesseling, C.G., Karssenberg, D., Burrough, P.A., and van Deursen, W.P. 1996. Integrating dynamic environmental models in GIS: The development of a dynamic modelling language. *Transactions in GIS*, 1(1): 40–48.

Westervelt, J.D. and Hopkins, L.D. 1999. Modeling mobile individuals in dynamic landscapes. *International Journal of Geographic Information Systems*, 13: 191–208.

Xiao, N., Bennett, D.A., and Armstrong, M.A. 2007. Interactive evolutionary approached to multiobjective spatial decision making: A synthetic review. *Computers, Environment, and Urban Systems*, 31: 232–252.

Yuan, M. 1999. Representing geographic information to enhance GIS support for complex spatiotemporal queries. *Transactions in GIS*, 3(2): 137–160.

Yuan, M. and Stewart Hornsby, K. 2008. *Understanding Dynamics of Geographical Domains*, CRC Press, Boca Rotan, FL.

28 Geographic Data Structures

May Yuan

CONTENTS

28.1 INTRODUCTION

Finding an appropriate representation is a critical step to problem solving. Winston states, "Coarsely speaking, a representation is a set of conventions about how to describe a set of things. A description makes use of the conventions of a representation to describe some particular things" (Winston, 1984, p. 21). It provides a means to describe, communicate, generalize, store, and analyze information. A good representation "makes important things explicit," while exposing "the natural constraints inherent in the problem" (Winston, 1984, p. 24). Traditionally, geographers use maps as the primary form of representing geographic data. Hartshorne (1939, p. 249) even argues, "if his problem cannot be studied fundamentally by maps—usually by a comparison of several maps—then it is questionable whether or not it is within the field of geography."

Advances in quantitative analyses and computer technologies have promoted many other ways to represent geographic data, such as spatial interaction matrices, fractals, and other complex mathematical models (for examples and studies in these models, see Representation of space and representation of space/time in Fotheringham and Wegener, 1999). The selection of information to be represented and the choice of a representational scheme are often driven by the purpose of the analyses, although they might also be based on available data, or on an abstraction of the actual phenomena being represented (Mark, 1986). In recent years, geographic visualization has emerged as dynamic and multimedia displays of geographic data, probing new insights to intrinsic spatial distributions of geographic variables that cannot be portrayed in a 2D, static map environment.

Since a representation provides both conceptual and computational foundations for processing, integrating, analyzing, and visualizing geographic data, the representation chosen for a geographic phenomenon has a profound impact on interpretation and analysis. In a GIS, the kinds of geographic information that can be encoded, computed, and visualized largely depend upon the embedded representation schemes. As such, geographic representation is an important subject in the study of theories and concepts in developing the next generation of GIScience and GIS applications (UCGIS, 1996).

There are many levels of spatial data representation in a GIS, ranging from computer-encrypted codes of 0s and 1s to models that are closer to the human conceptualization of reality. While detailed implementation of these representations is only significant to technical professionals, all advanced GIS users should understand spatial data objects, data structures, and data models to have a good grasp of how the system works, its applications, and limitations. *Spatial data objects* are digital representations of real-world entities, and they are the basic data unit that users can manipulate and analyze in a GIS database. They carry data about both the geometric and the thematic properties of represented entities at a certain level of abstraction. *Spatial data structures* refer to methods for organizing spatial data with emphases on efficiency on storage and performance (Franklin, 1991). Furthermore, *spatial data models* are high-level data structures that focus on "formalization of the concepts humans use to conceptualize space" (Egenhofer and Herring, 1991, p. 229). A spatial data model can be implemented in different spatial data structures, but a spatial data structure must conform to the properties in a corresponding spatial data model. Consequently, spatial data models determine the essence of what geographic information can be represented and how it is represented in a GIS. The following sections in this chapter provide essential discussions on representing spatial data in terms of spatial data models, spatial data structures, incorporating time into spatial representation, and trends and challenges in representing spatial data. There are many other excellent references on these subjects. Puequet (1984, 1994) provides analytical comparisons and syntheses of spatial data models. Samet (1990a,b) offers a comprehensive introduction to spatial data structures and their applications. Egenhofer and Golledge (1998) collect many thought-provoking articles in spatial and temporal reasoning providing new insights into space–time integration in geographic information systems. Peuquet (2002) comprehensively overviews the representation of space and time in philosophy, cognitive science, and geographic information science. In addition, Yuan et al. (2004) summarize needs for extensions to geographic representation as one of the 10 research challenges in Geographic Information Science. Interested readers should also consult standard work in information storage and retrieval, such as Date (1995) and Ullman (1980) for in depth discussions on various general data structures and relational data models that are commonly applied to handle attribute data in GIS databases.

28.2 SPATIAL DATA MODELS: OBJECT AND FIELD REPRESENTATIONS

In general, humans perceive the geographic world as a set of discrete entities and continuous fields. Discrete entities, often simply referred to as objects, are distinguishable by their independent and localized existence; their relative permanence of identity, attribute, and shape; and their manipulability. They are identified as individuals before any attributes they may possess (Couclelis, 1992). On the other hand, a continuous field describes the spatial variation of a geographic variable in a space–time frame. Most fields are scalar fields that only have a single value of the geographic variable at every point in the space–time frame. When a geographic variable is directional (such as wind), it may form a vector field to provide a distribution of directions with or without values in the frame (Goodchild, 1997a,b). Both objects and fields are represented well in topographic maps (Figure 28.1) where discrete entities like towns, roads, and lakes are marked by points, lines, and polygons, and a continuous field of topography is portrayed by contours and levels of shades.

However, currently spatial data models cannot accommodate both object and field perspectives in a single model. The constraint is mainly due to the incompatibility of

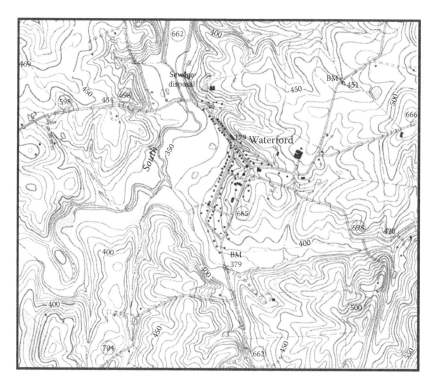

FIGURE 28.1 **(See color insert following page 426.)** An example of USGS topographic quadrangle maps showing both objects (roads, buildings, etc.) and fields (contours).

the spatial conceptualizations of objects and fields that require distinct sets of data objects to portray geographic space. Spatial data models designed to represent discrete entities use points, lines, and polygons as geometric primitives to form spatial data objects that depict spatial characteristics of the entities. Such an object-view of the world describes reality as an empty space containing a combination of conceptual, primitive, and compound objects. It emphasizes individual entities and has no constraints on space being exhausted (Frank and Mark, 1991). In essence, the idea is comparable to Newton's atomic view of the world, in that, space exists regardless of whether or not there are objects to occupy the space (Couclelis, 1982). Object representation is most appropriate for geographic entities for which boundaries can be well defined, and since points, lines, and polygons do not normally exist in nature, they are practical for engineering works or administrative and property lines, such as well locations, highways and streets, states and counties, and land parcels.

In contrast, the field perspective is comparable to Einstein's relative view of the world, where space is not a container populated with objects as in Newton's model, but a "plenum" characterized by a ubiquitous field. Since a field is spatially continuous, a representation of the field must have a value or be able to imply a value at every location. As quoted in Couclelis (1992, p. 70):

> There is no such thing as empty space, i.e. space without field. Space-time does not claim existence on its own, but only as a structural quality of the field (Einstein, 1960, *Relativity*, p. 155).

Unlike object representation, there are no identified things in a field representation. Rather, geographic things (or features) emerge through spatial and temporal aggregation of spatial units for which raw or derived values are within a threshold in a single field or multiple fields that form the basis for image interpretation and pattern recognition.

Although it is impossible to fully represent a continuous space in a digital world, several spatial data models are designed to provide various degrees of approximation. The commonly used models include regular spaced points (lattice), irregular spaced points, contours, regular cells (raster or grid), irregular triangular networks (TIN), and polygons (Goodchild, 1997a,b). Regular or irregular points are most popular for field surveys and weather observatory networks. Contours apply a set of isolines to show the spatial pattern of a geographic variable, such as elevation, temperature, or chemical concentration. Contours are very effective representation for visualization, but they are ineffective for digital computation compared to other field-based models. Point- or line-based models cannot represent a field completely because values of the focused geographic variable are only available at the locations of points or lines. In order to create a continuous surface that covers the entire field, spatial interpolation is necessary to transform point- or line-based models to area-based models, that is grid, TIN, or polygons (Bourrough and McDonnell, 1998). There are many spatial-interpolation methods, and each of them has different assumptions on spatial distributions. Lam (1983) provides an excellent discussion on commonly used spatial-interpolation methods.

Area-based field representations divide a space into a finite number of smaller areas. While area-based models are complete field representations, the value within each small area is set to a constant, assuming that there is no spatial variation within each of the small areas. Consequently, the size of these small areas determines the amount of information and the degree of detail in a field that an area-based model can represent. A field representation is regular if its smaller areas are of the same area and geometry; otherwise, it is irregular. Square is the most commonly used geometry to construct a regular field model, named *grid*. It can be directly related to the data acquired by remote sensing technology, especially useful to acquire large-scale geographic data. However, it is geometrically impossible to represent the spheroidal Earth (Chapters 2 and 4) with a single mesh of uniform, rectangular cells (Dutton, 1983). Other geometries, especially the triangular or hexagon mesh, will provide better coverage for the spheroidal Earth, but converting squared cell-based data captured by satellites to triangular or hexagonal meshes will inevitably introduce uncertainty.

Nevertheless, irregular triangular meshes, known as *Triangulated Irregular Network* (*TIN*), are particular effective to represent surfaces that are highly variable and contain discontinuities and breaklines (Peuker et al., 1978; Figure 28.2). It is effective because a TIN connects a set of irregularly spaced significant locations, each of which defines a point where there is a change in the surface. For example, all neighboring points of the peak of a mountain are downhill; all neighbors of a point along a stream, except for the downstream point, are uphill. These significant points form nodes of triangles, while linear features (such as streams or ridges) and boundaries of area features with constant elevation (such as shorelines) frame triangle edges. Alternatively, a field can be represented by a set of irregular polygons, also known as *irregular tessellation* (Frank and Mark, 1991). This approach is similar to thematic mapping where a value in each area represents an average, total, or some other aggregate property of the field within the area. Every point in the space–time frame of interest must lie in exactly one polygon, except for the ones on boundaries.

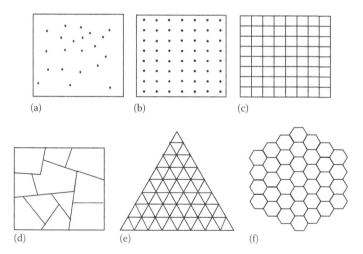

FIGURE 28.2 Examples of field representations: (a) irregular points; (b) regular points; (c) a grid; (d) polygons; (c) a triangular mesh; (f) a hexagonal mesh.

There is no overlapping polygon, and all polygons together must exhaust the frame. Typical geographic fields represented by irregular polygons include vegetation cover classes, soil types, and climate zones. Another field model of irregular polygons is *Thiessen polygons*, also known as *Dirichlet* or *Voronoi polygons*. Thiessen polygons are derived by first connecting data points to triangles via Delaunay triangulation (the same procedure to form a TIN) and then linking perpendicular bisectors of these triangles (Figures 28.3 and 28.4). Thiessen polygons are often used in meteorology

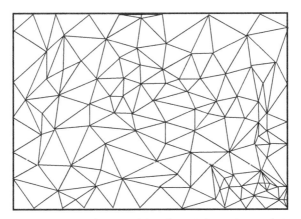

(a) Triangular irregular network based on a Delauney Triangulation

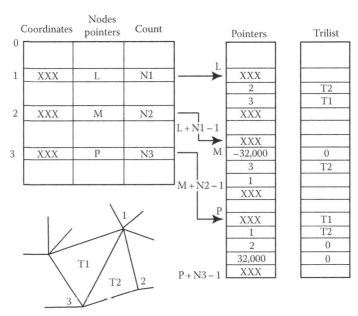

(b) Data structure of a TIN (detail)

FIGURE 28.3 The data structure of Triangulated Irregular Networks. (Reproduced from Bourrough, P.A. and McDonnell, R.A., *Principles of Geographical Information Systems*, Oxford University Press, New York, 1998, p. 66.)

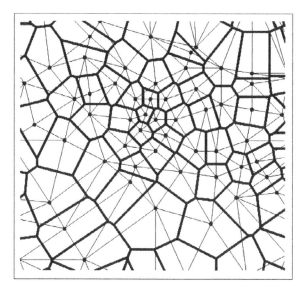

FIGURE 28.4   An example of Thiessen polygons (thick lines) and Delaunay triangles (thin lines).

and hydrology as a quick method for relating point-based data to space under the assumption that data for any given location can be taken from the nearest observatory station. However, the size of Thiessen polygons is highly related to the spatial distribution of data points, and a data point may represent unreasonably large area within which all locations are given the same value as the one at the data point. Consequently, the model may represent the field well in an area where sample points are appropriately dense for the variable under study, but it may overlook spatial variation in the sparsely sampled area.

Together, object and field representations reflect two complementary views of the geographic world. There are geographic things that can be easily identified as individual objects, and there are properties of the land that can only be described through continuous fields of attributes. These fields serve as substrates upon which objects are built through human activities (such as roads or ranges) or natural processes (such as rivers or islands). While current spatial data models are either object-based or field-based representation, most commercial GIS provide functions to convert data from one model to another with a certain degree of compromising accuracy. Nevertheless, there are geographic phenomena (such as wildfire and precipitation) that have both object-like and field-like properties. Object or field representation alone cannot fully capture the intrinsic properties of these phenomena, and a combined approach that uses raster data to represent field-based attributes (such as precipitation intensity) in a given object (such as a convective storm) can better capture the spatial variability of properties within the object (Yuan, 2001). A study of representing convective storms in Oklahoma suggests that a combined object-field approach can enrich the semantics of storm characteristics to be embedded in GIS databases (McIntosh and Yuan, 2005). Similarly, Cova and Goodchild (2002) also advocate for a hybrid approach to geographic representation that can link geographic objects (such as view sheds) to locations in a field. The intricate perception of fields and objects is also discussed in Galton (2004) from the perspective of human cognition. An integrated model that supports both object and field perspectives will facilitate a better modeling of reality. Much can be learned from entertainment technology that has developed representation enabling the interactions between moving objects and their background fields.

28.3   SPATIAL DATA STRUCTURES: VECTOR AND RASTER DATA

Object and field representations outline the conceptual models of reality. Implementation of the models relies upon appropriate data structures. Vector and raster are two major types of spatial data structures. Vector data are based on Euclidean geometry with *points*, *lines*, and *polygons* as primitive 0-, 1-, and 2-dimensional objects, accordingly. Raster data consist of only one spatial object type: *cells*. Since geographic things are clearly identified in an object representation, they can only be modeled by vector objects to describe their location and geometry. As to field representation, spatial data models may use vector data for regular or irregular points, contour lines, TIN (points and lines), and Thiessen or irregular polygons, or raster data for grids to represent spatial distributions of geographic properties.

28.3.1   Vector-Based Structures

In a vector geographic information system, primitive spatial data objects are points, lines, and polygons located by Cartesian coordinates in a spatial reference system. These simple geometric primitives indicate static locations and spatial extents of geographic phenomena in terms of XY coordinates at a certain level of abstraction. A point object only marks the location of a geographic entity, such as a well, by a pair of XY coordinates. A line object shows the location and linear extent of a geographic entity, such as a river, by a series of XY coordinates. Furthermore, a polygon object depicts the location and 2-dimensional extent of a geographic entity, such as a county, also by a series of XY coordinates. However, a polygon has its first XY pair the same as its last to ensure a closure. These simple geometric primitives constitute many early vector data sets in computer cartography, in which data input is the primary consideration in structuring spatial data, and little data manipulation is performed after the data have been input into the system from maps (Peucker and Chrisman, 1975). *The spaghetti data structure* is the representative of spatial data structures with simple points, lines, and polygons (Figure 28.5). In this data structure, every data object is independent of the others with no regard for connectivity or adjacency. A point shared by two lines or a boundary shared by two adjacent polygons is to be encoded twice in a spaghetti data structure. This not only results in data redundancy, but also impedes error checking and data analysis.

The spaghetti data structure was quickly replaced by spatial data structures that incorporate connectivity among spatial objects, that is, *topology* (Dangermond 1982). Topological data structures distinguish two types of points: (1) *nodes* that are endpoints on a line (often referred to as *from-node* and *to-node*) and (2) *points* that mark

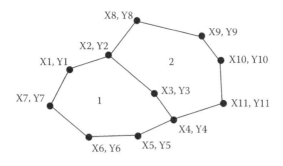

Polygon 1	Polygon 2
X1, Y1	X2, Y2
X2, Y2	X8, Y8
X3, Y3	X9, Y9
X4, Y4	X10, Y10
X6, Y6	X11, Y11
X7, Y7	X4, Y4
X1, Y1	X3, Y3
	X2, Y2

FIGURE 28.5 An example of the spaghetti data structure. Note that the points along the common boundary are recorded twice.

locations. Points located on a line also describe the shape of that line. For polygon data, topological data structures encode *polygon adjacency* by noting the polygons on both sides of a boundary line as the left-polygon and right-polygon in relation to the direction of the line. With these arrangements, it is easy to discern connected lines and adjacent polygons. If two lines have a common node, then they are connected. Likewise, if two polygons are the left- and right-polygons of a line, then they are adjacent.

There are several topological data structures, such as the GBF/DIME (Graphic Base File/Dual Independent Map Encoding), POLYVERT (POLYgon ConVERTer), and TIGER (Topologically Integrated Geographic Encoding and Referencing System). Both the GBF/DIME and TIGER systems were developed by the U.S. Bureau of the Census to automatically encode street networks and census units for the 1970 and 1990 censuses, respectively. The GBF/DIME system is centered on linear geographic entities, such as streets and rivers, and uses straight-line segments as the basic data objects to represent these linear entities. For each line segment, it encodes from-node, to-node, left-polygon (left-block), right-polygon (right-block), left-address, and right-address. Nodes are to be encoded twice when they connect two lines, and so do polygons when a polygon is on the right side of a line but on the left side of another (Figure 28.6). Nevertheless, such redundancy allows automated checking for data consistency (Puequet, 1984), and is later applied to the "chains" or "arcs" structure underlying POLYVERT and many modern vector GIS data models such as DLG (USGS Digital Line Graphs), SDTS (Spatial Data Transfer Standard), and the polygon layers in commercial systems such as ARC/INFO.

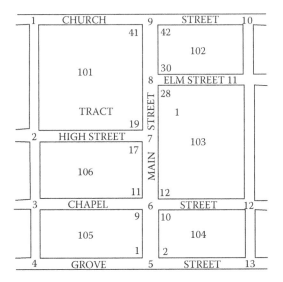

CENSUS ADDRESS CODING GUIDE RECORDS

STREET	TRACT	BLOCK	LOW ADDR	HIGH ADDR
Main	1	102	30	42
Main	1	103	12	28
Main	1	104	2	10
Main	1	105	1	9
Main	1	106	11	11
Main	1	101	19	41

DIME STREET SEGMENT RECORDS

STREET	NODE START	NODE END	TRACT LEFT	BLOCK LEFT	TRACT RIGHT	BLOCK RIGHT	LOW ADDR	HIGH ADDR
Main	5	6	1	105	1	104	1	10
Main	6	7	1	106	1	103	11	17
Main	7	8	1	101	1	103	19	28
Main	8	9	1	101	1	102	30	42

FIGURE 28.6 An example of the DIME file structure. (Reproduced from Peucker, T.K. and Chrisman, N., *Am. Cartographer*, 2(1), 59, 1975. With permission.)

Although DIME incorporates topology through line segments, it is ineffective to assemble the outline of a polygon. Moreover, it is laborious to update line segments and retrieve polygons (Peucker and Chrisman, 1975). Alternatively, POLYVERT uses *chain* as its basic spatial object. A chain, like a line segment, has two nodes at its ends, but, unlike a line segment, it may consist of many shape points. Thus, a single chain can sufficiently reference the boundary between two polygons in POLYVERT, rather than a series of line segments as in DIME. POLYVERT organizes geometric and topological information into three tables of points, nodes, and chains, and a polygon list and polygon-chain list (Figure 28.6). Since chains are the basic spatial objects, the chain table references to all other tables and the polygon list. Each chain has an identifier, references to the point table to obtain coordinates for its shape points, references to from- and to-node identifiers in the node table for node coordinates, and

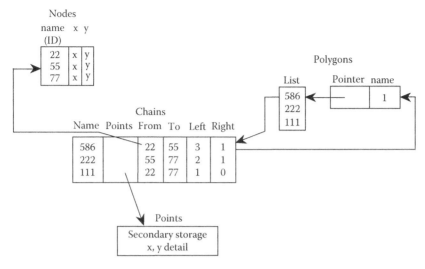

FIGURE 28.7 An example of the internal representation of the POLYVERT chain file. (Reproduced from Peucker, T.K. and Chrisman, N., *Am. Cartographer*, 2(1), 59, 1975. With permission.)

references to left- and right-polygon identifiers to individual polygons in a polygon list. The polygon list has built-in pointers, each of which directs a polygon to a list of all chains that make up that polygon. Such a data structure enables effective retrieval of polygons in two ways: (1) by adjacency from left- and right-polygons in the chain table; and (2) by polygon identifiers from the polygon-chain list.

The data structure used in POLYVERT (Figure 28.7) is similar to the design of the TIGER system. While the GBF/DIME encodes only straight lines and requires assembling of lines to retrieve a polygon, the TIGER system improves the census encoding by allowing storing curve lines (i.e., *chains*) and directly retrieving polygons of census units (Marx, 1986). To ensure that there is no duplication of features or areas, the TIGER system adapts the theories of topology, graph theory, and related mathematics to define the location and relationship of streets, rivers, railroads, and other features to each other and to the census survey units. Additional data objects are incorporated in the TIGER system to further differentiate topological significance (Table 28.1). An *entity point* is used to identify the locations of point features, such as wells and towers, or to provide generalized locations for areal features, such as buildings and parks. A *complete chain* has both topology (marked by its end nodes) and geometry (characterized by its shape points). Complete chains form polygon boundaries and intersect other chains only at nodes. When complete chains form a closure, they form a *GT-polygon* (geometry and topology polygon). GT-polygons are elementary polygons that are mutually exclusive and completely exhaust the surface. Chains that are not associated with polygons (i.e., have no left and right polygons) are *network chains*.

In total, the TIGER system consists of 17 record types (each of which forms a table) to encode geometric coordinates and attribute information for spatial data objects.

TABLE 28.1

A Summary of Spatial Data Objects Used in the TIGER System

	Point (0-Cell)	Line (1-Cell)	Polygon (2-Cell)
Topology	Node	Complete chains or network chains	GT-polygons
Non-topology	Entity point		
Attribute	Shape point		

Source: TIGER/Line, Technical documentation, 1999, http://www.census. gov/geo/www/tiger/tiger99.pdf.

Among the 17 record types, Type 7 has coordinates of entity points for landmark features, Types 1, 2, 3, and 5 record coordinates and geographic attributes of complete chains, Type I provides the linkages among complete chains and GT-polygons, and Type P contains coordinates of entity points for polygons. Common data items, such as TLID, FEAT, CENID, and POLYID, can be used to relate tables of the 17 record types. Although the TIGER system is structurally complex, it provides a rich spatial data framework for census and socioeconomic mapping.

Many more complex spatial data objects have been introduced to account for diverse geometric and topological properties of geographic phenomena. For example, the spatial data transfer standards (SDTS), developed in 1992 by the U.S. Geological Survey (USGS), contain many additional spatial objects, such as rings, universal polygons, and void polygons (SDTS document, http://mcmcweb.er.usgs.gov/sdts/ standard.html). When more and more complex spatial objects become necessary to satisfy the need for representing diverse geographic phenomena, topological structures that associate objects from various tables and lists create significant overhead for data management and retrieval. Non-topological data structures have recently regained popularity, for they can provide faster drawing speed and editing ability, allows overlapping and noncontiguous polygons, and require less disk storage. Most importantly, they enable one-to-one mapping between spatial data objects and geographic entities so that spatial data structures can be fully integrated with non-spatial database management systems (DBMS). For example, the state of Hawaii consists of eight islands. Using a topological data structure, it takes eight independent polygon objects, each of which has the attribute state as Hawaii (resulting in eight records). Alternatively, a non-topological data structure can associated one attribute record to a single data object of a *multipolygon* to represent the state of Hawaii. Chapter 27 in this book provides additional discussions on TIGER files.

Non-topological data structures are considerably simple compared to topological data structures. Every data object has its own stand-alone table or list to encode identifiers and coordinates. Since there is no topology embedded, there is no need to link tables in any way. However, topology is critical for spatial data manipulation and analysis. In order to fulfill the need for topology, advancement in computing power has made it possible to incorporate appropriate functions to build topology

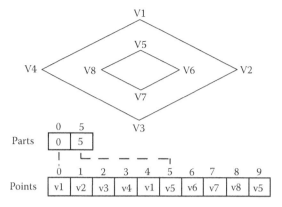

Polygon record contents

Position	Field	Value	Type	Number	Byte order
Byte 0	Shape Type	5	Integer	1	Little
Byte 4	Box	Box	Double	4	Little
Byte 36	NumParts	NumParts	Integer	1	Little
Byte 40	NumPoints	NumPoints	Integer	1	Little
Byte 44	Parts	Parts	Integer	NumParts	Little
Byte X	Points	Points	Point	NumPoints	Little
Note: X = 44 + 4 + NumParts.					

FIGURE 28.8 An example of the shape file structure. Here is a polygon with a hole and eight vertices. Parts and points indicate the number of parts and vertices (shape points) that constitute each part, linked by the address numbers or identifiers of the vertices. Polygon record contents detail X and Y coordinates and relevant structural information for the object. (Courtesy of Environmental Systems Research Institute, Inc., Redlands, CA.)

on the fly (i.e., build it when it becomes necessary). ArcView shape files and Spatial Database Engine (SDE), both developed by Environmental Systems and Research Institute (ESRI, Redlands, California) are examples of the modern non-topological data structures (Figure 28.8).

28.3.2 RASTER-BASED STRUCTURES

Unlike vector data, raster data are regular arrays of cells without explicit associations with XY coordinates except for one of its corner cell (usually the south-west corner). While a cell is easier for programming purposes to be identified by row and column number, the XY coordinates can also access it. In simple raster structures, there is a one-to-one relation between data value, pixel, and location. Therefore, all data values can be stored as a simple array with a Meta file indicating the numbers of rows and columns, cell size, projection and coordinate systems, and minimum values for X and Y coordinates (Figure 28.9). However, this is a storage-hungry approach because it uses the disk space for the entire array regardless of data distribution. Demand for large data storage can also degrade data processing performance. To

```
ncols       270
nrows       476
xllcorner   391253.1875
yllcorner   4064188.25
cellsize    3
NODATA_value −9999
```

−9999 −9999 −9999 −9999 −9999 −9999 2321.5 2321.295 2320.653 2319.938 2319.385 .....
−9999 −9999 −9999 −9999 2321.5 2321.5 2321.5 2321.093 2320.492 2319.851 2319.341 .....
−9999 −9999 2321.5 2321.5 2321.5 2321.5 2321.421 2320.977 2320.449 2319.905 2319.438 .....
−9999 −9999 2321.5 2321.5 2321.5 2321.5 2321.327 2320.94 2320.492 2320.024 2319.595 .....
−9999 −9999 2321.5 2321.5 2321.5 2321.5 2321.281 2320.964 2320.588 2320.179 2319.777 .....

FIGURE 28.9 An example of simple raster data structures.

effectively increase data processing performance and reduce the demand for data storage, two issues involving raster data structures need to be addressed: (1) compression methods: how to store the data, and (2) scan order: how to scan the data in an array. Since geographic phenomena often show a certain degree of spatial autocorrelation, it is common to have blocks of cells in a raster array with the same data value. For example, when raster structures are used to represent an area, all cells of the area will have the same value. Such properties are the basis for many compress and scan-order methods. Commonly used compression methods include chain codes, run-length codes, block codes, and quadtrees. Commonly used scan orders include row, row-prime, Morton and Peano–Hilbert. Quadtrees can be used as both compression and scan-order methods.

Compression methods aim to reduce the demand for data storage. *Chain codes*, also known as *Freeman-Hoffman chain codes*, store raster cells based on directions along linear features, especially on boundaries (Freeman, 1974). The method uses a number scheme (1–8) to represent each direction in a clockwise fashion. The chain-code method is particularly effective for storing region data because only the cells of region boundaries are encoded. It is also very useful for raster-to-vector conversion (i.e., *vectorization*) of region features. However, that the chain code method only stores relative locations of boundaries makes spatial operations difficult without reconstruction of a full grid. The *run-length coding* method is another simple yet effective data compression method (Figure 28.10a). It groups cells of the same value row by row and encodes these cells by a beginning cell, an end cell, and an attribute value. A general format for the code is (Row, (Min_Column, Max_Columm, Attribute)). The run-length coding method is particularly useful when there are just a few classes (attribute values), but is ineffective for coding continuous variation when each cell tends to have a unique value. The *block coding* methods extend the run-length codes to a two-dimensional space by grouping cells of the same value into square blocks. A general format for the block code is (Row and Column of the origin, Extent, Attribute), where "the origin" is usually the bottom left of the block and "Extent" is the block size in terms of the number of cells (Rosenfeld, 1980). The block coding method is most effective to store grids with large regions of classes where a region can be indexed into a few big blocks. However, like the run-length coding method, block codes are not suitable for grids with high spatial variations.

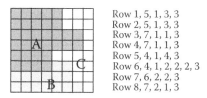

Row 1, 5, 1, 3, 3
Row 2, 5, 1, 3, 3
Row 3, 7, 1, 1, 3
Row 4, 7, 1, 1, 3
Row 5, 4, 1, 4, 3
Row 6, 4, 1, 2, 2, 2, 3
Row 7, 6, 2, 2, 3
Row 8, 7, 2, 1, 3

(a) Run-length codes: A = 1, B = 2, and C = 3

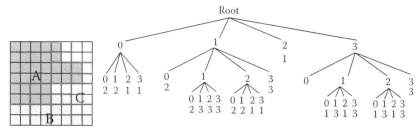

(b) A quadtree: A = 1, B = 2, C = 3

FIGURE 28.10   Examples of (a) run-length codes and (b) quadtree codes.

Another very popular coding method is the *quadtree*, which is also a scan-order method (Figure 28.10b). Distinguished from all the other coding methods above, quadtree is a hierarchical data structure, in that it is based on the principle of recursive decomposition of space (Samet, 1990a). The quadtree data structure divides a grid into four quadrants (NW, NE, SW, and SE), and will further divide a quadrant to four sub-quadrants if the quadrant is not homogeneous (i.e., contains only one attribute value). As a result, the quadtree method is only applicable to grids with both the numbers of rows and columns equal to a power of $2(2^n)$. A quadtree has a *root node*, which corresponds to the entire grid, and *leaf nodes*, which associate with the attribute values and quadrants without further divisions. In the worst-case scenario, all leaf nodes correspond to single cells. Nevertheless, the quadtree method is probably the most studied spatial data structure by both computer scientists and GIS scientists. It provides effective data access and spatial operations at multiple levels of resolutions (Mark and Lauzon, 1984; Samet, 1990b).

The scan-order methods are mainly concerned with performance in terms of data processing (Goodchild, 1997a,b). The *row* and *row-prime* methods scan one row at a time (Figure 28.11a and b), but the row-prime method reverses every other row. The compression methods used with the row-prime scan, also known as *Boustrophedon scan*, can achieve a greater degree of compression because adjacent cells are more likely to have the same value. However, like the row-scan method, the row-prime method only scans in a one-dimensional fashion, so that neighboring cells in an array do not always close to each other in a scan. Two-dimensional scan methods, such as *Morton* and *Peano–Hilbert*, put neighboring cells in a cluster by spatially recursive shapes (Figure 28.11c and d). The Morton method repeats a Z-like shape with four neighboring cells as a unit and repeats the shape at all levels. Unfortunately, the complexity of the Morton scan does not guarantee a better compression. Goodchild

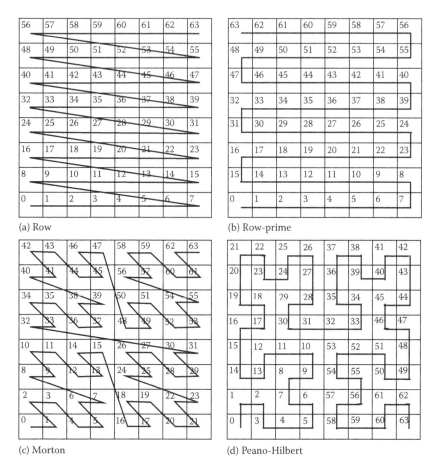

FIGURE 28.11 Examples of (a) row, (b) row-prime, (c) Morton, and (d) Peano–Hilbert scan order methods. Numbers represent the scan order in each method. (Modified from Bourrough, P.A. and McDonnell, R.A., *Principles of Geographical Information Systems*, Oxford University Press, New York, 1998, p. 57.)

and Grandfield (1983) show that the row-prime method in general produces better results than the Morton scan. Alternatively, the Peano–Hilbert method, also known as *Pi-order scan* or *Peano curve*, has a basic U-like shape that repeats at all levels, and this method generally gives the best results (Goodchild and Grandfield, 1983).

28.4 REPRESENTATION OF TIME IN A GIS

Geographic worlds are neither static nor planar. The above discussions on spatial data modeling and data structures overlook the need to handle dynamic or volumetric phenomena. Incorporation of temporal components into a spatial representation is not a trivial task because time has distinct properties from space. Timestamping is the most popular approach of incorporating temporal components into relational database management systems (RDBMS) and GIS (Yuan, 1999). For example in

1993

County	Population	Avg. income
Nixon	17,000	20,000

1994

County	Population	Avg. income
Nixon	20,000	19,800
Cleveland	35,000	32,000

1995

County	Population	Avg. income
Nixon	20,900	21,000
Cleveland	35,000	32,000
Oklahoma	86,000	28,000

(a) Time-stamped tables (Gadia and Vaishnav, 1985).

Stock	Price	From		To	
IBM	16	10-7-91	10:07 am	10-15-91	4:35 pm
IBM	19	10-15-91	4:35 pm	10-30-91	4:57 pm
IBM	16	10-30-91	4:57 pm	11-2-91	12:53 pm
IBM	25	11-2-91	12:53 pm	11-5-91	2:02 pm

(b) Time-stamped tuples (rows): an ungrouped relation
(Snodgrass and Ahn, 1985).

Name	Salary	Department
[11, 60] John	[11, 49] 15 K [50, 54] 20 K [55, 60] 25 K	[11, 44] Toys [45, 60] Shoes
[0, 20] U [41, 51] Tom	[0, 20] 20 K [41, 51] 30 K	[0, 20] Hardware [41, 51] Clothing
[0, 44] U [50, Now] Mary	[0, 44] U [50, Now] 25 K	[0, 44] U [50, Now] Credit

(c) Time-stamp values (cells): a group relation
(Gadia and Yeung, 1988). [11, 60] represents a period
starting at T_{11} and ending at T_{60}.

FIGURE 28.12 Timestamp methods used to incorporate time into relational databases. (Reproduced from Yuan, M., *Trans. GIS*, 3(2), 141, 1999. With permission.)

RDBMS, timestamping has been applied to tables (Gadia and Vaishnav, 1985; Figure 28.12a), individual tuples (Snodgrass and Ahn, 1985; Figure 28.11b), or individual cells (Gadia and Yeung, 1988; Figure 28.12c). Correspondingly, GIS data are timestamped with layers (the snapshot model in Armstrong, 1988; Figure 28.13a), attributes (the space–time composite model in Langran and Chrisman, 1988; Figure 28.13b), or spatial objects (the spatiotemporal objects model in Worboys, 1994; Figure 28.13c).

The snapshot model presents the simplest way to incorporate time into spatial data by a set of independent states. Since the method stores a complete state for every given snapshot, it encounters problems of data redundancy and possible data inconsistency, especially in dealing with large data sets. Alternatively, space–time composites and spatiotemporal object models explicitly represent information about

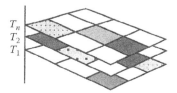

(a) Time-stamped layers (Armstrong, 1988).

Poly id	T_1	T_2	T_3	T_4
1	Rural	Rural	Rural	Rural
2	Rural	Urban	Urban	Urban
3	Rural	Rural	Urban	Urban
4	Rural	Rural	Urban	Urban
5	Rural	Rural	Rural	Urban

(b) Time-stamped attributes (columns): Space–time composites
(Langran and Chrisman, 1988).

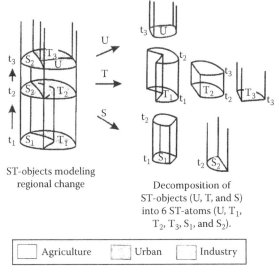

ST-objects modeling
regional change

Decomposition of
ST-objects (U, T, and S)
into 6 ST-atoms (U, T_1,
T_2, T_3, S_1, and S_2).

| ☐ Agriculture | ☐ Urban | ☐ Industry |

(c) Time-stamped space–time objects: the spatiotemporal
object model (Worboys, 1992).

FIGURE 28.13 Timestamp methods used to incorporate time into GIS databases.
(Reproduced from Yuan, M., *Trans. GIS*, 3(2), 144, 1999. With permission.)

changes. Both eliminate the problems of data redundancy and inconsistency to a
degree. However, the space–time composite model has problems keeping spatial
object identifiers persistent because updating space–time composites can cause
fragmentation of existing spatial objects (Langran and Chrisman, 1988). While the
spatiotemporal object model is able to maintain spatial object identifiers, it, as in all
timestamping approaches, has difficulty representing dynamic information, such as
transition, motion, and processes. Additional change-based data models are proposed

by Hazelton (1998) and Kelmelis (1998) to capture changes of geographic features in a 4-dimensional space–time Cartesian space. Nevertheless, such a change-based or timestamping approach uses geometrically indexed methods that "make the coordinate system of the layer into the primary index of the spatial representation" (Raper and Livingstone, 1995, p 360), not the process itself. As result, change-based data models are incapable of supporting spatiotemporal queries about information on the dynamic characteristics of geographic processes, such as movement, rate of movement, frequency, and interactions among processes. Geographic representation "*must deals with actual processes, not just the geometry of space-time*" (Chrisman, 1998, p. 91). Kelmelis (1998) propounds the use of process dynamics, temporal extent, and causal propagation as the basis for linking space and time in geographic and process modeling.

Recent development in GIS representation has emphasized dynamic processes. These models include Smith et al.'s (1994) modeling and database system (MDBS), Peuquet and Duan's (1995) event-based spatiotemporal data model (ESTDM), Raper and Livingstone's (1995) geomorphologic spatial model (OOgeomorph), and Yuan's (1996) three-domain model. MDBS takes a domain-oriented approach to support high-level modeling of spatiotemporal phenomena by incorporating semantics into data modeling, specifically for hydrological applications. However, MDBS is designed for hydrological data processing and modeling; it is not designed for representing or managing temporal information in GIS. On the other hand, ESTDM is conceptually simple and easily adaptable to other raster-based systems to represent information about locational changes at pre-defined cells along the passage of an event. While the model has shown its efficiency and capability to support spatial and temporal queries in raster systems, it will require a substantial redesign for use in a vector-based system (Peuquet and Duan, 1995). On the other hand, OOgeomorph is a vector-based system; it is designed to handle point data of timestamped locations. Its ability to handle spatial objects of higher dimensions and its applicability to systems other than geomorphology are not yet evident. The three-domain model provides a framework that extends both the space–time composite model and ESTDM model to enable representing histories at locations as well as occurrences of events in space and time, but it needs to be refined and tested.

The development of geographic research in the connections between space and time began with Hagerstrand's (1967) *Time Geography*, Langran's (1992) *Time in Geographic Information Systems*, Engerhofer and Golledge's (1998) *Spatial and Temporal Reasoning in Geographic Information Systems*, to Wachowicz's (1999) *Object-oriented Design for Temporal GIS*. Geographers have come a long way in the study of spatiotemporal constraints and representations. A clear trend directs to the use of processes to integrate space and time, and the application of object-oriented concepts to map real-world entities, including events and processes, to spatial data representations. Information about change is important, but it is more valuable to be associated with the real-world entities and processes that experience or cause the change. Identity-based change can offer new insights to entity interactions in a space–time frame, and therefore, it can serve as a foundation for spatiotemporal knowledge representation and reasoning (Hornsby and Egenhofer, 2000). A further success of temporal representation requires incorporation of models of human cognition of dynamic geographic phenomena to facilitate spatial analysis of the phenomena in a computer environment (Yuan, 1999; Mennis et al., 2000). Recent research has made important progresses in modeling of human activities (Kwan, 2004; Miller, 2007), lifelines (Mark and Egenhofer, 1998; Stefanidis et al., 2003) and spatiotemporal ontologies (Grenon and Smith, 2004).

28.5   CHALLENGES IN SPATIAL DATA REPRESENTATION

Harvey (1969, p. 191) identifies that, "The whole practice and philosophy of geography depends upon the development of a conceptual framework for handling the distribution of objects and events in space." He further cites from Nystuen (1963) that the conceptual framework should be appropriate for "(i) stating spatial distributions and the morphometric laws governing such distributions, and (ii) examining the operation of processes and process laws in a spatial context" (Harvey, 1969, p. 191).

The developments in GIS and GIScience have brought out renewed calls for such a conceptual framework through enforcing powerful and robust spatial data representation that can embrace diverse geographic objects and events and support the analysis of various spatial relations. The debate on field versus object representations evolves to the need for a better understanding of human cognition for representing spatial data (Coucleclis, 1992), and recently, the idea has been explored (Peuquet, 1994, Yuan, 1996, 1997, Mennis et al., 2000). The return of non-topological data structures from topological data structures allows a better integration of geometry and attribute data, to better match human perception the world. Although topology is still critical for spatial analysis, fast progresses in computer processing speed and random access memory enable quick construction of topological relations when necessary. Research on temporal GIS has progressed from change based to process-focused studies that seek space and time as integral compartments of geography, rather than two independent dimensions.

All the progresses have not only enriched the conceptual and theoretical basis in geographic information science but have also improved the usefulness of the geographic information systems. However, as the world grows to be more and more integrated yet complex through the development of technologies, geographic representation becomes more and more challenging. Emerging research issues reflect the need for a heterogeneous and massive, and distributed geospatial data, for example.

Heterogeneous data can include data of multiple formats, dimensions, and resolution. In parallel to the development of multimedia technology, geographic data have recently embraced a variety of large images and video and audio data. Integration of these new forms of data within a GIS database is non-trivial, for the current spatial data models are incapable of handling data of multiple forms. Consequently, query support is limited, if any, to file retrieval (i.e., access individual video files by keywords or some indexing scheme), rather than retrieve information based on an object. For example, a full integration should enable retrieval of all related video or audio clips when a user clicks a lake object on a GIS data layer, such as video clips about land use change around the lake or an audio explanation of the lake's history. These clips are created dynamically from larger video and audio files through the selection of data objects.

Data of multiple dimensions are critical for 3D and temporal applications. Currently, GIS data are two dimensional. Three-dimensional visualization techniques cannot fully support 3D GIS because a true 3D application requires information that can only be derived from analyzing 3D topological relationships beyond simple visualizing of the data volume. This is because topological integrity forms the basic operations to manipulate and analyze data in 2D, 3D, or 4D GIS (Egenhofer and Herring, 1991; Egenhofer and Al-Taha, 1992; Egenhofer and Mark, 1995; Hazelton, 1998). For example, a GIS must have capabilities to compute information about adjacency in the vertical dimension to answer a 3D query for areas where sandstone lies upon shale layers to identify areas of landslide potential.

Moreover, geospatial data have grown at a phenomenal rate as a result of advanced remote sensing and survey technology over the last decade. Yet, despite the massive amount of data coming on a daily basis, the utility of GIS technology in scientific research is considerably limited because the information implicit in GIS data is not easy to discern. This generates an urgent need for new methods and tools that can intelligently and automatically to transform geographic data into information and, furthermore, knowledge. The need is in part, for a broader information technology in knowledge discovery databases (KDD) that aims to extract useful information from massive amounts of data in support of decision making (Fayyad et al., 1996). Robust spatial representation with effective data structures is the key to facilitate geospatial knowledge discovery because a GIS cannot support computation on the information that it cannot represent. Geospatial knowledge cannot be synthesized when geographic phenomena cannot be fully embraced in a GIS.

The development of internet technology presents another challenge to spatial data representation because the trend promotes the use of distributed data, and furthermore, distributed computing. When data are distributed at multiple sites, the integration and usability of these data in an application depends upon both structural and semantic interoperability. A common representation for spatial data object specifications is critical to ensure effective communications among different data sets and systems. The Open GIS consortium with members from industry and academia has proposed an object-oriented spatial data model to serve the need for interoperability. The model provides a common base for developing open GISs, that is, data are interoperable at multiple GIS platforms. Its use for integrating distributed geospatial data on the internet is yet to be investigated. While the term "distributed data" signifies that data reside at multiple sites, the term "distributed computing" emphasizes sending functions to multiple data sites for computation and returning results from the individual sites to compile the final result. Distributed computing is best suited for intranet applications with large data residing at multiple computers so that getting all data to a single computer is inefficient and, sometimes, impractical. Similar to issues related to distributed data, data interoperability is critical to the success of distributed computing. A data model that can provide a common structural and semantic basis for data communication will significantly facilitate both distributed data and distributed computing applications.

Geographic data are by nature heterogeneous, massive, and distributed, and modern information technology has further promoted the complexity of geographic data. The idea of a digital Earth challenges GIS scientists to consider geographic data in all three complex issues. A truly developed digital Earth should contain geographic data of all kinds and at multiple resolutions to provide the user all possible perspectives to the current state as well as the history of the Earth at different levels of detail. Its data will be unprecedentedly massive and from multiple distributed sites in countries all over the world. The digital Earth challenge of spatial data integration and visualization cannot be overstated. While the digital Earth idea has sprouted many research questions, representation is at the heart of the fundamental issues that need to be addressed to ensure its success. The developments in spatial and temporal data models and data structures have built a foundation to represent geographic data for the digital Earth. There are plenty opportunities to empower a dynamic and ubiquitous spatial data representation to substrate the database for a digital Earth.

REFERENCES

Armstrong, M. P. 1988. Temporality in spatial databases. *Proceedings: GIS/LIS'88* 2:880–889.

Bourrough, P. A. and McDonnell, R. A. 1998. *Principles of Geographical Information Systems.* New York: Oxford University Press.

Chrisman, N. R. 1998. Beyond the snapshot: Changing the approach to change, error, and process. In M. J. Egenhofer and R. G. Golledge, eds. *Spatial and Temporal Reasoning in Geographic Information Systems.* New York: Oxford University Press, Chapter 6, pp. 85–93.

Coucleclis, H., 1982. Philosophy in the construction of geographic reality. In P. Gould and G. Ollson, eds. *A Search for Common Ground.* London, U.K.: Pion, pp. 105–140.

Coucleclis, H., 1992. People manipulate objects (but cultivate fields): Beyond the raster-vector debate in GIS. In A. U. Frank, I. Campari, and U. Formentini, eds. *Theories and Methods of Spatio-Temporal Reasoning in Geographic Space.* Berlin, Germany: Springer Verlag, pp. 65–77.

Cova, T. J. and Goodchild, M. F. 2002. Extending geographical representation to include fields of spatial objects. *International Journal of Geographical Information Science* 16(6):509–532.

Dangermond, J. 1982. A classification of software components commonly used in geographic information systems. In *Proceedings of the U.S.-Australia Workshop on the Design and Implementation of Computer-Based Geographic Information Systems*, Honolulu, HI, pp. 70–91.

Date, C. J. 1995. *An Introduction to Database Systems*, 6th edn. Reading, MA: Addison-Wesley Publishing Company, Inc.

Dutton, J. 1983. Geodesic modelling of planetary relief. In *Proceedings, AutoCarto VI*, Ottawa, Canada.

Egenhofer, M. and Al-Taha, K. 1992. Reasoning about gradual changes of topological relationships. In *Theory and Methods of Spatio-Temporal Reasoning in Geographic Space*, Pisa, Italy, A. Frank, I. Campani, and U. Formentini (eds.) Lecture Notes in Computer Science, Vol. 639, Springer-Verlag, pp. 196–219, September 1992.

Egenhofer, M. J. and Herring, J. R. 1991. High-level spatial data structures for GIS. In D. J. Maguire, M. F. Goodchild, and D. W. Rhind, eds. *Geographical Information Systems, Volume 1: Principles.* Essex, U.K.: Longman Scientific & Technical, pp. 227–237.

Egenhofer, M. and Mark, D. 1995. Naive geography COSIT'95, Semmering, Austria; In A. Frank and W. Kuhn (eds.). *Lecture Notes in Computer Science,* Vol. 988, Springer-Verlag, pp. 1–15, September 1995.

Fayyad, U., Djorgovski, S. G., and Weir N. 1996. Automating the analysis and cataloging of sky surveys. In M. Fayyad et al. eds. *Advances in Knowledge Discovery and Data Mining*. Boston, MA: AAAI Press/MIT Press, Chapter 19.

Fotheringham, A. S. and Wegener, M. 1999. *Spatial Models and GIS: New Potential and New Models*. London, U.K.: Taylor & Francis.

Franklin, Wm. 1991. Computer systems and low-level data structures for GIS. In D. J. Maguire, M. F. Goodchild, and D. W. Rhind, eds. *Geographical Information Systems, Volume 1: Principles*. Essex, U.K.: Longman Scientific & Technical, pp. 215–225.

Frank, A. U. and Mark, D. M. 1991. Language issues for GIS. In D. J. Maguire, M. F. Goodchild, and D. W. Rhind, eds. *Geographical Information Systems, Volume 1: Principles*. Essex, U.K.: Longman Scientific & Technical, pp. 147–163.

Freeman, H. 1974. Computer processing of line-drawing images. *ACM Computing Survey* 6(1) (March 1974), 57–97.

Gadia, S. K. and Vaishnav, J. H. 1985. A query language for a homogeneous temporal database. In *Proceedings of the ACM Symposium on Principles of Database Systems*, Portland, OR, pp. 51–56.

Gadia, S. K. and Yeung, C. S. 1988. A generalized model for a relational temporal database. In *Proceedings of ACM SIGMOD International Conference on Management of Data*, Chicago, IL, pp. 251–259.

Galton, A. 2004. Fields and objects in space, time, and space-time. *Spatial Cognition and Computation* 4(1): 39–68.

Goodchild, M. F. 1997a. Representing Fields, *NCGIA Core Curriculum in GIScience*, http://www.ncgia.ucsb.edu/giscc/units/u054/u054.html, last revised August 12, 2000.

Goodchild, M. F. 1997b. Quadtrees and scan orders, *NCGIA Core Curriculum in GIScience*, Unit 057, http://www.ncgia.ucsb.edu/giscc/units/u057/u057.html, last revised October 23, 1997.

Goodchild, M. F. and Grandfield, A. W. 1983. Optimizing raster storage: An examination of four alternatives. In *Proceedings, AutoCarto 6*, Ottawa, Canada, Vol. 1, pp. 400–407.

Grenon, P. and Smith, B. 2004. SNAP and SPAN: Towards dynamic spatial ontology. *Spatial Cognition and Computation* 4(1): 69–104.

Hagerstrand, T. 1967. *Innovation Diffusion as a Spatial Process*. Chicago, IL: University of Chicago Press.

Hartshorne, R. 1939. *The Nature of Geography*. Chicago, IL: University of Chicago Press.

Hartshorne, R. 1939. *The Nature of Geography*. Lancaster, Penn.: Association of American Geographers.

Harvey, D. 1969. *Explanation in Geography*. London, U.K.: Edward Arnold.

Hazelton, N. W. J. 1998. Some operational requirements for a multi-temporal 4D GIS. In M. J. Egenhofer and R. G. Golledge, eds. *Spatial and Temporal Reasoning in Geographic Information Systems*. New York: Oxford, pp. 63–73.

Hornsby, K. and Egenhofer, E. 2000. Identity-based change: A foundation for spatio-temporal knowledge representation. *International Journal of Geographical Information Science* 14(3):207–224.

Kelmelis, J. A. 1998. Process dynamics, temporal extent, and causal propagation as the basis for linking space and time. In M. J. Egenhofer and R. G. Golledge, eds. *Spatial and Temporal Reasoning in Geographic Information Systems*. New York: Oxford, pp. 94–104.

Kwan, M.-P. 2004. GIS methods in time-geographic research: Geocomputation and geovisualization of human activity patterns. *Geografiska Annaler, Series B: Human Geography* 86(4):267–280.

Lam, N. S. 1983. Spatial interpolation methods: A review. *American Cartographer* 10:129–149.

Langran, G. 1992. *Time in Geography*. London, U.K.: Taylor & Francis.

Langran, G. and Chrisman, N. R. 1988. A framework for temporal geographic information. *Cartographica* 25(3):1–14.

Marx, R. W. 1986. The TIGER system: Automating the geographic structure of the United States census. *Government Publications Review* 13:181–201.

Mark, D. M. and Egenhofer, M. J. 1998. Geospatial lifelines. In O. Günther, T. Sellis, and B. Theodoulidis, eds. *Integrating Spatial and Temporal Databases*. Dagstuhl-Seminar Report No. 228, http://timelab.co.umist.ac.uk/events/dag98/report.html Schloos Dagstuhl, Germany.

Mark, D. M. and Lauzon, J. P. 1984. Linear quadtrees for geographic information systems. In *Proceedings: IGU Symposium on Spatial Data Handling*, Zurich, Switzerland, August 20–14, pp. 412–431.

McIntosh, J. and Yuan, M. 2005. A framework to enhance semantic flexibility for analysis of distributed phenomena. *International Journal of Geographic Information Science* 19(10):999–1018.

Mennis, J. L., Peuquet, D., and Qian, L. 2000. A conceptual framework for incorporating cognitive principles into geographical database representation. *International Journal of Geographical Information Science* 14(6):501–520.

Miller, H. J. 2007. Place-based versus people-based geographic information science. *Geography Compass* 1:503–535.

Nystuen, J. D. 1963. Identification of some fundamental spatial concepts. *Papers in Michigan Academia of Science, Arts, and Letters* 48:373–384.

Peucker (now Poiker), T. K. and Chrisman, N. 1975. Cartographic data structure. *The American Cartographer* 2(1):55–69.

Peucker (now Poiker), T. K., Flower, R. J., Little, J. J., and Mark, D. M. 1978. The triangulated irregular network. In *Proceedings of the DTM Symposium, American Society of Photogrammetry and American Congress on Survey and Mapping*, St. Louis, MO, pp. 24–31.

Peuquet, D. J. 1984. A conceptual framework and comparison of spatial data models. *Cartographica* 2(2):55–69.

Peuquet, D. J. 1994. It's about time: A conceptual framework for the representation of temporal dynamics in geographic information systems. *Annals of the Association of American Geographers* 84(3):441–462.

Peuquet, D. J. 2002. *Representation of Space and Time*. New York: The Guilford Press.

Peuquet, D. J. and Duan, N. 1995. An event-based spatiotemporal data model (ESTDM) for temporal analysis of geographical data. *International Journal of Geographical Information Systems* 9(1):7–24.

Raper, J. and Livingstone, D. 1995. Development of a geomorphological spatial model using object-oriented design. *International Journal of Geographical Information Systems* 9(4):359–384.

Rosenfeld, A. 1980. Tree structures for region representation. In H. Freeman and G. G. Pieroni, eds. *Map Data Processing*. New York: Academic Press, pp. 137–150.

Samet, H. 1990a. *The Design and Analysis of Spatial Data Structures*. Reading, MA: Addison Wesley Publishing Company.

Samet, H. 1990b. *Applications of Spatial Data Structures*. Reading, MA: Addison Wesley Publishing Company.

Smith, T. R., Su, J., Agrawal, D., and El Abbadi, A. 1993. Database and modeling systems for the earth sciences. *IEEE* 16(1) (Special Issue on Scientific Databases): 33–39.

Snodgrass, R. and Ahn, I. 1985. A taxonomy of time in databases. In *Proceedings of ACM SIGMOD International Conference on Management of Data*, Austin, TX, pp. 236–264.

Stefanidis, A., Eickhorst, K., et al. 2003. Modeling and comparing change using spatiotemporal helixes. In *Proceedings of the Eleventh ACM International Symposium on Advances in Geographic Information Systems*, New Orleans, LA: ACM Press.

UCGIS, 1996. Research priorities for geographic information science. *Cartography and Geographic Information Systems* 23(3):115–127.

Ullman, J. D. 1980. Principles of database systems, Computer Science Press, Potomac, MD.

Wachowicz, M. 1999. *Object-Oriented Design for Temporal GIS*. London, U.K.: Taylor & Francis.

Winston, P. H. 1984. *Artificial Intelligence*, 2nd edn. Reading, MA: Addison-Wesley.

Worboys, M. F. 1994. A unified model of spatial and temporal information. *The Computer Journal* 37(1):26–34.

Yuan, M. 1996. Modeling semantic, temporal, and spatial information in geographic information systems. In M. Craglia and H. Couclelis, eds. *Progress in Trans-Atlantic Geographic Information Research*. Bristol, PA: Taylor & Francis, pp. 334–347.

Yuan, M. 1997. Knowledge acquisition for building wildfire representation in geographic information systems. *The International Journal of Geographic Information Systems* 11(8):723–745.

Yuan, M. 1999. Representing geographic information to enhance GIS support for complex spatiotemporal queries. *Transactions in GIS* 3(2):137–160.

Yuan, M. 2001. Representing complex geographic phenomena with both object- and field-like properties. *Cartography and Geographic Information Science* 28(2):83–96.

Yuan, M., Mark, D. M., Egenhofer, M. J., and Peuquet, D. J. 2004. Extensions to geographic representations. In R. B. McMaster and E. L. Usery, eds. *A Research Agenda for Geographic Information Science*. Boca Raton, FL: CRC Press, LLC, Chapter 5.

29 Spatial Analysis and Modeling

Michael F. Goodchild

CONTENTS

29.1 INTRODUCTION

In the previous chapters, we have seen how a wide variety of types of geographic data can be created and stored. Methods of digitizing and scanning allow geographic data to be created from paper maps and photographs. Powerful computing hardware makes it possible to store large amounts of data in forms that are readily amenable to manipulation and analysis using the routines stored in powerful software. Thus, the stage is set for a discussion in this chapter of the real core of GIS, the methods of analysis and modeling that allow us to examine data, to solve specific problems, and to support important decisions, using the capabilities of hardware, software, and data that together compose a GIS.

Chapter 27 introduced the various ways in which geographic phenomena can be represented in digital form. A road, for example, can be represented by recording an appropriate value in a swath of cells in a raster representation. With a cell size of 1 m, the swath corresponding to a major four-lane highway might be as much as

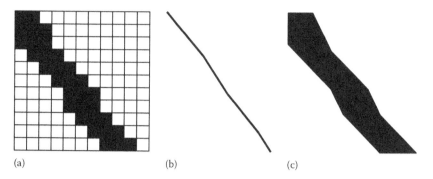

FIGURE 29.1   Three alternative representations of the same road: (a) a swath of raster pixels, (b) a vector centerline, and (c) a vector area.

50 or even 100 cells wide. Alternatively, the road might be represented as a single line, or *centerline*, in a vector database. In this case, its location would be recorded by specifying the coordinates of a series of points aligned along the road's center. Such centerline databases are now very commonly used in applications like vehicle routing and scheduling, and by sites such as www.mapquest.com or maps.google.com that offer to find the best routes between pairs of places, and are used daily by millions of people. Finally, the road might be represented as an area, encased by its edges (we use the term *cased* to describe this option, and also call it a *double-line* representation). Figure 29.1 shows the three options. Which of these is used in a given case depends on the nature of the application, on the limitations of the available software, on the origins of the data, and on many other factors.

This chapter is about turning data into useful information. Analysis and modeling can make calculations that are too tedious to do by hand, and by doing so provide numbers that are useful for many kinds of applications. They can be used to reveal patterns and trends in data that may not be otherwise apparent. They might also be used by one person or a group to draw the attention of another person or group, as might occur in a courtroom where geographic data are used by one side to make a point. Finally, the results of analysis and modeling can be used to provide the information needed to make decisions. In all of these cases, GIS is the engine that performs the necessary operations under the guidance of the user who issues the necessary instructions.

29.1.1   ORGANIZING THE POSSIBILITIES

A very large number of methods of analysis and modeling have been devised over the years, many dating from well before the advent of GIS, when calculations and measurements had to be performed by hand. The ability to process large amounts of data has quickly led to a rapid explosion in the list, and today it is virtually impossible to know about every form of spatial analysis. The developers of GIS software often provide thousands, and thousands more are added by specialized companies and individuals (e.g., the toolbox provided with ESRI's ArcGIS 9.2 provides over 500 distinct functions, in addition to the hundreds of operations that can be called from its drop-down menus). So, one of the most daunting aspects of analysis and modeling with GIS is simply keeping track of the possibilities.

One way to organize methods of analysis and modeling is by the data types on which they operate. There are operations that work on discrete objects—points, lines, and areas—and operations that work on phenomena conceptualized as continuous fields. At a different level, and following the concepts introduced in Chapter 27, it is possible to separate methods of analysis of vector data sets from those that operate on raster data sets, since there are very few instances of operations that require input of both kinds. Another possibility is to see every operation as a transformation, taking some kind of input and producing some kind of output, and to organize methods on this basis. The structure adopted here is a little of all of these, but is based primarily on popularity: since there are so many possibilities, it is most important to understand the ones that are used most often, and to leave the less popular ones to further study. Each method is described in terms of the problem it attempts to solve, the inputs required, and the outputs that it generates. Both raster and vector operations are covered.

Section 29.2 deals primarily with points, and the following section (Section 29.3) with areas. Section 29.4 looks in detail at rasters, using the framework of *cartographic modeling*, and includes methods of analyzing digital elevation models (DEMs). Finally, Section 29.5 examines methods for optimization and design, where the objective is to use GIS to find the best solution to problems. A more complete discussion is available in the text by Longley et al. (2005), and an advanced compilation is provided by De Smith et al. (2007).

29.2 METHODS OF ANALYSIS FOR POINTS

We begin with the simplest kinds of geographic objects. Suppose we have records of each of the customers of an insurance agent, including each customer's location. Perhaps these originated in an address list, and were subsequently converted to coordinates using the process known as *geocoding* or *address-matching* (Chapter 22). Plotted on a map they might look something like Figure 29.2. Several questions

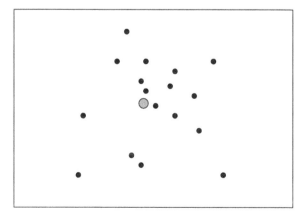

FIGURE 29.2 The locations of the customers of an insurance agent, shown in relation to the agent's office location, and ready for analysis.

might occur to a market researcher hired by the insurance agent to study the state of the business and recommend strategies. For each, there is a straightforward type of GIS analysis that can be used to provide the necessary information.

29.2.1 DISTANCE

First, we might ask how far each customer is located from the agent's location. The agent might be interested in knowing how many customers do their business with the agent rather than with some other agent located closer to the customer. What is the average distance between each of the customers and the agent, and are there areas near the agent where advertising might increase the agent's market share? All of these questions require the ability to measure the distance between points, a commonly used function of a GIS.

To measure the distance between two points we need to know their locations in (x, y) coordinates. Coordinate systems were introduced in Chapter 2, and many of them can be used to measure distances. The UTM coordinate system, for example, is based on measurements of coordinates in meters, so distances calculated in this coordinate system are easily understood. State plane coordinates are often expressed in feet. In these cases the distance between two points (x_A, y_A) and (x_C, y_C), representing the locations of the agent and a customer, is given by

$$D = \sqrt{(x_A - x_C)^2 + (y_A - y_C)^2}$$

This equation is derived from Pythagoras's famous theorem, and is thus also known as the Pythagorean distance, or the length of a straight line between the two points.

The Pythagorean distance applies to points located on a plane, not on the curved surface of the Earth, and so should be used only when the points are close together. A good rule of thumb is that the study area should be no more than 500 km across, because at that distance errors due to ignoring the Earth's curvature begin to approach 0.1%, even with the best-designed coordinate systems.

It is important to recognize that this problem of the Earth's curvature is not resolved by using latitude and longitude coordinates. Suppose we record location using latitude and longitude, and plug these into the Pythagorean equation as if they were y and x, respectively (this is often called using *unprojected* coordinates, and is also what happens if we specify the so-called Plate Carrée or Cylindrical Equidistant projection). In this case, the coordinates are measured in degrees, so the value of D will also be in degrees. However, one degree of latitude is not the same distance on the Earth's surface as one degree of longitude, except exactly on the Equator. At 32° North or South, for example, the lines of longitude are only 85% as far apart as the lines of latitude, and that percentage drops all the way to zero at the poles. Therefore, the equation cannot be applied to latitude and longitude. Instead, we should use the equation for distance over the curved surface of the Earth, otherwise known as the Great Circle distance, because the shortest path between two places follows a Great Circle (a Great Circle is defined as the arc formed when the Earth is sliced through the two points and through the center of the Earth). If the point locations are denoted by (ϕ_A, λ_A) and (ϕ_C, λ_C), where ϕ denotes latitude and λ denotes longitude, then the distance between them is given by

$$D = R\cos^{-1}\left[\sin\phi_A \sin\phi_C + \cos\phi_A \cos\phi_C \cos(\lambda_A - \lambda_C)\right]$$

where R denotes the radius of the Earth, or approximately 6378 km. Note, however, that this formula assumes that the Earth is a perfect sphere, ignoring the substantial flattening at the poles and other distortions (Chapter 2).

There are many other bases for measuring distance, because it is often necessary to allow for travel that must follow streets, or avoid barriers of one kind or another. Therefore, the actual distance traveled between two places may be much more than either of these formulae would predict. Two of these cases are dealt with later in this chapter: when the path followed is the one that minimizes the total cost or travel time over a surface (a case of raster analysis, see Section 29.4), and where the path follows a network of links with known lengths or travel times (a case of network analysis, see Section 29.4).

29.2.2 BUFFERS

Instead of asking how far one point is from another, we might turn the question around and identify all of the points within a certain distance of a reference point. For example, it might be interesting to outline on a map the area that is within 1 km of the agent's location, and subsequent sections will cover several applications of this concept. In GIS, the term *buffer* is used, and we say that the circle created by this operation constitutes the 1 km buffer around the agent. Buffers can be created for any kind of object—points, lines, or areas—and are very widely used in GIS analysis. Figure 29.3 shows buffers for each of these types of objects.

By finding a buffer, and by combining it with other information using the methods discussed in this chapter, we could answer such questions as: How many customers live within 10 km of the agent's location? What is the total population within 10 km of the agent's location (based on accurate counts of residents obtained from the census)? Where are the people with the highest incomes within 10 km of the agent's location?

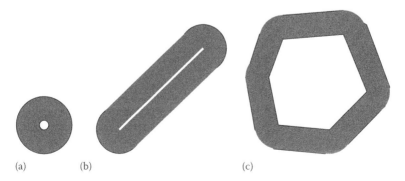

(a)   (b)   (c)

FIGURE 29.3 Buffers for (a) point, (b) line, and (c) area objects.

29.2.3 POINTS IN POLYGONS

The so-called point in polygon operation is another key feature of GIS analysis. Suppose we have a map of the tracts used by the census to publish population statistics, such as the map shown in Figure 29.4. In form, the map looks much like a map of states, or counties, or voting districts, or any of a number of types of maps that divide an area into zones for the purpose of counting and reporting. The point in polygon operation allows us to combine our point map with this map of areas, in order to identify the area that contains each point. For example, we could use it to identify the census tract containing each of the customer locations in our point data set. By counting customers in each tract, and comparing the totals to the known populations of each tract, we could get interesting data on market penetration by tract. Alternatively, we could join the attributes of the containing tract with the attributes of the customer, to see how neighborhood characteristics might relate to individual characteristics.

The point in polygon operation is actually very simple to execute, and so the method will be described briefly. One of the points in the figure will be used as an example. A line is drawn from the point, in this case diagonally upward. The number of intersections between this line and the boundaries of each polygon is counted. In the example, there is one intersection with the boundary of Polygon A, two with the boundary of Polygon B, and two with the boundary of Polygon C. The polygon that contains the point is the only one with an odd number of boundary intersections

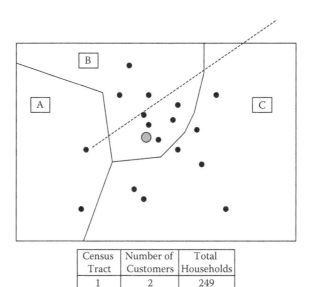

Census Tract	Number of Customers	Total Households
1	2	249
2	8	377
3	7	401

FIGURE 29.4 By using the point in polygon operation to identify the census tract containing each of the customer locations, it is possible to compare customer counts to other attributes of the tracts, such as the total number of households.

(and it will always be true that exactly one polygon has an odd number—all other polygons will have an even number).

One useful way to think about the result of the point in polygon operation is in terms of tables. The result of the point in polygon operation could be expressed in terms of an additional column for the census tract table shown in Figure 29.4, recording the number of points found to lie in that polygon, as a preliminary to computing the local market penetration. It is applications like these that make the point in polygon operation one of the most valuable in vector GIS.

29.3   ANALYSIS OF AREAS

Section 29.2 looked at one method that combines points and areas (or *polygons*, a term commonly used interchangeably with areas in vector GIS, since areas are normally represented as polygons as discussed previously). This section focuses specifically on operations on areas, again from the perspective of vector GIS.

29.3.1   MEASUREMENT OF AREA

One of the strongest arguments for the use of GIS, as compared with manual methods of analysis of information shown on maps, is that computers make it easier to take measurements from maps. The measurement of area is in fact the strongest argument of this type. Suppose we need to measure an area from a map, such as the area of a particular class of land use. Perhaps our task is to measure the amount of land being used by industry in a specific city, and we are given a map showing the locations of industrial land use, in the form of appropriately colored areas. In a medium-sized city, the number of such areas might be in the thousands, and so there would be thousands of measurements to be made and totaled. However, even the measurement of a single area is problematic. Manually, we would have to use one of two traditional methods, known as dot counting and planimetry. Dot counting proceeds by overlaying a prepared transparent sheet covered by dots at a known density, and counting the number of such dots falling within the area of interest. To get a reliable estimate the density of dots needs to be high, and the dots need to be small, and it is easy to see how tedious and inaccurate this task can be. Planimetry proceeds by using a crude mechanical device to trace the outline of the area—the result is read off a dial on the instrument, and again the process is tedious and error-prone. In short, manual methods are frustrating, and expensive in terms of the time required to obtain even a poor level of accuracy.

By contrast, measurement of area is extremely simple once the areas are represented in digital form as polygons. A simple calculation is made for each straight edge of the polygon, using the coordinates of its endpoints, and the calculations are summed around the polygon. Any vector GIS is able to do this quickly and accurately, and the accuracy of the final result is limited only by the accuracy of the original digitizing.

The world's first GIS, the Canada Geographic Information System, was developed in the mid-1960s precisely for this reason. A large number of detailed maps had been made of land use and other land properties, with the objective of providing accurate measurements of Canada's land resource. However, it would have taken decades, and a large workforce, to produce the promised measurements by hand, and the results would have been of disappointing accuracy. Instead, a GIS was developed, and all of the maps were digitized. In addition to straightforward measurement of area, the system was also used to produce measurements of combined areas from different maps, using the operation of polygon overlay, another important vector operation that is covered in the next section.

29.3.2   POLYGON OVERLAY

Figure 29.5 shows a typical application of polygon overlay. The two input maps might represent a map of land use, and a map of county boundaries. A reasonable question to ask of a GIS in a case like this is, "how much land in County A is in agricultural use?" To answer the question we must somehow compare the two maps, identifying the area that is *both* in County A and in agricultural land use, and measure this area. Polygon overlay allows this to be done. As a method, its history dates back to the 1960s, though efficient software to perform the operation on vector data sets was not developed until the late 1970s.

Figure 29.4 shows the result of the overlay operation, in the form of a new data set. Each of the polygons in the new data set represents a unique combination of the inputs, and the boundaries of the new polygons are combinations of the lines from the two inputs. The operation is actually reversible, since it is possible to recover both of the input data sets by deleting some of the attributes of the new polygons and merging polygons with the same remaining attributes. For example, by deleting land

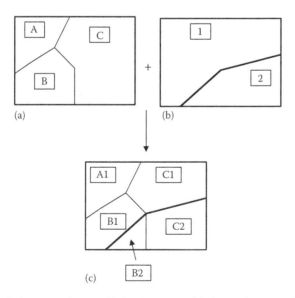

FIGURE 29.5   Polygon overlay, combining (a) a map with three polygons and (b) a map with two polygons to create (c) a map with five polygons. The attributes of the new map combine the relevant attributes of the input maps.

use attributes we recover the county data set. In other words, no information is lost or created in polygon overlay; instead, the information is simply rearranged.

The new data set makes it easy to answer questions like the one we started with. To determine the amount of land in County A in agricultural land use, we simply identify all polygons in the new data set with those attributes, compute their areas, and total. Polygon overlay has a myriad of uses, of which this kind of query is among the simplest. One popular use has to do with population statistics. Consider that a water supply agency wishes to estimate consumption in one of its service areas, and needs to know the total number of households in the area. The census provides statistics on the total numbers of households, but for areas defined by the census, not those used by the water supply agency. Therefore, a simple way to estimate households is to overlay the service areas on the census areas, and measure the areas of overlap. Then the household counts from the census are apportioned based on these areas, to obtain estimates that are often remarkably accurate. For example, if one census area has 1000 households, and 50% of its area lies in Service Area 1, and the remaining 50% lies in Service Area 2, then 500 households are allocated to each of these service areas. When all census counts are allocated in this fashion, the service area counts are summed.

Polygon overlay is an important operation for vector GIS, but it has two distinct versions. The examples discussed in this section have been of fields (as defined in Chapter 28), such that each point on each map lies in exactly one polygon, and the overlay produces a similar map. The overlay in this version is conducted on an entire map or *layer* or *coverage* at a time (these terms all have similar meaning in this context). However, an overlay can also be used for data that represent the *discrete object* view of the world. In this case, the input maps can be such that a point can lie in any number of polygons, including zero. This view of the world makes no sense for maps of counties, or land use, but it certainly makes sense for maps of potentially overlapping phenomena like forest fires, or zoning ordinances. In this case, a polygon overlay proceeds one polygon at a time, and is used to answer questions about individual polygons, such as "what zoning ordinances affect Parcel 010067?" Some vector GISs support only the field approach (e.g., versions of ArcInfo up to 7), and some (typically those with strong roots in CAD software) support only the discrete object approach, while others can potentially support both approaches.

In addition, it is important to recognize the differences between overlay in vector GIS and overlay in raster GIS. The latter is a much simpler operation, and produces very different results, for different purposes. The discussion of raster analysis below includes the essentials of this version of the overlay operation.

29.4   RASTER ANALYSIS

A raster GIS provides a very powerful basis for analysis, that is similar in many respects to the capabilities of other software that also rely on raster representations. For example, some of the raster operations described here will be familiar in concept to people who regularly use software for processing digital photographs or scanned documents, or for processing the images captured by remote sensing satellites. Vector data sets are similar in form to those used in computer-assisted design software (CAD) and in drawing software, but it is unusual to find comparable methods of analysis in these environments.

In raster GIS, the world is represented as a series of *layers*, each layer dividing the same project area into the same set of rectangular or square cells or *pixels*. This makes it very easy to compare layers, since the pixels in each layer exactly coincide. However, it also means that the ability to represent detail is limited by the pixel size, since any feature on the Earth's surface that is much smaller than a single pixel cannot be represented. Smaller features can be represented by reducing the pixel size, but only at the cost of rapidly increasing data volume.

Each layer records the values of one variable or attribute, such as land use or county name, for each pixel. The recorded value might be the average value over the pixel, or the value at the exact center of the pixel, or the commonest value found in the pixel. If there are many attributes to record (the census, e.g., reports hundreds of attributes for each county in the United States), then separate layers must be created for each attribute. However, some raster GISs allow a more efficient solution to this particular problem, in which a single layer is used to record county ID for each pixel, and a related table gives the many attributes corresponding to that ID (see Figure 29.6).

An excellent framework for thinking about raster analysis was provided some years ago by Dana Tomlin, who coined the term *cartographic modeling* (Tomlin, 1990). In this scheme, there are four basic types of operations (there is some variation in the terms in different versions of the scheme):

- *Local* operations, which process the contents of data sets pixel by pixel, performing operations on each pixel, or comparing the contents of the same pixel on each layer
- *Focal* operations, which compare the contents of a pixel with those of neighboring pixels, using a fixed neighborhood (often the pixel's eight immediate neighbors)

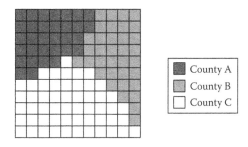

County	Population	Tax Rate (%)
A	1,998	5.5
B	3,941	6.0
C	15,227	6.75

FIGURE 29.6 By using raster cells to store the IDs of polygons with multiple attributes in a related table it is possible to store many attributes in a single raster layer.

- *Zonal* operations, which perform operations on zones, or contiguous blocks of pixels having the same values
- *Global* operations, which are performed for all pixels

Examples of each type of operation are given in the following sections.

29.4.1 LOCAL OPERATIONS

The simplest kind of local operations occur when a single raster layer is processed. This is often done in order to apply a simple reclassification, such as

All areas of Soil Classes 1, 3, and 5 are suitable building sites for residential development; all other areas are not.

This could be operationalized by reclassifying the soil class layer, assigning 1 to all pixels that currently have Soil Class 1, 3, or 5, and 0 to all other pixels. Another might be

Find all areas where the average July temperature is below 25°C.

Again, this could be achieved by reclassifying a layer in which each pixel's attribute is that area's average July temperature.

This kind of reclassification, using simple rules, is very common in GIS. Other local operations on a single layer include the application of simple numerical operations to attributes, for example, the July temperature layer might be converted to Fahrenheit by applying the formula

$$F = 9°C/5 + 32$$

GISs with well-developed raster capabilities, such as Idrisi, allow the user access to a wide range of operations of this type that create a new layer through a local operation on an existing layer.

Other local operations operate on more than one layer. The raster equivalent of a polygon overlay is an operation of this type, taking two or more input layers, and applying a rule based on the contents of a given pixel on each layer to create a new layer. The rules can include arithmetic operations, such as adding or subtracting, as well as logical operations such as

If the average July temperature is above 20°C and soil type is Class 1 then the pixel is suitable for growing corn, so assign 1, otherwise assign 0.

However, notice how different this is from the vector equivalent. The operation is not reversible, since the new layer does not preserve all of the information in all of the input layers. Instead of reorganizing the inputs, the raster overlay creates new information from them.

29.4.2 FOCAL OPERATIONS

Focal operations produce results by analyzing each pixel in relation to its immediate neighbors. Some of these operations work on all eight immediate neighbors, but others focus only on the four neighbors that share a common edge, and ignore the four diagonal neighbors. Sometimes we distinguish these options by referring to moves in the game of chess—the eight-neighbor case is called the *queen's case*, and the four-neighbor case is called the *rook's case* (see Figure 29.7).

Among the most useful focal operations are the so-called convolutions, in which the output is similar to an averaging over the immediate neighborhood. For example, we might produce a new layer in which each pixel's value is the average over the values of the pixel and its eight queen's case neighbors. Figure 29.8 shows a simple instance of this. The result of this operation would be to produce an output layer that is smoother than the input, by reducing gradients, lowering peaks, and filling valleys. In effect, the layer has been *filtered* to remove some of the variation between pixels, and to expose more general trends. Repeated application of the convolution will eventually smooth the data completely. Convolutions are very useful in remote sensing; where they are used to remove noise from images (see Chapter 20).

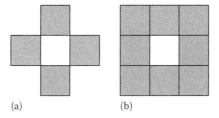

(a)   (b)

FIGURE 29.7 Two definitions of a raster cell's neighborhood: (a) in the rook's case, only the four cells that share an edge are neighbors, but (b) in the queen's case, the neighborhood includes the four diagonal neighbors.

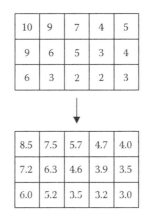

FIGURE 29.8 Application of a convolution filter. Each pixel's value in the new layer is the average of its queen's case neighborhood in the input layer. The result is smoother, picking up more of the general trend. At the edge, where part of the neighborhood is missing, averages are over fewer than nine pixels.

Of course, averaging can only be used if the values in each pixel are numeric and measured on continuous scales. Therefore, this operation would make no sense if the values represent classes of land, such as soil classes. However, in this case, it is possible to filter by selecting the commonest class in the neighborhood, rather than by averaging. Note also that special rules have to be adopted at the edge, where cells have fewer than the full complement of neighbors.

If the input layer is a DEM, with pixel values equal to terrain elevation, a form of local operation can be used to calculate slope and aspect. This is normally done by comparing each pixel's elevation with those of its eight neighbors, and applying simple formulae (given, e.g., by Burrough and McDonnell (1998)). The result is not the actual slope and aspect at each pixel's central point, but an average over the neighborhood. Because of this, slope and aspect estimates always depend on the pixel size (often called the distance between adjacent *postings*), and will change if the pixel size changes. For this reason, it is always best to quote the pixel size when dealing with slope or aspect in a GIS.

Slope and aspect can also be used to estimate the pattern of surface water flow over a DEM, a very useful operation in determining watershed boundaries and other aspects of surface hydrology. Each pixel's elevation is compared to those of its eight neighbors. If at least one of the neighbors is lower, then the GIS infers that water will flow to the lowest neighbor. If no neighbor is lower, the pixel is inferred to be a pit, in which a shallow lake will form. If this rule is applied to every pixel in a DEM, the result is a tree-like network of flow directions (Figure 29.9), with associated watersheds. Many more advanced versions of this simple algorithm have been developed, along with a range of sophisticated methods for studying water flow (hydrology), using GIS (see Maidment and Djokic (2000)).

Another powerful form of local operation forms the raster equivalent of the buffer operation and also supports a range of other operations concerned with finding routes across surfaces. To determine a buffer on a raster, it is necessary only to determine which pixels lie within the buffer distance of the object. Figure 29.10 shows how this works in the cases of a point and a line. But suppose we make the problem a little more complex, by asking for pixels that are within a certain travel *time* of a given point, and allowing travel speed to be determined by a new layer. Figure 29.11

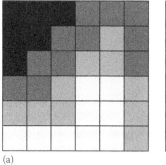
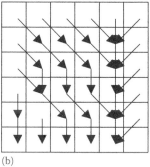

(a) (b)

FIGURE 29.9 (a) A digital elevation model, and (b) the result of inferring drainage directions using the simple rule "If at least one neighbor is lower, flow goes to the lowest neighbor."

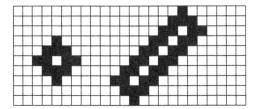

FIGURE 29.10 The buffer operation in its raster form, for a point and a line (indicated by the white cells).

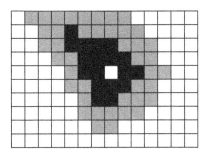

FIGURE 29.11 Illustration of a spreading operation over a variable friction layer. From the central white cell, all cells in the black area can be reached in 10 min, and all cells in the gray area can be reached in 20 min.

shows such a layer, and the result of determining how far it is possible to travel in given numbers of minutes. The operation is also known as *spreading*. Of course, this assumes that travel is possible in all directions, but this would be appropriate in the case of an aircraft looking for the best route across the Atlantic, or a ship in the open ocean.

This method is used very frequently to find the best routes for power transmission lines, highways, pipelines, and military vehicles such as tanks; in all of these cases, the assumption of travel in all directions is reasonably accurate. Origin and destination points for the route are defined. A raster GIS is used to create a layer of travel speed, or *friction*. Because this is not uniform, the best route between the origin and destination is not necessarily a straight line. Instead, the best route is modeled as a series of moves in the raster, from each cell to the most appropriate of its eight neighbors, until a complete route of least total time or least total cost is found (Figure 29.12). There have been many applications of this GIS method over the past three decades by highway departments, power utilities, and pipeline companies.

The final function considered here is the so-called viewshed operation. This also works on a DEM, and is used to compute the area that can be seen from a specified point, by a person at that point. When more than one point is specified, the GIS could be used to determine the area visible from at least one point, or in some military applications, it might be useful to determine the area invisible from all points.

The viewshed operation is often used in planning. For example, if a new clearcut is proposed by a forest management agency it might be useful to determine whether

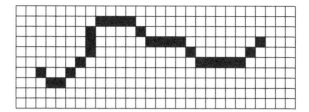

FIGURE 29.12   Finding the least-cost path across a variable-cost layer from a given origin to a given destination, using queen's case moves between raster cells. This form of analysis is often applied to find routes for power lines, pipelines, or new highways.

the scar will be visible from certain key locations, such as public roads. If a new paper mill is proposed, it might be useful to determine how visible it will be from the surrounding areas. One of the complications of the viewshed operation as implemented on a DEM is that it is difficult to incorporate the effects of tree cover on visibility, since a DEM records the elevation of the ground surface, not the top of the tree canopy. In addition, it is wise to be careful about the representation of the observer, since far more can often be seen by someone standing on a tower than by someone standing on the ground. Many GIS implementations of the viewshed operation simply assume an observer of average height standing on the ground.

29.4.3   ZONAL OPERATIONS

Rasters represent everything in pixels, so line objects and area objects are modeled as collections of pixel values, with no explicit linkage between them. Therefore, some operations that are very straightforward in a vector GIS, such as the measurement of area described in Section 29.3.1 are quite awkward in raster. If we simply count the pixels assigned to Class 1 on a soil layer, for example, we get the total area of Class 1, not the separate areas of the individual patches of Class 1. Zonal operations attempt to address this, by focusing on operations that examine a layer as a collection of zones, each zone being defined by contiguous pixels of the same class. Zonal operations can be used to measure the areas of patches, or their perimeter lengths, and the results are returned as new layers in which each pixel is given the appropriate measure, evaluated over the zone of which the pixel is a part.

29.4.4   GLOBAL OPERATIONS

Finally, global operations apply a simple manipulation to an entire layer, usually returning a single value. Examples include operations to determine the average of all pixel values, or the number of pixels having a specific value, or the numbers of pixels having each unique value, or the total area having each unique value. For example, a global measurement of area on a soil layer with six classes might return six areas, plus the total area. Global operations produce summary statistics, so they are often the last operation in a sequence.

29.5   OPTIMIZATION AND DESIGN

Many applications of GIS analysis provide information that can be used to make decisions, but some methods come much closer to recommending decisions directly. These are the methods that focus on optimization, by finding the answers that *best* address a problem, and methods that focus on design. Some of the earliest of these were advocated by Ian McHarg, then Professor of Landscape Architecture at the University of Pennsylvania, who developed a method of overlaying maps to find the best locations for new developments (McHarg, 1969). Each map to be overlaid represented some specific issue, such as preservation of agricultural land, or cost of construction, so that when all maps were overlaid the best location would be the one with the least impact or cost. McHarg developed this as a manual method, using transparent sheets, but it is easily automated in a raster GIS. The method of optimum route selection discussed in Section 29.4.2 is very similar in concept to McHarg's.

Today, there are many methods of optimization and design available in GIS, in addition to the one already discussed. Many are concerned with optimum routing and scheduling of vehicles operating on a road network, a type of problem that is almost always implemented in vector GIS. In the simplest instance, the GIS are used to find the shortest path through a road and street network between a user-specified origin and destination. Millions of people use this kind of analysis daily when they access Web sites to request driving directions between two street addresses, and the same kind of analysis is possible in the in-vehicle navigation systems that are becoming increasingly common.

Much more sophisticated versions of the same basic idea are used by school bus authorities to design the best routes to pick up children and take them to school, by parcel delivery companies to design routes for their drivers, and by utility companies that schedule daily work orders for their maintenance staff. The term *logistics* is often used to describe this application of GIS.

Another class of optimization problems is termed *location–allocation*. Here the issue is to find sites for activities that serve geographically distributed demand. Examples are retail stores, which must locate centrally with respect to their customers, schools and hospitals, fire stations and ambulance depots, and a host of other services. In all cases, the objective is to find locations that are best able to serve a dispersed population. The problem is known as location–allocation problem because the best solution involves both the *location* of one or more central facilities, and the *allocation* of demand to it, in the form of service areas (though in some cases this pattern of allocation is controlled by the system designer, as in the case of school districts, but in other cases it is determined by the behavioral choices of customers). In other versions of the problem, there is no *location* as such, but instead an area must be divided into the best set of districts, for such purposes as voting, or the assignment of territories to sales staff.

29.6   SUMMARY

It should be clear by now that a vast number of options are available in the form of methods of spatial analysis and modeling implemented in GIS. Any good GIS provides many of these, and more are available in the form of add-ons from other vendors, agencies, or individuals. In the case of one popular GIS, ArcGIS from ESRI, the basic package contains many operations in its menus and toolbox; others are available from ESRI in the form of additional special-purpose modules; more are available from other vendors to build on the power of ArcGIS in specific applications; and the package provides tools that allow the sophisticated user to add even more. If your favorite GIS appears not to permit a certain kind of operation, it is certainly worth hunting around, probably on the Web, to see if a suitable add-on is available.

Many GISs also allow operations to be combined into *scripts*, written in special languages such as Visual Basic or Python. Some allow this to be done through a visual interface, with operations represented as icons that can be linked and combined by the user. All of these options have greatly extended the power of GIS to solve problems, and to provide the kinds of information people need to make effective decisions.

Certain aspects should be considered in any application of analysis and modeling using GIS. First, is the analysis more suited to a raster or a vector approach? While many operations are possible in both, this chapter has shown how there can be subtle differences, and how some operations are much more difficult in one form than the other. Second, what level of geographic detail is needed? Most GIS operations give different results depending on the level of geographic detail of the inputs (e.g., the pixel size in a raster operation), so it is important that the level of detail be adequate. Third, what accuracy is achievable? All results from a GIS will be uncertain to a degree, because all inputs are uncertain, so it is important to determine whether accuracy is sufficient. Finally, how should the results be presented? GIS is fundamentally a graphic tool, and the images and pictures it produces can have enormous impact if they are well designed. On the other hand, poor graphic design can easily undermine the best presentation.

REFERENCES

Burrough PA and McDonnell RA (1998) *Principles of Geographical Information Systems.* New York: Oxford University Press.

De Smith MJ, Goodchild MF, and Longley PA (2007) *Geospatial Analysis: A Comprehensive Guide to Principles, Techniques and Software Tools.* Winchelsea, U.K.: Winchelsea Press. Available online at www.spatialanalysisonline.com.

Longley PA, Goodchild MF, Maguire DJ, and Rhind DW (2005) *Geographic Information Systems and Science.* Chichester, U.K.: Wiley.

Maidment DR and Djokic D (eds) (2000) *Hydrologic and Hydraulic Modeling Support with Geographic Information Systems.* Redlands, CA: ESRI Press.

McHarg I (1969) *Design with Nature.* Garden City, NY: Natural History Press.

Tomlin CD (1990) *Geographic Information Systems and Cartographic Modeling.* Englewood Cliffs, NJ: Prentice Hall.

30 Spatial Data Quality

Joel Morrison and Howard Veregin

CONTENTS

30.1 INTRODUCTION

Data describing positions, attributes, and relationships of features in space are termed spatial data. Events throughout history document the role of spatial data in influencing the outcome of key events. Almost any planning activity, movement of people, vacation travel, exploration, military campaign, or real-estate transaction relies upon spatial data. Some of these activities form the cornerstones of our current economic systems.

Prior to the electronic revolution that ushered in the information age, most spatial data users were continually seeking to increase accuracy and relevance of data—to seek more data, and data of higher precision, currency, and positional accuracy. The emphasis focused upon the x, y position of the data, with the quest for attributes receiving somewhat less attention.

Now, users of spatial data currently face, for perhaps the first time, a situation in which technology enables the potential mapmaker to routinely collect and use data with quality exceeding that needed and/or requested by users. In a digital world, the spatial data situation has changed so that data producers are collecting and developing data, structuring spatial data into data sets, and adding enhanced attribute information. The fact that these digital data sets may also be easily distributed allows other data producers to add attributes to features, add new features, and integrate the results with other digital data sets. A given data set may therefore be the product of a number of data producers. The acceleration of the data collection process creates numerous data sets of varying quality that duplicate Earth features at differing resolutions and which compete to satisfy users' needs. These readily available distributed data files are easily accessible electronically. Moreover, they can often be easily visualized using geographic information systems technology. (Throughout this chapter, the term visualization will be used to indicate the graphic rendering of spatial data, either on a monitor or in printed map form.)

These developments point to the fact that the cartographer's control over the new end product is much less than was previously the case during preparation of the printed map. In the era of the printed map, the cartographer controlled the data that was portrayed on the map and how it was symbolized, and was ultimately responsible for the success of the product. He or she, through experience, developed a sense of appropriateness for data use. Relying upon that sense, the cartographer evaluated the quality of each data source, and rendered the contents of the map accordingly, putting an invisible "stamp of approval" on the map.

The result of today's change in technology means that there are many more players involved in creating and using spatial data. In the current context, the data producer and the data user are often unrelated. The producer collects and develops data. The cartographer must ask: How do I describe the quality of the spatial data collected so that it can be assessed against the demands of customers? How do I select among competing spatial data sets? What criteria do I use to make this decision? Is it a decision based on the known quality of the spatial data, or on the availability or ease of accessibility of the data? After these decisions are made, the user must impart to the visualization, indications of the quality of the spatial data. Traditionally, cartographers have not faced such questions, and in fact, today we do not know how to answer them in our current digital environment. Oddly, we are even hampered by the new technology in creating concern about this aspect of visualization, because most digital renderings appear to be very precise and accurate.

30.1.1 IMPORTANT NEW QUESTIONS

Two aspects of the current situation deserve our attention. First, the user—not the cartographer or the producer—now must decide which, among competing data sets, to use for a specified visualization or analysis. To aid the user in making this decision, what spatial data quality information should the producer include in a spatial data set? How important is spatial data quality in deciding which spatial data sets to use?

Second, because computer renderings of digital data appear so precise and accurate, regardless of their inherent accuracy, there is a need to devise methods to indicate the quality of the data sets used in analyses and visualizations. There are generally few inherent clues in a well-rendered visualization to indicate the quality of the data. In the digital era, the analog methods that cartographers used to select and render data sets are no longer sufficient. Ideally, users who download digital files and create visualizations should indicate the quality of the data sets used. However, given the ease with which data sets can be augmented, transformed, and integrated, the quality of the end product may be quite difficult to assess. Moreover, methods to visualize spatial data quality are not well established. Therefore, the spatial data user—whether naive or sophisticated—while empowered to utilize spatial data sets for visualizations, may at the same time be required to make decisions about the underlying data, the consequences of which may be unknown.

To add to this state of affairs, geographic data sets are increasingly being shared, exchanged, and used for purposes other than those intended by the producer. The user can select from many data sets that were obtained from a variety of sources, by a variety of methods, by multiple data producers at different scales. Many of these data sets can be easily downloaded and manipulated in geographic information systems, and in the absence of statements about the quality of the data, the visualization of each data layer may appear to be precise and accurate. Generally, the user will find that data from different sources, when conflated and integrated, will not perfectly agree; geometry and attributes representing the same Earth features may not exactly match. The user will have no logical way to resolve these differences in the absence of statements about data quality. To an extent, this situation was present in the era of analog cartography. In addition, in those rare instances of conflated data in analog technology, the cartographer needed information about spatial data accuracy before assuming responsibility for using the data. In the absence of other information, the decision was often based on the reputation of the data producer. Today, many data producers are so new to the marketplace that they have not yet established reputations, which makes this assessment difficult.

Today's empowered user, acting as a cartographer, must select the most appropriate data sets for use in a given situation. How can a spatial data user and the creator of visualizations assess the quality dimensions of a given data set? What constitutes a statement of spatial data quality? How can the myriad of electronic data producers be required to include statements of spatial data quality with their data sets?

30.1.2 INITIAL RESPONSES TO NEW QUESTIONS

Society has begun to react to such uncertainties. One reaction has been the development and adoption of the ISO9000 family of standards (ISO, 1987), and the creation of certification organizations worldwide to evaluate compliance with ISO9001, that part of the standards against which certification can be obtained.

Since its initial development in 1987, the ISO9000 family has been revised several times. The current standards include ISO9000:2005 (Quality Management Systems—Fundamentals and Vocabulary), ISO9001:2008 (Quality Management Systems—Requirements), and ISO9004:2000 (Quality Management Systems—Guidelines for Performance Improvements). There are also industry-specific implementations, in which the general ISO9000 standards are tailored to specific industries and organizations.

The ISO9000 family—or more specifically ISO9001:2008— provides a set of standardized requirements for quality management, the concept whereby an organization has formal control over its production processes. Quality management refers to the processes and procedures that are implemented in order to meet customers' expectations, needs, and requirements while also improving performance in pursuit of these objectives.

The ISO9000 standard stresses control of the production processes, including defining measurable objectives, regular monitoring and auditing procedures, upstream quality-control procedures, documentation of procedures, and adequate understanding of customer requirements and expectations. The focus is on the management of the production process to achieve quality, rather than on downstream checking of the finished product.

ISO9000 has a strong customer focus, related to understanding customer needs and expectations and ensuring that these needs and expectations are met. This customer focus implies that quality targets will be strongly influenced by real or anticipated customers, rather than by *a priori* principles about how accurate or reliable a product could be or should be. Since quality is costly in terms of time and resources, one implication is that no more quality will be engineered into a process than is required to meet customer needs. This raises many interesting questions: How do spatial data producers adapt to this philosophy without a thorough knowledge of the spatial data quality of a given data set and the needs of their consumers? How does a producer assess quality needs when customers are unable to articulate their needs? What is the responsibility of the producer in documenting data quality for users? What aspects of quality are necessary and sufficient to describe and communicate the quality of a given digital data set?

ISO9000 represents a stark contrast to the traditional operating assumptions familiar to most analog cartographers. For most of recorded history, professional cartographers have tried to present the most accurate maps that they could produce given the working constraints under which they labored. The reason for this operating assumption was partly due to the fact that printed maps were multipurpose products. Very few widely disseminated printed maps could afford to serve a single purpose. The cartographer could never be certain of all of the uses for which a given map would be used. Maps had to be as accurate as possible to prevent inappropriate use. Electronic technology now encourages the production of unique products to satisfy single user needs, and this is reflected in the ISO9000 philosophy.

Today, a given visualization and the data sets on which it is based may have very specific purposes and quality characteristics. The knowledge required to assess the quality of digital spatial data extends well beyond assessment of the positional accuracy and an occasional reliability diagram of the analog world. Unfortunately, many of the traditional spatial data production institutions have not yet fully interpreted these changes in user needs. They still produce in the digital environment to the exacting positional standards of the analog data era, ignoring other, equally important, aspects of spatial data quality.

30.2 SPECIFICATION OF SPATIAL DATA QUALITY

Spatial data quality can be defined by three dimensions. First is the definition of the elements of spatial data quality. A second component consists of specification of easily understood indices for each of the elements defined in this first dimension. These metrics may accompany a digital data set and must have meaningful recognition by the community of spatial data users. The third part of the specification concerns the problem of presenting the known data quality in visualizations. The bulk of the remarks in this chapter will be aimed primarily at the first two parts. There is a parallelism in the developments relating to parts one and two, while research into topics included in part three has been accomplished in a more independent arena.

30.2.1 DEFINITION OF SPATIAL DATA QUALITY

In analog cartography, spatial data quality was synonymous with the quest for positional and attribute measurement accuracy. National Map Accuracy Standards were devised and implemented by most national mapping agencies (Thompson, 1987). For larger-scale work, standards were generated by local governments who were responsible for collecting and displaying spatial data (American Society for Photogrammetry and Remote Sensing, 1990). With the advent of the information age and digital cartography, mapping organizations lost the tight control they once enjoyed over these activities. In this new age, National Map Accuracy Standards are not only of little use, but are really irrelevant to most mapping.

30.2.1.1 Initial Work on Spatial Data Quality in the United States

During the 1980s, mapping experts in several parts of the world began research and collaborations in efforts to standardize different aspects of digital spatial data. Most of these efforts included a concern for spatial data quality, and more precisely for a definition of spatial data quality. One of the most comprehensive efforts was undertaken in the United States, where, in 1982, a National Committee on Digital Cartographic Data Standards (NCDCDS) was established under the auspices of the American Congress of Surveying and Mapping (ACSM). Over a five-year period this committee deliberated and produced a report entitled, "A draft proposed standard for digital cartographic data" (Moellering, 1987). One of the four sections of this report was devoted to digital cartographic data quality. This perhaps represents the first comprehensive statement on spatial data quality in the electronic age. Quoting from the report's statement of spatial data quality (Moellering, 1987):

> The purpose of the Quality Report is to provide detailed information for a user to evaluate the fitness of the data for a particular use. This style of standard can be characterized as "truth in labeling," rather than fixing arbitrary numerical thresholds of quality. These specifications therefore provide no fixed levels of quality because such fixed levels are product dependent. In the places where testing is required, several options

for different levels of testing are provided. In this environment, the producer provides the quality information about the data and the user makes the decision of whether to use the data for a specific application.

In the Moellering report, the specification of the components for reporting data quality is divided into five sections: lineage, positional accuracy, attribute accuracy, logical consistency, and completeness (Moellering, 1987). After 1987, a modified version of the proposed standard for the exchange of spatial data created by the Moellering committee was accepted by the National Institute of Standards and Technology as the Federal Information Processing Standard—173 (National Institute of Standards and Technology, 1994).

30.2.1.2 International Work on Spatial Data Quality

The work of the data quality subcommittee of the NCDCDS in the United States gained considerable worldwide acceptance. Cartographers in South Africa (Clarke et al., 1992), the United Kingdom (Walker, 1991), and Australia adopted aspects of the five elements embodied in the NCDCDS report and the NIST standard.

In the early 1990s, the Comité Européen de Normalisation/Technical Committee 287 (CEN/TC287) became one of the first international (regional) standards bodies to begin to focus on standards for geographic information to aid portability of geographic data within Europe (Comité Européen de Normalisation, 1992). Spatial data quality was one of its initial work items. Work on spatial data quality continues within this organization.

Beginning its work in 1991, The Commission on Spatial Data Quality of the International Cartographic Association accepted the five elements of the NCDCDS as important aspects of spatial data quality. The Commission in its deliberations also considered several other potential aspects of spatial data quality and reached consensus on two additional elements: semantic and temporal accuracy. In the Commission's book *Elements of Spatial Data Quality* (Guptill and Morrison, 1995), all seven aspects are explained.

In 1997, the International Organization for Standardization (ISO) established a new technical committee, TC211, geographic information/geomatics, to specifically devise a set of international standards for geographic information. Through an agreement of cooperation, the work of CEN/TC287 was taken into account in the development of the more global ISO standards. Experts working in CEN/TC287, as well as experts from the United States that had worked on the NIST standard, and experts from the ICA Commission, combined to ensure a continuity of ideas in the work of ISO/TC211.

Within ISO/TC211 five working groups (WG) were initially established and among the tasks identified for working group 3 (WG3, geospatial data administration) were work item 13 (NO13, geographic information—quality), work item 14 (NO14, geographic information—quality evaluation), and work item 15 (NO15, geographic information—metadata). Also of relevance to this discussion is work item 4 (NO4, geographic information—terminology) included under the tasks assigned to working group 1 (WG1, framework and reference model). ISO TC211/WG3/NO13 divided the aspects of spatial data quality into two categories: (1) data quality elements and

(2) data quality overview elements. Five data quality elements and three data quality overview elements were included in the Draft International Standard (ISO, 2001).

The publications of TC211 relevant to spatial data quality include ISO 19113 (Geographic Information—Quality Principles), ISO 19114 (Geographic Information—Quality Evaluation Procedures), and ISO 19115 (Geographic Information—Metadata). Work by ISO/TC211 is ongoing, with additional working groups being formed and others disbanded, and updated versions of documents (see http://www.isotc211.org/ for recent status). This work is important to all levels of spatial data users at international, national, regional, and local levels. It defines common terminology that enables all users to communicate with one another about spatial data. It also establishes standards that ease the exchange and international use of spatial data. Finally, it defines the elements of metadata that should accompany any data set to enable the wise use of the data. Digital spatial data sets are routinely used for purposes that differ from the producer's intended use; it is only through the accompanying metadata, and specifically through the metadata about data quality, that a potential user can make a rational decision on the use of a data set.

30.2.1.3   The Work of ISO/TC211/WG3/NO13—Quality Principles

The work of the ISO/TC211/WG3/NO13 on quality principles represents a comprehensive and definitive statement about the elements of spatial data quality. It combines the major threads of previous work done in various parts of the world over the past two decades. The following discussion therefore relies heavily on that work.

The ISO/TC211 project team working on quality principles found it necessary and convenient to define a model of data quality. Figure 30.1 is a modification of the model initially created by the project team.

A verbal description of the data quality model begins with a data set that is an identifiable collection of related data. These data represent real or hypothetical entities of the real world (which are characterized by having spatial, thematic, and temporal aspects.) The process of modeling the potentially infinite characteristics of real–world entities into an ideal form for making these entities intelligible and representable is defined as the process of abstracting from the real world to a universe of discourse. The universe of discourse is described by a product specification, against which the quality content of a data set, or specific parts of it, can be tested. The quality of a data set depends on either the intended use of the data set or the actual use of the data set: both data producers and data users can assess the quality of a data set within the confines of the data quality model. A data producer is given the means for specifying how well the mapping used to create the data set reflects its universe of discourse; data producers validate how well a data set meets the criteria set forth in its product specification. Data users are given the means for assessing a data set derived from a universe of discourse identified as being coincident with (some or all) requirements of the data producer's application. Data users can assess quality to ascertain if a data set can satisfy their specific applications, as specified within their user requirements (ISO 19113). This model is compatible with ISO terminology that defines quality as the "totality of characteristics of a product that bear on its ability to satisfy stated and implied needs" (ISO, 2001).

The ISO/TC211 project team's model in Figure 30.1 makes an important distinction. It acknowledges that data users commonly have differing needs or requirements from

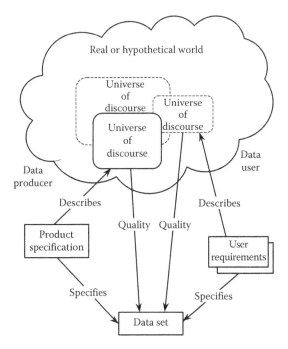

FIGURE 30.1   Modified ISO/DIS 19113 model of data quality.

the specifications used by the data producers in creating a data set. Therefore, both data producers and data users bear a responsibility for assessing the quality of a data set.

30.2.2   QUANTITATIVE ELEMENTS OF SPATIAL DATA QUALITY

The ISO/TC211 project team defined five data quality elements subject to quantitative assessment and three data quality overview elements that are nonquantitative in nature (Figure 30.2).

30.2.2.1   Completeness

The first recognized element of spatial data quality is completeness, described as the "presence and absence of features, their attributes and their relationships" (ISO, 2001). Completeness is applied to both features and attributes and can be measured by commission (the inclusion of features and/or attributes not found in reality) and by omission (the exclusion of features and/or attributes found in reality).

Completeness has been generally agreed by all working groups to be an element of spatial data quality. The FIPS 173 states,

> The quality report shall include information about selection criteria, definitions used and other relevant mapping rules. For example, geometric thresholds such as minimum area or minimum width must be reported.... The report on completeness shall describe the relationship between the objects represented and the abstract universe of all such objects. In particular, the report shall describe the exhaustiveness of a set of features. (National Institute of Standards and Technology, 1994).

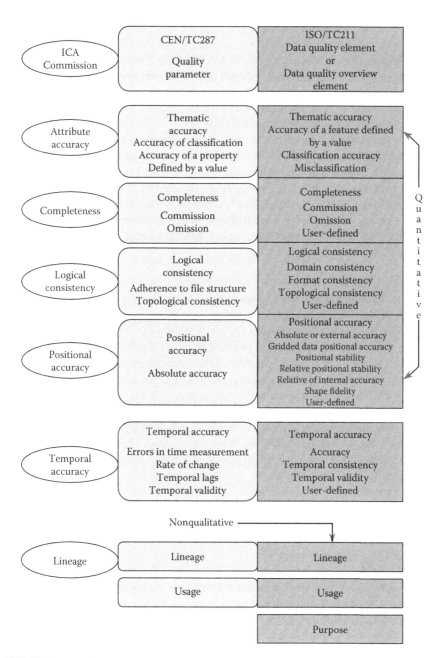

FIGURE 30.2 Crosswalk of spatial data elements adapted from ICA Commission, CEN/TC287, to ISO/TC211/WG3/NO13.

Professor Kurt Brassel and his associates at the University of Zurich expanded on the FIPS 173 definition of completeness in Chapter 5 of the book *Elements of Spatial Data Quality* as: "Completeness describes whether the entity objects within a dataset represent all entity instances of the abstract universe...the degree of completeness describes to what extent the entity objects within a dataset represent all entity instances of the abstract universe" (Guptill and Morrison, 1995).

In practical terms, many map users never considered completeness when analog maps were the only type available. The tacit assumption was that all instances of existing features—as appropriate for the scale of the map—were depicted. Now, it is no longer necessary for that to be the case and a relevant question about a digital file is to ask: What percentage of feature instances are in reality encoded in the data set? Are all culverts listed, or only those over 4 m in length? What percentage of those culverts over 4 m in length are claimed to be encoded by the metadata of the data set? Are all farm ponds encoded, or only those over 5 acres in surface extent? Are all water towers listed, or only those in active use over 30 ft in height? Completeness is now an integral part of spatial data quality, and a critical concern to digital spatial data users.

30.2.2.2 Logical Consistency

A second data quality element is logical consistency, which describes the "degree of adherence to logical rules of data structure, attribution, and relationships" (ISO, 2001). This element is also defined in FIPS 173 as follows: "logical consistency shall describe the fidelity of relationships encoded in the data structure of the digital spatial data" (National Institute of Standards and Technology, 1994). Logical consistency describes the number of features, relationships, or attributes that have been correctly encoded in accordance with the integrity constraints of the feature data specification. Tests can be either graphic or visual, and topological. Conceptual consistency, domain consistency, format consistency, and topological consistency are all testable for omission and commission.

Kainz states in Chapter 6 of the book *Elements of Spatial Data Quality*, "According to the structures in use, different methods may be applied to test for logical consistency in a spatial dataset." Kainz "gives an overview of the modeling of spatial data and relationships and their mathematical background," and follows that discussion with a section on "consistency rules derived from the characteristics of the underlying mathematical structure and a collection of tests that can be performed to test a dataset or scene for any logical inconsistencies present" (Guptill and Morrison, 1995).

Spatial data in digital form present a different set of concerns for the spatial data user. From the physical properties of entities and from human use of these entities a set of logical relationships can be deduced. For example, when two highways cross each other there is an intersection, an underpass, or an overpass. All drainage flows from higher to lower elevations. A given area cannot be simultaneously land and water. If water, it cannot have a railroad or a road without a bridge. Since digital data are often stored by feature layer, changes or updates to one digital data set layer will affect other layers of data. One cannot add a dam to a data layer without affecting another data layer. Automated checks for these logical relationships can reveal inconsistencies in a data set.

The logical consistency element of spatial data quality measures the internal consistency of the data in the various data layers of a data set.

30.2.2.3 Positional Accuracy

Positional accuracy has a longer history of concern by spatial data producers and users, and in the electronic age still represents a major concern. The ISO/TC211 project team states very simply that positional accuracy shall describe the "accuracy of the position of features" (ISO, 2001). However, the aspects of positional accuracy mentioned are numerous. According to the FIPS 173, "The quality report portion on positional accuracy shall include the degree of compliance to the spatial registration standard" (primarily latitude and longitude) (National Institute of Standards and Technology, 1994). It includes measures of the horizontal and vertical accuracy of the features in the data set. It must consider the effects on the quality of all transformations performed on the data, and report the results of any positional accuracy testing performed on the data.

ISO/DIS 19113 refers to absolute (or external) accuracy, relative (or internal) accuracy, and gridded data position accuracy. It also mentions positional stability and shape fidelity. Even though more work has probably been done on positional accuracy than on any other single element, much remains to be researched.

30.2.2.4 Temporal Accuracy

During the 1980s, as more digital spatial data began to become available, researchers began to devote more attention to temporal aspects of spatial data quality. The ISO/TC211 project team states that temporal accuracy shall describe the "accuracy of the temporal attributes and temporal relationships of features" (ISO, 2001). Temporal accuracy includes errors in time measurement, temporal consistency, and temporal validity.

The ICA Commission also accepted temporal information as an element of spatial data quality. Temporal information describes the date of observation, type of update (creation, modification, deletion, and unchanged), and validity periods for spatial data records. Prior to its specification as an element of spatial data quality, aspects of the temporal dimension were contained in each of the other elements of spatial data quality as defined in FIPS 173. Guptill states that in the use of digital spatial data and geographic information system technology, "users will collect, combine, modify, and update various components of spatial information. This will occur in a distributed, heterogeneous environment, with many parties participating in the data enterprise. In such an environment, access to reliable information about the temporal aspects of the spatial data is paramount" (Guptill and Morrison, 1995).

To realize the implications of the temporal dimension, spatial data users need only think about producing a map of all of a county's fence lines that have ever existed through history on the one hand, and a map of the current fence lines on the other. In analog topographic mapping it was accepted that the features shown on a given quadrangle might never have actually existed at the same time in reality (but did exist sometime within the approximate average of four years that it took to produce a topographic map). Today, when selecting a file of digital spatial data for use in the preparation of a visualization, the data user must have some indication of the dates of validity for a digital feature in order to produce a map. Time tags on all digital features are therefore necessary, and the availability and reliability of those time tags reflect on the spatial data quality of a data set.

30.2.2.5 Thematic Accuracy

The fifth data quality element is thematic accuracy. Thematic accuracy describes the "accuracy of quantitative attributes and the correctness of non-quantitative attributes and of the classifications of features and their relationships" (ISO, 2001). It is important to note that thematic accuracy includes the accuracy of the classification and/or misclassification (Chapter 21). It can be argued that the use of reliability diagrams in analog technology to display attribute accuracy is the analog equivalent of this element. However, it is now deemed crucial to the use of digital data to specify accuracy of the value assigned for any single attribute tied to an Earth position as well as the accuracy of the classification of that attribute.

These five elements are relatively consistent among the work of the ICA Commission, the CEN/TC287, and ISO/TC211 (Figure 30.2). The work by Salge on semantic accuracy, included as Chapter 7 in the ICA Commission's work (Guptill and Morrison, 1995), and the textual fidelity referring to the accuracy of spelling and the consistency of abbreviations in CEN/TC287, was not accepted by ISO/TC211/WG3/NO13.

30.2.3 Nonquantitative Elements of Spatial Data Quality

The ISO/TC211 project team also defined three nonquantitative spatial data overview elements. These are lineage, usage, and purpose. This is an expansion on previous work, since the ICA Commission, like the FIPS 173 standard, only defines lineage, while the CEN/TC287 defines both lineage and usage.

30.2.3.1 Lineage

The ISO/TC211 project team defines lineage as a description of the history of a data set and, in as much as is known, a recounting of the life cycle of a data set from collection and acquisition through compilation and derivation to its current form (ISO, 2000a). Lineage contains two unique components: (1) source information describing the parentage of a data set, and (2) processing history describing transformations of the data set, including the processes used to maintain it, whether continuous or periodic (ISO, 2000a). The FIPS 173 states, "The lineage portion of a quality report shall include a description of the source material from which the data were derived, and the methods of derivation, including all transformations involved in producing the final digital files" (National Institute of Standards and Technology, 1994). The description is to include the dates of the source material and the dates of ancillary information used for updates, with reference to the specific control information used, and the mathematical transformations of coordinates used in each stage from the source material to the final product.

Clarke and Clark state in Chapter 2 of *Elements of Spatial Data Quality* that, "Lineage is usually the first component given in a data quality statement. This is probably because all of the other components of data quality are affected by the contents of the lineage and *vice versa*." They admit that "The ultimate purpose of lineage is to preserve for future generations the valuable historical data resource" (Guptill and Morrison, 1995). Clarke and Clark list four components that provide the contents of a quality statement on lineage, and a format is suggested for the recording of lineage data for digital data files.

30.2.3.2 Usage

Usage, according to the TSO/TC211 project team, describes "the application(s) for which a data set has been used. Usage describes uses of the data set by the data producer or by other, distinct, data users" (ISO, 2001).

30.2.3.3 Purpose

According to the ISO/TC211 project team, purpose "shall describe the rationale for creating a data set and contain information about its intended use." The project team adds a footnote that a data set's intended use is not necessarily the same as its actual use. Actual use is described using the data quality overview element usage (ISO, 2000a).

Figure 30.2 gives a summary of Section 30.2, the specification of spatial data quality, and the identification of its elements. Three of the more important studies on spatial data quality are easily compared and in general, there is much agreement among the three.

30.3 SPATIAL DATA QUALITY AND ITS MEASUREMENT

The second necessary part of spatial data quality—easily understood and implemented methods for measuring and stating spatial data quality in terms of its defined elements—is within the scope of work of the ISO/TC211/WG3/NO14. Much less work has been devoted to specification of spatial data quality and its measurement. The ICA Commission presented a descriptive framework by Veregin and Hargitai in Chapter 9 of its book (Guptill and Morrison, 1995), and in addition, several individual researchers have addressed the problem.

Veregin and Hargitai attempted to define a metric for the seven elements of spatial data quality as defined by the ICA Commission. It was exploratory by its very nature but offered an outline of a logical schema for data quality assessment in the context of spatial databases. The schema is based on two concepts: "geographical observations are defined in terms of space, time, and theme. Each of these dimensions can be treated separately (but not always independently) in data quality assessment" and "the quality of geographic databases cannot be adequately described with a single component." The multidimensional aspect of spatial data quality is readily accommodated in the framework.

Aalders and Morrison in Chapter 34 of the *Geographic Information Research: Trans-Atlantic Perspectives* emphasize that in the information age three competing possibilities exist for a potential data user when a data set is selected for the delineation of a set of features or a region of the world (Craglia and Onsrud, 1999). These are (1) the quality of the data, (2) the availability of the data, and (3) the ease of accessibility. A conscientious data user needs to know what spatial data quality information is required, and how to impart that quality information in the visualization being prepared. Although their work gives examples, no actual metrics for the specification of spatial data quality are introduced.

Hunter's (1999) article, "New tools for handling spatial data quality: Moving from academic concepts to practical reality," focuses on such metrics. Five examples are presented: (1) tracking of feature coordinate edits and their reporting in visual data quality statements, (2) testing and reporting the positional accuracy of linear features of unknown lineage, (3) simulating uncertainty in products derived from Digital Elevation Models (DEMs), (4) incorporating uncertainty modeling in vector point, line, and polygon files, and (5) reporting data quality information at different levels of geographic information systems database structures.

Simley (2001) has written about the practical application of ISO9000 techniques to cartography. Based on the philosophy of ISO9000, Simley has apprised cartographers of tools available in the ISO9000 arena that could prove useful in increasing and maintaining the quality of spatial data. These same tools could be used for the assessment of positional and thematic accuracy and logical consistency. They are compatible with the ISO/CD 19114 (see discussion below).

30.3.1 THE WORK OF ISO/TC211/WG3/NO14

A project team working within ISO/TC211 under the direction of Dr. Ryosuke Shibasaki has accomplished the most comprehensive work on the derivation of evaluation metrics for the elements of spatial data quality. The two project teams that created ISO/DIS 19113 and ISO/CD 19114 worked closely together to maintain compatibility.

ISO/CD 19114 "recognizes that a data producer and a data user may view data quality from different perspectives" (ISO, 2000a). Each may set different conformance quality levels. The ISO/CD 19114 project team developed a quality evaluation process that is a sequence of steps that can be followed by both a data set producer and a data set user to produce a quality evaluation procedure that yields a specific quality evaluation result. Figure 30.3 illustrates the ISO/CD 19114 project team's data quality evaluation process flow.

For each different test of a data set required by a product specification or user requirement, a data quality element, data quality sub-element, and data quality scope, in accordance with ISO/DIS 19113, must be identified. Next, a data quality measure is identified for each test, and a data quality evaluation method is selected. A data quality evaluation method can be either direct or indirect. Direct methods are applied to the data specified, while indirect methods use external information to determine a data quality result. Direct evaluation may be accomplished by automated or nonautomated means (ISO, 2000a). Finally, the data quality results must be reported as metadata compliant with ISO/CD 19115 (ISO, 2000b).

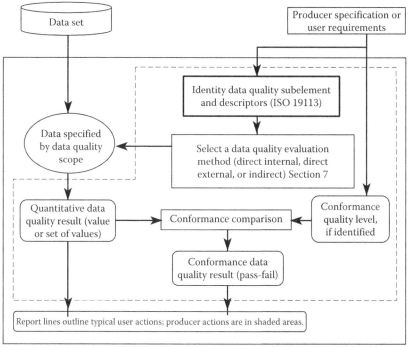

Dotted lines outline typical user actions; producer actions are in shaded areas.

FIGURE 30.3 Data quality evaluation process flow (as per draft dated May 31, 2000).

30.4 SPATIAL DATA QUALITY AND ITS VISUALIZATION

The ability of researchers to focus on the visualization of spatial data quality has changed dramatically with the tools available to cartographers from the electronic revolution. The use of computers has freed the visualization of spatial data from serving both as a database and as a visual display. The database is now separated and the need to retrieve precise information from the visual display is no longer necessary. Until the information age, cartographers have not been able to separate their research energies between database creation and visualization creation. Rather independently of the work reported above in Sections 30.2 and 30.3, several groups of scholars have been working on the graphic presentation of spatial data quality. This work is not as structured as the ISO work reported in Sections 30.2 and 30.3.

Visualization of spatial data quality was addressed in an early research initiative of the National Center for Geographic Information and Analysis (NCGIA) (see Beard et al., 1991). In addition, several recent book chapters discuss the current state of research in this area. Examples include, "Visualisation of uncertainty in geographical data" (Drecki, 2002), and "Communication and use of spatial data quality information in GIS" (Devillers and Beard, 2006).

30.5 CONCLUSIONS

Spatial data producers and spatial data users are forced to consider spatial data quality as a result of the electronic revolution. During the age of analog maps, when a few organizations tightly controlled collection, maintenance, and visualization of spatial data, it was possible to provide spatial data to users based upon established reputations built over many years. Products were static and users could not alter the information on printed maps. In the information age, electronic production requires systematization of specifications for spatial data quality so that potential users may select data sets appropriately. Because all individuals are now potential spatial data users and consumers, they have both a need and a right to know the quality of the data.

In response, over the past 25 years, communities of interested individuals throughout the world have worked to systematize processes for evaluating spatial data sets. Continued international efforts within the ISO emphasize the importance of the initiative, and the substantial agreements as to the definition of spatial data quality components. The next step, still in its preliminary stages, is the development of a series of metrics to measure each element of spatial data quality. The ISO standards have specified a process flow that will enable metrics to be selected and reported to potential data users, but the metrics themselves are still not systematized. In addition, more work in the area of visualization of spatial data quality is needed, as relatively few of the techniques that have been explored are now in actual usage within organizations that produce or consume spatial data. Although much work remains, the last 25 years has seen more systematic progress in the creation of knowledge about spatial data quality than was attempted in the previous century.

We should continue to expect great progress in this arena. Morrison and Guptill (1995) outline the future direction of this line of inquiry. They point to directions in need of further investigation and speculate about the use and transfer of digital spatial data in tomorrow's electronic world. In addition, there has been much recent research on the topic of spatial data quality. Several edited volumes have appeared in the last decade, including *Spatial Data Quality* (Shi et al., 2002), and *Fundamentals of Spatial Data Quality* (Devillers and Jeansoulin, 2006), among others. Some of the latest research has been presented at a series of international conferences on spatial data quality, with the latest of these, the 6th International Symposium on Spatial Data Quality, to be held in 2009. Earlier work includes *Accuracy of Spatial Databases* (Goodchild and Gopal, 1989), which was developed out of the first research initiative of the NCGIA. Several NCGIA technical publications on spatial data quality also resulted from this initiative.

Unfortunately, the fact that this work is still so new to the spatial data mapping community means that there are few readily available sources of information about how to collect and use information about digital spatial data quality. This presents another challenge that can only be resolved through further research and collaboration.

REFERENCES

American Society for Photogrammetry and Remote Sensing, Committee for Specifications and Standards, ASPRS accuracy standards for large maps, *Photogrammetric Engineering and Remote Sensing*, 56(7), 1990, 1068–1070.

Beard, K. M., Buttenfield, B. P., and Clapham, S. B. (eds.), Visualization of spatial data quality, National Center for Geographic Information and Analysis Technical Paper 91-26, Santa Barbara, CA, 1991.

Clarke, D. G., Cooper, A. K., Liebenberg, E. C., and van Rooyen, M. H., Exchanging data quality information, *A National Standard for the Exchange of Digital Geo-Referenced Information*, Chapter 2, National Research Institute for Mathematical Sciences, Pretoria, Republic of South Africa, 1992.

Comité Européen de Normalisation, Electronic Data Interchange in the field of Geographic Information—EDIGEO, Prepublication of CEN/TC 287, AFNOR, 1992.

Craglia, M. and Onsrud, H. (eds.), *Geographic Information Research: Trans-Atlantic Perspectives*, Taylor & Francis, London, U.K., 1999.

Devillers, R. and Beard., K., Communication and use of spatial data quality information in GIS, in Devillers, R. and Jeansoulin, R. (eds.), *Fundamentals of Spatial Data Quality*, ISTE, London, U.K., 2006, pp. 237–254.

Devillers, R. and Jeansoulin, R. (eds.), *Fundamentals of Spatial Data Quality*, ISTE, London, U.K., 2006.

Drecki, I., Visualisation of uncertainty in geographical data, in Shi, W., Fisher, P., and Goodchild, M.F. (eds.), *Spatial Data Quality*, Taylor & Francis, New York, 2002, pp. 144–163.

Goodchild, M. F. and Gopal, S., *Accuracy of Spatial Databases*, Taylor & Francis, New York, 1989.

Guptill, S. C. and Morrison, J. L. (eds.), *Elements of Spatial Data Quality*, Elsevier Science, Oxford, U.K., 1995, 202p.

Hunter, G. J., New Tools for handling spatial data quality: Moving from academic concepts to practical reality, *Journal of the Urban and Regional Information Systems Association*, 11, 1999, 25–34.

ISO, ISO 9000—1987 Quality Management and Quality Assurance Standards, International Standards Organization Copyright Office, Geneva, Switzerland, 1987.

ISO, ISO/TC211/WG3/Editing Committee 19114, Final text of CD 19114, Geographic Information—Quality Evaluation Procedures, International Standards Organization Copyright Office, Geneva, Switzerland, 2000a.

ISO, ISO/TC211/WG3/Editing Committee 19115, Final text of CD 19115, Geographic Information—Metadata, International Standards Organization Copyright Office, Geneva, Switzerland, 2000b.

ISO, ISO/Dis 19113, Draft International Standard, Geographic Information—Quality Principles, International Standards Organization Copyright Office, Geneva, Switzerland, 2001.

Moellering, H. (ed.), A draft proposed standard for digital cartographic data, National Committee for Digital Cartographic Standards, American Congress on Surveying and Mapping Report #8, 1987.

National Institute of Standards and Technology, Federal Information Processing Standard Publication 173 (Spatial Data Transfer Standard Part 1, Version 1.1), U.S. Department of Commerce, 1994.

Shi, W., Fisher, P., and Goodchild, M.F. (eds.), *Spatial Data Quality*, Taylor & Francis, New York, 2002.

Simley, J., Improving the quality of mass produced maps, *Cartography and Geographic Information Science*, 28(2), 2001, 97–110.

Thompson, M. M., *Maps for America*, 3rd edn., U.S. Department of the Interior, U.S. Geological Survey, U.S. Government Printing Office, Washington, DC, 1987.

Walker, R. S., Standards for GIS data, *Proceedings of the Third National Conference and Exhibition*, The Association for Geographic Information, Birmingham, U.K., 1991.

31 Cartography and Visualization

Robert B. McMaster and Ian Muehlenhaus

CONTENTS

31.1 INTRODUCTION TO BASIC CARTOGRAPHY

Once you have spent considerable time collecting, verifying, and editing your spatial data, your next step is to craft a powerful map that will clearly and succinctly communicate the patterns and idiosyncrasies of the data set. This chapter will review the processes you must go through in order to do this.

Traditionally, cartography has been defined as both the "art and science" of creating maps. However, this binary is an artificial one. It presupposes that art and science are independent and separable components in mapmaking. In reality, cartography is best thought of as a "craft" combining knowledge from both graphic design theory and mathematics. In order to become a stellar cartographer, one needs to hone, expand, and synthesize his or her skills in both of these realms—they are not separable. Thus, before we continue, we recommend that you consider yourself an artisan—cartography is something that you will only master via continual education

611

and practice. This chapter will present you with a basic overview of the rudimentary information necessary to design an intelligible map and to jumpstart, or refresh, your knowledge. Throughout the text, we provide references to more nuanced readings concerning the topics covered here.

Cartography is thousands of years old. Maps predate written language as a form of communication. For millennia, cartography has been used to enhance our understanding of the earth—its shape, distances, directions, and areas. In fact, flattening the spherical earth into a multitude of two-dimensional shapes that preserve one or more of the aforementioned attributes was first done by the Greeks. Although this chapter does not attempt to provide a historical overview of the cartographic discipline, it should be noted that today's marvelous technologies that allow us to accurately depict locations within centimeters are the culmination of thousands of years of cartographic development. We are but the tip of the iceberg, and, as with any knowledge, the tip is dependent on its base to stay above the water.

31.2 MAP PURPOSE

The first thing one must do as a cartographer is identify the purpose of your map. Before one begins symbolizing any data, before even laying the map out on the page (or more likely computer screen), take a moment and jot down who you expect will be reading the map—the intended audience—and what the purpose of the map is. Although this may seem rudimentary, it is not. In fact, having this written down when you begin designing the map will save you much consternation later on, when we begin generalizing and symbolizing. *The intended audience and map purpose must guide the map creation in order for you to communicate effectively.*

Map types can be broken down into two broad categories—*general-* and *special-purpose* maps. General-purpose maps are those that provide a geographical base and have a functional rationale. For example, a city map would be considered "general purpose." You can use the city map to find directions from your house to a local rummage sale, or you can utilize it to determine distances between your house and the nearest fire department. You cannot use the map to gauge the median incomes of different neighborhoods nor the odds that your backyard is sitting atop a radon hotspot. The purpose of such maps is to provide people reference data of some sort. A map of the world may show the location of all countries, and perhaps even different countries' flags, which is great for referencing. However, it will not allow you to compare different countries' stockpiles of uranium. Some of the most detailed general purpose maps are topographic maps, which depict the basic location of features on the surface of the earth, including transportation, hydrography, cultural features, and elevation. Topographic maps are like "geographic dictionaries." They emphasize geographic position and location, as well as planimetric accuracy.

However, not all maps are primarily concerned with position, location, and the accuracy of base map data. Indeed, *special-purpose maps* are interested in showing the distribution of a theme, such as characteristics of a population or land use. Beginning in the nineteenth century, cartographers, geographers, and statisticians began developing different methods of visualizing thematic data over base maps, including the choropleth, graduated symbol, dot, isarithmic, and dasymetric techniques (among several others). This was a revolutionary idea—to spatially illustrate data that had previously only existed in chart form. However, the usefulness of these techniques quickly became apparent, particularly for state governments attempting to make sense of the increasing amounts of data they were collecting about their populations and territories.*

As opposed to "dictionaries," special-purpose maps might best be considered as "geographical essays." Like a good snapshot, special-purpose visualizations can quickly tell a story without the need for words. By taking tabular data and illustrating it over a map via location, an audience is able to interpret the data more quickly and readily than if they were to look at numbers. However, a special-purpose map's readability and communicative power as an essay is only as good as its articulation; this is where cartographic design becomes extremely important. The rest of this chapter will review the five key realms of thematic (special purpose) map design that you should pay particular attention to when designing maps—scale, generalization, symbolization, data classification, and map design. Notice that these five realms combine both design and math (art and science). In order to become a proficient and successful thematic cartographer, you must practice the *craft* of melding these five realms together seamlessly.

31.3   CARTOGRAPHIC SCALE

The reason we project the Earth onto a flat piece of paper is to look at a particular part of the Earth's surface in greater detail—something that is impossible to do with a globe due to size constraints. Thus, the first thing any cartographer must do is to decide at what scale they want to present the data. Scale is simply the mathematical relationship between the map distance and the commensurate earth distance. This relationship is frequently represented via *representative fraction* (e.g., 1:500,000 where one unit of distance on a map equals 500,000 of those units in the real world). Map scales are often broken down into large scale (e.g., ~1:10,000) or small scale (e.g., ~1:1,000,000).† Large-scale maps will contain less mapped area; since the scale is large relative to the size of the Earth, such a map will be zoomed in on the Earth's surface. This is for the best if you hope to show a lot of detail on the ground (e.g., create a topographic map). However, this is not useful if you hope to show spatial patterns and themes across a wider area of land. In this case, you will want to use a small-scale map (small because its representation is tiny compared to the real proportions of the Earth). Small-scale maps show more area of the Earth allowing for a larger realm of comparison, but they do so at an expense—the level

---

* A fascinating history of the development of thematic cartography may be found in Robinson's (1982) *History of Thematic Cartography*. Also of note is MacEachren's (1979). "The evolution of thematic cartography: A research methodology and historical review" found in the *Canadian Cartographer* 16(1): 17–33.

† Many people get confused by large- and small scale when they look at representative fractions, often presuming that the fraction with a bigger number must be a larger scale. A simple way to keep this straight is to ask yourself which number is bigger—one-ten-thousandth or one-millionth? Since the numbers represent fractions, it becomes obvious that 1:10,000 is much larger than 1:1,000,000.

of detail small-scale maps provide may be unacceptable for purposes of location and distance evaluation or any other general-purpose uses.*

To determine the scale at which you create your map, you must take into account three things. First, recall the purpose of the map—which you have written down somewhere hopefully. Second, think about the intended audience—do they know an area well enough to be able to interpret an extremely large-scale map without any broader reference? Will the scale you choose include all areas affecting the theme of your map and of interest to your audience? (The last thing a cartographer wants to do is incense his or her audience by zooming in so far that they crop some culturally sensitive or relevant areas off the map.) Once you have determined the scale at which you would like to create your map, you must determine whether you have the appropriate data for mapping at that scale. For example, if you want to make a population map of the ethnic Hungarian diaspora in Central Europe, you may want to zoom in on Hungary and the Romanian Carpathians. Perhaps you envision mapping the percentage of ethnic Hungarians residing in the Transylvanian part of Romania by province. Yet, when you go to look at your data, all you have is the percentage of Hungarians residing in each country (e.g., Romania equals 1.4 million). Thus, you must create a map at the scale of the data—a map that shows both Hungary and Romania in their entirety. Such a representation would be completely ridiculous for the purpose of the map; so perhaps you are better off mapping the Hungarian minority in Slovakia and Serbia as well, zooming out even further to show ethnic-Hungarian populations throughout Central Europe.

31.3.1 SCALAR DEPENDENCE OF DATA GEOMETRIES

Geographic phenomena take the form of four different geometric shapes—points, lines, polygons, and volumes.† Spatial features in the real world can be represented using these geometries. However, determining which of these geometric features to use will change depending on the scale of your map. Geometry is scale dependent! For example, a country house shown on a large-scale map of a French village (e.g., 1:3000) should be drawn as a polygon—a building footprint. Yet, the same house on a map showing a section of the French province (e.g., 1:50,000) would be represented by a point. One problem with geographic information systems (GISs) today is that they do not change the geometry of geographic data automatically depending on the scale of the map. Thus, point data remains point data no matter how far one zooms in on the village. Vice versa, point data remains point data no matter how far one

---

* Of note to those using GIS, beware of the scale at which you decide to create your thematic map to make sure it parallels the spatial properties of your data! An increasing problem within modern GISs is the mixing of disparate scales (as database layers) without accounting for differential generalization, detail, and accuracy between and among these layers. The initial scale from which your data were obtained should always be included with the metadata to avoid confusion. *You should never create a larger-scale map than the scale at which your data was collected and compiled, no matter how tempting it may be!*

† Additionally, with the advent of computer cartography, a fifth type of geographic data, temporal, has been studied; though, the temporal characteristics of geographic data are often ignored on paper maps. There are notable exceptions to this rule; however, including Minard's now famous map of Napolean's march to Russia and back.

zooms out, even to the global scale. In fact, if the scale makes showing an individual house difficult, and if the house in question is not integral to the purpose of the map, the house should be generalized off the map entirely.

31.4   DATABASE AND CARTOGRAPHIC GENERALIZATION

The scale of your map will dictate to what extent your map will be generalized; the geometry of your data will determine what types of generalization can be used. The key thing to note here is that every map is generalized—since a map is merely a simplified representation and model of reality, some aspects of reality (typically many) must be trimmed or cut from the representation. Generalization is the simplification, de-emphasis, exaggeration, and enhancement of certain spatial data to make the message of your map clear. The level of generalization necessary is always affected by a map's scale. Unfortunately, many cartographers do not practice extreme prejudice when generalizing and attempt to cram too much data and detail on a map. Thus, the first thing any cartographer should do before they begin to generalize is again ask—what is the purpose of the map and who is the map audience? If the purpose of the thematic map is merely to show broader spatial patterns, then the accuracy of the base map itself may not be of primary concern, but data accuracy will be important. If the purpose is to show exactly where certain thematic variables occurred, but not necessarily quantify the intensity of these variables, then the data may be simplified but the base map must be accurate enough to show relatively precise location. As for the audience, if a map audience comprised mature map readers who are familiar with the data being mapped, perhaps less generalization will be more useful for interpretation. However, if a map audience comprised the general public or a younger audience that is less familiar with the data you are mapping, generalization is of paramount importance to declutter and emphasize the thematic data.

Map generalization can be broken down into two types—*database generalization* and *cartographic generalization*. The first method of generalization involves the filtering of cartographic information inside the database itself, and does not include any operation to improve graphical clarity. This process is never visual but occurs through the application of internal database operations. Perhaps, for instance, one has a database that was created at a scale of 1:25,000. However, for the purposes of the mapping project, this is too detailed. Thus, one might use database generalization to regenerate and filter the same data for a smaller scale, say 1:50,000. This operation would be conducted within the database itself.

The second method of generalization is extremely visual. *Cartographic generalization* is the process of modifying map features to make them aesthetically pleasing and clear at a reduced scale of representation. McMaster and Shea (1992) have defined cartographic generalization as "the process of deriving, from a data source, a symbolically or digitally encoded cartographic data set through the application of spatial and attribute transformations." Central to cartographic generalization are the map purpose and the intended audience. Obviously, if you are designing a map for airplane pilots, too much generalization might be a problem. On the other hand, if you are designing a map for school children, including all of the geographic features found on an aeronautical chart would make the map overly complex and take away from the message you are trying to communicate.

There is no such thing as "perfect" or even "optimal" generalization. However, there are six fundamental philosophical objectives that should inform and drive all types of generalization:

1. Reducing the complexity of the map/database
2. Maintaining the spatial accuracy of map data
3. Maintaining the attribute accuracy of map data
4. Maintaining the aesthetic quality of the map
5. Maintaining a logical hierarchy of map elements
6. Consistently applying the rules of generalization

There are 12 fundamental transformations that can be used to generalize the visualization of spatial data on a map (see Figure 31.1, which illustrates these generalization operations and shows the result of each type of generalization at a 50% scale reduction).

Simplification involves the elimination of unnecessary information. *Line smoothing* shifts and eliminates nodes along the path to improve aesthetic quality. This can be done manually or via mathematical splining and averaging methods. *Merging* is crucial for conglomerating two parallel lines at a reduced scale into one (e.g., two banks of a river into one line representing the entire river). *Exaggeration* and *enhancement* are both operators that actually accentuate certain features in order to preserve their significance under scale reduction. This may be important if a bay or other detailed feature such as the shape of a building's courtyard becomes accidentally simplified or lost due to scale change. Related to exaggeration is the critical operation of displacement, where features are purposefully shifted apart to prevent coalescence under scale reduction. The aforementioned operationalizations are primarily for line features, but others exist for point and areal features as well. *Aggregation* fuses multiple point features together (e.g., individual trees) to create an areal feature (e.g., forest). Along the same lines, *amalgamation* fuses several smaller polygons into one larger polygon. Whereas the *collapse* operator does the opposite, reducing areas into point features. *Refinement* and *typification* are operations that create a smaller-scale symbolic representation of a set of features.

In light of the complexity of taking into account all six of the above-mentioned keys to successful generalization, it comes as no surprise that in recent decades numerous attempts have been made to automate generalization within GISs themselves. This is a tall order for a computer program, particularly on the aesthetic front, and there is still much progress to be made before these operations will be efficient and effective in GISs.

31.5 MEASUREMENT LEVEL OF SPATIAL DATA

To determine how best to symbolize our spatial data—both via geometric symbolization (point, line, or polygon) and graphic manipulation (the manipulation of

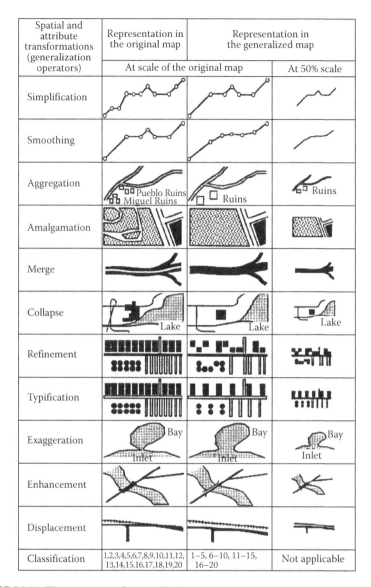

FIGURE 31.1 The processes of generalization.

the visual variables, to be discussed in the next section)—we must first analyze the measurement level of the data. Cartographic data can be categorized into the four established measurement levels of *nominal, ordinal, interval,* and *ratio.*

At their most basic, data can be measured at the nominal level. Nominal data is merely the classification of data as something—labeling. Data measured at the nominal level are completely qualitative, meaning they cannot be compared to one another mathematically. Examples of nominal-level data include land-use data (e.g., recreational, housing, commercial, industrial), race (White, African-American,

Asian), or ground coverage (forest, grassland, human-made). Like comparing apples and oranges, nominal-level data are unique in that all classes lie at the same level and cannot be compared to one another based on their own inherent attributes. This fact is crucial to remember when symbolizing this type of data, as one must not accidentally impart value or rank on nominal attributes—we will discuss this further in the next section.

The next level of measurement, *ordinal*, involves a ranking (or "order"). Ordinal-level data allow for differentiation between higher and lower levels. There is a presumed order in the data, but the level of difference between these rankings *is not* measurable. Examples of ordinal-level data would include a map showing the location and the rating of hotels—from one-star to five-star. It is impossible to quantify the difference between a one-star hotel and a five-star hotel, but most people would inherently choose a five-star hotel over the alternative (finances permitting, of course).

Interval- and *ratio-level* data are similar in nature and separable from other levels of measurement due to the fact that they are metric; that is, interval and ratio levels involve true numbers that can be used for purposes of statistical measurement and comparison. The difference between these two levels of measurement is dependent on the meaning of "zero." Interval-level data do not have a true, absolute zero-point. Ratio-level data do. Interval-level data are generally found when dealing with raw measurements socially constructed by humans. Ratio-level data include any type of derived data (e.g., any time there is a percentage, there is an absolute zero), as well as data that cannot be negative. These helpful guides do not make determining the difference between interval- and ratio-level data any easier, though. Take, for example, elevation. This concept would seem to have an absolute zero. Moreover, it would seem to be completely natural—not a component of human contrivance. We can define elevation as being associated with intervals, not absolutes. Where is the zero? Sea level. What is sea level? Why did we decide that sea level would be zero? It is an arbitrary zero. Other examples of interval data are temperature (excluding Kelvin) and time (unless you calculate your workday based on the Big Bang). As already mentioned, ratio-level data includes any derived numbers (which are "rates"), as well as any other type of data with an absolute zero. Age is ratio data. You cannot be negative years old. Precipitation is also ratio data; though, we arbitrarily choose how to measure rainfall (inches or centimeters), you cannot have negative rainfall.

These four levels of measurement can be broken down into two broader data types—qualitative and quantitative data. Nominal-level data are qualitative; they cannot be compared to one another. The three other types of data levels are quantitative; some type of statistical comparison is possible. Before you begin symbolizing your data, it is crucial that you analyze the level of measurement of the thematic data you are mapping. As will be illustrated in the next several sections, if you fail to do so there is a good chance that you will use an incorrect method of symbolization and confuse your map reader. Thus, please add the level of measurement to the list where you wrote down the map purpose and intended audience. It will come in handy.

31.6 FUNDAMENTAL VISUAL VARIABLES

Unlike natural phenomena, which can occur volumetrically, paper and computer screens constrain our ability to represent spatial data in three-dimensional form. (Although, as will be discussed, virtual and Web 2.0 technologies are changing our ability to "visualize" and interact with the third and fourth dimensions.) Thus, cartographers have traditionally identified three types of cartographic symbols—point, line, and area symbols. (Volumes are generally represented with skewed area symbols appearing three dimensional to the human eye.)

The use of visual variables allows us to "design" and manipulate the geometric properties of points, lines, and polygons to represent certain data variables. For well over 100 years, cartographers have been adapting and manipulating different visual variables to create thematic maps. Only in the past few decades, however, did cartographers begin to systematically categorize the different visual variable techniques that can be used, and associate these techniques with different types of data levels. Bertin is recognized as the first person to point out the prime visual variables with his influential book *The Semiology of Graphics* (1983). Many others have added to Bertin's original visual variables and modified his original definitions to better suit cartographic endeavors. The visual variables reviewed here and in Figure 31.2 are contrived from the work of DiBiase et al. (1992) and Muehrcke and Muehrcke (1992).

The key thing to note is that *depending on what level of measurement your data represents, certain visual variables should never be used*! Using our analogy of

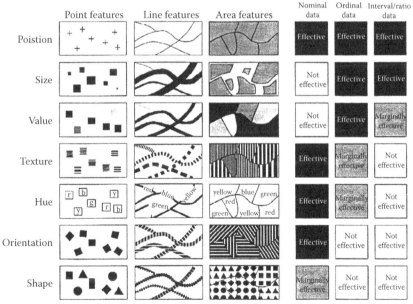

The visual variables and their effectiveness in signifying the three
levels of measurement of data (after Bertin [1983]).

FIGURE 31.2 The visual variables.

maps as communication, geometric map symbols are best thought of as the words comprising the graphic message; visual variables act as the grammatical framework structuring the words. For example

> The grammatical does fails not map of make if just visual to as established follow a use one the and order communicator rules will make one if does structure use variables English as sense correctly not.
>
> Just as English does not make sense if the communicator does not follow the established rules of grammatical structure, a map will not make sense if one fails to use and order visual variables correctly.

The first indented quote makes no sense to anyone. Yet, it contains the exact same wording as the quotation following it. The same kind of confusion confronts map readers when a cartographer uses cognitively inappropriate visual variables and symbolization for the data they are attempting to represent.

Although we often think of literacy as only concerning spoken and written languages, from birth humans also learn graphical literacy (Dondis, 1973). Although we can all read graphics, we are not natural communicators. Few of us are lucky enough to have formal training in graphic design. Thus, we need training and practice to effectively harness the grammar—or rules of visual variables and hierarchy—to effectively communicate spatial data. MacEachren (1995) and Slocum et al. (2008) both provide great overviews of how and when to effectively utilize different visual variables. The key thing to note is that the type of visual variable you use, regardless of the geometry of your symbol, must correlate with the data's level of measurement, otherwise confusion is likely to result.

31.7 CHOOSING A REPRESENTATION FOR YOUR THEMATIC MAP

Once you have determined the geometric properties you will use to map your spatial data—point, line, or polygon—your data's level of measurement, and you understand what type of visual variables work with this data, you must choose how to represent these data on your map (Figure 31.3). How does a cartographer choose what type of thematic map to create? Cartographers have been researching the effectiveness of

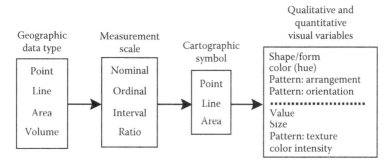

FIGURE 31.3 Geographic data type, measurement scale, cartographic symbol, and the qualitative and quantitative visual variables.

different map types for well over 50 years. Different physiological tests have resulted in a variety of results, some more conclusive than others. Perhaps one of the simplest and most effective models, however, came from the cognitive field. MacEachren and DiBiase (1992) argue that spatial phenomena can be arrayed along two axes: *continuous-discrete* and *abrupt-smooth*. Depending on the spatial nature of the phenomena for which we have collected data, some methods of symbolization and visual variable manipulation are more appropriate than others. Not only are they more appropriate, they will better communicate the data. The reason for this is that the representation of the data on the map will closely mimic the distribution of the measured phenomena in the real world.

An example of the typology of mappable phenomena is provided in Figure 31.4. Each of the boxes provides an indication of the typical data type found along the continuum as well as a specific symbolization type (below the box). For instance, the number of government employees is both abrupt and discrete and would require a graduated symbol. Government employees are abrupt, because the number of employees is contingent upon strictly demarcated political boundaries. This data is discrete because the samples occur in specific places, there are no government employees dotting the landscape; they are typically found in government buildings. Alternatively the average farm size is both continuous and smooth and would normally utilize the isopleths method. (The different methods of mapping will be discussed in upcoming sections.) As Figure 31.4 demonstrates, because farms dot the

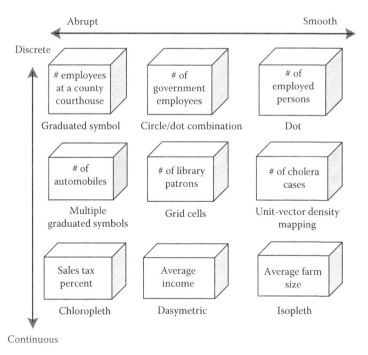

FIGURE 31.4 A topology of mappable data and mapping types. (From DiBiase, D. et al., *Cartogr. Geogr. Inf. Syst.*, 19(4), 201, 1992.)

landscape and sizes can change fluidly, the nature of this data is polar opposite from that of government employees.

Again, a GIS user creating a map must clearly understand the data and the type of phenomena it represents. More importantly, he or she must realize what visual variables and cartographic symbols are appropriate. The misuse of symbolization results in significant errors in map reading and interpretation, easily derailing the utility of the data one has spent time and money collecting. Thus, it is imperative that we review the common types of thematic symbolization found in Figure 31.4 so that you have a better understanding of when and where it is appropriate to use each. The forthcoming thematic mapping techniques—*dot, graduated, choropleth,* and *isarithmic*—have been developed over the past 100 years and are proven to effectively transmit information clearly and concisely.*

31.7.1 Dot and Graduated Symbol Methods

Two common point techniques used to map discrete data include the dot and graduated symbol methods of thematic representation (Figure 31.5). Dot mapping provides an excellent visual technique for viewing the clustering, dispersion, linearity, and general pattern of a distribution and is often applied to both population and agricultural data. The cartographer must determine both the size of the dot itself, and the unit value, or how many items equals a single dot (e.g., 100 acres of wheat harvested per dot). By taking into consideration the location of environmental or social constraints on wheat production (e.g., wheat is not grown in cities, nor where there is enough precipitation to grow a higher value crop such as corn), one can disperse the dots in relevant areas to more accurately show the distribution of wheat production.

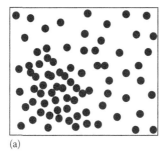 

(a)   (b)

FIGURE 31.5 The (a) dot and (b) graduated symbol mapping technique.

---

* Many other types of thematic maps also exist, but these four are the most commonly used ones. Flow maps, for example, can be used to create qualitative or quantitative representations of exchanges between two or more places—something that is difficult to do with the techniques reviewed here. However, flow maps are often not created, because they can be extremely time-consuming to make in modern GIS systems. For a review of other symbolization and thematic map types, including methods for portraying multivariate data, see the introductory books by Dent (1996), Robinson et al. (1995), or Slocum et al. (2008).

The graduated symbol method is excellent for data that are large in number, but close in space. For instance, when mapping the total shipping tonnage brought into ports on the eastern seaboard, the graduated symbol technique is ideal due to the density and size of the data points. Proportional symbols are unique in that their size is directly proportional to their data values. The proportions used to create the sizes can differ depending on the circumstances. There are three choices for proportional symbol creation. *Mathematical proportioning* creates a direct mathematical ratio between a data point's value and the size of the symbol compared to other data symbols. *Perceptual proportioning* skews the sizes of symbols to take into account humans' incapability of accurately estimating the size of areas.* Finally, *range-graded* proportional circles limit the number of proportional circles used on a map. A data set's values are broken down into different ranges and a unique circle size is used to represent all data values falling within an individual range. This way, a map can be made less complex, as perhaps you have 150 data points, but only five differently sized circles on the map, with each circle representing a range of data.

31.7.2 CHOROPLETH AND ISARITHMIC METHODS

For continuous data, two mapping techniques are readily available in most GISs—choropleth and isarithmic mapping (Figure 31.6).† The choropleth method involves applying value or color intensity to enumeration units (census tracts, counties, states,

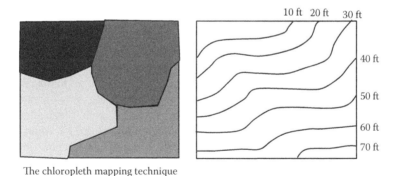

The chloropleth mapping technique

FIGURE 31.6 The chloropleth and isarithmic technique.

---

* A typical method of perceptual scaling is the Flannery method. This option is available in most GISs. Flannery surmised that humans tend to underestimate the size of larger areas, and came up with a formula for creating perceptually accurate areas. Although this formula was the standard method of proportional scaling for several decades, over the past 10 years the Flannery method has been largely debunked, because it fails to take into account the impact of neighboring symbols on one's perception of size.

† The dasymetric technique is often desirable over both choropleth and isarithmic methods, but is rarely used in modern GISs due to software limitations and cartographer time constraints. Similar to what can be done with a dot map, this method masks off areas where data points cannot possibly occur and derives choroplethic values based on these logical areas, not default enumeration units.

nations) based on some statistical value. The higher an enumeration unit's data value, the darker or more saturated the color value. Fundamental to every choropleth method are the concepts of *data standardization* and *classification*.

All choropleth data must be standardized. We repeat a choropleth map may never—*ever*—be used to map count data. If one maps raw data using the choropleth method, the visualization will suffer from an inherent areal bias. Not all enumeration units are the same size; thus, some enumeration units will naturally have more count data than others simply due to their areal extent. For instance, Texas and California have greater populations than Rhode Island or Connecticut. This should not be a surprise—Texas and California have huge areas compared to the other two states. If you standardize the data by area, however, Connecticut and Rhode Island are far more populated when it comes to the number of people per square kilometer. If you are interested in comparing the raw number of people living in states, you should use proportional symbols.

The isarithmic mapping technique involves lines of constant data value, such as elevation or temperature. These maps are constructed through the "interpolation" of point distributions, with each point having a unique data value. By weaving a pattern of constant data values around these points, a trained map reader is able to visualize a surface with peaks and dips in data values. Many types of data can be depicted using isarithmic maps, including population densities, rainfall amounts, and average incomes.

Isarithmic mapping can be further broken down into two types of representation depending on the nature of the data a cartographer is working with—*isometric* and *isoplethic maps*. *Isometric data* are those actually recorded at positions on the Earth's surface, for example, precipitation (collected and measured in individual containers), barometric pressure (collected by different barometers), temperature (collected at weather stations), and depth to bedrock (collected in different places). Thus, each data point represents one individual measurement that one's lines of constant value can weave around. The *isoplethic technique* is used to map data collected via enumeration unit (e.g., at the county or state level). As with choropleth data, isoplethic data are always standardized by area or some other attribute of the enumeration unit (e.g., number of persons). Thus, isoplethic data would include such things as population density, per capita income, and average gasoline price. Unlike isometric data points, whose locations are known, isoplethic data requires the construction of a point around which to weave lines of constant value. Thus, when mapping isoplethic data, a cartographer must first create a centroid in every enumeration unit. For example, if a cartographer working for a city government wants to map car thefts by neighborhood, he or she will first collect the number of car thefts recorded for each neighborhood, standardize this by the number of cars owned in each neighborhood, then find the geometric centroid of every neighborhood, and weave lines of constant value around these data points. If this process sounds complicated, it really is not. Yet, it is more involved in many ways than creating an automated choropleth map, which is one reason cartographers often mistakenly use choropleth maps to map continuous-smooth data when an isoplethic map would better represent the data.

31.8 NECESSITY AND RULES OF DATA CLASSIFICATION

Due to the fact that humans cannot differentiate among many different color values, and that cartographers often have multiple enumeration units that need shading, choropleth maps should use classified data. (Please note, however, that range-graded proportional symbols and other types of cartographic visualizations are also equally predicated on the classification of data.) This section will explain what classified data is and review a handful of classification techniques that are useful for illustrating the nature of your data and communicating the message you desire.

Data classification is the mathematical organization of different data points to a limited number of classes (or groups). For example, one may want to create a grayscale map of European wind power based on the number of kilowatt hours each country produces per person. With well over 30 countries to shade using the choropleth method, a unique shade of gray for each country would make the value and order of different countries' production hours unintelligible. Thus, the cartographer would "classify" or group like valued countries together and represent their data ranges with one shade. Countries with higher production per person would be shaded extremely dark. Countries with lower production would be shaded a very light gray. Data classification is the process of mathematically determining which countries get what shade of gray.

Different classifications of exactly the same data can result in significantly different visualizations and interpretations. This makes it imperative that careful consideration be given to each of the following methods of classification on a case-by-case mapping basis. When classifying data, the cartographer must keep in mind several considerations:

1. The purpose and intended audience of the map. Does the classification method chosen enhance the map's purpose or accidentally cover up the information that is trying to be conveyed? Are the number of classes selected (e.g., seven) suitable for the cognitive level of the audience (e.g., five year olds) or will the map be too difficult to interpret?
2. The range of data. Does the classification system encompass the full range of data, including both the minimum and maximum values?
3. No overlapping values or vacant classes. Each data value must fall in one class only. Vice versa, class created must have at least one data value fall within it.
4. The number of classes must be great enough to avoid sacrificing the accuracy of the data, but not be so numerous as to impute a greater degree of accuracy than is warranted by the nature of the collected observations. For example, if you are mapping the 13 provinces and territories of Canada, you should not have 10 classes (i.e., more than one class per two enumeration units) nor only two classes (i.e., dividing the 13 units into only two groups).*

---

* One useful technique for determining the number of data classes is the *Sturges Rule*. This rule states that the number of data classes (or x) should fall between $2^n < x < 2^{n+1}$. Thus, for the 50 states of the United States, the two values of n would be $(2^5 = 32) < 50 < (2^6 = 64)$. For 50 observations, the value x falls between 2^5 and 2^6, and either five or six classes would be needed.

5. Divide the data set into reasonably equal groups of observations.

6. Have a logical mathematical relationship if possible (Robinson et al., 1995).

31.8.1 DIFFERENT CLASSIFICATION METHODS

There are many methods and techniques for placing data values down into different classes. We will review some of the most common and rudimentary ones available in nearly all GISs. At the top of the illustration in Figure 31.7 is a set of 20 numbers arrayed along a value line from 10 to 85. The data values range from 11 to 84—quite a wide range for only 20 different values!

It is good practice to look at the dispersion of your data values along the histogram before doing anything else. Notice how the two highest values are outliers at 82 and 84; there are no data points falling in the 70s at all. The next thing cartographers must do is to figure out how many classes they want to create. Let us suppose that the audience is an atypical one: adults with training in map comprehension! Using *Sturgis's rule* (see Footnote 8), we have decided that five classifications are ideal for our data set.

Beneath this histogram are three different methods for classifying data. Each class has a number underneath it representing how many data values fall into each category. Many GISs will allow you to view a histogram and select from a variety of classification methods. The three most common classification methods are *equal interval*, *quantiles*, and *natural breaks*.

An *equal interval classification* assumes equal distance, or "range," between the class breaks. In this instance, the equal distance has been set at 15 units, which would yield class breaks of

Class 1:	10–25
Class 2:	25–40
Class 3:	40–55
Class 4:	55–70
Class 5:	70–85

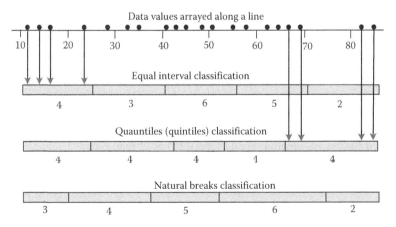

FIGURE 31.7 Data classification technique using equal intervals, quantiles, and natural breaks.

Alternatively, a quantiles classification puts an equal number of observations in each class. Thus, if there are 100 observations in a data set, and the user desires five classes (or in this case "quintiles"), 20 observations will be placed in each category, regardless of whether the class breaks apart observations that are very close to one another or, conversely, forces some that are far apart together. In the example provided in Figure 31.7, a quantiles classification is calculated (20 data values/5 classes = 4 observations per class), yielding class breaks of

Class 1:	10–25
Class 2:	25–41
Class 3:	41–52
Class 4:	53–66
Class 5:	66–85

Notice that the range of the fifth class has been extended quite a bit, and our outliers in the 80s will now be shaded the same color as several of the values in the 60s.

Finally, a natural breaks classification may be calculated. Here, the user selects the maximum "breaks" along the number line, or where the significant gaps appear in the data set. The idea is to minimize the internal variation of the data set, while maximizing the variation among the classes. Although the breaks are typically determined through graphical methods—number lines, histograms, and frequency curves—George Jenks created an "optimal classification method" that used an algorithmic approach to determine these ideal breaks (Slocum et al., 2008). For the sample data set, a possible series of breaks (found by inspecting the number line) might be found at 20, 38, 52, and 75, yielding classes of

Class 1:	10–20
Class 2:	20–38
Class 3:	38–52
Class 4:	52–75
Class 5:	75–85

Many other classification methods exist, including those using nested means, standard deviations, and area-under-the-curve calculations. The references listed throughout this chapter provide greater details on most of the techniques used by geographic information specialists and cartographers. The key thing to take away is that the user of GIS must be aware of both the classification methods provided by the GIS application and the limitations of those methods. The worst thing that a cartographer can do is simply select a classification method based on convenience or what looks "right," as this rarely leads to an effective or meaningful representation of the data.

31.9 TYING IT ALL TOGETHER: GRAPHIC DESIGN, HIERARCHY, AND FIGURE GROUND

We have reviewed what goes into preparing and shaping the spatial data one is mapping. We have reviewed scale, generalization, visual variables, symbolization, and data classification. What remains to be done is to present this information in a clear

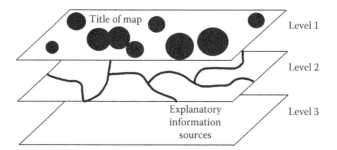

Level 1

Level 2

Explanatory
information
sources

Level 3

FIGURE 31.8 Figure–ground relationships and the visual hierarchy.

and easily received manner that will help map readers avoid any misunderstanding. This section will give a brief overview on map design, visual hierarchy, and figure ground (Figure 31.8).

Again, before beginning to design the layout of a map, the purpose and the audience should be the first thing that pops into the cartographer's mind. What is the map attempting to communicate? What visual style will best reflect the rhetoric of the data or your message? It is imperative that the cartographer present the mapped elements in a logical manner. Not necessarily logical to the cartographer but to the eventual map viewers.

A map should never be presented as a uniform graphical plane, but one that visually emphasizes certain features and deemphasizes others. This is particularly true when it comes to allowing a map reader to reference what it is they are supposed to be looking at and what is merely tangential information. The information and the symbolization of primary importance should be promoted and the supplemental information of tertiary worth subdued graphically. The concept of figure ground pertains to the visual differentiation between the prominent features on a map (figures) and the background components (grounds). Dent (1996) systematizes the approach to the creation of effective figure–ground relationships on maps by arguing that one must establish a hierarchy among the features that are to be mapped. This hierarchy in turn can be matched with an established set of cognitive visual levels (see Figure 31.8).

The differentiation of map levels follows an orderly progression. The actual thematic symbolization (e.g., graduate circles, choropleth values, dots) should be positioned at the highest visual level. These should be emphasized so that they become the figure. Certain map elements that are instrumental in helping map readers decipher the information should also be prominently displayed at this level (e.g., the title, legend, and cartographic labels). The ground should follow at a lower level, with the base map (e.g., political boundaries, cultural and physical features, and land–water boundaries) deemphasized but still available should someone need to reference the location or relationship among several data points. All other cartographic information should even be further subdued so that it does not take away from the spatial information on the map; such information may include map credits, a north arrow, graticule, logos, or similar items.

Although GISs have become incredibly helpful in almost all facets of cartographic map symbolization, their increasing ubiquity has also resulted in a plethora of poorly designed and illegible maps. GISs will not automate map design. Before a map goes to press, or to the Web, one should always double-check with Dent's hierarchy to make certain that the following information visually jumps out in the following order: (1) thematic data; (2) crucial map elements; (3) base map data; and (4) all other data. When maps muddle this order and confuse figure ground, the purpose and message of the map becomes difficult to discern.

31.10 TECHNOLOGICAL DEVELOPMENT AND ONLINE CARTOGRAPHY

Technology has always played a role in the evolution of cartography, particularly cartographic methods (Monmonier, 1985, 2002). Recent advances in computer hardware and software, coupled with high-speed Internet access and the advent of Web 2.0, have changed cartography more rapidly than any other technological developments since the dawn of thematic cartography (MacEachren, 1995; Goodchild, 2007). These technological innovations have had a great impact on the purpose of cartography, shifting it from merely being a tool of spatial communication to a tool of spatial exploration. A new term to describe this expanded capability is *geographic visualization*—or *geovisualization*.* What is this new visualization? According to MacEachren and Monmonier (1992)

> In the context of scientific visualization, 'to visualize' refers specifically to using visual tools (usually computer graphics) to help scientists explore data and develop insights. Emphasis is on purposeful exploration, search for pattern, and the development of questions and hypotheses. Advances in scientific visualization are changing the role of maps and other graphics as tools of scientific investigation.

DiBiase (1990) argues that technological development has ushered in a whole new realm of possibilities with mapping. He breaks mapping down into to two broad uses: *visual thinking* and *visual communication*. Maps have always been about the communication of spatial data; we have drummed this fact sufficiently already; so let us examine what is meant by "visual thinking."

Up until recently, maps were static. One could not interact with them. Obviously this is no longer the case. Due to the fact that maps are becoming user-interactive and are also linked to spatial databases via network technologies, maps can now be used for data mining. DiBiase (1990) hypothesized, and research has begun to support, that this new use for maps may be even more powerful than their original role in communication. Maps help humans develop knowledge visually. Moreover, maps are not merely created for public presentation; one can create maps using GIS for private data exploration. DiBiase (1990) and MacEachren (1995) both view this as a major paradigm shift. Cartographers are increasingly adjusting their research to focus on dynamic cartography and geovisualization technologies (Figure 31.9).

---

* For a comprehensive survey of some recent research activities in this area, see the special issue of the journal *Computers and Geosciences* (Vol. 23, Number 4) edited by MacEachren and Kraak (1997).

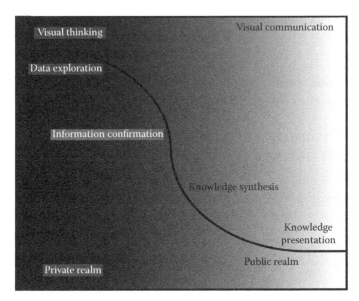

FIGURE 31.9   Shifts in cartographic communication. (From DiBiase, D., *Earth Miner. Sci. Bull.*, 59(2), 13. With permission.)

31.11   NONVISUAL VARIABLES

One of the more interesting advancements in the area of visualization is the potential for four-dimensional cartography. Unlike the "static" maps discussed so far, dynamic maps include the temporal dimension, which is normally visualized through animations. In such a dynamic world, the visual variables based off of Bertin's original work are insufficient. DiBiase et al. (1992) proposes that dynamic mapping offers at least three new modes of cartographic expression:

1. Animation: the illusion of motion created from a sequence of still images
2. Sonification: the representation of data with sound
3. Interaction: the empowerment of the viewer to modify a display

They also propose a revised set of dynamic variables that are useful for understanding animation. These include

1. Duration. Duration is the number of units of time that a scene is displayed. This scene duration may be used as a design variable—longer scenes allow for a more thorough study of a distribution.
2. Rate of change. The rate of change is a proportion, m/d, where m is the magnitude of change in position and attributes of entities between scenes and d is the duration of each scene.
3. Order. Order is the sequence in which the scenes are presented.

Ironically, chronological order is not always the ideal method for exploring a spatial distribution with animation. These three dynamic variables can be used for a variety of visualizations, including *emphasizing location*, *emphasizing an attribute*, and *visualizing change*. Modern GISs and online mapping applications are increasingly incorporating these dynamic capabilities into their software.

31.12   CONCLUSION

The user of a GIS must first be aware of the purpose of the map they are creating and keep in mind the nature of the map's intended audience. Second, a GIS user must be aware of the cartographic capabilities of the system they are using. Rather than take a cookie cutter approach to making a map, they should experiment and explore with the symbology by using different visual variables and exploring different thematic approaches to representing the data. Only through experimentation and tinkering with the arsenal of cartographic techniques and methods available will a GIS user create an intelligible map for his or her audience. Far too frequently, GIS users lack the background in cartography to adequately and usefully present the data they have spent considerable time and energy analyzing. The worst option is to default to customized approaches packaged within a GIS. Only through taking into consideration all of the cartographic steps and concepts outlined herein, and addressed in more detail in the recommended readings scattered throughout the previous pages, can one be assured of creating a useful and enlightening thematic map.

REFERENCES

Bertin, J. 1983. Semiology of Graphics. Translated by W. J. Berg. Madison, Wisconsin: University of Wisconsin Press.

Dent, B. D. (1996). *Cartography: Thematic Map Design*. Dubuque, IA: WCB Publishers.

DiBiase, D. (1990). Visualization in the earth sciences. *Earth and Mineral Sciences Bulletin* 59(2): 13–18.

DiBiase, D. et al. (1992). Animation and the role of map design in scientific visualization. *Cartography and Geographic Information Systems* 19(4): 201–214, 265–266.

Dondis, D. A. (1973). *A Primer of Visual Literacy*. Cambridge, MA: The Massachusetts Institute of Technology.

Goodchild, M. F. (2007). Citizens as sensors: The world of volunteered geography. *GeoJournal* 69(4): 211–221.

Jiang, B. (1996). Cartographic visualization: Analytical and communication tools. *Cartography* 25(2): 1–11.

Lloyd, R. (2000). Understanding and learning maps. In *Cognitive Mapping: Past, Present, and Future*. R. Kitchin and S. M. Freundschuh, eds. London, U.K./New York: Routledge, pp. 84–107.

MacEachren, A. M. (1979). The evolution of thematic cartography: A research methodology and historical review. *The Canadian Cartographer* 16(1): 17–33.

MacEachren, A. M. (1994). *Some Truth with Maps: A Primer on Symbolization and Design*. Washington, DC: Association of American Geographers.

MacEachren, A. M. (1995). *How Maps Work: Representation, Visualization, and Design*. New York: Guilford Press.

MacEachren, A. M. and M.-J. Kraak. (1997). Exploratory cartographic visualization: advancing the agenda. *Computers and Geosciences* 23(4): 335–343.

MacEachren, A. M. and M. Monmonier. (1992). "Introduction," Special issue on Geographic visualization, *Cartography and Geographic Information Systems* 19(4): 197–200.

McMaster, R. B. and K. S. Shea (1992). *Generalization in Digital Cartography*. Washington, DC: Association of American Geographers.

Monmonier, M. S. (1985). *Technological Transition in Cartography*. Madison, WI: University of Wisconsin Press.

Monmonier, M. S. (2002). *Spying with Maps: Surveillance Technologies and the Future of Privacy*. Chicago, IL: University of Chicago Press.

Muehrcke, P. C. and J. O. Muehrcke (1992). *Map Use: Reading, Analysis, and Interpretation*. Madison, WI: JP Publications.

Robinson, A. H. (1982). *Early Thematic Mapping in the History of Cartography*. Chicago, IL: University of Chicago Press.

Robinson, A. H. et al. (1995). *Elements of Cartography*. New York: John Wiley & Sons, Inc.

Slocum, T. A. et al. (2008). *Thematic Cartography and Geographic Visualization*. Upper Saddle River, NJ: Pearson Prentice Hall.

32 Carrying Out a GIS Project

Rebecca Somers

CONTENTS

32.1   INTRODUCTION

This chapter provides guidelines for implementing a geographic information system (GIS). It describes the steps, activities, and issues involved in carrying out a GIS project. In some respects, developing a GIS can be daunting—particularly for big projects. There are so many decisions to be made and so many tasks to do and to coordinate. In other respects, GIS implementation can seem deceptively simple—particularly if one makes simple assumptions. This discussion will put GIS implementation in perspective and help readers understand the GIS implementation process and get started.

32.1.1   GIS PROJECTS AND PROGRAMS

GIS efforts can be classified as projects or programs, ranging from small efforts to large, complex undertakings. A project is a one-time effort, developing a GIS that will serve a specific limited-term use and then be abandoned. Examples of such projects might include the performance of an environmental analysis, the production of maps for a survey, the development of a long-range land used plan, or the design and development of a park. As can be inferred from these examples, however, GISs do not usually disappear when the designated project is over. The GIS developed to plan and construct a facility such as a park or community could then remain as an ongoing management tool for that facility. Permanent, ongoing GISs are usually termed programs. Although they may start small, and may even start out as a project, the goal of these programs is to develop a lasting system or asset that will facilitate the organization's work. A GIS project or program may be small and simple, involving limited software, data, and users; it may be large and complex, involving myriad data sets, applications, and users and complex systems and databases; or it could fall anywhere in between. Today, many GISs are integrated with organizations' other systems, such as enterprise resource planning, computer-aided design and drafting, work management, and electronic document management systems—providing organizations with integrated spatial data management systems.

32.1.2   TYPICAL IMPLEMENTATION APPROACHES

Different types of organizations implement geospatial data and technology differently. The GIS components and implementation approach used by any organization depend on its specific needs, but can by characterized by organization type. See Part V of this book for details related to the use of GIS in various sectors.

Most local-government GISs, for example, are multipurpose, comprehensive, and enterprise-wide systems. They are designed to serve all of the organization's geospatial data-handling needs, integrate all its spatial data, and make the data readily accessible to all users and departments. The data are usually large scale, based on high-accuracy parcel information, although local governments also use some other smaller-scale, generalized data (FGDC, 1997). The implementation approach used by most local governments is a centrally managed, highly coordinated effort that maximizes efficiency and minimizes redundancy and incompatibilities in data and systems. Leveraging GIS efforts and assets is crucial to local governments because many employees, contractors, and citizens use the same data and these systems can be very expensive.

Utilities also develop enterprise-wide geospatial data systems, and the integration extends beyond GIS technology. These organizations typically interface or integrate their geospatial data and systems with facilities management, work order management, customer information, facilities modeling, and other operational data systems. Much of their critical information is high-accuracy, high-resolution data, although they have smaller scale data, particularly because some utilities cover very large geographic areas. Utility GIS implementations are also usually extensive, expensive efforts.

Private sector organizations, on the other hand, take a wide variety of approach to GIS, depending upon their business operations, extent, and geospatial data needs. Obviously, small companies would develop smaller, usually more focused, GISs than would larger companies. However, even for larger companies, approaches vary. Some develop various degrees of integrated, enterprise-wide systems that help integrated data for functions such as marketing and distribution. Others develop several independent GIS projects or programs within the company. For example, the GIS resources developed in professional services firms, such as engineering companies, are usually intended to serve specific projects or clients. The data and applications developed for one project may have no relation to those developed for another project and never will. In this type of situation, GISs are implemented independently, focusing only on each project's needs. This independent development of specific GISs within an organization is a business-tools approach. This approach is also used by other types of companies that develop limited GISs in different areas such as marketing, customer service, and facilities planning. In these cases, even if the GISs do cover the same geographic areas, the costs of developing them are usually low, business units are independent, and the costs of coordination may outweigh the savings achieved by that coordination. However, in situations where the organization finds that it is starting to spend significant resources developing and supporting independent GISs, it may take a service-resource approach to GIS development. In this model, GIS support services and perhaps data resources are developed and made available to the operating units. This approach benefits the operating units and the company as a whole, but does not involve the extensive coordination, comprehensive planning, and enforced standards that the enterprise approach does. In any event, the GIS implementation model must match the organization's business model or implementation will be difficult and prone to failure (Somers, 1998).

32.2 THE GIS IMPLEMENTATION PROCESS

Although organizations differ in their approaches to coordinating their GIS implementation activities and in the resultant components and characteristics of their systems, they follow the same basic implementation process. It is a structured process that ensures that the GIS will meet users' needs.

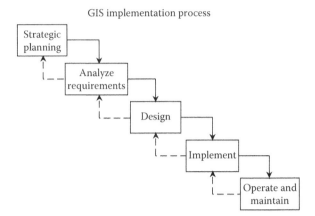

GIS implementation process

FIGURE 32.1   The GIS implementation process. (From Somers, R., *GIS Implementation and Management Workbooks*, Somers-St. Claire, Fairfax, VA, 1993–2008. With permission.)

The process involves five basic phases:

- Planning—strategic planning; establishing the scope and purpose of the GIS and developing an approach and general plan
- Analysis—determining specific GIS requirements
- Design—developing the system, data, and organizational design of the system
- Acquisition and development—acquiring software, hardware, and data and crafting them into the organization's specific system
- Operations and maintenance—putting the system into use, operating it, and maintaining the system and data

This is a form of common system development processes and is illustrated in Figure 32.1. It includes feedback loops from each step because information gained in one task may require going back to a previous step and doing more work on task or modifying its approach. For example, analysis of users' needs in Step 2 may necessitate going back to the planning step and reexamining assumptions and goals or adjusting budget plans. The steps in the process may be repeated in phases at higher levels of detail as the system is built out (Somers, 1996; 1993–2008).

The following discussion provides an overview of these tasks. It usually takes the viewpoint of a GIS program of moderate to high complexity, so that all points can be covered. The tasks discussed are crucial for GIS programs of this type. They also apply to simpler and smaller GIS projects and programs, but can be more easily accomplished for those types of efforts.

32.3   PLANNING A GIS PROJECT AND GETTING STARTED

Planning is the first step in developing a GIS project or program of any size or type. Careful planning lays a solid foundation for all the subsequent steps of GIS implementation and helps an organization or individual avoid costly mistakes.

For a GIS program, the first step is strategic planning. The organization must determine its basic needs and purpose for implementing a GIS, develop the vision and mission for the program, establish goals, and identify the key steps to achieving them (Somers, 2007). For a limited GIS project or very small program, there is often the temptation to skip strategic planning. Whatever the size of the GIS effort, and whether the process is technically called strategic planning or not, it is necessary to address some basic questions before moving ahead.

First, the scope and nature of the GIS must be defined. Will it be a one-time project or an ongoing program? Will it be used for all geospatial data handling in the organization or for only a specific subset of functions, such as mapping? Will most people in the organization use it or will users be limited in numbers or job function? Will this GIS implementation be part of a larger GIS effort? Will geospatial data and technology be integrated with other data and systems? Will GIS significantly change the way the organization does business or will its impact be limited? These types of questions help establish the scope and character of the GIS—Whether it will be a simple limited work tool confined to specific tasks, an enterprise-wide infrastructure that organizes and integrates spatial data and changes the way the organization operates or something in between. The scope of the GIS guides all the remaining planning and implementation tasks.

In addition to specific planning tasks, the planning phase also includes identification and education of participants. The scope of the GIS determines who should be involved in its development. For example, if the GIS will be an enterprise-wide system serving many different types of uses and applications in many departments, then representatives from those areas must be involved. They represent the end-user perspective and their involvement will ensure that all requirements are met. In addition, large-enterprise GISs often require the involvement of top management to provide support. Another perspective to consider is the skills needed to carry out GIS implementation. The scope and nature of the GIS effort will determine the types of background and expertise required to implement and operate it. Required skills might be added through education and training, hiring, or obtaining outside assistance. For GIS efforts of even modest size, it is usually advisable to form a team and assign a leader. Even if things appear simple at this stage, they will become more complex in the course of developing the GIS and a project manager will be necessary.

In order to work effectively, most GIS implementors and future end users require some GIS education. Providing organized GIS educational opportunities in the planning phase is especially important for large projects with many diverse participants. Each participant must acquire an appropriate level of GIS knowledge so they can participate effectively in the planning and analysis activities that lie ahead. This may involve internal briefings as well as participation in outside GIS events. In any case, it is very important that education be tailored to individuals in terms of their backgrounds, how they will be using the GIS, and how they will participate in the GIS development process.

Another key component of GIS planning is developing resource and benefit estimates. Again, the scope of the GIS largely determines these initial estimates. The nature of the uses, the data, and the system provide indicators of costs. Specific cost details cannot be known yet, but general estimates can be developed based on these factors: the number of users; the type of planned GIS use—professional GIS, desktop, or web-based systems; the type and amount of application development, customization, and integration that will be required; and the amount and type of data. These factors will provide indications of general cost levels—whether the GIS will cost hundreds, thousands, or millions of dollars. Likewise, the scope and nature of the GIS provide general indicators of expected return on investment (ROI). Moderate benefits will be gained from map creation applications; higher benefits can be gained from data and map maintenance, depending on the volume; and the highest levels of benefits can be realized from data access, manipulation, and analysis. Together, these estimates can provide a broad picture of the costs, benefits, and time required to implement the GIS.

Planning a large complex GIS may take weeks or months. For these types of projects, a formal GIS strategic planning approach is required. Planning a smaller, simpler effort may take only a few days or less and may be done through an informal, yet still analytical method. In any case, however, questions concerning the purpose and goals of the GIS, the scope of the effort, the approach, the participants and their roles, and resource requirements must be asked and answered before moving on (Somers, 1996; 2007).

32.4   ANALYZING REQUIREMENTS

Although people usually have general ideas of what they need from GIS and the planning task addresses those needs, specific requirements analysis is needed to provide the necessary information for effective GIS implementation. In this task, the future uses of GIS and current geospatial data handling situation are examined in analytical detail. The goal is to identify the functional and data needs of the GIS applications, as well as the organizational environment.

32.4.1   BUSINESS PROCESSES AND GIS FUNCTIONS

First, all the business processes that involve geospatial data and will use GIS in the future must be identified. The GIS scope defined in the planning task guides the identification of these processes and users. If the GIS is being developed to serve only one function, this step will be easy. It is far more complex in large organizations where GIS will be used extensively, and it is common to miss future GIS users when their use of or need for geospatial data is not apparent today. They may be using non-graphic forms of geospatial data and not realize it. It is important to identify as many of the GIS's future users as possible, even if they may not come online for a long time. Identifying their future needs and building those considerations into the GIS design and implementation will ensure that the system can be expanded to accommodate them when the time comes.

A business process analysis approach is very useful in analyzing GIS requirements. This is a structured approach that examines each current (and planned) operation and maps out the steps and the data involved. The method focuses on the purpose, inputs, outputs, and steps of the process and the data used to perform it. In some methods, these are called use cases. Tasks and decisions performed by individuals and work units are mapped out. Specific databases and data items, formats, sources, characteristics, and flows are included. Links to other work processes and systems are also identified. In the course of the analysis, desired changes are also included, based on the incorporation of GIS into the process and other changes the organization may be planning. The goal is to define future GIS-based work processes. This business process analysis method provides all the components that describe each user's GIS needs. Another benefit of this approach is that it prevents the mere transposition of current manual operations into GIS. It helps users and organizations make the best use of the power of geospatial data and technology.

Finally, the organizational environment will present conditions, constraints, and opportunities that affect the requirements and business process analysis. For example, the organization may have standards that must be followed, regulatory requirements to meet, set mandates and operating procedures, or politics that affect the way it does business and the way GIS must fit that business. These conditions must be factored in. They may affect individual work processes as well as the overall requirements for the GIS.

32.4.2 PERFORMING THE ANALYSIS

This analysis is accomplished through user interviews and work sessions. If the planned GIS is small and the end user is the implementor, this task may be relatively simple, but it must still be approached using analytical methods in order to produce the required information. The individual must take a close, critical look at how he or she intends to work and beware of assumptions about data and software. He or she must also be sure to identify the individual work processes and planned uses of the GIS. They may involve different data and functionality requirements.

It is more common that there are many future GIS users involved in many different business processes in different parts of the organization. They have different viewpoints, different missions and work processes, and different needs in terms of geospatial data and processing tools. In this situation, the GIS implementor or analyst works with each individual user or group to perform the business process analysis. The analyst needs skill and experience in business process analysis to do this effectively. This is also where users' GIS education, accomplished in the planning task, becomes important. They need to understand the basic tools that GIS can provide to their work in order to effectively communicate with the analyst to jointly design their future GIS work process and to ensure that all their requirements are addressed.

32.4.3 ANALYSIS RESULTS

The requirements analysis results in a clear, documented specification of the detailed GIS needs of the future GIS users as well as organizational support factors. There is a description and diagram of each future GIS work process and its functional and data needs. Many of these will become applications. Any constraints, opportunities, or problems associated with individual work processes, user groups, or the organization as a whole are also identified.

The requirements analysis also provides additional information about the expected benefits and costs of the GIS. Benefits, such as decreases in work time or staff levels, increases in levels of service, and improved data and decision support, are identified in the course of analyzing each work process and the improvements that GIS will bring to it. Analysis of the work processes also provides a better idea of cost information based on data and functionality requirements, as well as the number of users in each group. The detailed information on benefits and costs can then be used to develop detailed expected ROI information.

Some of this information derived from the work process analysis will contribute to the technical implementation of the GIS; some will be used in the project and GIS management components. All of these products are necessary to build an effective GIS, but they are just pieces—they are not sufficient to go ahead with GIS implementation. They must be put together in the next step.

32.5   DESIGNING SYSTEM COMPONENTS

The design task involves putting the components together: determining the characteristics and combination of software, hardware, data, processes, and people that will be required to meet the organization's GIS needs. The challenge is to meet the organization's overall goals for GIS and the specific needs of the diverse users and applications, while developing an integrated and effective design. Developing this design is a crucial step prior to obtaining and implementing any GIS components.

32.5.1   SYSTEM ARCHITECTURE

The system architecture is a representation of the conceptual system design. It depicts the infrastructure of the system—the main components and how they are related to each other. For example, it would show the major databases and data storage, major applications or groups, system interfaces, user access interfaces, and data creation processes. It would show the relationships among these components and the major routes by which data enters the system and by which users access the system data and functionality. The architecture would indicate which arrangements will be used, such as client/server, centralized DBMS, service-oriented architecture, web interfaces, and federated GIS networks. There are a wide variety of architectural options, so an organization can develop a specific architecture that meets its specific needs most effectively.

32.5.2   DATA AND DATABASE DESIGN

Data are the most important aspect of a GIS and should be given primary consideration. Case studies and industry experience indicate that organizations generally spend the largest portion—as much as 80%—of their GIS budgets on data. Accordingly, the largest portion of effort and consideration should be spent on the data that will become part of the GIS—rather than the disproportional attention many people devote to technology. A GIS is, after all, just a tool to better use and maintain spatial data, so system design should focus on the data needed to do work and how they are to be handled.

GIS data and databases are discussed in previous chapters. (See Chapter 25 on Fundamentals of GIS and Chapter 26 on Geographical Data Structures.) At this point in the GIS development process, it is up to each individual or organization to define their specific GIS database. The data requirements of individual users and work processes are accommodated in an integrated design. This may be one database, a distributed database, or even data residing on different systems with access provided through service-oriented or federated architecture. In enterprise GIS, the goal is usually to develop one version of a shared database that meets all users' needs with minimum redundancy and maximum usefulness and accessibility. The design addresses several aspects of the data.

- Data characteristics are defined to suit the combined system requirements. Each entity is described in terms of type and appearance, format, accuracy or resolution, attributes, links to other GIS entities and other databases, volume, source, maintenance responsibility and standards, and distribution and access.
- Data relationships are identified and described through data models, indicating relationships at the entity level (see Chapter 26).
- Data access and handling requirements are described, ensuring that each user and application will have access to needed data in the required form (see Chapter 25).
- Relevant temporal aspects are identified. These will support applications and data management functions such as time series analysis, planning, backup, archive, and retrieval.
- Metadata are identified at the appropriate level (e.g., feature class or entity level) and in terms of how they will be used.
- The landbase or basemap is defined, based on users' needs. Content, accuracy, and maintenance procedures for this data set will be crucial to most applications and users. This data set usually involves some of the largest creation or conversion costs, but also may provide some opportunities for cost-effective acquisition and/or sharing.

For a large organization developing an enterprise GIS, the database design can be a complex process.

32.5.3   SOFTWARE AND APPLICATIONS

The operational needs of the business processes and of the overall database support environment are examined to derive the required functions of the GIS software and applications. GIS software is discussed in Chapter 28 on GIS hardware and software. As with the database design, this is the point at which an individual or an organization identifies their specific software needs to support their specific applications and environment. This is not yet the point to select software, but to develop a comprehensive description of what will be needed in terms of functionality. There are several important aspects to consider.

- Architecture. What types of tools are needed to support the desired architecture? Granted, this can become somewhat of a chicken-or-egg or necessarily iterative process. The architectural vision determines the software, but the software determines what architectural is possible.

- Applications support. What functions will the applications perform and what basic GIS software tools will be needed to support applications development?
- Data support. What functions and features are needed to support the data design through creation, operation, use, and maintenance activities?
- Data access. What types of data access tools will be needed by users?
- Data integration. What are the data and systems integration support requirements?
- Performance. What are the performance requirements for applications and for other aspects of system operation?

Large organizations implementing enterprise GISs try to minimize the number of different GIS software packages that they implement. Software compatibility is important not just for data sharing, but for system support as well. Therefore, they seek to develop a comprehensive set of system specifications that will help them choose a suite of products that will meet all users' needs. A single user has more latitude in choosing whichever software package will most closely match their specific application needs, but may want to consider outside compatibility and standards for the purposes of future data sharing (and any such needs should have arisen in the planning and the requirements analysis tasks).

Applications will be developed using the chosen software, or perhaps purchased. Either way, the applications software must meet the users' needs while fitting into the overall architecture and software environment.

32.5.4 Integration with Other Systems and Databases

Historically, most GISs have been stand-alone systems. Today, GIS users and developers recognize the need to interface GIS with other systems and data. Requirements, uses, and tools for the full integration of geospatial data and technologies with other data and systems are becoming more prevalent. An individual's or organization's needs for data and systems interfaces and integration are derived in the planning and requirements analysis tasks. At this point, those needs must be examined and defined in terms of the system design—specifically which data or aspects of the GIS must interface or integrate with other data and systems.

A small GIS may truly be a viable stand-alone system. Larger organizations, such as local governments and utilities, however, usually need to integrate their GISs with systems and databases that support functions such as appraisal, emergency response, facilities maintenance, work management, or customer support.

32.5.5 Management Components

Along with the technical aspects of the GIS design, the management aspects must be designed also. Standards and procedures for database development, data maintenance, data management, system support and management, user support, system governance, and project management and coordination must be developed. The particulars will be based on the character and specifics of the GIS design and the environment and users that are to be supported. It is necessary to develop the management design at this stage because it affects other aspects of design as they affect it. For example, if a certain data set, as initially designed, is unsupportable, it must be redesigned now. Another reason that the management components must be designed now is that some of them will be needed soon in the process.

32.5.6   Determining Resource Requirements and **ROI**

What will it cost to develop a GIS and how long will it take? These are probably the most commonly asked questions. The answer for every organization is different. The resource requirements for any particular GIS depend on the organization's needs. So, as with all other GIS components, the resource requirements are derived from the planning, analysis, and design steps.

Initial cost estimates for the GIS are established in the planning task. Then in the analysis and design tasks, details are collected that provide additional information needed to calculate costs. The combined costs for the development and operation of the GIS include the following components:

- Hardware purchase, upgrade, and maintenance
- Software purchase, development, enhancement, and upgrade
- Software support
- Application development and maintenance
- Systems integration
- Data purchase, license, conversion, collection, and creation
- Database development
- Data maintenance and enhancement
- Data preparation
- Quality control and assurance
- Training—initial and ongoing
- On the job learning
- Recruiting and hiring
- System maintenance and enhancement
- Staff
- Consulting and contracting services
- Management time

These development and operational costs are offset by the benefits that the GIS will provide. Tangible benefits include costs that can be avoided by using GIS to provide the needed data or functions and costs that can be reduced by performing tasks more efficiently with GIS. Depending on the organization, benefits may also include income and profit. Intangible benefits cannot be measured and, therefore, cannot be factored into the numerical part of the cost–benefit analysis, but they are often some of the most important benefits and should be identified. Such benefits may include better products, better service to citizens, and better planning or analysis results.

Once the costs and benefits have been identified, they must be transformed into a ROI analysis and an evaluation method that can be used by the organization. Some organizations may focus on total costs versus total benefits and payback periods. Others are interested in comparative measures such as the internal rate of return or the net present value that they can use to evaluate GIS investments with respect to other investments.

GIS implementation, operation, and use will also require human resources. The number of people, type and level of skills needed, and how they should be organized and managed to best support the system depends on the type of GIS that will be implemented. Likewise, the skills that users will need depend on their system use. Training, recruiting, and staffing requirements will be derived from these needs.

Time requirements must also be calculated. All the tasks and expenditures necessary to implement the GIS as designed must be scheduled for development in an implementation plan. As mentioned earlier, large enterprise-wide GISs can take years to complete. Small projects may be finished in weeks or even days.

32.5.7 Developing a Detailed Implementation Plan

Although a general plan is developed as part of the planning task, a detailed GIS implementation plan must be developed based on the detailed information identified in the requirements analysis and developed in the design task. The implementation plan spells out all tasks including data development, system acquisition and development, organizational development, and GIS management. It describes the tasks, schedule, and responsibilities for realizing all details specified in the requirements analysis and design tasks, and provides a roadmap and management tool for doing so.

A typical implementation plan would include several key components.

* GIS vision and scope
* Participants: roles, responsibilities, and organization
* GIS design: architecture, database design, applications, software requirements, hardware, requirements, and integration
* Tasks: data acquisition, creation, and/or conversion; system acquisition and development; application development; organizational development, including staffing and training; and task responsibilities
* Schedule: the schedule of tasks and milestones
* Budget
* Management procedures

The need for a detailed implementation plan and project manager is evident for large GIS projects that can take many years and millions of dollars to complete. But detailed planning is necessary even for small systems. Without a plan, significant money and time can be lost through mismanagement. Documenting even a simple plan helps ensure that all aspects are covered and that implementation proceeds as planned.

32.5.8 Design Results

There are several important products that result from the design phase:

* System architecture
* Database design, including data descriptions, data model, and metadata specifications

- Applications descriptions
- Data/application/user correlations
- System functionality description, focusing on software requirements
- Management and organization design and components
- Budget estimates for system implementation and operation and ROI analysis for entire system and major components
- Implementation plan

All of these components work together, so they must all be developed and completed together before moving on. If emphasis is given to some aspects while others are neglected, the GIS components can get out of sync, sowing the seeds for later problems.

In this five-phase GIS implementation process, the first three phases—strategic planning, requirements analysis, and design—may be considered the planning stage, as they culminate in the detailed implementation plan and actual implementation follows in the next two phases. (This aligns with the 10-stage GIS planning methodology discussed by Tomlinson (2003).)

Some implementation methods defer detailed requirements analysis until after the general system design is completed in the design phase, and couple application design with application and data development in the implementation stage. While this approach may permit the process to move more quickly and deliver some earlier applications, it runs the risk that the system design may not account for some important functionality, data, or integration discovered later in the detailed application analysis.

32.6 ACQUIRING AND DEVELOPING GIS COMPONENTS

Many people find the task of selecting GIS data and system components daunting—understanding and choosing among the many alternatives can become overwhelming. But if the work done in the requirements analysis and design tasks is thorough, then selecting, procuring, and developing GIS components should not be difficult. It may still be complex and time- and resource-consuming, but the decisions and tasks should be a matter of following the specifications and plans developed in the design phase.

GIS software packages provide the basic tools for input, editing, storage, maintenance, management, access and retrieval, manipulation, analysis, display and output of spatial data (see Chapter 28). These tools alone may satisfy some users' needs, and most applications will be built with these tools. So the challenge is to select the GIS package or software suite that best meets the organization's needs. A detailed specification is crucial in doing this. It provides the standard against which all alternatives will be evaluated. (Granted, as the GIS market continues to evolve and products differentiate themselves, choices have become more obvious in some cases. Still, any apparent "obvious" solution should be compared to the specifications or capabilities derived from the organization's requirements.)

Database implementation also presents many options. Depending on the type of data needed, the data sources available, and the costs associated with different sources and methods, an organization may choose to buy, license, collect, or convert the needed spatial data. And among these different sources and methods, there are many choices. As with software selection, the challenge is to find the most cost-effective alternative that meets the organization's needs, and the way to do this is to have a detailed data specification against which to compare alternatives. If the organization is developing or converting the data itself, then the detailed specification provides the guidelines for performing the operation.

A variety of system and data development activities may also be required to complete the GIS components. Major items may include applications and other software development, database design, and system installation and integration.

In all cases, the ROI of alternatives must be considered. While organizations and individuals may desire the most detailed, best quality data the cost may be prohibitive. Likewise, highly customized applications may be appealing but too expensive. The ROI of each component and of the system as a whole is an important factor in making implementation decisions.

32.6.1 PROCUREMENT METHODS

Depending on the organization and the GIS, any of a variety of methods may be employed for the acquisition of system components. Many public-sector organizations must follow formal request for proposal (RFP) procedures. The organization may supply required "boilerplate" content and procedures, and the GIS design and plan components comprise the technical specifications. Other organizations are free to acquire products and services through less-formal means. In either case, however, GIS products and services specifications should be thoroughly documented and the alternatives should be evaluated against those specifications and potential ROI.

A large, formal evaluation process usually involves evaluating written responses to the RFP specifications, and then meeting with short-listed vendors to conduct a more detailed evaluation. Evaluation criteria include not only the vendor's response to the specifications and requirements, but their demonstrated ability to meet the organization's needs, experience, track record, costs, and other factors identified in advance. Implementors of small GISs may not have the resources or opportunities to conduct such an in-depth evaluation, but should still screen vendors according to documented requirements, and then spend extra time looking closely at a small set of viable alternatives.

Managing GIS product and services contracts effectively is as important as making the right vendor selection. Clear contract operating procedures must be developed, along with a plan and schedule. Reports must be regular and useful. Vendor responsibilities and deliverables must be clearly specified. And the client's responsibilities must be also specified and agreed upon.

32.6.2 DATABASE DEVELOPMENT

For many organizations, database development will be the biggest part of their GIS effort. Therefore, to obtain cost-effective and timely data development, many organizations contract out the creation and conversion of their data. Local governments, for example, often contract out their basemap (planimetric, topographic, geodetic control, and digital orthophoto) creation and parcel data conversion. Utilities often do the same for their landbase and facilities data. Other types of organizations acquire GIS as a tool to handle the data they create or obtain in the course of their operations or projects. In these cases, there may be little or no basemap development. Still other organizations, particularly those performing "business GIS" applications buy or license much of their data (Somers, 1996). Chapter 25 discusses GIS data development. It is up to the organization to tailor basic spatial data input processes to suit their specifications.

Whatever the method or source for data development, quality assurance must be given prime attention. Quality control requirements should be built into the RFP or data development specifications; the vendor must respond to them, assuring the quality of the data throughout the delivery process; and the user must take responsibility for verifying and maintaining data quality (see Chapter 30).

32.6.3 PUTTING THE GIS INTO OPERATION

Except for limited projects, putting a GIS into operation is a fairly lengthy process that requires careful management. An organization has work that it must continue to do without interruption or slowdown while the GIS is being implemented. And for large GISs, the implementation process can take months or years.

It is usually advisable to conduct GIS implementation in phases. In large GIS efforts, there is simply too much data and software to implement all at once. Deciding which data to develop first and/or which users to supply with system access first relies on the analysis done in the design task. That analysis should have revealed factors such as which applications and users need which data sets and software functionality and what the costs and ROI of those applications will be. These factors, in addition to other organizational priorities and opportunities, will reveal advantageous starting points. Organizations often choose to implement those parts of the system that require the least amount of time, money, and effort in order to get early, cost-effective benefits from the GIS. The demonstration of early GIS use can often satisfy many users and managers and build support and resources for further development.

For more complex applications and aspects of the system, it is often advisable to develop a pilot project before proceeding with full implementation. The pilot project comprises a representative, yet relatively small, set of the data, system, applications, and procedures. It gives the GIS implementors and users the opportunity to evaluate the software, data, and procedures and make necessary changes before committing full funding and effort.

In addition to planning phased development and conducting pilot projects, there are other considerations and methods for introducing GIS operations into the organization, including developing test systems, staging databases, operating parallel systems, managing switchovers, and other proven information technology (IT) implementation methods. In most cases, putting a GIS into operation must be viewed and managed as a process, not an event.

32.6.4 IMPLEMENTATION CONTRACTORS

As can be seen from the discussion so far, planning, designing, and implementing a GIS can be a complex task, requiring a great deal of technical knowledge as well as management skill. In order to simplify the process, some organizations hire implementation contractors after they have done their initial strategic planning. This approach puts the responsibility for technical knowledge and tasks, task coordination and management, and ultimate system design and implementation in the hands of one contractor, with the owner organization retaining project management control. While this removes the need for the system owner to have the technical skills to perform the tasks, it can also remove some of the options and decisions. The relief from technical requirements may be a benefit, but the relief from decision making can be a mistake. So if this route is taken, the contract must be structured so that the full set of alternatives is considered at each stage; the client is fully informed of the requirements findings, alternatives, and potential outcomes; and the client participates fully in the decision making. Another alternative contracting approach is to have consultants perform individual stages of the implementation process—planning, requirements analysis, and design—with full review and decision points—and even contract limits—at the end of each stage.

32.7 MANAGING GIS OPERATIONS

As the GIS project moves from development into operations, the main goals become maintaining the asset and supporting users. A GIS has a life cycle, as does any system or facility. Throughout the phased development, as well as once the system is in full operation, the system components must be maintained, added to, and phased out when necessary. There may frequently be new technology, data, applications, and users, and the cycle of planning, analyzing, designing, and implementing must be continued for effective incorporation of new components and maintenance of existing ones.

In large GIS environments, many of the GIS operation tasks are similar to IT management tasks: user support, trouble shooting, training, system management, network management, data management, vendor support coordination, configuration management, system backups, and so forth. And, in fact, many organizations are now incorporating GIS management into their IT operations. However, due to some of the special characteristics of spatial data and technologies, certain aspects of GIS system management require special attention.

While all IT managers are concerned with data management, this responsibility presents additional challenges to GIS managers. Spatial data represent a very expensive and valuable asset. Spatial data and technologies are also powerful tools. Therefore, such matters as data stewardship, data maintenance standards and procedures, metadata, and data access and security are vitally important (Somers, 1996). Developing equitable and manageable standards and procedures for funding ongoing GIS system operation, maintenance, development, and data access is also an important aspect of GIS management.

32.7.1 GIS ORGANIZATION AND GOVERNANCE

In an enterprise GIS, organizing and working with participants is an important aspect of implementing and managing the GIS. The participants have many different viewpoints and needs for the GIS. Correctly identifying and working with them is important for system planning and requirements analysis, as well as for ongoing system operation.

A common model used by most organizations implementing GIS is a multitiered committee model that appropriately involves the participants in different aspects of the system and facilitates coordination and communication. This model is represented in Figure 32.2.

In this model, users are organized into three groups. An executive committee provides policy guidance and support to the GIS; a technical committee provides the input and technical guidance for development and operation of the GIS; and a users group (often formed once the GIS is operational) provides a forum for user discussion and input. The GIS Manager coordinates these groups in addition to managing the GIS implementation and operation (Somers, 2001; 1993–2008).

This model also embeds the governance of the system—how it is managed, how users and stakeholders participate in planning, how data is maintained and by whom, and how decisions are made.

32.7.2 SYSTEMS OPERATIONS

One of the main responsibilities of the operational stage of the GIS life cycle is ensuring that the system is operational, available, and reliable. As mentioned earlier, this involves many traditional IT tasks adapted to the GIS environment including system administration; data administration and database management; data access, dissemination, and distribution; system and data security; and system enhancement, upgrade, and migration (Somers, 2001).

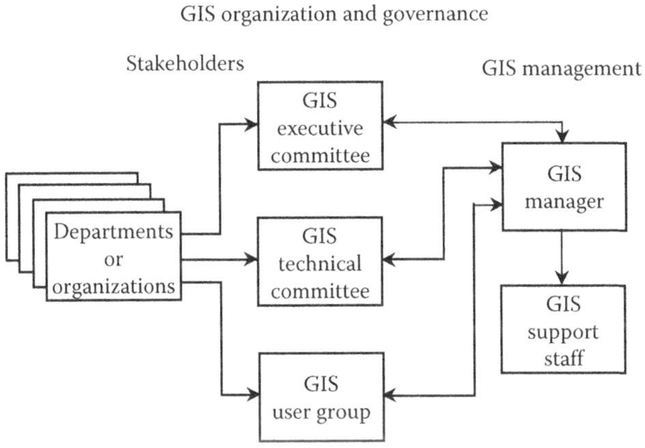

FIGURE 32.2 GIS organization and governance. (From Somers, R., *GIS Implementation and Management Workbooks*, Somers-St. Claire, Fairfax, VA, 1993–2008. With permission.)

32.7.3   DATA MAINTENANCE

As discussed earlier, data is the most important asset of the GIS. During database design, maintenance procedures and responsibilities are established. This includes the identification of data stewards and the development of data standards. During the operational phase of the system, these procedures, standards, and responsibilities are put into place and become operational. Actually, as data is first converted these processes are initiated, as the data must be kept up-to-date to be useful.

32.7.4   USER SUPPORT AND TRAINING

Users are trained as they begin to use the system and user support and continued training are important ongoing activities. The GIS must support the users' business processes. In order to do this, the users must be adequately trained in a timely manner, and provided with effective means of support. This may include a GIS help desk or GIS help integrated with the IT help desk. It may involve placing GIS representatives in users departments. It includes ensuring that the system, applications, data, standards, and procedures are well documented and that that information is readily accessible. Some information may be build into the system applications or data. The bottom line is that adequate support must be provided so that the users' work is enhanced by GIS, not impeded.

32.7.5   POLICIES AND PROCEDURES

Effective policies and procedures form the foundation for the successful implementation, operation, and management of a GIS. They document the decisions made and the standards that have been set and inform everyone during all phases of system development and use. They facilitate communication and provide a mechanism for coordinating activities. The most important aspect is that sound policies and procedures ensure a reliable system and data.

Policies cover the GIS goals, participants' roles and responsibilities, data management principles, data access and distribution guidelines, cost distribution, and system governance. Procedures provide details on system operation, system standards, data development, data standards, data maintenance, data access and distribution, planning procedures, system change procedures, quality control, and user support.

32.8   GIS IMPLEMENTATION CHALLENGES AND SUCCESS FACTORS

Several factors present GIS development challenges, but are crucial to successfully developing a large GIS project or program.

- Effective planning. A formal strategic planning methodology is required for large complex projects.
- Requirements-based GIS development. GIS software and data will be big investments. It is crucial to understand the organization's entire set of

requirements and specifications before selecting or developing data and software. Mistakes caused by hasty or ill-informed decisions or assumptions can be very costly.

- Skillful GIS leadership and management. A large project with many participants requires leadership as well as good project management.
- Upper management support. Large projects that involve many different user groups or departments and will entail significant resources require upper management understanding and support.
- Starting small. Demonstrating meaningful results early in the process and for minimum expenditures has been a key success factor for countless GIS projects.
- User involvement, coordination, and communication. Effective structures, policies, and procedures must be developed for organizing, coordinating, and communicating with the user community. Involving users in the design of the system is also crucial for ultimate GIS acceptance and adoption (Eason, 1988).
- Education and training. Ensuring that GIS implementors and users and organizational managers have adequate GIS information and training is crucial to successful GIS development, adoption, and use.
- Change order management. A GIS project can get swamped with constant requests for additions and changes, particularly because as participant's GIS knowledge and exposure grows, so do their demands. Effective change order management is crucial to keeping it on track.
- Managing risk. Assessing organizational risk at the outset of the project and developing implementation methods that contain and minimize it increase the chances of project success (Croswell, 1991).
- Politics. Successful navigation of the organizational political waters is necessary to gain and maintain support for the GIS.

Small GIS projects and programs have fewer or smaller challenges, but still require attention to a few key points in order to be successful.

- Planning and analysis. Many people undertaking small GIS projects feel that planning and analysis are not necessary. They believe they know what they need from a GIS and how to go about building it. By the same token, however, if they do have a clear idea of what they need, then planning and analysis should be fast and easy. But these tasks are still necessary to ensure that all important aspects are covered and that small, seemingly obvious decisions do not lead to big problems.
- Following the implementation process. Likewise, it is important to follow all the steps of the GIS implementation process. The tasks may be easy and the process may go quickly for small systems, so it should not be a burden. But it is a safeguard in making GIS decisions and investments.
- Data quality. Developers of small systems often have limited resources and end up obtaining or developing inadequate data. Furthermore, they invest additional money and time in those data. Data are an investment, and it is important to look ahead and ensure that data that solve immediate needs

have longevity. Again, the steps and methods of the GIS implementation process help implementors do this.

- Avoiding unnecessary expenditures. The flip side of underinvesting in GIS is overinvesting. Limited GIS projects may be able to make due with limited GIS data and software. However, some users spend more money on more sophisticated software and data than they need and more time learning and using those systems than they would have spent on simpler systems. Again, the analyses done in the GIS implementation process will help implementors of such projects select appropriate GIS components (Somers, 1993–2008).

REFERENCES

Croswell, P. 1991. Obstacles to GIS implementation and guidelines to increase opportunities for success. *Journal of the Urban and Regional Information Systems Association.* URISA, Washington, DC, 3(1): 43–56.

Eason, K. 1988. *Information Technology and Organizational Change.* Taylor & Francis, London, U.K.

Federal Geographic Data Committee. 1997. *Framework Introduction and Guide.* FGDC, Washington, DC.

Somers, R. 1993–2008. *GIS Implementation and Management Workbooks.* Somers-St. Claire, Fairfax, VA.

Somers, R. 1996. How to implement a GIS. *Geo Info Systems,* 6(1): 18–21.

Somers, R. 1998. Developing GIS management strategies for an organization. *Journal of Housing Research,* 9(1): 157–178.

Somers, R. 2001. *Quick Guide to GIS Implementation and Management.* URISA, Park Ridge, IL.

Somers, R. 2007. GIS strategic planning: Evaluating and selecting approaches. *URISA Annual Conference Proceedings.* URISA, Park Ridge, IL.

Tomlinson, R. 2003. *Thinking about GIS.* ESRI Press, Redlands, CA.

33 Geographic Information Science and Society

Robert B. McMaster and Francis Harvey

CONTENTS

33.1 INTRODUCTION

Whereas as recently as the 1990s maps could be considered to be ubiquitous, by the end of first decade of the twenty-first century, geographic information instead has become ubiquitous, accessible to a large part of our society. This is true at least in many parts of the world—with other parts just a few years behind and in some cases even leapfrogging ahead. Society, institutions, and people are involved and affected in myriad ways—influences, impacts, and institutions are all issues. Geographic information system (GIS) and society research has both examined these developments and played a role in helping to conceive geographic information technologies.

At its root, GIS and society scholarship considers how geographic information technologies impact people and organizations and, inversely, how people and organizations use the technologies. This more broadly also examines how interactions between technologies and society and GIS enable societal transformations. Maps for every state and urban area are commonplace, but instead of free gasoline station maps found everywhere 30 years ago in North America, maps are now available with a few mouse clicks and are found in advertisements, newspapers, flyers, etc., at every turn and every day. But more changes in how people access geographic information are already being written on the wall. However, a backlash against the limitations of displaying maps on the miniature displays of cell phones, handheld global positioning systems (GPSs), and in-car navigation systems suggest that the business of making street atlases is going to stay, just in a different way than before. The map atlas and online navigation will coexist for sometime to come.

In the area of consumer GIS, frequently branded neogeography (neogeo), profound transformations are taking place as scores of entrepreneurs work toward developing the next "killer application" that will take the world by storm. The growth of neogeo comes from the availability of low-cost GPS receivers, high-quality hardware and software for consumers and professional applications, developed on pioneering work in GIScience. Integration with existing geographic information and infrastructures allows consumer applications to reach economies of scales that hold prices to extremely low levels and make GIS capabilities available to many people. Without GIScience, these recent developments would not be happening, yet we are aware from recent studies of GIS that technological abilities neither define nor lead societal developments. The interactions between society and technology in specific situations are quite complex. We do see a number of issues, some, especially accuracy and scale, already described in the 1980s, that are key to the use of geographic information technologies and thus a fundamental part of GIScience studies of GIS and society (Sheppard, 1995; McMaster et al., 1997; Curry, 1998; Sheppard et al., 1999; Pickles, 2004). The renewed concern about accuracy in consumer GIS applications points to their centrality and the awareness of their importance for GIS and society research.

We consider three key areas of GIS and society research in this chapter. Section 33.2 begins with an overview of this chapter and five perspectives on GIS and society activities. Section 33.3 focuses on Public Participation geographic Information Science (PPGIS) covering both theoretical and applied work, pointing out the contributions made in these areas, and the important roles of fundamental research. After this, we turn to examine the more recent developments of volunteer geography and consider the potential and issues, particularly accuracy, that it involves. Section 33.6 turns to research on organizations and disciplines that develop and use GIS. Work in this area engages with the foundations of GIS and society (data provision, coordination, collaboration) and also takes into consideration the development of GIS as a professional field of employment with its related societal consequences and potential. Section 33.7 concludes this chapter with a summary of these issues and an outlook on the potentials for further theoretical and applied work.

33.2 ISSUES AND PERSPECTIVES IN GIS AND SOCIETY RESEARCH

Connecting the studies of these various GIS and society issues are several theoretical threads. Theoretical work in GIS and society has gone through a variety of phases; however, the seven themes that emerged from the NCGIA initiative 19 in 1996 remain relevant: (1) the social history of GIS as a technology; (2) the relevance of GIS for community and grassroots perspectives and lifeworlds; (3) issues of privacy, access to spatial data, and ethics; (4) the gendering of GIS; (5) GIS, environmental justice, and political ecology; (6) GIS and the human dimensions of global change; and (7) alternative kinds of GIS.

The 1996 list, for its continued relevance, has been refined by further work. A workshop on GIScience and Technology held at Ohio State University in 2002 (Marble) identified these six society-orientated research areas:

- Geospatial data availability: its sources and influences
- GIS&T workforce studies
- Conditions associated with the adoption of GIS&T-based approaches
- Spatial understandings, or the cognitive ability of people to work with spatial data
- Cross and longitudinal studies of the use of GIS&T
- Improved tools for the societal evaluation of GIS&T activities

As the discussions and engagements evolved, a series of broader issues around the relationship between GIS and society emerged. In the recently published book, *A Research Agenda for Geographic Information Science* (2004), it was proposed that research into the interrelationship between GISs and society addresses several broad questions (Elmes et al., 2004), including

- In what ways have particular logics and visualization techniques, value systems, forms of reasoning, and ways of understanding the world been incorporated into existing GIS techniques, and in what ways do alternative forms of representation remain to be explored and incorporated?
- How has the proliferation and dissemination of databases associated with GIS, as well as differential access to these databases, influenced the ability of different social groups to utilize this information for their own empowerment?
- How can the knowledge, needs, desires, and hopes of noninvolved social groups adequately be represented as input in a decision making process, and what are the possibilities and limitations of GIS technology as a way of encoding and using such representations?

33.2.1 Five Perspectives on GIS and Society

As reflected in the 2004 research agenda book, work in GIS and society has taken five different approaches, including: the critical social theory perspective, the institutional perspective, the legal and ethical perspective, the intellectual history perspective, and the public participation perspective. We discuss each of these in the following text.

33.2.1.1 Critical Social Theory Perspective

A *critical social theory perspective* is concerned with underlying issues of power and limitations in the ways that populations, locational conflict, and natural resources are represented within current GISs and the extent to which these limits can be overcome by extending the possibilities of geographic information technologies;

with the ways in which the nature of and access to GIS simultaneously marginalize and empower different groups in society with opposing interests; and with questions of how the evolution of geographic information technologies reflects both societal structures and priorities as well as the practices of those who develop and utilize the technologies (Elmes et al., 2004; Harvey et al., 2005; Harris and Hazen, 2006; O'Sullivan, 2006).

The social theory approach asks the basic question, "Does the use of geographic information systems and other geospatial technologies benefit society?" As an example, if government agencies have rich sources of spatial information that can be used in decision making—for instance, spatial data for the siting of an industry within a community—and the community itself does not have such information, and perhaps more importantly the power (mapping software) to use such information—does this differentially empower the government and disempower the "society" being impacted? Although this is less of a problem with the slow democratization of spatial data and mapping than before, it is clear that those with rich sources of spatial data and powerful analysis software still have an advantage in making complex decisions.

A major theme of this perspective is that of protecting the confidentiality of people (National Research Council, 2007). How can we make social data available for basic and applied research in public health, criminology, housing, and racial segregation studies while, at the same time, ensuring the confidentiality of the individual herself or himself? As pointed out in the NRC study, "The increased availability of spatial information, the increasing knowledge of how to perform sophisticated scientific analyses using it, and the growth of a body of science that makes use of these data and analysis to study important social, economic, environmental, spatial, and public health problems has led to an increase in the collection and the preservation of these data and in the linkage of spatial and nonspatial information about the same research subjects" (2007, vii). An excellent example of this problem lies in the field of public health analysis where much of the data is only released at a course county-level, whereas methods for identifying clusters and analysis require finer-grained data, most optimally down to the individual. The release of such individual health records is obviously not desirable and perhaps unthinkable.

33.2.1.2 Institutional Perspective

The *institutional perspective* is concerned with the implementation of GIS by institutions; with the costs and benefits associated with implementation, the equity of the distribution of these costs and benefits among individuals and social groups, and the ongoing coordination and maintenance of GI by institutions. It also addresses the development of theories, tools, and techniques for determining the impact of GIS on policy decisions and on expectations about the agencies implementing them and with their impact on interactions between agencies, on citizens' relationships with government agencies, and on people's beliefs and actions with regard to the use and the management of land (Elmes et al., 2004).

The institutional perspective looks at the effects of geospatial data and technologies on the operation, economics, and social interactions when a GIS is implemented. It is clear, for example, that when a large government agency, such as a state department of transportation, implements a GIS it will have enormous impacts on institutional structures and decision making. As a separate GIS or data analysis unit is set up, the agency needs to consider the cost-benefit of using GIS, the education and the training of possible employees, and the relationships with other units with the agency. Within the Minnesota Pollution Control Agency, for example, which utilizes GIS for myriad purposes, decisions were also needed on access to very sensitive sources of information. The U.S. DOT regulates which data from its national inventory of bridge status can be provided.

33.2.1.3 Legal and Ethical Perspective

A *legal and ethical perspective* is concerned with the changing institutional processes and pricing mechanisms governing access to spatial data with the proliferation of proprietary spatial databases; with how these changes are rooted in governmental and legal regulation; with the ethical implications of these changes; and with possible legal remedies (Elmes et al., 2004). The growing use of GIS and geospatial data and technologies has resulted in a concomitant growth in legal considerations. Who is responsible for errors in spatial databases should a tragedy occur such as a shipwreck that results in oil spills or loss of life? What are the intellectual property issues with spatial databases and computer algorithms? Are spatial databases legal documents?

Related issues involve privacy and the surveillance of society. The acquisition of rich databases on all aspects of our personal lives is now commonplace, and used for applications in geodemographic analysis and marketing. Despite governmental controls, the acquisition and the use of such private data seems to be expanding at an alarming rate. What are the surveillance concerns with high-resolution imagery, or even the lower-resolution imagery of Google Earth and other such sites? More recently, there has been an attempt to create a "code of ethics" for GIS by the Urban and Regional Systems Association (URISA). As articulated on the URISA Web site, "This Code of Ethics is intended to provide guidelines for GIS professionals. It should help professionals make appropriate and ethical choices. It should provide a basis for evaluating their work from an ethical point of view. By heeding this code, GIS professionals will help to preserve and enhance public trust in the discipline" (http://www.urisa.org/about/ethics).

The code consists of four primary categories, including obligations to society, obligations to employers and funders, obligations to colleagues and the profession, and obligations to individuals in society. For example, the "obligations to society" section states, "The GIS professional recognizes the impact of his or her work on society as a whole, on subgroups of society including geographic or demographic minorities, on future generations, and inclusive of social, economic, environmental, or technical fields of endeavor. Obligations to society shall be paramount when there is conflict with other obligations" (http://www.urisa.org/about/ethics). The code suggests, "Do the best work possible; Be objective, use due care, and make full use of education and skills; Practice integrity and not be unduly swayed by the demands of others; Provide full, clear, and accurate information; Be aware of consequences, good and bad; Strive to do what is right, not just what is legal." One can argue about the exact wording of such codes, but the existence suggests that the discipline has evolved to a certain state of maturity.

When using geospatial technologies to evaluate environmental justice, for instance, myriad ethical issues must be considered such as the release of maps depicting the very complex and in many ways uncertain relations between hazardous materials and sensitive populations. How will such visual representations affect housing values, insurance rates, and perhaps even transportation patterns?

33.2.1.4   Intellectual History Perspective

A fourth approach to understanding GIS and society is the *intellectual history perspective*, which is concerned with tracing the evolution of geographic information technologies; with the dynamics through which dominant technologies are selected from a variety of potential geographic information technologies at critical points in time; and with the societal, institutional, and personal influences governing these selection processes; and with the question of whether and why productive alternative technologies have been overlooked (Elmes et al., 2004). While some have argued that much of what we have today resulted from research and development in the military-industrial complex, others can point to a richer history that involved actors from the private and governmental sectors. For instance, the concept of topology in spatial databases comes out of the careful work of the U.S. Census in the 1970s in building the DIME files. This, and the work of the Harvard Laboratory for Computer Graphics and Spatial Analysis, institutionalized this particular data model (what morphed into the georelational model) over other possibilities. By the early 1990s, certain paradigms in GIS so dominated that other approaches could not fully succeed. For example, the quadtree-based Spatial Analysis package (SPANS) produced by Tydac Corporation, which provided a method of variable cell-size raster processing, simply could not gain a foothold in the market and failed. In the context of environmental justice, it was the original study by the United Church of Christ Commission for Racial Justice (1987) that initiated the use of geospatial technologies to study the relationship between toxic waste sites and minority populations.

33.2.1.5   Public Participation Perspective

Finally, the *public participation GIS perspective* is concerned with how a broader effective use of GIS by the general public and by community and grassroots groups can be attained with the implications for empowerment and marginalization within such groups using GIS (Sieber, 1996; Elmes et al., 2004).

33.3   PUBLIC PARTICIPATION GEOGRAPHIC INFORMATION SCIENCE

One of the most vibrant areas of GIScience, public participation GIS, has developed research programs and applied projects around the world. Examples integrate researchers, practitioners, and citizens in myriad ways. For example, a recent project at Syracuse University, "The Syracuse Hunger Project," utilized GIS and mapping technologies to understand and help solve hunger-related issues in the city and to deal with nutritional inequalities. The researchers determined the location and operating hours of food pantries, the number of meals served, statistics on participation in government food aid programs, and census data about income, race, and housing quality. The study used spatial approaches to better understand the pattern of hunger, and where resources might be targeted. PPGIS research has made important advances in its engagement with issues for GIS and society research.

This example shows the possible relationship between geospatial technologies and social problems and raises a series of vexing questions on what broadly is called GIS and society. Some of these questions include: Who has access to the data and software? Are certain parts of society empowered, or disempowered, by the use of GIS?

Research on these issues turns to examine the fundamental interactions and processes by which decisions are made and how to best develop technologies to support democratic decision-making. Work by Tim Nyerges and Piotr Jankowski with numerous other contributors has empirically tested the ways that GIS technologies influence decision-making and pointed to improvements (Jankowski and Nyerges, 2008).

As we think about the access to and the use of spatial information by a public, however that is defined, the emerging field of public participation GIS will be very important. Recent research has identified a series of models by which the public can gain access to spatial information, maps, and in some cases analysis itself (Leitner et al., 2000). These six models include

1. Community-based (in-house) GIS
2. University/community partnerships
3. GIS facilities in universities and public libraries
4. "Map rooms"
5. Internet map servers
6. Neighborhood GIS center

It is obvious that the most promising of these six is the Internet map server, where there have been a multitude of sites created over the recent five years, perhaps one of the best known of which is "Google Earth." Such map servers are emerging at many scales, from the local (community-based mapping systems to look at neighborhood characteristics) to the state-level (sites showing the distribution of natural resources) to global-scale sites such as Google Earth. Each of these six models has their own distinct advantages and disadvantages. As noted in a paper by Leitner et al. (2000), each these models of delivery can be differentiated along several dimensions, including communication structures, nature of interaction with GIS, location, stakeholders, and the legal and ethical issues.

33.4 MATURATION OF PUBLIC PARTICIPATION GIS

A fundamental question in the maturation of PPGIS is what do we mean by a public?

When the Minnesota DNR creates and distributes geospatial wetlands data, are these not a form of public GIS? When the Bureau of the Census distributes digital versions of boundary files and statistical information, is this not public? The real meaning here is a public participation GIS provides access at a more grassroots level, a level of communities, of neighborhoods, of even the individual.

For the purpose of this discussion, PPGI Science represents a body of knowledge that parallels GIScience, but with specific concerns toward the broader utilization of these technologies by all involved people, especially the proverbial average citizen, often represented through community groups, grassroots organizations, and other such entities. What, then, are the fundamental questions that are emerging? Some might include

- What technologies are most appropriate for community groups? How does the technological expertise get maintained in communities? How do community groups gain access to the appropriate data?
- What models of access are most appropriate?
- What are the methods of localized knowledge acquisition that are most appropriate? What are new methods—Primary survey work?
- How will these technologies fundamentally change the political/social structures of community groups?
- What forms, and new forms, of representation are best suited for public participation work?
- How do neighborhood groups deal with issues of scale (e.g., relationship with municipal and state regulations)?
- Is there a fundamental difference in the cognition of space and spatial principles among community groups?

33.5 VOLUNTEER GEOGRAPHIES

The availability of user-friendly yet customizable Web services for online mapping has changed the ways that people can access and work with geographic information. Ultimately, this has utterly reconfigured the broader relationship between societal knowledge of place, geographic information resources, and the roles and uses of maps. Coupled with the boom in low-cost GPS equipment, which makes it very easy to collect locational data, what is becoming known as *volunteer mapping* has seen amazing growth in the last 10 years. Volunteered geographic information includes a broad range of digital data including geotagged images online, Wikipedia entries with geographic coordinates and more specialized place descriptions found in Wikimedia, mashups produced with Google Earth, NASA WorldWind or similar virtual earth and mapping applications, to volunteer efforts to create public-domain geospatial data layers, such as OpenStreetMap (OSM). The possibilities for creating user-generated content are endless.

While many of the potentials of volunteer mapping dovetail with PPGIS applications, it is actually distinct as the volunteer aspects focus more on producing geographic information and developing infrastructures, often commercially oriented, although some of the more interesting applications develop counter to governmental restrictions on the availability of GI. In fact, as in the case of OSM, volunteer mapping is actually producing alternative data sets to government sanctioned, funded, and controlled data. As the acquisition and/or licensing costs may be exorbitant for most people, volunteer mapping has attracted a great deal of interest in European countries, where the costs of street and basic topographical data can be very high (map of OSM coverage).

Taking off from this basic description of volunteer geographies, a useful division comes by distinguishing volunteer geographies underwritten by government bodies, on the one hand, and commercial grassroot organized collection and provision of geographic information on the other. Until now, most virtual geographic information fits the latter category, although examples of WaterShed Watches supplementing citizen-collected data with geographical location information and then mapped certainly exist. Governments and nongovernmental organizations already offer data for creating browseable and interactive maps of their data. More prevalent and the driving force behind much of volunteer geographic information are the multitude of citizen group and commercial applications. Examples abound from around the world use these online mapping packages to powerfully visualize their data to support citizen group activities, for example, the Neighbors Against Irresponsible Logging (NAIL), documented online. More popular examples include Wikimapia, Flickr, and scores of other volunteer geographic information applications.

A number of companies are now developing ways to take on roles in commercially orientated social networking and connect them to specific hardware platforms. For example, the application Twinkle, available on Apple's iPhone, allows contacts to follow updates about your location and brief messages that you wish to share. Loopt, another application, allows you to find out where people on a contact list are and what they are up to. Twinkle and Loopt are just two of the seven social networking applications for the iPhone that allow people to track each other in various ways. While there are certainly some privacy concerns to be considered, at the moment these applications are all based on an opt-in licensing model that requires each user to specifically indicate how they want to make their locational information available to a list of contacts and the general public. These are becoming part of location-based social networking, expected to have worldwide revenues of $3.3 billion by 2013.

While sophisticated analysis is seen to be less important than revenue-sharing models for successes in this area, the ability to locate accurately is a larger concern, as well as for other volunteered geographical information. Goodchild (2007) describe inaccuracies of up to 60 ft in the Santa Barbara, California area. Others have described even more significant discrepancies. The problems are further complicated by the lack of information available to help users of these online applications assess accuracy. At present, visible location differences between imagery and data for features, for example a road edge, are the main indicators people have to rely on. With more and more people coming to rely on volunteer mapping data sources, the maintenance of accuracy and easily understood information about its accuracy will become increasingly important.

33.6 ORGANIZATIONS AND PROFESSIONS

GIS and society research has also examined the settings and conditions for the creation and the management of geographic information. Studies in this area have accompanied the development of GIS from even before the acronym was coined. For example, Urban and Regional Information Systems (URISA) founder Edgar Horwood wrote about data-sharing issues back in the 1960s. Since then a number of researchers have advanced work in this area working on improving the underlying institutional and legal issues as well as critical engagements with the assumptions about organizations and data sharing. As the rapid growth of GIS led to discussions and tensions with related disciplines surrounding the professionalization of GIS careers, a number of people began examining and developing resources and guidelines for the nascent profession. The GIS Certification Institute (GISCI), creating standards for professional GIS practice, is an outcome of these activities.

Through the 1970s and into the 1980s, the GIS applications boom for public administration purposes continued. Notable in its merging of organizational issues and technological capabilities is work on what has become known as the Multipurpose Land Information System (National Research Council, 1980, 1984; McLaughlin, 1984). Organizational studies came along with software and technology development (Chrisman et al., 1984; Chrisman, 1987; Warnecke, 1987; Tulloch Jr. et al., 1996) and became milestones in improving and at times revisiting the frameworks guiding organizational developments. A key resource at the time, William Huxhold's *An Introduction to Urban Geographic Information Systems* (Huxhold, 1991) marks the de facto ascendency of GIS as a profession in its own right. The National Center for Geographic Information Analysis began at this point, among other research activities, on an initiative on the use and the value of geographic information (I-4) followed later by an initiative on sharing geographic information (I-9), which became key meetings in developing this area of research during the 1990s.

During the 1990s, research on organizational issues shifted through these activities to produce a body of work of lasting influence (Azad and Wiggins, 1995; Campbell and Masser, 1995; Onsrud and Rushton 1995; Petch and Reeve, 1999). Often drawing on Rogers' work on diffusion (Rogers, 1983), this research produced the first broad empirical studies of the development of GIS in organizations. A shift in focus paralleling the rise of networked computing, the Internet, and, ultimately, the now nearly ubiquitous World Wide Web, led to a shift at the end of this decade to research examining the interorganizational relationships that Horwood and others had pointed to (Nedovic and Pinto, 2000, 2004; Harvey, 2003; Harvey and Tulloch, 2006). Data sharing was engaged at large scales now as the development of spatial data infrastructures (SDI) (National Research Council, 1994; Federal Geographic Data Committee, 1997). Research has become both empirically and theoretically broad with great significance. For example, Wehn de Montalvo (2003) analyzed data-sharing behavior among South African government agencies from a social psychological perspective.

During the same period, beginning in the early 1990s, the growth of GIS activities in government and other sectors had become so significant, that researchers began to focus on issues related to professionalization and the relationships to other disciplines (Obermeyer, 1994; Obermeyer and Pinto, 1994). The dynamics of growth were taking hold and the impact of this work would never take hold in the manner of a manifesto, but became part of a discussion in URISA and related academics about the need to refine a profession of GIS to retain the intellectual core and the vibrancy of the field. Whether GIS is a profession or discipline seems to have been largely answered by developments in the last five or so years, but this differentiation to existing fields, frequently with strong disciplinary identities, goes on. This takes place in various ways around the world. In the United States, the GISCI, with now over 2000 certified GIS professionals, has become a key actor in developing GIS as a profession. GIScience issues, particularly accuracy in its relevance to professional practice and for related fields such as surveying, continue to support research in this area. Of significance for both academic and professional discussions have been work on ethical issues and the often understudied roles of mapping and geographic information issues (Chrisman, 1992; Rundstrom, 1993; Crampton, 1995; Pickles, 1995, 2004; Obermeyer, 1998).

33.7 SUMMARY

The ubiquity of GI in most parts of the world and its growing significance elsewhere continues to put GIS and society researchers before challenges and opportunities. The speed of these changes is astonishing, and we often struggle to keep up with GI-based applications that a few years ago would have come across as science-fiction dreams lifted from the pages of futuristic comics. Fortunately, there is a sound basis and continuation of activities. While the vibrancy of the activities is of great relevance for GIS and society work, we should not lose sight of their origins and continue to follow core areas of GIScience work on PPGIS and organizations.

As this chapter shows, any new technology undoubtedly will have an impact on the society that it is embedded in. Conversely, the society itself has helped to determine the shape and the success of that technology. GISs are now, more than ever, having a significant impact on most societies. Publicly available mapping Web sites, low-cost GPSs, vehicle navigation systems, and myriad other mobile mapping technologies (mobile devices and services using digital maps populated with rich attribute information such as restaurants, hospitals, schools, and retail), coupled with the wide dissemination of free spatial data have changed the way many humans now live. But the opportunities come with costs including differential knowledge that impacts decision-making, a loss of "spatial" privacy, and the increase in spatial surveillance such as the cameras now installed in urban areas at key locations. Research in maps, GIS, and society must continue to look at all aspects of GIS on our society, and in particular how these sophisticated technologies can help "society" understand the complex social and natural environment in which they live.

REFERENCES

Azad, B. and L. L. Wiggins (1995). Dynamics of inter-organizational geographic data sharing: A conceptual framework for research. *Sharing Geographic Information.* H. J. Onsrud and G. Rushton, eds. New Brunswick, NJ, Center for Urban Policy Research: 22–43.

Campbell, H. and I. Masser (1995). *GIS and Organizations. How effective are GIS in practice?* London, Taylor & Francis.

Chrisman, N. R. (1987). Design of geographic information systems based on social and cultural goals. *Photogrammetric Engineering and Remote Sensing* **53**(10): 1367–1370.

Chrisman, N. R. (1992). Ethics for the practitioners of geographic information systems embedded in "real world" constraints of guilds, professions and institutional sponsorship. *Proceedings of the GIS/LIS 92,* San Jose, CA, ACSM-ASPRS-URISA-AM/FM.

Chrisman, N. R. et al. (1984). Modernization of routine land records in Dane County, *Wisconsin: Implications to Rural Landscape Assessment and Planning*, URISA Professional Papers Series, Washington, DC.

Crampton, J. (1995). The ethics of GIS. *Cartography and Geographic Information Systems* **22**(1): 84–89.

Curry, M. (1998). *Digital Places. Living with Geographic Information Technologies.* New York: Routledge.

Elmes, G. et al. (2001). GIS and Society. University Consortium on Geographic Information Science white paper.

Elmes, G. et al. (2004). GIS and society: Interrelation, integration, and transformation. In *A Research Agenda for Geographic Information Science.* R. B. McMaster and E. L. Usery, eds. Boca Raton, FL: CRC Press, pp. 287–312.

Federal Geographic Data Committee (1997). *Framework Introduction and Guide.* Washington, DC: Federal Geographic Data Committee.

Goodchild, M. F. (2007). Citizens as sensors: the world of volunteered geography. *GeoJournal* **69**(4): 211–221.

Harris, L. and H. Hazen (2006). Power of maps: (Counter) Mapping for conservation. *ACME: An International E—Journal for Critical Geographies* **4**(1): 99–130.

Harvey, F. (2003). Knowledges and geography's technology—Politics, ontologies, representations in the changing ways we know. *Handbook of Cultural Geography.* K. Anderson, M. Domosh, S. Pile and N. Thrift, eds. London, Sage Publications: 532–543.

Harvey, F. et al. (2005). Introduction, special issue on critical GIS. *Cartographica* **40**(4): 1–4.

Harvey, F. and D. L. Tulloch (2006). Local government data sharing: Evaluating the foundations of spatial data infrastructures. *International Journal of Geographical Information Science* **20**(7): 743–768.

Huxhold, W. E. (1991). *An Introduction to Urban Geographic Information Systems.* New York: Oxford University Press.

Jankowski, P. and T. L. Nyerges (2008). Geographic information systems and participatory decision making. In *The Handbook of Geographic Information Science.* J. P. Wilson and A. S. Fotheringham, eds. Oxford, U.K.: Blackwell Publishing, pp. 481–493.

Leitner, H. et al. (2000). Modes of GIS provision and their appropriateness for neighborhood organizations—Examples from Minneapolis and St. Paul, MN. *Journal of the Urban and Regional information Systems Association* **12**(4): 45–58.

Marble, D. (2002). *Geographic Information Science and Technology in a Changing Society.* Washington, DC: Batelle Memorial Foundation.

McLaughlin, J. D. (1984). The multipurpose cadastre concept: Current status, future prospects. *Seminar on the Multipurpose Cadastre: Modernizing Land Information Systems in North America.* B. J. J. Niemann, ed. Madison, WI: Institue for Environmental Studies, University of Wisconsin-Madison, pp. 82–93.

McMaster, R. et al. (1997). GIS-based environmental equity and risk assessment: Methodological problems and prospects. *Cartography and Geographic Information Systems* **24**(3): 172–189.

National Research Council (1980). *Need for a Multipurpose Cadastre.* Washington, DC: National Academy Press.

National Research Council (1984). *Modernization of the Public Land Survey System.* Washington, DC: National Academy Press.

National Research Council (1994). *Promoting the National Spatial Data Infrastructure through Partnerships.* Washington, DC: National Academy Press.

National Research Council (2007). *National Land Parcel Data: A Vision for the Future.* Washington, DC: The National Academies Press.

Nedovic-Budic, Z. and J. K. Pinto (2000). Information sharing in an interorganizational GIS Environment. *Environment and Planning B: Planning and Design* **27**(3): 455–474.

Nedovic-Budic, Z., J. K. Pinto et al. (2004). GIS database development and exchange: interaction mechanisms and motivations. *URISA Journal* **16**(1): 16–29.

O'Sullivan, D. (2006). Geographical information science: critical GIS. *Progress in Human Geography* **30**(6): 783–791.

Obermeyer, N. (1994). GIS: A new profession? *Professional Geographer* **46**(4): 498–503.

Obermeyer, N. (1998). Professional responsibility and ethics in the spatial sciences. In *Policy Issues in Modern Cartography*. F. Taylor, ed. Oxford, U.K.: Elsevier Science, pp. 215–232.

Obermeyer, N. J. and J. K. Pinto (1994). *Managing Geographic Information Systems*. New York: Guildford Press.

Onsrud, H. and G. Rushton (1995). Sharing Geographic Information: An Introduction. *Sharing Geographic Information*. H. Onsrud and G. Rushton, eds. New Brunswick, NJ, Center for Urban Policy Research: xiii–xviii.

Pickles, J., ed. (1995). *Ground Truth. The Social Implications of Geographic Information Systems. Mappings: Society/Theory/Space*. New York: Guilford Press.

Pickles, J. (2004). *A History of Spaces. Cartographic Reason, Mapping, and the Geo-Coded World*. New York: Routledge.

Reeve, D. and J. Petch (1999). *GIS, Organisations, and People. A Socio-technical Approach*. London, Taylor & Francis.

Rogers, E. M. (1983). *Diffusion of Innovations*. New York: The Free Press.

Rundstrom, R. A. (1993). The role of ethics, mapping, and the meaning of place in relations between Indians and Whites in the United States. *Cartographica* **30**(1): 21–28.

Sheppard, E. (1995). GIS and society: Towards a research agenda. *Cartography and Geographic Information Systems* **22**(1): 5–16.

Sheppard, E. et al. (1999). Geographies of the information society. *International Journal of Geographical Information Science* **13**(8): 797–823.

Sieber, R. (2006). Public participation geographic information systems: A literature review and framework. *Annals of the AAG* **96**(3): 491–507.

Tulloch, D. L. et al. (1996). Comparative study of mulitpurpose information systems developmentss in Arkansas, Ohio, and Wisconsin. *GIS/LIS' 96 Proceedings*, Denver, CO, ASPRS/AAG/URISA/AM-FM.

Warnecke, L. (1987). Institutional research needs to maximize geographic information development. *International Geographic Information System Conference (IGIS) Symposium*, Arlington, VA, NASA.

Wehn de Montalvo, U. (2003). In search of rigorous models for policy-oriented research: A behavioral approach to spatial data sharing, *URISA Journal* **15**(APA 1): 19–28.

Part V

Applications

34 Biophysical and Human-Social Applications

Debarchana Ghosh and Robert B. McMaster

CONTENTS

34.1 APPLICATIONS OF GIS

Geographic Information Systems (GIS) have been used to support a wide range of applications. This chapter discusses data sources, analytical approaches, and examples of applications in both biophysical and human-social areas. The sophistication of GIS use in the application areas has been increasing over the past decade. Early on, GIS were used primarily for mapping distributions, such as the location of car thefts, or of toxic waste sites. Now, however, GIS are utilized for advanced types of both analysis and display, such as modeling transportation flows and in selecting idealized "routes" for deliver trucks, or locations for police and fire stations. This chapter is divided into two sections—biophysical applications and human-social applications. Although this division might seem somewhat arbitrary, as explained in Chapter 28, spatial data is either "field" or "object." Natural resource data, such as forest stands, drainage basins, or vegetation covers, tend to be continuous (field). Urban-social data, such as population, industry, and transportation, tend to be discrete (object). One must be careful, however, as many applications might very well require both types of data—transportation and terrain. An excellent review of applications may be found in a recently published year 2000 issue of the *Journal of the Urban and Regional Information Systems Association* (Vol. 12, No. 2). In this issue, there are comprehensive reviews of GIS and crime, emergency preparedness and response, public health, transportation planning and management, water resources, and urban and regional planning.

34.2 BIOPHYSICAL APPLICATIONS

Numerous disciplines in the biophysical arena such as ecology, forestry, soil science, and land use planning have been concerned for some time with the management of natural resources at different scales over space and time. A major goal has been to manage natural areas in a sustainable manner or to preserve remaining wilderness areas. Historically, however, resource planners employed management models that were nonspatial in nature such as the forest service's FORPLAN model. More recently, over the past 10–15 years, resource managers have recognized the benefits of using GIS to support the spatial analysis of natural resources. Crain and McDonald (1984) present three evolutionary stages of the development of GIS from land inventory to land analysis to land management purposes. The initial phase involves developing a spatial database that can eventually be used in the final phase, involving more sophisticated analytical capabilities as well as predictive modeling and decision making abilities.

GIS have been used to better understand natural ecosystems and ecological processes in order to conserve or restore them. Additionally, these ecosystems have been impacted to various degrees by human activity as populations grow and economic activity and development increases. Thus, GIS have been instrumental in monitoring, evaluating, and resolving environmental impacts. GIS provide planners and managers with the capability to test different alternatives to proposed actions in order to assess the possible impacts and outcomes of an action. "What if" scenarios can be modeled by spatial analytical capabilities of GIS and help to support improved decision-making. Geographic visualization further enhances our ability to explore potential outcomes of different scenarios.

An important consideration in the development of biophysical applications is the scale of analysis. The issue of scale is a fundamental question for geographers but also to other scientists in the biophysical arena such as ecologists. For example, how does scale influence the understanding of spatial pattern and ecological processes? GIS is able to support analysis at a variety of scales and, depending on the availability of data, can integrate data at disparate scales from a variety of sources (e.g., global positioning system [GPS], various governmental data, commercial data). The purpose of this section is to describe basic data sources and analytical operations useful for biophysical applications and then to discuss more specific applications in order to illustrate the benefits of GIS for such applications.

34.2.1 DATA

Biophysical applications involve the use of data from a variety of sources including remotely sensed imagery (e.g., satellite imagery and aerial photographs), field data collected by GPS and other technology, and digital data and hardcopy maps available from local, state, and federal government agencies such as land use/land cover data, species distributions, wetlands, toxic release inventory (TRI) data, forest resource inventories, climatic and weather data, hydrologic and geologic data, geo-demographic data, and terrain representations. For example, many natural resource applications involve the use of satellite imagery such as Landsat thematic mapper data to map vegetation cover and monitor forest health and landscape change. The Natural Resources Conservation Service uses GPS and GIS to map soils and related data in order to monitor soil and water quality and to recommend soil conservation options to landowners. Many state and local government agencies also have a variety of data useful for biophysical GIS applications (e.g., Minnesota's Land Management Information Center and Department of Natural Resources). A variety of environmental data are becoming very accessible in digital form via the Internet. For example, the U.S. Environmental Protection Agency highlights spatial data and applications at their Web site http://www.epa.gov/epahome/gis.htm including an Internet-based mapping application called "Maps on Demand" (http://www.epa.gov/enviro/html/mod/index.html). This system allows users to map environmental information such as Superfund sites, hazardous waste releases, water discharge permits, and tax incentive zones for brown fields using the Envirofacts Data Warehouse. Overall, most applications involve the use of a wide variety of data from different sources at different scales and GIS is able to integrate these data sources for use in a variety of applications.

A key characteristic of most natural landscape features such as soils, vegetation, and climate is that they exhibit a great deal of natural variation and uncertainty with respect to boundary location. For example, symbolizing soil boundaries as sharp clean lines on a map is inconsistent with the great degree of heterogeneity that may actually occur on a local scale (Burrough and McDonnell, 1998). Miller (1994) provides a useful discussion of the challenges faced when mapping very dynamic natural phenomena like plant and animal species distributions. Users of such data must be aware of possible limitations of such data (e.g., data quality, accuracy and/or precision, sampling density) depending on the purpose of a particular application. The development of data standards and metadata are helping to improve the use of such data, as are advancements in data processing and analysis. Researchers in the area of geographic visualization are also developing techniques for visualizing data accuracy and uncertainty.

34.2.2 SPATIAL ANALYSIS AND MODELING

One of the fundamental aspects of GIS is the ability to undertake a variety of spatial analytical operations that can assist in modeling complex biophysical processes. The basic analytical capabilities of GIS have already been covered in Chapter 29. In addition to these analytical capabilities (e.g., reclassification and overlay analysis), quite sophisticated analytical methods include the incorporation of spatial pattern analysis, geostatistical techniques, and accuracy assessment and error modeling operations. For example, Klopatek and Francis (1999) discuss various spatial pattern analysis methods that are useful for landscape ecology, which focuses on the relationship between spatial patterns and ecological processes at a variety of scales. Specific analytical techniques include point and patch analysis such as nearest-neighbor methods, join-count analysis, and lacunarity analysis. Griffith and Payne (2000) and Goovaerts (1997) describe various geostatistical techniques useful for natural resource evaluation and biophysical applications such as kriging, the use of stochastic simulation for assessing spatial uncertainty, and understanding the influence of spatial autocorrelation on spatial analysis. Digital elevation models can be used for a variety of terrain analysis such as developing viewsheds and delineating watershed regions. Understanding the influence of data and modeling accuracy on biophysical applications is another key analytical area in GIS. Mowrer et al. (1996) include examples of GIS-based applications that examine various spatial accuracy assessment approaches in natural resources management and environmental sciences. In order to illustrate how various spatial data sources and analytical techniques can be used to support biophysical applications using GIS, some examples are provided in Section 34.2.3. It is important to note that these are only a few of many possible applications in the broader range of biophysical applications that exist today.

34.2.3 EXAMPLES OF BIOPHYSICAL APPLICATIONS OF GIS

GIS can support spatial analysis and decision making in a wide variety of biophysical applications including landscape ecology and conservation planning (Haines-Young et al., 1993; Scott et al., 1996), natural resource management (e.g., forest and wildlife management) (Morain, 1999), and environmental modeling at various scales (e.g., hydrologic, atmospheric, soils, and ecological modeling) (Goodchild et al., 1996). The need to explore the impact of human activity on the environment also involves the integration of geodemographic data and analysis techniques, and more recently the incorporation of participatory democracy or public participation GIS in an effort to incorporate localized knowledge in order to understand how technology serves society. The examples discussed in this section illustrate how GIS can be used in forest management and bioresource conservation planning.

34.2.3.1 GIS and Forest Resources Management

GIS is being used by the federal, state, and local forest agencies as well as private forestry companies as spatial decision support tools to monitor and manage public and private forest lands. More specifically, GIS can be used to manage a variety of activities that fall under the "multiple uses" notion such as timber harvesting activities, wildlife habitat, livestock grazing, watershed management, recreational activities, mineral extraction, and the protection of rare and endangered species, and archaeological sites. GIS management models include those that are descriptive, predictive, and prescriptive (Green, 1999). Descriptive models are the most common and involve some form of suitability analysis, for example, determining the optimal site for recreational trail development based on a set of key spatial criteria or deciduous forest focal specie (Manton, 2005). Predictive models allow users to test "What if" scenarios based on a set of spatial variables (e.g., predicting the movement of a toxic gas plume given certain meteorological conditions) while prescriptive models support a full range of spatial decision-making capabilities that allow managers to assess the different management actions. The following applications include examples of these different management models.

Timber harvest planning and analysis often involves the use of GIS to integrate site-specific forest stand inventory data along with terrain data, sensitive species, and roadless areas to evaluate the environmental impact of proposed timber sale on national forest lands. GIS can be used to develop a harvest plan, evaluate potential environmental impacts, and assess alternatives to the plan. A key aspect of such a study is to protect roadless areas by using data such as Forest Service mapped roads, Roadless Area Review and Evaluation II (RARE II) mapping, and roadless areas data produced by the Sierra Biodiversity Institute. Such analysis often incorporates optimization modeling and proximity analysis (e.g., how to most efficiently skid logs to the nearest logging road and minimize the erosion and the sedimentation of streams and other environmental impacts). Figure 34.1 shows an example of a timber harvest suitability map indicating high, medium, and low suitability of stands. The spatial

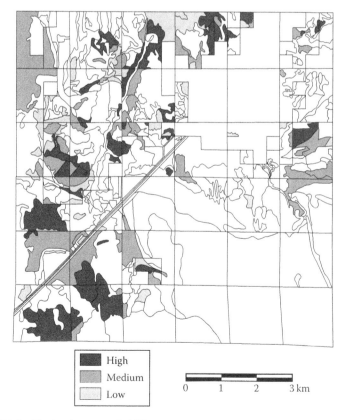

FIGURE 34.1   Timber harvest suitability map.

model that generated this map was designed to assess the suitability of stands in the study site for pulpwood management based on certain forest inventory criteria.

It consisted of three sub-models: pulpwood cover type, biological, and economic sub-models. Fourteen source maps were used including cover type, physiography, cover size, cover density, age, diameter at breast height (i.e., the diameter of a tree at 4.5 ft above ground level), basal area per acre (i.e., the sum of basal areas, the cross-sectional area of a tree trunk at breast height, of individual trees on the acre), site index (i.e., an expression of forest site quality based on the expected height of dominant trees at a specific age), mortality, damage, volume per acre, cords per acre (a cord is equivalent to a pile of wood measuring 128 ft³), stand acreage, and roads. A detailed discussion of model development is given in McMaster (1999). Assessing the impact of data quality on the outcome of such analysis is another important concern. For example, forest stand data could be tested against satellite images or orthophotos or field collected data to ground truth the data. The influence of resolution or model specification on harvest decisions can also be assessed. McMaster (1997, 1999) has examined the impact of varying resolution and attribute accuracy on pulpwood harvest decisions.

The applications of GIS in forest resource management can also include planning and policies (e.g., multiple uses, new recreational sites, management of soil erosion, and local participatory policies) (Cerda, 2004; Robiglio, 2005; Carver, 2006; Zhang, 2006; Ward, 2007; Zandersen, 2007; Baskent, 2008), watershed analysis (Korets, 2008), wildlife habitat management (e.g., determining optimal elk calving areas, roe deer habitats, and northern spotted owls) (Olson, 2004; Vospernik, 2008), forest pests and diseases (e.g., predicting areas that are most susceptible to spruce bud worm and gypsy moth infestations) (Meentemeyer, 2008), forest fires (Choung, 2004; Xu, 2006; Karpov, 2007; Kaloudis, 2008), and the modification of forest landscape structure, affecting the balance of biodiversity (Girvetz, 2003; LeMay, 2008; Miyamoto, 2008).

For example, LeMay (2008) described how GIS combined with ground data, and remotely sensed imageries were used to etch out the spatial pattern and characteristics of specie's stand including their composition, age, height, crown closure, and productivity. In forest fire management research, GIS is often used with other statistical techniques to identify the relationships between forest fires and potential environmental factors. Xu (2006) used forest inventory data and remotely sensed data to create a risk map for forest fire in Jilin province of China. First, the principal component analysis was used to understand the relationships between forest fire potentials and environmental factors. Second, the classifications of these factors were performed within GIS, generating three maps: a fuel-based fire risk map, a topography-based fire risk map, and an anthropogenic-factor fire risk map. These three maps were then synthesized to generate the final fire risk map. Third, the linear regression results showed that the most important factor contributing to forest fire ignition was topography, followed by anthropogenic factors.

34.2.3.2 Conservation GIS and Gap Analysis

Another major biophysical application area centers on preserving and maintaining biodiversity and endemism using GIS. Applications in this area can involve different strategies, for example, a species approach often focusing on endangered species or a more holistic habitat approach on a broader scale. Gap analysis is a program of the National Biological Service interested in determining the conservation status of land cover types, vertebrates, and habitats on a regional scale. Those species not adequately represented in the nation's existing conservation spaces are referred to as conservation "gaps." Gap analysis does not focus solely on rare or threatened species but also on "ordinary" species represented in biodiversity reserves. A key assumption of this approach is that terrestrial vertebrates and vegetation cover can serve as indicators or surrogates for biodiversity in a region. Sources of data used in this approach include Landsat thematic mapper imagery, agency and museum collection records, background knowledge of species' ranges, and previously generated hardcopy and digital maps (e.g., vegetation cover, land use/land ownership data). A key focus of gap analysis is to improve upon the inadequacies of traditional species distribution maps by using a variety of data sources to develop more accurate representations of species distributions. Butterfield et al. (1994) illustrate the gap analysis approach to map the breeding distribution patterns of 366 species of vertebrates. At present, the administration and the scale of the gap analysis program is at the state or regional level (Scott et al., 1996). The gap analysis program Web site (http://www.gap.uidaho.edu) highlights other research and applications of this approach to preserving biodiversity. GIS is ideal for undertaking gap analysis because it is a useful means of handling large volumes of data from disparate sources.

In some instances, GIS has been used to manage and monitor individual species that are recognized to be rare or endangered and in need of special attention (e.g., old growth redwood forests and grizzly bears in the United States). For example, GIS can be used to evaluate the success of grizzly bear recovery zones designated by the U.S. Fish and Wildlife Service (i.e., areas where the population can increase to a more sustainable level) by using grizzly bear siting data, roadless area data, and National Park Service area data (The Conservation GIS Consortium, 1999). MacKinnon and Wulf (1994) illustrate the use of GIS to design protection areas for the threatened giant pandas in China. Lipow (2004) also develops gap analysis approach to evaluate whether the genetic resources conserved in situ in protected areas are adequate for conifers in western Oregon and Washington (USA) (Lipow, 2004).

The future of GIS in supporting biophysical applications is sure to grow and improve as advancements are made in data and analytical procedures and as research continues to improve our understanding of how scale influences ecological process and how they are modeled.

34.3 HUMAN-SOCIAL APPLICATIONS

Our society is faced with many complex human-social issues that involve the use of large amounts of data in an effort to understand and evaluate the spatial dynamics of a problem so effective policies and solutions can be developed. GIS provide planners and decision makers with the ability to identify at-risk populations using a vast array of geodemographic data and spatial analytical tools. Private businesses also use GIS to support market research and real estate sales. One can find an increasing number of applications of GISs to human-social problems, including geodemographic analyses, crime assessment, environmental justice studies, and public health issues.

34.3.1 Data

Human-social applications involve the use of census data, land use/land cover information, transportation networks, public health data, property boundaries and geographic base files like TIGER, and zoning. The backbone of much of the spatial analysis for human-social applications are the data sets provided by the census, including the TIGER files and accompanying statistical data. Academic researchers as well as planners, decision makers, and community organizers utilize census data to assess a multitude of population characteristics, including those on housing, poverty, race/ethnicity, income, and transportation, and increasingly wish to analyze temporal trends. In contrast to some biophysical data such as soils and vegetation that tend to be more variable, some human-social data are more precise, such as cadastral data and property boundaries. Users of such data are very concerned with accuracy and precision since errors can be costly and involve liability issues. Increasingly, data from the public health sector are being provided, albeit at rather coarse resolutions due to confidentiality constraints. For instance, in the State of Minnesota one can obtain public health data only at the county-level, unless confidentiality is maintained; this is normally only possible for research purposes. Thus the type of address-specific data that geodemographic analysts utilize for marketing purposes would be impossible to acquire for health studies. Another key source of data for human-social applications involves transportation information. The Federal Department of Transportation, along with all states, maintain detailed data on transportation routes, flows, and conditions. One specific type of transportation data are the Traffic Analysis Zone (TAZ) data sets for urban areas, where census data on where people work are reaggregated into zones used to predict traffic flows.

34.3.2 Spatial Analysis and Modeling

GIS has facilitated the rapid growth of geodemographic analysis, including geomarketing, and many forms of population analysis in general, by integrating together the wealth of population data that are now spatially referenced with the powerful spatial-analytic capabilities of GISs. Examples of these spatial analyses include the assessment of environmental justice/racism at multiple spatial scales (regional, urban, community); the calculation of segregation indices and the evaluation of urban poverty, and the identification of concentrated poverty; the development of neighborhood indicators, including a multitude of economic and social measures based on population data; and the spatial/temporal analysis of census data.

Increasingly, researchers are utilizing GISs and census geographic base files for historical geodemographic analyses. For instance, after the 1990 census, it was possible to document the changes in geodemographics between 1980 and 1990, using the 1990 TIGER files. A common application was mapping the change in minority populations between the two periods. It should be noted, however, that researchers are constrained mostly to two or three decades of temporal analysis with the availability of only pre-1970 digital files.

Suitability analysis can be useful in both human-social and biophysical applications to determine the optimum location of some phenomena given a set of criteria. For example, where is the best site to locate a school or day care given certain geo-demographic and other relevant criteria.

34.3.3 EXAMPLES OF HUMAN-SOCIAL APPLICATIONS OF GIS

34.3.3.1 Racial Segregation Studies

One specific use of temporal census boundary files and data is in the computation of segregation indices. Although a multitude of segregation indices have been developed over the past 10 years, nearly all work applies such indices to 1990 data. The application of such indices is normally applied for the purposes of identifying areas of concentrated poverty, and to identify the income isolation that many groups experience.

34.3.3.2 Environmental Justice Research

Environmental justice studies, which rely extensively on census data attempt to determine if certain "at risk" populations (minority, young, old, disabled, poor) bear a disproportionate exposure to environmental chemicals, are only able to assess the conditions from 1990 onward. Increasingly, however, researchers have argued that examining the historical perspective of toxic sites is crucial in such work. The basic question, of course, is which came first—the toxic site or the "at-risk" population? Proving or disproving environmental inequity requires "intent," which can only be determining with careful painstaking historical reconstructions of both the industries themselves and the associated "at risk" populations. Did, for instance, a sub-population begin to migrate to certain spaces and a toxic industry then purposefully targeted that space for a factory? Or did the subpopulations themselves, because of a complex series of social-economic issues, cluster around a site due to depressed land values? Intent and causality can only be determined through historical reconstructions. Multiple studies utilize GISs to assess the effect of both spatial scale and data resolution on the results of environmental justice (McMaster, 1997). Figure 34.2 shows the results of an analysis of environmental inequity in Minneapolis. On this map, the percent of the population in poverty is depicted in relation to the 38 TRI sites in Minneapolis. One can also perform simple types of spatial analyses here, where Figure 34.3 illustrates the same data, but where the TRI sites have been buffered to 1000 yards in order to ascertain whether low-income populations live "close" to TRI sites in Minneapolis.

34.3.3.3 Crime Analysis

Crimes have location or spatial relevance and hence have a geographical element attached to them. It is here that GIS functionalities can play an important role in crime assessment. Typical GIS functions like overlay, mapping, and the display of multiple data sets can be effective in understanding the underlying causes of crime and help law enforcements to identify crime "hot-spots," make better decisions, target resources, and formulate strategies as well as for tactical analysis (e.g., crime

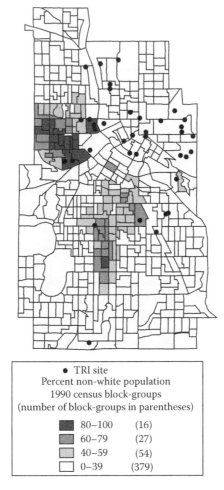

FIGURE 34.2 Overlay of TRI sites on the distribution of percent non-white population in Minneapolis, Minnesota.

forecasting, geographic profiling). The GIS applications in crime analysis can be broadly divided into three groups: (1) mapping of crime clusters and hot-spots (Loukaitou-Sideris, 1999; Vogt, 1999; Block, 2000; Rengert, 2001; Vann, 2001; Boba, 2003; Randerson, 2004; Levine, 2006; Pain, 2006; Brower, 2007; Lockwood, 2007; Markovic, 2007), (2) spatiotemporal analysis of crimes and their related factors (Hirschfield, 1995; Ratcliffe, 2002; Egilmez, 2003; Brunsdon, 2007), and (3) crime forecasting (Ouimet, 1999; Corcoran, 2003; Gorr, 2003; Brower, 2007).

34.3.3.4 Public Health and Epidemiological Studies

The field of public health and epidemiology identifies, monitors, and maintains health information of human population. Recently, it is more and more relying on spatial solutions to track the spatiotemporal spread of infectious diseases, delineate exposure areas and environmental correlates, identify hotspots of chronic diseases,

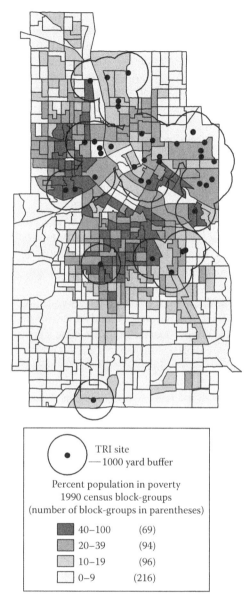

FIGURE 34.3 Proximity of low income population to TRI sites in Minneapolis, Minnesota.

estimate rates of deaths from injuries, or locate areas with poor health facilities. GIS with its powerful spatial analytical capabilities is making substantive contribution in this field at a highly practical level. Important questions such as "Where is the best place for the new clinic?," "Where is the highest disease mortality?" "Where is the disease cluster?," are inherently spatial and readily addressable using current GIS and statistical tools for analyzing health data (Gatrell and Loytonen, 1998).

One can point to a plethora of studies that are utilizing GIS functionalities for public health analysis. The applications can be broadly divided into four categories: (1) environmental health (Cuthe et al., 1992; Beck et al., 1994; Glass et al., 1995; Crabbe et al., 2000; Jacquez and Greiling, 2003; Jacquez et al., 2005; Schroder, 2006; Vanwambeke et al., 2006), (2) communicable and infectious disease prevention and control (CDC, 1996; Kitron, 1997; Schlenker, 1999; Ruiz et al., 2004; Prince et al., 2005; Trooskin et al., 2005; Zou et al., 2006; Leblond et al., 2007; Ozdenerol et al., 2008), (3) chronic disease (Rogers, 1991; Smith, 1995; Lewis-Michl, 1996; Kulldorff, 1997; Zeman, 1997; Peleg, 2000; Heineman, 2001; Reynolds, 2001, 2002; Jacquez and Greiling, 2003; Albert, 2004; Brody, 2004; O'Leary et al., 2004; Poulstrup, 2004; Rushton et al., 2004; Bunnell, 2006; Marcilio, 2006; McElroy et al., 2006; Vinnakota, 2006; Yu, 2006; Hsu, 2007; Kingsley, 2007; Matthews, 2007; Wagner, 2007; Lian et al., 2008; Viel, 2008; Wu, 2008), and (4) health care planning, assessment, access, and surveillance (Gordon and Womersley, 1997; Casey, 1999; Cheatham, 1999; Rushton, 1999; White and Cerny, 1999; Wall and Devine, 2000; Brabyn and Gower, 2004; Luo, 2004; Sherman et al., 2005; Shuai et al., 2006). The following paragraphs will briefly discuss some of the important GIS applications in each of the four categories mentioned above.

By combining spatial data on morbidity, mortality, demographic, and environmental risk factors, GIS technologies are directly used for understanding diseases or disease risks related to environmental exposure. One study modeled disease risk and identified the at-risk cohort for further evaluation on the basis of exposure to radioactive iodine (Henriques and Spengler, 1999). The investigators, using appropriate data sets, created a map that identified communities with elevated risk and therefore would benefit from a screening program. The Agency for Toxic Substance and Disease Registry (ATSDR) was using this map to educate people in the identified regions about their eligibility for the screening program. GIS with its efficient capability to store, manage, link, and analyze spatial data from disparate sources (U.S. Census, Environmental Pollution Agency, and TRI) could advance in developing predictive model for exposure to lead poisoning (Wartenberg, 1992; CDC, 1997). Wartenberg provided a risk map of exposure to lead poisoning, on the basis of which, local health department could prioritize its lead-screening programs, while minimizing travel time and expenses. Recently, in response to the need of spatiotemporal disease diffusion models or risk assessment models, a STIS—a Space Time Information System within a GIS environment have been developed to visualize and analyze disease rates simultaneously through space and time (Jacquez and Greiling, 2003; Jacquez et al., 2005). Another common trend in environmental health studies is to couple GIS, statistical, and mathematical models to produce cutting-edge research methodologies (Devine and Louis, 1994; Prince et al., 2005; McElroy et al., 2006; Schroder, 2006; Vanwambeke et al., 2006). In one such investigation, multi-level analysis was combined with GIS to study the environmental risk determinants of dengue infection in Thailand (Vanwambeke et al., 2006). The location of a person's house, the surrounding environment, including irrigated fields and orchards were important determinants for dengue infection in Thailand.

The knowledge of spatial distribution of communicable diseases or infectious diseases along with their association with environmental and anthropogenic risk factors is essential to understand the dynamics of disease transmission. GIS with its exploratory spatial data analysis functions is very effective in understanding the complex relationships between disease outcome, associated vectors and reservoirs, hypothesized risk factors, and human population. GIS mapping can provide new insights into both disease ecology and appropriate intervention strategies, thereby enhancing our ability to prevent and control communicable and infectious diseases. For example, public health officials could apply GIS technologies to prevent and control vaccine-preventable diseases, namely, hepatitis A by directing immunization efforts to geographic clusters or disease "hot-spots" with significantly high numbers of children with no immunization (Schlenker, 1999). In a similar study, clusters of people with hepatitis C were identified along with their demographic characteristics. This resulted in targeted prevention and educational campaigns to raise awareness of hepatitis C risk factors (Trooskin et al., 2005). GIS was also coupled with a mathematical model to investigate the spread of another communicable infection, SARS, in a close community (Prince et al., 2005). The application of geospatial technologies, including GIS, remote sensing, and GPS also plays an important role in understanding, controlling, and identifying risk factors of vector-borne infectious diseases. GIS is answering questions such as where do emerging disease arise? What is the spatial distribution of a disease? What are the local and regional risk factors of disease transmission? There are several examples where GIS can prove very effective in controlling efforts of mosquito-borne infectious disease, namely, LaCrosse encephalitis (Kitron, 1997), malaria (Beck et al., 1994), or identifying potential risk factors for West Nile Virus (Ruiz et al., 2004; Zou et al., 2006; Ozdenerol et al., 2008). Epidemiologists have also used GIS in tick-borne Lyme disease studies, where they considered geographic areas providing habitat for the tick vector and animal hosts as high-risk areas for disease transmission (Glass et al., 1995; CDC, 1996).

Public health officials can respond to the challenges of chronic diseases in their communities by using GIS technology. An example of such GIS applications is to target communities with relevant health education and health promotion messages. Another common usage of GIS functionalities in chronic disease is to identify geographic clusters of significantly high number of disease cases than the surrounding local areas and investigate possible associations between socioeconomic, behavioral, environmental, and other fac tors. Cancer is one such chronic disease, where GIS is extensively used to identify clusters (Smith, 1995; Kulldorff, 1997; Peleg, 2000; Heineman, 2001; Jacquez and Greiling, 2003; Albert, 2004; Rushton et al., 2004; Jacquez et al., 2005; Hsu, 2007; Kingsley, 2007; Matthews, 2007; Lian et al., 2008) and target areas or communities in need for screening, education, and testing. GIS is also applied to identify risk factors for cancer incidences, namely, pesticides, toxins, aquifers, and environmental factors. (Lewis-Michl, 1996; Reynolds, 2001, 2002; Brody, 2004; O'Leary, 2004; Poulstrup, 2004; Bunnell, 2006; Marcilio, 2006; Yu, 2006; Wagner, 2007; Viel, 2008; Wu, 2008). Recently spatial modeling in public health is combining spatial statistics with GIS to better understand disease etiology. McElroy and others presented an innovative method to identify geographic disparities in the early detection of breast cancer in Dane County, Wisconsin using GIS and Bayesian statistics (McElroy et al., 2006). In another similar investigation of cancer mortality, association rule mining was combined with GIS to reveal the spatial patterns of socioeconomic inequality in cancer mortality in the United States and identify regions that need further attention (Vinnakota, 2006). Mapping heart disease mortality, another important chronic disease in the United States, is also useful in directing public health activities and identify racial and ethnic disparities.

The use of GIS has significantly increased the abilities of public health agencies and personnel to ensure access to health care, planning, and surveillance. Health officials can evaluate and improve health care access by assessing the spatial or geographic relationship between the location of potential clients and available services (Rushton, 1999; White, 1999; Brabyn and Gower, 2004; Luo, 2004; Sherman et al., 2005). Maps created in GIS are used to determine the market area for local and state health departments (Gordon and Womersley, 1997; Casey, 1999) and estimating the proportion and the location of people without health insurance (Cheatham, 1999). In recent years with the development of more sophisticated geospatial technologies, GIS has accelerated the use of mapping as an epidemiological tool by automating the production of disease maps and by enabling users to do spatial referencing and the manipulation of the data on which the maps are based (Wall and Devine, 2000). As a result, federal, state, and local health agencies have begun to incorporate GIS-based approaches into their systems for surveillance of a number of health outcomes such as cancer, birth defects, and injuries along with other spatial information (Wall and Devine, 2000). In a related example, a pilot system is developed to integrate real-time surveillance and GIS in order to enhance West Nile virus dead bird surveillance in Canada (Shuai et al., 2006). This real-time system has streamlined, enriched, and enhanced national West Nile virus dead bird surveillance in Canada by improving the productivity and the reduction of operation costs.

34.4　CONCLUSIONS

There is a growing literature on most of these topics, but is found in dispersed sources. One can often find much of the literature in the cognate disciplines (e.g., public health), and not in the core GIS literature. Readers interested in seeing other examples of GIS applications are encouraged to read the trade magazines of *GeoWorld* and *GeoSpatial Solutions*, as well as the literature published by the vendors themselves.

REFERENCES

Albert, S. M. 2004. Cancer screening among older women in a culturally insular community. *Preventive Medicine* 39(4):649–656.

Baskent, E. Z. 2008. Developing and implementing participatory and ecosystem based multiple. *Forest Ecology and Management* 256(4):798–807.

Beck, L. R., M. H. Rodrigues, S. W. Dister, A. D. Rodrigues, and E. Rejmankova. 1994. Remote sensing as a landscape epidemiologic tool to identify villages at high risk for malaria transmission. *American Journal of Tropical Medicine and Hygiene* 51:271–280.

Block, R. 2000. Gang activity and overall levels of crime: A new mapping tool for. *Journal of Quantitative Criminology* 16(3):369–383.

Boba, R. 2003. Crime analysis through computer mapping. *Forensic Science International* 136:1–1.

Brabyn, L. and P. Gower. 2004. Comparing three GIS techniques for modelling geographical access to general practitioners. *Cartographica* 39(2):41–49.

Brody, J. G. 2004. Breast cancer risk and historical exposure to pesticides from wide-area. *Environmental Health Perspectives* 112(8):889–897.

Brower, A. M. 2007. Spatial and temporal aspects of alcohol-related crime in a college town. *Journal of American College Health* 55(5):267–275.

Brunsdon, C. 2007. Visualising space and time in crime patterns: A comparison of methods. *Computers Environment and Urban Systems* 31(1):52–75.

Bunnell, J. E. 2006. Possible linkages between lignite aquifers, pathogenic microbes, and. *Environmental Geochemistry and Health* 28(6):577–587.

Burrough, P. and R. McDonnell. 1998. *Principles of Geographic Information Systems*, New York: Oxford University Press.

Butterfield, B., B. Csuti, and J. Scott. 1994. Modeling vertebrate distributions for gap analysis. In *Mapping the Diversity of Nature*, ed. R. Miller, pp. 53–68. London, U.K.: Chapman & Hall.

Carver, A. D. 2006. Applying spatial analysis to forest policy evaluation: Case study of. *Environmental Science & Policy* 9(3):253–260.

Casey, N. M. 1999. Active patients attending the North Hill Child Health clinic, Akron, Ohio, 1997. *Journal of Public Health Management and Practice* 5(2):51–52.

CDC, Centers for Disease Control and Prevention. 1996. Lyme disease: United States, 1996. *Morbidity and Mortality Weekly Report* 46(23):531–535.

CDC, Centers for Disease Control and Prevention. 1997. Screening young children for lead poisoning: Guidance for state and local public health officials, Atlanta, GA.

Cerda, J. P. 2004. Using geographic information systems to investigate uncertainty in forest planning. *Forestry Chronicle* 80(2):262–270.

Cheatham, M. 1999. Percent of adults without health insurance in Ingham County, Michigan. *Journal of Public Health Management and Practice* 5(2):53–54.

Choung, Y. 2004. Forest responses to the large-scale east coast fires in Korea. *Ecological Research* 19(1):43–54.

Corcoran, J. J. 2003. Predicting the geo-temporal variations of crime and disorder. *International Journal of Forecasting* 19(4):623–634.

Crabbe, H., H. Ron, and M. Nuria. 2000. Using GIS and dispersion modeling tools to assess the effect of the environment on health. *Transaction in GIS* 4(3):235–244.

Crain, I. and C. McDonald. 1984. From land inventory to land management: The evolution of an operational GIS. *Cartographica* 21(2 and 3):40–46.

Cuthe, W. G., R. K. Tucker, E. A. Murphy, R. England, E. Stevenson, and J. C. Luckardt. 1992. Reassessment of lead exposure in New Jersey using GIS technology. *Environmental Resources* 59:318–325.

Devine, O. and T. Louis. 1994. Empirical Bayes estimators for spatially correlated incidence rates. *Environmetrics* 5:381–398.

Egilmez, S. F. 2003. Evaluation of the crime rate influencers using multivariate analysis. *Forensic Science International* 136:13–14.

Gatrell, A. and M. Loytonen. 1998. GIS and health research: An introduction. In *GIS and Health*. London, U.K.: Taylor & Francis, pp. 3–16.

Girvetz, E. 2003. Decision support for road system analysis and modification on the Tahoe. *Environmental Management* 32(2):218–233.

Glass, J. E., B. S. Schwartz, J. M. Morgan, D. T. Johnson, P. M. Noy, and E. Israel. 1995. Environmental risk factors for Lyme disease identified with geographic information systems. *American Journal of Public Health* 85:944–948.

Goodchild, M., L. Steyaert, B. Parks, C. Johnston, D. Maidment, M. Crane, and S. Glendinning. 1996. *GIS and Environmental Modeling: Progress and Research Issues*. Fort Collins, CO: GIS World Books.

Goovaerts, P. 1997. *Geostatistics for Natural Resources Evaluation.* New York: Oxford University Press.

Gordon, A. and J. Womersley. 1997. The use of mapping in Public Health and planning health services. *Journal of Public Health Management and Practice* 19(2):139–147.

Gorr, W. 2003. Introduction to crime forecasting. *International Journal of Forecasting* 19(4):551–555.

Green, K. 1999. Development of the spatial domain in resource management. In *GIS Solutions in Natural Resource Management*, ed. S. Morain. Sante Fe, NM: OnWord Press, pp. 5–15.

Haines-Young, R., D. Green, and S. Cousins. 1993. *Landscape Ecology and GIS.* London, U.K.: Taylor & Francis.

Heineman, E. 2001. Geographic information system for the Long Island Breast Cancer Study. *Epidemiology* 12(4):435.

Henriques, W. D. and R. F. Spengler. 1999. Locations around the Hanford nuclear facility where average milk consumption by children in 1945 would have resulted in an estimated median iodine-131 dose to the Thyroid of 10 Rad or higher, Washington. *Journal of Public Health Management and Practice* 5(2):35–36.

Hirschfield, A. 1995. GIS and the analysis of spatially-referenced crime data—Experiences. *International Journal of Geographical Information Systems* 9(2):191–210.

Hsu, C. E. 2007. A spatial-temporal approach to surveillance of prostate cancer. *Journal of the National Medical Association* 99(1):72.

Jacquez, G. M. and D. A. Greiling. 2003. Geographic boundaries in breast, lung and colorectal cancers in relation to exposure to air toxics in Long Island, New York. *International Journal of Health Geographics* 17(2:4):1–22.

Jacquez, G. M., D. A. Greiling, and A. M. Kaufmann. 2005. Design and implementation space–Time intelligence system for disease surveillance. *Journal of Geographical Systems* 7(1):7–23.

Kaloudis, S. 2008. Design of forest management planning DSS for wildfire risk reduction. *Ecological Informatics* 3(1):122–133.

Karpov, A. I. 2007. GIS-based computer code for the evaluation of forest fire spread. *Thermal Science* 11(2):259–270.

Kingsley, B. S. 2007. An update on cancer cluster activities at the Centers for Disease. *Environmental Health Perspectives* 115(1):165–171.

Kitron, U. 1997. Spatial analysis of the distribution of LaCrosse encephalitis in Illinois, using a Geographic Information System and local and global spatial statistics *American Journal of Tropical Medicine and Hygiene* 57:469–475.

Klopatek, J. and J. Francis. 1999. Spatial pattern analysis techniques. In *GIS Solutions in Natural Resource Management*, ed. S. Morain. Sante Fe, NM: OnWord Press, pp. 17–40.

Korets, M. A. 2008. GIS as a tool for identification of forest water protection areas. *Contemporary Problems of Ecology* 1(3):353–355.

Kulldorff, M. 1997. Breast cancer clusters in the Northeast United States: A Geographic Analysis. *American Journal of Epidemiology* 146(2):161–170.

Leblond, A., A. Sandoz, G. Lefebvre, H. Zeller, and D. J. Bicout. 2007. Remote sensing based identification of environmental risk factors associated with West Nile disease in horses in Camargue, France. *Preventive Veterinary Medicine* 79(1):20–31.

LeMay, V. 2008. Estimating stand structural details using nearest neighbor analyses to link ground data, forest cover maps, and Landsat imagery. *Remote Sensing of Environment* 112(5):2578–2591.

Levine, N. 2006. Crime mapping and the CrimeStat program. *Geographical Analysis* 38(1):41–56.

Lewis-Michl, E. L. 1996. Breast cancer risk and residence near industry or traffic in Nassau and Sulfolk Counties, Long Island, New York. *Archives of Environmental Health* 51(4):255–265.

Lian, M., D. B. Jeffe, and M. Schootman. 2008. Racial and geographic differences in mammography screening in St. Louis City: A multilevel study. *Journal of Urban Health-Bulletin of the New York Academy of Medicine* 85(5):677–692.

Lipow, S. R. 2004. Gap analysis of conserved genetic resources for forest trees. *Conservation Biology* 18(2):412–423.

Lockwood, D. 2007. Mapping crime in Savannah—Social disadvantage, land use, and violent. *Social Science Computer Review* 25(2):194–209.

Loukaitou-Sideris, A. 1999. Hot spots of bus stop crime—The importance of environmental attributes. *Journal of the American Planning Association* 65(4):395–411.

Luo, W. 2004. Using a GIS-based floating catchment method to assess areas with shortage of physicians. *Health and Place* 10:1–11.

MacKinnon, J. and R. D. Wulf. 1994. Designing protected areas for giant pandas in China. In *Mapping the Diversity of Nature*, ed. R. Miller. London, U.K.: Chapman & Hall, pp. 127–142.

Manton, M. G. 2005. Modelling habitat suitability for deciduous forest focal species—A sensitivity analysis using different satellite land cover data. *Landscape Ecology* 20(7):827–839.

Marcilio, I. 2006. Residence near a municipal solid waste incinerator and cancer risk: An analysis using a geographic information system (GIS). *Epidemiology* 17(6):S479–S479.

Markovic, J. 2007. GIS and crime mapping. *Social Science Computer Review* 25(2):279–282.

Matthews, B. A. 2007. Colorectal cancer screening among midwestern community-based residents. *Journal of Community Health* 32(2):103–120.

McElroy, J. A., P. L. Remington, R. E. Gangnon, L. Hariharan, and L D. Anderson. 2006. Identifying geographic disparities in the early detection of breast cancer using a geographic information system. *Preventive Chronic Disease* 3(1):A10.

McMaster, R. B. 1997. Examining the impact of varying resolution on environmental models. Paper read at Proceedings, GIS/LIS, October 1997, at Cincinnati, OH.

McMaster, S. 1999. Assessing the impact of data quality on forest management decisions using geographical sensitivity analysis. In *Geographic Information Research: Trans-Atlantic Perspectives*, eds. M. Cragalia and H. Onsurd. London, U.K.: Taylor & Francis, pp. 477–495.

Meentemeyer, R. K. 2008. Early detection of emerging forest disease using dispersal estimation. *Ecological Applications* 18(2):377–390.

Miller, R. 1994. *Mapping the Diversity of Nature*. London, U.K.: Chapman & Hall.

Miyamoto, A. 2008. The influence of forest management on landscape structure in the cool-temperate forest region of central Japan. *Landscape and Urban Planning* 86(3–4):248–256.

Morain, S. 1999. *GIS Solutions in Natural Resource Management*. Santa Fe, NM: OnWord Press.

Mowrer, H., R. Czaplewski, and R. Hamre. 1996. Spatial accuracy assessment in natural resources assessment and environmental sciences. In *Second International Symposium*. Fort Collins, CO: USDA, Forest Service, 728p.

O'Leary, E. S. 2004. Pesticide exposure and risk of breast cancer: A nested case-control. *Environmental Research* 94(2):134–144.

O'Leary, D., A. Marfin, S. Montgomery, K. AM., J. Lehman, and B. Biggerstaff. 2004. The epidemic of West Nile virus in the United States, 2002. *Vector Borne Zoonotic Disease* 4:61–70.

Olson, G. S. 2004. Modeling demographic performance of northern spotted owls relative to forest habitat in Oregon. *Journal of Wildlife Management* 68(4):1039–1053.

Ouimet, M. 1999. Crime mapping and crime prevention. *Canadian Journal of Criminology-Revue Canadienne De Criminologie* 41(3):414–417.

Ozdenerol, E., E. Bialkowska-Jelinska, and G. N. Taff. 2008. Locating suitable habitats for West Nile Virus-infected mosquitoes through association of environmental characteristics with infected mosquito locations: A case study in Shelby County, Tennessee. *International Journal of Health Geographics* 7(12).

Pain, R. 2006. "When, where, if, and but:" Qualifying GIS and the effect of streetlighting on crime and fear. *Environment and Planning A* 38(11):2055–2074.

Peleg, I. 2000. The use of a geographic information system (GIS) to identify small. *Gastroenterology* 118(4):3930.

Poulstrup, A. 2004. Use of GIS and exposure modeling as tools in a study of cancer. *Environmental Health Perspectives* 112(9):1032–1036.

Prince, A., X. Chen, and K. C. Lun. 2005. Containing acute disease outbreak. *Methods of Information in Medicine* 44(5):603–608.

Randerson, J. 2004. Mapping the path of crime epidemics. *New Scientist* 182(2447):20–20.

Ratcliffe, J. H. 2002. Aoristic signatures and the spatio-temporal analysis of high volume. *Journal of Quantitative Criminology* 18(1):23–43.

Rengert, G. F. 2001. Mapping crime: Principle and practice. *Professional Geographer* 53(2):298–300.

Reynolds, P. 2001. A case-control pilot study of traffic exposures and early childhood leukemia using Geographic Information Systems. *Bioelectromagnetics* 5:S58–S68.

Reynolds, P. 2002. Childhood cancer and agricultural pesticide use: An ecologic study in California. *Environmental Health Perspectives* 110(3):319–324.

Robiglio, V. 2005. Integrating local and expert knowledge using participatory mapping and. *Forestry Chronicle* 81(3):392–397.

Rogers, M. Y. 1991. Using marketing information to focus smoking cessation programs in specific census block groups along the Buford highway corridor, DeKalb County, Georgia, 1996. *Public Health Management and Practice* 5(2):55–56.

Ruiz, M. O., C. Tedesco, T. J. McTighe, C. Austin, and U. Kitron. 2004. Environmental and social determinants of human risk during a West Nile virus outbreak in the greater Chicago area, 2002. *International Journal of Health Geographics* 3:8:2–11.

Rushton, G. 1999. Methods to evaluate geographic access to health services. *Journal of Public Health Management and Practice* 5(2):93–100.

Rushton, G., I. Peleg, I. Banerjee, G. Smith, and M. West. 2004. Analyzing geographic patterns of disease incidence: Rates of late-stage colorectal cancer in Iowa. *Journal of Medical System* 28(3):223–236.

Schlenker, T. L. 1999. Incidence rates of Hepatitis A by ZIP code area, Salt Lake County, Utah, 1992–1996. *Journal of Public Health Management and Practice* 5(2):17–18.

Schroder, W. 2006. GIS, geostatistics, metadata banking, and tree-based models for data analyzing and mapping in environmental monitoring and epidemiology. *International Journal of Medical Microbiology* 296(1):23–36.

Scott, J., T. Tear, and F. Davis. 1996. *Gap Analysis: A Landscape Approach to Biodiversity Planning.* Bethesda, MD: American Society for Photogrammetry and Remote Sensing.

Sherman, J. E., J. Spencer, J. S. Preisser, W. M Gesler, and T. A. Arcury. 2005. A suite of methods for representing activity space in a healthcare accessibility study. *International Journal of Health Geography* 4:24.

Shuai, J. P. B., P. Sockett, J. Aramini, and F. Pollarl. 2006. A GIS-driven integrated real-time surveillance pilot system for national West Nile virus dead bird surveillance in Canada. *International Journal of Health Geography* 5(1):17.

Smith, T. J. 1995. Differences in initial treatment patterns and outcomes of lung cancer. *Lung Cancer* 13(3):235–252.

Trooskin, S. B., J. Hadler, T. St. Louis, and V. J. Navarro. 2005. Geospatial analysis of hepatitis C in Connecticut: A novel application of a public health tool. *Public Health* 119:1042–1047.

Vann, I. B. 2001. Crime mapping and its extension to social science analysis. *Social Science Computer Review* 19(4):471–479.

Vanwambeke, S. O., B. H. van Benthem, N. Khantikul, C. Burghoorn-Maas, K. Pranat, and L. Oskam. 2006. Multi-level analyses of spatial and temporal determinants for dengue infection. *International Journal of Health Geography* 5(1):5.

Viel, J. F. 2008. Dioxin emissions from a municipal solid waste incinerator and risk of invasive breast cancer: A population-based case-control study with GIS-derived exposure. *International Journal of Health Geographics* 7(1):5.

Vinnakota, S. A. N. S. L. 2006. Socioeconomic in equality of cancer mortality in the United States: A spatial data mining approach. *International Journal of Health Geography* 5(1):9.

Vogt, S. 1999. Geographic information systems—Crime-mapping, a breath of fresh air. *Kriminalistik* 53(12):821–823.

Vospernik, S. 2008. Modelling changes in roe deer habitat in response to forest management. *Forest Ecology and Management* 255(3–4):530–545.

Wagner, S. E. 2007. Characterizing spatial and environmental determinants of prostate. *Epidemiology* 18(5):S155–S156.

Wall, P. A. and O. J. Devine. 2000. Interactive analysis of the spatial distribution of disease using a geographic information system. *Journal of Geographical Systems* 2(3):243–256.

Ward, K. 2007. Transformation of the oak forest spatial structure in the Minneapolis/St. Paul metropolitan area, Minnesota, USA over 7 years. *Landscape and Urban Planning* 81(1–2):27–33.

Wartenberg, D. 1992. Screening for lead exposure using Geographic Information System. *Environmental Research* 59:310–317.

White, E. 1999. Geographic studies of pediatric cancer near hazardous waste sites. *Archives of Environmental Health* 54(6):390–397.

White, G. F. and K. C. Cerny. 1999. Client demographics (age abd gender) at low-income clinics in Austin/Travis County, Texas 1995-1996. *Journal of Public Health Management and Practice* 5(2):47–48.

Wu, K. S. 2008. Relationships between esophageal cancer and spatial environment factors. *Science of the Total Environment* 393(2–3):219–225.

Xu, D. 2006. Mapping forest fire risk zones with spatial data and principal. *Science in China Series E-Technological Sciences* 49:140–149.

Yu, C. L. 2006. Residential exposure to petrochemicals and the risk of leukemia: Using geographic information system tools to estimate individual-level residential exposure. *American Journal of Epidemiology* 164(3):200–207.

Zandersen, M. 2007. Evaluating approaches to predict recreation values of new forest sites. *Journal of Forest Economics* 13(2–3):103–128.

Zeman, P. 1997. Objective assessment of risk maps of tick-borne encephalitis and Lyme borreliosis based on spatial patterns of located cases. *International Journal of Epidemiology* 26(5):1121–1130.

Zhang, H. M. 2006. Quantifying soil erosion with GIS-based RUSLE under different forest. *Science in China Series E-Technological Sciences* 49:160–166.

Zou, L., S. N. Miller, and E. T. Schmidtmann. 2006. Mosquito larval habitat mapping using remote sensing and GIS: Implications of coalbed methane development and West Nile virus. *Journal of Medical Entomology* 43(5):1034–1041.

35 Local Government Applications: Toward E-Governance

Zorica Nedović-Budić

CONTENTS

35.1 DIFFUSION OF SPATIAL TECHNOLOGIES IN LOCAL GOVERNMENTS

Local governments are created to take care of the general health, the welfare, and the prosperity of their citizens and communities. In the United States, they are often referred to as "creatures of the state," as they receive their powers through state laws that expressly define local government functions (the so-called Dillon rule), or through the approval of local charters and the establishment of greater self-government and control (the so-called home-rule). Examples of local government concerns include regulating physical growth of the community; preventing fire hazards; providing infrastructure for roads, water, and sewer; stimulating economic development; minimizing crime; building schools; and securing publicly accessible recreational opportunities. In doing so, the local governments channel a variety of interests and goals toward decisions and policies that take into consideration and try to protect public interest.

Computerized information systems have been adopted since the 1950s as a way to improve government performance in serving and protecting the public interest by increasing government's efficiency, effectiveness, and accountability. In the late 1980s, the availability of user-friendly and affordable geospatial technologies, including geographic information systems (GIS), global positioning systems (GPS), and remote sensing (RS), has prompted their intensified adoption by local governments (Masser and Onsrud, 1993). Geospatial information technologies have been met with great enthusiasm, given their suitability to handle local government data, over 70% of which can be referenced by location (O'Looney, 2000).

Since the late 1980s, local governments have invested heavily in geospatial technologies, GIS in particular. A survey by Public Technology Inc. indicates that by the early 2000s GIS have become integral resources in various local functions, including urban planning, public works, financial, public safety, and economic development (Public Technology, Inc. 2003). Both early (Budić, 1993; Croswell, 1991; French and Wiggins, 1990) and recent GIS studies (Caron and Bédard, 2002; Haithcoat et al., 2001; Huffman and Hall, 1998; ICMA, 2002; Kreizman, 2002; NACo, 1999; Norris and Demeter, 1999; Warnecke et al., 1998) provide useful information about the diffusion and the use of geospatial technologies in the United States. Over the course of two decades, almost all large cities and counties in the United States started to use GIS. RS and GPS have not been as prevalent, although they are important sources of spatial data for local government applications, digital orthophotography at the county level in particular. However, the availability of satellite images of resolution 1 m and higher may change this practice, and increase the reliance of local governments on satellite data.

Along with other information technologies employed at the local level, geospatial technologies are contributing to the emerging phenomenon of E-governance (Greene, 2001). They are becoming an essential part of inter- and intra-governmental processes to support local functions, public service, and decision making at the operational, management, and policy levels (Georgiadou et al., 2006). They also represent the building blocks in the development of regional, state, and national spatial data infrastructures (NSDI) (Nedović-Budić and Budhathoki, 2006; Harvey and Tulloch, 2006).

35.2   EXAMPLES OF LOCAL GOVERNMENT APPLICATIONS

Geospatial technologies (GIS/GPS/RS) are applied to a wide range of local government functions. The scope of those functions varies across more than 87,000 of the local government entities accounted for in the United States (over 3000 counties; about 35,000 municipalities and townships; over 13,000 school districts, and 35,000 special districts) (U.S. Bureau of the Census, 2008, Web site). The main local government functions are health and safety; public works; recreation and culture; urban development; administration; finance; and management (Huxhold, 1991) (Table 35.1).

Health and safety applications rely primarily on emergency services—fire and police protection. These two are the key local-government functions, and, in many cases, the backbone of local government usage of geospatial technologies. Street network files coupled with accurate addresses and routing software allow for prompt response to emergency calls. With increased occurrences of natural disasters (e.g., flooding, earthquakes, and storms) and incidences of terrorism, geospatial technologies have also become a prominent tool for disaster mitigation and evacuation planning.

Figure 35.1 displays results of a spatial query on the properties that may be affected by 100 year flooding. General community health applications are also common as shown in Figure 35.2 with environmental factors hypothesized as related to the extent

TABLE 35.1
Functions of Local Government

Health and Safety	Public Works	Recreation and Culture
Building codes	Building and structures	Cable television services
Disaster preparedness	Forestry services	Community events
Fire services	Transportation	Historic/cultural preservation
Health services	Utilities	Recreational facilities/parks
Police services	Planning	Education
Safety services	Engineering	
Sanitation	General	

Urban Development	Administration	Finance
Business regulation	Equipment maintenance	Budget administration
Economic development	Information services	Property assessment
Environmental amenities	Records management	Revenue collection
Harbor management		

Housing	Management	
Land-use control/mapping	Coordination of services	Districting
Neighborhood preservation	Elections	Land-records management
Planning	Legislative activity	General
Demographic analysis	Long-range planning	
General	Policy development	

Source: Huxhold, W.E., *An Introduction to Urban Geographic Information Systems*, Oxford University Press, Oxford, U.K., 1991.

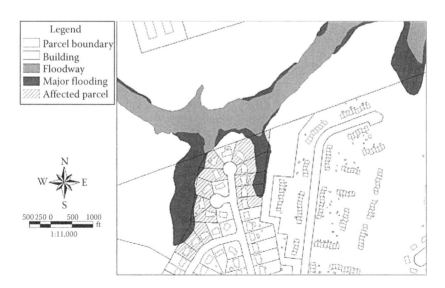

FIGURE 35.1 Properties affected by flooding. (Courtesy of PC ARC Info Training Kit, ESRI, Inc.)

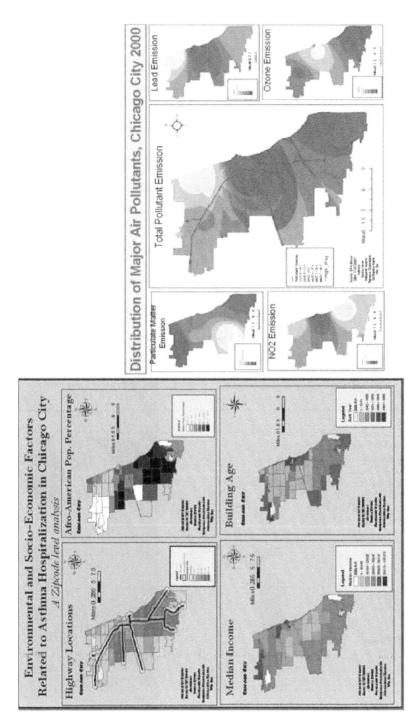

FIGURE 35.2 (See color insert following page 426.) Occurrence of asthma and its causes, City of Chicago, IL. (Courtesy of U.S. Census Bureau; UIUC, UP418 GIS class project.)

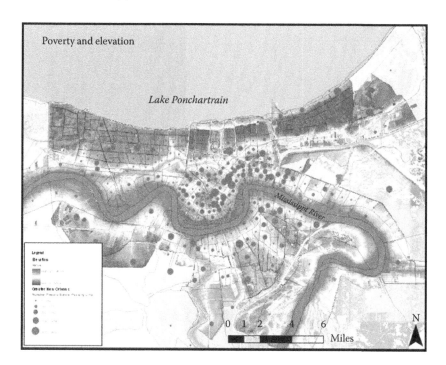

FIGURE 35.3 **(See color insert following page 426.)** City of New Orleans–Presence of population in poverty. (Courtesy of USGS; U.S. Census Bureau; UIUC, UP418 GIS class project.)

and the spatial distribution of asthma cases in the city of Chicago. In the emergency context, geospatial technologies have played important supporting role in responding to both 9/11 (New York, 2001) and Katrina (New Orleans, southern Louisiana, 2005) crises (Adam et al., 2006; Butler, 2005). Figure 35.3 relates the results of GIS overlay analysis of ground elevation and the population in poverty by census block group in the city of New Orleans.

Urban development applications span most of the planning departments' functional areas including neighborhood and economic development, land use and comprehensive planning, and environmental protection. Figure 35.4 shows an analysis of street pattern in two neighborhoods in the city of Urbana, IL—one traditional grid pattern in the city's historic district and one curvilinear suburban pattern in a more recently developed area. Figure 35.5 displays properties in the city of Champaign, IL, tax increment financing (TIF) district, and information on the changes in their equalized assessed value (EAV) after the program's 10 year and 17 year period. Figure 35.6 displays population density data for the Cities of Champaign and Urbana, IL, by census block group, providing the basis for further socioeconomic and urban change analyses. Figure 35.7 shows the City of Champaign proposed land use map that was featured in the city's 2002 comprehensive plan update.

Public works applications are usually performed by engineering units and often entail datasets with higher degree of positional accuracy than those used by planners.

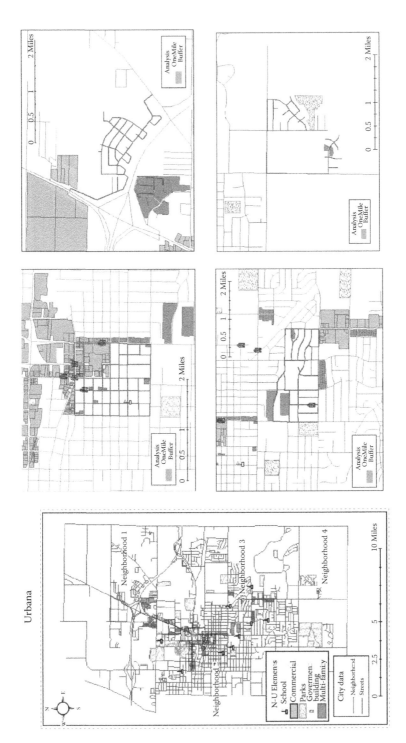

FIGURE 35.4 (See color insert following page 426.) Street pattern analysis, City of Urbana, IL. (Courtesy of U.S. Census Bureau TIGER files and city of Urbana; UIUL, UP418, GIS class project.)

FIGURE 35.5 **(See color insert following page 426.)** Tax increment district (TIF) performance analysis, City of Champaign, IL. (Courtesy of City of Champaign, IL; UIUC, UP418, GIS class project.)

The maintenance of water utility features (Figure 35.8) and the display of street network characteristics (Figure 35.9) are two examples of engineering base data.

Administration and *management* applications focus on inventorying and allocating local government resources, providing public access to those resources, performing political functions, and creating long-range plans and policies. Figure 35.10 is an example of using geospatial technologies for creating voting districts.

Recreation- and *culture*-related applications focus on providing public amenities and facilities in the areas of recreation, education, and culture. Geospatial technologies can be useful in planning and managing those recreational amenities and

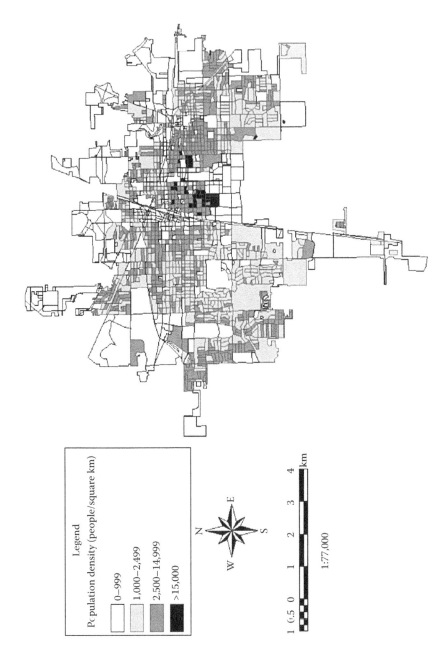

FIGURE 35.6 Population density by census block group. (Courtesy of U.S. Census Bureau; UIUC, UP318 GIS class project.)

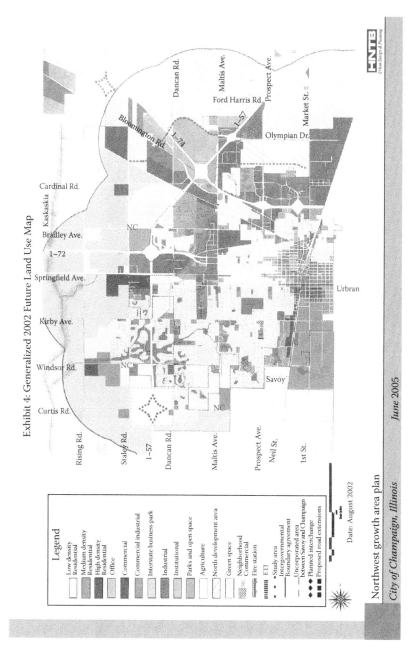

FIGURE 35.7 **(See color insert following page 426.)** Proposed land use map, City of Champaign, 2002 Comprehensive Plan update. (Courtesy of City of Champaign, IL.)

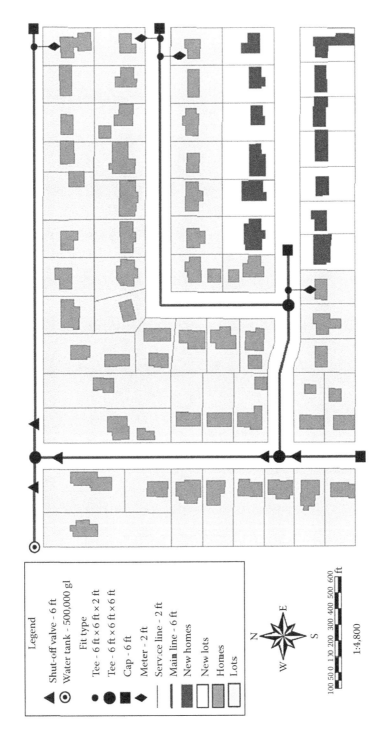

FIGURE 35.8 Utility mapping. (Courtesy of ArcInfo 8.0 sample/exercise data-set, ESRI, Inc.)

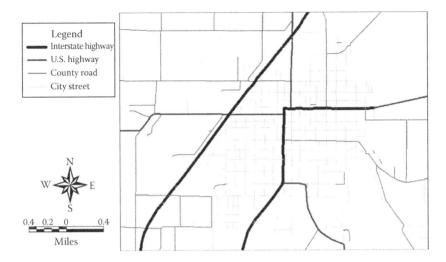

FIGURE 35.9 Street network. (Courtesy of UIUC, UP318, GIS class project.)

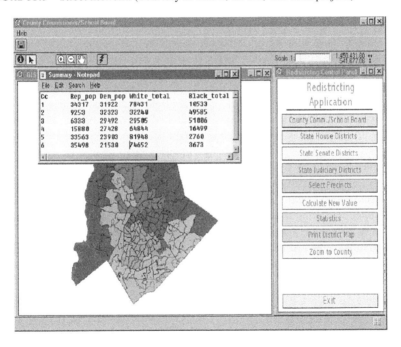

FIGURE 35.10 **(See color insert following page 426.)** Election districts generated by GIS districting function. (Courtesy of Macklenburg County, North Carolina, GIS Department.)

facilities. For example, street network files can be used to derive service areas of each recreational area in a city (Figure 35.11). This could also be done by using distance by air that is less precise representation, but useful for a general understanding of the population capture area and the distribution of parks (Figure 35.12). Land cover data derived from satellite images are used to understand the vegetation patterns between

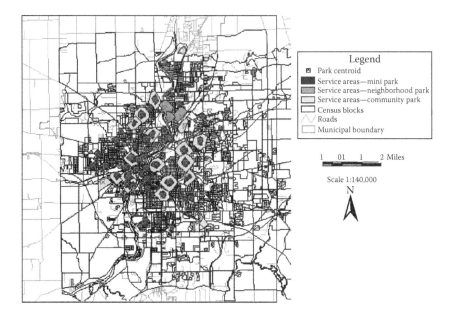

FIGURE 35.11   (See color insert following page 426.) Delineation of service areas for mini, neighborhood, and community parks. (Courtesy of Illinois Department of Natural Resources; UIUC, GIS-based Illinois Recreational Facilities Inventory Project (G-IRFI).)

existing recreational sites and to explore the landscape patterns that may provide for connectivity between the sites. Streams, abandoned railroads, bike trails, and utility corridors are targeted as potential connections (Figure 35.13). Finally, for educational purposes, Figure 35.14 illustrates a suitability analysis of school location and accessibility to walking students in the City of East St. Louis, IL.

Finally, property assessment is probably the most fundamental local government function in the area of *finance* for which geospatial technologies are becoming extensively applied. Deriving from various legacy land records management systems and computerized mapping software, the geospatial technologies applications in property assessment allow already sophisticated computerized-assisted mass appraisal (CAMA) systems to take advantage of spatial representation and visualization tools (see Figures 35.1 and 35.5). Many localities are currently displaying their property information on the Internet, allowing for the display and the search of parcel and tax assessment data. An example from Honolulu City and County, Hawaii, is provided in Figure 35.15.

35.3   BUILDING A LOCAL DATABASE

Federal Geographic Data Committee (FGDC, Web page) has, through a survey of geospatial data producers and users at all levels and across public, private, and nonprofit sectors, identified the following seven layers for which there is a common and recurring need: geodetic control, orthoimagery, elevation, transportation, hydrography, governmental units, and cadastral information. These are also developed or

Urbana Park District Service Areas

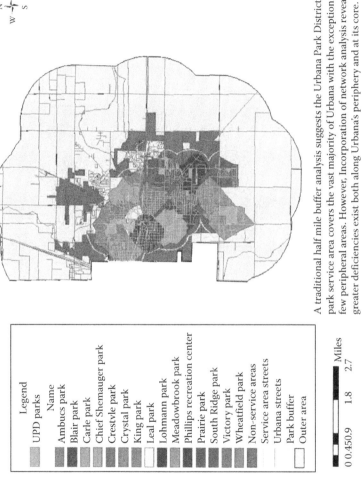

Legend

UPD parks

Name
Ambucs park
Blair park
Carle park
Chief Shemauger park
Crestvle park
Crystal park
King park
Leal park
Lohmann park
Meadowbrook park
Phillips recreation center
Prairie park
South Ridge park
Victory park
Wheatfield park
Non-service areas
Service area streets
Urbana streets
Park buffer
Outer area

0 0.45 0.9   1.8   2.7 Miles

A traditional half mile buffer analysis suggests the Urbana Park District park service area covers the vast majority of Urbana with the exception of a few peripheral areas. However, Incorporation of network analysis reveals greater deficiencies exist both along Urbana's periphery and at its core.

FIGURE 35.12 (See color insert following page 426.) Park service analysis, City of Urbana, Illinois. (Courtesy of UP418, UIUC GIS class project, City of Urbana Park District.)

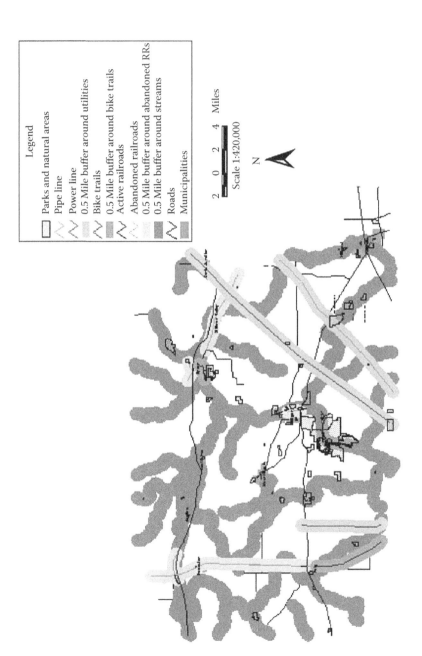

FIGURE 35.13 **(See color insert following page 426.)** Connectivity analysis for regional recreational/open space, City of Rockford, Illinois. (Courtesy of Illinois Department of Natural Resources; UIUC, GIS-based Illinois Recreational Facilities Inventory Project (G-IRFI).)

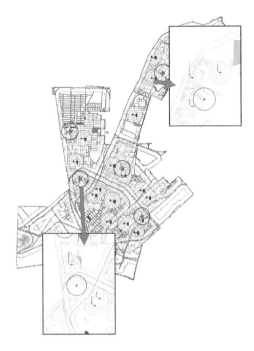

FIGURE 35.14   (See color insert following page 426.) Suitability analysis of elementary schools' accessibility to walking students, East St. Louis, IL. (Courtesy of U.S. Census Bureau, City of East St. Louis; UIUC, ESLARP project and UP418, GIS class project.)

FIGURE 35.15   (See color insert following page 426.) Parcel mapping, City and County of Honolulu, Hawaii. (Courtesy of City of Honolulu, HI, GIS, http://gis.hicentral.com/.)

used at the local level. As reported in a national survey by Warnecke et al. (1998), over 40% of the local governments sampled have the following components in their geospatial database:

1. Roads
2. Hydrology
3. Political/administrative boundaries
4. Cadastral/land records
5. Land use/zoning
6. Elevation
7. Digital imagery
8. Geodetic control

Geodetic control, parcels, street centerlines, and planimetrics (e.g., hydrography, building outlines) are the most common local base maps. These maps are complemented by natural/environmental features, land use, and other community relevant data. Data conversion and GIS database development consume substantial resources, particularly in the initial stage of system implementation. Database investments often exceed a million dollar figure, and rise exponentially with an increase in the accuracy of the map (Antenucci et al., 1991). The database costs represent a considerable portion—up to 80%—of the total system development. The conversion of cadastral boundary and other information from paper format (plat books) to digital files is probably the most extensive effort, particularly for large jurisdictions striving for high data resolution and accuracy. This process is complex and has many technical and organizational aspects. Manuals such as the one developed for Massachusetts' local governments and Urban and Regional Information Association's membership are useful guides to ensure that parcel mapping is properly managed and results in high-quality database (Parrish, 1999; University of Massachusetts, 1999).

Geographic information databases are developed from multiple sources, including paper maps, digital files, and field inventory (i.e., primary data collection). Depending on cost-effectiveness, database developers have a choice of several methods to employ (Montgomery and Schuch, 1993; Ngan, 1998):

1. Digitizing
2. Keyboard entry
3. Scanning (raster)
4. Automatic conversion (vectorization)
5. Surveying (GPS)
6. Coordinate geometry (COGO)
7. Aerial photography/digital orthophotos
8. Satellite RS
9. Computer-aided design (CAD) files (drawings)
10. Light detection and ranging (LIDAR)
11. Third party data (digital line graphs—DLG; topological integrated geographic encoding and referencing—TIGER; digital elevation model—DEM; community data; commercial databases, etc.)

Data conversion may be completed in-house or outsourced to contractors (albeit with care, in order to ensure the building of in-house technical capacity). In addition to data conversion, establishing a geodatabase includes database design, implementation, and management (Yeung and Hall, 2007). Other relevant aspects have to do with understanding of user needs and organizational processes and functions that the database and the system serve, as well as the adherence to spatial data standards. Finally, geospatial data that is dynamic and time-sensitive requires a continuous update and maintenance.

Whenever possible, it is most efficient to use existing ("third party") data already available in digital form. The development of spatial data infrastructures (SDIs) in the United States and worldwide have started to facilitate searching and accessing spatial data from many sources, producers, and depositories (Masser, 2005). SDIs rely on Internet-based data clearinghouses or portals that feature data or metadata on a variety of themes and geographic areas, as well as of diverse scale and resolutions. For the United States, the national portal is called Geospatial One Stop (http://gos2.geodata.gov/wps/portal/gos). Geospatial data also come in a variety of formats including drawing, proprietary GIS formats, and image files. Table 35.2 provides a selected list and summarizes the information on various geospatial-data formats (Lazar, 1998; Theobald, 2007). For further discussion of data acquisition and development methods, see Chapters 17 and 28.

35.4 MANAGING LOCAL GOVERNMENT GEOGRAPHIC INFORMATION RESOURCES

Local government adoption of geospatial technologies requires various organizational and management activities. Geospatial technology implementation may be pursued in a single department, shared by multiple departments, or corporate (enterprise-wide). Huxhold and Levinsohn (1995) differentiate three types of organizational placements: (a) operational—within an existing functional unit (e.g., planning, tax assessment, public works/utilities, information systems, data processing, or GI/GIS departments); (b) administrative—as a separate functional unit focused on managing information resources, directly responding to top management; and (c) strategic—as a service unit to government top management and administration. Regardless of the placement, effective government information systems integrate data at operational, management, and policy levels. Policy-level information is the most general, used to address organization-wide issues; management information is used for the efficient and the effective allocation of government resources; and the most detailed operational-level information supports the daily delivery of services to the public, for example, the issuance of building permits, pavement repairs, and tax assessment. The key to successful use and the management of geospatial technologies and information resources is that database development and maintenance becomes internalized into daily organizational processes.

Geospatial technologies and information resources can also be distributed across organizational boundaries to include multiple local governments and nonprofit groups, or to involve private sector partners (O'Looney, 2000). Interorganizational

TABLE 35.2
Digital Geographic Data Formats

Format	Description
AutoCAD drawing exchange format (DXF)	Vector format that has become the de facto standard for transfer of data between different CAD systems; not well suited for transferring attribute data. AutoCAD drawing is *.DWG.
ARC/INFO export format (e00)	Vector format intended to transfer coverages or GRID data and some third-party products.
ArcView shape file format	Openly published, this vector format is available for use by other GIS vendors. It consists of three types of files: main files (SHP), index files (SHX), and dBASE tables (DBF) for storing attributes.
ETAK	Etak, Inc.'s format for storing street networks.
ERDAS	GIS (*.gis) and Imagine (*.img, *.igw; raw *.raw) files
GRID, SDE raster	ESRI (spatial analyst) raster format
GIRAS	The USGS's geographic information retrieval analysis system (GIRAS) provides land use and land cover data.
MapInfo interchange format (MIF/MID)	MIF files store vector graphical information; MID files store attribute data.
MicroStation design file format (DGN)	Openly documented (except for some newer and product-specific extensions) vector format used by Bentley's MicroStation CAD software. The format does not store attribute data, but can store links to relational database records.
MicroStation IGDS	Interactive graphic design software—CAD export format.
Intergraph raster file	*.CIT or *.COT
DLG	Vector format used by the USGS; providing data on transportation, hydrography, contours, and public land boundaries. To a lesser extent, used by other federal and state government agencies.
USGS DEM	USGS's digital elevation model (DEM)—a raster grid of elevation values.
Triangulated irregular networks (TIN)	Represent surfaces using a network of irregularly shaped triangles that are continuous and nonoverlapping; they are typically based on point-based elevation values. TIGER/line format
	The topologically integrated geographic encoding and referencing (TIGER) format is used by the U.S. Census Bureau to distribute vector and attribute data from its TIGER database (census boundaries and data).
Spatial data transfer standard (SDTS) format	Standard format (used by the USGS, other federal agencies in Australia and South Korea) designed to support all types of vector and raster spatial data, as well as attribute data. The topological vector profile (TVP) and the raster profile are the implementation of subsets of the SDTS.

TABLE 35.2 (continued)
Digital Geographic Data Formats

Format	Description
Spatial archive and interchange (SAIF) format	The national spatial data exchange standard in Canada, supporting vector, raster, and attribute data.
National transfer format (NTF)	The national spatial data exchange standard in the United Kingdom supporting vector, raster, and attribute data.
Vector product format (VPF)	Vector format with attribute data that is part of the NATO DIGEST standard. It is used by the digital chart of the world (DCW).
Tagged image file format (RIFF or TIF)	Raster format frequently used for imagery; GeoTIFF is an extension of TIFF that includes georeferencing information.
Joint photographic experts group (JPEG) or JPG) format	Compression format for storing full color and gray scale images

Sources: Lazar, B., External data sources and formats, in Hohl, P. (ed.), *GIS Data Conversion Strategies, Techniques, and Management*, Chapter 9, OnWord Press, Albany, NY, 1998, 179–204; Theobald, D.M., *GIS Concepts and ArcGIS Methods*, 3rd edn, Conservation Planning Technologies, Fort Collins, CO, 2007.

geographic information activities are stimulated by existing interdependencies between various organizations, but are also challenged by the complexity of the relationships (Nedović-Budić et al., 2004). To illustrate this complexity, Fletcher (1999) proposes four levels of interoperability: global, regional, enterprise, and product; three types of interoperability: institutional, procedural, and technical; and three dimensions of interoperability: horizontal, vertical, and temporal (Figure 35.16).

An important factor for achieving interoperability and multi-participant developments of geospatial technologies is sharing and easy access to geospatial information (Harvey and Tulloch, 2006). Sharing geospatial information is believed to promote more effective use of organizational resources and cooperation among involved organizational entities (Nedović-Budić and Pinto, 2001; Onsrud and Rushton, 1995). Obstacles to data sharing are numerous, including both technical and nontechnical factors, and socio-technical approach is recommended as a way to deal with both and balance them (Reeve and Petch, 1999). On the data side, for example, it is very hard to resolve the varying needs for scales and accuracy that local governments located in the same region may have. The scales of geographic data layers used by local governments range from 1:500 to 1:24,000. The accuracy requirements range from sub-centimeter to dozens of feet and the temporal scale (i.e., the frequency of update) can also vary greatly across the local organizations. On the institutional side, the challenges also persist, but are tackled by the knowledge of practices that lead to successful sharing (Johnson et al. 2001; Nedović-Budić et al., 2004; Tulloch and Harvey, 2007).

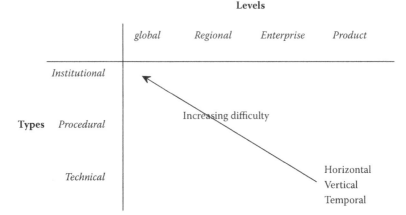

FIGURE 35.16   Interoperable system dimensions. (Courtesy of Fletcher, D.R., URISA publication Enterprise GIS, 1999.)

35.5   TOWARD E-GOVERNANCE

The advances in information and communication technologies in the 1990s, Internet in particular, have facilitated the access and the exchange of geospatial information. Local governments now add Internet GIS (or Web-GIS or Web mapping; Harder, 1998; Peng and Tsou, 2003) as their standard application to enable open access to local geospatial resources and increase the efficiency and the effectiveness of their public services. Tax assessment information and parcel maps are the common feature of those sites, with other data layers being readily available for search, viewing, and, in some cases, overlay and annotation. Examples abound at municipal, county, and regional levels. Following are a few: Metro GIS—seven-county Minneapolis-St. Paul (Minnesota) metropolitan area (http:// www.metrogis.org/); City and County of Honolulu, Hawaii (http://gis.hicentral. com/); San Diego (City/County) Geographic Information Source (http://www.san-gis.org/); Guilford County, NC (http://gcgis.co.guilford.nc.us/guilford_new/); City of Miami, FL (including 3D viewing with Virtual Earth; http://www.miamigov. com/GISWebPortal/pages/default.asp).

These interactive GIS Web sites and databases that they build upon also represent the building blocks of the regional, state, and NSDI.* While the integration of geospatial information resources at the metropolitan level is not easy to accomplish (Knaap and Nedović-Budić, 2003), those regions are the key unit for addressing many societal issues—from urban growth and environmental degradation to disaster recovery and national security. The importance of building a sustainable and useful NSDI became particularly apparent after the events of 9/11 (CAD DIGEST Web

---

* The United States was among the first countries to embrace the idea of building the NSDI about 15 years ago. The main impetus was given by President Clinton's Executive Order 12906 issued in April of 1994 (FGDC Web site) and Office of Management and Budget's (OMB) Circular A-16 and E-government Act of 2002 (FGDC Web site; U.S. OMB Web site).

site; GIS monitor Web site) and hurricane Katrina (UCGIS Web site). To enhance the emergency preparedness across the localities, the U.S. Department of Homeland Security is maintaining the First Responders initiative for training, equipment, and management (http://www.dhs.gov/xfrstresp/).

These technological and institutional developments bear many civilian benefits by democratizing the access to geospatial information and facilitating its use by a broad spectrum of interested individuals, groups and organizations from public, private, and nonprofit sectors. Local planning, for example, is coming online (Yigitcanlar, 2006). The term public participation GIS (or PPGIS) and the related activities have become common across the developed and developing societies (http://www.ppgis. net/). PPGIS provides support for public involvement in policy and decision making processes at various levels on many issues (Schlossberg and Shuford, 2005). The visualization of geospatial information is an important element of understanding and resolving societal and community problems (Hopkins and Zapata, 2007; van den Brink et al., 2007). This concept has extended from data access toward non-institutionalized developments of geospatial databases that tap into the enthusiasm and the energy of ordinary citizens for collecting, capturing, and contributing data to a common source. The so-called volunteered geographic information (VGI) has already started to have an impact on the perceptions and practices related to the management of geospatial resources, primarily in terms of reconceptualizaton of the notions of information producers and users (Budhathoki et al., 2008). Probably the most well-known VGI exercise is the OpenStreetMap worldwide map project (http://www.openstreetmap.org/).

35.6   CONCLUSION

This chapter presents the trends in the adoption of geospatial technologies in local government settings, a cross-section of local government functions and their corresponding applications, and the major sources and issues in building a local government geospatial information database. The local governments are the most massive user community for geospatial technologies. At the same time, they are the most challenging environments for introducing the new technologies. The implementation of geospatial technologies in local government settings is ridden by obstacles of both technical and organizational nature. Finances, staffing, and organizational conflict are the strongest impediments to the successful introduction of geospatial technologies. On the technical side, data development costs, data formats, and integration are the main implementation challenges.

Local governments, however, are also the most important community of current and potential users of geospatial technologies and the investments into their modernization are worthwhile. Most decisions affecting the quality of life and the livability of local settlements are made at the local level, where the new geospatial technologies can make the largest impact. Under the circumstances, those technologies have become a constituent element in the establishment of E-government and increased accessibility of data to a variety of public, private, and nonprofit entities.

REFERENCES

Adam, B., Boyd, K., and Harvey, C., Hancock county EOC support: Mapping the hardest-hit county post-katrina, Paper 1864 in *Conference Proceedings of 26th Environmental Systems Research Institute (ESRI) User Conference*, San Diego, CA, August 7–11, 2006. http://gis.esri.com/library/userconf/proc06/papers/papers/pap_1864.pdf.

Antenucci, J.C., Croswell, P.L., Kevany, M.J., and Archer, H., *Geographic Information Systems: A Guide to the Technology*, Van Nostrand Reinhold, New York, 1991.

Budhathoki, N.R., Bruce, B., and Nedović-Budić, Z., Reconceptualizing the role of the user of spatial data infrastructure, *GeoJournal*, 72, 2008, 149–160.

Budić, Z., GIS use among southeastern local governments, *Journal of Urban and Regional Information Systems Association*, 5(1), 1993, 4–17.

Butler, H., *Katrina Imagery Warehouse: The Inside Story*, Directions Magazine, 2005. http://www.directionsmag.com/article.php?article_id = 2008&trv = 1.

CAD DIGEST—The reading room for computer aided design, *9/11—The Technical Side*. http://www.caddigest.com/subjects/wtc/index.htm.

Caron, C. and Bédard, Y., Lessons learned from the case studies on implementation of geospatial technologies, *Journal of Urban and Regional Information Systems Association*, 16, 2002, 17–36.

Croswell, P.L., Obstacles to GIS implementation and guidelines to increase the opportunities for success, *Journal of the Urban and Regional Information Systems Association*, 3(1), 1991, 43–56.

Federal Geographic Data Committee (FGDC—http://www.fgdc.gov), Framework overview. http://www.fgdc.gov/framework/

Fletcher, D.R., The interoperable enterprise, in N.R. Von Meyer and S. Oppman (Eds.), *Enterprise GIS*, Chapter 2, Urban and Regional Information Systems Association, Park Ridge, IL, 1999, pp. 7–16.

French, S. and Wiggins, L., California Planning agency experiences with automated mapping and geographic information systems, *Environment and Planning B*, 17(4), 1990, 441–450.

Georgiadou, Y., Rodriguez-Pabón, O., and Lance, K.T., Spatial data infrastructure (SDI) and E-governance: A quest for appropriate evaluation approaches, *Journal of Urban and Regional Information Systems Association*, 18, 2006, 43–55.

GIS Monitor, *GIS Articles and Data Links on Attacks*. http://www.gismonitor.com/articles/comment/092601disaster_articles.php.

Greene, R.W., *Open Access—GIS in e-Government*, ESRI Press, Redlands, CA, 2001.

Haithcoat, T. Warnecke, L., and Nedović-Budić, Z., Geographic information technology in local government: Experience and issues, in *The Municipal Year Book 2001*, International City/County Management Association, Washington, DC, 2001.

Harder, C., *Serving Maps on the Internet—Geographic Information on the World Wide Web*, ESRI Press, Redlands, CA, 1998.

Harvey, F. and Tulloch, D., Local-government data sharing: Evaluating the foundations of spatial data infrastructures, *International Journal of Geographical Information Science*, 20, 2006, 743–768.

Hopkins, L.D. and Zapata, M. (Eds.), *Engaging the Future: Forecasts, Scenarios, Plans and Projects*, Lincoln Institute of Land Policy, Cambridge, MA, 2007.

Huffman, L.A. and Hall, G., Use of innovative technology applications in local government, *Special Data Issue*, No. 1, International City/County Management Association, Washington, DC, 1998.

Huxhold, W.E., *An Introduction to Urban Geographic Information Systems*, Oxford University Press, Oxford, U.K., 1991.

Huxhold, W.E. and Levinsohn, A.G., *Managing Geographic Information System Projects*, Oxford University Press, New York, 1995.

International City/County Management Association (ICMA), *2002 Survey*. http://www.icma. org, 2002.

Johnson, R., Zorica, N.-B., and Kathy, C. (with J.K. Pinto), Lessons from practice—A Guidebook to Organizing and Sustaining Geodata Collaboratives. Reston, VA: Geo Data Alliance.

Knaap, G. and Nedović-Budić, Z., *Assessment of Regional GIS Capacity for Transportation and Land Use Planning*, report to Lincoln Institute for Land Policy, HUD, and U.S. DOT, University of Maryland and University of Illinois @ Urbana-Champaign. 2003. http://www.urban.uiuc.edu/faculty/budic/W-metroGIS.htm

Kreizman, G./Gartner, *U.S. Public-Sector GIS Survey: Key Issues and Trends*, Market Analysis, December 2002. http://www4.gartner.com/Init

Lazar, B., External data sources and formats, in P. Hohl (Ed.), *GIS Data Conversion Strategies, Techniques, and Management*, Chapter 9, OnWord Press, Albany, NY, 1998, pp. 179–204.

Masser, I., *GIS Worlds—Creating Spatial Data Infrastructures*, ESRI Press, Redlands, CA, 2005.

Masser, I. and Onsrud, H., *Diffusion and Use of Geographic Information Technologies*, Kluwer Academic Publishers, Dordrecht, the Netherlands, 1993.

Montgomery, G.E. and Schuch, H.C., *GIS Data Conversion Handbook*, GIS World, Fort Collins, CO, 1993.

National Association of Counties (NACo). http://www.naco.org

Nedović-Budić, Z. and Budhathoki, N.R., Technological and institutional interdepen-dences and SDI–the Bermuda square? Editorial, *International Journal of Spatial Data Infrastructure Research*, 1, 2006, 36–50. http://ijsdir.jrc.it/

Nedović-Budić, Z. and Pinto, J.K., Organizational (Soft) GIS interoperability: Lessons from the U.S., *International Journal of Applied Earth Observation and Geoinformation*, 3, 2001, 290–298.

Nedović-Budić, Z., Pinto, J.K., and Warnecke, L. GIS database development and exchange: Interaction mechanisms and motivations, *Journal of the Urban and Regional Information System Association*, 16, 2004, 15–29.

Ngan, S., Data models, collection considerations, and cartographic issues, in Hohl, P. (Ed.), *GIS Data Conversion: Strategies—Techniques—Management*, Chapter 8, Onword Press, Santa Fe, NM, 1998, pp. 161–178.

Norris, D.F. and Demeter, L.A., Information technology and city governments, in *The Municipal Year Book 1999*, International City/County Management Association, Washington, DC, 1999.

O'Looney, J., *Beyond Maps: GIS and Decision Making in Local Government*, ESRI Press, Redlands, CA, 2000.

Onsrud, H.J. and Rushton, G., *Sharing Geographic Information*, Center for Urban Policy Research, New Brunswick, NJ, 1995.

Parrish, K., *Digital Parcel Mapping Handbook*, Urban and Regional Information Systems Association (URISA), Park Ridge, IL, 1999.

Peng, Z.R. and Tsou, M.H., *Internet GIS—Distributed Geographic Information Services for the Internet and Wireless Networks*, John Wiley & Sons, New York, 2003.

Public Technology, Inc., *National GIS Survey Results: 2003 Survey on the Use of GIS Technology in Local Governments*, 2003. http://www.pti.org

Reeve, D. and Petch, J., *GIS Organisations and People—A Socio-Technical Approach*, Taylor & Francis, London, U.K., 1999.

Schlossberg, M. and Shuford, E., Delineating "Public" and "Participation" in PPGIS, *Journal of Urban and Regional Information Systems Association*, 16, 2005, 15–26.

Theobald, D.M., *GIS Concepts and ArcGIS Methods*, 3rd edn. Conservation Planning Technologies, Fort Collins, CO, 2007.

Tulloch, D. and Harvey, F., When data sharing becomes institutionalized: Best practices in local government geographic information relationships, *Journal of Urban and Regional Information Systems Association*, 19, 2007, 51–59.

U.S. Census Bureau, *The 2008 Statistical Abstract* (112th edn.), Table 415, US Government Printing Office, Washington, DC, 2008. http://www.census.gov/compendia/statab/tables/08s0415.pdf

United States Office of Management and Budget (OMB), *Circular A-16 Coordination of Geographic Information, and Related Spatial Data Activities; Circular A-76 E-government; A-119 Federal Participation in the Development and Use of Voluntary Consensus Standards and in Conformity Assessment Activities; A-130 Management of Federal Information Resources.* http://www.omb.gov.

University Consortium for Geographic Information Science (UCGIS), *GISc Resources for Hurricane Katrina.* http://www.ucgis.org/kartina.

University of Massachusetts, Office of Geographic Information and Analysis, *Parcel Mapping Using GIS—A Guide to Digital Parcel Map Development for Massachusetts Local Governments*, Commonwealth of Massachusetts Executive Office of Environmental Affairs, Massachusetts Geographic Information System (MASSGIS), 1999.

Van den Brink, A., van Lammeren, R., van de Velde, R., and Däne, S., *Geo-Visualisation for Participatory Spatial Planning in Europe—Imagining the Future*, Wageningen Academic Publishers, Wageningen, the Netherlands, 2007.

Warnecke, L., Kollin, C., Beattie, J., Lyday, W., and French, S., *Geographic Information Technology in Cities and Counties: A Nationwide Assessment*, American Forests, Washington, DC, 1998.

Yeung, A.K.W. and Hall, G.B., *Spatial Database Systems: Design, Implementation and Project Management*, Springer, Heidelberg, Germany, 2007.

Yigitcanlar, T., Australian local governments practice and prospects with online planning, *Journal of Urban and Regional Information Systems*, 18, 2006, 7–17.

36 Geographic Information Technology in State Governments of the United States

David Mockert

CONTENTS

36.1 INTRODUCTION

States play an important role in coordinating the Geographic Information (GI) and GI Technology (GIT) produced at the local (municipal and county) levels for use at the state and federal level as well as the commercial use of geospatial data. The use and the application of geospatial technology at the state level has moved from an agency-specific business function to play a more enterprise-wide role throughout state governments. Where GIT historically grew out of development for managing resources in specific business areas such as Departments of Transportation (DOTs) and Departments of Natural Resources (DNRs), the use of GIT is being applied in almost every function of state government. This chapter provides an overview of GI/GIT applications, governance, coordination activities, data-sharing initiatives, and trends among the nation's 50 states.

713

36.2 APPLICATIONS

The use of geospatial technology within the nation's 50 state governments has become pervasive throughout most areas of government, including transportation, health and family services, natural resources, state land, and agricultural asset management. State government applications also are expanding in human and social services, public health, public safety, criminal justice, emergency management, economic development, and growth management. One critically important area that is applying GIT to support its business is in state emergency management planning and operations (Table 36.1).

States and state agencies initially used GIT independently to automate mapping processes or conduct rudimentary analyses. More recently, GIT has been applied to all levels of activity within state organizations, from policy analysis and development to planning, management, operations, regulation, adjudication, licensing, leasing, and other applications. Geographic Information System (GIS) also supports business and financial functions in government, including the analysis and the optimization of revenues and expenditures, collections and particularly in resource allocation. More and more state agencies apply GIT as an essential tool to manage business processes and deliver public services in a more efficient, effective, and equitable manner. GIT is becoming an important component of government Information Technology (IT) modernization and "reengineering" initiatives within and among individual agencies. While GIT benefits are typically measured in terms of individual agencies and governing functions, the trend is clearly toward multiple agencies within a state government now using GIT within an enterprise-wide approach. Some larger states, such as New York and California, have over 20 separate agencies using some form of GIT. As stated by Wendy Rayner, New Jersey's Chief Information Officer, "every agency, if they know it or not, needs GIS capability to make decisions." A review of some key state government applications and experience is provided in the following text.

36.2.1 EMERGENCY MANAGEMENT

The application of GI in the area of homeland security and emergency management grew significantly following the events of 9/11 and in response to hurricanes Rita and Katrina in 2006. Applications for situational awareness and disaster response management are significantly improved using visualization technologies provided by GIS. The application involves real-time analysis and data management activities, including the integration of data from multiple sources. Disaster response has been an important impetus for GIS activities and funding in some states. For example, while GIS efforts were previously underway in many states, agencies expanded their GIS efforts in response to major disasters including the Midwest floods, west coast fires, and the intense hurricane season of 2006. These disasters, particularly 9/11, caused other states to expand GIS efforts to better prepare for emergencies. GIS use for emergency planning and preparedness has also expanded to almost all states by 2008. Based on regional concerns, GI is being applied to all types of natural and man-made disasters such as flooding, hazardous materials/radiological incidents, earthquakes, wild fires, pandemic outbreaks, tornados, and hurricanes.

TABLE 36.1
GIS Use in State Government (1995)

No. of States	Function of State Government
General State Government: Administration, Planning, Finance, Revenue, Asset Management	
13	Revenue, including property taxation
13	Census data center
12	State planning
9	Budget, finance, comptroller, state property management
4	State surveyor, cartographer, geographer
3	Library
1	Banking regulation
Environmental/Natural Resources	
49	Water—quantity, quality, rights, or drinking
42	Wildlife, game, fish, or biological resources
39	Geological survey
30	Waste management, including solid, low level
29	Air quality
27	Forestry
27	Agriculture
24	Oil/gas/mining regulation and reclamation
22	Public lands management
22	Parks management
20	National heritage program
18	Coastal resources
12	Energy
Cultural Resources	
19	Historic preservation
14	Archaeology
1	Other—museum
Infrastructure	
50	Transportation
9	Utility regulatory commissions
Human Services	
25	Health (primarily epidemiology)
6	Social services
5	Employment security and labor
3	Education
Other	
24	Public safety, emergency management, and military
20	Economic development
20	Community and local affairs

Shortly after the September 11, 2001, terrorist attacks, the National Geospatial Intelligence Agency (NGA) was given a new mission to begin collecting critical geospatial datasets within the United States to support homeland security applications. In response to the new mission, NGA developed a product called Homeland Security Infrastructure Program (HSIP) Gold. The intention was to create a common set of geospatial data layers identifying the locations of critical infrastructure throughout the United States. There are over 300 layers in HSIP Gold, which includes everything from political boundaries to chemical facilities, hotels to Internet service provider locations, and school buildings to water bottling stations. State Emergency Management Agencies (EMAs) are collecting and using these HSIP Gold data layers to support their planning, response, and recovery efforts.

The growing severity and the cost of disasters over the past two decades resulted in an increased need for and the use of better technological capabilities to support emergency response and recovery. Almost as important, GIT is applied by state emergency management organizations to better prepare for and mitigate the impact of disaster events. State EMAs and others are broadening their GIT applications as a result. Many of these activities have been conducted in coordination with the Department of Homeland Security (DHS) and Federal EMA (FEMA).

36.2.2 PLANNING, GROWTH MANAGEMENT, AND ECONOMIC DEVELOPMENT

The planning function has long used GI/GIT because it inherently requires the integration and the analysis of many, often disparate types and formats of data. However, political support for planning is often cyclical, and can directly impact the resources available, the support for and the use of GIT. Some of the earliest state GIS applications can be traced back over 30 years by state planning agencies to assist in land use planning, often stimulated by federal funds. For example, federal support helped develop the Maryland Automated GI (MAGI) system, one of the first state GIS initiatives in the country that was used in statewide land use planning and specifically to prepare the state comprehensive land use and development plan. Mississippi's Automated Resource Information System (MARIS) used similar funding to become one of the earliest users of satellite imagery and one of the first state GIS programs to be authorized by statute (http://www.maris.state.ms.us/). Early state planning statutes and initiatives in Connecticut, Minnesota, New York, North Carolina, and Ohio also stimulated land use data systems, and became leading drivers for GIS.

Several land-planning applications have emerged over time. Minnesota, Ohio, and South Dakota conducted land suitability and capability analyses for specific regions or counties using geoprocessing tools. States have identified "critical" land areas warranting protection using GIS since the 1970s. Farmland preservation and the protection of shore lands, rivers, or wetlands were early applications, such as by New York's Adirondack Park Agency and the California Coastal Commission. Recent attention has been given to protecting drinking water supply, evidenced by the Great Lakes Initiative—and wildlife habitat, such as through the U.S. geological survey (USGS) Gap Analysis Program, which is a nationwide, state-by-state project to inventory, evaluate, and preserve biological diversity. In addition, The Nature

Conservancy's Natural Heritage Program, initiated in 1974, provided an early focal point in each state for the management of biological data inventories that can be used with GIS (http://www.ncnhp.org/). As a result of more public referenda to use state funds to acquire land for public use, states increasingly use GIS to prioritize areas for purchase. For example, voters in Florida recently approved $3.2 billion for land acquisition, with GIS used to inventory potential sites, set criteria for site selection, and monitor acquisition efforts.

Among the strongest state GI/GIT legislative directives to date are state planning or growth management laws. At least 25% of the states adopted planning laws that have legitimized and increased such usage within state governments, and also often in local governments, such as Georgia, Maine, New Jersey, Vermont, Rhode Island, and Washington. In a few states, funding and assistance were provided to develop or assist in preparing local plans in accordance with state goals, often including GIS and related data development. For example, Vermont's regional planning councils were equipped with GIS hardware and software to help localities, and Georgia's Regional Development Centers were given significant data roles for GIS. Other states have also provided funding for GIS use for local and regional planning, including New Hampshire and Utah.

Some states are using GI/GIT in support of economic development, to attract businesses to locate facilities in their states. Some states receive significant political support for this application; such as to attract businesses and optimize siting decisions. For example, Illinois, North Carolina, and others used GIS to develop proposals for the superconducting super collider project in the late 1980s. South Carolina is long recognized as a leading GIS user for economic development. The state developed natural resources, transportation, utilities, and socioeconomic data with GIS specifically to attract businesses, with successes such as the Bavarian Motor Works (BMW) manufacturing plant. Similarly, in 2006, Indiana was able to win the bid for the location of Honda Motor Company's latest manufacturing plant due in part to the IndianaMap (http://www.in.gov/igic/temp/IndianaMapROI.pdf).

Several states have adopted "smart growth" initiatives that require increased GIS use accordingly. In addition, the U.S. Department of Housing and Urban Development provided GIS software and support for use by its 50 state counterparts and grantees. This is an important trend for the future because almost by definition, planning encourages the coordination and the integration of often disparate and sometimes conflicting data to analyze conditions, develop scenarios, and plan for the future. This can, in turn, encourage multiple agencies to develop compatible data. This attention to data can further stimulate coordination and commonality among agencies within individual governments, and incent coordination with the private business sector.

36.2.3 Environment and Natural Resources

Authority, roles, and responsibilities for Environmental and Natural Resources (ENR) are traditionally shared by federal, state, and local governments, with each level having a leading role for specific ENR functions. States have numerous ENR responsibilities, some of which were among the earliest and currently are some of the most developed applications utilizing GI in state government today. Early uses were initiated to better manage various individual natural resources for which states have stronger official responsibility than the federal government, such as forestry, water, and wildlife.

Water resources management was one of the earliest and is one of the strongest uses of GIS in states. For example, the Texas Water Oriented Data Bank that was authorized by the legislature in the 1960s to coordinate response to droughts and floods was responsible for stimulating some of the earliest uses of GIS among the states. Water usage and analysis was also one of the first GIS uses in other states, such as Florida, Idaho, Iowa, Massachusetts, Utah, and Wyoming. For example, Idaho's Department of Water Resources (DWR) began to use satellite imagery and geoprocessing to analyze and monitor irrigated agriculture as early as the mid-1970s, later extending to other applications. Other states similarly used GIT for water quantity, a critical function for western states, including appropriation and use management. However, with the 2006–2008 drought in the southeastern United States, water quantity also has become an important focus for all states, both for ground and surface water. A list of state water agencies can be found at the following Web site: http://www.waterwebster.com/state_framebottom.htm.

State Forestry Organizations (SFOs) were among the earliest GIS users to help accomplish their missions, including planning for and managing state-owned forests, conducting statewide forest inventories, mitigating and controlling fires, managing pests, and providing technical assistance for and conducting oversight over private forested lands. One of the largest SFOs, Washington's DNR (WDNR) uses GIS to help manage over 3 million acres of land held in trust. WDNR has one of the largest GIS installations of any state agency in the country. Other states with strong forestry GIS usage include Alaska, California, Maine, Minnesota, and Wisconsin. GIS also has been used extensively to manage non-forested public lands and other natural resources.

Environmental protection is a growing application in states, due to both state initiative and federal direction and funding. Both state and federal environmental legislation has stimulated the development of GIS and data for many applications over the years. For example, Illinois and Kentucky used federal funding for some of their earliest GIS activities to determine lands unsuitable for coal surface mining. Ohio's Environmental Protection Agency (EPA) was the first state environmental agency to start using GIS in 1975, with both state and U.S. EPA (USEPA) funding. USEPA and state support for GIS has grown across the country during the last two decades. GIS use has grown for individual media such as water, air, and waste management, but also to integrate the use of data for "multimedia" decision support and management. Recent environmental GIS applications help integrate and streamline processes, such as one-stop permitting, and public access to data about environmental conditions, increasingly available through the Internet.

GIT is also used significantly in the area of state fish and wildlife management. The National Biological Information Infrastructure (NBII) is a broad, collaborative program to provide increased access to data and information on the nation's biological resources.

36.2.4 INFRASTRUCTURE: TRANSPORTATION AND UTILITIES

State governments have several important infrastructure responsibilities, including the management of highway, transit, and other transportation facilities and services, and the regulation of utilities. Most early transportation and utility users initiated the use of Computer-Aided Drafting or design (CAD) or mapping (CAM) systems to automate the preparation of facility or highway construction drawings and maps. Among utilities, they became known as Automated Mapping/Facilities Management (AM/FM) systems.

State DOTs/highways were among the earliest, and continue to be some of the largest state government users of GIT, initially to automate various large-scale manual drafting and mapping processes (CAD/CAM), and more recently, to apply analytical capabilities for the analysis of conditions and to improve planning and management processes. DOTs initiated CAD/CAM systems beginning in the 1960s, as in New York and Pennsylvania, with most penetration in other states during the 1980s. Virtually all state DOTs had some type of CAD/CAM activity by 1990. Some mapping installations in DOTs have been the largest technology configurations of any GIS-related systems in state governments.

Highway safety and pavement management were early areas benefiting from analytical GIS capabilities, though these programs often were developed separate from CAD/CAM operations within DOTs. Early examples include Arizona's accident location identification and surveillance system, which was developed in 1970 and was the first statewide digital spatial database in Arizona, and Ohio's state accident identification and recording system. Ohio DOT (ODOT) retrieved and displayed accident-related information for over 112,000 miles of roads in the state, and GIS capabilities were later expanded for use in road inventory, pavement and bridge management, skid resistance, and other applications. Wisconsin's DOT was perhaps the first in the country to apply GIS as part of a comprehensive information management approach to fulfill operational, management, and planning needs, and several other DOTs have adopted similar approaches.

Interagency GI/GIT coordination has grown in state governments over time, but DOTs often operate separately from other agencies. However, DOTs often work with counterparts in other states to address similar needs, with more attention to common GIT issues across states. For example, various interstate DOT committees address related issues, the Transportation Research Board has sponsored studies about GIS in the 50 DOTs, and a national GIS in Transportation (GIS-T) conference has been held annually since 1987. Federal transportation direction, funding, and technical assistance has stimulated GIT use in DOTs and other transportation organizations, including the nation's over 300 Metropolitan Planning Organizations (MPOs). Today, it is generally known that all state DOTs and almost all MPOs use GIT.

While states do not traditionally operate utilities, it is important to recognize that they have regulatory authority over those operating within their borders. Some Public Utilities Commissions (PUCs) use GIS for both internal and regulatory purposes, such as to monitor service areas and analyze demographics. GIS has recently been applied to help manage competition among utilities to ensure the public is adequately served. Early innovators included North Dakota and Ohio, with other states emerging as active users including Kansas, Mississippi, and New Hampshire.

36.2.5 AGRICULTURE AND HEALTH SERVICES

Agricultural GIS applications also are expanding in states, including the evaluation of various environmental conditions, crop inventories and yield forecasting, and pest infestation and crop disease monitoring. The use of GIT for land and soil management is common among most states.

For example, in the state of Florida, the agriculture industry employs approximately 645,000 people and has an overall economic impact estimated at $87 billion each year. Supporting this important economic engine while safeguarding the public is the stated mission of the Florida Department of Agriculture & Consumer Services (FDACS). The department ensures the safety and the wholesomeness of food through rigorous inspection and testing programs. Department personnel monitor fruits and vegetables for pesticide residues, investigate incidents of food-borne illness, and regularly inspect grocery stores and food-processing plants to make sure they are clean and safe. The department is also Florida's lead agency for consumer protection. To help manage its business, FDACS has developed a geospatial data repository (GDI) for spatially managing both regulated production facilities and personnel throughout the state. Integrated location data for employees and regulated entities allows management to locate (visually mapping) field inspection personnel to make best use of their time and resources.

Researchers, public health professionals, policy makers, and others use GIS to better understand geographic relationships that affect health outcomes, public health risks, disease transmission, access to health care, and other public health concerns. GIS is being used with greater frequency to address neighborhood, local, state, national, and international public health issues. The Center for Disease Control (CDC) has information on state health information (National Center for Health Statistics) at this location http://www.cdc.gov/nchs/gis.htm.

36.2.6 OTHER APPLICATIONS

Planning, ENR, and transportation were the early, and continue to be the dominant, uses of GIS in states. Other applications emerged in innovative agencies during the 1980s, but did not penetrate most states until recently. Growth in the quality and the availability of data and technology has stimulated additional state government functions and agencies to apply GIS. As shown in Table 35.1, several states now apply GIS for public safety and emergency management; health, human social services; cultural resources, and other uses. Other public safety functions also use GIS, such as the California Department of Justice, which uses GIS for narcotics tracking. In addition, as mentioned earlier, modernized communications systems with (E911) capability increasingly use GIS to optimize response.

While adopting GIT later than other parts of state governments, various state human and social services, labor and education agencies started using GIT in the 1980s. Table 36.1 shows that most state GIS use has been in epidemiology, largely because ENR data can be analyzed with data about the occurrence of diseases to understand environmental implications on people. Additional applications are emerging as well. For example, New York's Department of Social Services pioneered social services GIS applications to support the state's neighborhood-based strategy for services delivered to children. South Carolina has used GIS to target services for children based on the locations of poverty, crime, teenage pregnancy, infant morbidity, and other conditions. Many states use demographic and migration data to help locate health care and other facilities and services. State labor departments also use GIT to geographically analyze labor statistics and trends to target and locate services, provide job search and placement for welfare recipients, and supply input for transportation and land use planning. Some state education departments have used GIS for similar program planning and management applications, particularly to ensure the equitability of resource allocations. GIS applications are expected to grow dramatically in the twenty-first century, particularly in public safety, and human and social services, because these are among the most costly functions of state governments, and because continually improving data and technology can be deployed for these applications in a very cost-effective manner.

36.3   GI COORDINATION ACTIVITIES IN STATE GOVERNMENTS

States have been providing coordination of geospatial data through data clearing-house initiatives spurred on by the Geospatial One Stop (GOS) and the National Spatial Data Inventory (NSDI). The USGS has recently realigned its spatial programs into a National Geospatial Program Office (NGPO), bringing the National Map, GOS, and the Federal Geographic Data Committee (FGDC) into a single program office. With the creation of the NGPO, the essential components of delivering the NSDI and capitalizing on the power of place will be managed as a unified portfolio that benefits the entire geospatial community (Shanley, 2007).

Policy support for coordinated geospatial data activities is growing within states across the country, and GIS is increasingly viewed as a way to manage, present, and use data from several sources. Several internal and external factors have influenced the development and the use of GIT in states over time. Public and governmental needs and actions increasingly seem to drive GIT and related data development. Practitioners and researchers alike recognize that GIT support, use, and utility depend on political, institutional, financial, and human factors in addition to technical capabilities. Depending on conditions within individual governments, these factors may be obstacles, facilitate efforts, or in some cases, result in a combination of both influences (Figure 36.1).

Societal and government trends affect GIT in several ways. Early experimentation and development was facilitated by increasing public consciousness, societal concern, and, particularly, federal government activism, though such action is cyclical and varies in direction and emphasis over time. Strong federal government involvement in domestic matters during the 1960s led to legislation and programs providing incentives, assistance, and funding for state governments, such as land use planning discussed earlier. Governmental activism also was reflected in the "environmental movement" as numerous significant and long-lasting federal and state environmental laws and institutions were established. While funding support for planning has diminished, many state ENR GIS applications continue to be aided by federal ENR agencies in the Departments of Agriculture, Commerce and Interior, and the USEPA

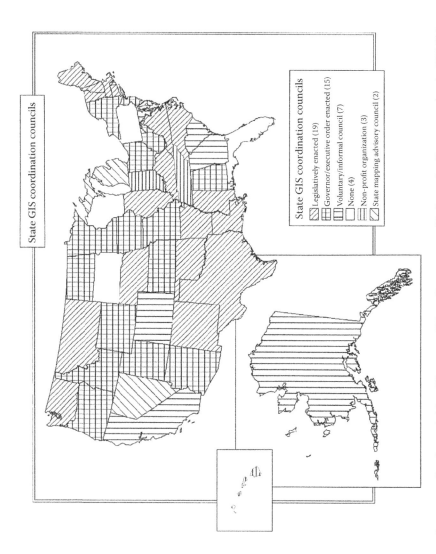

FIGURE 36.1 Map of state coordination activities. (Courtesy of Texas Department of Information Resources, Austin, TX, February 21, 2008.)

as discussed earlier. Federal government activism also is exhibited in direct transfer of technology, data and assistance to states. The National Aeronautics and Space Administration (NASA) funded the investigation and the use of satellite imagery and image processing in states during the late 1970s and early 1980s. The USGS and the FGDC provide aid to states through the CAP grant program (see Chapter 37). State government direction and actions also stimulated GIT development and use. States and localities initially responded well to federal funding, but during the 1980s began to resist federal preemptions, regulations and "unfunded mandates" that accompanied some legislation. At the same time, many new federal roles were transferred or "devolved" to states, which helped stimulate states to improve their institutional capacity and technical capabilities. States have become increasingly active in many areas addressed by the federal government, as well as additional areas such as economic development. State initiative in these areas created opportunities for new GIT applications as described earlier. At the same time, as governing issues continue to cross local government boundaries, state governments are increasingly required to address local issues, such as land use planning. These trends and state activism create fertile ground for new GIT applications at the state level as described earlier.

GIT conditions also are influenced by political, institutional, and financial factors within individual governments. Moyer and Niemann concluded that these factors are one of the least understood, least discussed, and most important aspects of GIT (Moyer and Niemann, 1994). There are inherent difficulties in modernizing government and adopting innovation, such as the diffusion of GIT and other integrating information systems. Leadership support, policy direction, new institutional arrangements, and stable funding are typically required to develop enterprise approaches and to maximize benefits across multiple agencies. However, the implementation of internal organizational, personnel, and financial changes are often resisted even when leadership direction and support exists. These factors increasingly impact GIT over time. Although limited investigation has been conducted about these factors, it is clear that several agency level and statewide GI/GIT initiatives have been abandoned over the years due to these factors.

Funding is a particularly crucial factor in states, since limited resources can severely stymie GIT development, use, and coordination. Justifying GIT expenditures is difficult because anticipated benefits are often hard to document and long term in nature, and not necessarily quantifiable in economic terms because savings are often created by reducing the duplication of efforts and improving cooperation between agencies and governments. This problem has diminished to some extent in recent years, but remains in many government organizations. But even when accepted, if GIT is not funded in the long term as a regular component of an agency's business processes and budget, then continual efforts are required to justify work and expenditures, thus detracting from actually applying the technology to an agency's needs and effectively coordinating plans, efforts, and data with others.

While internal governmental factors, and particularly bureaucratic and financial conditions, can stymie state adoption and hinder the full realization of GIT potential and benefits, it is well recognized that technological developments stimulate such usage. Better and cheaper GIT tools, and more available and useful data, are strong drivers for technology adoption. However, government data are highly influenced by the internal institutional conditions discussed earlier—resulting in most databases being highly fragmented. Data are usually contained within individual programs and agencies, and not managed or financed as a strategic enterprise-wide resource. Combining and integrating data from differing sources is challenging at best due to data incompatibilities and other technical factors, but also because data coordination usually is not a high priority of political or administrative leaders. As a result, there is limited support and funding to address these issues or to develop standards, data coordinating centers, or other mechanisms to provide access and share data resources.

36.4 STATEWIDE GIS GOVERNANCE

There are a number of different approaches to GIS coordination at the state level. Many state programs were developed organically from the "ground up" rather than the "top down," where entrepreneurial GIS users often jumpstarted GIS to provide a particular business need. Numerous stories reveal how internal GIS advocates broke barriers between agencies to coordinate efforts and share data, often without the knowledge or the support of their superiors. Previous efforts coordinated hard copy mapping through State Mapping Advisory Committees (SMAC), but GIS use often emerged in different agencies. Concern with duplicative GIS activities led to broader attention to GI in the 1990s. In general, the maturation of coordination approaches in states has meant an evolution in the conceptualization, the use, and the range of technologies, data, applications, and participants. As a result, state government authorizing direction, interorganizational groups, and coordinating staffs are beginning to address GI/GIT in a holistic manner. While most attention is on GIS, focus is growing to coordinate remote sensing and Global Positioning System (GPS) activities, as well as GI residing in tabular databases. Over time, previous SMAC and other related interagency efforts are being subsumed conceptually and organizationally within broader GI/GIT approaches. Enterprise-wide GI/GIT approaches and institutionalization is evidenced in several ways in state governments, and provides evidence of continuing maturation of statewide GI/GIT approaches.

Authorizing directives are increasingly adopted in the form of statutes and executive orders that directly or indirectly promote GI/GIT development and coordination. The most common purpose of these directives is to sanction GI/GIT coordination, and often to authorize GI/GIS coordination groups, or studies. Some directives authorize related offices, databases, or funding for statewide or narrower missions, such as natural resources management, environmental protection, or growth management. Some directives specifically address GI or "GIS data" in a modification of open record laws to provide cost recovery, while others authorize GI/GIT assistance to local governments. While more GI/GIT directives are adopted over time, few are comprehensive, such as to fund a coordination program, or establish oversight to ensure multiagency data commonality or compliance with direction.

States began establishing broadly focused interagency GI/GIT coordination groups in the mid-1980s, now generally known as "GI Councils" (GICs). The level of activity of these GICs varies considerably from state to state and the degree to which a group operates officially can also change over time. Some states have very cyclical GI/GIT coordination histories, including the establishment of new groups after others have ceased operations. For example in the state of Wisconsin, where the Wisconsin Land Information Board (WLIB) was sunset through legislation in 2005 and a new coordination body, the Wisconsin Geographic Information Coordination Council (WIGICC) was created in 2008. However, state GI/GIT groups have generally experienced an increase in authorization, membership participation, strength, resources, and direct or implied responsibility and influence over the direction of GI/GIT among their agencies.

Today, 47 states have at least one group with some degree of official stature (see Figure 36.1). Some states have a policy-level group advised by one or more technical group, sometimes focused on GIS, GPS, base mapping, standards or other issues. Group participants can include representatives of virtually all state government functions, sometimes with legislative participation. Members increasingly represent additional sectors, particularly localities, but also federal agencies, regional organizations, academic institutions, Indian tribal governments, utilities, and others. Broadening participation typically expands focus from interagency coordination to also address the needs of other GI/GIT users operating within a state. Moreover, increasing involvement by multiple sectors often facilitates many forms of GI/GIT partnerships.

Increasing incidence of statewide GI/GIT coordination entities is a third example of state government institutionalization. Specifically, the role of Geographic Information Officer (GIO) is becoming more common within states (Idaho, Indiana, Minnesota, Missouri, Montana, and Wisconsin). These entities serve as a focal point within the state government, and can range from less than the full time effort of one individual to an office with over 30 staff. They typically complement, chair, and/ or staff GI/GIT groups discussed earlier, and can have several GI/GIT development and coordination responsibilities. According to NSGIC, as of 2008, every state has designated an official GIS coordinator (Table 36.2).

TABLE 36.2
Incidence and Authorization of State GI/GIS Coordinators

Authorized		Unauthorized			
Year	Number	% of Total Authorized	Number	% of Total Unauthorized	Total Coordinators
1985	10	59	7	41	17
1988	15	52	14	48	29
1991	30	75	10	25	40
1994	31	77.5	9	22.5	40
1995	33	80.5	8	19.5	41
2008	46	92	4	8	50

36.5   STATE GI/GIT COORDINATION ROLES, RESPONSIBILITIES, AND ACTIVITIES

Most statewide GI/GIT coordination efforts are conducted by a combination of groups, coordination entities, and others' work. They have varying roles, and conduct several different activities according to the nuances of individual states. However, facilitating communication and coordination among GIS users is an initial and continuing goal in virtually all states. One of the most significant differences is that some statewide GI/GIT entities primarily coordinate activities while GIT work is housed in individual agencies. In other states, a statewide entity may provide GIT services for agency or external clients, acting as a "service bureau." With either approach, more and more groups and entities serve in statewide policy and planning roles, with or without official direction or oversight in this regard. For example, mature groups and staffs may lead, develop, and adopt plans, policies, procedures, guidelines, and standards; prioritize and implement statewide data layers; monitor agency GI/GIT activities; and/or provide access to statewide data resources. Information about GI/GIT coordination activities is increasingly available on Web sites, but also in annual reports, directories, newsletters, and other media. State GI/GIT groups and/or entities can have one or more of the following roles and activities:

- Serve as a clearinghouse concerning activities, projects, and plans about GI/GIT in state agencies and possibly other entities, including the provision of directories, guides, annual reports, newsletters, and other materials with regularly updated information
- Provide data clearinghouse, access, and dissemination functions for data indexed and possibly maintained in a state GI/GIS database, and perhaps provide customized data searches, manipulation, and interpretation to meet user needs
- Develop and implement data and metadata policies, guidelines, standards, and procedures to encourage data commonality and sharing, including accuracy and scale requirements to meet overall state needs
- Promote collaborative planning for future data development and other work, including helping prioritize, coordinate, and gather resources to develop and maintain data
- Fund local initiatives for data creation and sharing activities
- Synthesize input from various entities to prioritize common data and other needs, gather resources to accomplish these needs, and carryout data development and/or acquisition plans
- Develop data, sometimes with general appropriation or collaborative interagency funding to ensure data is useful for more than one purpose, project, or agency
- Provide contract GIT services for state agencies and others
- Staff GI/GIT coordination and user groups
- Hold GI/GIT conferences and meetings to facilitate information exchange
- Provide GIT educational services for state agencies and others

In many respects, state coordination groups and entities serve as collaborators for GI/GIT across several sectors. For example, most of the states have been active supporters of the annual "GIS day" first held on November 19, 1999, which was sponsored by National Geographic Society, the Association of American Geographers, and Environmental Systems Research Institute. GIS day was initiated to increase awareness about how GIS has made substantial contributions to society, with particular focus on educating children. Through the efforts of state GI/GIT coordinators, about 30 governors signed proclamations officially designating GIS day in 1999. Many open houses, demonstrations, and other activities have been held on these days in various places across the country.

36.6   STATEWIDE DATA-SHARING INITIATIVES

The effective use of GIT increasingly requires accurate and quality data. Statewide data are a very valuable resource not only for state agencies but also for substate entities, such as local governments and regional organizations, federal agencies, private companies, and other organizations. State GI/GIT direction, roles, and approaches vary considerably by state, particularly concerning data initiatives.

Most states are, to some degree, developing clearinghouses and standards to provide access to GI and associated metadata. To date, all of the states have some form of GI clearinghouse (http://www.columbia.edu/acis/eds/outside_data/stategis.html) with most linked to the FGDC and the NSDI. State approaches differ in that some focus on providing access to metadata, while others also provide access to actual data resources in an integrated format. For example, California, Florida, and Virginia have distributed clearinghouses with metadata available from a central location, and state agencies are responsible for managing and maintaining their own data holdings. Alternatively, states such as Indiana, Maryland, North Carolina, and Utah provide direct access to data often developed by several agencies and projects.

While most state coordination efforts strive to provide access to data or metadata, states differ significantly concerning their data roles, specifically whether or not, and the degree to which they act at a statewide, multiagency level to determine, manage, fund, develop, implement, maintain, and/or distribute statewide data. Most data roles in the past, and in some states today, are conducted at a programmatic level to meet individual functional or agency needs. However, state GI/GIT coordination groups and entities increasingly recognize that independent agency data activities often result in duplicative, incompatible, costly, and sometimes conflicting data. States vary considerably in the direction, approach, long-term commitment, and programs developed to address increasing demand for multipurpose, statewide data resources for use with GIS.

Some options include the following:

State as owner: Certain datasets are needed by many agencies and are too complicated and costly for one agency to manage. In these cases, a state entity with central GI/GIT roles may officially manage the production and the maintenance of a certain dataset, and "owns" and disseminates the data. Data requiring large capital investments and strict standards are candidates for state ownership, such as digital orthophotos and land cover (see framework discussion in the following text).

State as coordinator: A common role is to coordinate and help manage data that are primarily the responsibility of others. A central GI/GIT entity may initially create a dataset, and then oversee future data development and maintenance by one or more other entities. For example, the state may develop a seamless transportation dataset that is maintained by local governments and/or regional organizations using state standards and guidelines. In this case, the state collects updates at predetermined intervals and integrates them at the state level. The state may financially support various aspects of this data role, including distribution.

State as catalyst: State government can encourage and/or support one-time data development, and provide standards and guidance for future work, but have a diminished role regarding maintenance and updates. This is applicable for specialized data that are not a leading statewide, multiagency data priority.

States with GI/GIT entities conducting data development and management roles differ in terms of their choice of datasets, scales, and currency. For example, states with small areas but relatively large populations may develop data at the 1:5,000 scale, while larger, rural states may focus efforts at the 1:24,000 scale. Most data efforts typically focus on "base mapping" or "foundational" or "framework" data that can be used with GIS for several purposes. FGDC worked with representatives of several sectors during the late 1990s to determine that certain datasets are most often needed for GIS applications and to derive additional data (http://www.fgdc. gov/framework). Seven discrete data themes were identified as "framework" data, including geodetic control, digital orthoimagery, elevation, highway center lines, hydrography, governmental units, and cadastral data. States have generally adopted this approach, but with unique variations such as adding or further specifying categories of data according to individual state priorities.

All seven data themes are useful, but digital orthoimagery has been a recent and significant data resource for use with GIS. Digital Orthophoto Quadrangles (DOQs) are generally scanned and correct aerial photographs that combine the detail of a photograph with the spatial properties of a map. DOQs typically have spatial resolutions of 1 m (1:12,000 or better) and are produced from either black-and-white or color infrared photos. With the rapid adoption of DOQs as base imagery by states, DOQs are assuming roles similar to those of the original 1:24,000 scale, 7.5-min quadrangle maps produced by the USGS. DOQs are being produced nationwide under the auspices of the National Digital Orthophoto Program (NDOP) and other independent efforts. Several federal agencies participate in NDOP, which is coordinated by a federal interagency body that assigns priorities, obtains funding from contributing agencies, and organizes input data for DOQ development. NDOP endeavors to complete DOQ production in the United States and continue to support the products through maintenance and data updates, including adaptation to new data sources, products, and production methodologies as they become available.

While NDOP provides standardized data across states, state progress, resources, and approaches at developing other data resources vary considerably, depending on both statewide and agency-specific efforts. A state's overall data resources for use with GIS depend on applicable statewide data initiatives, but also data developed by functional agencies to meet their individual missions, such as natural resources or highway management. However, an increasing number of states have received funding directly from their legislatures for statewide data development efforts, such as Florida, Kansas, Texas, and Utah. One of the most "data rich" states is Minnesota, particularly because its legislature has specifically committed up to $5 million per year for the development of natural resources data for use with GIS through a dedicated fund.

Some examples of statewide data development programs are described in the following text. These examples illustrate the variety of tools states employ to better distribute their data. While the role of states in GIS data development and services is rather straightforward, the degree of creativity applied to these services is very high. The examples cited are not meant to illuminate the "best" programs or describe what should be done with spatial data. They are intended to show what solutions states apply in developing and distributing spatial data.

Florida: Maximizing data availability and access. Emphasis in this high growth state is on making access to data as organized and easy as possible, with strong use of standards and metadata. GIT is distributed among several agencies represented and a wide range of geographically referenced data is available through the Florida Data Directory (FDD). Florida has had a statewide approach to GI/GIT since 1985 when the Growth Management Data Network Coordinating Council was formed to help meet the state's growth management needs. It evolved to become the Florida GI Board (GIB), authorized by a statute in 1996. Like most states, GIB is aided by technical groups. However, Florida is one of the few states also having a GI/GIT group that comprises federal agency representatives; the Federal Intrastate Interagency Coordinating Committee. This group helps strengthen GI/GIT coordination between Florida and the federal government. Florida's approach to data access is also unique in that FDD more closely follows the search tools available in library referencing schemes. Primarily metadata driven, entities contributing data provide metadata to the metadata collector portion of FDD where it is cataloged and verified. The automated library system cataloging server then provides access to contributed data through direct data downloads or links. Users search for data via catalog indices or through text queries. Searching the category list, for example, the user selects the theme of interest (i.e., forestry) and the server provides information derived from the metadata. Detailed summary metadata is provided, including data description, citation, dates, format information, and a graphic showing the regions covered. FDD is an excellent example of proper metadata application and the ease of data access. Planned FDD enhancements include an Internet-based interactive map interface through which users can define regions and select data. In addition to FDD, the University of Florida's Department of Urban and Regional Planning operates another type of data distribution hub, which is based on spatial data packaging. The Geo-Facilities Planning and Information Research Center (GeoPlan) manages and distributes the Florida Geographic Data Library (FGDL). The FGDL distributes satellite imagery, aerial photographs, tax data, and other data for use with GIS. Data are organized by county and other political boundaries and are updated frequently.

Texas: *Building foundational data.* As one of the nation's largest states with 254 counties—the only one with over 160 counties—Texas has many GI/GIT challenges. One of the earliest issues identified by statewide GI/GIT coordinators was the lack of statewide data resources. As a result, Texas's coordinating group (now named the Texas GI Council [TGIC]), organized a multiyear, seven layer, cost-sharing effort to develop much of the framework data at 1:24,000 scale in one large program; the Strategic Mapping Program (StratMap—http://www.stratmap.org). The legislature provided funding for the four-year program to develop DOQs, elevation contours, hydrography, transportation, boundaries, digital elevation models (DEMs), and soil surveys statewide where possible. TGIC also developed partnerships with federal, local, and regional entities to raise required funds, providing many leveraging opportunities and benefits. The limited availability of state funds combined with the goal of completing seven data layers made the success of the multiagency and multilevel partnerships even more imperative. TGIC will regularly maintain and provide access to the data. StratMap, while still ongoing, shows that states can form partnerships with entities ranging from small local governments to huge federal agencies interested in common mapping goals and regions. The unique and effective position of states between local and federal entities is strongly evidenced in Texas in this effort to develop and implement such a large cooperative data development program.

Vermont: *A large-scale solution for a small state.* This New England state covers only 9609 sq. miles, and has a history of state government activism in land planning issues, coupled with strong municipal governance and few county government roles. The General Assembly (legislature) first authorized statewide GI activities in 1988 through the state's Growth Management Act, and has provided direction since then. It created the Vermont Center for GI (VCGI) in 1992, and established it as a public, not-for-profit corporation in 1994 (http://geo-vt.uvm.edu/). VCGI is developing and implementing a comprehensive strategy to develop and use GI/GIT, including statewide base layers. A statewide digital database based on 1:5,000 scale orthophoto coverage was initiated in 1988, before the federal 1:12,000 orthophotos program described earlier (NDOP) was developed. VCGI's Web site provides access to the state's many online data holdings. Available data include building locations, elevation contours, soils, feature codes, well locations, electric transmission corridors, and 1:5,000 roads, which can be used for many applications. Such rich data resources often are not available at the statewide level, and particularly at this scale. In addition, VCGI is helping the Vermont Enhanced 9-1-1 Board implement statewide E911 service (http://www.state.vt.us/e911/). A spatially referenced statewide address database was developed using GIS and GPS, and now is used for additional applications (Westcott, 1999). VCGI also uniquely developed an approach for the U.S. Census Bureau to incorporate Vermont's street centerline network data into the Bureau's Topological Integrated Encoding and Referencing (TIGER) database, thus improving both the positional and attribute accuracy of TIGER data in Vermont.

Wisconsin: *State-sponsored statewide and local database development.* A unique legacy of support for parcel-based land information systems and expertise in satellite remote sensing has enabled Wisconsin to have unparalleled programs and data resources. The components of Wisconsin's statewide GI/GIT approach include the Wisconsin Land Information Program (WLIP) and a GIS service center located in the Department of Administration, a State Cartographer's Office, the Wisconsin Land Information Clearinghouse (WISCLINC) and the Wisconsin Initiative for Statewide Cooperation on Land Cover Analysis and Data (WISCLAND)—(http://www.sco.wisc.edu/wisclinc/). Unique among the states, WLIP provides grants to counties to facilitate land information mapping and record keeping, funded by fees assessed for all land deed transfers in the state. The legislature created the WLIP and the Wisconsin Land Information Board (WLIB) to govern it in 1989. Most collected funds (two-thirds) are retained by counties for land record documents, mapping, and related salaries, while the remaining one-third are state-managed, including the provision of local grants and administrative support. The result is a local/state partnership that mixes local knowledge and responsibility with state standards and support. WISCLAND is a separate public/private cooperative effort to develop, maintain, and distribute statewide GI, including several "framework" datasets including DOQs, elevation models, and hydrography. Unique compared to other states, WISCLAND includes a multipurpose land cover database created by processing satellite data derived from 30 m Landsat imagery. The data are classified according to a schema designed to meet the state's needs, yet are compatible with established systems such as the Anderson classification. Additional data often not available in states include wetlands, land use, digital soil surveys, and floodplains. WISCLINC is the state's clearinghouse that provides access to much of the state's GI.

36.7 CONCLUSION

Over the past two decades, states have greatly expanded the application of GI into all areas of government. States have also coordinated and institutionalized GI/GIT approaches and activities within their organizations from a business area–specific tool to become more of an enterprise function. States employ more than just "smaller" versions of federal GIT; in fact, in some respects, states have a wider range of applications and more advanced institutional arrangements than other sectors. In addition, state GIT activities have an increasing influence upon federal, local, and other organizations. This is largely because overall state activism has increased, while much federal direction and funding that supported state and local GIT activities in past years has decreased. Another important trend is that the notion of a "state" is expanding beyond state government as an entity to also encompass other organizations with interests and efforts for the area defined by a state's geographic borders (referenced by the Great Lakes Initiative). States increasingly provide external organizations with a focal point and opportunities to help develop and conduct GIT coordination and activities across several sectors via the state coordination councils. These initiatives promise to enhance GIT usage, maturation, and coordination, as well as data development, quality, maintenance, and access within states and other sectors. As a result, improved government efficiency, effectiveness, and equity will be the result—along with multiple benefits for the public they serve.

REFERENCES

Arizona Department of Transportation, *Feasibility Report and Implementation Plan for Application of a Computer-Aided Design and Drafting (CADD) System*, Arizona Department of Transportation, Phoenix, AZ, 1984.

Demers, L. et al., *Survey of GIS Use by Public Utility Commissions*, New York State Department of Public Service, Albany, NY, 1995.

Moyer, D. D. and Niemann, B., Institutional arrangement and economic impacts, *Multipurpose Land Information Systems: The Guidebook*, vol. 17, Federal Geodetic Control Subcommittee, Coast and Geodetic Survey, National Oceanic and Atmospheric Administration, Silver Spring, MD, 1994, pp. 1–28.

National Academy of Public Administration, *Geographic Information for the 21st Century: Building a Strategy for the Nation*, National Academy of Public Administration, Washington, DC, 1998.

National Research Council, *Promoting the National Spatial Data Infrastructure through Partnerships*, National Academy Press, Washington, DC, 1994.

Shanley, L., *GIT Governance: State Models and Best Practices*, Wisconsin Geographic Information Office, Madison, WI, 2007.

Warnecke, L., *Geographic Information/GIS Institutionalization in the 50 States: Users and Coordinators*, National Center for Geographic Information and Analysis, University of California, Santa Barbara, CA, 1995.

Westcott, B., GIS and GPS—The backbone of Vermont's statewide E911 implementation, *Photogrammetric Engineering and Remote Sensing*, 65(11), 1999, 1269–1276.

37 National, International, and Global Activities in Geospatial Science: Toward a Global Spatial Data Infrastructure

John J. Moeller and Mark E. Reichardt

CONTENTS

37.1   INTRODUCTION

The need for geospatial information and technology as a tool to help enable a better understanding of the relationship of people, places, and things is now well recognized. Many nations have well established programs for mapping, surveying, and other geographic information. Other nations have been moving toward the improved coordination and integration of geospatial information and technology. However, in the early 1990s these coordination efforts took on a new level of importance as the need for data sharing became evident and technological revolutions in Internet communications and geospatial tools became a part of the building blocks of the information society. The global geospatial information community recognized the opportunity for collaboration among national and regional initiatives to achieve a broad global network and to extend these initiatives into a global infrastructure.

Many countries recognized the importance of geographic information in place-based decisions that address the needs for sustainable development identified in international activities such as the Kyoto protocol and the Rio and Johannesburg Agenda 21 Summits. The 1990s also saw emerging capabilities in geographic information systems, information technology, and communications technology that called for an improvement in the way in which geographic information was collected, managed, made accessible, and used. These changes in technology were occurring at the same time as a general trend of governments to incorporate e-government processes into their business practices and accompanied a growing awareness of the need to address the social, economic, and environmental dimensions of issues as part of decision-making. In the early 2000s, a renewed focus was placed on national security and terrorism, climate change, and on monitoring and mitigating the impact of anthropogenic influences on global ecological systems. A recent example initiative is the Global Earth Observation System of Systems (GEOSS) that is bringing global earth observing systems and spatial data infrastructure together to connect the producers of environmental data and decision-support tools with the end users of these products. GEOSS has a 10-year implementation plan that runs through 2015.

National security, homeland defense, disaster response, and trans-boundary social and environmental issues have been major drivers in the increased use of geospatial science and technology by governments in the past few years. However, the potentially most significant development affecting the use and influence of geospatial information is the explosion of mass-market geospatial capabilities for citizens. Geospatially oriented consumer tools are fueling an explosion of creative applications that integrate geospatial capabilities into day-to-day activities of citizens.

The future promises a continued evolution of the role and nature of Spatial Data Infrastructures (SDIs) and the promise of the realization of a global infrastructure with a set of policies, standards, practices, technologies, and relationships to facilitate the flow of geospatial data and information across government organizations and between all sectors and levels. This chapter will provide an overview of the development of national infrastructures with case examples from some nations and will briefly discuss the emerging Global Spatial Data Infrastructure (GSDI).

37.1.1 National Defense and Security

National defense and security issues remain as high priority considerations for many nations. In the early 2000s, national defense was coupled with homeland security as a primary driver for many increased uses of geospatial and related technologies. The increase in the terrorism threat in many places around the globe stimulated defense, intelligence, and security uses and applications. In the period since early 2000, there has been a significant increase in the development and implementation of defense and security SDIs not only in the United States, but also in many other nations. These SDIs are rapidly incorporating international and industry consortium standards such as those from the International Organization for Standardization (ISO) and the Open Geospatial Consortium (OGC). They are also moving toward compatibility through standards for interoperability with other SDIs. National defense and homeland security are often drivers for investments in geospatial science and technology. For example, in the United States, the National Geospatial-Intelligence Agency has been a major partner in the development of open standards and specifications that are benefiting not only the National System for Geospatial Intelligence but also SDI and GSDI activities around the world. While there are still many barriers to overcome in developing and adopting standards and protocols for information sharing within the national defense and homeland security community and between that community and others, much progress has been made.

ELEMENTS OF A SPATIAL DATA INFRASTRUCTURE

Spatial data infrastructures (SDIs) provide a basis for spatial data discovery, evaluation, and application, and include the following elements:

- *Geographic data*: the actual digital geographic data and information.
- *Metadata*: the data describing the data (content, quality, condition, and other characteristics). It permits structured searches and comparison of data in different clearinghouses and gives the user adequate information to find data and use it in an appropriate context.
- *Framework*: includes base layers that will probably differ from location to location. It also includes mechanisms for identifying, describing, and sharing the data using features, attributes, and attribute values, as well as mechanisms for updating the data without complete recollection.
- *Services*: to help discover and interact with data.
- *Clearinghouse*: to actually obtain the data. Clearinghouses support uniform, distributed search through a single user interface; they allow the user to obtain data directly, or they direct the user to another source.
- *Standards*: created and accepted at local, national, and global levels.
- *Partnerships*: the glue that holds it together. Partnerships reduce duplication and the cost of collection and leverage local/national/global technology and skills.

Source: GSDI Association Brochure, available at: http://www.portal.gsdi.org/files/?artifact_id=313.

As defense and security related programs move forward, their progress is paying dividends and producing benefits in many nondefense communities by stimulating new technologies, advanced applications, broadly useful open standards, and commercial activities that generate jobs and other economic opportunities. Technological advances many stimulated by defense and security, are giving us the capability to collect, process, and use vast amounts of spatial information to address and solve a wide variety of issues in ways not previously possible.

37.1.2 TRANS-BOUNDARY SOCIAL AND ENVIRONMENTAL ISSUES

The concept of sustainable development calls for a pattern of resource use that aims to meet our present human needs while preserving the ability of future generations to meet their own needs. It calls for a balance of economic, social, and environmental issues so that each is sustainable now and into the future. The world's population continues to grow. Many nations are seeking to attain a higher standard of living for their citizens placing an increasing strain on a number of natural and ecological systems.

Many environmental issues often have accompanying social impacts that cause disruption/breakdown of social systems, result in human suffering, or create conflicts among groups of people.

Natural disasters have devastated many parts of the globe in recent years. These may or may not be connected to Global Climate Change, but their destructive impact is unquestioned. Whether it is the tsunami in the Asia-Pacific, hurricane Katrina flooding in New Orleans, wildfires in the western United States or the severe droughts experienced in Australia and China in 2006, these events cause human suffering and damage to the natural environment.

The production and consumption of energy is a continuing issue around the globe. Developed and wealthier societies have traditionally consumed the largest percentage of the world's energy production. With many nations' seeing an increase in their standard of living to include increasing automobile use the demand for energy resources is increasing and pushing consumer cost upward. The economic stakes in energy resources and production are enormous and contribute to many global tensions and conflict between nations and regions of the world.

One of the most significant problems facing a large segment of the world population is adequacy of their food and water supplies. While the world has not experienced famines on the scale of those in China in the late 1950s and early 1960s, famines have caused untold human suffering and have resulted in millions of deaths. Approximately 1.1 billion people have no access to improved drinking water (World Health Organization and UNICEF 2005). As we move forward, the issue of providing food and clean sources of water for the most needy around the globe will continue to be a problem. The loss of lands from crop production, competition for food sources for energy production, growing populations, and uncertain climatic conditions all contribute to the problem.

Geospatial technologies have been used to help address the planning for, response to, and mitigation of issues for many of the areas cited above. However, as studies have shown, that existing geospatial data and technology have not been used as extensively or as effectively as they could be. Recommendations have pointed to the need for increased data sharing, improved availability of easy to use and interoperable tools, and the need for greater awareness of existing resources (data, technology, human, and financial) which can be made available for use. SDIs that meet multiple needs are frequently lacking and their enhancement is often pointed to as a desired goal for the future.

SDIs that capitalize on Internet technology and the development of interoperable geoprocessing systems and technology are helping communities and nations actually implement place-based decision processes. The modern world has produced many marvelous capabilities, but these capabilities have also given man the ability to cause profound changes in the environment, in economic forces, and in social interaction. These changes must not occur in a vacuum without consideration of their impact on the well-being of our citizens and on future generations and their opportunities for a sustainable future. The improved understanding of place and the affects of actions within that place hold the promise that communities and societies will use geospatial science to provide information in a way that addresses issues to achieve long-term sustainability.

37.1.3 Geospatial Capabilities for Citizens

We have often heard the phrase "Act locally–think globally." This has been an idea for some time in the geospatial community but until recently was simply an idea. The Internet has created a reality in which we exist and work in both a local and a global context. We are linked across oceans by communications and information in such an extensive fashion that a person can be part of a local group at one moment and then part of a global group in the next, without having changed their physical location. Thus, the concept of place and the interconnectedness of our actions take on new dimensions. This concept has been given a new reality by the advances within geospatial technologies that have made them more useful and usable by citizens without specialized training or knowledge of geospatial tools or techniques. Until recently, mass-market geospatial capabilities were essentially limited to hard copy maps. A person without an in-depth knowledge of geospatial processing found it difficult to perform geospatial visualization or reasoning.

The Internet explosion, distributed networks, wireless communications, and other new tools have spurred radical changes in the use of geospatial capabilities. Geospatially oriented consumer tools such as Google Earth, Virtual Earth, MapQuest, Yahoo! Maps, in-car navigation systems, location-enabled mobile phones, and hand-held GPS devices are fueling an explosion of creative applications that integrate geospatial capabilities into day-to-day activities of citizens. A significant benefit of this technology revolution is that many more people understand the value and importance of geospatial information. Traditionally, geospatial uses and approaches focused on single disciplines or technologies. Now they are more multidisciplinary, enabling people to think about or use information in more ways. Today, more people access digital technology as part of their daily lives in the style that works best for them. Geospatial capabilities are a growing part of the tools they commonly use.

While citizens often do not need to consider the ramifications of a single local action on the global environment, they often need to consider its impact in relation to the actions within their neighborhood. Geospatial technologies, as a part of the information technology platform, are available to citizens in their homes, give the ability to tie into larger networks, and to see and better understand these interconnections. Local information is part of larger regional and global information sets and all components of this connected system must be healthy for the whole system to be at its best. Thus, it is important that connections be built and maintained from the local to the global in a way that preserves the perspectives and sovereignty of local input while advancing worldwide sustainability. The key role in achieving this balance of interdependence is through the efforts of nations in the development and use of geospatial information. As spatial data/information infrastructures develop, it is becoming apparent that many types of geographic information are important for global thinking and action. Science and technology are helping create new knowledge and understanding that can advance our collective ability to solve local and ultimately global problems if we can share what we know. Data and information are infinitely sharable commodities. Sharing does not diminish an individual's own store of data and knowledge. Conversely sharing enriches the collective storehouse. National SDIs are being improved to increase the access and availability of spatial data and to enrich the data and knowledge wealth of that nation. If these national efforts can continue to coalesce around a few critical linking mechanisms, we will likely see great strides forward in our ability to "Act locally—think globally" and to take advantage of the growing wealth of geospatial capabilities for citizens.

37.2   LOCAL TO INTERNATIONAL REQUIREMENTS FOR SPATIAL DATA INFRASTRUCTURES

The needs and requirements for geospatial content and services vary from nation to nation. Much of this variation is associated with the cultural, legal, and policy background of a given nation or region of the world. The trend toward more significant and more effective incorporation of geospatial capabilities into information technology programs and architectures is continuing to grow worldwide. Nations continue to seek to achieve better use of existing technology, to better capitalize on investments that produce geospatial data, and to understand emerging technologies that provide geospatial capability.

The technical components of a SDI appear to be of general agreement and a solid framework of standards for technical interoperability has emerged through the work of organizations such as the ISO Technical Committee 211 Geographic Information/ Geomatics, OGC, and other multinational standards organizations and national standards bodies. Many nations see the benefits that can be derived from the use of geospatial data and technology and the understanding of the need for interoperability is likewise growing. While most successful endeavors have a well-defined policy base, policy and information security still remain as challenges for many SDI and geospatial-information sharing efforts.

37.2.1   Policy Frameworks

From a policy perspective, there are several key issues that drive the development of local, national, and regional efforts to improve the availability and use of geospatial information. The key policy drivers appear to be

- The implementation of electronic government initiatives
- The need to improve national or homeland security; technology modernization
- The need to achieve better government performance and coordination of government functions
- The desire to address growing problems of sustainability of economies and ecological productive capability

A wide-reaching policy initiative is the Infrastructure for Spatial Information in Europe (INSPIRE). INSPIRE is a European Union initiative designed to advance the creation of a European spatial information infrastructure. This infrastructure will provide improved user access to interoperable geospatial information and services from a variety of sources at the local to global levels, and in support of a range of users and needs. A European Parliament and Council of the European Union Directive and Implementing Regulations will establish a policy framework within which Member States will develop their spatial information infrastructures in such as way as to be compatible and useable in a community and trans-boundary context. This ambitious initiative not only builds upon a consistent policy framework, but also recognizes the need for extensive coordination among the different parties who will implement and use the infrastructure.

The significant level of effort being invested to establish a policy base for INSPIRE is required to reach multinational agreement and exceeds efforts in other parts of the world. However firmly established polices also exist in countries such as Canada, the United States of America, and Australia. The policy framework for each nation is different but has been incorporated into government practices at the national level and has implementation mechanisms that reach into local levels of government. While there are differences in the level of national guidance to other levels of government these and other efforts with well-established policy bases and standard practices have transitioned smoothly during changes in government or administration parties.

37.2.2   Common Standards and Practices

It is frequently said that the world runs on standards. Like other areas of data, information and technology sharing, common standards, and "best practices" improve our ability to publish, discover, access, and use geospatial information. The work of consensus standards development organizations such as ISO TC211 and the OGC has produced standards that provide a platform for implementing interoperable geospatial architectures and infrastructures. With these recognized standards in place, many commercial and open source technology providers are offering products that conform to these standards. While additional standards and implementation procedures are still needed, the existing geospatial standards framework is enabling sharing and reuse of data and services in ways that were previously not available and are helping communities, governments, and nations to address the geospatial requirement of critical issues as well as everyday business needs.

Common data standards, and encodings and metadata are enabling cooperating organizations and jurisdictions to collect, describe, maintain, and share geospatial information, with greater potential for consistency and applicability of use. Many SDI initiatives at the national and regional levels emphasize a common set of standards based core or "framework" geospatial data that provide broad geographic coverage with accurate base information from which other themes of geospatial information may be added to support a range of decision making needs from the local to national or regional levels.

37.2.3 COLLABORATION, SHARING AND PRIVACY, AND SECURITY PROTECTION

The effective use and implementation of an SDI—whether it is corporate, local community, national, or global in scale—is dependent upon collaboration and sharing. As we have learned, no one organization is the steward of all of the geospatial data that it requires to carry out its mission/business functions. Collaboration is necessary whether an organization works with other agencies or companies to share data for a specific proprietary process, or whether they are operating in the public domain. Effective organizations have demonstrated the ability to operate in a collaborative manner with partners in a business process to create information resources that can be shared to enhance the use of common network services on the Web.

The world is rapidly becoming networked. A critical requirement for improving performance in today's environment is the ability to collaborate with others. The increase in data sharing, use of common services, and greater levels of collaboration bring increased concerns about the protection of privacy, rights, and security of proprietary or classified information. The geospatial community like many other segments of the information society is still seeking best practice solutions for the protection of privacy and security. The resolution of privacy and security concerns is an example of the convergence of geospatial and mainstream Information and Communication Technology (ICT) communities. As the geospatial community becomes increasingly integrated with other elements of the information and communication infrastructure, capabilities and practices will be reused for addressing related problems. Solutions exist within other parts of the ICT community that may provide answers to the privacy and security issues of the geospatial community. In many instances, this is already happening. Experiences in several instances suggest that there are two keys to successful integration of technologies and achieving the desired balance of openness/sharing and privacy/security protection. These keys are: the adoption of clear and consistent policies for sharing and security and the use of standards based technologies, services, and processes.

37.3 CASE EXAMPLES—LOCAL TO NATIONAL LEVEL

The following sections discuss the progress of SDI development and implementation around the globe. These are just a few case examples that provide some perspective regarding the similarities and differences between initiatives, and the implications associated with establishing a global SDI necessary to collaborate on issues of great environmental, social, and economic importance.

37.3.1 THE UNITED STATES OF AMERICA

A major milestone related to today's U.S. National Spatial Data Infrastructure (NSDI) occurred in 1953, when the U.S. Office of Management and Budget issued Circular A-16 on Surveying and Mapping to create a federal process for the assessment and coordination of surveying and mapping functions being performed across the federal government. Through a revision of this circular in 1990, the Federal Geographic Data Committee (FGDC) was established as a formal coordination mechanism with the objective of building a "national digital spatial data resource" (United States Office of Management and Budget 1990).

A major report (National Research Council 1993) solidified the concept of the NSDI. This report, along with a strong interest in the reform of the federal government by the Vice President's National Partnership for Reinventing Government, resulted in the endorsement of the NSDI concept. This ultimately led to Presidential Executive Order 12906 in April 1994 (United States Executive Office of the President 1994), calling for more aggressive national efforts to advance the NSDI. In the 1990s, the FGDC stakeholders established a core set of essential NSDI components comprising data content and metadata standards; a nationwide set of framework data; metadata to help inventory, advertise, and intelligently search for geographic data sets; a NSDI clearinghouse that allows for catalog searches across multiple geospatial data servers on the Internet; and strong partnerships with the public and private sectors to advance the NSDI. Through advancements in each of these components, and a strong U.S. policy for low-cost access to federal information, significant progress has been achieved toward fulfillment of the NSDI's executive order.

Significant progress has been made in the promotion of data content standards and metadata practices. In addition to the development of a nationwide standards based "Framework" of core geospatial information, the FGDC has endorsed some 23 data content standards for use nationwide, and has 14 standards in various stages of development through the FGDC committee process (FGDC 2008). The FGDC has worked with federal, state, local, and tribal stakeholders to implement a national metadata standard that is now in broad use at the federal level and by other communities throughout the nation to help inventory, advertise and access geospatial data. Today, the www.geodata.gov portal provides a discovery and access service for geospatial data, services and other online resources through a distributed network of NSDI clearinghouse nodes located across the nation. The geodata.gov portal was advanced under a "Geospatial One Stop" program; one of 24 priority e-government initiatives approved by the U.S. Office of Management and Budget (OMB) in 2001.

Through the geodata.gov, visitors may access metadata, geospatial, and related services made available by federal, state, local, tribal, and other contributors.

To help stimulate and accelerate community adoption of NSDI best practices, the FGDC continues to provide annual grants to state, local, and tribal governments and other implementing organizations. As of June 2008, the NSDI CAP has provided over $18M in funding to more than 600 NSDI-related projects (Robinson 2008).

In 2007, the FGDC in conjunction with the OMB launched a Geospatial Line Of Business (LOB) program that helps to understand, leverage, and optimize the collective resources of the federal government in geospatial information acquisition, maintenance, and application across the federal agencies. With the goal of reduced costs and improved service, the LOB works to coordinate policy effort that includes approximately 30 participating federal departments and agencies. To better partner with nonfederal stakeholders in the NSDI, a National Geospatial Advisory Committee (www.fgdc.gov/ngac) comprising nonfederal committee members was established in early 2008 to provide a platform for discussion on the NSDI geospatial program, policy, and management issues.

37.3.2 CANADA

In 1999, the GeoConnections program was launched with a $60M CAN government funding investment to accelerate the development of the Canadian Geospatial Data Infrastructure (CGDI) through partnerships and matching investments from all levels of government, with the private and academic sectors. GeoConnections, now in its second phase, is funded through 2010 to "to maintain, operate, and expand the CGDI and, in particular, to support its use in decision making on public safety and security, public health, environment and sustainable development, and matters of importance to Aboriginal peoples" (GeoConnections 2006).

The CGDI is the mechanism by which the Canadian federal government is enhancing access to geospatial information and in turn accelerating the development of knowledge based economic activities. Coordinated through the GeoConnections initiative, which reports through Natural Resources Canada, development of the CGDI has included advancements in the following fundamental areas:

1. *Access*: Through the *GeoConnections Discovery Portal*, providing national electronic access to digital geospatial or geographic information held by public agencies, greatly expanding access, and usage. A GeoGratis portal was also established as a mechanism for suppliers to post geospatial data for free distribution.
2. *Data*: Through *www.GeoBase.ca*, a common national framework for geospatial information, making information easier to use and by coordinating the establishment of thematic data standards based on user needs assessment of the targeted communities.
3. *Standards*: Participation in the development of and adoption of national geomatics and metadata standards, as well as international standards through participation in ISO and the Open Geospatial Consortium, making information easier to use.

4. *Partnerships*: Federal, provincial, and local government partnerships, as well as partnerships with the private sector have been established to improve CGDI availability and usefulness to decision makers, leading to greater efficiencies in cost-sharing activities (production, management, and distribution).
5. *Supporting policy environment*: Key geospatial data sharing arrangements between federal, provincial, and territorial ministries and additional policy and best practice recommendations to improve the accessibility and use of information by the public and private sectors to acquire add-value, commercialize, and use this government information.

The CGDI vision statement revised for phase II of the GeoConnections program, acknowledges the successful build-out of a core CGDI infrastructure, and places emphasis on the access and application of geospatial information for decision-making across a range of business and societal issues:

> "To enable access to the authoritative and comprehensive sources of Canadian geospatial information to support decision-making" (GeoConnections 2005).

As with the United States, funding opportunities for projects geared toward sustainable activities grounded on solid understanding of user needs are major portions of the CGDI program thrust.

37.3.3 THE NETHERLANDS

In recent years, important steps for further development of the Dutch NSDI have been taken. The Minister of Housing, Spatial Planning and the Environment (VROM) is the coordinating minister for geo-information in the Netherlands and introduced several policy measures:

- Boosting innovation and knowledge development in the professional geo-field with a substantial subsidy to the Space for Geo-Information (RGI) program
- Founding the Council for Geo-Information (GI Council), with strategic advisory duties
- Forming Geonovum, a new foundation of public sector parties, created with support from the Ministry of VROM and others
- Arranging for formal consultation with industry, the geo-profession and the academic world
- Defining the legal framework for, and implementing, Authentic Registers of key geospatial information

The above activities and policies have, together with the European INSPIRE directive, helped increase commitment in the professional field, and have resulted in successful and ambitious projects with enthusiastic support. The establishment of Authentic Registers represents a major policy undertaking.

In 2008, the next essential step in the Netherlands was taken by the approval of a well-supported and consistent approach and implementation strategy for the development of the NSDI by the Minister of VROM, called GIDEON. GIDEON implementation will produce a national facility for location-specific information. The Ministry of VROM is directing the GIDEON implementation, and reports to parliament. The GI Council has a coordinating role.

GIDEON acknowledges the paramount role of Authentic Registers of key geospatial information and its government has mandated the use by public parties. This is an important point, as government organizations will be responsible for accuracy, currency, and overall quality of their geospatial data that should increase recurring use of authentic data for a range of public needs. This significant legal and policy position should also lead to greater harmonization and interoperability with other geospatial information and related technologies concerning public programs.

Geonovum, as the executive committee of the Dutch NSDI, plays a key role in the implementation of GIDEON. Geonovum devotes itself to providing better access to up-to-date geo-information in the public sector. Geonovum develops and manages the geo-standards necessary to implement GIDEON. In its role as enabler of the NSDI, Geonovum connects public sector managers with geospatial information professionals.

37.3.4 COLOMBIA

The following Colombia Spatial Data Infrastructure (ICDE) overview is summarized from an SDI "Cookbook" developed by the Global Spatial Data Infrastructure Technical Working Group (2000) and from Gómez and Arias (2008) and Arias (2007).

> The Colombian Spatial Data Infrastructure (ICDE) is defined as the set of policies, standards, organizations, and technology working together to produce, share, and use geographic information in Colombia in order to support national sustainable development.

SDI efforts in Colombia have evolved over the years, largely because of government mandates to respond to a variety of municipal to national programmatic needs. In the early 1990s, a series of laws and decrees were issued for Colombian agencies to develop information systems to manage the environmental, geologic, demographic, and mapping data for the nation. The Agustin Codazzi Geographic Institute (IGAC), the official producer of national information regarding cartography, geography, agrology, and cadastre, formulated a strategy based on the concept of NSDI to support the evolution of geospatial information management. This strategy has been implemented through a series of key national and international actions as summarized below:

- 1996: A high-level team drafted a set of government guidelines on information, emphasizing the need for key policy to manage geospatial information as a strategic national resource. The concept of the ICDE was formed because of this activity.
- 1997: Legislation required municipalities to develop plans to define and regulate land use.

- 1997: An Inter-institutional committee (National Committee for Geographic Information) involving 13 government agencies was set up to promote and coordinate SDI activities, including development of a set of national geospatial data and metadata standards.
- 1998: The Colombian government prioritizes the establishment of an alliance between Colombia and the United States, the "Environmental Alliance for Colombia" (Pastrana 1998), aimed at the promotion of technical, scientific, managerial, informational, financial, and political cooperation for the knowledge, conservation and sustainable development of Colombian natural resources.
- 2000: IGAC convened an International Symposium on Geospatial Metadata with the participation of representatives from the US FGDC, OGC, and several NSDIs. The Symposium led to the creation of the Permanent Committee on GIS Infrastructure for the Americas (PC IDEA), and the formalization of methodology for establishing the Colombian National Spatial Data Clearinghouse.
- 2000: The main Colombian producers and users of geographic data, the National Oil Company and the Coffee Growers Federation, agreed to jointly develop and implement the ICDE with the government agencies. Four clearinghouse nodes were implemented.
- 2001: Colombia hosted the 5th Conference of the Global Spatial Data Infrastructure, assumed the Presidency of GSDI and helped consolidate this organization as a nonprofit association.
- 2002: The ICDE National Project was officially accepted by the National Planning Department (DNP) and was included as a permanent part of the National Budget.
- 2003: IGAC delivered the Geographic Metadata Management Software, implemented the SDI and metadata curriculums in their educational programs and documented about 105,000 metadata records. Results motivated the design of the Colombian Geoportal architecture.
- 2004: The ICDE coordination and development was recognized as a function of the National Mapping Agency (IGAC).
- 2006: The Colombian Space Commission (CCE) was created to advance a range of national strategic goals. Given the many common societal, economic, and environmental goals of the CCE and the ICDE, the IGAC successfully proposed the inclusion of the ICDE as a priority of the CCE because of the obvious synergies to be realized among technologies, data sources, geo-processing, and applications.
- 2007: The second generation of the Colombian Clearinghouse, the National Geospatial Information Directory, was developed offering services such as: metadata catalog, static maps, geographic visualization, thematic queries, specific applications, data downloading and interoperability services, institutions from the education, environmental, risk management and land management, geology, and hydrocarbons sectors have focused investments on GIS implementations based on this SDI architecture.

From a research and development perspective, the IGAC continues to assess different solutions to provide access to ICDE services, through an architecture that allows using, updating, and creating new functionalities and delivering software packages that can be implemented by institutions with limited budgets. Toward this end, the IGAC software development team has been implementing OGC specifications, GIS and Web mapping services, making the most of available free open source tools.

Capacity building activities like education and applied research are ongoing as well as contributions to develop the SDI components in the Andean Community (CAN), Cuenca del Plata Committee (Argentina, Brazil, Bolivia, Uruguay, and Paraguay). Capacity building is also ongoing with the Pan American Institute for Geography and History (PAIGH) and the Permanent Committee on Spatial Data Infrastructure for the Americas (PC IDEA), the creation of a Latin American Metadata Profile with PAIGH, the U.S. National Geospatial-Intelligence Agency (NGA), ISO, and the InterNational Committee for Information Technology Standards (INCITS).

Colombia is focusing the next generation of ICDE to provide sound geospatial information products and services to the Global Earth Observation System of Systems (GEOSS), as the most advanced way of understanding our planet to achieve sustainable development.

37.3.5 AUSTRALIA

Since 1996, the Australia New Zealand Land Information Council has been promoting the concept of an Australian Spatial Data Infrastructure (ASDI). The ASDI provides a technical and administrative framework that will allow spatial data of all types to be discovered, accessed, correlated, and analyzed, thereby extending its usefulness and maximizing the community benefit from the government investment in the data.

ASDI Vision Statement:
Australia's spatially referenced data, products, and services are available and accessible to all users.

Source: ASDI Internet Framework Technical Architecture, 2003.

ANZLIC Vision Statement:
Australia's and New Zealand's economic growth, and social and environmental interests are underpinned by quality spatially referenced information.

Source: ANZLIC, Strategic Plan 2005–2010, Milestone 5: National Framework Data Themes, not dated. Available at www.anzlic.org.au/get/2442847451.pdf?hilite=strategic+plan+ 2005–2010.

A key part of the ASDI is the Australian Spatial Data Directory (ASDD) that currently provides a mechanism for discovering and accessing existing spatial data, thereby avoiding duplication. The ASDD is a clearinghouse that provides online access to the ASDI; it may also include product development tools. To make the ASDD work, metadata standards have been developed and are being widely implemented. Recently the ANZLIC metadata standard has been updated to conform to ISO19115 and is now a profile within the ISO structure. Additionally, various tools are being developed to support the entry and management of the ASDD and a pilot project commenced in early 2008 to replace the existing ASDD software with the GeoNetwork open source metadata tool. Additional work to provide further capabilities to GeoNetwork is planned including providing catalog services.

ANZLIC has developed a set of core policies and guidelines as "best practice" to promote sound implementation of the ASDI nationwide. Policy documents produced by ANZLIC and available on their Web site address topics such as metadata capture and maintenance, custodianship, and data management guidance related to privacy, pricing, and access to sensitive information. ASDI compliance criteria have been developed and evaluated in jurisdictions. In 2001, ANZLIC policy and guidance was leveraged to formalize a Data Access and Management Agreement to provide consistent access to data sets managed by the Australian National Land and Water Resources Audit and its partners nationwide. In 2001, in addition and in line with the ANZLIC best practices views, the Australian Government implemented a Spatial Data Access and Pricing Policy that is administered by the Office of Spatial Data Management.

A major focus area of the 2005–2010 ANZLIC Strategic Plan is the definition and implementation of an accurate, current nationwide framework of geospatial data themes. Similar in nature to the framework or "core" data sets specified by other SDI initiatives, the ASDI framework envisions the development of nationwide themes related to cadastre, geodetic network, imagery, topography, address, administrative boundaries, and place names. Together these themes would serve as a core data set to which other geospatial themes can be oriented for a range of decision-making requirements from local to national levels. Much of this fundamental data framework has now been developed and maintained by PSMA Australia Ltd., an unlisted pubic company wholly owned by the State, Territory, and Australian Governments. The fundamental data is acquired from the jurisdictions and integrated by PSMA to provide national coverage. This data is available through commercial resellers and is being increasingly used to support government and commercial sector business activities. The Geo-coded National Address File (G-NAF) is becoming the basis of a national address management framework, being developed to provide consistency in address related activities across all jurisdictions. As of this writing, focus was being placed on the development of a National Elevation Data Framework that will model the landform and coastal seabed in support of a range of critical environmental and climate topics such as coastal inundation and water security.

ANZLIC and the ASDI are undergoing review and evaluation through a number of mechanisms including the Intergovernmental Committee on Surveying and Mapping (ICSM), who employed a paper to provide an updated view of the ASDI. A public private partnership in developing the ASDI was seen as the way forward, and the work being undertaken by PSMA Australia to link the land information agencies of each jurisdiction together is one capability that will add to the ASDI.

Progress may readily be seen at the local to state levels where standards based SDI and Enterprise Architecture best practices have been leveraged heavily to implement interoperable Web-based services that broadly serve the needs of government and citizens. In Western Australia for example, Landgate, the state's land authority, and Western Australian Land Information System (WALIS), have implemented the Shared Land Information System (SLIP) enabler. The SLIP service was initiated to: (1) simplify and speed government access to and application of information in decision making, (2) minimize costs of administering and disseminating information, and (3) maintain data integrity by enabling broad and direct user access to information to the information held by authoritative custodians. This program links the information assets of 19, local, state, and federal agencies, enabling broad user access to hundreds of data sets including land and other relevant geospatial themes via standards based Web services. A range of standards based SLIP services enable users to organize and apply this wealth of information in a location context for improved decision making. The SLIP program broadly supports a range of land use management programs including commercial diversification, land valuation, land registration, infrastructure planning, mining, and emergency management services (Armstrong 2008) (Figure 37.1).

Government agencies at all levels are presently investigating a range of options for better coordinating spatial data capabilities across all agencies to improve service delivery, policy development, and internal business processes. Business needs from the traditional spatial communities such as natural resource, emergency management, and climate change are being supported by increasing interests in using spatial infrastructures to support activities of the government.

37.3.6 CATALONIA, SPAIN

The Spanish NSDI, "Infraestructura de Datos Espaciales de España" (IDEE) (www.idee.es), follows the decentralized federated structure of the Spanish Government: national, regional, and local. It connects in a transparent and seamless way geospatial catalogs, gazetteers, and services distributed in servers all around Spain.

The ideas of the INSPIRE initiative and the standards of OGC and ISO/TC 211 make this possible. This standards framework includes: a Spanish Core Metadata standard (NEM for "Núcleo Español de Metadatos"), a minimum set of ISO19115 metadata items; a Spanish Gazetteer Model (MNE for "Modelo de Nomenclátor de España"), a common schema for gazetteers compliant with ISO19112; and a Spanish Data Model for geographic vector data (MDE for "Modelo de Datos Español") based on ISO19107 and GML. All are based on consensus and approved by the Spanish National Geographic High Council.

Catalonia's SDI was the first to be established in Spain, setting a model for other regions and for the IDEE.

In 2002, the government of the autonomous region of Catalonia (Spain) began the IDEC project (SDI of Catalonia). IDEC began as a collaboration between the Cartographic Institute of Catalonia (ICC), two departments of the regional government—the *Generalitat* and the Department of Land Policy and Public Works—and the Secretary of the Information and Telecommunications Society (STSI) of the Department of Universities, Research, and the Information Society. The first year of IDEC was devoted to planning and creation of a collaborative framework. The following year, the institutional agreements were made regarding general

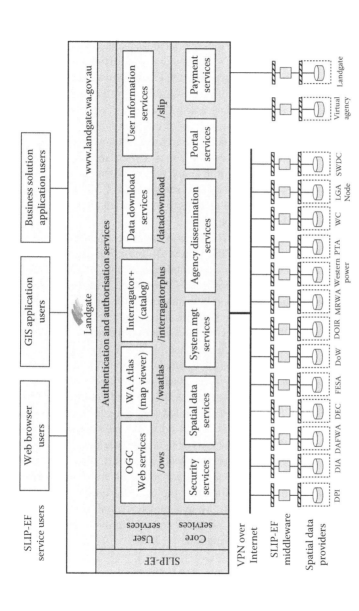

FIGURE 37.1 Shared land information platform architecture. (Data from Armstrong, K. GSDI-10 Technical Session 6.1. Shared and information platform—A cost effective spatial data infrastructure supporting sustainable development. 2008. Available at http://www.gsdi.org/GSDI10/papers/TS6.1paper.pdf. With permission.)

concepts, architecture, and technologies. Implementation began in 2004. The first goal was data sharing within the departments of the *Generalitat*, followed by similar data sharing goals with local governments.

IDEC offers several services as part of its geoportal, the most important of which is the multilingual Catalog Server, with more than 18,000 records of metadata available (53,000 in both Spanish and English), describing data available from over 70 providers. Metadata is also available for services and schemas. The Viewer allows the user to access more than a dozen WMS servers from different providers who together provide about 125 layers of geodata. OGC standards-based services that provide Web-based query and delivery of vector-based and raster-based data are also active. Data themes such as environment, coastal information, transportation, etc., have proved useful to users involved in a variety of other initiatives.

One example is the EUROSION Project, a European initiative funded by the EC to promote better management of the coastal zones. Others include UNIVERS, a regional initiative to connect the WMS of the university Departments in Catalonia to share land information and other geospatial information; SITCAT, a corporate governmental GIS; and LOCAL, a project that aims to incorporate the municipalities in the Information Society and e-government.

An important 2008 study titled "The Socio-Economic Impact of the Spatial Data Infrastructure of Catalonia" (by Pilar Garcia Almirall, Montse Moix Bergadà, Pau Queraltó Ros Universitat Politècnica de Catalunya, Centre of Land Policy and Valuations and Max Craglia (Editor), European Commission Joint Research Centre, Institute for Environment and Sustainability) quantified the costs and benefits of IDEC. The study indicates that "the total investment to set up the IDEC and develop it over a 4 year period (2002–2005) was recovered in just over 6 months…. Wider socio-economic benefits have also been identified but not quantified. In particular, the study indicates that Web-based spatial services allow smaller local authorities to narrow the digital divide with larger ones in the provision of services to citizens and companies."

As described above, SDIs have become an essential ingredient in overall national information infrastructures. The cases of the United States, the Netherlands, Canada, Spain, and Australia represent a few of the SDI initiatives now underway. Many of these efforts have common characteristics, a strong national priority for spatial data as a national resource, the desire to reduce duplication of effort, the need to promote policy and "best practice" guidance to address collaboration and sharing, emphasis on common standards for interoperability, and the need for unique public–private partnerships at all levels to advance SDIs. Many nations are establishing spatial catalogs based on internationally accepted practices and standards to inventory, advertise, and promote access to spatial information.

37.4 REGIONAL AND GLOBAL SPATIAL DATA INFRASTRUCTURES—A COMMON LANGUAGE TO TRANSCEND BORDERS

Geospatial data provides information that can help decision makers at all levels make sound decisions about critical issues such as natural or man-made disasters to everyday matters. SDIs are designed to enable these decision makers and others to share and apply critical geospatial information and services programs across organizations, political, and administrative boundaries. SDIs help improve collaboration within the geospatial community and facilitate communication and interaction with other communities through interoperability of data, services, technology, and processing. Interoperability and supporting standards have provided a platform for the implementation of SDIs reaching from the local to national levels to share and apply critical geospatial information vital for improved decision-making. Similarly, the concept of Regional and Global Spatial Data Infrastructures has arisen as a result of the need to address the broad issues that tend to transcend national borders.

The United Nations (UN) Conference on Environment and Development held in Rio de Janeiro in 1992 called for action to reverse environmental deterioration, and for the establishment of a sustainable way of life to carry us into the twenty-first century. To achieve these goals, Agenda 21 of this conference called out measures to help reduce pollution, deforestation, and to address reversing other adverse environmental activities. Geospatial data was identified as a critical resource for decision-making. Similarly, geospatial data and technology have been identified as critical to the success of many other global or multinational initiatives. Military activities, disaster planning and response, earth monitoring systems, disease prevention, and endangered species management are but a few of the issues for which geospatial information and technology has played an important role. In response to the growing need, for broader multinational collaboration on environmental, economic, and social issues, regional and multinational SDI efforts have emerged, with a focus on the issues unique to a broad area. Regional SDI initiatives have continued to grow in almost all parts of the world. Examples include the Permanent Committee on Geospatial Infrastructure for Asia and the Pacific, the Permanent Committee on Geospatial Data Infrastructure for the Americas (PCIDEA) and the Pan-American Institute of Geography and History (PAIGH) in the Americas, the Infrastructure for Spatial Information in Europe (INSPIRE), and several regional initiatives that support SDI for Africa.

The Permanent Committee on Geospatial Infrastructure for Asia and the Pacific is a forum for nations from Asia and the Pacific to cooperate and contribute to the development of a regional and a global geographic information infrastructure and to share experiences and work together on matters of common interest. It aims to maximize the economic, social, and environmental benefits of geographic information.

Within the Americas, two multinational organizations PCIDEA and PAIGH promote the development of interoperable SDIs that can form a hemispheric network that contributes to the GSDI. Initiatives are underway to advance harmonized metadata profiles and other common implementations of SDI standards for interoperability.

In Africa, a number of organizations are working to assist in the development of compatible SDIs to facilitate access and use of geospatial information for a wide range of needs. Chief among these organizations are the United Nations Agencies, the Regional Center for Training in Aerospace Surveys, the African Association of Remote Sensing of the Environment, and EIS-Africa.

The European Union Nations are engaged in the most structured initiative for a multinational SDI. In March of 2007, the European Parliament and the Council of the European Union, issued a directive establishing an Infrastructure for Spatial Information in the European Community (INSPIRE). The initiative is intended to create a European spatial information infrastructure that delivers integrated spatial information services to users for discovery and access of spatial or geographical information from a wide range of sources. INSPIRE is also intended to reach from the local level to the global level, in an interoperable way for users to include policy-makers, planners, and managers at European, national, and local level, and the citizens and their organizations.

As we face increasing environmental stress on a global scale, nations have recognized that there are pressing needs to better monitor the Earth's systems. The Group on Earth Observations was formed in response to calls for action by the 2002 World Summit on Sustainable Development and by the G8 (Group of Eight) leading industrialized countries. It was recognized that international collaboration is essential for exploiting the growing potential of Earth observations to better understand Earth systems and to support decision making for Sustainable Development. GEO is a voluntary partnership of governments and international organizations that is coordinating efforts to build a Global Earth Observation System of Systems, or GEOSS.

GEO is truly international in scope with membership including 73 governments and the European Commission, in addition to 51 intergovernmental, international, and regional organizations. Of particular note is the linkage of GEOSS and SDI. The GEO 10-Year Implementation Plan Reference Document advocates the use of existing SDI components in areas such as geodetic reference frames, common geographic data, standard protocols, and interoperable system interfaces. The European Commission provides a Co-Chair of GEO and has offered INSPIRE as a contribution to GEOSS. The Commission's Joint Research Centre is leading European efforts to link GSDI and GEOSS and has provided leadership in promoting the coherent and interoperable development of these initiatives through the adoption of standards, protocols, and open architectures.

In the early 1990s, the concept of SDIs at national levels began to take hold in various countries around the globe. As a result, there was a growing recognition that a global network was emerging that offered the promise of extending these initiatives into a global infrastructure. In September 1996, the first Global Spatial Data Infrastructure Conference was held in Bonn, Germany. The participants of that event established a dialogue and a framework that has matured into a robust GSDI that is able to support issues ranging from sustainable development, emerging economies, improved decision-making, the needs of the information society, and the resolution of problems of extreme poverty affecting many around the world. There have now been 10 GSDI conferences, with the results of each conference helping to provide the additional detail and focus needed to promote a global infrastructure capable of supporting transnational and global issues. A GSDI steering committee was formed and grew to include, representatives from all continents. In recognition of the need for a more stable organization the GSDI Association was put in place with the adoption of the association's bylaws in July 2003. The GSDI Association now carries forward the leadership for the vision and goals of GSDI through an organization, which is "dedicated to international cooperation and collaboration in support of local, national, and international spatial data infrastructure developments that would allow nations to better address social, economic, and environmental issues of pressing importance."

GSDI is defined as "… the policies, organizational remits, data, technologies, standards, delivery mechanisms, and financial and human resources necessary to ensure that those working at the global and regional scale are not impeded in meeting their objectives…" (http://www.gsdi.org/).

The GSDI offers significant potential to guide the development of compatible national and regional infrastructures that offer nations the opportunity to collaborate on the significant issues that affect the broader transnational and global community. To better understand the GSDI, it is appropriate to look at the concepts it is built upon and some of the advancements that have been achieved to date by nations and organizations working collaboratively to achieve international consensus on data, standards, dissemination, policy, and resources. This section summarizes some examples of activities that have contributed toward the accomplishment of the Global Spatial Data Infrastructure.

With local, national, and regional activities as the basis for action, the concepts of GSDI were drawn from the collective experience of the participants in the first several GSDI conferences. These concepts were a reflection of many different SDI interests and have remained relatively stable over time. They have formed a framework for development of many of the resolutions of GSDI conferences and activities of the organization:

- *Open and extensive involvement of interested parties*—the necessity to have the involvement and support of decision-makers and others from business, government, and academia. The GSDI Association strongly promotes open involvement and use of the capabilities of SDIs at all levels.
- *Collaboration and communication with organizations*—a successful GSDI requires the many international activities that are involved in different aspects of GSDI development to communicate, coordinate, and collaborate. The geospatial community has developed a culture of collaboration over the years and has done much to incorporate cooperation and collaboration into many of its projects and programs. Other communities have also recognized the need for information sharing and for cooperative working relationships. As GSDI and the geospatial community continue to reach out to others, its experience will be valuable in helping address problems faced in many areas around the globe.
- *Education and research*—these are needed to expand knowledge and awareness and to demonstrate the use and application of geospatial data and technology. GSDI promotes educational activities and operates several small grant programs to help promote GSDI awareness and SDI implementation.
- *Standards*—it is essential to have a family of international standards to provide the technical foundation for GSDI. These standards should include geospatial and other information technology standards. The ISO Technical Committee 211 Geographic Information/Geomatics and the Open

Geospatial Consortium have developed many of the standards necessary for GSDI to operate. The critical standards for SDI implementation have been identified and will be promoted by GSDI as a requirement for best practice SDI implementations.

- *Sharing of benefits*—as successful practices, pilot projects, prototypes, and demonstrations are shared, and organizational and technical best practices are promoted and greater benefits are obtained. The GSDI Technical Working Group has released a "GSDI Implementation Guide," which will serve as a global reference guide for SDI developers. This document provides significant detail regarding core data categories, accuracy, and resolution; the standards and best practices for data documentation, discovery, and access; case studies and other services necessary for nations to assure globally compatible SDIs. This document includes the internationally agreed upon standards, specifications, and proven practices to promote consistency and ease of data sharing across national borders.
- *Core GSDI components*—the SDIs generally include common core components in order to find, access, share, and use data between local, national, and global levels and across organizational and jurisdictional lines.

The GSDI is recognized to include the following core components:

- *Metadata and other standards*—ISO standards are now in place for geospatial metadata and a growing number of ISO Compliant Metadata Profiles are being developed. The Open Geospatial Consortium (OGC) is a global standards organization that promotes the development and use of open system standards for geoprocessing and related technologies. OGC standards are now in place to provide a technical baseline for SDI implementation. The work of ISO Technical Committee 211 and the OGC along with that of National and Multi-National Standards Development Organization has enabled the implementation of standards-based interoperable SDIs around the world.
- *Core/base data and other geospatial data*—include basic sets of commonly used data as well as other geospatial data resources. The International Steering Committee for Global Mapping (ISCGM) created in response to Agenda 21 is leading the production of a global data set (global map) containing elevation, vegetation land use, drainage systems, transportation networks, and administrative boundaries at 1 km resolution. As of June 2008, approximately 180 nations are participating in this effort. The global map will provide a consistent set of base mapping information that can be used as a building block by SDIs around the world.
- *Clearinghouse/data portal network*—distributed networks that contain collections of metadata and data that can be searched through common standards based user interfaces. Catalog and registry services are being hosted by many nations in the GSDI community.
- *Web services*—standards-based Web applications for exchanging data or performing other functions ranging from simple transactions to complex

business processes. As standards for interoperability become more fully implemented SDIs are moving rapidly to Web-services environments. GSDI promotes the use of Web-services as part of the evolution of technologies that support SDI development.

- *Policies/legal frameworks*—guidelines and rules by which a particular SDI and its resources are administered. Policies and legal frameworks differ greatly around the world and it is not the intent of the GSDI to drive policy, rather to offer a collaborative environment for member nations to discuss, understand, and potentially resolve differences so that appropriate action can be sought nationally. Compatibility of policy frameworks is an area that may potentially become a major factor in determining the degree to which data can be shared on a global basis.
- *Partnerships*—organizational and other working relationships for communication coordination and collaboration to include activities to reduce duplication, share skills, leverage technology and data investments, and conduct joint projects. GSDI has been working with many organizations to address issues related to funding, human resources, education and other of the core components of GSDI. Partnerships with organizations such as the World Bank, the United Nations, National Mapping Organizations and professional societies/consortia that advance the collection and use of geospatial information have proven beneficial to helping the entire community move forward in a way that benefits our global society.

Indeed, the contributions of government and private programs have yielded key elements of the GSDI—many of which have become part of the overall GSDI reference environment—needed to help gain compatibility at a transnational and global level. However, much more work needs to be accomplished to address the remaining technology, policy, and resource issues that are limiting the implementation of the GSDI. The linkage of GSDI to global initiatives such as GEOSS will help show the value of SDIs as a resource in addressing critical trans-boundary issues. However, as important as SDIs are for a problem area such as earth monitoring, their true value is as a ready and available resource of data, technology, and services for many issues.

The GSDI concepts have proven to be sound over time and in many different locations and political/economic situations. As technology has advanced, the core components of the GSDI are becoming more and more feasible to implement and are enabling many innovative approaches for developing, sharing, and using geospatial information and technology. Standards are in place for many geospatial technologies and service interfaces. More and more commercial products are using those standards. However, organizational barriers remain a significant hurdle. While technology changes rapidly, organizational policy and culture change slowly. Thus, the concepts of openness, collaboration, communication, education, and sharing benefits are even more important in the future, as nations continue the progress of the past 10 years.

37.5 SDI TRENDS

With basic technical interoperability now a reality and with awareness of SDI concepts and benefits growing worldwide, we see increasingly successful efforts to implement SDIs, as described in the examples above. The overall trend appears to be toward continuous expansion of and interoperability between SDIs. We can also discern other trends by looking at the evolution of ICT markets as well as the evolution of political, social, business, cultural, and legal frameworks in which SDIs—and information systems in general—reside. Considerable challenges and opportunities face us as our "information ecosystem" evolves.

Some of the key opportunities lie in emerging technology trends:

- Computing capability and Internet bandwidth are growing along with Web service architectures and platforms for grid computing, service chaining, and workflow management.
- That growth supports improved Web-based modeling, such as climate modeling, as well as improved decision support systems in domains such as traffic management and disaster management. Inputs to models and decision support systems can come from diverse sources, with standard interfaces "hiding" the diversity of processing services and data formats.
- The proliferation on the Web of sensors and transducers (building environmental sensors, water quality sensors, webcams, satellite imaging, vehicle sensors, cell phones, cameras, etc.) and new sensor Web standards make it easier to access, task, fuse, and use sensor information, adding a new dimension to SDI.
- Cooperation between standards organizations and affiliated trade associations is enabling unprecedented convergence of technologies and communication between application domains. An example is Building Information Models (BIM), which give a wide range of "Architecture, Engineering, Construction, Owner, Operator" (AECOO) stakeholders unprecedented access to full life-cycle information about buildings and capital projects such as airports and roads. The ability to fuse BIM and other "built" information with geospatial information to create an urban three-dimensional model allows improved support for urban planning, public safety, transportation and logistics, retail site planning, and a range of other important functions.
- Semantics and ontologies are leading to "intelligent networks" in which queries yield better results, and in which queries need not be as specific as before. This result is being achieved by two separate methods that are frequently seen as competing with each other, but with great potential to be leveraged collectively for greater search capability:

  1. Standard metadata (with pointers to online data and services) registered in catalogs that implement ISO metadata and OGC Catalog Services standards make possible spatial searches of unprecedented power.
  2. The search engine approach provides a "brute force" method of finding data, arguably without structured metadata and catalogs.

Apart from searches, semantic issues are critical because different communities of interest such as property assessors, climatologists, and emergency response officials use different vocabularies. SDI standards from ISO and the OGC, as well as standards from the World Wide Web Consortium and other SDO are critical to address this need. As the Semantic Web evolves, it will be coevolving with the Geospatial Semantic Web.

Some of the major business trends affecting SDI include

- Major Internet companies are offering advertising-driven geo-information offerings for the masses. Published mapping application programming interfaces such as Google's KML, which is now an OGC standard, make it easy for even beginning developers to create "mash ups" that display data on maps or Earth images.
- In some mash ups, social networks and other Web sites, it is now easy for Web site visitors to contribute data. Examples include: local reports on the progress of recent wildfires in San Diego; the USGS "Did you feel it?" earthquake surveys; the Flicker Web site that lets users "geotag" photographs; the use of GeoRSS to provide spatial tags to RSS feeds; and OpenStreetMap, a free, user-editable map of the whole world.
- An extended geospatial data e-commerce framework is emerging, with technical, policy and "best practice" approaches that support access control, security, rights management, pricing, and provenance tracking. As the information network becomes more dense and inescapable, concerns about sovereignty become more important. Individuals, businesses and governments are becoming more focused on requirements for control of access to data and services. Not surprisingly, groups have formed in the OGC to address digital rights management, privacy, and security as they relate to geospatial information. Their work will have an enduring influence on the future of our industry, and, it seems certain, on the future of society. Communities of interest need to define their own policy guidance on standards and versions that they employ as a technology baseline, and they need to define rules and best practices related to digital rights management, privacy, liability, and security. Different domains often have very different requirements.
- In developing areas of the world, mobile devices and networks offer alternative means of extending the SDI to support end-user decision-making. Access to the Internet as well as reliable sources of power are still problematic in many areas of the developing world, but the growth of mobile devices relative to landline telephones offers a potential new avenue for Internet communications in developing regions. For instance, in Africa, the number of cell phone owners eclipsed the number of wire line phone subscribers in 2001 (Butler 2001), and as of 2007, over 85% of Africa's 280 million telephone subscriptions were wireless (International Telecommunications Union 2008). While device costs, subscription costs, and access to adequate bandwidth are continuing challenges in developing nations, wireless appears to be the dominant channel for access to the SDI.

## 37.6	SUMMARY

The essence of SDIs is to help improve decision making through better use of geospatial capabilities and information that exist from the local to global level. SDIs are advancing at the local, national, regional, and global levels to encourage collaboration and sharing for place-based decision-making. An emerging structure is developing through global and regional SDI organizations to help organizations establish and implement SDIs that serve not only local to national needs, but also those broader issues that will require collaboration across borders.

SDIs must remain adaptive to the rapidly changing technology environment so that innovative technologies, business models, and policy arrangements can be leveraged quickly to extend SDI capabilities. This must be done with care to maintain backwards compatibility so that government, businesses consumers, and citizens at every stage of SDI development can leverage the benefits of a global SDI.

We have entered an age in which we cannot help but see that we are all connected through our shared dependence on the Earth and its natural systems and limitations. In the area of SDI, we have an opportunity to learn much from each other about coevolving our social technologies—government, education, business, law, etc.— with our physical technologies to meet our shared needs.

ACKNOWLEDGMENTS

The authors wish to thank the following for their work in providing content for this chapter: Sylvain Latour, Director, GeoConnections Program, 615 Booth Street, Ottawa, Canada, ON K1A 0E9; Ben Searle, General Manager, Government Office of Spatial Data Management, GPO Box 378, Canberral ACT, 2601 Australia; Lilia-Patricia Arias Duarte, Head of Geographic Information Research and Development Center, Agustin Codazzi Geographic Institute (IGAC), Carrera 30 #48–51 Oficina CIAF 103 Bogotá, Colombia; W.C. (Watse) Castelein (MSc), Geonovum, Barchman Wuytierslaan 10, Post box 508, 3800 AM, Amersfoort, the Netherlands.

REFERENCES

African Association of Remote Sensing of the Environment, http://www.itc.nl/aarse/.
ANZLIC, Strategic Plan 2005–2010—Milestone 5: National Framework Data Themes, not dated. Available at www.anzlic.org.au/get/2442847451.pdf?hilite=strategic+plan+2005–2010.
Arias, L.-P. IGAC response to the *Survey National Spatial Data Infrastructure Assessment*, (survey designed by J. Crompvoets, A. Bregt, L. Grus, T. Delgado, A. Rajabifard, and B. van Loenen) November 2007.
Armstrong, K. GSDI-10 Technical Session 6.1 Paper entitled Shared land information platform—A cost effective spatial data infrastructure supporting sustainable development, 2008. Available at: http://www.gsdi.org/GSDI10/papers/TS6.1paper.pdf.
Butler, R. Cell phones may help "Save" Africa, July 2005. Available at: http://news.mongabay.com/2005/0712-rhett_butler.html.
Coleman, D.J. and McLaughlin, J. GSDI Theme Paper #1, Defining Global Geospatial Data Infrastructure (GGDI): Components, stakeholders and interfaces, 1997.
EIS-Africa, http://www.agirn.org/agirn/eis_africa.php.
Federal Geographic Data Committee, *A Strategy for the National Spatial Data Infrastructure*, Federal Geographic Data Committee, Reston, VA, 1997.

Federal Geographic Data Committee, FGDC *Endorsed Standards and Standards in Development* web pages, 2008. Available at www.fgdc.gov.

GeoConnections, Vision document, 2005, http://www.geoconnections.org/publications/tvip/Vision_E/CGDI_Vision_final_E.html.

GeoConnections, Mapping the future together online: Annual report 2005–2006: Laying the groundwork, 2006, http://www.geoconnections.org/publications/reports/ar/05-06_AR_E.pdf.

Geonovum, National Spatial Data Infrastructure Executive Committee homepage, 2008. Available at http://www.geonovum.nl.

Gómez, I.-D. and Arias, L.-P., The evolution of the National Spatial Data Infrastructure in Colombia: A high level strategy to support policy formulation and spatial information management in the context of the Colombian Space Commission, 2008. Available at: http://www.gsdi.org/GSDI10/papers/TS2.4paper.pdf.

Group On Earth Observations, The Global Earth Observation System of Systems (GEOSS) 10 Year Implementation Plan, February 2005. Available at: www.earthobservations.org/about_geo.shtml.

GSDI Bylaws, www.gsdi.org/Association%20Information/bylaws%20and%20organization%20chart/GSDIBylaws_17April04.pdf.

GSDI Conference Proceedings and Resolutions, Chapel Hill, NC, http://www.gsdi.org/gsdi-Conferences.asp.

Infrastructure for Spatial Information in Europe, http://www.ec-gis.org/inspire/.

International Telecommunications Union, ICTs in Africa: Digital divide to digital opportunity, 2008. Available at: http://www.itu.int/newsroom/features/ict_africa.html.

National Research Council, *Toward a Coordinated Spatial Data Infrastructure for the Nation*, National Academy Press, Washington, DC, 1993.

Pan American Institute of Geography and History, http://www.ipgh.org/.

Pastrana, A., *Formal Announcement of Alianza Ambiental por Colombia*, meeting held in Washington, DC, October 1998.

Permanent Committee for Geospatial Infrastructure for Asia and the Pacific, www.pcgiap.org/.

Permanent Committee on Geospatial Data Infrastructure for the Americas, http://www.cp-idea.org/nuevoSitio/indice.html.

Pilar, G.A. et al., M. Craglia, ed., JRC Scientific and technical report titled. The socio-economic impact of the spatial data infrastructure of Catalonia, 2008. Available at: www.ec-gis.org/inspire/reports/Study_reports/catalonia_impact_study_report.pdf.

Regional Center for Training in Aerospace Surveys, http://www.rectas.org/.

Rhind, D., GSDI Theme Paper #2, Implementing a Global Geospatial Data Infrastructure (GGDI), 1997.

Robinson, M., Interview regarding the U.S. Federal Geographic Data Committee Cooperative Agreements program, June 2008.

Rodriguez, A., An NSDI for Spain, *GIM International*, 2005. Available at: http://www.gim-international.com/issues/articles/id404-An_NSDI_for_Spain.html.

The Economist, *Does Not Compute*, November 8, 2007.

United States Executive office of the President, Coordinating geographic data acquisition and access: The National Spatial Data Infrastructure, Executive Order 12906, April 1994. Available at: www.archives.gov/federal-register/executive-orders/pdf/12906.pdf.

United States Office of Management and Budget, Circular Number A-16 Revised: Coordination of surveying, mapping, and related spatial data activities, Washington, DC, 1990. Available at: www.whitehouse.gov/omb/circulars/a016/a016.html.

World Health Organization and UNICEF, *Water for Life: Making It Happen*, World Health Organization, Geneva, Switzerland, 2005. Available at: www.who.int/entity/water_sanitation_health/waterforlife.pdf.

38 Private Sector Applications

*John C. Antenucci, Robin Antenucci,
and J. Woodson Smith*

CONTENTS

38.1 THE PARADIGM BROADENS, SHIFTS, AND BLURS

The application of geospatial information technology in a commercial setting is quite different today than at the beginning of the decade. Although the business drivers remain the same, for example, efficiency, speed, cost to market, the accessibility of products and services, reduced overhead, and higher profits, there has been a broadening of how a commercial entity relies on the related geospatial technology and the underlying data.

Historically, the private sector use of spatial information technology such as geographic information systems (GIS), global positioning system (GPS) devices and data, and satellite and aerial imagery (and photography), among others, was more often than not, analogous to public sector usage. Some of these functions included general (but automated) mapping of features and events, the correlation of data from multiple sources using location as a common denominator, addressing the logistics of access (or delivery) using location and routing-based algorithms, and the visualization of information using geographic two- and three-dimensional projections of data.

Similarly, the initial applications of spatial information followed the trace of public-sector agencies that pioneered the use of GIS and remote sensing (RS)—initially in a broad range of natural resource, environment, and land use planning applications. Subsequently, the granularity and the scale of practical applications included parcel and tax information and facility and infrastructure management. As a consequence, it was not unexpected that early commercial uses of geospatial information systems were led by private entities with a stake (or supporting role) in forestry; oil, gas, and agricultural; land development and real estate; and water, wastewater, electric, gas, and telecommunication utilities.

Moreover, as a general rule, the early users of GIS in the private sector mirrored the public-sector approach to spatial information by acquiring systems and software, converting or building databases, developing applications, and building, inside the institution, a management and staff component that was responsible for the technology and the delivery of solutions to other parts of the organization. The use of outside experts to design systems, convert data, build applications, etc., was comparable to the practices of the public sector.

Neither the public sector nor the private sector has abandoned this approach to spatial information systems—one that might be characterized as "institution-based" geospatial information capability. However, given more recent developments in the private sector, the servicing of geospatial needs of the public and private sector has broadened the paradigm to include an "external-based" geospatial information capability.

This development both shifts the source to Web-based and embedded functionality and blurs the distinction between the private sector benefiting (i.e., profiting) from the use of externally based geospatial technology but also supplying it to other users for a fee or for free. And, even the supply of the capability for free to the end user is motivated by the profit incentive satisfied, in part, by third parties (e.g., Web-based advertisements/advertisers).

38.2 TECHNOLOGY AND INFORMATION INTEGRATION YIELD PROFITS

Profits generated through higher revenue, reduced cost, accelerated earnings, and competitive positioning and market share remain a basic motivating influence of the commercial sector to embrace a new "workflow" or information source. Changes to workflow or the business process within an organization are frequently met with skepticism and even hostility. Commercial ventures are frequently hesitant to tread on new ground if there is not a precedent and an experience set that establish the economic value of such a change.

Despite the fact that for the better part of 40 years the commercial sector has been constantly refreshed with more potent and less expensive technology, organizations continue to demonstrate hesitancy when approached with the idea of changing or "reengineering" their business process to leverage a new technology or the adoption of one. The introduction of the GPS, GIS, and RS techniques has not been an exception to the rule.

GIS may have had, perhaps, the higher barrier to entry in the commercial marketplace given its parentage in academia and in the government sector. GPS, originally developed for national defense by commercial entities, has established itself as a broad-based utility for the commercial sector. RS technologies, developed for the U.S. space and defense programs, have, with the deployment of commercial high-resolution satellites, become a pervasive source of spatial information.

38.3 COMMERCIAL APPLICATIONS

Coupling two or more of these technologies into a commercial workflow represents a significant challenge to organizational dynamics in the private sector. The integration requirements associated with the technologies have not always been trivial from a technical standpoint. Likewise, the integration of the derivative information has been equally a challenge both from a technical viewpoint and from a perspective of absorption into the analytical needs and information flow of organizations.

Now the private sector (as well as public sector and private users) have a range of choices on the degree to which an organization utilizes institution-based versus external-based geospatial information capability. In many ways, the external-based capability, whether Web-based or embedded, eliminates considerable initial investment and the commitment of calendar time in drawing on (or exploring) the benefits of geospatial capabilities.

Moreover, the end user of the capability may not be the private sector who remains the beneficiary from a profit standpoint by supplying the Web-based or embedded capability to a third-party user as part of its effort to present itself or sell its products or services.

The discussion that follows presents an overview of both models. First, the description of the ever-increasing, near-ubiquitous access to geospatial information from a myriad of sources and for a range of applications that are near breathtaking in their novelty and creativity—as well as those that have a more traditional heritage. Second, a revisit and an expansion of geospatial applications that are more institution-based in their construct. And here, the distinction begins to blur all the more—as institution-based geospatial initiatives begin to incorporate, embed, and rely on external functionality and data.

As the case studies in the following text exemplify, the successes are real and not so unusual once an appreciation of the individual technologies is developed. GPS is frequently relied upon as a baseline and as a vehicle for accumulating and registering data for a particular geography. Remotely sensed data in the broadest definition of the phrase provides a broad range of techniques for the collection, the analysis, and the portrayal of data. GIS finds itself used as a source of collateral data, as a mechanism for aggregating data (and their respective positions), and serving as both an analytical tool and a visualization media.

And, on the other hand, the benefits are more than monetary and the utility nearly unconstrained, where a feature can be associated with some referencing system (Ardila, 1996).

38.3.1 EXTERNAL-BASED APPLICATIONS: WEB AND EMBEDDED ACCESS TO GEOSPATIAL INFORMATION

In many ways, the most dramatic impact on the private sector has come from the "consumerization" of spatial data and technologies—a consumerization that has been made possible by still relatively new and evolving communications media (i.e., the Internet and mobile communications).

Perhaps the earliest widespread use of spatial data on the Internet took the form of sites displaying maps for navigation (MapQuest, Yahoo! Maps, Google Maps, and others) or weather information (www.weather.com, www.wunderground.com).

Google Maps, Google Earth, Virtual Earth (Microsoft), and Yahoo Maps are a few of the commercial mapping platforms that have greatly expanded and transformed the user experience to geospatial information. By providing basic and advanced versions, the suppliers have evolved a geospatial platform that can be embedded in Web sites or other applications as geo-referencing or visualization tools, or, expanded by third-party developers, serve as a "platform" for more advanced or specific geospatial applications.

Each of these platforms relies, in part, on a number of commercial data suppliers, many of these businesses having preceded the platform supply companies. The suppliers of maps, roads and centerlines, parcels, and other map feature data include companies such as Boundary Solutions, Europa Technologies, GDT, Lead Dog Consulting, Tele-Atlas, and a host of country-specific data suppliers such as IMTCAN (Canada) and ZenRin (Japan). Much of the source material is derived from government records. The mapping bases for these commercial data sources have migrated from government sources (e.g., NASA) to include commercial satellite imagery providers such as CNES/Spot, Digital Globe, and Terra Metrics, and government source aerial imagery typically acquired through contracts with commercial firms.

Sources and types of data have generally expanded over time, and many of these sites now combine relatively high-resolution satellite imagery, terrain modeling, and scanned topological maps, in addition to the basic road maps that they initially offered. Today, much of the importance of the mapping sites is derived from the other online applications (or "mashups") that use them as base maps.

GPS has become almost commonplace in the private sector. As receivers have become smaller and less expensive, consumer uses of GPS have expanded. Many cell phones offer GPS functions, for example, and dash-mounted GPS units for automobiles are increasingly common. Likewise, GPS units for recreational and fitness uses have entered the market; in the 1990s, it was a novelty to perform GPS mapping using a bicycle, but, today, GPS units are marketed specifically to bicyclists and runners as training tools (often incorporating other sport-specific functions like heart-rate monitors or cadence sensors).

A huge variety of applications draw on online mapping and/or GPS technologies:

- Customer care—Plug-in maps are almost ubiquitous additions to Web sites for stores, restaurants, etc. Typically linked to one of the online mapping services, they give customers the ability to plan a route to the establishment.
- Recreation/fitness—A number of Web sites (e.g., www.mapmyride.com, www.mapmyrun.com, www.gmap-pedomoter.com, www.motionbased. com) draw on data from Google Maps and Microsoft Virtual Earth to enable mapping, saving, and sharing routes for running, hiking, and bicycling. Routes can also be downloaded to or uploaded from GPS receivers (see Figure 38.1). Then there is the new sport of "geocaching," a form of treasure-hunting in which GPS coordinates are used as clues and GPS receivers are used to find a cache.
- Safety—A number of devices, for example, RoamEO, The Pet Detective, Garmin Astro, and others, use GPS (sometimes with a handheld receiver and sometimes in combination with a cell phone) to track a pet and tell the owner where the pet is. Similar devices are available for tracking children, cars, or children in cars. Also catering to the desire for safety is the Web site www.criminalsearches.com, which enables search for criminal records by geography.

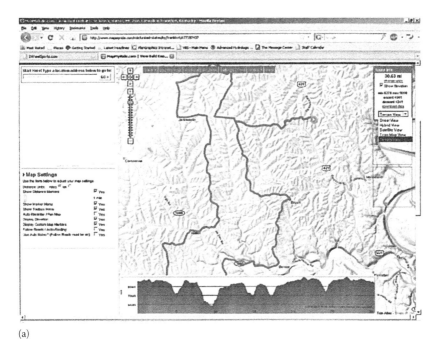

(a)

FIGURE 38.1 (See color insert following page 426.) Screenshots courtesy of www. mapmyride.com: One route shown against three different base maps—(a) shaded digital relief map.

(*continued*)

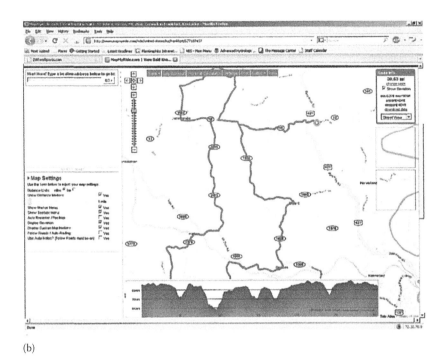

(b)

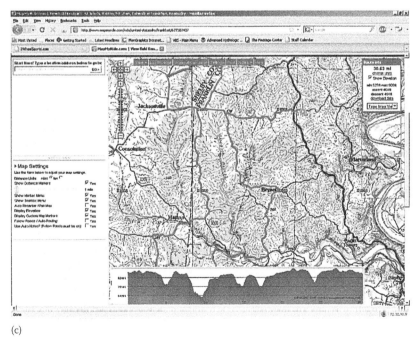

(c)

FIGURE 38.1 (continued)   (See color insert following page 426.) Screenshots Courtesy of www.mapmyride.com: One route shown against three different base maps—(b) state highway map, (c) USGS topo map. (Screenshots from www.mapmyride.com)

- Real-time traffic flow—Taking functionality a step or two beyond that of standard automobile GPS units, Dash Navigator (www.dash.net—"the first two-way, Internet-connected GPS navigation system") is a subscription-based GPS navigation service that delivers real-time traffic flow information based on data received from the dashboard units in subscribers' cars. The network, in effect, informs itself—in other words, the network is effectively its own data.
- Real estate—The Web site Zillow (www.zillow.com) searches for and displays a variety of real estate data—houses for sale and houses recently sold, sortable by attributes such as number of rooms, size, or price—against a base map.
- Hobbies—Online photo albums like Picasa or Flickr make it possible to add locational information to photos and view thumbnails as an overlay to a base map.

What do these and other applications have in common?

1. They hide the technology. The user sees only the typically simple and easy-to-use interface, with no idea of, or need for understanding, the underlying databases and applications.
2. They provide little information about their data. Granted, the typical user may not care to know, for example, what the resolution of the satellite imagery on a mapping Web site is; but, nevertheless, the lack of even basic meta-data—just when was this satellite image taken, anyway?—does limit their utility to less-casual users.
3. Web sites that use geography as a tool for aggregating data from disparate sources (criminal records, say, or housing prices) on the national level are subject to wide variations in data sources at the local level. Mapped search results on a Zillow or a criminal searches Web site could be very accurate in some geographical areas and very misleading in others.

38.3.2 Revenue and Profit Origins of External-Based Applications

Who makes money from these private applications of spatial data and technology, and how?

38.3.2.1 Data Owners/Developers

The mapping sites use data under license from companies like NAVTEQ and TeleAtlas, who compile, research, and update the databases of maps, imagery, and attributes that form the sites' basic content. (Since NAVTEQ and TeleAtlas are the two main sources of data for all the mapping sites, it follows that the basic information available from any mapping site is much the same.)

38.3.2.2 Application or Interface Providers

Two basic business models bring cash into their coffers, and many sites use both. The business model most frequently used is based on advertising; this business model can be simple (e.g., fee based on presence versus time) or sophisticated (e.g., algorithms that serve up ads based on search terms or the location of the mapped area, and advertisers may be charged only if the user "clicks through" a given ad). The other business model is based on subscriptions; in many cases, a free subscription may give users a limited degree of functionality, while a "premium" paid subscription brings added functionality and perhaps fewer ads.

38.3.2.3 Advertisers

J.P. Wanamaker, many years ago, is said to have complained that "Half the money I spend on advertising is wasted; the trouble is I don't know which half." One wonders to what extent any advances in technology have changed that essential predicament. Regardless, advertisers charge fees for placement on pages where the application interfaces are presented, and the fees are part of the income stream from the site, and, at the same time, the advertisers, presumably are accruing an incremental increase in business, revenue, and profit from the exposure.

38.4 INSTITUTIONAL-BASED COMMERCIAL APPLICATIONS

The number of commercial institution-based applications of GIS, GPS, and other geospatial technologies and information approaches the length of a string. The utility is as broad and deep as the types of business processes that make up contemporary commerce. A less than rigorous review of recent periodicals yielded a substantial number of representative examples of commercial geospatial applications. A sample of the visualization and transportation applications follows.

38.4.1 VISUALIZATION

38.4.1.1 Informed Sales and Marketing

Ohio-based Windstream Yellow Pages provides more than 375 directories for telephone companies in 35 states totaling 8 million copies each year. In addition, Windstream processes more than 100,000 yellow page ads and 18,500 community and information pages for its directories on a yearly basis. The key challenge for Windstream Yellow Pages has been to compile demographic, commercial, and geographic data and then transform it into functional business development tools for its sales force.

Communication analysts within Windstream developed an application to create maps showing the distribution boundaries of the directories and to depict demographic and market trends by location. The map product transformed extensive and sophisticated data into a visually intuitive sales collateral. The reliance on the geospatial platform provided the analyst with a doubling of output and a broader and customized set of products for the sales force (http://www.mapinfo.com/case-studies/communications).

38.4.1.2 Efficiencies for Map Production

The Ordnance Survey of Great Britain recently upgraded its geospatial platforms to provide enterprise-wide capabilities for management, planning, coordination, and control of data capture and production activities. The Ordnance Survey is Great

Britain's national mapping agency that sells and licenses a range of commercial data products. The agency's geographic database describes more than 440 million individual features, with a change rate of more than 5000 updates per day. The survey's large-scale data holdings will be managed in a centralized geospatial database, and a standards-based interface will integrate field- and office-based editing to ensure consistency in products and the ability to develop new ones (www.intergraph.com/assets/pressreleases/2007/48158.aspx).

38.4.1.3 Shipping Communications with a Map

P&O Nedlloyd is one of the world's largest suppliers of point-to-point container shipping services. The sales division of P&O Nedlloyd sells containerized transportation in different locations in Europe driving bonuses to the sales force for filling empty cargo space. Conversely, the sales force is penalized if containers are sold in a location where no empty containers are available. Nedlloyd's underutilized asset was one central database in which the whole logistical process is stored. The challenge successfully met was the development of an application that would present the information on a map in a manner where the sales force could utilize the store of information. Utilizing a two-digit zip code for all of Europe provided the visual context for the geographical analysis on the location of empty containers versus the location of the potential shipper. The application included a capability to publish these maps via the Internet to the sales force, making the dynamics of the supply and the demand of empty containers readily available (http://www.geodan.com/markets/private-sector/po-nedlloyd/).

38.4.1.4 Photomaps Boost Fieldwork Productivity

Geologic VR, LLC, a geosciences firm specializing in exploration, monitoring, and visualization services, devised a method that significantly improves the efficiency and the accuracy of geologic fieldwork by making a radon survey essentially paperless. The firm created a GIS for the project area by obtaining aerial photos and combining them with digital geologic and topographic maps (see Figure 38.2). Surface lineaments were identified from the photography and included in the database.

Fieldwork included the use of mobile GIS devices with built-in GPS receivers capable of real-time differential correction. In addition, digital cameras equipped with photo mapping software permitted the geo-referencing of photographs (www.esri.com/news/arcuser/0807/uraniumaps/html).

The three technologies were used simultaneously to identify the location of the lineaments in the field, take photographs of the preselected radon detector sites, and used GPS to collect location and the digital camera to record the bar code of the detector and plot sample. Team members were able to obtain between 36 and 54 detectors per day, and most of the time was spent traversing terrain rather than setting up sample stations. The results of the day's work were downloaded and mapped daily on return to the base camp.

The reduction of field time relative to the number of samples was a major benefit. More importantly, there was the near elimination of the need to transcribe field notes. Field scientists concluded, though, that the greatest benefit was the ability to retrace and relocate particular observation points in the field, an exercise where a mere meter in error may result in a substantially different reading and result.

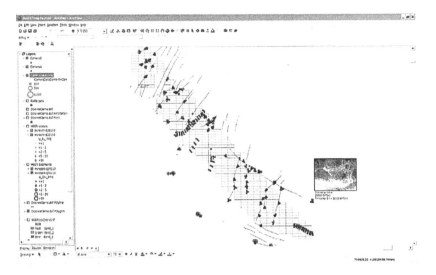

FIGURE 38.2 **(See color insert following page 426.)** Index map to lineament structures and radon detector readings with embedded image metadata. (From Bobbitt, R. and Johnson, C., Boosting fieldwork productivity with photomaps, *ArcUser Online by ESRI*, July–September 2007. With permission.)

38.4.2 TRANSPORTATION

38.4.2.1 Australia's Largest Road

EastLink is Australia's largest road project and will deliver Melbourne's second fully electronic toll way. Approximately 45 km in length, the freestanding road connects eastern freeways to the city's southeast. With a continual stream of incoming data related to design, survey, environment, construction, community feedback, and planning, the communication and the retention of information were identified as major issues. An innovative multistage GIS strategy was implemented, including the translation of computer-aided design (CAD) data and the development of Web-enabled applications. With limited resources, the main focus of the geospatial system targets the requirements of the environment, community, property, safety completions, and tunnels groups (see Figure 38.3).

The benefits were real and multiple and included the savings of approximately $50,000 per year by avoiding outsourcing of services to external suppliers. The use of geospatial data also provided multiple efficiencies, including the improvement, by 80%, in locating and collating information and the speed in data collection, validation, and updating erroneous data entry and the benefits associated with accurate data (www.esri/news/arcnews/summer07articles/developing-australias.html).

38.4.2.2 Vehicle Recovery

Green Flag is the UK's third largest assistance group providing emergency support and rescue to the automotive industry. Green Flag, a subsidiary of the Royal Bank

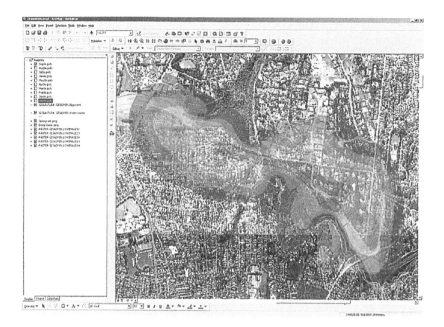

FIGURE 38.3 **(See color insert following page 426.)** Water table impact zones resulting from tunneling for the East Link Project, Melbourne, Australia. (From Developing Australia's Largest Road Project; ArcNews online Summer 2007. With permission.)

of Scotland, partners with independent recovery agents, regional garages, and repair specialists to respond to calls for Green Flags members rather than maintaining its own fleet of recovery staff and vehicles.

In a two-year IT re-engineering program, Green Flag has paid particular attention to its call center application for logging help calls and dispatching vehicles. The aim is to reduce roadside waiting to a maximum of 35 min. As the mapping application is an important part of the mission-critical call center solution matching the affected motorist with an available, in-network, recovery agent or garage, Green Flag adopted a mainstream IT approach to integrating the mapping application into a relational database using a set of spatially extended development tools (middleware) for building a map interface and mapping application. The mapping application provides capabilities, to be used by the call center, for the estimation of the journey distances for the validation of costs associated with a particular incident. Moreover, Green Flag advertises that it will compensate the stranded driver £10 for any member who must wait more than 60 min after a call is logged.

The application solution has resulted in improved customer services, faster decision-making by the call center, reduced maintenance costs by retiring legacy GIS and investing in a spatially enable database, and achieving reliability and resilience in supporting more than 100 uses in disparate locations and tens of thousands of customers (www.agi.org.uk/pooled/articles/BF_TRAINART/view. asp?Q = BF_TRAINART_159646).

38.4.2.3 Improving Transportation Infrastructure

Louisiana timed managers (LTM) was selected to minimize the duration of Louisiana's transportation infrastructure model for economic development (TIMED). TIMED is the largest transportation program in Louisiana's history and includes plans to widen 536 miles of highway, and the construction of three new bridges. The project is budgeted at $4.7 billion.

GIS was initially used by LTM to assemble data in preparing its bid and schedule in competition for the project. Subsequent to the award, a GIS Internet site was established to allow environmental engineers to readily access geospatial information without becoming expert in the technology. Field biologists were equipped with mobile data-collection platforms utilizing GIS and GPS for the collection and the posting of environmental data, among other uses. The capability was used to support decision-making in the identification and the evaluation of proposed highway alignments.

The implementation provides real time links to documentation resulting in the analysis and the incorporation of CAD-based data into the GIS and the establishment on interoperability between the GIS and the CAD-based data for use by the design engineers. Similarly, the system was used to facilitate environmental documentation and the application of required permits.

38.4.3 Agriculture and Natural Resources

38.4.3.1 Environment and Banana Production

In Colombia, South America, banana production is an important part of the economy, accounting for 21% of all agricultural products. Banana exports to the United States from Columbia represent approximately 18% of banana imports from Columbia (comtrade.un.org), are valued at more than $400 million each year, and are 10% of the world's banana exports (www.fao.org/corp/statistics/en/). The banana crop is vulnerable to a damaging fungus known as the Sigatoka fungus that can destroy a major portion of each annual harvest if not controlled properly.

Banana plantations managed by growers such as Pole and Chiquita have traditionally used ground-based flagging crews to guide crop dusting aircraft. Social and environmental scientists find this method hazardous to the flagging crews, particularly from physical contact and the inhalation of fungicide. In addition, the technique is imprecise, environmentally unsound, and an inefficient use of costly aircraft and pilot time.

In 1995, a group of growers began to apply GPS technology to improve their spraying programs. The components of the system include a differential GPS (DGPS) receiver, GPS, and differential correction antennas mounted above the cockpit, an indicator light bar mounted in the cockpit, and a moving map display. Spray boundaries were mapped and digitized for each banana crop lot in each grower's plantation. Hazards such as power lines and water bodies were also mapped. A GIS database was developed to track data on each lot sprayed such as the chemical used, percent of coverage, etc. (Ardila, 1996).

With the new system, as the pilot approaches the lot to be sprayed, the indicator light bar displays the plane's location relative to the path to be sprayed through an interconnection between the DGPS and the digital map. LEDs on the indicator light bar change color as the area to be sprayed is approached. Red indicates "do not spray," changing to orange to indicate "prepare to spray," and when the plane is over the target lot, a green light on the display tells the pilot to "release the chemical spray." At the end of the flight line, the display changes back to red. If the pilot nears a power line or water body, a warning is flashed on the light bar notifying the pilot to change course.

The system has provided a number of significant benefits to the growers' operations, including

- Flexibility—The pilots no longer have to wait on flaggers to relocate ground markers. Pilots can also review their flight lines prior to landing and make course or spraying corrections without having to wait until the next mission.
- Reduced operational costs, as the flaggers are no longer needed. The traditional, manual system costs growers more than U.S.$150,000 per year. The incorporation of digital cockpit maps and GPS reduced system costs by more than 80% per year.
- Precise control—Data stored in the moving map display accurately records information about the spraying operation that is used to determine if the spraying missions are being carried out properly by the contracted pilots.
- Environmental protection—The digital cockpit map display allows the pilot to know if a no spray area such as a water body is being approached. Eliminated exposure of workforce to direct contact with airborne fungicide.

38.4.3.2 Geospatial and the Lumber Industry

Weyerhaeuser Corporation of Tacoma, Washington, has integrated GIS technology with GPS and RS technologies in identifying potential stands of merchantable timber throughout the Pacific Northwest United States (Needham, 1999).

With millions of acres of private forestlands to assess for the potential procurement of softwood and hardwood logs, Weyerhaeuser is using RS techniques to rapidly identify tree types and stand age characteristics to avoid the costly process of using field crews to complete this work. In the Pacific Northwest, foresters use classified satellite imagery to identify tree stands consisting of primarily mature deciduous and coniferous hardwoods. These stands are mapped and, using the company's GIS system, combined with land ownership data, roadways, and other base data. The information is downloaded to laptop PCs, equipped with GIS software, and integrated with GPS that are installed in company trucks. Procurement foresters use the GPS unit to navigate to potential procurement sites and investigate the area from the ground to obtain more detailed information about the timber stand and site.

These technologies have enabled Weyerhaeuser to more quickly and accurately prioritize areas for procurement. This means that they can develop their pricing strategies more quickly and begin negotiations with the landowners sooner than competitors. By having tree stands identified and mapped, the company is also able to develop better strategic plans for future purchases and react more quickly to changing market demands for various types of timber. With the GIS provided in the truck, staff can record more detailed information about particular stands, and, using the GPS devices, can more quickly navigate to the potential sale sites.

From pest control to harvest schedules, the use of GIS integrated with remotely sensed data is giving Weyerhaeuser the tools needed to stay competitive and maintain profit margins.

38.4.3.3 GIS and Remote Sensing Improves Oil Exploration

In its constant efforts to explore the earth for petroleum resources, the U.S.-based company, Texaco, has implemented GIS and RS technologies, to assist in its exploration activities (Lyle, 1999).

Satellite imagery is combined with surface and subsurface geology data, well locations, and topographic data to identify viable sites for exploration. The imagery is analyzed and used to map lineaments and surface feature characteristics that may indicate the presence of hydrocarbons. These tools help the company narrow the focus from huge land areas with potential oil fields to smaller areas that, through the analyzed data, indicate viable sites for on-the-ground exploration and sampling. This technique of analyzing imagery saves the company hundreds of hours of staff time by enabling geologists to more rapidly identify potential exploration sites across large land areas. The wide spectrum of collected data types and imagery allows staff geologists to "see" the site and gain perspective on the relationships of surface and subsurface features.

Once an area has been targeted for exploration, the company uses GIS to develop models for the drilling sites reflecting not only geological but also social and environmental considerations. In the field, workers collect detailed information about the site such as digital photos of outcrops, video strips of topography, and field observation notes. GPS units are also used in the field to identify and precisely locate seismic lines and core sample locations. This information is combined with base information in the GIS and is used to form a complete picture of the potential exploration target area. In the GIS, 3D representations are generated for the potential drill sites. These generalized data are subsequently imported to CAD tools and used to build detailed site plans for the location of the well rig machinery, roadways, and other drill site infrastructure.

The GIS tools in use at Texaco have given the company improvements in the integration and the coordination of activities at potential drill sites, better and substantive data for the evaluation and the analysis of sites, savings in time and money, and a means to collect and store critical corporate information that remains in the corporate database and is unaffected by changes in key personnel. By optimizing their operations with technology-driven processes, Texaco has realized benefits in both dollar savings and staff efficiencies.

38.5 FINANCIAL ADVANTAGE FAVORS INTEGRATION

The adoption by a commercial enterprise of any or all the three technologies—GPS, GIS, and RS—would be driven by an exception of both tangible and intangible benefits. As characterized in *Geographic Information Systems: A Guide to the Technology* (Antenucci et al., 1991), a formal structure for examining benefits of technology implementation identifies five distinct benefits:

- Type 1—Quantifiable efficiencies in current practices, or benefits that reflect improvements to existing practices
- Type 2—Quantifiable expanded capabilities, or benefits that offer added capabilities
- Type 3—Quantifiable unpredictable events or benefits that result from unpredictable events
- Type 4—Intangible benefits or benefits that produce intangible advantages
- Type 5—Quantifiable sale of information, or benefits that result from the sale of information service

In the preceding case studies, at least three benefit types are apparent. Types 1, 2, and 4 are apparent, and the potential for achieving Types 3 and 5 benefits exists. Clearly, the actions of a commercial enterprise to take on a new technology, to undertake the integration of several technologies, and to change the workflows and business practices to leverage the potential for cost savings, higher productivity, and higher profits, are typically quantifiable.

The use of GPS and mobile GIS to guide spraying operations reduced costs in terms of reduced fuel consumption, lower costs of fungicide, and lower costs of ground staff. And, it does not take much imagination to reflect on the intangible sociological and human health benefits that accrue when the operation eliminated exposing "flag bearers" to direct aerial spraying. It is not much of a stretch to suggest that the companies may have avoided a currently unquantified liability in the event that the exposure to the fungicide results in financial claims for illness and death. Carrying this example one step forward, the fact that the companies can systematically document the effectiveness of specific fungicides in specific fields under known conditions creates the opportunity that they may sell the resultant information to others (a Type 5 benefit).

An expedited design and build of the Melbourne electronic freeway may decrease the carbon footprint of a geography beset with ozone depletion and symptoms of climate change.

Green Flag not only serves to minimize the unquantifiable pain and the rage of a stranded motorist but also likely reduces the hazard to and caused by the incapacitated automobile or lorry—and mitigating a potential liability to its parent organization's insurance subsidiaries.

Texaco might see secondary income from the information collected as well fields are jointly developed and production and regulatory compliance information is shared. Similarly, an oil company may find the information base created for one purpose to be invaluable in the event of a catastrophic failure of equipment, fires, or natural disasters.

Returning to the initial premise, justification for the technology in the commercial sector will more often than not be established against expectations of productivity enhancements, new capabilities, and the generation of information from, heretofore, disjointed sets of data. As a consequence, commercial enterprise can establish measurement systems that facilitate a determination of return on investment made in the introduction and the integration of the selected technologies.

REFERENCES

Antenucci, J.C. et al., *Geographic Information Systems: A Guide to the Technology*, Chapman & Hall, New York, 1991.

Ardila, M.J., Precision Farming, supplement to *GPS World Magazine*, Advanstar Communications, Cleveland, OH, 1996.

Bobbitt, R. and C. Johnson, Boosting fieldwork productivity with photomaps, *ArcUser Online*, July–September 2007, http://www.esri.com/news/arcuser/0807/uraniumaps.html.

Developing Australia's Largest Road Project; *ArcNews Online* Summer 2007, http://www.esri.com/news/arcnews/summer07articles/developing-australias.html.

Lyle, D., GIS in exploration, *Hart's Oil and Gas World*, June, 1999.

Needham, S., *Personal Communication*, Weyerhaeuser Corporation, Tacoma, WA, 1999.

39 Geospatial Solutions for Utility Infrastructure: A Summary of the Use of GIT in Utilities Today

Mary Ann Stewart and Kathryn Hail

CONTENTS

39.1   INTRODUCTION AND BACKGROUND

The use of geospatial technology in utilities—gas, electric, water and wastewater, pipeline, and telecommunications organizations—has a well-established history. A groundbreaking experiment between what was then called Public Service Company of Colorado and IBM in the late 1960s is considered the genesis of the discipline that came to be called automated mapping/facility management, or AM/FM. Dubbed the "Cheyenne Project" because of its trial site in the Wyoming service territory, the success of this effort spurred additional activity and attracted entrepreneurial individuals to establish private consultancies to service growing demand for this revolutionary approach. Utilities of all types would embrace the technology to drive their network-related business applications.

In the 1980s, hardware migrated to desktops and software capabilities dramatically expanded. Consequently, almost every utility in North America and most local government agencies have either implemented geospatial information technology (GIT) systems, or are in the process of doing so. The information in this chapter is the result of an annual effort to track the adoption of geospatial technology by the Geospatial Information & Technology Association (GITA).

Founded in 1978 and chartered as a nonprofit association in 1982 as AM/FM International, the association provided the emerging AM/FM field a forum for education and information exchange. Its name was changed to GITA in 1998 to recognize the emergence of a broader technology niche incorporating geographic information system (GIS), global positioning system (GPS), remote sensing, and other related technologies now being used to solve complex network and infrastructure issues in utilities.

For the past 9 years, a report has been prepared from the responses of utility organizations to various questions about the use, costs, and GIT implementation approaches of utilities in five vertical markets. The 2008 edition, called the geospatial technology report, features responses to a Web-based series of questions from over 400 utilities. Table 39.1 shows the breakdown of participants according to their vertical market.

No private sector companies or providers of geospatial products or services are allowed to participate; the report is for users of the technology only. As such, it is the only study to independently survey utility-based technology users for detailed

TABLE 39.1
Statistical Summary of Study Participants Sector/General Info

	Electric	Gas	Telecom	Water	Pipeline
Survey participants	88	34	11	39	26
Low meters/ customers	122	1	500	1	4
High meters/ customers	8.6 million	13 million	7 million	18 million	1 million
Low square miles	20	1	45	5	54
High square miles	600,000	400,000	1 million	1 million	41,000

project information across the full spectrum of GIT users. It represents a unique collective "snapshot" of how utilities are using the technology, the most important applications in each vertical market, and other financial and benchmarking trends based upon nine years of research.

39.1.1   2008 GEOSPATIAL TECHNOLOGY REPORT: RESULTS, TRENDS, AND GENERAL OBSERVATIONS

A multitude of technologies have reached the point of viability for enterprise-wide implementation, resulting in real opportunities for the creation of the open spatial enterprise. Interoperable solutions based on open standards, Web access to spatial information, and open relational database technology are enabling the transformation to geodata-based environments. All sectors are increasingly participating in mobility applications and the use of wireless and GPS technology.

Commercial off-the-shelf (COTS) land-based datasets, increased the availability of imagery, and price wars among hardware providers have lowered the cost of entry for many potential GIT users to levels that enable even the smallest of enterprises to realize benefits and payback in shorter cycles. The continued increase in demand for GIT is driving what appears to be increased spending in utility sectors. Spending may be focused in specific technology areas driven by acute needs such as security, regulatory compliance, and utilities policy changes.

39.1.2   WORKFORCE ISSUES

In general, across all industry sectors, the most common educational background for an employee is a 4 year college degree, and the most relevant academic disciplines are GIS and engineering. Electric and gas utilities depart somewhat from this generalization in that they have a substantial number of geospatial employees with only high school degrees.

Respondents indicated the number of employees in each organization actively engaged in the use of geospatial technology. As many as 6000 employees may use the technology in a telecommunications company, as many as 11,300 employees in a public sector organization, as many as 2500 in an electric utility, and as many as 1000 in a pipeline company. A maximum of only 600 employees use the technology in the water, wastewater, and storm water industry. Gas utility data was unavailable. Employees in all fields may use the technology anywhere from 1% to 100% of their time, with average use in the range of 23%.

39.1.3   BENCHMARKING

Metrics were collected to quantify costs per unit of landbase or facilities, and ranges of cost are presented by each industry sector. Table 39.2 illustrates survey participants' maintenance cycles. In some sectors, the percent growth of facilities is also quantified. Updating the GIS landbase data on a continuous basis is a costly undertaking, and many organizations defer this activity to control costs. The electric sector gravitates toward 1 week maintenance cycles, gas between one-week to

TABLE 39.2
Facilities Maintenance Cycle by Industry Facilities Maintenance

	Electric (%)	Gas (%)	Telecom (%)	Water (%)	Pipeline (%)
Daily	16	8	0	37	0
Weekly	29	15	30	25	0
Monthly	33	46	60	4	8
Longer	22	31	10	33	92

one-month cycles, telecommunications toward one-month cycles, and water toward greater-than-one-month cycles. Most sectors utilize a 1 month maintenance cycle; however, pipeline overwhelmingly benchmarks on longer than 1 month cycles. The water sector is a bit unusual in having strong percentages both at daily-to-weekly and longer-than-a-month cycles.

39.2   IMPLEMENTATION APPROACHES

In an enterprise approach to implementation, geospatial data are stored in a corporate data repository, and applications support a wide range of corporate business processes and interfaces. In a departmental approach, geospatial data and applications are undertaken at a departmental level. These data and uses include, land and right-of-way data management, maps and records, engineering, and operations. A service organization approach involves a specific GIS group within the company, which provides GIS services on demand to various users in the organization and sometimes outside the organization.

Enterprise deployments have remained somewhat steady across all industries, as have the percentage of departmental and service approaches. Table 39.3 displays implementation approaches according to vertical markets. Electric, gas, and pipeline utilities show a decrease in enterprise implementations since 2006, while water utilities show

TABLE 39.3
Implementation Approaches of Study Participants Sector/Approach

	Electric (%)	Gas (%)	Telecom (%)	Water (%)	Pipeline (%)
Enterprise	64	55	56	63	29
Departmental	27	25	33	30	43
Service group	9	20	11	7	29

Note: The geospatial technology report referenced throughout this chapter has been published by GITA since 1998 and contains detailed information on the completeness, the complexity, and the direction of GIS projects being implemented at nearly 500 infrastructure-based organizations. For more information on the geospatial technology report, please visit http://www.gita.org/gtr.

an increase. A relatively consistent message across all industries is that more mature organizations are building solutions that fit their existing business process rather than trying to make technical changes and organizational changes simultaneously.

39.3 MULTIPLATFORM GIS IMPLEMENTATIONS

A variety of factors influence the decision to implement GIS across the enterprise, including the lower cost of entry from a software perspective and the advancement of "open GIS (OGIS)" via data standards. From OGIS standards to pipeline open data standard (PODS—an independent database modeling initiative applicable to gas and liquid gathering, transmission, and distribution pipeline systems), and more. The ability to exchange data across platforms enables users to buy the right application for the job, even if it means running more than a single GIS platform. Approximately 60% of all participants were using more than a single platform, and over three quarters of those were sharing data among systems.

Survey respondents were also asked to indicate which of the various relational databases are in use in their companies. Oracle users accounted for about 40% of the total responses, indicating that more utilities are moving to Oracle as the relational database of choice. Just under a third of respondents reported using SQL server, a significant increase from the previous year. Another 20% reported using Access as their primary database. There is still use of multiple relational databases in utilities, especially in large organizations, but the degree of such use is unreported. The use of GIS-dedicated mobile computers has also increased steadily since the question was initially included in the survey three years ago. The use of mobile GIS technology is most common in the telecommunications sector, with all respondents having some mobile applications. On the other end of the mobile technology spectrum, the pipeline and public sectors reported the lowest level of use. Overall, 84% of respondents in all sectors reported some mobile use, a significant increase from the 70% reported in the previous year's survey.

39.4 PROJECT COSTS

Project costs are broken down in three ways. There is a category for annual budget information for most cost categories, providing valuable information regarding upcoming spending patterns. Expanded categories provide levels of detail for software and applications, hardware only (servers and workstations), data, and implementation services. The data category is further broken down to describe facilities and landbase data in terms of conversion and data purchase, detailing COTS data and data purchased from noncommercial sources. The availability of the new budget information presents contrasts in spending patterns. Several sectors have very high 2007 budgets for applications software maintenance, with electric participant budgets at 78% of cumulative spending and telecommunications budgets at 132% of cumulative spending. The remaining sectors' 2007 budgets for applications software are also high, ranging from 9% to 44% of cumulative spending for maintenance. Higher still are telecommunications budgets for integration services at 80% of cumulative spending, perhaps indicating a permanent increase to annual budgets for the sector.

39.5   LANDBASE MANAGEMENT

Business drivers for GIT implementation vary according to the disciplines described in this report. Most utilities implement GIT to manage a large network of geographically dispersed distribution or collection facilities. They focus on the data directly related to those networks—whether they are pipes or wires—and the related equipment. More utilities are recognizing that landbase data empower the GIS to physically locate the assets and enable effective applications. This is resulting in a trend of increased sophistication in utility landbase data. Public sector GIS users are more focused on the land fabric itself, along with a variety of nonutility land features, including cadastral (property, easements, right of way, and tax and boundary data), planimetric (man-made visible features such as streets, curbs, paved surfaces, and fences), topographic (elevations, viewsheds, and hydrology), and demographics (data about population makeup, fire hazards, retail locations, traffic counts, crime statistics, known offenders, etc.). Accordingly, this is the sector indicating greatest aggregated spending on landbase, followed by the electric sector. Public sector has high 2007 budgets for landbase maintenance, at 25% of cumulative expenses. Electric utility 2007 budgets are high for new data acquisition, at 34% of cumulative expenses for COTS data and 31% for non-COTS data. The gas sector follows these two in cumulative spending and shows high 2007 budgets for the maintenance of non-COTS landbase, at 39% of cumulative spending.

39.6   LANDBASE ACCURACY

A primary factor in the cost of landbase data is the level of positional accuracy. Participants were generally unwilling to spend their GIT project budget on 6-in. levels of accuracy but all sectors had increased the use of 2 ft accurate landbase from last year's survey. Further indication that all sectors are continuing to increase their level of landbase accuracy is the decrease in the use of 50 ft accuracy. The use of diverse level of accuracy is a reflection of GIS users varying their landbase accuracy levels based on geography, the density of population, and their distribution networks.

39.6.1   LANDBASE SOPHISTICATION

The majority of all sectors uses a mix of raster and vector landbase data. The drive to automate processes and integrate redundant systems across the enterprise helps cost-justify more intelligent and sophisticated landbase qualities, such as connected street centerline networks and seamless coordinate systems. It also helps cost-justify the use of commercially available raster imagery as a backdrop. Improved landbase enables operational and customer service departments to realize more benefits. When dispatch and supervisory control and data acquisition (SCADA) use the same landbase as engineering and construction, the organizations can improve their bottom line. Although including raster data to complement vector data is widespread among this year's survey participants, there is a continued decline in those relying on raster data alone. The gas utility sector showed greatest use of vector-only landbase (32%), followed by electric at 24% and telecom at 22%.

39.7 FACILITIES MANAGEMENT

The primary business driver for implementing GIS at utilities is the automation of the many activities surrounding the design, the engineering, the construction, the operation, and the maintenance of their facility networks. The majority of costs are incurred with activities ranging from initial conversion of the network data to the ongoing maintenance and the distribution of the data. The primary benefits or returns on investment (ROIs) are realized through implementing user applications in these areas. The overall sophistication of the facilities data model has been increasing in areas where benefits can be realized and decreasing in areas where it proves more cost-effective to integrate with legacy or new lower-cost systems. The tools available to users for data input are becoming easier to use and the availability of budget dollars for outsourcing is decreasing, resulting in more in-house efforts. Using raster imagery in conjunction with vector facility data is low in cost, deferring some conversion costs to an incremental approach.

The current report shows that the greatest geospatial budgets were high in the electric sector, at 42% of cumulative spending for facilities maintenance. The gas sector also showed high budgets for maintenance, at 42% of cumulative spending. The telecommunications sector showed budgets at 84% of cumulative spending for new facilities data and 52% of cumulative spending for data maintenance, possibly showing the beginning of an increase in telecommunications sector budgets.

Prioritizing areas for GIS application implementation is a major concern to strategic planners so the system can begin to realize return on investment as soon as possible. Planners recognize high return areas, so system data conversion and application development costs are scheduled and incurred based on these groups.

39.8 MAPS AND VIEWS

Before it was commonly referred to as GIS, the technology began as automated mapping and was then expanded to include facilities management (AM/FM). As a digital map is easier to update than redrawing a paper map, generating various maps and views remains the primary application area for GIS. As with previous indications of steady levels of implementation in this mature application area, there is little indication of change other than moderate overall increases in implementation in the water and telecommunication sectors. There is a significant number of design and development stage projects in the electric, telecom, and water sectors, which should result in a jump in implementations in future years' results.

39.9 OPERATIONS AND MAINTENANCE

In earlier studies, most applications associated with this area reported a lower than 60% implementation rate in any industry. Current results show a decrease in most applications reporting less than 50% implementation, with the exception of increased implementations of compatible units and cost estimating in the electric sector and of hydrant inspection and maintenance and one-call applications in the water sector. There are many areas with a large percentage of projects in design or development, including valve or manhole inspection and maintenance, hydrant inspection and maintenance, and relating leaks to mains/services at water utilities; street maintenance in the public sector; work sketches, design engineering, and as-builts in the electric sector. Generally recognized as the logical, single point-of-entry for new facilities data across most industries in a post-conversion implementation, increases here correspond with the high top 10 ratings for mobile computing and pen-based systems in utilities as they increase work sketch and as-built applications. Electric and gas companies have very strong percentages of implementations in most or all application areas in this category. Telecommunications utilities have a high percentage of projects in design or development for design engineering, compatible units, cost estimating, and work print plots, indicating that this sector is gradually making a move into the work order area.

Public sector and gas utilities have implemented many projects using GPS capabilities, while telecommunications, gas, and electric companies have implemented many field viewers. There are a significant number of projects in the development or design stages, including CD/ROM applications in the public sector, and GPS, database access, and wireless applications in the telecommunications sector.

Among the lowest implementation percentages across all sectors, miscellaneous systems include applications that are considered part of corporate marketing and financial systems, such as customer targeting, business geographics, competitive analysis, and tax applications. Most segments are low and decreasing, perhaps because the required functionality is more cost-effective to implement on other linked systems. Tax polygon reporting was implemented to some extent by public sector, gas, electric, and telecommunications utilities. Gas utilities have also implemented consumer target marketing. Telecommunications have a number of projects at the design and development stage in consumer target marketing, business geographics, competitive analysis, and tax polygon reporting. The gas sector has design and development projects in competitive analysis.

Interfacing GIS facilities data with other corporate systems involves sending data to other systems, importing data into the GIS from other systems, or both. Systems that create or maintain specific data types are the best source of data for other systems. They eliminate redundant data entry and maintenance and reduce the potential for discrepancies among systems used for decision support. Information technology operates at a higher level of efficiency when users of various types share data rather than re-create them, and as data become more homogeneous and readily shared, this activity should increase in many of these areas.

Interfaces with node reduction/segment reduction have increased in electric utilities, as have interfaces for load flow analysis, pressure analysis, and valve area isolation analysis for gas companies. Water companies have implemented many flow analysis and pressure analysis projects and have a significant number of projects in design or development for valve area isolation analysis. Activity tends to be low in the public sector, with the exception of a good number of engineering analysis projects already implemented or in design or development.

39.10   WORK AND MATERIALS MANAGEMENT

These projects reflect the potential for real construction cost savings, as design engineers are enabled with "what-if" scenarios for optimizing designs from the labor and materials cost side.

Electric utilities have the greatest percentage of implementations in this category at just above 50% for compatible units and work management. Gas companies also have a significant number of implementations, implementing 19% for compatible units and 29% for work management. Water and telecommunication companies are comparable in implementation with 12% for compatible units and 33% for work management and 29% for compatible units and 29% for work management, respectively.

39.11   CUSTOMER INFORMATION SYSTEMS/ TROUBLE/OUTAGE MANAGEMENT

Electric utilities are the sector with the greatest percentage of implementations in this area, particularly for window or link with GIS, trouble call-taking, voice response unit, outage analysis, and incidences displayed. Telecom companies also have a good number of outage analysis applications, while gas companies have implemented many automated meter reading and computer-aided crew dispatch projects. Water utilities have a substantial number of projects at the design or development stage for outage analysis and automated meter reading.

39.11.1   LINKS TO BACK-OFFICE SYSTEMS

Interfaces here include financial, human resources, SCADA, workflow automation, forecasting and planning, and document management. The greatest percentage of SCADA/EMS/distribution automation implementations are found at electric and water utilities. Telecom, gas, and electric companies have a high implementation of document management systems. Telecoms have a relatively high percentage of forecasting/planning and document management projects in the design and development stage, and the public sector has a high percentage of document management and workflow automation projects in the design and development stage.

39.12   TOP 10 APPLICATIONS AND TECHNOLOGIES

The Geospatial Technology Report (GTR) surveys the top 10 applications and technologies of current geospatial usage by field. Data issues dominate a list of the top three issues faced by utility companies, followed by upgrades/migrations and regulatory compliance issues. Vendor replacement emerged as a new important issue on this list. Currently, there appears to be rising interest in asset management. Gas and pipeline utilities indicated an increased focus on regulatory compliance. Mobility and wireless communications continue to be important technologies for utilities.

39.12.1   ELECTRIC UTILITY INDUSTRY

Trouble call/outage analysis maintained the No. 1 spot in applications from the previous year's survey. Data maintenance moved up to second, trading places with asset management at fifth place. Conversion/data capture made the list at tenth place. Condition-based asset maintenance, automated notification, transmission GIS, and land management were nominated as new applications categories.

As in past years, the focus of the top 10 technologies is on the collection, the transmission, and the display of data—mobile applications maintained the No. 1 position, with system integration coming second and wireless in third. Digital orthophotography entered the list at tenth place. Service-oriented architecture (SOA), LiDAR, and Linux were nominated as new technologies categories.

Top 10 applications—Electric

1. Trouble call/outage analysis
2. Data maintenance
3. Engineering work order design
4. Engineering analysis
5. Asset management
6. Work management
7. Field automation/workforce automation
8. Mobile mapping
9. Customer Information System (CIS) integration
10. Conversion/data capture interface

Top 10 technologies—Electric

1. Pen/mobile computing
2. System integration
3. Wireless
4. GIS software choices
5. GPS
6. Internet/intranet
7. Document management/workflow
8. Windows/Windows NT/Windows XP
9. Data storage/warehousing
10. Digital orthophotography

39.12.2 GAS UTILITY INDUSTRY

Regulatory compliance rose to the top of applications ranked, replacing work management. This is a fairly dramatic shift and more likely reflects strong interest in regulatory compliance systems rather than a cessation in the second-ranked work management applications. Network analysis, routing applications, and CIS integration all rose in rank, while asset management and one-call did not make the top 10 list, indicating interest in these applications may have peaked and is now diminishing. Business development was recommended as an application category for next year's survey.

Top 10 technology results were similar to last year, with wireless access rising to tie for first place with Internet/intranet technologies. Data access, mobile computing, and GPS remain the focus for tools that gas utilities are leveraging to improve their bottom lines. This list shows a dichotomy of technology and business practice issues, with data quality possibly residing in both camps. Nonproprietary programming rose to the top 10 list. Interoperability was recommended as a technology category for next year's survey. Landbase issues were recommended as a top three issues category for next year's survey.

Top 10 applications—Gas

1. Regulatory compliance
2. Work management/process automation
3. Network analysis/simulation
4. Routing/location-based applications
5. CIS integration with AM/FM/GIS
6. Replacement maintenance monitoring/management
7. Construction design
8. Leak detection/management
9. Mobile data collection/viewing/access
10. Facility maintenance—monitoring and management

Top 10 technologies—Gas

1. Internet/intranet tie with wireless access
2. Data access
3. Pen/mobile computing
4. GPS
5. Spatial data in standard relational database management system (RDBMS)
6. Document management/workflow
7. Data exchange/OGIS
8. Digital orthophotography
9. Satellite imagery
10. Nonproprietary programming object linking and embedding (OLE)/ Component Object Model (COM)

39.12.3 Pipeline Industry

For the top 10 applications list, regulatory compliance took first place. The integration of facility and operational/maintenance data held strong at the second place. Risk management rose from seventh to third place. Routing applications held at fourth place. SCADA dropped from first to fifth place. The top 10 technologies list shows OGIS holding its rank at first place. Technology for solving data issues was popular, with survey data capture in second place, followed by data integration in third place, data maintenance tools in fourth place, field data collection in fifth place, and data sharing in sixth place.

Top 10 applications—Pipeline

1. Regulatory compliance—integrity management regulations
2. Integration of facility/operational/maintenance data
3. Risk management
4. Pipeline/transmission line routing
5. SCADA
6. One-call
7. Reporting
8. Field operations/reporting
9. Outage management
10. Environmental management

Top 10 technologies—Pipeline

1. OGIS—Integrated Spatial Analysis Techniques (ISAT)/PODS
2. Survey data capture
3. CIS, in line inspections (ILI), and soils data integration
4. Data maintenance tools
5. Field computing/field data collection
6. Data sharing
7. Internet/intranet
8. Integration software
9. Electronic document management
10. Distribution data input

39.12.4 TELECOMMUNICATIONS INDUSTRY

The current top applications list shows only nine rankings due to the frequency of ties for places. It is typical for applications with the highest return on investment to rank highest here, as well as those applications addressing current business trends or those leveraging new technologies. Engineering design, data conversion, work management, and system integration show highest rankings. The top technology list shows the same top priorities as last year's list. Data exchange/OGIS and applications integration have moved up significantly. As with the top applications list, there are only nine rankings due to the frequency of "ties."

Top 10 applications—Telco

1. Engineering design tie with data conversion
2. Work management
3. System integration
4. GIS database management (tie with asset management)
5. Field computing and collection
6. Facility maintenance tie with data management
7. Core GIS migration

   8. Facilities modeling
   9. Investigation/litigation support

Top 10 technologies—Telco

   1. GIS software
   2. Internet/intranet
   3. Data exchange/OGIS
   4. GPS tie with applications integration
   5. Pen/mobile computing
   6. Document management/workflow tie with system integration
   7. Satellite imagery tie with wireless
   8. Data access
   9. Object-oriented software development tools

39.12.5   WATER AND WASTEWATER INDUSTRY

Many of the top 10 applications and their rankings are very similar to the past year's list. Noteworthy changes include core GIS migration rising to No. 2 on the list and work management dropping to No. 8 on the list. There were additional applications not making the top 10 that deserve mention. These are facility maintenance, facility model integration, distribution water quality/security and safety, and investment/litigation support.

Most of the top 10 technologies are also similar to last year's list. Noteworthy changes include the appearance of data storage/warehousing and document management/workflow on the list. Other frequently mentioned technologies not making the top 10 include wireless, object-oriented software development tools, and data access/portability.

Top 10 applications—Water/wastewater

   1. GIS database maintenance
   2. Core GIS migration
   3. Data management
   4. Data conversion
   5. Field computing and collection
   6. System integration
   7. Asset management
   8. Work management
   9. Engineering design
  10. Replacement forecast planning

Top 10 technologies—Water

   1. GIS software
   2. GPS

3. System integration
4. Applications integration
5. Pen/mobile computing
6. Internet/intranet
7. Data storage/warehousing
8. Document management/workflow
9. Satellite imagery
10. Digital orthophotography

The geospatial technology report referenced throughout this chapter has been published by the geospatial technology report since 1998 and contains detailed information on the completeness, the complexity, and the direction of GIS projects being implemented at nearly 500 infrastructure-based organizations. The preparation of the 2009 edition of GITA's geospatial technology report will begin in the fall, with publication scheduled for January 2009. For more information on the geospatial technology report, please visit http://www.gita.org/gtr.

Index